LUBRICATION

LUBRICATION

Raymond C. Gunther

Registered Professional Engineer, formerly Lubrication Consultant, Engine Builder's Division, Mobil Oil Corporation, and Officer-in-Charge, Refrigeration School, Army Service Forces Training Center

Chilton Book Company

PHILADELPHIA NEW YORK LONDON

Published in Philadelphia by Chilton Book Company
and simultaneously in Ontario, Canada,
by Thomas Nelson & Sons, Ltd.
ISBN 0-8019-5526-2
Library of Congress Catalog Card Number 70-153134
Designed by Carole L. DeCrescenzo
Manufactured in the United States of America

Dedicated to my wife,
ALMIRA DOAK GUNTHER

CONTENTS

PREFACE

The practice of lubrication is as old as the wheel and axle, but the *science* of lubrication gradually developed from those early years during which the production of goods by hand was slowly replaced with manufacture by machine, and the production of power by the water wheel changed to generation by the steam engine. Scientific progress in industry and transportation is dependent upon and has demanded a corresponding progress in lubrication methods and materials. New and more critical parameters have been created in the form of operating temperatures and pressures that challenge the ingenuity of man to discover new lubricants and lubricating methods. The selection and application of lubricants, once a perfunctory matter, have become more precise and demanding.

The purpose of this book is to present the subject of lubrication as a practical tool required in the operation and maintenance of frictional machinery. As such, it is designed for use by personnel who are responsible for the selection, application, handling, and maintenance of lubricants and lubricating systems. Included in this category are plant engineers, plant lubrication engineers, salesmen of lubricants and lubricating devices, supplier engineers, lubrication supervisors, general lubrication personnel, and students of technical institutes and industrial schools. Its contents are designed to serve as a reference source for many practical lubrication subjects.

The text material includes the fundamental categories related to lubrication science:

1. Chemical and physical description of lubricants of various kinds and states, including their constituents, properties, and specifications.
2. Bases of lubricant selection.

3. Methods of lubricant application.
4. Care and handling of lubricants and operation of lubricating systems.
5. Basic frictional components—bearings, gears, cylinders, and ways.
6. Mechanics of machinery.
7. Description of the major types of machinery.
8. Lubricants in nuclear power plants.

The contents of this book are based primarily on the experience of the author as a lubrication engineer and consultant in various fields, including industry and transportation, and to some extent on intensive research into pertinent experimental fields.

Appreciation is expressed to those equipment manufacturers for the many illustrations provided to supplement the text material.

RAYMOND C. GUNTHER
Westgate Hills
Havertown, Pennsylvania
October 1970

LUBRICATION

1

INTRODUCTION

The study of lubrication recognizes an ever present force tending to interfere with the motion of bodies. This force, called *friction,* is interposed between contacting surfaces of two bodies and resists motion of one body relative to the other. Friction may be reduced in magnitude, but never eliminated. This fact precludes the possibility of establishing perpetual motion within the sphere of the earth's influence at temperatures above absolute zero (−459.6°F).

In some forms, friction is destructive to the rubbing surfaces of engineering materials, such as iron, steel, brass, and babbitt, an action called *wear.* In any form, friction causes energy to be wasted. Ultimately, it is dissipated as heat, causing a temperature rise of the bodies involved.

In many instances friction is useful. In a brake it is the force that causes a vehicle to slow down and stop. It is also the force known as traction, as between a tire and the road surface. Locomotives get greater traction for drawing heavy loads upgrade when sand, an abrasive, is applied to increase the friction between the metal tires of their driving wheels and the rails.

Lubrication is the science of reducing friction by the application of a suitable substance between the rubbing surfaces of bodies having relative motion. Substances employed to reduce friction are called *lubricants.* Because friction wastes energy and tends to destroy rubbing surfaces, its influence must be minimized where motion is desired. Lubrication therefore becomes highly important in maintaining mechanical equipment at maximum operating efficiency and prolonging its useful life.

While the definition of a lubricant includes any substance capable of counteracting the force of friction, practical applications demand the employment of only those possessing inherent characteristics compatible with sound economic practices. Consequently, comparatively few substances are selected for a majority of lubrication requirements.

FORMS AND FUNCTIONS OF LUBRICANTS

Lubricants may be gaseous, liquid, plastic, or solid. They serve to (a) reduce the power required to initiate and maintain motion of a part, (b) minimize wear of rubbing surfaces, and (c) control, within reasonable limits, operating temperatures developed by friction. In some instances the lubricant acts as a seal to prevent the entrance of contaminants into the frictional region. Some lubricants serve as rust preventives. Some materials are self-lubricating.

BASIC FRICTIONAL COMPONENTS

There are three basic types of frictional components employed to support mechanical motion: *bearings, gears,* and *cylinders.* The degree of protection against the destructive forces of friction afforded any type of component depends on the qualities of lubrication applied. Such qualities include (1) suitability and adequacy of lubricant selected, (2) effectiveness of the method and interval of lubricant application, and (3) the amount of preventive maintenance devoted to the lubrication system and the frictional component.

It is axiomatic that lubrication is a poor substitute for a favorable mechanical condition. A lubricant in some instances may act as a temporary cushion against the effects of improper design, misalignment, excessive loading, etc., but it should never be depended on to provide permanent protection under circumstances involving mechanical defects. Where favorable mechanical conditions exist, the following rule will always apply, regardless of the type of frictional system: *Effective lubrication depends on the proper application of an adequate amount of correct lubricant at the time it is needed.*

The importance of lubrication is universal. Modern civilization depends on a thin film of lubricant to carry on essential mechanical processes necessary for its existence. Without this lubricating film, all mechanical equipment would cease to function due to the force of friction; power generating stations, transportation systems, manufacturing plants, mines, quarries and all processing facilities would come to a complete standstill. Even so, in spite of all the progress made in the field of lubrication, much of the energy produced in the world is lost through friction.

LUBRICATION CATEGORIES

Lubrication considerations vary somewhat according to the commercial field or other endeavors toward which they are directed. Classified on the basis of enterprise, the subject of lubrication can be divided into seven general categories: (1) industrial, (2) automotive, (3) marine, (4) railroad, (5) aviation, (6) space, and (7) nuclear.

While many lubrication practices are common to all of the above, deviations in some instances are necessary to satisfy the operating conditions peculiar to any one of the individual categories.

1. INDUSTRIAL LUBRICATION. The scope of industrial lubrication is extremely broad. It embraces all mechanical equipment related to the various types of manufacturing and industrial processing plants, public utilities, mines and quarries, research and development laboratories, food and beverage processing plants, cold storage warehouses, amusement facilities, printing establishments, and all other industrial installations. Because of its diversification industrial lubrication presents many complex problems.

2. AUTOMOTIVE LUBRICATION. In terms of lubricant consumption, automotive lubrication is the largest category. This is because of the large number of automobiles and trucks in use and the way they consume lubricants. The number of kinds of lubricants required in this category, however, are relatively few when compared with industrial requirements. Automobiles, trucks, tractors, buses, contractors' and agricultural equipment, and industrial shop and yard vehicles are included in the automotive lubrication category.

3. MARINE LUBRICATION. Ships when at sea are isolated from their major sources of mechanical parts. Any break-down of prime movers or running gears may prove serious. Many mechanical components related to the marine trade are especially designed to meet the peculiarities of ship maneuverability. These peculiarities must be considered in planning a successful marine lubrication program. This group includes all vessels from large ocean-going ships to small outboard motor boats, including ferries, tugs, lake and river craft, fishing boats, racing boats, and naval vessels including submarines.

4. RAILROAD LUBRICATION. As in the marine trade, railroad lubrication poses many problems of its own. While a number of these problems are associated with rolling stock, components of the right-of-way, such as signal equipment, switches, and curves, also manifest their own peculiar lubrication problems. In addition to rolling stock and right-of-way equipment, maintenance and repair facilities are included in the railroad lubrication category.

5. AVIATION LUBRICATION. Aircraft lubrication is one of the most recent of the categories listed above. Because of the rapid change in design of power plants, controls, and auxiliary equipment, resulting in progressive increases in speed, power range, and maneuverability, the field of aviation promises to continue in a highly fluid state. Aircraft operation is carried on under a wide range of environmental conditions. They are exposed to extreme variations in temperature, the limits of which are governed by their zone or zones of operation (torrid, temperate, and frigid), season, and altitudes.

Such changes in environment must be successfully coped with if safe protection of airborne mechanical equipment is to be maintained. Lubricant selection is not the sole factor of complication; methods of application and lubrication system safeguards are also involved. All powered aircraft are included in this group, whether heavier or lighter than air—helicopters, airships, commercial and military planes of all types, and private airplanes.

6. SPACE VEHICLES. The lubrication problems in space lie in the environment of low pressure, vacuum, and temperature extremes. Natural radioactivity in space does not seem to present any lubrication problem, but evaporation and outgassing of liquid and many solid lubricants under high vacuum conditions require other than conventional lubricants to perform satisfactorily. Gaseous lubricants and self-lubricating materials appear to be the answer in some but not all cases.

7. NUCLEAR ENERGY SYSTEMS. The effects of radiation on hydrocarbons and many synthetic lubricants limit their use in nuclear power plants to relatively short periods of service. While much of the equipment, such as power-generating components, may be considered conventional, it is the frictional surfaces located within the influence of nuclear reaction that require special attention.

SCIENCE OF LUBRICATION

Practical lubrication is not an exact science. Although a given mechanical system portrays maximum performance under a set of constant operating conditions with a lubricant possessing specific qualifications, including fluidity and slipperiness, the selection of such a lubricant is not always feasible. The reason for this is that the operating factors, such as speed, load, and temperature, are not always constant. A lubricant that produces maximum results under a given set of operating conditions may not do so when one or more of those conditions have changed. Lubricant selection, then, should be made on the basis of obtaining optimum performance under all predictable operating conditions, giving priority to the lubricant which will perform best under the most usual conditions.

COMPROMISE FOR OPTIMUM RESULTS. Where a number of different kinds of machines are involved, even the latter generality is subject to some modification. Otherwise, the number of kinds of lubricants required for a given machine or an entire industrial plant may reach unmanageable proportions. It is thus obvious that the practical selection of lubricants for, say, a large industrial plant becomes a compromise in trying to obtain the best possible mechanical performance with the least number of lubricants. That is not to preclude the use of the ideal lubricant where the service of the frictional component warrants exacting considerations. But lubricant minimization should not be carried to a degree tending to jeopardize the protection of any frictional part.

The selection of a proper lubricant for a given requirement is of prime importance, but it is not the only consideration in the correct lubrication of a frictional part. Good lubrication also involves methods of application and care of the lubricant and lubrication system.

2

FRICTION

Friction opposes motion regardless of the shape, size, or nature of the bodies in contact. A journal rotating in a bearing, water flowing through a pipe, an airplane traveling through the air, or a wheel rolling over a road—all are examples of frictional conditions.

The principal causes of frictional development appear to be (a) molecular attraction between bodies in physical contact, (b) interlocking of opposing surface irregularities (asperities) present on all surfaces, and (c) surface waviness.

The amount of friction developed between two contacting surfaces depends on the following factors.

1. Type of frictional system
2. Pressure exerted between contacting bodies
3. Nature of the rubbing surfaces
4. Condition of the rubbing surfaces

The degree to which friction develops is indicated by

1. Amount of heat generated
2. Rate of surface wear of solid bodies
3. Power required to initiate and maintain motion
4. Ratio of power output to power input of any mechanical system
5. Degree of pressure drop along a given length of conduit during fluid flow

KINDS OF FRICTION

There are several kinds of friction. Included are sliding or solid friction, rolling friction, fluid friction, boundary friction and mixed-film friction.

Sliding friction is the force that resists relative motion between sliding solid bodies whose opposing surfaces are clean and dry, i.e., nonlubricated. Any two dry solid bodies in physical contact sliding over each other, such as iron on iron, iron on brass, and steel on babbitt, are examples of sliding friction. Dry sliding friction represents friction of the highest magnitude, generally.

Rolling friction is the force that resists relative motion between two solid bodies when one or both roll over the surface of the other. A ball or wheel rolling over a hard surface, such as a floor or road, is an example of rolling friction. Pure rolling friction represents friction of the least magnitude. Friction is theoretically nonexistent between perfect rolling bodies.

Fluid friction is the force that resists the flow of liquids or gases. Such a force opposes the sliding action, one over the other, of the numerous molecular layers that compose the fluid mass. Two solid bodies bearing on each other and having relative motion, whose opposing surfaces are fully separated by a fluid film, represent fluid frictional conditions. Fluid friction also exists between the molecular layers of a fluid (liquid or gas) flowing through a conduit. The magnitude of fluid friction is determined and reported in terms of viscosity, which varies for any fluid with temperature and pressure.

Boundary friction is the force that resists relative motion between two solid bodies whose opposing surfaces are wetted by a lubricant, but whose surfaces are barely separated by the lubricant film or adsorbed layer of contaminant. Protection of rubbing surfaces under boundary conditions is provided by such characteristics of slipperiness or adhesiveness as the lubricant may possess. Lubricant films with a magnitude of one or more molecular layers are conducive to the development of boundary friction. Fluidity of the lubricant creating the film is not an important factor under conditions defined here.

Mixed-film friction is the force that resists relative motion between two solid bodies whose opposing surfaces are partially separated by a full fluid film. Mixed film friction differs from boundary friction only in the degree with which a sparse film of lubricant is capable of maintaining a state of separation between opposing surfaces. Any increase in pressure caused by additional loading, or shock, sustained by the moving body tends to convert mixed-film friction to boundary friction. Under favorable operating conditions, however, mixed-film friction tends toward the development of fluid friction. Frictional magnitude under mixed-film conditions depends on the surface properties and the fluidity of the lubricant, and in some cases on the *lubricity* of the fluid. A hand-oiled bearing is representative of mixed-film frictional conditions after most of the periodically applied lubricant has drained from the rubbing area. Continued draining of the lubricant from the bearing, without replenishment, will ultimately lead to a condition of boundary friction and thence to nearly dry sliding friction. The latter usually occurs when surface temperatures rise sufficiently to drive all of the remaining fluid from the frictional area.

STATIC AND KINETIC FRICTION

The force required to initiate motion of a body at rest is called *static friction*. The force required to continue motion of a moving body is called *kinetic friction*. Static friction is always greater than kinetic friction for the same bodies under similar conditions. For this reason, it takes less effort to keep a moving body in motion than it does to initiate motion of the same body at rest. Newton's First Law of Motion, related to inertia, in some cases applies here, i.e., bodies at rest tend to remain at rest while bodies in motion tend to remain in motion.

FRICTION VS MECHANICAL EFFICIENCY

The forces of friction prevent any machine from producing an amount of work equivalent to the amount of energy put into it. The energy relationship of any machine may be expressed by the equation

$$\text{work output} = \text{work input} - \text{friction}$$

From this expression, the mechanical efficiency of any machine can be derived:

$$\text{percentage efficiency} = \frac{\text{work input} - \text{friction}}{\text{work input}} \times 100$$

or

$$\text{percentage efficiency} = \frac{\text{work output}}{\text{work input}} \times 100$$

and

$$\text{friction horsepower} = \text{horsepower input} - (\text{percentage efficiency} \times \text{horsepower input})$$

CONVERSION OF FRICTION TO HEAT ENERGY

Frictional energy is ultimately converted to thermal energy. The action of energy conversion obeys the *First Law of Thermodynamics* and the universal law governing the *Conservation of Energy*. These laws state:

First Law of Thermodynamics. *When thermal energy is transformed to mechanical energy, a definite amount of thermal energy disappears to emerge in like amount as mechanical energy; and, conversely, when mechanical energy is transformed to thermal energy, a definite amount of mechanical energy disappears to emerge in like amount as thermal energy.*

Law of Conservation of Energy. *Energy may be converted from one form to another, but it cannot be created or destroyed.*

In any machine, part of the input energy is converted from mechanical energy to frictional energy and thence to thermal energy. Since all other forms of energy may be converted to it, thermal energy is considered to be energy in its lowest form.

A simple gear train having an input of 100 hp and an output of 98 hp is said to have an efficiency of $^{98}/_{100} \times 100 = 98\%$. The difference of 2% represents energy lost as friction, which in turn is converted to thermal energy. The latter is borne out by the subsequent temperature rise that occurs in all mechanical work systems.

The amount of thermal energy developed in this gear train may be determined by the equation.

$$\begin{aligned}\text{Btu/min} &= \text{frictional horsepower} \times \text{mechanical equivalent}\\ &= F \times J\end{aligned}$$

where

$$\begin{aligned}\text{F} &= \text{frictional horsepower}\\ &= \text{horsepower input} - (\text{efficiency}\\ &\quad \times \text{horsepower input})\end{aligned}$$

Therefore

$$F = 100 - (98\% \times 100) = 2\text{hp}$$

and

$$J = 42.4 \text{ Btu/min/hp}$$

J represents the Joule equivalent, or the equivalency of thermal energy to mechanical energy. One horsepower is equivalent to work done in lifting 33,000 lb to a height of one foot in one minute. One British thermal unit is the amount of heat necessary to raise the temperature of 1 lb of pure water 1°F, as from 59°F to 60°F.

$$1 \text{ hp} = 33{,}000 \text{ ft-lb/min}$$

or

$$\begin{aligned}1 \text{ hp} &= 33{,}000 \text{ ft-lb/min}\\ 1 \text{ Btu} &= 778 \text{ ft-lb}\\ 1 \text{ hp} &= 42.4 \text{ Btu/min}\end{aligned}$$

To convert mechanical energy to thermal energy, multiply horsepower by 42.4 Btu. Therefore,

$$\begin{aligned}\text{Btu/min} &= 2 \text{ hp} \times 42.4 \text{ Btu}\\ &= 84.8 \text{ Btu/min, frictional heat}\end{aligned}$$

SLIDING FRICTION

Machine parts that develop sliding friction include a journal rotating in a fixed plain bearing, guides, ways, skids, thrusts, crossheads, and gears. Friction reaches its highest magnitude under dry sliding conditions. The reason for this is twofold:

First: Unwetted parallel surfaces permit maximum area contact when two solids are pressed together, a condition under which molecular attraction attains its greatest influence. The bonding force of molecular attraction increases with better surface smoothness because the area of actual contact becomes proportionately greater. Johannsen blocks are difficult to pull apart or slide across each other because of their high degree of surface smoothness. It it were possible to obtain perfect smoothness, any two such surfaces would adhere to each other permanently and (probably) would require the same force to separate them as it would to tear apart the solid material of which they were created. It is conceded, however, that a large amount of the force here is due to external pressure (atmospheric).

Second: The surfaces of all engineering metals, regardless of their "finish," consist of numerous asperities. The size of such asperities, both as to density and degree of irregularity, governs surface smoothness, or roughness. Surface irregularities are both jagged and wavy (Fig. 3-4). The jagged phase is composed of innumerable projections and depressions that cover the entire surface area. The waviness phase is developed by a series of waves that present the surface with an irregular contour. When two such surfaces slide over each other, a confliction of asperities occurs, tending to block the passage of one projection over another located on the opposing member.

If forced to slide on each other, the projections of the hardest member will shear its opposing projections. Shearing action becomes more aggravated as pressure between the two surfaces is increased. If the pressure is great enough, spot or local temperatures will reach a level high enough to cause the metal to soften. Plastic flow occurs when local temperatures reach the point of metal softening, permitting welding or bonding of the two surfaces. Seizure results as relative motion continues, followed by wiping of the bearing metal.

It is often necessary, or in some cases convenient, to prepare new bearing surfaces by a process known as "running-in." The purpose of the running-in process is to develop a higher degree of smoothness between opposing surfaces. This is accomplished by allowing the more prominent projections of one surface to shear off, or otherwise smooth out, the more prominent projections of its opposing surface during a period when the host machine is operated with or without a light load. A reduction in friction, improved operation, and longer life of rubbing parts is obtained by running-in newly assembled mechanical equipment.

To prevent bonding due to localized heating during the shearing action, the running-in lubricant may be compounded with an antiweld agent, such as sulphur or chlorine. An oily agent may also be added to relieve the additional frictional force generated by initial close fits, or virgin surface conditions, or both. By using sulfurized or chlorinated fats, both antiweld and oiliness characteristics can be imparted to the running-in lubricant. The use of such agents often results in reducing the running-in time compared with the time required with noncompounded mineral oils.

Waves are relatively wide irregularities, having a crest-to-crest dimension of $\frac{1}{32}$ in. or greater and a depth from crest to base of less than 0.0002 in. Roughness (finely spaced minute projections) dimensions are on the order of ten-millionths of an inch for highly polished surfaces up to more than 200 millionths of an inch for relatively smooth surfaces.

The laws of dry sliding friction are given in Table 2-1.

Table 2-1. GENERAL LAW OF FRICTION

Condition	*Sliding Friction*	*Rolling Friction*	*Fluid Friction*
Load	Approximately proportional to the pressure between the sliding surfaces	Proportional to the load	Nearly independent of the pressure
Speed	Nearly independent of speed. Varies slightly with different ranges of velocity. Less at high velocity	Nearly independent of speed	Proportional to relative velocity at lower speeds and to the square of velocity at higher speeds (above 1600 ft/min)
Surface condition	Decreases with greater surface smoothness	Dependent on the condition of surfaces. Decreases with smoothness of surfaces	Dependent on the degree of smoothness, or roughness. Distortion of fluid layers occurs with pronounced surface roughness
Nature of materials	Varies with different materials and with different combinations of materials. Greater with soft materials	Dependent on the nature of materials. Decreases with surface hardness	Independent of the nature of materials
Area of contact	Nearly independent of area of contact		Directly proportional to the contact area
Size of rolling member		Varies inversely as the radius of the rolling member	
Viscosity of fluid			Varies with viscosity (fluidity) of similar fluids, increasing as viscosity increases

ROLLING FRICTION

Theoretically, a ball establishes point contact with its bearing surface. A roller establishes line contact. Under pure rolling conditions, friction would not exist, since a point or a line has no area. Applied loads, however, tend to flatten the rolling member and also indent the bearing surface, or stator. Deformation of opposing members establishes area contact. The area of contact increases with greater loading and softer materials. Consequently, ball and roller bearings are made with hard steel. As a ball or roller rolls over a surface, it forms a band along its path, the width of which depends on the amount of flattening incurred on the ball, or the length of the roller. The bearing metal displaced as a result of indentation forms a temporary mound, or "swell," immediately in front of the rolling member, the front being the direction of rolling. An exaggeration of this condition is shown in Fig. 2-1. Rolling friction, then, is the sum of the resistances offered by the mound and the ever-present surface asperities. Some sliding action also occurs due to slippage of the rolling member in its constant effort to lift itself over the obstructions. Mathematically, the number of rotations a ball would make when traveling around a circle would be equal to the circumference of the circle divided by the circumference of the ball, assuming the circle and center line of the ball were congruent. Because of slippage, the distance that a rolling member of a ball bearing or roller bearing travels around its race is slightly greater than the mathematical value.

The difference between static and kinetic friction for rolling bodies is

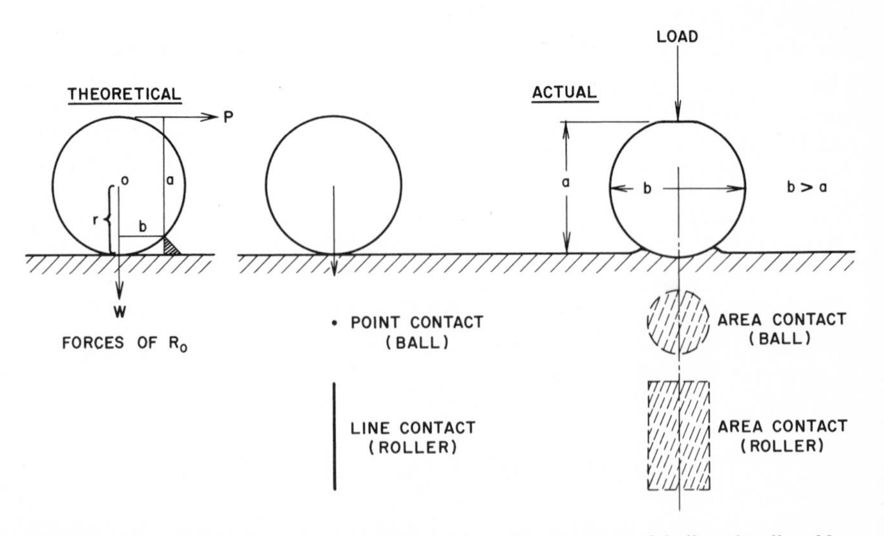

Figure 2-1. Theoretical and actual (exaggerated) surface contact of ball and roller. Note distortion of rolling element and surface (stator) under load. Resistance due to friction (R) equals the coefficient of friction (f) times the load (w) divided by the radius (r), or $R = fw/r$.

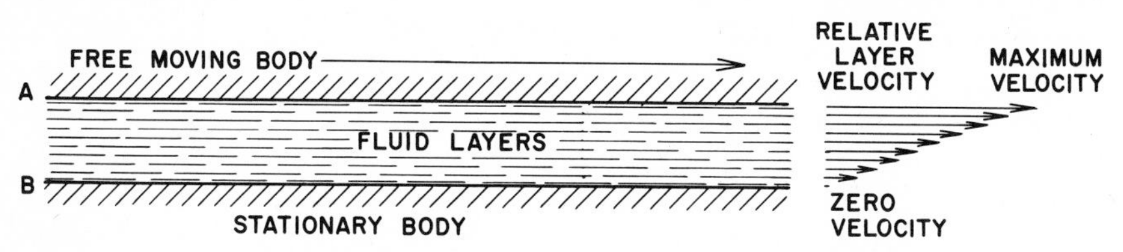

Figure 2-2. Demonstration of fluid friction portraying condition within a liquid interposed between a moving and a stationary surface. Velocity of layer at *A* is that of moving body. Velocity of layer at *B* is zero. Free body develops greater rate of velocity at *A* with liquids that have high fluidity (low viscosity) than with those of low fluidity (viscous).

slight, much less than that for sliding friction. Some general laws of rolling friction are given in Table 2-1.

FLUID FRICTION

The fluidity of a liquid (or a gas) is an indication of the force required to cause each layer of molecules to slide over its companion layers. Fluid friction is the force that resists the sliding action of the numerous molecular layers that compose the liquid mass (Fig. 2-2). Liquids that do not flow readily, such as castor oil, cylinder oil, and glycerin, have relatively high fluid friction. Rapid flowing liquids, including water, spindle oil, and alcohol, are fluids with relatively low fluid friction.

The action of the several molecular layers in sliding over each other is called *shear*. The speed with which they slide is known as the *rate of shear*. The rate of shear within a fluid mass governs its fluidity. The fluidity of a liquid is expressed in terms of viscosity. Viscosity is the measure of liquid fluidity, i.e., the measure of molecular friction within the fluid mass. The relationship of viscosity and fluidity can be expressed as follows:

$$\text{viscosity} \propto \frac{1}{\text{fluidity}}$$

Thus fluidity decreases as viscosity increases. Viscosity, then, is the measure of fluid friction within the mass. The laws governing fluid friction relative to bearings are given in Table 2-1.

COEFFICIENT OF FRICTION

Friction, being a force that resists motion, is measurable. Its value is equal to the counterforce required to initiate or maintain motion. Because friction varies with any change in condition associated with moving bodies,

Table 2-2. COEFFICIENT OF FRICTION FOR VARIOUS FRICTIONAL SURFACES

Frictional Conditions	*Coefficient of Friction f (Average)*
Fluid friction	0.001–0.005
Rolling friction:	
Ball	0.002
Roller	0.004
Sliding friction:	
Dry	0.15 –0.40
Boundary	0.080–0.10
Mixed film	0.020–0.08

the counterforce will also differ in each case. The value of frictional force remains constant only when all the prevailing conditions remain constant. When all the conditions except the load (i.e., the pressure between opposing surfaces) remain the same, the force of friction will also vary, but in constant relationship to the load.

The coefficient of friction also varies according to the kind of friction involved, as shown in Table 2-2. Comparing f for the various types of frictional systems establishes their relative efficiency.

SLIDING FRICTION. If a weight W of 10 lb were placed on block A (Fig. 2-3), and a pull of 6 lb, as determined by a spring scale, were required to initiate motion, then the static friction, being equal to the counterforce (pull), would also be 6 lb. If a drop in pulling force occurred, say to 4 lb, by continuing the block in motion, it would indicate that kinetic friction is equal to the pull of 4 lb. Any increase in weight placed on block A would require a correspondingly greater pulling force, provided that all other conditions remained the same.

By increasing the weight W to 20 lb, the respective values of static and kinetic friction would increase correspondingly to 12 and 8 lb. While the

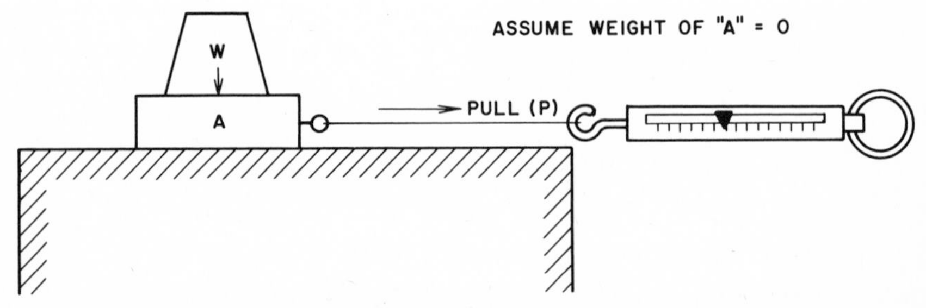

Figure 2-3. Apparatus for measuring coefficient of friction, f ($f = P/W$).

amount of pull varies with the load, their relationship remains constant as demonstrated below:

$$\text{ratio } \frac{\text{pull (lb)}}{\text{load (lb)}} = \text{constant}$$

Namely,

kinetic friction

$$\text{1st case: } \frac{\text{pull}}{\text{load}} = \frac{\text{friction}}{\text{weight}} = \frac{4}{10} = 0.4$$

$$\text{2nd case: } \frac{\text{pull}}{\text{load}} = \frac{\text{friction}}{\text{weight}} = \frac{8}{20} = 0.4$$

static friction

$$\text{1st case: } \frac{\text{pull}}{\text{load}} = \frac{\text{friction}}{\text{weight}} = \frac{6}{10} = 0.6$$

$$\text{2nd case: } \frac{\text{pull}}{\text{load}} = \frac{\text{friction}}{\text{weight}} = \frac{12}{20} = 0.6$$

The ratio of the force required to overcome friction to the load (pressure) imposed between the surfaces of opposing bodies is called the *coefficient of friction.* The coefficient of friction can be determined for any frictional system, whether it be sliding, rolling, or fluid friction. It is constant for any specific set of conditions, but it manifests the effect of any change in those conditions, as expressed by appropriate laws of friction (Table 2-1).

The coefficient of friction, f, can be determined in several ways, including the direct pull method (Fig. 2-3), or by use of the inclined plane (Fig. 2-4).

The relationship of the various forces present in the direct pull method is expressed by the equation

$$f = P/W$$

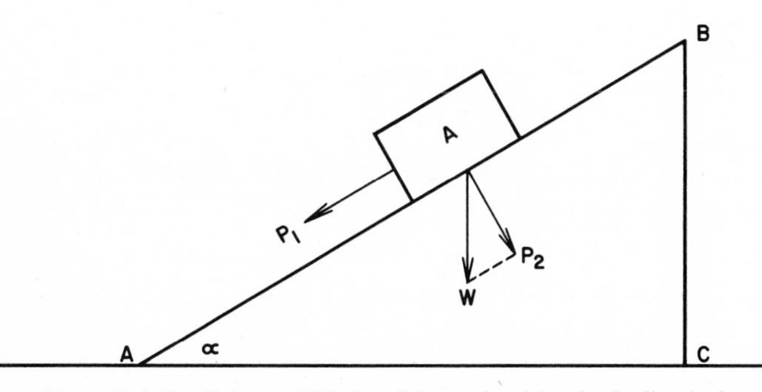

Figure 2-4. Coefficient of friction f determined by the inclined plane ($f = \tan a$).

and under constant frictional conditions and varying loads,

$$P = f \times W$$

where

$P =$ force in pounds (pull) normal to the surface, to overcome friction (lb)
$W =$ weight of movable body (lb)
$f =$ coefficient of friction

The coefficient of friction is an indication of frictional conditions between two rubbing surfaces, decreasing as the force of friction diminishes. Gravity supplies the pulling force in the inclined plane method of determining the coefficient of friction. Block *A* (Fig. 2-4) resting on the inclined plane surface *BA*, exerts force in two directions: P_1 and P_2. P_1, parallel to *BA*, is the force tending to cause block *A* to slide down the inclined plane as indicated by the arrow, and P_2 is perpendicular to the sliding surface *AB*. The sum of P_1 and P_2 equals the resultant force *W*, the total weight of block *A*. The relative force exerted by either P_1 or P_2 depends on the slope represented by the angle α. The force P_1 increases while P_2 decreases correspondingly as the angle α becomes greater. The distance *BC* also increases as the angle α becomes greater. Therefore, the force P_1 varies as the distance *BC*, reaching its maximum value when $\alpha = 90°$ and its minimum value of 0 when *AB* is parallel to *AC*. This is the same as saying that P_2 varies as the distance *AC* relative to *BC*. It follows that the forces P_1 and P_2 are functions of the relationship of the distances *BC* and *AC*, expressed by

$$\frac{P_1}{P_2} = \frac{BC}{AC}$$

Trigonometrically,

$$\frac{BC}{AC} = \tan \alpha$$

Therefore,

$$\frac{P_1}{P_2} = \tan \alpha$$

But P_1 equals the amount of force required to cause block *A* to slide, which force is provided by the pull of gravity. And P_2 equals the pressure exerted between the sliding surfaces at the time sliding began. P_2 then is the actual weight of block *A* bearing on the surface *AB* when the angle α becomes great enough to cause sliding. Therefore P_2 is the load.

By definition, the coefficient of friction *f* is the ratio of pull in pounds to the load in pounds, or

$$f = \frac{\text{pull}}{\text{load}} = \frac{P_1}{P_2}$$

Therefore

$$\frac{BC}{AC} = \frac{P_1}{P_2} = f = \tan \alpha$$

So that

$$f = \tan \alpha$$

when α becomes great enough to cause block A to slide on the surface BA in the direction of P_1.

The coefficient of friction varies with different materials or combinations of materials (Table 2-3). This means that the value of f will depend on the materials which compose block A and the sliding surface BA in the inclined plane (Fig. 2-4), or the corresponding parts shown in Fig. 2-3. For example, steel on steel has a dry sliding friction (kinetic) coefficient of about 0.50, whereas f for cast iron on cast iron is approximately 0.15. Steel on cast iron has a coefficient of friction of 0.25.

ROLLING FRICTION. It is easier to roll a barrel on its sides than to slide it up-ended. The effort to roll a large barrel is less than that required to roll a small barrel, providing the parameters are constant for both. In the case of any rolling element, ball or roller, once set rolling it would continue to roll indefinitely if certain conditions are met. Those conditions include

(a) Rolling elements are perfect circles
(b) Rolling way surfaces are level
(c) Rolling surfaces are perfectly smooth
(d) Rolling elements and surfaces are incapable of being deformed or indented
(e) Rolling occurs in a vacuum

The reason for the phenomenon of perpetual rolling in the case of perfect parameters is that, theoretically, rolling elements do not make *area* contact

Table 2-3. COEFFICIENT OF FRICTION OF SOME MATERIALS IN SLIDING CONTACT

Bronze on bronze	0.20
Cast iron on bronze	0.21
Cast iron on brass	0.20
Hard steel on hard steel	0.42
Mild steel on mild steel	0.57
Hard steel on babbitt	0.34
Mild steel on bronze	0.34
Mild steel on copper lead	0
Wood on wood	0.25 to 0.50
Metal on oak	0.50 to 0.60
Leather on cast iron	0.56

with their rollways. A ball makes *point* contact while the roller makes *line* contact. Where there is no area contact, friction cannot develop. Friction is the barricade on the road to perpetual motion.

Theoretical parameter values are not applicable under practical conditions. There are no perfect conditions, and if by chance one or more developed, practical operations would soon eliminate them. Antifriction bearings are therefore not frictionless because

1. The weight of the roller and the weight on it cause some deformation of the rolling element and the raceway. Sliding *friction* develops, which in the case of ball bearings includes the effect of spin.
2. The rolling element is in physical contact with certain parts of the bearing assembly that (a) keeps the individual rolling elements separated, (b) guides the elements around their periphery, and (c) maintains the position of separators within their locating race. Sliding friction develops in all cases.
3. *Spin* of balls at high rotating speeds to develop additional sliding friction.
4. *Hysteresis* losses.
5. Losses due to *shearing* of oil film, or of excess oil usually present.
6. *Windage,* a function of velocity of revolution.

Note that no reference is made to friction due to the act of rolling. Any parameter identified as "coefficient of rolling friction" is therefore a misnomer or refers to the effect of friction associated with a *system.* Even if the load on a rolling bearing approaches zero, the very act of motion develops friction with some measurable value. Because the force is a combination of sliding, hysteresis, shearing action, and windage, the measurable value is *torque,* or the force necessary to promote rolling motion. Torque of a rolling bearing is therefore finite.

The major factors that govern the torque magnitude are

(a) Size of bearing
(b) Type of bearing—roller (tapered, spherical)
—ball (radial, thrust, etc.)
(c) Geometry—relative sizes of parts
(d) Speed
(e) Load
(f) Lubricant—type, fluidity, quantity, etc.

These factors decide the parameters that compose the system.

To determine the theoretical force necessary to sustain pure rolling, only the parameters representing load and size of rolling member are considered, so that

$$F_r = W/r$$

where

F_r = force of rolling
W = load (lb)
r = radius of rolling element (in.)

If the values of W and r are applied to a rolling element, free of any of the frictional hindrances listed for rolling bearings above except deformation of contact surfaces, the theoretical equation must be modified to meet practical conditions. Surface deformation caused by rolling motion takes the form of (1) a depression in the raceway and (2) a flattening of the roller at the point of contact. The material displaced by the depression builds up to form a mound in the direction of rolling and to a lesser extent behind the rolling element. The tendency on the part of the rolling element is to shift its load from the line of centers to a line normal to the new bearing area or forward mound, which is the temporary high point of the rolling surface. The distance between the load line and the line normal to the mound depends on the elastic properties of the materials and the load magnitude. Rolling cannot be accomplished by a given rolling element on more than one plane simultaneously. Therefore, while the rolling element continues to roll along a depressed line parallel to the original raceway, it also tends to climb the mound. It never succeeds in leaving the maximum depth of the depression, however, because the displaced material continuously moves as a wave ahead of the rolling element. Rolling thus occurs only on the one plane, i.e., the one of greatest depth, while sliding occurs on every increment of surface area that forms the contact curvature of the mound. In a ball bearing, the curvature takes the form of the ball, concave spherical, while roller bearings form a straight concave surface.

The force F required to initiate or sustain motion is a function of the vertical distance of the crest of the mound from the load line, or eccentricity ε, i.e., according to the balance of moments,

$$F \times r = W \times \varepsilon$$

where

r = radius of the rolling element
W = load

or

$$F = \frac{W\varepsilon}{r}$$

which is the theoretical equation modified by the effect of deformation.

According to *Coulomb's Law*, the value of ε is the coefficient of rolling

Table 2-4. COEFFICIENTS OF FRICTION OF VARIOUS ROLLING BEARINGS

Bearing	*Coefficient of Friction*
Self-aligning ball	0.0010
Cylindrical roller with flange-guided short rollers	0.0011
Thrust ball	0.0013
Single-row deep-groove ball	0.0015
Tapered and spherical roller with flange-guided rollers	0.0018
Needle	0.0045

A. Palmgren, *Ball and Roller Bearing Engineering,* 3rd ed., SKF Ind.

friction. Transposing the value of ε becomes the coefficient of rolling friction f_r, or

$$f_r = \frac{F}{W/r} = \frac{F_r}{W}$$

If F is in pounds and r is in inches, then

$$F \times r = \text{torque (in.-lb)}$$

Note in the basic equation $F = W/r$ that the ideal rolling force varies (a) inversely to the radius of the rolling element and (b) proportionally to the load. It is nearly independent of the speed, which is not a factor in the equation. However, speed does exert some influence, as shown by tests, on the order shown below for a single-row self-aligning ball bearing at constant load:

$$250 \text{ rpm-}f_r = 0.001$$
$$1250 \text{ rpm-}f_r = 0.00125$$
$$1750 \text{ rpm-}f_r = 0.00133$$

The effect of speed is not great compared with the effect of load. However, the value of f_r does vary with the type of rolling bearing (Table 2-4).

3

WEAR

Wear is the transformation of matter by use. Mechanically, wear results in diminishing the dimensions of machinery parts. It is caused by the forces of attrition, abrasion, and erosion. Mechanical wear is the most important of four types of wear that convert useful materials to other substances, usually useless. The other three are chemical wear (e.g., corrosion, rust), bacteriological wear (e.g., decay and rot), and electrolytic wear (e.g., metal removal and pitting). Like friction, all four of these processes can serve useful purposes, such as, respectively, dimension control, etching with acids, cheese making, and electroplating. Although chemical, bacteriological, and electrolytic wear are usually less significant, they are involved in many instances and must be dealt with.

When useful material is converted to useless material by any of the four processes mentioned, particularly by mechanical and electrolytic wear, it may change in form but not in state. The *fines,* or debris, resulting from mechanical wear are the same material as that of the original mass, such as a bearing, from which they came, or an oxide of such materials. Metal removed from, say, a gear by electrolysis is, in many cases, deposited at some other location as nearly the same material as the gear. Such substances become useless because of change of form.

Also classified as wear is the destruction caused by fracture, whether from a sharp blow or from overloading, or even by confined expansion space due to elevated temperature.

When two metal surfaces rub hard on each other, producing heat, there is an exchange of surface metal, and this is called *cold-welding*. Cold-welding occurs when a bond developes between the tips of surface asperities of two surfaces when a pressure-temperature combination exists in great enough magnitude to cause plastic flow. The transfer of metal by welding is a form of surface destruction.

The remaining cause of destruction is material fatigue, usually inherent in metals after long periods of exposure to repeated stress. A rolling element, such as a ball or roller of a rolling bearing, moving along the arc of its race under maximum pressure followed by another ball or roller in continuous rotation, is an example of conditions conducive to metal fatigue, both in the element and in the race.

Regardless of the cause and degree of destruction, when a mechanical component is no longer capable of giving service to the extent required, it is said to be "worn out" and must be replaced. It is therefore necessary for lubrication personnel not only to provide the means of retarding wear by attrition, abrasion, corrosion, erosion, and electrolysis, but also to prevent sudden mechanical failure by timely replacement of components based on intelligent prediction of metal fatigue according to service life.

WEAR CONSIDERATIONS

From an industrial standpoint, wear is associated with and primarily caused by the relative motion of surfaces in physical contact. The efforts of the lubrication art must be directed toward maintaining separation of wear-potential surfaces and preventing the development of other destructive forces supplemental to friction that prematurely end the service life of mechanical parts. Thus the subject of wear must necessarily embrace all mechanisms apropos of motion that tend to render frictional components useless, except deliberate destruction, error in design, or metallurgical deficiencies. Wear must, therefore, be considered a complex process as evident by the various mechanisms that cause it, any combination of which may be exerted simultaneously.

MECHANISMS OF WEAR

Pure wear may be defined as the loss of material caused by the rubbing (sliding) of two bodies together unaided in any way other than their adhesion at or adjacent to asperity junctions, including abrasion caused by the debris resulting from adhesion. The mechanism of pure wear is harmful because the debris that results either becomes an abrasive that will remove additional surface material or "irons out" to fill the depressions and form part of and render smoother the existing surfaces without gross loss of material during the initial phase of sliding. Both actions are functions of the relative hardness of the surfaces in contact and form the basis of two kinds of wear mechanisms, *adhesive wear* and *abrasive wear*.

As stated previously, wear is a complex process often occurring as a result of some combination of the following mechanisms:

1. Adhesive wear
2. Abrasive wear
3. Corrosive wear
4. Fatigue
5. Fretting
6. Miscellaneous minor wear mechanisms (electrolysis, erosion, solubility)

1. ADHESIVE WEAR. Frictional surfaces are not continuous planes that define perfect smoothness, but are made up of numerous alternate asperities and depressions of various height and depth. When two such surfaces are rubbed together, they bear on the tips of the asperities, resulting in high local pressures and cold welding. The result of such metallic junction is a shearing process as relative motion continues along a parallel path. Wear occurs according to the nature of opposing surfaces which governs the strength of the junction and subsequent wear pattern. Adhesion (cold welding) and subsequent shearing of the asperities at the interface or within the bulk of either or both surfaces is the most fundamental type of wear.

If a hard metal surface is rubbed against a softer one, the asperities of the harder metal will plow furrows in the softer metal. The metal particles displaced from the furrows will often become transferred to the harder metal. In some cases, opposing asperities become so strongly welded that shearing does not occur at the junction interface, but causes grain displacement (*galling*), creating minute pits or craters in the softer metal. If mutual shearing occurs without adhesion, a self-smoothing process is initiated, providing that the sheared fines can be kept to microscopic size or flushed

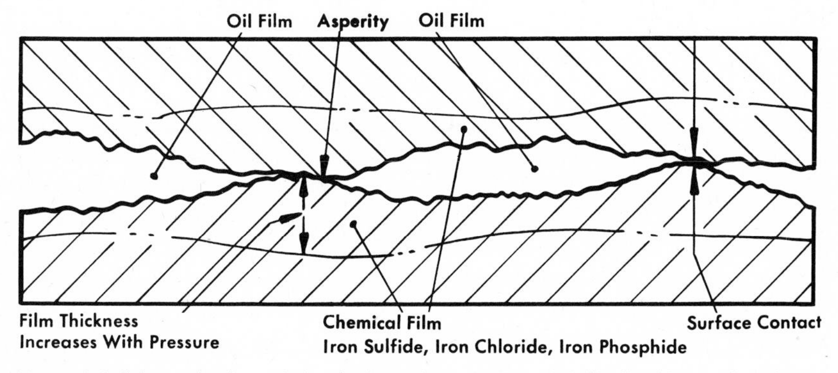

Figure 3-1. Schematic view of chemical reaction of EP rock drill oil with metal surfaces. *Courtesy, Henry A. Pohs, Gardner-Denver Co.*

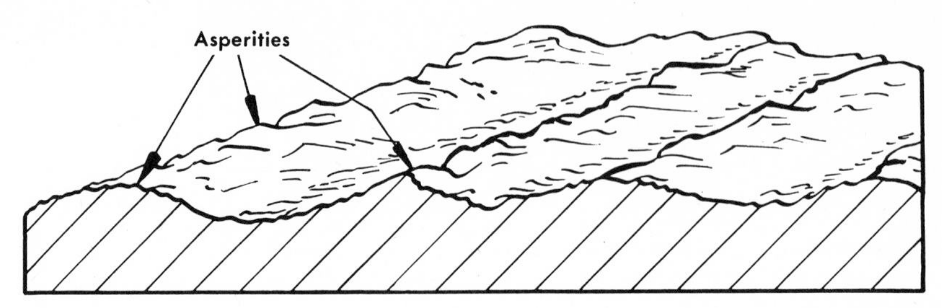

Figure 3-2. Enlarged schematic side view of a machined metal surface. *Courtesy, Henry A. Pohs, Gardner-Denver Co.*

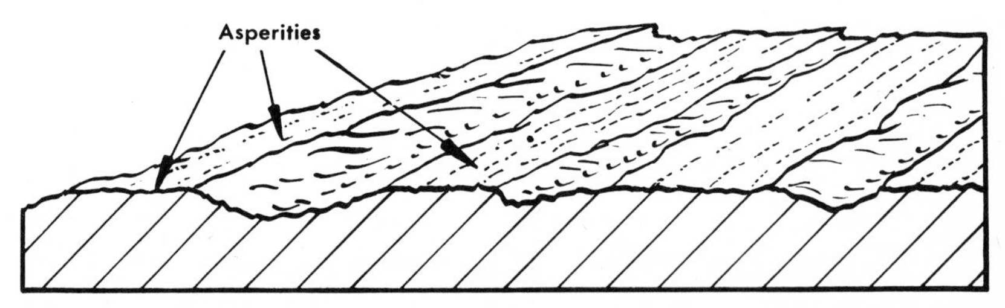

Figure 3-3. Same enlarged side view as Fig. 3-2 after grinding to provide a surface smoother than the machined surface. *Courtesy, Henry A. Pohs, Gardner-Denver Co.*

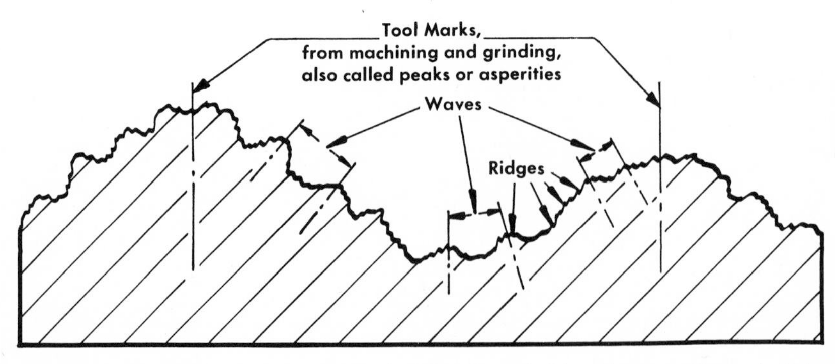

Figure 3-4. Exaggerated sketch of typical surface irregularities. *Courtesy, Henry A. Pohs, Gardner-Denver Co.*

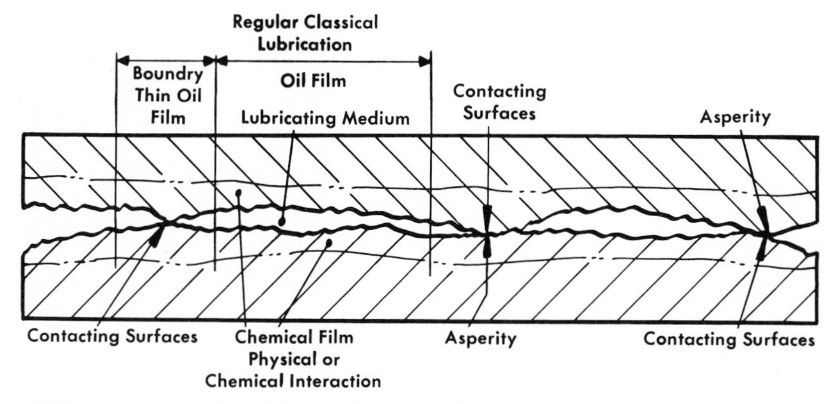

Figure 3-5. Two surfaces in contact, illustrating main ideas in lubrication theory. View simulates a pair of 8 rms microinch surfaces magnified 5000 times. *Courtesy, Henry A. Pohs, Gardner-Denver Co.*

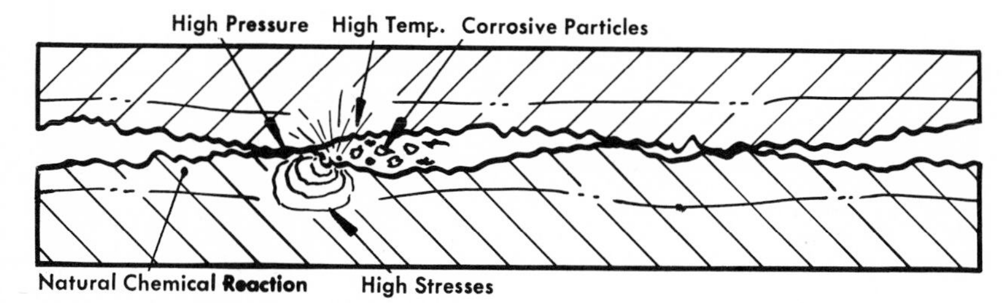

Figure 3-6. Schematic view of corrosion contribution to the phenomenon of wear. *Courtesy, Henry A. Pohs, Gardner-Denver Co.*

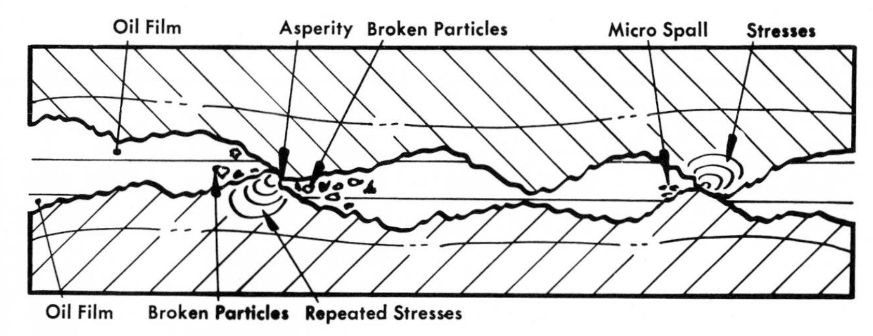

Figure 3-7. View of fatigue contributions to wear. These failures are due to repeated overstressing. *Courtesy, Henry A. Pohs, Gardner-Denver Co.*

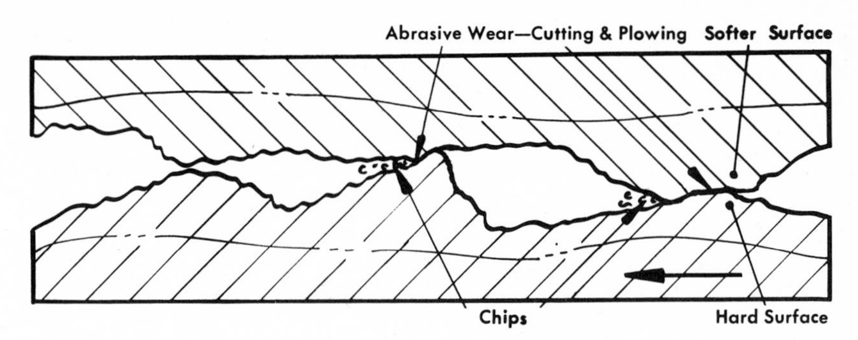

Figure 3-8. Abrasive wear as developed between hard and soft surfaces. *Courtesy, Henry A. Pohs, Gardner-Denver Co.*

from between the surfaces as they are produced. The process is accomplished during "break-in" periods involving such equipment as internal combustion engines and gears with the use of lubricants containing antiweld additives, such as sulfur, chlorine, and phosphorus. Some authorities believe there is evidence that adhesion at the interfaces is the major cause of friction on clean (nonlubricated) surfaces.

2. ABRASIVE WEAR. Abrasion occurs when a particle of hard material (oxides) is allowed to rub the surface of a softer material. Abrasive particles may accumulate by the shearing action occurring between two rubbing surfaces. The particles may come from the harder material or from a softer material containing hard surface particles. Abrasive materials may also enter the rubbing area by various means from an external source.

3. CORROSIVE WEAR. This type of surface deterioration is the result of chemical attack. Acid vapors within the surface environment or acids developed during oxidation of lubricating oils are sources of corrosive agents that cause etching. The decrease in surface area leads to losses in bearing area and causes a general weakening of the supporting component, be it

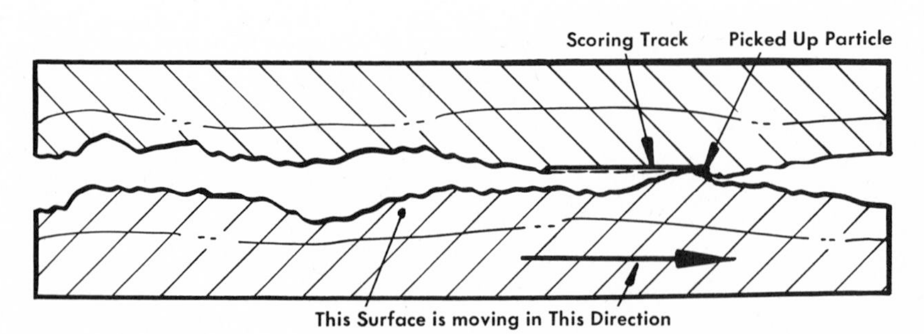

Figure 3-9. Scoring and picking up phenomenon, which leads to seizure. *Courtesy, Henry A. Pohs, Gardner-Denver Co.*

gear tooth or bearing. Corrosion can often be eliminated, especially when the cause is due to the lubricant. The use of oils containing suitable additives, such as corrosion inhibitors, is discussed in Chapter 10. Corrosion of frictional parts and associated components can also be reduced by adherence to a selected oil change program.

Corrosive vapors in the atmosphere may be isolated from a bearing by the blanketing effect of the lubricating oil. Another form of corrosion is caused by the oxidation of most metal surfaces. However, unless constantly disturbed, this type of corrosion is retarded by the oxide layer developed during the oxidation process, which acts as a self-protecting film.

4. FATIGUE. Cold drawn steel wire will eventually break at the point where it is subjected to bending a number of times. The stresses developed during the successive bending operations finally weaken the wire structure to the point of fatigue, causing breakage. All mechanical components subject to changes in stress, particularly those in which the stress causes structural deformation, will eventually reach a point of fatigue and will fracture, or break. As more fully discussed in Chapter 13, bearing life is a function of stress magnitude and frequency. This leads to bearing life predictions in operating hours based on load and speed (rpm). The bearing life factor changes with any change in either or both of these factors.

In sliding bearings repeated loading and unloading often causes surface failures in the form of insidious cracks followed by a major breakdown of the supporting structural mass. Repeated deformation due to stress cycling usually affects fatigue initially in the subsurface region of the bearing followed by surface blemishes of various dimensions. This means that the insidious cracks mentioned above are not a forecast of what will happen beneath the surface but a warning of what has already happened in that region.

5. FRETTING. Fretting, often referred to as false Brinelling, is a type of corrosion occurring on the surfaces of two metals in contact (Fig. 3-10). It is usually caused by an almost imperceptible motion resulting from vibration. It causes deterioration of the surfaces at the area of contact by the formation of debris, mostly oxides. The answer to the fretting problem is not lubrication, but lies either in the elimination of relative motion with respect to the opposing surfaces, or in plating the surfaces with materials that will resist the corrosive action of fretting because of their low oxygen absorption rates.

Metals that absorb oxygen at substantially high rates, such as stainless steel and aluminum, are highly susceptible to fretting corrosion. However, fretting also occurs with some nonoxidizing materials, so that the term "fretting corrosion" appears to be a misnomer.

6. MISCELLANEOUS WEAR MECHANISMS. In addition to the wear mechanisms already discussed, there are others that occasionally appear, electrolysis, erosion, and solubility. In some instances, the activity of such

Figure 3-10. Wear due to fretting corrosion.

mechanisms is readily identified while in other cases damage is difficult to trace.

Electrolysis. Surface damage by electrolysis takes the form of pitting and sometimes etching. It is caused by the flow of eddy currents originating as leakage from a main electrical power source, or from some process generating static electricity. The loss of metal due to the metal plating activity accompanying electrolysis is usually concentrated in an area through which the electrical current passes on its way to a ground terminal. Surface damage by electrolysis is often erroneously diagnosed as chemical corrosion, or galling.

The author investigated the cause of pit-like blemishes on a nichrome-steel shaft operating on a high-speed printing press. The journal bearing was lubricated by a grease which the operators believed to contain an acid causing damage by corrosion. The cause was found to be a static electrical current generated by the paper sheets sliding on each other in a dry atmosphere. The problem could have been solved by installing an air conditioning system (humidifier for winter and dehumidifier for summer) to maintain a relative humidity of about 60%. Instead it was solved by grounding the frame at a point near the source of static current.

Leakages from power sources require more serious attention because they result from short circuiting, which may cause lethal shock and/or result in fire. Currents of this kind must be held within the circuits designed to carry them by proper insulation and grounding. Electrolysis often occurs on antifriction bearings, rolling elements of rolling bearings, and gears.

Erosion. The removal of surface material by erosion is the result of incessant impingement of particles of matter. The damage occurs in a form similar to that of sand-blasting, but may also have the appearance bordering on systematic abrasion. Erosion often appears on high-speed gears whose teeth plough through a lubricating oil bath or which are lubricated by an oil spray whose particles continuously impinge on the tooth surfaces traveling at extremely high speed. A lubrication system designed by the author to reduce or eliminate gear tooth erosion is described in Chapter 11. It involves lubricant misting along with a refrigerant to achieve the dual purpose of gear case cooling.

Where erosion is caused by a limited number of separated solid particles continuously present, the surface roughness may be characterized by deeper cavities, especially where local spots are repeatedly impinged upon.

Solubility. Some metals are soluble in oils under certain circumstances. The relationship of oil to copper in the presence of refrigerants is described in the discussion of "Copperplating" in Chapter 21.

SURFACE RESPONSE TO WEAR

The field of lubrication deals generally with solid surfaces, usually metallic. Plastics, rubber, and wood also account for numerous frictional parts, such as gears and bearings. Some of the materials used are soft and ductile, while others are hard and in some cases brittle. A number of them possess such properties that fall somewhere between the two extremes.

Components that comprise a kinematic pair may be of similar materials, such as steel on steel, as in most gears. However, journal and thrust bearings in general are made up of dissimilar metals, such as steel on bronze and steel on babbitt. As described in Chapter 12, many bearings are made by combining one metal with another, either as a mixture or a matrix. Worm gears are made with both the worm and wheel of similar metals, but more often the worm is steel and the wheel is bronze. Cast, malleable, and ductile irons are also used, particularly for gears. Journal materials differ generally only in the type of steel used and metallurgical treatment, although other materials are used for special purposes.

Thus there are many combinations of opposing frictional surfaces. All of such surfaces must perform smoothly and efficiently and must have an extended service life. Each combination exhibits individual frictional and wear characteristics. Some combinations when rubbed together dry (non-lubricated) portray a coefficient of friction that is relatively high while others show lower values. The same may be said of the wear rate for various material combinations that form a given pair.

While the resistance to wear is interdependent on surface hardness, practical considerations involving abrasives tend more toward softer, or even elastic bearing materials. It is for good reason then that the selection of hardened steels for gears and journals on the one hand and bronzes and babbitts for bearings on the other has become universal practice.

Except for strength and in many instances ductility and flexibility, many other surface properties diminish in importance with correspondingly greater hydrodynamic film thickness and effective oil filtration. This is not so, however, under conditions of dry boundary or mixed film lubrication, including "startup" and "shutdown" intervals under nonhydrostatic operations. In the latter situations, adhesive wear particularly, and abrasive wear generally, develop, making it desirable to provide a combination of appropriate surface properties and lubricant characteristics that will maintain a low coefficient of friction and minimize wear.

Because friction has a high correlation to adhesive wear, the effort to reduce both must be directed toward the magnitude of shear stress and the region of shear relative to the rubbing surfaces under load. This may be done ideally by the selective pairing of (1) materials between which the adhesive bond is minimal and (2) lubricants that provide surface films of sufficient thickness to cushion low magnitude asperities on one hand and act as an anticohesive and oiliness agent on the other.

Wear is minimized when shear occurs at the point of adhesion or junction of opposing asperities. If the junction is weaker than either of the surface materials, then shear occurs in the region of the junction, which becomes the actual interface.

Wear is maximized when shear occurs in the mass of metal composing the surfaces. If the junction is stronger than both metals, shearing will occur in the mass of both, but to a greater extent in the weaker one. This encourages extensive wear on the metal, requiring a lower shearing stress. The latter is also true when the junction is stronger than only one of the opposing metals, except that the debris of the sheared metal will tend to "iron out" and adhere to the stronger metal surface or become loose debris.

In the case of opposing surfaces of the same metal, and in fact with any combination of relative stress values between metals and junctions, grain displacement during shear and adhesion will harden or soften the metals compared with their original states. The process is known as *work hardening* or *work softening,* respectively. Their mechanisms are described elsewhere in this text. In work hardening, shearing occurs generally in the weaker surface mass in combinations of dissimilar metals, but it occurs in the mass of both surfaces when similar metals are opposed to each other. This is because work hardening strengthens the junction and leads to a higher wear ratio.

Work softening weakens the junction. When the process causes both metals to become stronger than the junction, the results are similar to those described above for relatively weak junctions, i.e., wear is minimal.

Bearings, as described under the subject elsewhere herein, are variously prepared to combat the forces of friction and wear. A number of materials or combinations of materials are used in the construction of plain axial and radial type rotative bearings. In some instances, lineal (slider type) bearings are used to accomplish either or both of the latter advantages, depending on the priorities and demands of the service. To obtain the desired performance capabilities of the metals or combination of metals employed, they may be associated within the bearing construction either as (1) a nearly homogeneous mass (such as a bronze sleeve), (2) a matrix (in which a hard metal is supported in a soft matrix), (3) a heterogeneous mass (in which one or more different metals or other materials are interspersed in another), or (4) a laminate (consisting of two or more materials bonded together in an overlay sequence).

WEAR FACTORS

Adhesive wear is believed to be the major cause of friction. It is also the most fundamental type of wear. Wear is interdependent on surface hardness when considered in its basic forms of adhesion, abrasion, and surface fatigue. It is a function of applied pressure and relative velocity, increasing as either or both are increased. In this relationship, adhesive wear is correlated with surface roughness or the magnitude of asperities that exist on such surfaces. It follows then that friction causes wear, which appears to be paradoxical to the hypothesis that adhesive wear is a major cause of friction. The answer to this lies in the influence of cold-welded junctions of the contacting asperities formed during the rubbing process on both adhesive wear, and the retarding effect of motion, which in this case is defined as friction.

It was previously stated that the adhesive wear pattern was a function of junction strength versus surface metal strength. It was also implied that, by providing certain combinations of metals conducive to shear at the junction interfaces, minimum wear becomes a probability and often a fact. By introducing lubricants capable of weakening or preventing asperity juncture to the rubbing surfaces, adhesive wear can be retarded even under extreme boundary conditions.

Wear, then, can be a complex process involving cold welding or adhesion, subsequent shearing, ploughing, abrasion, corrosion of virgin metal, metal surface fatigue, and plastic deformation (work hardening). Of the four primary causes, adhesion is the most difficult to control and therefore the process mostly responsible for surface wear. It can be greatly retarded by (1) providing a sufficiently thick lubricating film through hydrodynamic or hydrostatic action, or (2) by introducing an antiweld agent by means of a lubricant capable of preventing the formation of strong junctions. Wear by adhesion alone is a long gradual process and does not account for the unexpected and early destruction of frictional surfaces that cause sudden mechanical failure. An adequate supply of filtered lubricant that will reduce the coefficient of friction, maintain safe temperature levels, provide a continuous hydrodynamic film of sufficient minimum thickness, and adhere to the frictional surfaces during periods of rest will not only retard adhesive wear, but also prevent corrosive and abrasive wear, the latter two by blanketing and flushing, respectively. Because fatigue is a function of continuous stress frequency, it poses little danger to other than rolling bearings.

Wear, friction, and thermal effects represent the three major factors toward which the role of lubrication is directed. The beneficial results of effective lubrication are expected to include (1) extended service life, (2) greater ease of relative motion and (3) maintenance of an optimum operating temperature resulting from a frictional system. The performance of a lubricant, then, is based on its ability to (1) retard wear, (2) reduce friction, and (3) control thermal effects. In addition to lubricant consideration, all three factors are influenced by bearing design, including length-to-diameter ratio involving side leakage in narrow bearings and journal deflection in wide bearings. Clearance ratio and surface finish also affect frictional performance of bearings, as determined by the dimensionless formula ZN/P, described in Chapter 9. The latter is a criterion of film thickness which is correlated with wear, the coefficient of friction, and the mechanical-to-thermal energy conversion pattern.

ACTUAL CONTACT AREA VS SWEPT AREA. All surfaces are composed of a series of asperities and depressions superimposed on a series of waves. The roughness of surfaces depends on the wave frequency and the magnitude of the peaks and valleys. When two surfaces are in contact unimpaired by intervening contaminant films of any kind, the weight of one is carried on those areas formed by junctions of those asperities and waves whose heights protrude above a line representing a nominal surface with a perfect plane. The actual area in dry, stationary contact is thus the sum of the areas of those peaks and ridges capable of juncture. The actual area of contact is therefore only a small fraction of the total swept area of the opposing surfaces. The remaining areas are a series of clearance gaps having various headroom magnitudes, ranging upward from about 100 angstroms. (One angstrom unit = 1×10^{-8} centimeter or 1×10^{-4} micron (μ).)

The surface profile is fragile, changing as the pressure between opposing surfaces increases and more so when relative motion occurs under any but ideal hydrodynamic or hydrostatic conditions.

The load being borne on the peaks, especially on those superimposed on the crest of the waves, tends to deform the supporting structure (asperities), causing elastic deformation whose area of contact is expressed in terms of diameter by the Hertz equation, in which the load and Young's Moduli for both surfaces are important functions. Elastic deformation is followed by plastic flow under increased load, during which cold-welding, better described as adhesion, occurs between areas in actual contact. As the load becomes greater, the subsequent squeeze action causes peaks of correspondingly lesser magnitudes to come together, thus increasing the actual area of contact. The result is a relatively smoother surface, accompanied by a temperature increase. Under extreme loading, cold-welding is converted to hot-welding in some local areas, so that what is accepted as a normal adhesion force is substituted by seizure when rubbing begins. On one hand, adhesion tends toward a more congruent interfacial contour over the entire contiguous area, while on the other hand seizure tends to leave this course and develop blemishes, which reduces the bearing area potentiality. Such seizure should not be interpreted as *scuffing* or *wiping*, which are respectively due to abrasion and distortion, during relative motion of opposing surfaces. Neither should it be confused with *pitting* or *cratering* due to displacement of granular metal by localized shock, which is analogous to the loosening of a small pebble from a concrete surface by something akin to hammer blows.

RUNNING-IN PROCESS

From the foregoing discussion we see that a fresh bearing surface should be carefully treated to avoid pronounced rupture of the asperities and avoid cleavages below the desired interface. This would cause permanent surface damage and interfere with the normal activity of the gradual adhesive process. The wear process becomes greatly aggravated when the preparation of surfaces is accomplished by severe machining operations that cause deep work hardening. In the latter case, the junction or weld will fail to shear at the interface, cause rupture well below the surface of the weaker metal and pull out work hardened sections, leaving deep blemishes. The metal so removed will form debris that will become abrasive, to the further detriment of the surface.

To avoid damage developed during initial stages of rubbing, virgin surfaces should not be subject to their full design load and speed capacities. With a procedure known as "running in" (wearing in or breaking in), work is performed under light loads, moderate speeds, and an adequate supply of lubricant. During the running in process, metals not subject to severe work-hardening will shear off at the junction interfaces, gradually reduce the magnitude of the asperities and retard wear. Surfaces that have been work-hardened during their machining process should be further prepared by annealing to relieve the hardening stresses and return the crystals to their uniform size. The running in should not be confused with a *polishing* operation. Polishing is a form of abrasive wear accompanied by elevated temperatures. The abrasive materials to be effective must have a higher melting point than the surfaces being polished.

Polishing does not usually develop what is known as a *Beilby layer*. This surface layer is formed by a combination of metal, metal oxide and in some instances crystallites of various materials, such as graphite and molybdenum disulphide, subject to alternate melting and solidification of the metallic debris. The presence of a Beilby layer requires greater pressure to bring about a strong weld at asperity interfaces so that any resulting surface blemishes will form at a lesser depth.

CAUSES OF SURFACE DAMAGE

Surface damage appears in different forms and patterns and varies from slight scratches to complete destruction by seizure or breakage. Some of the causes of surface damage are listed below:

1. Lack of lubricant
2. Incorrect lubricant
3. Misapplication of frictional part
4. Misalignment of frictional part
5. Metallurgical weakness
6. Initial surface roughness
7. Overloading
8. Overspeeding
9. Inadequate heat conduction
10. Incorrect clearance
11. Shock
12. Vibration
13. Insufficient backlash (in gears)
14. Abrasives
15. Electrolysis
16. Corrosive environment
17. Bearing warping
18. Loose bearing, or bond

Some of these causes apply to all frictional components, while others are inherent to specific groups, such as gears. Those that are common to ball and roller bearings are described in Chapter 13. Surface damage to gears and causes is covered in Chapter 14. Those that occur with plain bearings, including both rotating and sliding types, are given in Chapter 12.

Surface damage is identified by individualistic marking or pattern and described by a word that pictures the type of blemish or damage. For example, a *scored* bearing or gear tooth is one on which the surface has been cut by a rather sharp object, such as a hard protuberance (asperity) located on a mating surface. The action is known as scoring (Fig. 3-11).

Another kind of surface damage is *etching,* which is caused by a corrosive substance in contact with a frictional part. Etching is usually identified by a random and sometimes exotic pattern etched in the surface to various depths, depending on the strength of the corrosive and the length of contact time.

As an introduction to the nomenclature used in regard to wear and other kinds of surface damage, the terms most used are given here:

Pitting (incipient and destructive)	Seizure	Brinelling
Abrasion	Erosion	Crowding
Scoring	Fretting	Interference
Galling (mild and advanced)	Fatigue	Flaking
Spalling	Scuffing	Etching
Corrosion	Burning	Rolling
Wiping	Polishing	Peening
Gouging	Scraping	Featheredging
	Cracking	Rippling
	Fracture (breakage)	Scratching

Some of these terms are extensions or subdivisions of others that represent the more basic concepts. However, in the interest of being more selective in surface damage description some redundancy appears desirable. For example, scuffing may describe a series of scores or cuts appearing close to each other. It may also be described as mild scoring. Scoring is also identified as galling. This is not entirely correct in that scoring is a cutting action, while galling is defined as wear (or chafing) due to rubbing. Scratching and gouging are forms of scoring at low and high magnitudes, respectively. Lapping may follow the initial breaking off of original surface asperities to stabilize the resulting pitting action.

The major causes of frictional surface failure may be reduced to the five listed below:

1. Original surface deficiencies, i.e., roughness and physical damage due to careless handling or installation.

2. Mechanical damage during operation (shock, vibration, misalignment, foreign debris, particles of broken teeth in gears).

3. Metal fatigue as a result of (a) metallurgical deficiencies (softness, brittleness), (b) unsuitable materials, (c) depletion of life expectancy (espe-

Figure 3-11. Scoring of inner ring caused by "creep."

cially in rolling bearings), and/or (d) excessive overloading (pressure-speed factor).

4. Lubricant contamination by abrasives, corrosives, and/or advanced deterioration (neutralization number).

5. Lubricant deficiency (incorrect lubricant, insufficient lubricant), either from a sparse principal supply, ineffective application (misdirection, incorrect grooving of bearings, line blockage, etc.), or complete absence of lubricant (broken feed lines, pump failure, low oil level, neglect, etc.).

Any of these surface damage patterns may occur, but the basic ones that appear most frequently are

Pitting	Fatigue (surface and subsurface)
Abrasion	Plastic flow or wiping
Spalling	Fracture
Galling	Corrosion
Scratching	

All of these patterns are causes of *destructive wear.*

Most of the surface damage patterns occur in all frictional surfaces, while some, including erosion, featheredging, rippling, and rolling, are more common with gears.

Gears usually fail as a result of overloading, abrasives, improper mounting, or defective related parts, such as bent shafts, inadequate tooth contact, and worn or mis-aligned bearings. Exposing gear teeth to stresses that promote (a) continuous bending and (b) excessive compressive stresses causes scuffing, scoring, pitting, and galling and leads to gear failure. Bending fatigue results in fracture.

To eliminate many of the causes that lead to surface destruction requires appropriate remedies, which include these five:

1. Avoid overloading.
2. Assure correct mounting.
3. Supply an adequate amount of suitable and clean lubricant.
4. Operate within reasonable limitations.
5. Provide preventive maintenance service.

EFFECT OF LUBRICATION ON SURFACE DETERIORATION

Metallurgical qualities of frictional components are the responsibility of the metallurgist, while the dimensions of such parts depend on the designer. The physical condition of frictional surfaces is the combined responsibility of the manufacturer, installation mechanic, and maintenance personnel. Respectively, they control the smoothness, handling, and operational care of bearings, gears, etc. If a part is properly prepared in all respects, correctly installed, and judiciously operated, its service life expectancy will depend on lubrication considerations. Such considerations may be summed up by the golden rule of the lubrication art: *Effective lubrication depends on the proper application of an adequate amount of correct lubricant at the time it is needed and in a clean condition.*

Under ideal operating conditions, an antifriction bearing with a life expectancy of, say, 6000 hours at a given load and speed will perform satisfactorily for at least that number of hours. Gears last for years under ideal operating conditions when properly lubricated, even though they are subject to short periods of abnormal operation, such as overloading, shock, and vibration. A correct lubricant can combat some types of frictional surface abuse, but not continuously for long periods. There are certain deficiencies, however, that even the most selective lubricant cannot overcome.

Many of the surface deformities listed above are due to physical deficiencies and external forces that are completely independent of the lubricant. A number of them are the result of deficiencies in the process which they serve, while others are caused by design and metallurgical shortcomings.

Correct lubricants may retard the normal wear to which all moving frictional surfaces are vulnerable. Where hydrodynamic films are developed, the lubricant may combat abrasive wear if hard particle sizes remain below the minimum film thickness which they would have under normal wear conditions. Larger abrasive particles may be flushed from between opposing surfaces and prevented from compounding abrasive damage if the oil is in constant circulation. Oil will act as a cushion against reasonable shock and vibration to prevent surface blemishes from occurring. However, oil will also enter and become entrapped within any existing incipient cracks to aggravate surface blemishes by acting as a hydraulic agent under heavy loading.

Normal wear is the only cause of ultimate surface damage that a lubricant is able to combat. The lubricant minimizes surface damage in several ways:

1. Reduces static and dynamic friction by containing *polar* additives where circumstances dictate, such as under boundary or quasi-boundary (mixed-film) conditions.

2. Maintains flooded or *hydrodynamic* (or hydrostatic) films by having the proper viscosity.

3. Prevents welding under extreme pressure conditions by containing anti-weld agents.

4. Establishes strong lubricating films to reduce metal-to-metal contact through the use of *antiwear* additives that become active under higher rubbing temperatures developed by heavy pressures.

5. Introduces corrosion inhibitors to protect surfaces against chemical deterioration.

6. Dissipates heat when supplied and circulated adequately.

Lubricants are most effective when they are (1) kept clean by filtration, (2) freed from moisture by settling or centrifuging, (3) changed periodically, (4) analyzed regularly, and are (5) subject to a program of preventive maintenance.

4

SURFACES

Two case histories concerning friction and wear on the cylinder walls of an internal combustion engine illustrate an interesting phenomenon related to metal surface behavior and friction.

First Case. Some years ago a prominent automobile manufacturer, in an effort to combat friction and wear and increase engine life, honed the cylinder walls to effect a high degree of surface smoothness, described as micro-finish, or mirror finish. The result was not successful.

Second Case. During World War II, the engine of a captured German Stuka dive bomber was inspected by another prominent car manufacturer to evaluate certain physical aspects related to its excellent performance. One point of curiosity surrounded the condition of the cylinder walls, which had been subjected to tremendous forces when the pilot pulled out of a dive and abruptly climbed back to a safe altitude, the entire maneuver executed at very high speed. It was found that the originally smooth surfaces of the cylinder walls had been lightly sandblasted during its manufacture, and they were still in excellent condition.

The reason for the difference in the degree of performance obtained by these two surfaces, representing extremes of smoothness in one case and of roughness in the other, lies in the following discussions covering surface interactions, friction, wear, and lubrication practices.

SURFACE MATERIALS

The science of lubrication deals with *solid surfaces*. The surfaces involved are those exposed areas that guide the motion or receive and/or transmit the load for the masses forming the various frictional components of machinery. These loads are dynamic, so that such surfaces are exposed to or expose other surfaces to *relative motion.* Relative motion between surfaces in contact with each other precipitates either a sliding or rolling action, or a combination of both. Surfaces exposed to sliding or rolling motion are the objectives of lubrication. Examples of basic machine components with moving surfaces in contact are

Common bearings and journals
Ways and carriages
Crossheads and guides (bed plates)
Gear teeth, or worms, and their mating teeth
Cams and followers
Piston rings and cylinder walls
Balls and rollers and their races
Chain links and pins
Wire rope strands
Chains and sprockets
Valve stems and their seats

Most machine load-carrying surfaces with relative motion are made of various metals. Each kind of metal has its particular intrinsic properties, some of which are inherent, while others are imposed by special treatment or preparation. Many of them are alloys, fabricated by combining two or more kinds of metals or by adding various nonmetal elements to impose specific properties in either a basic metal or an alloy. Prominent alloys used to prepare friction surfaces include steels, bronzes, brasses, babbitts (white metal), hardened lead, cadmium-base, aluminum-base, and copper-lead. Cast iron and silver are also employed in the friction field.

Other metal surfaces are prepared on solids produced by compressing metal powder that is sintered at high temperature to fuse and develop a strong porous material. Porous bearings are of various composition, such as aluminum, copper and tin, and iron and copper.

There are some nonmetallic solids used on which friction surfaces are prepared. They include wood (lignum vitae, hard maple), stone, synthetic materials (Bakelite, nylon, Teflon), glass, rubber, and carbon graphite. A more detailed discussion of frictional materials is presented in Chapter 12.

SURFACE STRUCTURES

The frictional surfaces of machine components, act as a first-line support for the mass that forms the load-carrying components, such as those previously listed. There are four types. First, they may be prepared directly on the mass by appropriate machining to achieve the desired profile, in which case the surfaces possess all the properties of their host material. Gear teeth that are projections of the gear blank, cylinders, and ways are examples of such surfaces.

The second type is exemplified by the inside area of cylindrical inserts, such as liners, sleeves, and bushings, which present surfaces on materials that may or may not be the same as the machine member into which they are inserted. (The problem of compatibility of dissimilar metals arises where careful selection of insert materials for certain machine components is not practiced, especially in the case of pressed fits.) The surfaces of inserts will possess the properties of the material on which they are prepared.

A third type of surface is the exposed area of a thin strip of desired material, either laid on a backing shell to provide rigidity (bimetal), or applied directly on the component metal. In either case, the supporting material is made to conform to the design profile of the frictional surface. In some cases, a third material differing from both the surface material and backing strip is introduced between the two to form a trimetal lamination. The purpose of the third material is to provide additional properties (e.g., conformability) required by design considerations.

The *matrix*-formed surface is a fourth type, with characteristics that often differ from the individual materials that form the surface. Copper and lead are among the metals used in one variety, in which particles of lead are distributed within a copper matrix. In another variation, the lead forms a continuous network throughout the copper phase. Lead, tin, and antimony are other material combinations used in surface alloys of this type.

Surfaces behave to a large extent in a manner governed by the volume properties of the materials on which they form the exposed areas suitably prepared for their intended purpose. Some materials, such as cast iron, are hard and brittle, while others, like lead, are soft and pliable. Still others possess these properties in varying degrees, ranging between the two extremes. There are many other properties, however, in addition to the two mentioned above that govern the selection of a frictional material. What is said of the mass material may, in most cases, be said of their frictional surfaces. Thus, the laws governing the behavior of a particular volume also apply to its surfaces.

PROPERTIES OF MATERIALS

Properties of materials are associated with energy systems and describe their behavior according to the kind of energy involved. There are a number of kinds of energy, but the kinds shown in Table 4-1, along with their associated properties, are of particular interest here: electrical, chemical, thermal, and mechanical.

Magnetic energy is often and correctly described and listed as a separate kind of energy apart from electrical energy. It is listed under electrical energy here for logical and convenient reasons.

Table 4-1. PROPERTIES OF SURFACE MATERIALS

Electrical Properties	*Chemical Properties*	*Thermal Properties*	*Mechanical Properties*
Conductivity	Corrosion	Conductivity	Elasticity
Magnetic hysteresis	Rust	Specific heat	Elongation
Electrolysis	Fretting	Dimensional changes	Strength
Magnetism	Effects of	Expansion	Ultimate
	Acids	Contraction	Tensile
	Alkalis	State changes	Compressive
	Salts	Melting point	Shear
		Physical changes	Torsional
		Hardness	Roughness
		Brittleness	Ductility
			Brittleness
			Hardness
			Resilience
			Flexibility
			Bending
			Toughness
			Wear resistance
			Endurance (fatigue)
			Damping capacity
			Mechanical hysteresis
			Creep
			Machinability
			Impact resistance

The effect of one kind of energy in relation to another can be demonstrated by listing just a few of the changes in various properties of metals caused by thermal and other processes:

Many metals become brittle (or more brittle) at low temperatures, even to the point of shattering at cryogenic levels (−240°F and below).

There is a definite decrease in hardness and yield strength and an ultimate decrease in tensile strength of steels at elevated temperatures accompanied by a marked increase in elongation and reduction of area.

ELECTRICAL PROPERTIES

1. Superconductivity of thermal and electrical energies develops in certain metals at cryogenic temperatures.

2. Thermal conduction of certain metals, such as copper, at extremely low temperatures can be controlled by inducing a magnetic field about the conductor, a process used in magnetic refrigeration.

3. Metal surface deterioration results when subjected to electrolysis, especially in frictional components that act as a ground for uncontrolled electric currents.

CHEMICAL PROPERTIES

If steel is exposed to moisture, its surface will attract oxygen atoms, thus generating a chemical reaction to form a hydrated oxide ($Fe(OH)_2)_2$, or ferrous hydroxide, that subsequently converts to the complex substance *rust*.

Chemical interchange in the formation of the ferrous hydroxide is one of *ionization* involving both positively and negatively charged particles called *ions*. Conversion to rust continues as the chemical reaction penetrates deeper into the mass as each succeeding molecular layer becomes the new surface as a result of corrosion of the outer layers. Penetration is possible because of the loose structure of the rust already formed. Rust is one of a number of chemical effects that may occur on the surfaces of materials. The susceptibility of a metal to respond to chemical reactions, such as rust and etching, depends on its *corrodibility*. Oxides also tarnish the surfaces of such metals as brass, bronze, and copper. Coatings of sulfides, chlorides, and nitrides also form on certain metal surfaces.

ADSORBED FILMS. Some substances, including oil, grease, and water, tend to form a film over a metal surface. This is in accordance with polar theory, described in Chapter 6. Such adsorbed surface films, or layers, develop a film so firm that they appear to have been adsorbed by the metal surface, and some films can displace other substances present on the surface. Adsorbed films tend to interfere with a chemical reaction that one metal would otherwise develop when brought in contact with another.

THERMAL PROPERTIES

EXPANSION. All metals react to temperature changes up to their melting points by changing lineal, areal, and cubical dimensions. They expand with temperature increase and contract with any decrease. The rate at which they expand or contract is given in terms of a *coefficient of expansion,* which is a rate of dimensional change per degree change in temperature (Table 4-2). The coefficient of expansion is not based on any particular unit of measurement. The result of any calculation, therefore, will be expressed in terms of the unit employed.

Recalescence is a property of steels which means that at certain critical periods in cooling red hot steel, a spontaneous glow appears during which the temperature remains constant or actually rises, i.e., the metal becomes visibly hotter, after which it resumes its normal rate of cooling. The reverse is true during heating steel when the rise in temperature becomes temporarily retarded. These phenomena occur at the point in cooling or heating known as the *critical temperature* for a given steel, or more accurately, the *critical range.*

Table 4-2. COEFFICIENT OF LINEAR EXPANSION FOR VARIOUS SOLIDS

	Per 1°C	× 5/9	=	*Per 1°F*
Aluminum	0.000023	× 5/9		0.0000127
Brass	0.000019	× 5/9		0.00001052
Copper	0.00017	× 5/9		0.00000933
Glass (average)	0.000009	× 5/9		0.000005
Iron, cast	0.000011	× 5/9		0.0000059
Iron, wrought	0.000012	× 5/9		0.0000067
Lead	0.00003	× 5/9		0.0000170
Nickel	0.000010	× 5/9		0.0000056
Platinum	0.000009	× 5/9		0.000005
Rock salt	0.00000846	× 5/9		0.0000047
Silver	0.000019	× 5/9		0.0000105
Silver, German	0.0000184	× 5/9		0.0000102
Steel (average)	0.0000123	× 5/9		0.00000685
Tin	0.000021	× 5/9		0.000012
Zinc	0.000029	× 5/9		0.000016

COEFFICIENTS OF CUBICAL EXPANSION FOR VARIOUS LIQUIDS/°F

Calcium chloride		Petroleum oil	0.00054
6% solution	0.000140	Sodium chloride	
40% solution	0.000253	20% solution	0.00023
Glycerin	0.000208	Sulfur	0.00127
Ice	0.00069	Turpentine	0.00054
Mercury	0.000101	Water	0.000105
Olive oil	0.00040		

MECHANICAL PROPERTIES

Elasticity

Elasticity is the ability of a material to be deformed and to return to its original dimension upon release of the deforming force. A rubber band will stretch as much as six times its original length without change in physical structure. Rubber is recognized for its great elasticity.

With some exceptions, metals possess elastic properties that vary according to their intrinsic structure. Engineering materials such as steel can be elongated within certain dimensional limits and return to the original length upon release of the force tending to deform them. A round 1-in.-diam bar of cold drawn steel (AISI C1040) can be elongated only about 17% in 2 in., whereas a hot-rolled low-carbon steel of the same size can be stretched 38% in 2 in., both steels returning to their original lengths.

The maximum stress that a given material can withstand and return to its original dimension is called the *elastic limit* of that material. Any additional stress causes a structural deformation within the material mass that permanently changes its size, both in length and cross section. The new fixed dimension is referred to as *permanent set,* usually identified as *specified set,* because it is related to the amount of stress that developed the strain corresponding to a given magnitude of set.

STRESS is the external force applied to a material, and it is measured in pounds per square inch (psi). It is determined by dividing the axial load by the original area of cross section, or

$$\text{stress} = S = \frac{P}{A_0} = \text{psi}$$

where

S = unit load (psi)
P = total load applied to test specimen (lb)
A_0 = area of original cross section of test specimen (sq in.)

STRAIN in a test specimen is the ratio of change in length, caused by stress, to the original length, or the amount of stretch or elongation in inches per inch.

$$\text{strain} = \varepsilon = \frac{AL_0}{L_0} \times 100 = \text{percentage change}$$

or

$$\varepsilon = \frac{L_f - L_0}{L_0} \times 100 = \text{engineering strain}$$

where

ε = unit strain (in. per in.)
AL_0 = change in original length, or L_0
L_0 = original length of specimen (in.)
L_f = final length of specimen (in.)

The equation applies to stresses due to both tension and compression, when the latter is determined on a test specimen that is short enough to preclude bending.

STRESS—STRAIN RELATIONSHIP. A typical relationship of stress to strain is shown in Fig. 4-1, in which stress values form the ordinate and strain the abscissa. It is called a *stress-strain diagram.* From this diagram several important parameters regarding the properties of materials can be established:

1. The stress-strain relationship between zero stress and some stress value *a* governed by the intrinsic properties of the test specimen is shown to be a straight line. This means that the amount of change in elongation of the specimen is proportional to the amount of stress applied up to the psi shown on the ordinate at a point corresponding to *a.* The range between 0*a* may be called the line of proportionality and *a* the *proportional limit,* because beyond *a* the line departs from the straight line relationship.

2. Between points *a* and *b* on the diagram, the line tends to curve

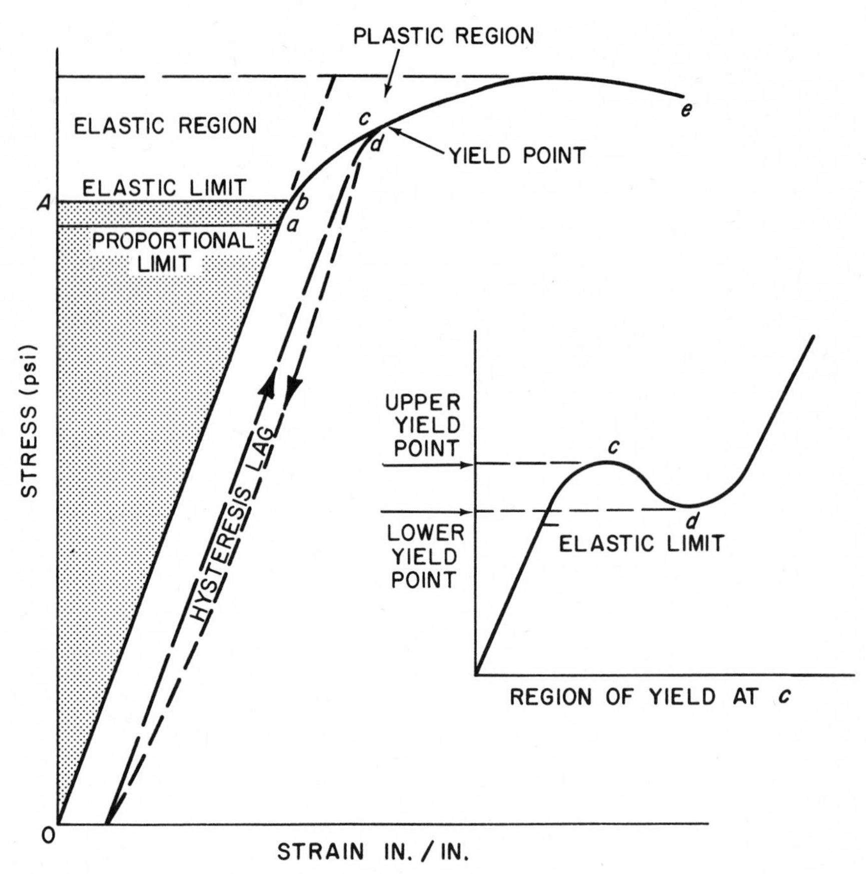

Fig. 4-1. Stress-strain diagram.

slightly. Point *b* signifies increased stress applied to the test specimen to a magnitude shown on the ordinate at a point corresponding to *b* on the curve. If it were found that, upon release of the load when the stress reached the value corresponding to *b*, the material returned to its original length but did not at any point beyond, it would be said that the stress value corresponding to *b* is the elastic limit for the material tested. The state of the material between 0 and *b* or below the elastic limit *b* is one of elasticity, and the range is called the *elastic region*.

3. Because on subsequent application and release of a load having a stress of any increment of value above that corresponding to *b* on the ordinate, the specimen did not return to its original length, the material is said to have entered a plastic phase. The stress-strain curve beyond the elastic limit *b* represents the *plastic region* for the material tested.

4. Above *b* the line curves sharply. Upon reaching the crest at *c*, it tends to dip but continues to increase strain rapidly, while the stress remains

constant. With some materials, the stress will (1) decrease slightly for a short period or (2) continue to increase slowly.

5. The stress corresponding to point c is the *yield point* of the material being tested, and being situated at the crest of the curve, i.e., at maximum stress, it is also called the *upper yield point.*

6. The point where the dip in the curve reaches its lowest stress value, which is at d on the curve in Fig. 4-1, is called the *lower yield point.*

7. For materials that show no abrupt change in slope, as shown between c and d, a parameter known as the *yield strength* is determined when the elongation has reached a stress value corresponding to a *specific* amount of permanent set. This is demonstrated by the line running parallel to the line of proportionality from c to an arbitrary point on the abscissa, and usually 0.20% is the specified set. The specified set is reported with the stress at the point where the yield strength is selected.

8. The curve continues from point d with a rapid increase in strain with little or no increase in stress until the latter drops abruptly to zero upon rupture of the specimen at point e.

As previously stated, the curve in Fig. 4-1 is hypothetical, but fairly representative of many types of steels. Because of differences in some properties, or even in property magnitudes, of some materials the slopes of the curve will vary or even be nonexistent, as in the case described for determining yield strength.

HOOKE'S LAW. Returning to line $0a$ of Fig. 4-1, it was previously implied that it is located within the elastic region. It was also stated that stress is proportional to strain within this range. This is the basis for *Hooke's Law,* which states:

In the region below the elastic limit of a material, the stress is directly proportional to the strain.

YOUNG'S MODULUS. The ratio of stress to strain below the proportional limit is described as the *modulus of elasticity* (E)

$$\text{or} \qquad E = \frac{\text{stress (psi)}}{\text{strain (in. per in.)}} \qquad \text{or} \qquad \frac{S}{\varepsilon}$$

POISSON'S RATIO. Under tension, a metal test specimen tends to contract laterally and expand longitudinally. During compression the tendency is toward lateral expansion and subsequent longitudinal contraction. The definition of strain denotes change in dimension due to stress so that the term may be applied to such changes in any appropriate direction. For *tensile* or compressive stresses, the directions of strain are longitudinal or axial (parallel to the applied load) and lateral or radial (perpendicular to the load). Within the elastic region ($0a$) on the stress-strain diagram (Fig. 4-1) the lateral strain (ε_r) and axial strain (ε_a) are proportional to each other and their ratio is constant for a given material and is expressed as *Poisson's Ratio* μ, or

$$\mu = \frac{\text{lateral strain}}{\text{longitudinal strain}} \quad \text{or} \quad \frac{\varepsilon_r}{\varepsilon_a}$$

For steel, $\mu = 0.333$; cast iron, $\mu = 0.250$; concrete, $\mu = 0.100$, approximately in each case.

Another constant within the elastic limit for metals is *shear,* which is the tendency for a load to deform the mass by causing the parallel layers to slide over each other. The shear constant is called the modulus of elasticity in shear, or *modulus of rigidity* G. Because all three elastic constants are involved in a shear situation, their relation to each other is shown in the equation

$$G = \frac{E}{2(1 + \mu)}$$

or

$$E = 2G(1 + \mu)$$

in which G and E are expressed in pounds per square inch.

Strength

If the load on the specimen is continually increased after entering the plastic region, the degree of deformation also continues. If carried to the ultimate capacity of the material, rupture occurs. The unit stress that is applied immediately before rupture is called the *ultimate strength* of the material tested. This is represented by *e* on the stress-strain diagram (Fig. 4-1).

If the load applied tends to pull the specimen apart, a state of tension exists, and the applied force is axial or perpendicular to the plane of stress. The ultimate strength of a material under tension is referred to as its *tensile strength.*

If the load applied tends to crush the specimen, a state of compression exists and, as under tension, the applied force is also perpendicular to the plane of stress. The ultimate strength of a material under the force of compression is known as the *compressive strength.*

When the load is applied parallel to the plane of stress during which the parallel layers of material tend to slide over each other or to be displaced in the direction of their line of contact, a shearing action is said to occur. The ultimate stress of a material under the latter conditions is called *shear strength.*

The ultimate stress that a test piece such as a shaft, of uniform cross section, can withstand when twisted or subjected to torque is its *torsional strength.* It produces a shear stress in the cross section of the shaft from zero at the center of the neutral axis to maximum at the outermost circumference T. The distance T multiplied by the twisting force L acting on the rod is the *moment of twist.*

Surface Roughness

Surfaces are those areas of a substance that are exposed to other substances. They may have any shape—flat, circular, smooth, rough, jagged, etc. Considering that flat surfaces have their center of axis at some infinite point or that circular surfaces can be "rolled out" to become flat, it will be assumed here that all surfaces have a straight contour and that any two surfaces in contact are parallel relative to each other. Any surface can be described with words or illustrated in a drawing. To do either, certain definitions must be recognized:

1. In cross section, a surface exists in profile, which is its contour either in part (given length of section) or as a whole.
2. The profile may be a series of closely spaced irregularities, or asperities, having height or depth relative to some straight base line. Such irregularities make up what is defined as *roughness.* Roughness, therefore, has dimensions which describe its degree.
3. The profile may also evidence a series of another kind of irregularity, best described as waves, or sine curves, whose peaks (crests) and valleys (roots) are spaced further apart than the roughness asperities, which are superimposed on the waves. This type of irregularity is known as *waviness.*
4. Seen from above, the predominant irregularities may show a pattern of direction relative to the area boundaries of the surface. An analogy of this would be a map view of the Allegheny Mountains, seen stretching from northeast to southwest relative to the North and South Poles. Such a pattern is called *lay,* and it may appear to be relatively straight, circular, swirling, or otherwise.
5. Dispersed among these irregularities may be found protrusions and/or indentations called *flaws.* Flaws may be in the form of peaks, ridges, scratches, pin holes, incipient cracks, etc.
6. Roughness of surfaces associated with bearings, gear teeth, cylinders, ways, etc., is measured in terms of microinches, i.e., millionths of an inch (1 microinch = 0.000001 inch). The numerical roughness value is an average of the vertical distances measured from the base line, which is the actual surface plane, to the crests and roots of the asperities.

This average dimension for a given length of section along the surface profile may be reported as the arithmetic average or as the root-mean-square (rms) average. By measuring all the deviations ($y_1, y_2, \ldots, y_n$) in height

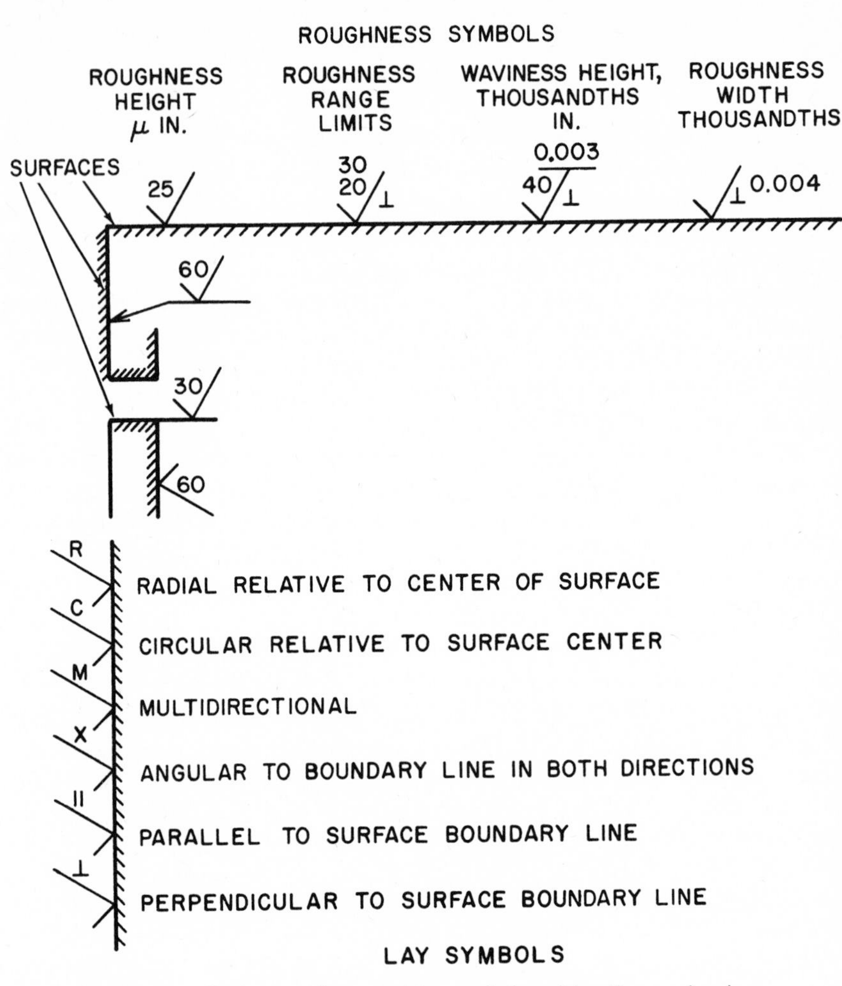

Fig. 4-3. Surface roughness and lay symbols and location on drawing.

and depth relative to the base line, as shown in Fig. 4-3, either of the averages may be determined as follows:

The arithmetic average is equal to the sum of the deviations (y) per given section divided by the number of deviations measured (n),

or $$\text{arithmetic average} = \frac{y_1 + y_2 + y_3 + y_4 + \cdots + y_n}{n}$$

The root mean square (rms) is equal to the square root of the sum of the individual deviations squared divided by the number of deviations measured (n), or $$\text{rms} = \sqrt{\frac{y_1^2 + y_2^2 + y_3^2 + \cdots + y_n^2}{n}}$$

Roughness values are usually given in microinches rms in the various handbooks. The base line is correctly placed if it is in the center of a curve representing the various deviations, as shown in Fig. 4-2.

Smoothness is a relative term, but with machining processes, a fine line is drawn between roughness and smoothness. The surface of a steel shaft smooth turned on a lathe may have a roughness of 180 to 220 μin. rms, which is relatively rough in terms of precision. The same metal polished to a superfinish of 3 μin. or less would be considered to have a high degree of smoothness by any standard. A fine human hair, in comparison, has a diameter of about 1200 μin.

The surfaces of all engineering metals, regardless of their finish, consist of numerous asperities. The size of such asperities governs surface smoothness, or roughness, which, in turn, is a function of the machining process. These processes include turning (as on a lathe), drilling, reaming, grinding, lapping, shaving, and, in some instances, scraping. To form a sharp surface, as for a knife edge, a honing and stropping process is used. Table 4-3 shows the range of roughness values reported in root-mean-square heights for various surface finishing methods.

WAVINESS. The profile of a surface is not a perfect plane but consists of a series of waves or undulations. The magnitude of waviness varies according to the care and methods used in the smoothing process. Waviness may also be due to surface deformation caused by unbalanced strains within the parent body, e.g., warping, uneven tightening of fasteners (bolts), and thermal stresses. Waves are relatively wide irregularities with a crest-to-crest dimension of 1/32 in. or greater, and a depth from base to crest of less than 0.002 in.

The American Standard of Surface Roughness, Waviness, and Lay

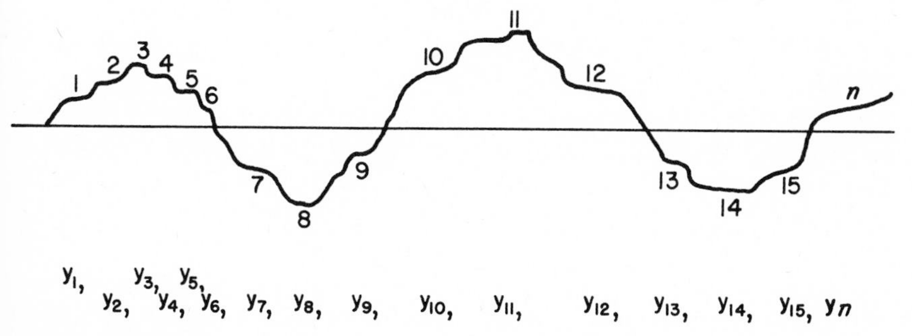

Fig. 4-2. Height and depth of surface asperities relative to a given plane, showing dimensions in microinches.

Table 4-3. RANGE OF SURFACE ROUGHNESS

Industrial Finishing Procedure	*Roughness Range (μin., rms)*
Turning, rough	63–1500
Milling, rough	63–1000
Shaping	32–500
Grinding, rough	32–200
Boring, reaming	32–63
Broaching	8–125
Commercial grinding	8–63
Finish grinding (precision)	4–16*
Honing and lapping	4–16*
Super finish	1–8*

* Peak roughness will vary from 4 to 7 times rms roughness.

specifies both roughness and waviness in terms of width and height of asperities as follows:

Waviness Width Rating—Waviness widths may be specified directly in inches.

Height Rating—The height of roughness or waviness is specified in one of the following terms:

Maximum peak-to-valley height
Average peak-to-valley height
Average (either rms or arithmetical) deviation from the mean surface.

(The type of height rating intended should be included in all specifications, etc.)

LAY SPECIFICATION. The direction of the visible surface marks is called the lay. The lay of a surface is specified by the lay symbol indicating direction of visible surface marks (Fig. 4-4).

FLAWS. The coefficient of friction for the same surface materials varies according to the surface roughness, being higher as the height and depth of the asperities become greater. When such surfaces are rubbed together for the first time (first pass), the peaks of one tend to shear the peaks of the other in a plowing action. If the surfaces are both ferrous materials and if the interfacial pressure is great enough, high spot temperatures accompany the shearing process, tending to cause welding of one peak to another. When this happens, continuous relative motion causes the weaker peaks to be pulled away from their respective surfaces, leaving flaws in the form of pits or galls.

In repeating the rubbing action by either a reciprocating or rotating motion, the asperities are gradually worn down, and this is accompanied by a diminishing frictional resistance, i.e., the coefficient of friction decreases.

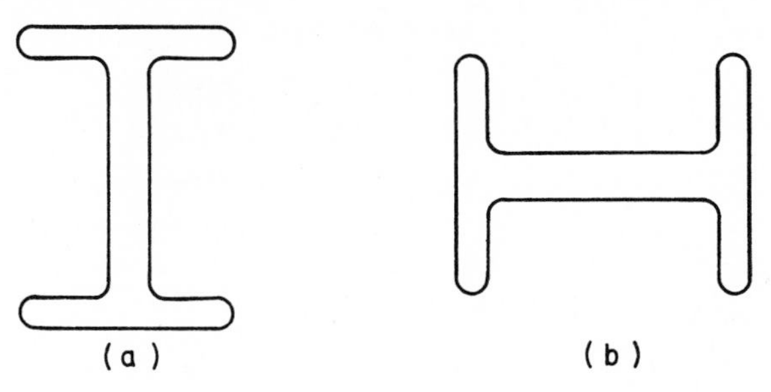

Fig. 4-4. Cross section of I-beam. (a) Web vertical, (b) web horizontal.

Each time the same surface areas cross each other is called a pass. By interposing a molecular film of an antiwelding lubricant (say, sulfurized sperm oil) between the opposing surfaces and subjecting them to a program of multiple passes, a wearing in or breaking in process will be carried on. If the process is allowed to continue, the surfaces will ultimately become highly polished, at which time their load-carrying capacity will increase. The increase occurs because

1. The coefficient of friction is reduced.
2. A more uniform boundary lubricating film up to 25 μin. thick may be established.
3. The actual area of surface contact is increased.

Manufacturers of gears, internal combustion engines, etc., break in their products in this manner, using a low-viscosity mineral oil compounded with an antiwear, antiweld additive. A magnetic drain plug and/or magnetic filter should be employed during the breaking in period to catch the iron and steel fines resulting from wear and to prevent their return to the friction areas with the lubricating oil. Most units incorporate a self-sealing drain plug that can be temporarily removed to clear the fines periodically without draining the oil.

Surface analyzers and profilers that measure and record roughness in microinches, rms, are used by many manufacturers of mech
ment.

The *polar moment of inertia J* of a surface is the sum of the separate moments (the distance of each elementary area multiplied by the square of its distance from the center of gravity of the surface),

$$J = \frac{\pi d^4}{32}$$

the polar moment of inertia for solid shafts or

$$\frac{\pi(d^4 - d_1^4)}{32}$$

the polar moment of inertia for hollow round shafts, where d and d_1 are outside and inside diameters, respectively.

The *modulus of rigidity for torsion,* then, is expressed as

$$E_t = \frac{TL}{\theta J}$$

where

T = twisting moments (in.-lb.)
L = length between cross sections between jaws (in.)
θ = degree of turns (radians)
J = polar moment of inertia

For a solid round rod of cold rolled steel with a length between jaws of 6 in. (L), and 0.8 in. in diameter (d), the author found by test the modulus of rigidity to be 11,000,000.

ELONGATION. Elongation is the amount of increase in permanent extension of a test specimen under tension near the area where rupture occurs when the stress is raised to an ultimate magnitude. It is expressed in percentage of original gauge length relative to the latter dimension, such as 26% increase in elongation in 2 in., or

$$\text{percentage elongation} = \frac{\text{final length} - \text{original length}}{\text{original length}} \times 100$$

Reduction in Area. Any increase in elongation of a material is accompanied by a reduction in cross-sectional area. The reduction in area is identified as the smallest area at the point of rupture relative to the original area of the test specimen. It is expressed by the equation

$$\text{percentage reduction of area} = \frac{\text{original area} - \text{final area}}{\text{original area}} \times 100$$

BENDING. A property denoting flexibility is described as bending, and the force required to flex a body is the *bending stress.* The tendency to bend is a function of the applied load, as on a beam, and the length of the span whose supports act as counterforces to the load.

The only way to reduce the bending quality of any given material, i.e., to increase its stiffness, is to increase its bulk or make appropriate modifications in its cross section relative to the load. For example, placing a structural beam, such as an I beam, on one of its flanges with the web in a vertical position provides greater stiffness than if the same beam is placed with its web horizontal (Fig. 4-2). The resistance to bending is a function of the modulus of elasticity.

To produce equilibrium in a beam, the internal moment of resistance must equal the external moment of bending, or

$$M = SI/C$$

where
$M =$ external bending moment (in.-lb)
$S =$ safe unit working stress on the outermost fiber of the beam at the section of beam related to M
$C =$ distance from neutral axis (center of a uniformly square, round, or rectangular beam) to the outermost fiber of the beam
$I =$ moment of inertia of the cross section of the beam.

By transposing the equilibrium equation, it can be written

$$\frac{M}{S} = \frac{I}{C}$$

The term I/C represents the *section modulus,* or the ratio of the moment of inertia of cross section to the distance from the neutral axis to the outermost part of the beam. The section modulus depends on the cross-sectional area distribution throughout the beam and not at all on the beam material, regardless of its composition (alloy) or heat treatment.

The section modulus is the capacity of a beam section to resist any bending moment applied.

Bending may be multiple or simple. Simple bends are usually referred to as one-way bends, e.g., gear teeth, which are a form of cantilever beam. Refer to Chapter 14 for further discussion of strength and wear on gear teeth.

Toughness

The capacity of a material to withstand hard usage, as from shock, impact, or other dynamic force, is called toughness. It inclines toward the plastic region of the stress-strain curve and represents the ability to absorb a high degree of external impact energy without permanent structural damage, or its ability to withstand high stress and deformation before rupture. An example of tough material is the metal used for the teeth of power shovel buckets.

The toughness of a given material can be evaluated through the use of the stress-strain curve by determining both its ultimate strength and strain at the point of rupture. Multiply these factors to obtain a *toughness index number:*

$$\text{toughness index number} = S_u \times \varepsilon_r$$

where

$$S_u = \text{ultimate strength}$$
$$\varepsilon_r = \text{strain at rupture}$$

The area under the stress-strain curve up to the point of rupture is an indication of toughness of a given material when related to the same area of a curve of another material.

RESILIENCE is the property of a material that allows it to return to its original shape or form upon release of an applied force that caused its deformation. It is related to elasticity in that the energy required to deform it is stored within the body mass. The internal force of resilience is equal to the force of deformation, while the mass structure remains in the elastic region, i.e., the force is at or below its elastic limit.

The resilience of a material is measured in terms of its *modulus of resilience* which is expressed in inch-pounds per cubic inch:

$$\text{modulus of resilience} = \frac{(\text{stress at proportional limit})^2}{2 \times \text{modulus of elasticity}}$$

or

$$U = \frac{S_p^2}{2E}$$

where

$$U = \text{modulus of resilience}$$
$$S_p = \text{stress at proportional limit (psi)}$$
$$E = \text{modulus of elasticity}$$

Comparative values of modulus of resilience may be shown as follows: steel, 21; wrought iron, 12.5; cast iron, 1.2.

The unit $S_p^2/2E$ is representative of the unit work required per unit volume, the latter being the area of the material multiplied by its length. This work may be graphically shown on the stress-strain curve by an area formed by a vertical line from A, Fig. 4-1, to the abscissa and the line $0a$ of the curve up to a. The work area is shaded on the diagram.

DUCTILITY. The ability of a material to withstand a high degree of permanent deformation prior to rupture is called *ductility.* It is also the capacity to be extended by beating (as with a hammer), drawn into wire (as through a die), bent, or otherwise worked. The measure of ductility is evaluated by the magnitude of elongation at the point of rupture as determined by the stress-strain curve. The greater the dimension of stretch, the more ductile the material.

CREEP is a property of metals associated with continuous loads for extended periods, and it is intensified by high temperatures. It is a tendency to stretch, especially with components made of such nonferrous metals as tin, lead, copper, and zinc and their alloys, such as the babbitts, bronzes, and brasses, and to a lesser extent with ferrous metals.

An important property of bearing materials, creep develops in a series of three phases, or time stages:

First stage. Deformation due to creep develops relatively fast during the initial period, and it gradually slows down when the stress tends to become uniformly distributed throughout the mass and eventually causes strain hardening.

Second stage. Creep continues to develop at a slow constant rate when the annealing affected by high operating temperature tends to balance the strain hardening effect.

Third stage. The creep rate increases during this stage when the effect of the annealing process gradually exceeds the influence of strain hardening to cause ultimate rupture.

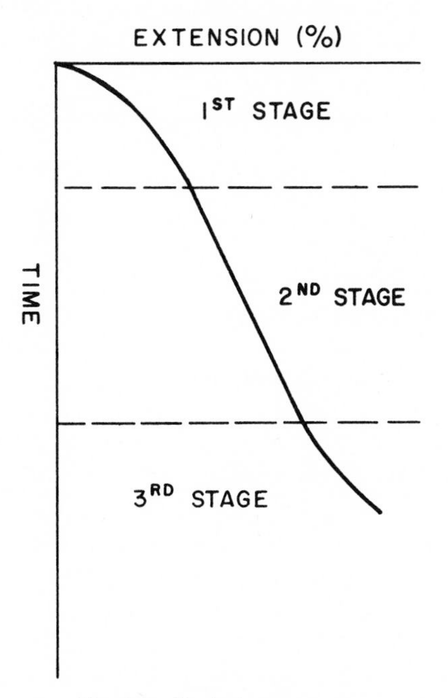

Fig. 4-5. Typical creep curve.

METAL FATIGUE. With few exceptions, such as a journal turning on its bearing, load-carrying components are subject to work cycles or alternate loading. These components may be static, as in the case of machine tool ways, or dynamic, in the form of a carriage that slides over the ways with a reciprocating action. Individual balls and rollers of antifriction bearings are exposed to an action that is alternately one of loading and unloading as they constantly pass in and out of the pressure zone. The bearings of crankshafts, connecting rods, and wrist-pins of internal combustion engines are all subject to alternating loads. Cams, gear teeth, and numerous indexing mechanisms are included in the category associated with cyclic loading.

Cyclic loading may be alternating, reversing, or repeating in form, and all are demonstrated in one or more of the examples mentioned above.

Surfaces exposed to cyclic loading transmit a constant wave of stress variations to their parent material that range from zero (in some cases) to a maximum value corresponding to load applied during any given cycle.

All metals have a life expectancy measured by the number of cycles of alternate loading and unloading they are capable of withstanding before *fatigue,* or structural deterioration, develops to the point of *fatigue failure.* For a given material, the number of cycles varies according to the magnitude of maximum loading and the cycle time period, which governs the amount and frequency of deformation experienced during each cycle.

For a given material, the number of cycles varies according to (a) the magnitude of average load (stress) and the (b) frequency of the cycle (time). The factors of stress magnitude and cyclic frequency govern the amount of deformation experienced by the same component under a given set of operating conditions. The life expectancy tends to increase when

1. The load is reduced at a given speed
2. The speed is decreased and the load remains constant
3. Both the load and speed are reduced

The life expectancy of materials is discussed in more detail in Chapter 13, Rolling Bearings.

Metal fatigue may be defined as the condition of a material after it has been subjected to a maximum number of alternate stresses regardless of the difference in stress magnitude between loaded and unloaded conditions. It is usually initiated by the formation of a crack, which continues to grow until the remaining cross section of material becomes too slender to transmit the stresses uniformly to the remaining mass. Rupture then occurs, or in the case of bearings, the rupture is described as a bearing failure due to metal fatigue.

The life expectancy of a material, which is usually reported according to the limiting stress corresponding to an appreciably large number of stress cycles prior to fatigue, may be described as its *endurance.* Endurance is usually measured in terms of average life expectancy in hours under specific conditions of load and speed (stress frequency). The life expectancy of a ball bearing, for example, may be given as 3800 hr at 1000 rpm and 2200 lb load.

MACHINABILITY. The property of a material associated with the ability to transform its shape by removing some of its mass by an appropriate machining process, such as cutting, planing, tapping, drilling, and broaching, is machinability. The nature of the basic material, its heat treatment, and intrinsic properties that oppose molecular layer separation, such as work hardening during the machining process in some cases, all contribute to the quality of machinability. It is usually reported by a *machinability rating* in percent that indicates the machining quality of a material compared with another whose ease or difficulty of machining is widely known. Ratings are based on the kind of metal (ferrous or nonferrous) and range from above 100% for materials that machine with relative ease (e.g., standard malleable iron, 120%) to about 25% for the most difficult materials (e.g., high carbon, high chromium tool steel, 25%).

WEAR RESISTANCE is that property of a material that is associated with the attrition of its surfaces when in rubbing contact with another body of the same or of a dissimilar material. The amount of wear is measured by determining weight losses of mechanical components (bearings, hydraulic pumps, etc.) after a given number of hours at a specified temperature and constant loading (see Chapter 3).

SURFACE ENDURANCE. A mutual function exists between the wear qualities and surface endurance of materials. *True wear* is the result of area friction and not of the abrasive action of sharp objects, whether they be microparticles or large bodies of appropriate hardness. The loss of surface materials by whatever means, except deliberate mutilation, is that which disappears as a result of productive usage and which will be identified as *practical wear,* or in the interest of brevity, wear, throughout this text. In so defining it, all the factors erroneously defined as functions of true wear must be considered for practical reasons not always academically recognized.

To accept all theories associated with surface attrition due to friction would be to define gouging, welding, and brazing with their subsequent tearing effect, and destruction caused by overheating and overloading, as forms of wear. Many more frictional components, including bearings, gears, and cylinders, are destroyed because of poor maintenance practices and carelessness than are replaced because of normal wear.

With the selection of the best materials for a given set of operating conditions and correct surface maintenance in or out of operation, an optimum service life can be obtained.

Hardness

Hardness is not a definite property of metals, relying as it does on a number of variables, such as metallurgical treatment, chemical composition,

Table 4-4. SCALE OF RELATIVE HARDNESS (MOHS' SCALE)

No.	Material	No.	Material
10	Diamond	5	Apatite
9	Corundum	4	Fluorite
8	Topaz	3	Calcite
7	Quartz	2	Gypsum
6	Feldspar	1	Talc

structure and other mechanical considerations. Because of its importance, it will be treated as a property here, associated with a specific material having definite specifications of ingredients, thermal treatment, etc. (e.g., steel, low carbon, SAE 1010, annealed).

Hardness is the quality of a material that keeps it firm or unyielding against impressed forces. It is apparent by resistance of a surface to plastic deformation. If the quality is judged on the basis of relationship, the degree of resistance to surface deformation of a diamond in comparison with other minerals is the criterion. The reciprocal of hardness is softness. The relationship is shown in the Mohs' scale, Table 4-4, in which a number denotes the relative hardness of various minerals headed by 10 for diamond, the hardest, and descending in value to 1 for talc, the softest. Other hardness tests, each with its special apparatus, are the Shore Scleroscope, Vickers, and Rockwell.

Such a hardness scale does not denote specific hardness magnitude and is analogous to a scale denoting thermal intensities as being extremely cold, very cold, cold, warm, hot, very hot and extremely hot instead of in terms of degrees Fahrenheit, Celsius, etc.

For practical purposes, the properties of materials must be expressed in precise values. Such values, particularly for metals such as steel, are determined using exacting laboratory techniques. The techniques used to determine hardness employ an indenting process in which a round or pointed element is forced against a surface and the degree of penetration measured. (See discussion on hardness tests, Chapter 13.) Hardness as determined by the Brinell testing apparatus is indicated by a number and is called *Brinell hardness;* the higher the number the greater the hardness. A material such as pure rolled aluminum, with a Brinell hardness of 23 to 44, is considered soft. Chromium steels (e.g., SAE 5150) are described as very hard, some of them having a Brinell hardness of 400 and higher, depending on the heat treatment.

Hysteresis

Hysteresis may be interpreted as the lagging of an effect behind its cause. As applied to metals, particularly the surfaces of rolling bearings, it is produced by cyclic loading below the elastic limit of the bearing materials. Hysteresis is evident under both tensile and compressive forces. It is the underlying phenomenon behind the damping capacity of a metal to dissipate energy without resulting in failure due to cyclic loading. This phenomenon is known as *mechanical hysteresis effect.*

The effect of hysteresis is simply demonstrated by stating that any elastic material upon being compressed returns to its original state at some time after the compressive force has been relieved. This return lag is the measure of hysteresis for any material. The mechanical energy involved during the process converts to thermal energy, resulting in a rise in temperature.

In the case of a ball bearing under load, the cyclic stresses that are rapidly applied by the succession of balls, one following quickly behind the other, are complex factors that cause some loss in the theoretical performance of an otherwise highly efficient rolling motion. Therefore, some of the energy loss in antifriction bearings must be accounted for by the loss of heat generated by the hysteresis effect.

The effect of hysteresis is demonstrated by the *hysteresis loop* shown in Fig. 4-1. The curve *abc* represents the application of stresses to a metal specimen, say steel, while the path *cde* represents the sudden release of that stress. Reapplication of the load, represented by the path *efg*, leaves a loop formed by lines *cde* and *efg*. This hysteresis loop tells an interesting as well as an important story, briefly stated as follows:

1. A large hysteresis loop accompanied by a rapid conversion of mechanical to thermal energy portrays a higher damping effect.
2. A high damping effect is conducive to better vibration absorption.
3. A better vibration absorption encourages longer service life to metals subject to cyclic loading.
4. The phenomenon per se is not necessarily associated with other metal properties, such as tensile strength or hardness.

MOLECULAR ATTRACTION. The phenomenon displayed by some solids in their tendency to adhere to other specific solids may be likened to the molecular attraction existing in liquids. Within the mass, the attraction exists in all directions, while at the surface the part of the molecule exposed to the external environment is left unsatisfied while continuing to possess an attraction to or for other compatible molecules. This phenomenon is a form of energy which at the surface is described as *surface energy*. The magnitude of surface energy may have a definite relationship to the enthalpy of the material and its state. However, the effect of surface energy of one material on another is considerably reduced or even eliminated by the presence of a film that may in any way appear between them.

5

MECHANICS OF MACHINERY

This text deals with lubrication as applied to practical machines, i.e., machines capable of performing useful work. Successful lubrication of a given machine depends on a correct analysis of the mechanical characteristics of each of its members. Therefore, the study of the science of machinery is essentially a part of the education of personnel responsible for recommendations concerning the lubrication of complex mechanical equipment. This chapter deals with definitions relating to the science of machinery, or mechanics.

MECHANICS

The branch of science which treats of motion, forces, and the effect of forces on rigid bodies is mechanics. When motion and forces relate to machine parts, the science is referred to as *mechanics of machinery.* Mechanics of machinery is concerned with the construction, operation, and efficiency of a machine and each of its parts. It is based on three fundamental laws; *Newton's Three Laws of Motion:*

First law. A body will remain at rest, or in uniform motion in a straight line, unless acted upon by an external force that compels change. This is the *law of inertia.*

Second law. Every motion or change of motion is proportional to the acting force and occurs in the direction in which the force acts.

Third law. Every action is opposed by an equal and opposite reaction.

The science of mechanics is divided according to the conditions under which a body exists as related to motion and forces. The principal divisions are *statics* and *dynamics.* Dynamics is further divided into two branches, *kinematics* and *kinetics*. Statics is the branch of mechanics that deals with forces and the effect of forces acting on rigid bodies at rest. Dynamics is the branch dealing with motion and the effect of forces acting on rigid bodies in motion. Kinematics is a branch of dynamics embracing the study of motion without regard for the forces that cause motion. Kinetics is the other branch of dynamics that deals with forces acting on rigid bodies in motion and the effect of such forces in altering such motion. Machine frames, crankshafts, connecting rods, gears, bearings, and pistons are examples of rigid bodies. The scope of the science may be broadened to include fluids (hydraulic and pneumatic) and flexible connectors (belts, chains, and ropes) which are employed to transmit power and motion.

MECHANISM

A mechanism consists of a combination of rigid members, one of which is fixed, so designed and connected that when motion is transmitted to one member, all others respond in a definite manner (Fig. 5-1). The purpose of a mechanism, as differing from a machine, is to modify or transmit motion. However, a machine may consist of one or any number of mechanisms. A complex machine is composed of a series of mechanisms, each performing a definite function that combines to accomplish a common purpose, i.e., the purpose for which the machine is intended.

Motion may be transmitted to mechanisms from several sources:

1. From a common source within the parent machine to any number of mechanisms.
2. From a source external to the parent machine.
3. From various convenient sources within the parent machine.
4. Relayed in turn from one mechanism to another.
5. By takeoff from a mechanism of one machine to a mechanism of a neighboring machine.
6. Utilizing all or any combination of the foregoing arrangements.

A mechanism is known in mechanics of machinery as a *kinematic chain*.

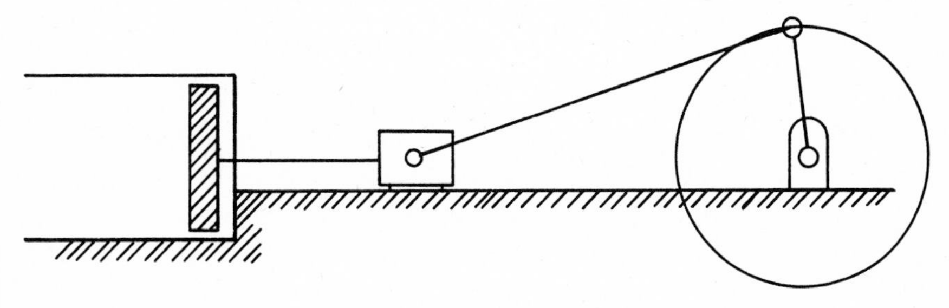

Fig. 5-1. Simple mechanism.

A kinematic chain is composed of (1) a fixed support or frame, (2) a driving component, (3) a coupler, and (4) a follower.

MACHINE

A *machine* is an apparatus consisting of a number of rigid members, each having a definite function, i.e., to modify or transmit motion, which together act to perform some useful work through the modification of energy. Machines vary in the number of mechanisms they require, ranging from the hand press employing a simple lever arrangement to an openside shaper having a number of complicated mechanisms (Fig. 5-2).

Machines consist of both fixed and moving parts. Fixed parts include those that remain stationary relative to other members of the same machine. They may or may not have motion relative to the earth. Stationary members act to support and/or guide moving parts and to lend strength to the whole assembly. Frames, braces, ways, guides, beds and some types of bearings are examples of stationary machine members. Moving parts, i.e., parts

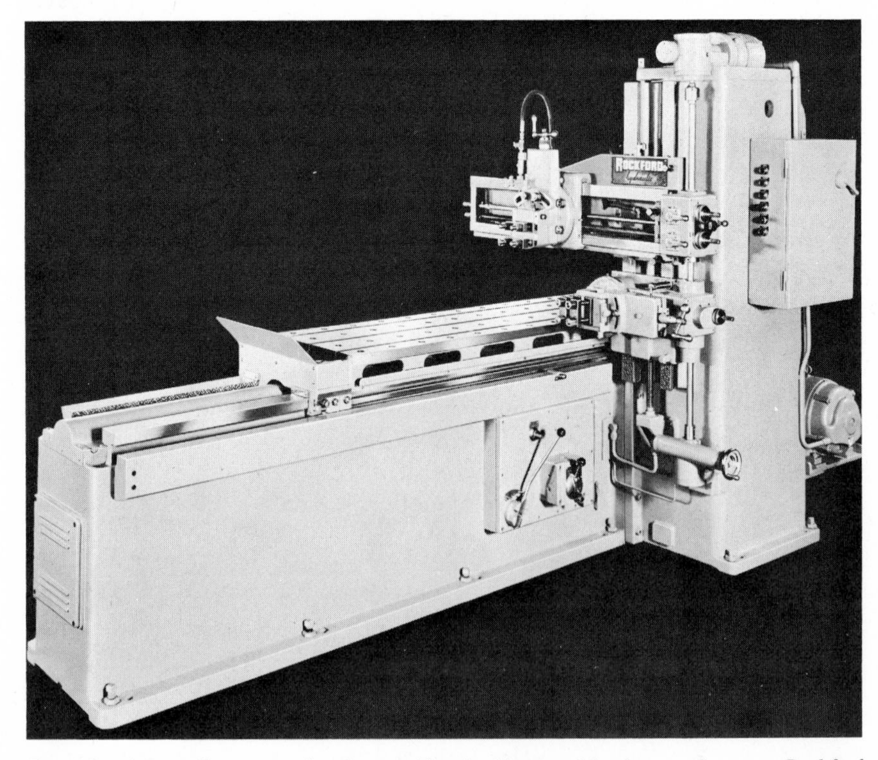

Fig. 5-2. An excellent example of a machine is this openside shaper. *Courtesy, Rockford Machine Tool Co.*

having relative motion as related to other joining members of the same machine, include gears, chains, wheels, sprockets, pulleys, rolls, rollers, cranks, pistons, levers, and journals.

Machines may be classified according to their major functions, the most common of which include—

Power generation. Steam turbines, gas turbines, hydraulic turbines, steam engines, internal combustion engines.

Industrial production. Paper machines, metal rolling mills, cement-making machinery, rubber mills, brick machines, presses, food canning machines, textile looms, shoe machinery, woodworking machinery, refrigeration equipment, linoleum-making equipment, die-casting machines, injection molding machines, foundry equipment, etc.

Metal machining. Lathes, boring mills, milling machines, broaches, drill presses, grinders, screw machines, and hammers

Material handling. Conveyors, elevators, pumps, hoists, cranes, derricks, and lift trucks

Transportation. Motor vehicles, railroad rolling stock, ships, and aircraft

Construction. Shovels, tractors, conveyors, pile drivers, and special earth-moving and road-laying equipment

Gas handling. Compressors, blowers, and fans

Mining machinery. Coal cutters, loaders, screens, crushers, air drills, shovels, and conveyors

Textile manufacture. Looms, winders, twisters, and sewing machines

Agricultural equipment. Tractors, combines, balers, conveyors, and pickers

Other. Laboratory equipment, chemical and pharmaceutical machinery, packaging equipment, office equipment, and household appliances

MOTION

A body in the act of changing its position relative to surrounding bodies is said to be in motion. A body free of motion relative to surrounding bodies is said to be at rest. All motion is relative, but the base of relationship must be plainly stated. Most problems relate to the earth as a base.

When a body moves over the earth's surface, its motion relative to the earth, or to any other body at rest on the earth's surface, is described as its *absolute motion.* A vehicle traveling over the earth's surface at a speed of 50 mph has an absolute motion in relation to the earth of 50 mph.

When two or more bodies move over the earth's surface simultaneously, the motion of any one is related to the motion of each of the others, i.e., they have *relative motion.* Consider two vehicles traveling over the earth's surface in opposite directions. One travels west at an absolute speed of 50 mph, and the other travels east at an absolute speed of 40 mph. Their relative motion to each other is 40 + 50 = 90 mph. If both vehicles were traveling in the same direction at the speeds given above, their relative motion would be 50 − 40 = 10 mph. Their absolute motion, however, would remain 50 and 40 mph, respectively.

Consider four parts of the same machine, one of which is the frame (part 4), attached to the floor of a building and supporting the others. Assume that the remaining three members move along parallel lines at a rate of 5 ft/min relative to the frame. Assume also that parts 1 and 2 travel in the same direction but opposite that traveled by part 3. The following conditions of motion exist:

(a) Parts 1, 2, and 3 are in absolute motion relative to the frame at a rate of 5 ft/min.

(b) The relative motion of part 1 in respect to part 2 is zero because both are moving at 5 ft/min in the same direction.

(c) The relative motion of part 2 in respect to part 1 is zero for the same reason given in (b).

(d) Parts 1 and 2 have relative motion in respect to part 3 because the latter travels in the opposite direction to parts 1 and 2.

(e) Part 3 has relative motion in respect to parts 1 and 2 for the same reason given in (d).

(f) Because parts 1, 2, and 3 are moving at the same rate relative to the frame but in opposite directions (i.e., parts 1 and 2, in respect to part 3), the relative motion of parts 1 and 2 in respect to part 3 is twice that of their absolute motion, or 10 ft/min.

(g) The relative motion of part 3 in respect to parts 1 and 2 is also 10 ft/min for the same reason given in (f).

PATH OF MOTION. A body in motion moves in a definite direction, which is a succession of points or positions it occupies in its course of travel. Allow the central-most particle of a moving body to represent the body mass; then that particle would describe an imaginary line through the several points to form a path of motion. A path of motion may be straight or curved.

To control a body in motion, it must be *constrained,* i.e., suitably connected to other bodies capable of directing such motion against influences that tend to alter motion adversely. Thus constrained, a body may be directed to move along a desired path in a prescribed manner and returned to its original position in a definite time. When a body, or mechanism has performed all its functions once, i.e., has passed through all its positions and returned to its original or starting position, it is said to have completed a *cycle of motion.* A *period* is the time necessary for a mechanism to complete one cycle of motion.

TYPES OF MOTION. Motion may occur along any path, either in a *plane* or in *space.*

When all the particles of a moving body remain parallel to a fixed plane, it is said to have *plane motion.* A block of wood sliding over a table top is an example of plane motion. If the block of wood follows a straight line, all its particles describe straight-line paths, and its plane motion is considered to be *rectilinear.* If all its particles describe identical curved paths while deviating to any degree from a straight line, its plane motion is *curvilinear.* A vehicle rounding a curve in a level road is an example of plane curvilinear motion. In either case, the action is one of *translating,* although the latter term is generally used to describe rectilinear motion.

Plane rotation exists in bodies in which all its particles move in parallel circular paths that remain at constant distances from a fixed axis, which is perpendicular to the fixed plane of motion. A flywheel rotating about a fixed shaft is an example of plane rotation, which is the usual type of motion of machine parts.

Plane motion may also be a combination of translation and rotation. A rotating body that also moves along a straight line described by its axis of rotation is a form of plane motion called *helical motion.* Any particle of a turning nut moving along its threaded bolt is an example of helical motion as is a screw threading its way through a wood board.

A body whose path of motion describes any series of spheres relative to any appropriate plane, and whose every particle remains at constant distances from a fixed point that also acts as a common center for all spheres described, is considered to have *spherical motion.* The centrifugal flyball governor is the classical example of spherical motion. The universal joint is another example.

Space motion is demonstrated by a projectile driven from its course by cross windage.

Motion may be *continuous, intermittent,* or *reciprocating.*

Continuous motion exists if a machine member does not stop or reverse during or between cycles. Rotating bodies such as pulleys and gears are examples of machine members usually subject to continuous motion.

If a machine member stops for a definite interval during each cycle, it is said to have intermittent motion. Indexing mechanisms, some cam arrangements, geneva wheels, etc., have intermittent motion.

A machine member that reverses direction during and/or between each cycle is considered to have reciprocating motion, and pistons are the most common examples.

DISPLACEMENT. Motion being defined as an act of a body during change of position infers that such a change is made with respect to some reference point. Therefore, it implies displacement of the body in respect to its original position or attitude of reference. Displacement may be *linear* or *angular.* The magnitude of linear displacement may be given in inches, feet, or miles, and it applies to bodies having rectilinear or curvilinear translation. Angular displacement is given in revolutions, degrees, or radians, and it applies to rotating bodies such as flywheels, pulleys, and gears. In addition to the magnitude, the *direction* of displacement is an essential factor.

SPEED. Speed is the rate of motion of a body. The term speed expresses magnitudes of distance and time without regard for direction. Speed is expressed as miles per hour, feet per minute, etc. Without reference to direction, the speed factor alone is not sufficiently complete for practical purposes.

VELOCITY. The term velocity is applied to moving bodies as an expression of (1) distance, (2) time, and (3) direction. When these three factors are given, the displacement of a moving body can be exactly determined for any interval of elapsed time after leaving its original position, or position of reference. When direction is accepted as being fixed, either by natural causes as in the case of falling bodies, or by standard practices, such as the clockwise or counterclockwise rotation of an electric motor shaft, the terms speed and velocity may be assumed to be identical by definition.

Velocity is an expression applied to the speed of a body traveling in a stated direction at any given instant only. It implies the distance the body would travel if it were allowed to maintain constant speed throughout the time expressed. It does not necessarily mean that the motion of the body must be maintained in the same direction and at the same speed throughout the entire time expressed. Neither is it limited to the time expressed.

This discussion has considered the velocity of bodies having linear displacement, or *linear velocity*. Linear velocity is expressed in distance unit per time unit.

Angular velocity expresses the direction and rate of speed of bodies having angular displacement, such as rotating pulleys. It is expressed in revolutions per minute, revolutions per second, radians per second, etc. 1 radian is an arc whose length equals the radius of the circle of which the arc forms a part, or 57.3°. 1 revolution equals 360° or 2π radians ($2 \times 3.1416 \times 57.3° = 360°$).

External diameter surface speeds of rotating bodies are often more important than the number of revolutions per unit of time in determining proper lubrication practices.

When a body moves at constant velocity, motion is said to be *uniform*. If the velocity of a body changes from instant to instant, its motion is non-uniform or *variable*.

ACCELERATION. Acceleration is the rate of change of velocity of a body in respect to time and may be expressed by the equation

$$\text{acceleration} = \frac{\text{change in velocity}}{\text{time interval}}$$

or

$$\text{acceleration} = \frac{\text{final velocity} - \text{original velocity}}{\text{time interval}} = \frac{v_f - v_0}{t}$$

then

$$\text{change of velocity} = \text{acceleration} \times \text{time interval}$$

$$\text{average velocity} = \frac{\text{original velocity} + \text{final velocity}}{2}$$

$$\text{final velocity} = \text{original velocity} + (\text{acceleration} \times \text{time})$$

$$\text{distance traveled} = \text{average velocity} \times \text{time interval}$$

where

$$\begin{aligned} v &= \text{velocity at any given time} \\ v_0 &= \text{initial velocity} \\ v_f &= \text{final velocity} \\ v_a &= \text{average velocity} \\ v_f - v_0 &= \text{change in velocity} \\ a &= \text{acceleration} \\ t &= \text{time} \\ d &= \text{distance traveled} \end{aligned}$$

then

$$at = \text{change in velocity}$$

$$v_f = v_0 + at$$

$$v_a = \frac{v_0 + v_f}{2} = \frac{2v_0 + at}{2} = v_0 + \frac{1}{2}at$$

$$d = t(v_0 + \tfrac{1}{2}at) = v_0t \times \tfrac{1}{2}at^2$$

where

$$v_0 = 0$$

then

$$d = \tfrac{1}{2}at^2 \qquad \text{(uniform acceleration)}$$

The equation $d = \frac{1}{2}at^2$ expresses the law that *the distance moved by a uniformly accelerated body starting from rest varies as the square of the time* (t^2).

Acceleration is expressed in terms of velocity change per unit of time for each similar unit of time that a body continues to change velocity at a uniform rate.

One example of both acceleration and deceleration may be shown in the study of a reciprocating pump driven by a crank through a connecting rod, crosshead, and piston rod. The crosshead starts ($v = 0$) from one end and moves at an accelerated rate until it reaches one-half its total travel distance of one stroke, at which point maximum velocity is reached. The instant the crosshead crosses the halfway point, its velocity starts to diminish until it reaches the other end of its stroke, whereupon velocity again equals zero. It may be said that acceleration during the first half of crosshead travel is positive and during the second half, negative.

A body whose speed increases with equal magnitude during each unit of time, i.e., second, minute, hour, etc., is described as having *uniform* acceleration. If the speed changes vary in magnitude during each unit of time, acceleration is said to be *variable.*

The classic example of uniform acceleration is a falling body moving through a frictionless atmosphere. Motion here is caused by the force of *gravity* acting on the body pulling it toward the earth. Acceleration produced by gravity on all bodies is uniform at a magnitude of 32.2 ft/sec/sec. (or 980 cm/sec/sec), which value is represented by the symbol g. Substituting g for a in the equation $v = at$, it becomes

$$v = gt$$

and

$$t = \frac{v}{g}$$

where g always equals 32.2 ft/sec/sec.

When a body falls freely from rest, then

$$d = \tfrac{1}{2} gt^2$$

and for falling bodies having initial velocity,

$$d = v_0 t + \tfrac{1}{2} gt^2$$

The velocity of a body falling from rest through any distance ℓ is determined by the equation

$$v = \sqrt{2gd}$$

Revolving bodies turn about an axis external to its own mass, whereas *rotating* bodies turn about their own axis. For example, the earth rotates upon its own axis, but revolves about the sun. The flyballs of a centrifugal governor revolve about the auxiliary shaft on which they are mounted. Where the distance between a revolving body and its external axis remains fixed the arrangement may be treated as one of rotation. If the distance is variable, as in the case of the flyball governor, other dynamic factors must be considered.

FORCE

The action or energy that changes or tends to change the state of motion of a body on which it acts is called *force.* A body at rest will remain at rest unless an external force acts on it to initiate motion. A body moving along a straight line will continue its straight line path unless an external force acts on it to change its direction. Also, a body will continue to move at constant velocity unless an external force acts on it to increase or decrease its speed.

Force has magnitude. The intensity of a force acting on a body, either at rest or in motion, governs the magnitude of change in the state of motion of that body. According to Newton's Third Law of Motion, the magnitude of change in the state of motion of a body is proportional to the intensity of the acting force.

WORK. Force being an action serves as the criterion that *energy* is present and being exerted. Therefore, energy is the impelling force that creates change in the state of motion of a body to do *work* through the action of force. Therefore, one of the functions of work is force. The second function of work is displacement, or the distance moved by the work element. The relation of the two functions is shown in the formula

$$\text{work} = \text{force} \times \text{distance}$$

or

$$W = f \times d$$

where, usually,

f = force expressed in pounds
d = distance expressed in feet or inches

Work being the product of force, expressed in pounds, multiplied by distance, expressed in feet, results in a combined value called foot-pounds. Work, then, is expressed in foot-pounds.

POWER. To determine the capacity of any work element to do work, a third function, *time,* must be considered. The capacity of a machine to perform useful work is called *power.* Power is the ability of a body to do work in a given period of time, or, as the rate of doing work. The relation of time (t) to force (f) and distance (d) is shown in the equation,

$$P = \frac{f \times d}{t}$$

where

t = time expressed in seconds or minutes
P = power expressed in foot-pounds per second or per minute

The conventional method of expressing power is by ratio of work done in foot-pounds per minute, as compared with an arbitrary value adopted as a standard, an example of which is horsepower (hp), 1 hp being equivalent to 33,000 ft-lb of work performed in 1 min.

To determine the power of any work system in terms of horsepower the following formula is employed:

$$\text{horsepower} = \frac{\text{force in pounds} \times \text{distance in feet moved in 1 min}}{33{,}000}$$

or

$$\text{hp} = \frac{f \times d}{33{,}000 \times t}$$

where

$$t = \text{time in minutes}$$

FORCE BEHAVIOR. In order that force may exist as an action, i.e., to abide by its definition given previously, some kind of opponent or opposing force must also exist. If this were not true, there would be no change in the state of motion of bodies either at rest or in motion. Newton's Third Law of Motion states that for every action, there is an *opposite* and *equal* reaction. Therefore a force defined as an action is always met with an opposing force, or *reaction*. Force, then, is one of a pair of equal, opposite, and *simultaneous* actions that impel change in the state of motion of bodies, or in some instances change the form of bodies themselves. An object resting on a table exerts a force downward due to the pull of gravity measured by the weight of the object. The action of this force is met by a reaction of an equal upward force of the table. The object may be considered as being under the action of two forces, one upward and one downward, both of which are equal, opposite, and simultaneous. However, the magnitude of force is in no way altered by the assumption of pairs of forces, i.e., action and reaction. If the above weighed 10 lb, it would act with a force downward of 10 lb against the table top. The table top would react with a force of 10 lb upward against the object. The action being equal, opposite, and simultaneous relative to the reaction, a total of 10 lb force would exist between the opposing surfaces of object and table. Note that the magnitude of force is equal only to that applied by the action and *not* the sum of the pairs.

Forces, as considered here, are actions that occur between physical bodies, one upon the other. Such bodies may be in direct physical contact or remote. A book resting on a table is an example of bodies in direct physical contact. The gravitative pull of the earth or a magnetic field are instances wherein forces exist between bodies remote from each other.

An *applied force* is one in which opposing bodies exert a *push* or *pull* effort. Where the effort tends to pull the parts of one body apart, the condition is one of *tension*. A weight suspended from an overhead rigid support by a rope is an example of a body (the rope) under tension. If the effort tends to push or squeeze the parts of a body together, a condition known as *compression* exists. In the example given above, the book resting on the table tends to compress the table legs. Any piece of metal acted on by the jaws of a pair of pliers, or a vise, is under compression.

Tension and compression are *internal forces* called *stresses* that result from an application of external forces on the body under stress.

A force may be *concentrated* or *distributed.* A concentrated force is one that acts against an area of a given foundation, the size of which approaches a point. A distributed force acts on an area of a given foundation large enough to spread the action beyond the limits of a point even to the extent of the entire surface of the foundation facing the action. A slender pole supporting a heavy object, and a rope suspending a weight, are examples of bodies subject to concentrated force. A concrete foundation supporting a steam turbine, and the floor of a warehouse over which the contents of the building are reasonably divided, are examples of distributed force. The distinction between concentrated and distributed force may be recognized by considering two positions of a cone, one in which the cone rests on its base and the second in which the cone is inverted to rest on a small area formed by cutting off a small portion of the apex.

To fully describe a force, all of its *characteristics* must be known. Characteristics of the force include *magnitude, direction, sense,* and *point of application.* Magnitude is expressed in total load as ounces, pounds, tons, grams, kilograms or any fraction of such measurements. It may also be expressed in pressure per unit area (psi). Direction may be horizontal, vertical, or at some angle relative to the horizontal or vertical. Direction may be more fully described by denoting whether the force acts upward, downward, right, left, clockwise, counterclockwise, outward, or inward, characteristics described as the sense of the force. The point of application is that point or area on a body to which an external force is applied by physical contact.

TRANSMISSIBILITY OF FORCE. Forces may be *coplanar,* i.e., when all forces of a system lie in the same plane, or *noncoplanar* when all forces of a system do not lie in the same plane.

Coplanar and noncoplanar systems may be further described as follows:

(a) Coplanar, parallel—When all forces of the same system are in the same plane and parallel.

(b) Coplanar, concurrent—When all forces of the same system are in the same plane and intersect at a common point.

(c) Coplanar, nonconcurrent—When all forces of the same system are in the same plane but do not intersect at a common point.

(d) Noncoplanar, parallel—When all forces of the same system are not in the same plane but are parallel.

(e) Noncoplanar, concurrent—When all forces of the same system are not in the same plane but intersect at a common point.

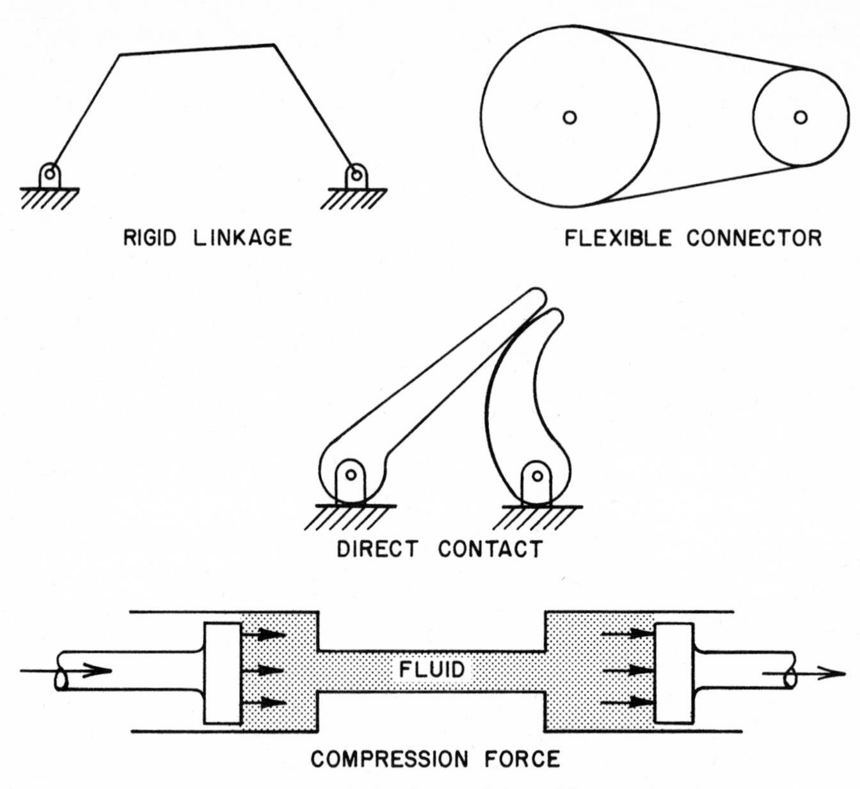

Fig. 5-3. Various methods of transmission of force and motion.

(f) Noncoplanar, nonconcurrent—When all forces of the same system are not in the same plane and do not intersect at a common point.

(g) Collinear—When all forces in a parallel system act along a single line of action.

The forces acting in machines having plane motion are usually in parallel planes. It is therefore standard practice to analyze the system as though all forces were located in the same plane, or if necessary appropriate forces may be analyzed separately as being in a plane perpendicular to the first plane.

THE KINEMATIC CHAIN

Machines are composed of one or more kinematic chains called mechanisms. As previously stated, kinematics is the branch of dynamics dealing with the study of motion without regard for the forces that cause motion. To understand the principles of mechanism, the relation of the various parts to each other must be recognized, especially as regards to motion. Kinematics of machines, then, is the name applied to the analysis of the motion, shape, methods of support, and control of such parts.

DEFINITIONS

Mechanism. A kinetic chain composed of a group of elements or links so related as to produce motion, having a definite geometrical pattern (Fig. 5-4). Kinematically, its purpose is to transmit motion from an input source to another mechanism, or point of production (output).

Link. An element, such as an arm, crank, rod, or frame, that constitutes the mechanism. Kinematically, it is represented by a *bar,* or line. A *rigid link* is capable of transmitting both tensile and compressive forces. *Tension links* are those represented by belts, ropes, or steel bands. *Compression links* transmit energy by either hydraulic (liquid) or pneumatic (gaseous) elements.

Linkage. A group of links and *pins,* or *contactors,* geometrically arranged to form a mechanism, or kinematic chain. Pins in the form of axles, shafts, bushings, etc., maintain permanent contact, while contactors, such as cams, gibs, crossheads, and rollers maintain sliding or rolling contact between adjoining elements. Linkages are identified by the number of bars that make up the assembly.

Pair. A pin or contactor that holds two adjoining links together in permanent or frictional contact. Pairs are identified with the selected geometrical forms and configurations to which they relate.

Pairing elements. The links or elements that form the pair or joint.

Lower pairs. Joints that allow area or surface contact, such as a full bearing and journal arrangement.

Higher pairs. Joints that permit line or point contact, such as balls or rollers in contact with other curved, or flat surfaces, such as some types of cams and wheels.

Closed pairs. Pairing elements that are held in contact by mechanical means, such as the full bearing and journal arrangement mentioned above.

Open pairs. Pairing elements that are held together by gravity, springs, etc.

Turning pairs. Pairs in which one element is allowed to rotate or oscillate in a single plane relative to the other element of the pair. A crank turning on its shaft constitutes a turning pair.

Sliding pair. A pair that allows sliding between the elements forming the pair, e.g., a combination of carriage and way.

Spherical pair. A pair that allows one element to move in any appropriate plane relative to its mating element. The flyball governor and universal joint are representative of the spherical pair arrangement.

Frame. The structural part of a machine that supports the moving

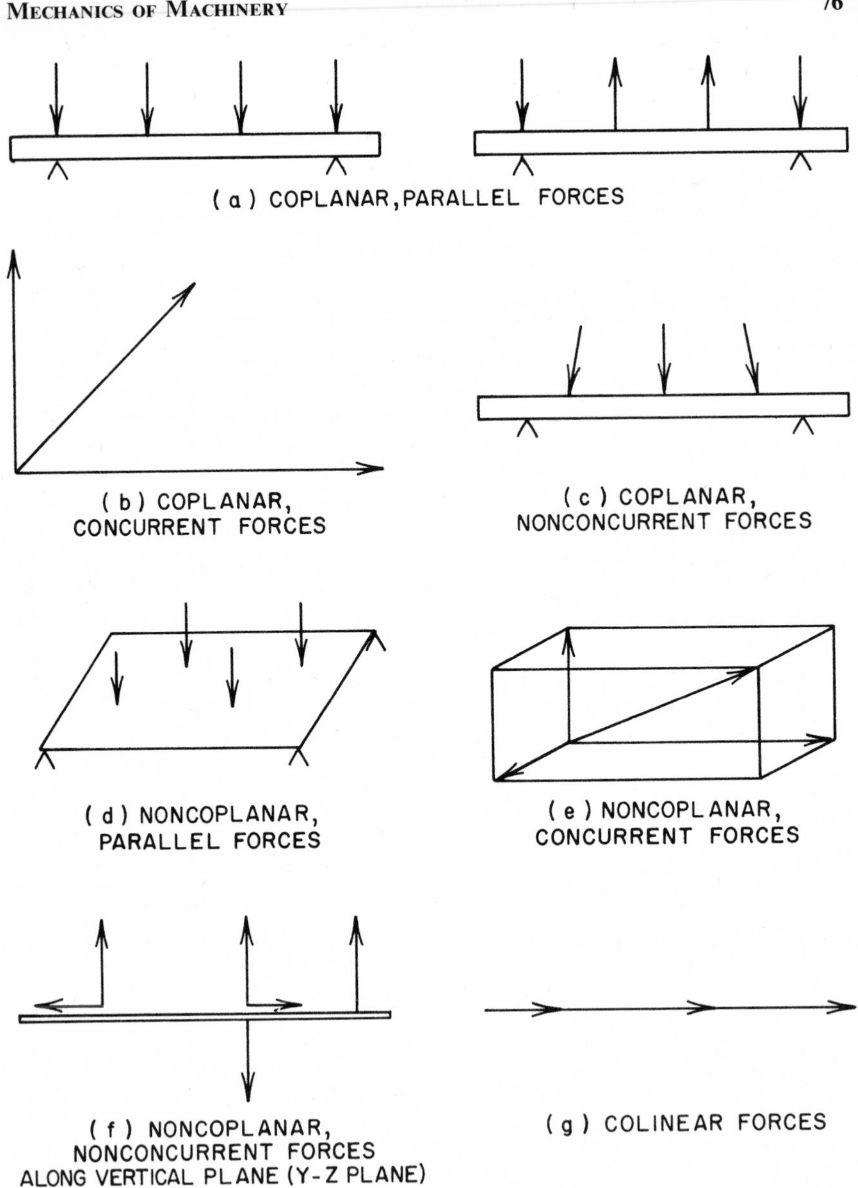

Fig. 5-4. Direction of forces.

mechanisms and upon which the required number of mechanisms are mounted in some orderly manner so as to achieve chronological progression of energy from the source to the point of production.

Bearing. Any surface on which a load is concentrated for support. Kinematically, it is a contacting surface between mating elements having relative motion. Bearings may be stationary or have motion relative to its mating element or to its frame.

Crank. A rigid arm that rotates about another element, such as a shaft or pin, to which it is fixed and whose axis is perpendicular and common to the crank.

Lever. A rigid arm that is fixed at a fulcrum and acted on at two other points by two forces tending to cause it to rotate about the fulcrum.

Fixed pair. Elements having their connecting surfaces in permanent contact with each other.

Instantaneous center. Primarily the concern of machine designers, and variously called *instant center.* It represents the axis of rotation of one element relative to its pairing element at an instant in time at which the two elements have no relative velocity, only absolute velocity. Instantaneous centers may be located anywhere within the plane of motion, i.e., on the links or at infinity, depending on the geometrical circumstances of the instant. The instantaneous center also becomes the *fixed center,* or *permanent center,* when the links referred to are mechanically connected. Therefore, there are as many fixed instantaneous centers as there are pairs.

Where it is necessary to determine the number of instantaneous centers for a mechanism having, say, four links, there are at least four fixed instantaneous centers. Referring to Fig. 5-5e, these fixed centers are located at A, B, C, and D. However, these represent only those links mutually connected, i.e., A combines links 1 and 2; B combines 2 and 3; C combines 3 and 4; and D combines 4 and 1. Other combinations include 1 and 3, and 2 and 4, making six combinations altogether. Thus, the number of instantaneous centers for any given linkage is a constant having the relationship

$$C_i = \frac{N(N-1)}{2}$$

where

C_i = number of instantaneous centers
N = number of links in the kinematic chain

for the chain in Fig. 5-5e, then,

$$C_i = \frac{4(4-1)}{2} = \frac{12}{2} = 6,$$

for a four-link mechanism

To locate the remaining instantaneous centers, designated E and F, the *Aronhold-Kennedy Theorem of Three Centers* will be demonstrated. This theorem states that *when any three bodies have relative plane motion, there are three instantaneous centers,*

$$N = 3 \qquad \text{or} \qquad C_i = \frac{3(3-1)}{2} = 3$$

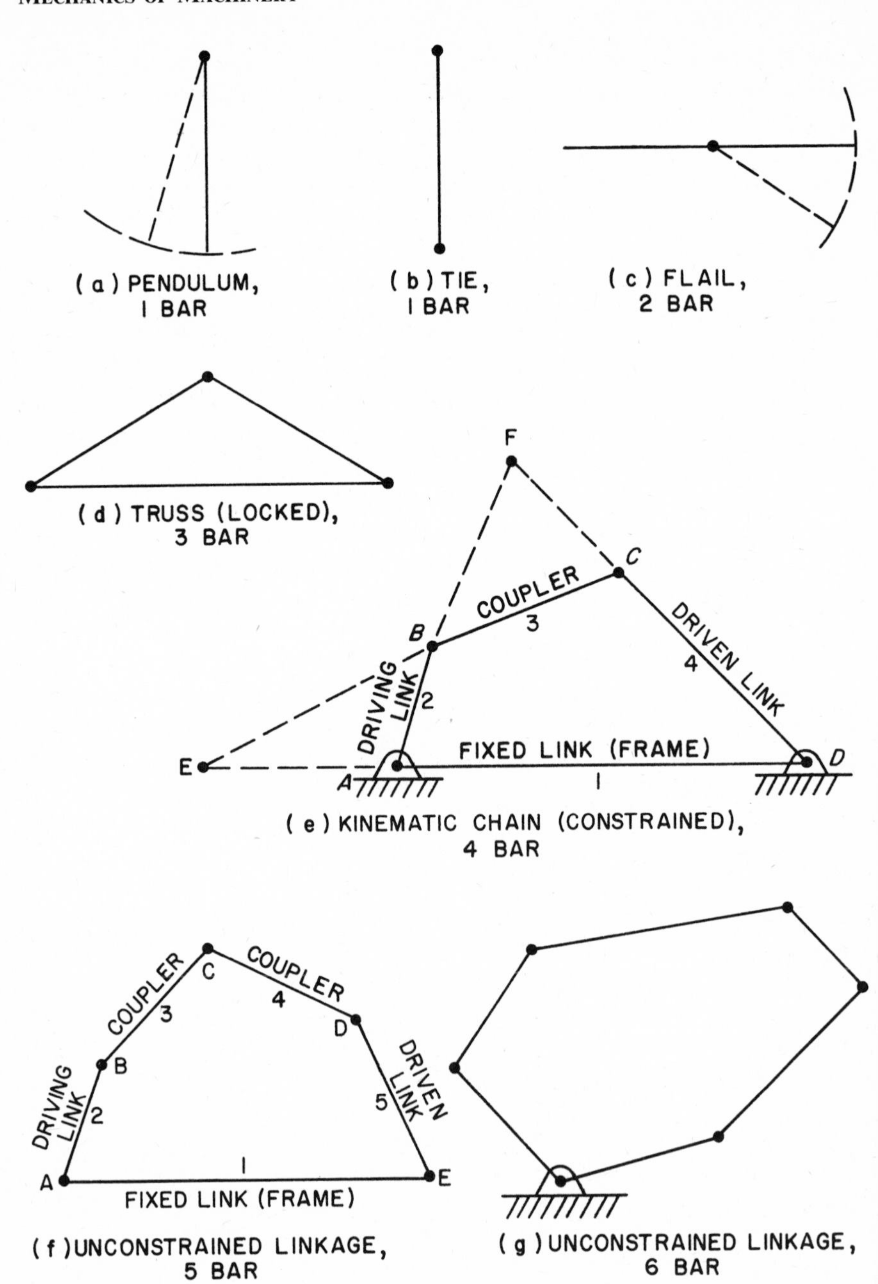

Fig. 5-5. Various linkages.

all of which lie on a straight line. This is demonstrated in Fig. 5-5(*f*), in which four of the instantaneous centers are known by reason of obvious fixed pairs. By extending a straight line *A*–*D* along link 1, to intersect with a straight line extended through *C*–*B* along link 3, the instantaneous center for the combination of links 1 and 3 is established at *E*. Again by extending a straight line *D* to *C* along link 4, and again *A* to *B* along link 2, the instantaneous center for the combination of links 2 and 4 is established at intersection *F*. Thus all six instantaneous centers for the kinematic chain as determined by the equation $C_i = N(N - 1)/2$ are established.

LINKAGES

Kinematics is concerned with linkages and their analysis relative to geometry, displacement, rotation, and motion, i.e., velocity and acceleration. The first concern is with the geometry of linkages, which is the relationship of lines and points that represent links and pairs that form the kinetic chain, or mechanism. The links in turn represent the actual static and dynamic components, such as the frame, arms, push or pull rods, bearings, gears, and cams, which are dimensionally developed as to conformation, size, and strength, after all kinematic solutions are determined.

Putting these lines and points together in various numbers and geometrical forms results in the hypothetical arrangements shown in Fig. 5-5. Before analyzing the linkages, however, consideration of pertinent kinematic laws is in order:

1. A kinematic chain is a *workable* mechanism in which all machine elements have definite motion relative to each other, i.e., if one of the elements is fixed and motion initiated in another, all points of the remaining elements will always move in definite constrained paths. Thus the configuration of all elements is controllable and will respond in the same manner during each cycle of performance.
2. If the length of one link in a chain is greater than the sum of the lengths of all other links, it is not a workable mechanism.
3. A kinetic chain is composed of (a) fixed link (e.g., frame), (b) driving link, (c) coupler, and (d) driven link.
4. Linkages may be *locked, constrained,* or *unconstrained.*
5. If a linkage contains pairing elements whose longitudinal center lines may become aligned in a manner as to form a *dead point* (such as the dead center position of piston-connecting rod-crank of an internal combustion engine), some force must be applied to prevent locking. This force may be supplied either by the momentum of the elements involved or from some external source, such as a flywheel or knockout rod to carry the pair through the dead point position.
6. The motions of mechanisms are the functions of the position of the pairs relative to each other and do not depend at all on their size or shape.

Referring to Fig. 5-5, the following arrangements are found:

(a) One link, fixed at one end—Useful only for purposes similar to that of a free-swinging punching bag or a pendulum.

(b) One link, fixed at both ends, acts only as a tie.

(c) Two links, or bars, connected to each other. Represents the principle of the flail used to thresh grain.

(d) Three bars attached to each other. This represents the *locked* linkage, or truss, in which all links must move as a solid unit. It is a ternary link having three constraints.

(e) Three-bar chain. Represents an oscillating arm that forms a higher pair with an eccentric driver or cam. This arrangement is also an open pair with gravity acting to maintain contact between the driving element a and follower b.

(f) Four-bar linkage is shown having (1) frame, (2) crank (driver), (3) coupler, and (4) another crank (follower). If crank 2 is turned relative to frame 1, then the motion and position of the coupler and driven crank is *constrained,* i.e., their relative position and motions are determinate during any increment of displacement. The four-bar linkage is the most practical of all mechanisms, and except for dead-point situations requires no other augmenting member or force to maintain kinematic control. A determinate mechanism can be developed if the mathematical relationship involving the number of pins and bars (constraints) is a constant having the value of 4. This is demonstrated in the discussion on four-bar linkages below.

(g) Five-bar linkage is an *unconstrained* linkage, i.e., the relative positions of all elements are not determinate. For example, if crank 2 is turned relative to frame 1, it cannot be predicted with any assurance that the driven links will always move in the same path about pin D during repeated cycles of the motion of arm 2.

(h) Six-bar linkage with certain modifications can be made constrained as was true with a three-bar linkage. Ordinarily, however, it is an unconstrained linkage. This latter is usually true with linkage having any number of bars, except those whose relationship conforms to the rule of constant $= 4$.

FOUR-BAR LINKAGE. Inasmuch as a mechanism may be identified on the basis of the number of bars or constraints, consider that a bar is a rigid link, several of which are necessary to form a mechanism. Constraint of each bar requires they be fastened in some manner to another bar. Being rigid and fastened, say by a pin, about which each bar can turn, the following situation develops:

(a) The distance between each pin located at opposite ends of the bar is fixed and equal to the length of the bar.

(b) The motion of the bar governs the motion of the pin connecting the driver to the driven element. It is the responsibility of the designer to have the driven element follow a desired path through every cycle.

(c) One bar is capable of fixing two pins.

(d) It is also true that three bars are necessary to connect three pins, five bars to connect four pins, seven bars to connect five pins, etc., in similar progression.

(e) It is obvious by the number of bars versus pins that such combinations will bring about either a locked, constrained, or unconstrained linkage. As mechanisms depend on constraint, only those combinations resulting in constraint can be of practical use.

As stated previously, the four-bar linkages are kinematically determinate. To determine the classification of a linkage having any number of pins, the following rules may be applied:

$$(2 \times \text{number of pins}) - (\text{number of constraints}) = \text{constant}$$

or

$$2N_p - N_c = \text{constant}$$

If the constant equals 3 the linkage is locked.
If the constant equals 4 the linkage is a practical mechanism.
If the constant is equal to or greater than 5 the linkage is constrained.

Thus a linkage having four bars and four pins, or $(2 \times 4) - 4 = 4$, is a constrained mechanism, a linkage having six pins and six bars, $(2 \times 6) - 6 = 6$, an unconstrained linkage, and in the case of three pins and three bars, then $(2 \times 3) - 3 = 3$, a locked linkage.

INVERSIONS. By fixing each link of a four-bar linkage in turn to a frame, a different mechanism can be developed. Transferring fixed links from one to another is called an inversion of the original linkage. There are as many possible inversions as there are links in the mechanism (Fig. 5-6). A classical example of a four-bar linkage inversion is such machine tools as the shaper versus the planer. By changing the fixed link of the shaper whereby the *tool moves* across the stationary work face, a planer is developed whereby the *work face moves* to allow the stationary tool to perform its work of metal removal.

Inversion of mechanisms may change their absolute motions but does not change the motion of the links relative to each other.

Other inversions are the lathe versus boring mill and conventional gear train (of two mating gears mounted on their own shafts) versus the epicyclic gear train (in which one gear rides around the other).

Links may be of any form providing the desired motion throughout the *chain of mechanism* is not subject to interference. Linkages can be designed to convert motion in the following ways:

1. Continuous rotation into continuous rotation
2. Continuous rotation into reciprocating motion
3. Continuous rotation into oscillating motion

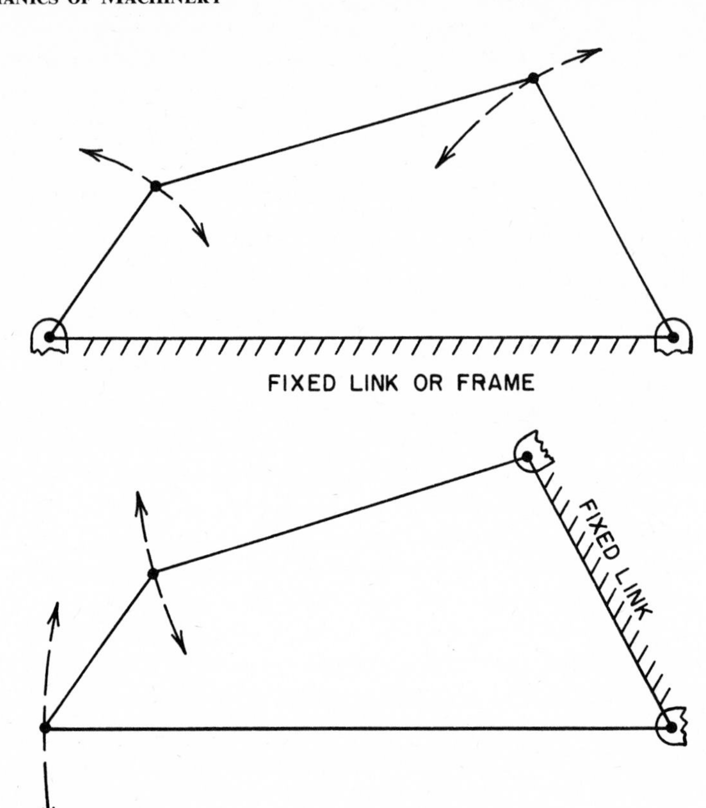

Fig. 5-6. Inversion of four-bar linkage.

4. Reciprocating motion into reciprocating motion
5. Reciprocating motion into continuous rotation
6. Reciprocating motion into oscillating motion
7. Oscillating motion into oscillating motion
8. Oscillating motion into continuous rotation
9. Oscillating motion into reciprocal motion

Chains may be *locked, constrained,* or *unconstrained.* A locked chain is one whose links are incapable of having relative motion in respect to each other. It can move only as one body and so has purely structural application, such as a triangular corner plate. The constrained chain is the basic linkage of all practical mechanisms and is also called a *kinematic chain.* With one link fixed and another set in motion, all other links will move respectively in the same fixed path. An unconstrained chain has no practical application, because with one link fixed, movement of the remaining links cannot be controlled or constrained to repeatedly follow fixed paths.

The simplest form of constrained chain is the four-link mechanism made up of a fixed link (frame support), driving link, connecting link, and follower, or driven link, the last three of which are allowed to turn about the axis (cylindrical) of their pairing elements in a parallel plane. Such links make up the kinematic chain commonly called a mechanism. A mechanism, then, is a kinematic chain having one link fixed. The four-link mechanisms include the four-bar and slider-crank mechanism.

By varying the relative position and linear dimensions of links of a four-bar mechanism and fixing each link in turn, numerous variations in motion (displacements) of driver and follower links can be developed. The necessity of identifying more specifically the function and motion of the various links within a given mechanism gives rise to additional nomenclature other than that mentioned. Thus a rotating link is called a *crank,* an oscillating link is known as a *beam* or *lever,* and a connecting or floating link is described as a *connecting rod.* Usually the fixed link acts as the supporting member or frame of the assembly.

Common types of four-link mechanisms include the drag link, slider-crank mechanism, scotch yoke, toggle, pantograph, parallel linkage, locomotive side-rod drive, and the Corliss valve gear.

A typical machine may be considered to possess certain basic mechanisms that provide (1) the power to drive it, (2) means for controlling speed, (3) ability to perform work for producing a desired result, and (4) necessary control and/or adjustment mechanisms. The source of power may be an electric motor, internal combustion engine, turbine, or line shafting. Transmission of power from source to the machine may be by direct connection, gear train, speed reducer or increaser, variable-speed unit, belt and pulley arrangement, or chain and sprocket drive. The journals of the driver and driven units are usually connected by a coupling of either the rigid or flexible type. The work-producing mechanism may be composed of a series of journals, bearings, gears, slides, chains, cams, etc., that are so connected and coordinated as to produce a desired result at a given rate. Controls are mechanical, electrical, electronic, pneumatic, hydraulic, or some combination of these to govern speeds, tolerances, size, color, etc., that may either stop the machine or make automatic adjustment to attain adopted standards of performance.

All such parts must be connected in some way to provide the necessary force and motion required by the work-producing mechanism and controls. Such connections may involve journals, bearings, gears, chains, cam, pistons, cylinders, slides, and ways that compose the pairing elements, the pairs, links, and mechanisms described previously. All pairs support motion and therefore require some kind of lubricant.

Lubrication personnel must, therefore, be familiar with the frictional characteristics of the various mechanical elements that comprise the entire machine or machines for which they are responsible. They should also be able to readily trace the flow of power of any given machine from its source to production mechanism.

6

TYPES AND PROPERTIES OF LUBRICANTS

Any substance that can reduce friction is a lubricant, but practical application demands the choice of only those that have inherent lubricating characteristics and that are compatible with sound economic practices. Consequently, comparatively few substances are selected for the majority of lubrication requirements.

Some of the more important properties variously required of lubricants (according to the operating conditions) are—

1. Proper fluidity or plasticity under conditions of operation.
2. Film strength commensurate with loads.
3. Chemical stability.
4. Lubricity or oiliness or slipperiness to the degree required by operating factors.
5. Adhesiveness to bearing surfaces.
6. Purity, i.e., freedom from contaminants that detract from the efficiency of the lubricant.
7. Noncorrosive characteristics.
8. Rustproofing capability.
9. Resistance to water-wash.
10. Resistance to foaming.
11. Good sealing properties.
12. Low rate of change in fluidity due to temperature fluctuation (small temperature coefficient of viscosity, or high VI).
13. Fire resistance.
14. Minimum volatility or out-gassing.
15. Resistance to the effect of nuclear radiation.
16. Availability at a reasonable cost commensurate with application.
17. Good emulsifying qualities.

Only a few such properties are present in a natural lubricant, but additional properties are attained by refining and the use of additives.

Classification by State

Lubricants may be gaseous, liquid, plastic, or solid. Their classification according to physical state includes materials and coatings that are self-lubricating. The additives listed under solids are usually not lubricants themselves but contribute important lubricating properties when added to an oil.

GASEOUS LUBRICANTS

- Air, helium, carbon dioxide, and others

LIQUID LUBRICANTS

Mineral oils (from petroleum crudes)
- Straight, or unadulterated
- Compounded with fixed oils or their derivatives
- Compounded with special additives
- Compounded with fixed oils or their derivatives, plus chemical additives, such as polymers and metal soaps

Fixed oils
- Animal (acidless tallow, oil, lard oils, etc.)
- Vegetable (castor oil, rapeseed oil, palm oil, etc.)
- Fish (sperm oil, porpoise jaw oil, etc.)

Synthetic fluids
- Silicones
- Silicate esters
- Phosphate esters
- Polyglycols
- Dibasic-acid esters
- Chlorofluorocarbon polymers
- Fluoro-compounds (fluoroesters)
- Neopentyl polyol esters
- Polyphenyl ethers

Soluble oils or compounds
- Mineral oil compounded with emulsifying agents
- Synthetic fluids compounded with emulsifying agents

Liquid metals

PLASTIC LUBRICANTS

Greases
- Soap-thickened mineral oils
- Soap-thickened mineral oils with special additives
- Soap-thickened synthetic fluids
- Yarn fibers saturated with fluid grease
- Semifluid grease
- Mineral oil thickened with nonsoap gelling agents

Synthetic grease (silicones)
Block grease (nonflowing at room temperature)
Petrolatum, or petroleum jelly

SOLID LUBRICANTS

Graphite
Molybdenum disulfide
Molybdenum diselenide
Mica
Talc, or soapstone
Vermiculite
Polymer films
Soft metal films
Boron nitride
Boron nitride (pyrolytic)
Borax
Silver sulfite
Cadmium iodide
Lead iodide
Tungsten disulfide
Titanium disulfide
Niobium sulfide
Niobium selenide
Tantalum sulfide
Tantalum selenide
Rhenium sulfide
Rhenium selenide
Lead carbonate
Lead monoxide
Wax

ADDITIVES

Metallic phosphates (zinc phosphate)
Metallic oleates (lead oleate)
Metallic chlorides
Metallic sulfides
Metallic stearates (lead stearate)
Metallic oxides
Metallic oxalates

GASEOUS LUBRICANTS

Gases used as lubricants include air, helium, nitrogen, and hydrogen. Because air is readily available it is used more than other gases. Gaseous lubricants at present are used mostly for lighter bearings, as for the inertial gyroscopes used for navigation. Heavier applications include the thrust bearings of some vertical turbines and the rotating mount of large telescopes. In gas-lubricated bearings, the coefficient of friction approaches zero.

An advantage of lubricating gases is that their viscosity is not greatly affected by temperature change, so that they perform efficiently through a wide range of applications. The viscosity of air under atmospheric conditions is 0.018 cP. Gases are helping to solve the lubrication problem where temperature extremes are encountered, as in the fields of space exploration and cryogenics. Since they are not adversely affected by radiation, gases are ideal hydrostatic lubricants for use in nuclear environments.

Practical use of gaseous lubricants considers those conditions capable of developing either a hydrodynamic or hydrostatic film. As true with liquid lubricants, a hydrodynamic gaseous film depends on the formation of a wedge of gas between the surfaces of bodies in relative motion, while a hydrostatic film is developed by external pressure. Gases do not possess nearly the amount of fluid inertia offered by liquids, and their inertia may be considered negligible. This leads to streamline or laminar flow even at high velocity, which makes them useful in high-speed bearings. A running in process, whereby the high spots of a bearing are worn off to develop a smoother surface, is possible and often applied to liquid-lubricated surfaces. The process cannot be applied to gas-lubricated bearings.

Gas-lubricated bearings must necessarily be precision built, especially those that are to operate at high temperatures and pressures. This means that surface roughness must be held to within 5 microinches rms. When allowed to rest, the surfaces of opposing members make physical contact. Some external means must therefore be provided during initial startup to protect such surfaces against wear. This becomes more important if startup is initiated under load. Under light load or no load, a rounding of the inlet edges may be sufficient for developing an instant film. Under relatively heavy loads, some means of relieving the load, such as retraction, should be furnished during the starting period. A bearing can be lifted from the rotor either mechanically or hydrostatically.

An essential requirement of gas-lubricated bearings is almost perfect alignment. Since perfection cannot be built into a bearing, it becomes necessary to supply some method of self-alignment during operation. This is accomplished by providing a cushioning arrangement such as a diaphragm on which a journal bearing is mounted. Thrust bearings are equipped with pivots with free motion in several directions so as to maintain a constant relationship between all areas of both stator and rotor. Such arrangements are known as *gimbals*.

Other advantages of using gas as a lubricant include its low friction, freedom of contamination, and nondrip properties (as encountered with oil and grease).

The coefficient of friction of a hydrodynamic film, developed on a thrust bearing of the Kingsbury type, has been found to be as low as $f = 0.001$. Hydrostatic films of air or oil developed by external pressure have an extremely low coefficient of friction, having been measured as low as 0.00000075 with oil. It could be expected to be even lower with air as the lubricant.

Where loads are light, the compressibility of air can be neglected for purposes of mathematical analysis. But heavy loading does have an effect on performance which should be considered when such calculations are made. Thus a compressibility effect that reflects any change in the volume of the air passing through the bearing should be considered. Otherwise, the conventional equations for noncompressible fluid may be used.

Because of its nondrip property, air is an ideal lubricant for certain bearings employed in the food, textile, and chemical industries, where contamination must be avoided. Air allows wide flexibility of speed, from less than 1 rpm under hydrostatic conditions to upwards of 100,000 rpm under hydrodynamic conditions.

Because the fluidity of a gas decreases (its viscosity increases) as its temperature rises, air-lubricated bearings will operate in extremely hot zones, with the temperature limited generally only by metallurgical considerations.

Hydrostatically lubricated bearings are supplied with lubricant from an external pressure source (Fig. 6-1). Pressurized lubricants can be introduced into the bearing either radially or longitudinally. The hydrostatic method of applying a lubricating film is usually used on bearing systems

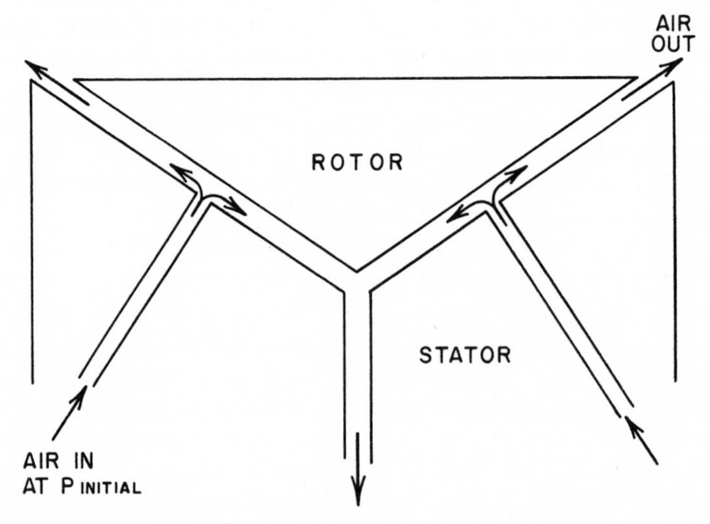

Fig. 6-1. Schematic cross section of hydrostatic (gas-lubricated) cone step bearing.

in which the relative motion in respect to stator and rotor is too low to develop a hydrodynamic film of the self-acting type.

For hydrodynamic conditions, the minimum film thickness for a thrust bearing can be determined by the equation $h_0 = \sqrt{0.066\mu V\ell/P_{av}}$ where h_0 = minimum film thickness, μ = absolute viscosity, V = runner velocity (mean), ℓ = length of shoe in direction of rotation, and P_{av} = average thrust pressure. Viscosity of air at 70°F and 29.69 in. Hg is about 0.018 cP.

LIQUID LUBRICANTS

Mineral Oils

The most important group of liquid lubricants comprises the *mineral oils* produced from petroleum crudes by distillation and a series of refining processes. Petroleum crudes are complex compounds of carbon and hydrogen called *hydrocarbons.* They are composed of 83 to 87% carbon and 11 to 14% hydrogen by weight. Sulfur, nitrogen, and oxygen may also be present in small amounts, their presence and content varying according to the source of the crude.

Crudes are classified according to their predominant hydrocarbons, of which there are four basic series: Paraffins, Naphthenes, Asphaltenes, and Aromatics.

The paraffins and naphthenes are the important hydrocarbons as related to lubricating oils because of their greater chemical stability and also because they represent the largest classes of hydrocarbons. Crudes whose predominant hydrocarbon compounds consist of paraffins are called *paraffinic base crudes.* Those in which naphthenes predominate are known as *naphthenic base crudes.* Crudes having about equal parts of paraffins and naphthenes are called intermediate, or *mixed base crudes.*

The districts within the United States in which the three base crudes generally occur are shown in the table.

Base Crude	*Typical Source*
Paraffinic crudes	Eastern district
Naphthenic crudes	Gulf Coast and California districts
Mixed base crudes	Mid-continental district

Crudes are not made up solely of paraffins or naphthenes but contain both in varying amounts. Pure paraffinic or pure naphthenic base crudes do not occur in nature. Because the chemical structure of crudes varies considerably, including those from the same district, their classification shown in the table is very general.

The United States Bureau of Mines in their publication RI 3279,

September 1935, entitled "Base of a Crude Oil" presents a more thorough method for classifying crudes. Divided into nine classes, the crudes are identified with the relative boiling points (fractions) of the two basic hydrocarbons—paraffins and naphthenes—that make up the crude in each case. The first term of each classification identifies the hydrocarbon having the lighter fraction, or whose boiling point falls within the range of 182°F to 527°F at a pressure of 29.92 in. Hg. The second term identifies the hydrocarbon having the heavier fraction or whose boiling point is within the range of 527°F to 572°F at a pressure of 1.6048 in. (40 mm) Hg. The following shows the nine basic classes of crudes presented by the Bureau of Mines:

1. Paraffin base, in which the distillates are paraffinic throughout.
2. Paraffin, intermediate base, in which the light fractions are paraffinic and the heavy fractions are intermediate.
3. Intermediate, paraffin base, in which the light fractions are intermediate and the heavier fractions are paraffinic.
4. Intermediate base, in which the distillates are intermediate throughout.
5. Intermediate, naphthene base, in which light fractions are intermediate and heavy fractions are naphthenic.
6. Naphthene, intermediate base, in which the light fractions are naphthenic and the heavy fractions are intermediate.
7. Naphthene base, in which distillates are naphthenic throughout.
8. Paraffin, naphthene base, in which light fractions are paraffinic and the heavy fractions are naphthenic.
9. Naphthene, paraffin base, in which the light fractions are naphthenic and the heavy fractions are paraffinic.

Paraffins have the general chemical formula C_nH_{2n+2}, with the number of hydrogen atoms *always* two more than twice the number of carbon atoms. The simplest member of the paraffin hydrocarbon family is methane, having the formula CH_4 arranged as follows:

```
     H
     |
  H—C—H
     |
     H
```

Ethane has the formula C_2H_6, or

```
     H  H
     |  |
  H—C—C—H
     |  |
     H  H
```

and propane, C_3H_8, appears as

```
    H  H  H
    |  |  |
H—C—C—C—H
    |  |  |
    H  H  H
```

The molecular formula for normal butane, C_4H_{10}, is constructed as

```
    H  H  H  H
    |  |  |  |
H—C—C—C—C—H
    |  |  |  |
    H  H  H  H
```

Note that all paraffinic molecules have their atoms linked chain-like together in a manner portraying saturation, i.e., all atoms are satisfied or joined to an atom of another element according to their valence. Paraffins are a chain series of hydrocarbons that are chemically stable, and more so generally than the naphthenes.

Naphthenes, another family of hydrocarbons with relatively good chemical stability, have the general formula C_nH_{2n}, or always having twice as many hydrogen atoms as carbon atoms. Their general construction takes the form of a ring of atoms, as shown below for cyclobutane, with the formula C_4H_8:

```
    H  H
    |  |
H—C—C—H
|   |  |  |
H—C—C—H
    |  |
    H  H
```

The aromatic hydrocarbons, with the general formula C_nH_n, are an unsaturated group composed of equal numbers of carbon atoms and hydrogen atoms and whose ring structure is shown for one member of the family, benzene, C_6H_6:

```
        H
        |
        C
      /   \\
H—C       C—H
    ||      |
H—C       C—H
      \   //
        C
        |
        H
```

As produced from the well, petroleum crudes are not composed of simple molecules of paraffinic, naphthenic, or aromatic structure, but may contain any number or all in combination. Lubricating oils produced under conditions that allow some unsaturated hydrocarbons to remain will be subject to the formation of oxy-materials, i.e., unsaturates that combine with free oxygen or other unsaturates to form sludges, gums and other undesirable products, especially when subjected to heat. Early reactions occur when poorly refined oils are subjected to normal operating temperatures (110 to 140°F) for extended periods or for much shorter periods at higher temperatures prevalent in internal combustion engines. Good lubricants are made, not found.

Fixed Oils

Fixed oils are fatty substances that are extracted from animals, vegetable matter, and fish. They are called fixed oils because they will not volatilize without decomposing. When cooled sufficiently, fixed oils become fats. The temperature of solidification varies with the kind of oil.

Fixed oils will combine with oxygen of the air and become gummy. The process is known as *drying*. Some fixed oils dry readily, while others combine with oxygen at a much slower rate. The former types are called

Table 6-1. SOME CHEMICAL CONSTITUENTS AND PROPERTIES OF FIXED OILS (AVERAGE)

Fixed Oil	*Specific Gravity at 60° F*	*Saybolt Viscosity at 100° F*	*Saponification Number*	*Iodine Number*	*Titer (°C)*	*Principal Fatty Acids*
Coconut	0.96					Acedic ($C_6H_{12}O_2$ to $C_{14}H_{28}O_2$)
Castor	0.964	1370	180	85	21	Ricinoleic, stearic
Cottonseed	0.921		19.5	110	2–3	Linoleic, oleic, palmitic
Corn		180	187	115		Oleic, linoleic, stearic
Lard	0.915	100–210	194	70		Oleic
Linseed	0.935	130	150	180	19	Linoleic
Mineral	Varies	Varies	None	12	Varies	Normally free
Neatsfoot	0.916	220	197	70	26.5	Oleic
Olive	0.915	190	192	85	20	Oleic, palmitic, linoleic
Palm	0.923		20	55	40	Palmitic
Palm nut	0.978		25	14	22	—
Peanut		330	191	101	40–45	Oleic, linoleic
Rapeseed	0.915	260	174	95	13	Erucic
Rapeseed (blown)		900–1000	206	60		—
Sperm	0.88	105	130	80	11	(Liquid waxes)
Tallow	0.95	225	194	65	45	Stearic
Porpoise jaw			25	49		Sovaleric
(unstr.)	0.926		14.4	77		

drying oils and are used in paint manufacturing, the most familiar of which is linseed oil. The slower drying types are known as *nondrying oils* and include the types associated with lubrication. The tendency of a fixed oil to dry is determined by its *iodine number*. The iodine number is the amount (percent) of iodine absorbed by an oil under controlled test conditions. The higher the iodine number, the better the drying qualities of the oil.

Fixed oils are composed of fatty acids and alcohols, the radicals of which are combined to form fatty esters. A means of determining free and combined fatty acids in a fixed oil is the saponification test. The relative values of various fixed oils in respect to fatty acids can be determined by their *saponification number*. The saponification number is the number of milligrams of potassium hydroxide taken up by 1 gram of oil under controlled test conditions. The higher the saponification number, the greater the net amount of base-consuming constituent present in the oil. Table 6-1 lists some of the more important fixed oils, showing average iodine and saponification numbers, and their Saybolt viscosities at 100°F.

The lubricity and emulsifying characteristics of some fixed oils make them appropriate for use as additives for mineral oils. Typical uses for several of the more suitable fixed oils are given below.

Fixed Oil	*Use*
Tallow (acidless)	An emulsifying agent for steam cylinder oils.
Lard oil	Metal-cutting oil, both straight and blended with mineral oil.
Castor oil	Blended with mineral oil to produce castor machine oil.
Rapeseed oil	Blown rapeseed oil is blended with mineral oil to produce an emulsifying lubricant used in marine service, e.g., stern tube lubricant.
Palm oil	Metal roll oil. Fluxing dip in tinplating of steel, either straight or blended with mineral oils. Also acts as a covering fluid to prevent molten tin from oxidizing.
Olive oil	Textile finishing and a yarn lubricant.
Peanut oil	Has been used blended with diesel engine fuel to lubricate injector pump plungers. Also as a blending agent in lubricating oils.
Sperm oil	An excellent lubricant used as an ingredient for metal-cutting oil bases, forming and drawing oils. Sulfurized sperm oil is an effective antiwear agent for lubricants and metal-cutting compounds.
Porpoise jaw oil	A stable fixed oil of high oiliness employed as a lubricant for watches, clocks, and control and recording instruments.
Fish oils	Menhaden, sardine, and herring oils are sometimes used in lubricating greases and are usually hydrogenated to eliminate their odor.

Castor, lard, tallow, rapeseed (blown), and sperm oils represent the more important lubricating types of fixed oils. They are very seldom used alone, but usually are employed as an additive to increase oiliness or promote emulsibility of mineral oils. Lard oil, acidless tallow oil, and degras (lanolin) are used principally with steam cylinder oils, where they lend emulsibility and heat-resistant qualities. Tallow is used in manufacturing greases. Lard and sperm oils are blended with mineral oils to produce metal-cutting oils. Rapeseed, when blown with air, increases viscosity considerably. It is used with mineral oil for lubricating stern tubes and other water-washed bearings. Palm oil is an excellent metal-drawing and roll oil. Porpoise jaw oil is used for lubricating watches and other small timepieces. Olive oil is employed in the combing of wool in the form of a soluble oil, known as wool oil. Corn oil is an excellent base for molybdenum disulfide, enabling the latter to adhere more readily to metal surfaces. Rosin oil is used in the manufacture of cold-set greases. Castor oil, like all other fixed oils, is an excellent lubricity agent. It has been used for lubricating racing cars, but it forms hard gummy deposits when exposed to heat, as will any other fixed oil.

Extracts of fixed oils, such as oleic acid, possess high oiliness properties and film strength characteristics. Oleic acid is used to improve the lubricating qualities of mineral oils and is particularly valuable as an additive in oils employed for boundary lubrication. Oleic acid, or elaine, as it is commonly called, or oils containing even small amounts of the fatty acid, should not be used in galvanized containers or applicators, with which it sets up a chemical action. About 1% of oleic acid is all that is required to be blended with mineral oils for maximum performance.

Mineral and fixed oils are considered *natural* substances. The fluids discussed in the next section are *synthetic.*

Synthetic Fluids

To meet the extraordinary demands placed on lubricants by the progress of modern technology, it is sometimes necessary to supplement the established mineral oils. It is also necessary to develop new kinds of lubricants when the products of petroleum crudes cannot even be considered because of their specific and related deficiencies. In the case of supplements, for example, hydraulic fluids exposed to ignitable environments, such as molten die-casting operations, welding production lines, and some steel mill operations (such as fettling), must be fire resistant as a safety precaution. While high quality petroleum oils render excellent service in most hydraulic systems where the chance of their starting a fire is remote, such as those employed on conventional machine tools, presses, elevators, and hoists, they are flammable and should not be used where any fire hazard exists. Environments requiring new and special types of lubricants include those with extremes of temperature, high vacuum, nuclear radiation, and chemical reducing effects.

Variations in certain properties of synthetic fluid lubricants, such as the viscosity and vapor pressure of silicones, can be made by changes in the size of certain molecules and by substituting one group of atoms for another. The more important synthetic fluids contain compounds that have been synthesized to obtain a desired intrinsic quality, particularly that of thermal stability and resistance to sustained ignition (Table 6-2). They include the polymers and esters, which embrace the acids, glycols, ethers, and silicones. Development in the field of synthetic lubricants continues in the race against further novel and more demanding physical, thermal, and chemical requirements.

The synthetic fluids most commonly used commercially are the silicones, polyglycols, phosphate esters, dibasic acid esters, and silicate esters.

SILICONES. Silicones are compounds synthetically formed by synthesizing certain constituents and organic groups, which are then polymerized to produce various products, which include fluid lubricants (oils), resins, and silicone rubbers (elastomers). It is the silicone oils that are of interest here and which will be discussed briefly considering the wide complexity of their formulation and preparation.

The relationship of the chemical constituents may be shown by their arrangement in the general structure shown below:

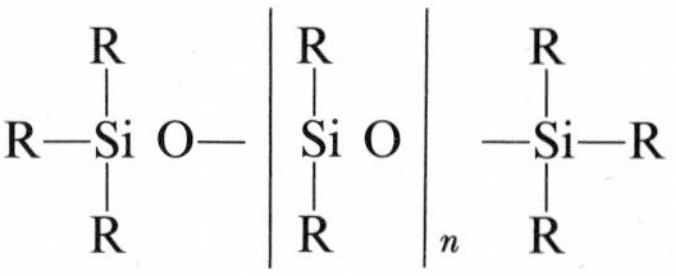

$$R-\overset{R}{\underset{R}{Si}}\ O-\left|\ \overset{R}{\underset{R}{Si}}\ O\ \right|_n-\overset{R}{\underset{R}{Si}}-R$$

in which the basic elements silicon and oxygen form the inorganic main structure and R the organic groups, which are attached to the silicon atoms. The organic groups are usually methyl or phenyl or some combination of both. With the substitution of R by the methyl (alkyl) group CH_3 the structure takes the repeating linear form presenting the polymer (compounds made up of many of the same type of continuously repeating chemical units) called dimethyl polysiloxane. The siloxane is one of a class of compounds

Table 6-2. SOME CHARACTERISTICS OF SYNTHETIC FLUID LUBRICANTS

Type	*Lubricating Capability*	*Oxidation Resistance*	*Thermal Stability*	*Viscosity Index*
Silicones	Poor	Good to approx. 450°F	Very good	180+
Polyglycol esters	Good	Poor	Fair	Up to 150
Diesters	Good	Fair	Very good	Up to 170
Fluorocarbons	Good	Excellent	Excellent	Low

composed of alternate silicon and oxygen atoms in their structure (linear or cyclic) and often containing methyl or phenyl type radicals as replacements for some or all hydrogen atoms:

$$
\begin{array}{cccc}
CH_3 & & CH_3 & \\
| & & | & \\
Si & -O- & Si & -O \\
| & & | & \\
CH_3 & & CH_3 &
\end{array}
$$

Having only two units, this structure represents a *dimer,* molecular weight of 162, which corresponds to a low viscosity of 0.65 cSt. Increasing the number of units by one in each case from two to nine, the resulting compounds would be identified by the words trimer (3), tetramer (4), pentamer (5), hexamer (6), heptamer (7), octamer (8), nomamer (9), and polymer (many). Silicones in which the units represent more than one type of repeating substituent are identified as *copolymers.* Dimethyl silicone lubricants are polymers. Phenyl, chlorophenyl, and fluoroalkyl silicone lubricants are copolymers. As the number of chemical units increases so does the viscosity, which ranges from 0.65 cSt at 77°F to over 1,000,000 cSt at 77°F. To prevent any further change in molecular weight as a result of condensation, the ends of the chain are blocked by the attachment of unreactive compounds, such as the trimethylsiloxy groups.

Another change in the silicone oil that occurs with any lengthening of the molecular chain is in the *surface tension,* which is relatively low when compared with that of mineral oils. For dimethyl silicones, the surface tension ranges from about 16 dyn/cm for very low viscosity fluids up to a maximum of about 21 dyn/cm at 50 cSt at 77°F.

Silicones behave like Newtonian fluids up to 1000 cSt and a shear rate of 10,000 sec^{-1}. Above these values dimethyl silicones depart Newtonian behavior to become non-Newtonian in nature, having reversible thixotropic properties, i.e., they return to their original apparent viscosity when the effect of shear is removed and the substance allowed to rest.

End-blocked silicone chains tend to maintain a somewhat flat viscosity-temperature slope, or high viscosity index. The small change in viscosity holds true over a wide range of temperature variation so that they remain usefully fluid at both high and low temperatures.

Dymethyl silicone fluids possess high chemical stability, because of their inertness in regard both to heat exposure and to contact with other substances that are otherwise reactive. Their incompatibility with most substances precludes their fortification by the use of the lubricant additives in general use. This characteristic necessitates that improvements in most directions must be built into the polymer as one of the substituents.

For example, because of the low surface tension and inertness of dimethyl silicones, they lack sufficient adsorption qualities and chemical interaction to form an adequate lubricating film under boundary conditions and therefore are not recommended for thin film lubrication. However, they possess many desirable qualities as hydrodynamic lubricants, including proper viscosity (as required), high shear resistance, antifoam characteristics, and relatively low volatility, besides chemical stability, high viscosity index and excellent heat resistance. Dimethyl silicone fluids are limited in their temperature of thermal decomposition, although it is quite high relative to mineral oils. The process of building-in specific improvements may begin with creating better thin-film lubrication and raising the thermal decomposition temperature.

The substitution of halogens, in the form of either chlorine or fluorine, into the molecular structure improves their lubrication value. Such substituents result in chlorophenyl-methyl and trifluoropropyl methyl silicone fluids, neither of which equal the lubrication performance of mineral oils fortified with lubricity and/or film strength additives. As measured on the Shell Four-Ball wear testing machine, a silicone fluid of the halogen-phenyl type compounded with an organo-metallic additive (made soluble by copolymerizing with dimethyl siloxane) has shown improved lubrication properties on steel-to-steel sliding surfaces.

PROPERTIES OF SILICONE OILS. The silicone family of lubricating fluids possesses many redeeming qualities that merit their use as lubricants and as hydraulic fluids and in high shear service, such as in torque converters.

Viscosity. The silicones increase in viscosity to over 1,000,000 cSt as their molecular weight is made greater. They can therefore be made available at numerous viscosity levels, but they demand, and economics requires, that only a limited number of silicone fluid lubricants at suitable viscosities be manufactured.

Volatility. Relatively low.

Flash point. Dependent on volatility (size of polymer), ranging up to 600°F for phenyl-methyl ratios of 1.30.

Thermal stability. Good to excellent.

Oxidation resistance. The silicones can withstand long service at high temperatures, i.e., they are highly resistant to heat and oxidation. Because their oxidation rates are low, they will maintain their desirable properties for long periods, even when exposed to oxidizing conditions. The deteriorating threshold temperatures vary with the type of silicone, ranging from about 600°F for dimethyl fluids to about 700°F for high phenyl methyl silicones under nonoxidizing conditions. In service, the temperatures should not exceed 300°F for dimethyl silicones and about 400°F for the high-phenyl fluids.

Hydrolytic stability. Silicones resist the action of water.

Shear. Dimethyl silicones show a low change in viscosity, even under extreme shear conditions. They are greatly superior to thickened mineral oils in this regard.

Viscosity-temperature coefficient (VTC). Determined by the formula

$$\text{VTC} = 1 - \frac{\text{viscosity at } 210^\circ\text{F}}{\text{viscosity at } 100^\circ\text{F}}$$

The VTC of silicones is low, ranging from about 0.60 for dimethyl fluids to approximately 0.90 for phenyl methyl (ratio 1.3:1) silicone fluids. This compares with a VTC of 0.892 for a mineral oil having a VI of 95. The dimethyl and low-phenyl methyl (ratio 0.05:1) fluids have a VTC ranging from 0.57 to 0.62, showing the excellent viscosity-temperature relationship of silicones.

Compressibility. They are compressible and remain fluid in the intermediate and higher viscosity ranges, but the viscosity increases to a high degree under heavy loads. Compressibility is measured by the percentage of reduction in volume based on atmospheric pressure (14.7 lb/in.^{2}) as compared with the applied pressure, or

$$\text{compressibility} = \frac{\text{volume (atm pressure)} - \text{volume (applied pressure)}}{\text{volume (atm pressure)}} \times 100$$

Boiling point. Depends on viscosity (size of polymer), but will not boil when the viscosity is above 50 cSt at 77°F under any pressure.

Freezing point. Dependent on viscosity, increasing as the viscosity increases and ranging from −8°F to below −140°F.

Specific gravity. Close to that of water, ranging from about 0.95 to 1.10.

Surface tension. Low compared with mineral oils, the latter showing values of 30 to 35 dyn/cm as compared with a maximum of 31 dyn/cm for the dimethyl silicones. The maximum surface tension is about constant for silicone fluids above viscosities of 50 cSt.

USES OF SILICONE OILS. The high shear resistance of silicones makes them excellent mediums for fluid type torque converters. Inertness and chemical stability assure their continued high-temperature bearing service for extended periods. The low viscosity-temperature coefficient displayed by some silicones is advantageous in meeting both high and low ambient temperature conditions and for requirements calling for minimum viscosity change with temperature. Their overall characteristics are conducive to their application as lubricants under thick-film sliding and rolling action. They show good results in the lubrication of ball bearings and slider bearings whose opposing surfaces are steel and bronze. They are excellent damping fluids.

PHOSPHATE ESTERS. An ester is a compound of an alcohol radical and an acid. An *ortho*-type acid is a class of acids which contains the same elements in different proportions, such as the orthophosphoric acid (H_3PO_4). *Tertiary* substances are formed by the substitution of all three hydrogen atoms of an ortho-compound by either an aromatic radical (aryl), such as phenyl, or an alcohol (alkyl) radical. By replacing the hydrogen atoms in phosphoric acid with an organic aryl (phenyl), or alkyl group (alcohol), or both, they become tertiary orthophosphate esters classified as triaryl phosphates, trialkyl phosphates, and alkyl aryl phosphates. Phosphate esters are substances whose characteristics are governed by organic radicals of molecular structure.

The structure of the original molecule of phosphoric acid may be represented by

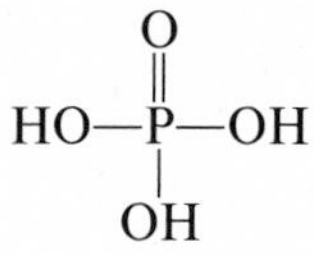

$$\begin{array}{c} \text{O} \\ \| \\ \text{HO—P—OH} \\ | \\ \text{OH} \end{array}$$

or H_3PO_4.

The structure would convert to the following for an orthophosphate ester:

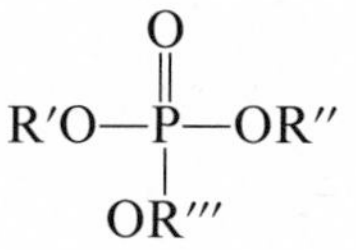

$$\begin{array}{c} \text{O} \\ \| \\ \text{R}'\text{O—P—OR}'' \\ | \\ \text{OR}''' \end{array}$$

if the hydrogen atoms are replaced by the organic radicals of the same or different groups represented by R. A *secondary* phosphate ester is formed when two of the R groups are alkyl or aryl and one is replaced by hydrogen. A *primary* phosphate consists of two hydrogen groups and one of either the aryl or alkyl group.

The tertiary phosphate esters are excellent wear agents and comprise the principal base ingredient of synthetic lubricants. They are widely used in the field of fire-resistant hydraulic fluids.

The properties of phosphate esters vary extensively according to the combination of the R groups and to the particular R group or groups attached to the phosphate moiety. Synthetic lubricants having required characteristics may be obtained by blending different phosphate esters.

The commercial types of phosphate esters have the desirability of being inherently fire resistant. They constitute the backbone of the fire-resistant hydraulic fluid requirement for the industrial, aircraft, and marine industries. The inherent fire-resistant property of phosphate esters should not be interpreted as nonflammability. They will burn under certain conditions.

In addition to their hydraulic capabilities as fire-resistant fluids, some tertiary phosphate esters serve as antiwear agents in mineral oils. Tricresyl phosphate, with a viscosity of 38 cSt at 100°F, is one of the more important lubricating additives.

Because of the large number of different tertiary phosphate esters, any property classification other than a general description would be impractical. The following property description of this class of synthetic liquids, therefore, applies generally to those in wide commercial use as hydraulic fluids and lubricants.

PROPERTIES OF PHOSPHATE ESTERS. The phosphate esters have certain properties which make them well suited to applications where fire resistance is important.

Fire resistance. Phosphate esters are inherently fire resistant but are not non-flammable. Their autogenous (self-produced) ignition temperature ranges from about 970°F to over 1200°F. Their hot manifold ignition temperature is usually slightly above the autogenous ignition temperature by 50 to 90°F. Flash tests with a high-pressure spray pattern ignited by an oxyacetylene torch show self-suppression of flash as the stream leaves the ignition zone, which implies a flammability rating of good to excellent.

Viscosity. A broad range of viscosities is available, from extremely low to very high.

Viscosity-temperature coefficient. Alkyl type esters have a VTC equal to that of petroleum oils, whereas the VTC of triaryl esters is lower. This property is not important in such services as hydraulic systems, in which temperatures remain fairly constant. The VTC here may be rated fair to good.

Thermal stability. Phosphate esters remain stable up to certain threshold temperatures. For general service, a limit of 300°F is established for many phosphate esters. For the service in which they predominate as fire-resistant fluids, the limit is sufficient to provide adequate protection for extended periods. Some triphenyl esters, however, have thermal decomposition temperatures of higher than 900°F, but because of the presence of catalysts and other disturbing influences involved in commercial systems, 900° should not be considered for practical use.

Antiwear characteristics. The polar characteristics of phosphorous compounds offer lubricating film development tendencies conducive to those qualities required to resist wear. One of the outstanding esters for good antiwear characteristics is tricresyl phosphate. Certain primary and secondary phosphate esters are excellent wear agents. Such esters include the dithiophosphates, particularly zinc dithiophosphate. They are used as additives to meet the severe lubrication requirements of the new hydraulic pumps for which selected MS-sequence-approved (petroleum-base) motor oils have been recommended.

Hydrolytic stability. This is the property of fluids that measures their resistance to degradation due to the presence of water. Degradation leads to the formation of insoluble sludges and excess acidity accompanied by changes in certain physical properties. Phosphate esters, particularly trimethyl phosphate, are susceptible to the hydrolizing process. In general, they may be rated from poor to good.

Volatility. Generally low.

Bulk modulus. Relatively low volume change under pressure, i.e., high bulk modulus, well above 300,000 psi usually. Where M_B is the bulk modulus, and p_1, p_2, v_1, and v_2 are the initial and final pressures and volumes, respectively,

$$M_B = \frac{p_2 - p_1}{(v_1 - v_2)/v_1}$$

Pour point. Depends on structure, from +20°F to less than −55°F.

Specific gravity. Relatively high, approaching or above that of water. This property must be considered in designing transfer equipment, such as pumps and fluid motors, to avoid the creation of a partial vacuum within the fluid mass, a condition known as cavitation. The specific gravity of the phosphate esters is in the range of 1.16 to 1.42.

Chemical solvency. Phosphate esters mix well with other substances, which makes them excellent additives for other synthetic fluids, petroleum oils, and greases, as well as for synthetic plastic lubricants. Such additives include tricresyl, triphenyl, and tributyl phosphate esters.

Resistance to radiation. The phosphate esters will decompose when exposed to most types of radiation. This also applies to phosphate ester additives, which do not in any way appear to be protected by the carrier. The effect is to form new materials (ionization), including acids, and to radically increase viscosity.

USES OF PHOSPHATE ESTERS. These synthetic fluids find use in a number of services where their fire-resistant and lubrication characteristics are advantageous and warranted. Such services include hydraulic systems and some of the lubricant systems of the following equipment:

Glass making machinery	Aircraft
Missile handling equipment	Heat-treating lines
Production welding lines	Air compressors
Die casting machines	Gas compressors
Underground mining equipment	Gas turbines
Some marine requirements	

The greatest fire hazard exists where fluid spray due to breaks in pressurized conduits causes contact with an open flame or hot surface.

Welders have been fatally burned when their torches inadvertently severed feed lines carrying non-fire-resistant lubricants.

Disadvantages. Phosphate esters cause deterioration of certain materials used for packings, seals, gaskets, hoses, accumulator bladders, flexible couplings, insulating lacquers, and ordinary paints. Materials such as natural and buna rubber and neoprene will swell and deteriorate when in contact with or close to phosphate esters. However, materials that are compatible with phosphate esters include butyl rubber, silicone rubber, nylon, epoxy, and Teflon; these can be safely used for the components mentioned. The author recommends against painting the interior surfaces of hydraulic systems and lubricant reservoirs regardless of the fluid employed, but there are special coatings available for the purpose. Their identification is best made by the supplier of synthetic fluids in all cases.

Like new petroleum oil, phosphate esters tend to loosen debris and clean the internal surfaces of systems through which they flow. This tendency can be disastrous if such debris clogs feed lines or enters pumps and fluid motors. Filtering helps if the debris remains in circulation, but new or used systems should be cleaned by solvent flushing of large systems. Initial system flushing should always be followed by a secondary flushing, using the fluid that will ultimately become the permanent medium. The fluids used in primary and secondary flushing must be thoroughly drained from the system and either discarded or appropriately filtered before being reused as a flushing medium.

POLYGLYCOLS. A glycol may be defined as a compound having intermediate properties between those of glycerin and alcohol; hence the name glyc-ol. Alcohol being monatomic and glycerine triatomic, the glycol compounds are diatomic. There are two important basic glycols, the simplest being ethylene glycol and the other, propylene glycol. Polymers of ethylene and propylene glycols are called by the common term *polyglycols,* or more specifically, polyalkylene glycol.

The reaction of an alcohol and propylene oxide produces a series of polyglycols that are insoluble in water. By the addition of ethylene oxide in the reaction, a series of water-soluble polyglycols result. Both types are used as synthetic lubricants in applications suitable to their individual properties.

The general formula for the polyglycol polymer is

$$RO-\left|-CH_2-\underset{\displaystyle R'}{\underset{|}{C}}H-O-\right|_n-R''$$

where R, R′ and R″ represent the particular substituent groups that determine the type of glycol produced. Such types include diol, triol, monoether, diether, ether-ester, and diester polyglycols. The molecular weight of the particular type produced depends on the value of *n*, which indicates the repeating magnitude of the bracketed constituents.

By substitution of R, R′ and R″ in the general structure, the following glycols would be produced:

Polyethylene glycol R, R′, and R″ = H (hydrogen)

$$HO-\left[-CH_2-\underset{\displaystyle H}{\underset{|}{CH}}-O\right]_n-H$$

Polypropylene glycol R and R″ = H
R′ = CH_3

$$HO-\left[-CH_2-\underset{\displaystyle CH_3}{\underset{|}{CH}}-O\right]_n-H$$

Mixed oxyethylene – oxypropylene glycol

$$HO-\left[-CH_2-\underset{\displaystyle H,\ CH_3}{\underset{|}{CH}}-O\right]_n-H$$

etc.

Both the water-insoluble and water-soluble types are used as lubricants. As mentioned earlier, water-solubility is established by the presence of ethylene oxide in the molecular structure.

PROPERTIES OF POLYGLYCOLS. In general, the properties of both water-insoluble and water-soluble polyglycols are comparable. Their viscosity-temperature coefficients are excellent, having viscosity index values up to 160. They are equal or superior to mineral oils in volatility, have good lubricating characteristics, and are compatible with many engineering materials, including rubber to some extent. They will dissolve paint. The polyglycols are generally compatible with additives. They rate fair to good in thermal stability and resistance to hydrolysis, but their thermal and oxidation stabilities can be improved by the addition of oxidation inhibitors up to about 400°F bulk temperature. Their flash point ranges between 400 and 500°F for lubricant types. Their degradation products are either volatile or soluble in the fluid mass, which is an advantage toward preventing the formation and accumulation of sludges. They are pumpable at −40°F, which appears to be their low-temperature limit for viscosities acceptable to bearing lubrication and hydraulic services. Polyglycols are made in a wide range of viscosities. Spontaneous ignition temperatures range between 640 and 750°F, according to viscosity and type of structure, reaching the maximum level with the mixed oxide types of monoethers. The water-soluble types are considered fire-resistant to the degree suitable for use in mine and marine service. The high cost of water-insoluble polyglycols prohibits their wide adoption as substitutes for petroleum-base lubricants in a number of requirements, including vacuum pumps, internal combustion engines, bearings, and gears. The lower cost of water-soluble polyglycols relative to the water-insoluble types makes the former attractive as fire-resistant hydraulic fluids in mines where the loss due to line breakage is an important factor.

WATER-SOLUBLE POLYGLYCOLS. The water-soluble polyglycols are commonly referred to as fire-resistant water-base fluids and are used extensively as hydraulic liquids. They are mixtures of polyglycols, ethylene glycol, water, and suitable additives. The water content varies from about 40 to 50% (the average being about 45%) by volume. Because of the presence of water, one of the additives is a corrosion inhibitor capable of preventing rust both in its liquid and vapor phase. The vapor phase protects the areas of the circulation system not immersed in the fluid mass. Polygylcols are true chemical solutions; they are therefore stable at normal operating temperatures during storage and service.

Since water is the fire-resistant ingredient in this type of fluid, it must be maintained at a specific volume level prescribed by the supplier. Thus, if the water content of the initial volume of the fluid is, say, 43%, it is essential that this level be held to within ±4% to maintain proper viscosity and effective fire-resistant qualities. The tendency is to lose water by evaporation during service. At temperatures of 120 to 130°F, the loss rate is not critical.

The viscosity of water-base glycols varies inversely with water content. The loss of water by evaporation therefore increases the viscosity of the fluid and also decreases its fire-resistant effectiveness. Water loss, determined by change in viscosity, must be constantly replaced by an equal volume of distilled or deionized water. Without water, glycols will burn.

Another loss encountered during operation in circulating systems is the depletion of the corrosion inhibitor in performing its rust-combatting activity in areas above the liquid line in its vapor phase. The amount of inhibitor loss is indicated by a drop in alkalinity of the fluid. When the alkalinity is reduced below a specified (by the supplier) point, it is necessary to add a sufficient amount of inhibitor to return its strength to the original level. The inhibitor and the amount to be added should follow the advice of the supplier.

Because of the higher specific gravity of water-base glycols and their noncompatibility, petroleum oils will float on top and can be removed by skimming. Most other synthetic fire-resistant fluids will settle to the bottom of water-base glycol reservoirs and can be drained off by the same method by which water is drawn from the bottom of mineral oil reservoirs. Care must also be taken to determine the compatibility and miscibility of water-base glycols from different sources. The suppliers involved should be contacted in this regard.

The amount of water to be added to water-base glycol fluids is based

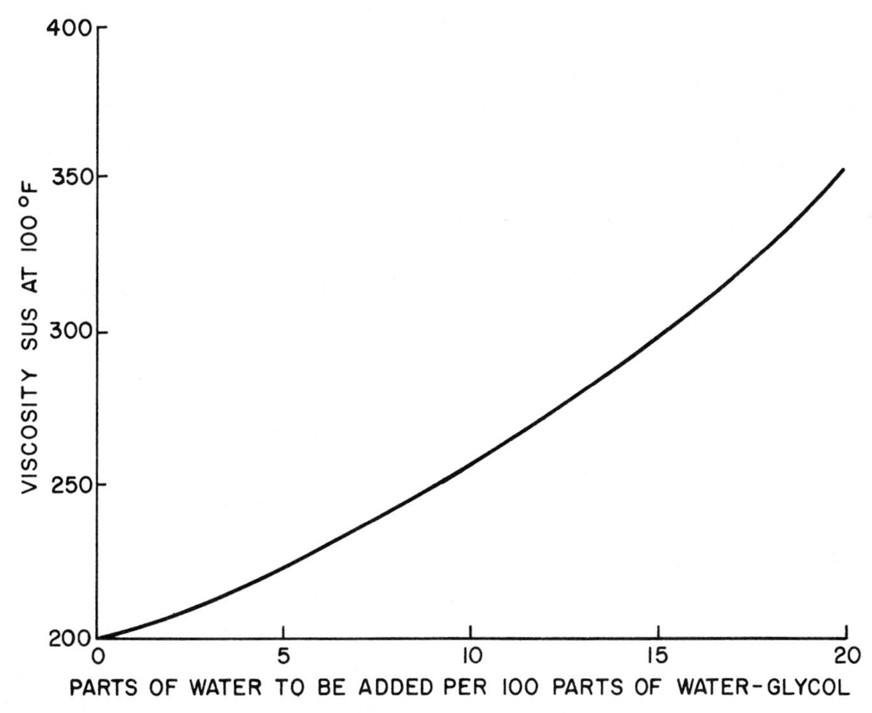

Fig. 6-2. Water conversion chart.

on the increase in viscosity as indicated on a water make-up chart wherein measured viscosity is plotted against the percentage of water required to return the solution to its original viscosity. The curve on the chart (Fig. 6-2) is a reference related to an individual glycol or any glycol of the same kind (brand) having a viscosity of 200 SUS at 100°F.

EXAMPLE: How many parts of water per 100 parts of water-base glycol should be added if the measured viscosity is found to be 300 SUS at 100°F.

Solution: Find the point of intersection of the curve with the line representing 300 SUS at 100°F. Drop vertically to the ordinate and find 15 parts. Therefore, 15 parts of distilled or deionized water per 100 parts of water-glycol fluid is required to reduce viscosity to the original viscosity of 200 SUS at 100°F.

Use. The largest field of use of the water-base glycols is that of fire-resistant hydraulic fluids in which balanced vane-type and in-line axial piston pumps are used. They are not suitable with pumps employing heavily loaded antifriction bearings, including gear or vane-type pumps, because of resulting rapid wear. Such pumps should be converted to plain (bushing) bearings if water-glycols are to be the hydraulic medium. Viscosity in the field of hydraulics is usually in the range of 150 to 300 SUS at 100°F.

DIBASIC ACID ESTERS. Dibasic acid esters are organic diesters formed by the reaction of a dibasic acid and an alcohol. Diesters developed as synthetic lubricants include in their formulation sebacic acid ($C_{10}H_{18}O_4$), azelaic acid ($C_9H_{16}O_4$), or adepic acid ($C_6H_{10}O_4$), all of which are long-chain dicarboxylic acids, long-branched primary alcohols, and long-chain monobasic acids, such as pelargonic acid reacted with glycols.

Castor oil is a glyceride and is the starting material for the C_{10} sebacic acid. Animal fats and oils make available C_9 azelaic and pelargonic acids. Adipic acid is available from petroleum hydrocarbon synthesis, e.g., cyclohexane or cyclohexanol.

The structure of dibasic acid esters is represented by the formula

$$R'{-}O{-}\overset{\displaystyle O}{\overset{\|}{C}}{-}R{-}\overset{\displaystyle O}{\overset{\|}{C}}{-}O{-}R'$$

where R′ represents the alcohol organic radical and R the carbon chain of the acid used in the formulation. A typical structure would incorporate the radical and acid carbon chain by substituting for R and R′:

$$C_8H_{17}{-}O{-}\overset{\displaystyle O}{\overset{\|}{C}}{-}C_8H_{16}{-}\overset{\displaystyle O}{\overset{\|}{C}}{-}O{-}C_8H_{17}$$

The esters are identified by their constituents, such as shown below for several synthetic lubricants and their typical viscosities at −40°F:

Di(2-ethylhexyl) sebacate	1450 cSt
Di(3, 5, 5, -trimethylhexyl) sebacate	(400 cSt at 100°F)
Di(2-ethylbutyl) azelate	500 cSt
Di(2-ethylhexyl) adipate	800 cSt
Di(2-ethylbutyl) adipate	220 cSt
Dipelargonate (of depropylene glycol)	960 cSt

The viscosity indexes of these esters range up to 180 (Dean and Davis).

PROPERTIES OF DIBASIC ACID ESTERS

Viscosity-temperature characteristics (VTC). The viscosity indexes of these synthetic lubricants are excellent, most of them being above 140, calculated according to the Dean and Davis method.

Specific heat. This is an important property considering their principal use in gas turbines where temperatures range from 300 to 400°F. Dibasic-acid esters of the MIL-L-7808 type are comparable with mineral oils of similar viscosity in thermal conductivity, but they have a greater capacity to absorb heat at room temperature. They have comparable or slightly lower specific heat values at the higher temperatures prevalent in gas turbines. Used oil appears to have a higher heat capacity than the same oil when new, due to the presence of combustion debris in the used oil. Their heat-absorbing capability is considerably above some other synthetic lubricants, such as the silicones and chlorofluorocarbons. The specific heat increases as temperature increases, so that the values of dibasic acid esters range from about 0.45 Btu/lb/°F at 75°F to about 0.62 Btu/lb/°F at 450°F. Mineral oils have specific heat values at those temperatures of approximately 0.38 to 0.64 Btu/lb/°F, respectively. For average conditions, a specific heat value of 0.43 Btu/lb/°F has been found by the author to give optimum results in calculating minimum mineral oil flow volume for circulating systems with proper cooling.

Pour point. One of the important characteristics of these diesters is their low pour point, ranging downward from −60°F to less than −100°F.

Thermal stability. Thermal stability of dibasic esters may be rated good. Tests of di(2-ethylhexyl) sebacate show that thermal decomposition does not occur below 525°F. With esters of secondary alcohols, decomposition begins at a lower temperature (about 450°F). However, the use of oxidation inhibitors has proved highly satisfactory. Dibasic esters possess fairly good resistance to hydrolysis.

Volatility. In general, the volatility of the dibasic-acid esters is low. For example, the percentage weight loss of di(2-ethylhexyl) sebacate in 168 hr at 150°F in a convection type oven was less than 0.05%. In this regard, they show a lower volatility for a given viscosity than equivalent mineral oils. Tests of those esters named above indicated weight losses under the foregoing conditions in the range of 0.05 to 0.07% adipate with the exception of di(2-ethylbutyl) adipate, whose loss was under 0.40%.

Flash point. Generally high. Most of them are in the range of 400 to 470°F.

Foaming characteristics. Similar to the foaming characteristics of hydrocarbon types of similar viscosity. Silicones are used as an antifoam agent in esters as well as in other synthetic lubricants and mineral oils.

Chemical solvency. Dibasic esters have a pronounced solvent effect on paints.

Lubricity. Highly refined straight mineral oils show an average static coefficient of friction value of 0.19. The same values of unfortified dibasic acid esters average 0.21 at the same temperature (room temperature).

Compatibility. Unlike the silicones, the dibasic esters accept most of the essential additives. They have a solvent effect on paint and rubber, but Buna N, silicones, and Teflon may be used in direct contact as gasket and seal materials with safety. They are practically immiscible with water.

Bulk modulus (compressibility). The bulk modulus of fluids increases with pressure and decreases with temperature, and its value is reported as pounds per square inch. The higher the value, i.e., the nearer it approaches 1.0, the less compressible is the substance. The dibasic acid ester, di(2-ethylhexyl) sebacate under 20,000 psi applied pressure shows a bulk modulus of about 325×10^{-3} at 77°F and 245×10^{-3} at 210°F. This compares with corresponding values for the same temperatures, but under 30,000 psi applied pressure of 365 and 300, respectively. Silicone oil shows a greater compressibility, having a bulk modulus corresponding to an applied pressure of 30,000 psi and temperatures of 77°F and 210°F of 260×10^{-3} and 205×10^{-3}, respectively. Mineral oil and dibasic acid esters show comparable bulk modulus values, particularly at the higher temperature.

Reaction to radiation. The effects of radiation on most synthetic oils are many-fold, including an increase in viscosity, foaming tendency, and acidity. The diesters are no exception. Radiation also decreases the flash point and the autogenous ignition temperature.

USES. Because of their good thermal stability, low volatility, heat transfer properties, and comparable lubrication and wear characteristics, the dibasic acid esters have been adopted commercially for use in lubricating gas turbines. While the diesters mentioned are included in this discussion to furnish examples of this class of lubricants, the chemistry of the esters is not confined to the few listed. The development of jet aircraft and gas turbines tends toward more severe thermal conditions with predicted bearing and bulk oil temperatures of 800 and 700°F in jet engines at speeds in the magnitude of four times that of sound. The answer appears to lie in the development of more complex esters as base stock, such as the tetra-esters mentioned in the discussion of gas turbines (Chapter 18).

Another use for the dibasic esters is for instrument oils, low-volatility greases, and special hydraulic oils.

ORTHOSILICATE ESTERS. One of the more recent synthetic fluids to enter the field of lubrication is a class of silicate esters. This class of fluids has the advantages of high-temperature properties combined with low-temperature characteristics. These properties result in a fluid having practical service viscosities through the range of −65°F to 400°F, which makes these fluids adaptable for hydraulic systems of supersonic aircraft, at which temperatures the corresponding viscosities are 2500 cSt and 2.5 cSt, respectively, as specified by Mil-H-8446, covering high temperature hydraulic fluids.

The structure of silicate esters, also called orthosilicate esters, is represented as follows:

```
            R
            |
            O
            |
R—O—Si—O—R
            |
            O
            |
            R
```

where R represents the organic groups, which may be similar or different. The orthosilicate esters being a reaction product of silicic acid, $Si(OH)_4$ and an alcohol or a phenol, R may be an alkyl or an aryl alcohol or mixtures of the two. Common types of silicate esters include the tetraalkyl, tetraaryl, and mixed alkyl-aryl orthosilicates. The organic groups may be further extended to include such substituents as chlorine, fluorine, nitrogen, and alkoxy groups.

The silicate ester structure shows a bond of silicon-oxygen-carbon esters, which may identify them variously as tetralkoxy-silanes, or tetraaryloxy-silanes. Thus the difference between the silicate esters, the tetraalkyl silanes, and the tetraaryl silanes is shown by the direct silicon-carbon bond of the silanes having different fluid properties.

Another group of compounds having many properties similar to those of the orthosilicate esters is composed of the *dimer silicates,* which also contain the silicon-oxygen-carbon bond. These compounds are identified as hexaalkoxy-disiloxane, or hexaaryloxy-disoloxanes. They have a more complex structure:

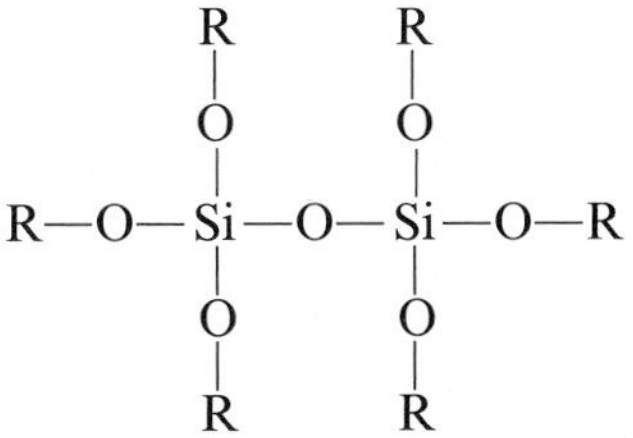

```
          R       R
          |       |
          O       O
          |       |
  R—O—Si—O—Si—O—R
          |       |
          O       O
          |       |
          R       R
```

in which R may be substituted for an alkyl or aryl group. They are called by the shortened name disoloxanes but are treated here under the general nomenclature of silicate esters. The properties of both orthosilicate and disoloxane compounds are determined by the kind and size of the organic groups and can be varied by group substitution or by blending different compounds.

Representatives of the silicate esters are listed in Table 6-3.

PROPERTIES OF SILICATE ESTERS. In addition to the wide range of serviceable fluidity shown for the esters listed in Table 6-5, they are available

Table 6-3. VISCOSITIES OF SILICATE ESTERS

	Viscosity (cSt)		
	−65°F	210°F	400°F
Tetra (2-ethylbutyl) silicates	207	1.6	
Tetra (2-ethylhexyl) silicates	1320	2.35	
Hexa (2-ethylbutoxy) disiloxane	2296		3.82
Hexa (2-ethylhexoxy) disiloxane	3047	4.08	
Tetra (2-butyl) silicate	57	0.95	

in many intermediate viscosities. Most of the alkyl types possess excellent viscosity-temperature properties. The viscosity index ranges from 150 to 200 and the lubrication characteristics are fairly good. Silicate esters have high thermal stability and low volatility. They favorably accept and are compatible with most additives. In the higher temperature service range, oxidation inhibitors are essential. Antiwear agents are required when they are subject to boundary lubrication conditions.

The specific heat of tetra (methylphenyl) silicate ranges from about 0.45 at room temperature (75°F) to 0.95 Btu/lb/°F at 475°F. Tetraethyl silicate has an average specific heat value of about 0.5 Btu/lb/°F for temperatures of 70 to 210°F.

The pour point of silicate esters ranges from −85 to −100°F, and their flammability characteristics may range from poor to fair, with autogenous ignition temperatures varying according to formulation from 455°F for tetraethyl silicate to 810°F for tetra benzyl silicate.

The silicate esters have a minimum effect on engineering materials, but will harden some types of rubbers when left in contact for protracted periods of time at higher temperatures.

The use of suitable additives, including rust inhibitors, film strength agents, and antioxidants, is practiced either to inhibit intrinsic properties as appropriate or to combat other factors, such as water, during their service life.

USES OF SILICATE ESTERS. Their major use is in the field of high-temperature hydraulic fluids, particularly in military aircraft. They also serve as heat transfer media, low-viscosity hydraulic fluids, airborne electronic system coolants, and low-viscosity bases for greases.

HALOCARBON COMPOUNDS. The commercially available chloro-fluorocarbon lubricants are low-molecular-weight polymers derived from chlorotrifluoroethylene, $CF_2Cl\text{-}CFCl_2$, having the typical structure in which all the hydrogen atoms have been replaced by the chlorine and fluorine atoms:

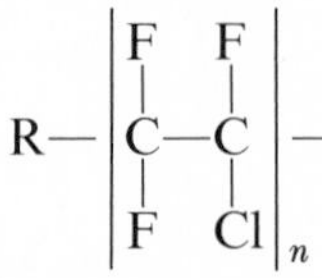

$$R-\left[\begin{array}{cc} F & F \\ | & | \\ C- & C \\ | & | \\ F & Cl \end{array}\right]_n-R'$$

where R and R′ represent the terminal groups of Cl, F or CF_2, and F, C, and Cl represent the fluorine, carbon, and chlorine constituents.

The replacement of hydrogen in alkyl hydrocarbons by chlorine alone results in an unstable product having strong chemical activity leading to highly corrosive, toxic compounds. The addition of fluorine as a substituent results in a more stable product. Thus the polymers of chlorotrifluoroethylene are highly stable and remain inert in the presence of acids, alkalies, and oxidizing compounds.

The inert qualities of chlorotrifluoroethylene polymers when maintained at temperatures below 500 °F (above which they will thermally decompose) make them highly resistant to oxidation.

Because of the halogens, which impart low surface tension (23 dyn/cm at 77 °F), these polymers possess good lubricity. However, their viscosity-temperature coefficient is high, resulting in a low viscosity index. Like many other halogenated hydrocarbon compounds, they are nonflammable. Fluorination is not easily achieved, but the short bond between the carbon atom and the atoms of both chlorine and fluorine results in a product not easily disturbed under other than abnormal conditions, such as temperature in the region above the point where depolymerization is initiated (500 °F).

The halocarbon lubricants vary widely in their effect on paints and rubber materials. They appear to have little effect on nylon. Because they are essentially noncorrosive, they can be used with most engineering materials under normal temperatures. They are not suitable as additives with either petroleum or synthetic oils due to their inertness (low solubility).

Chlorotrifluoroethylene oils are made with various molecular weights; the higher the molecular weight, the greater the viscosity of the oil. They are heavy (in weight) fluids, having densities in the region of 1.9 grains/cm^3 at 100 °F. They possess low pour points, as low as -100 °F with some fluids.

In the case of compatibility with plastic and rubber materials, lubricants, mineral or synthetic, vary in their effect, depending on temperature in many cases. With some materials compatibility cannot be achieved under any circumstances. The effect of incompatibility shows up in various ways, such as a change in volume of the sealing material, which may swell or shrink. Softening or hardening of the material is also possible. Complete destruction may be rapid or gradual. Oil discoloration may occur. The manufacturer of the lubricant is the best authority as to what materials can be safely used as seals, gaskets, packing, etc., with his lubricant, and also the conditions of use.

PROPERTIES AND USES OF CHLOROFLUOROCARBON LUBRICANTS. A summary of the properties of halocarbon fluids as lubricants emphasizes their utility:

(a) chemical stability
(b) essentially noncorrosive
(c) nonflammability
(d) excellent lubricity
(e) high density
(f) good dielectric strength

By reason of such characteristics, they are used as lubricants for mechanical equipment that handles highly reactive elements and compounds. These include the lubrication of oxygen compressor cylinders and piston rods and turbo-pump bearings of aircraft using oxidizing chemicals that promote combustion. The halocarbon oils also serve as insulating and spark-repressing fluids in electrical equipment. They are also used for a base stock for grease to give it a wide operating temperature range of between −40 to 400°F.

POLYPHENYL ETHERS. Polyphenyl ethers were developed to meet the severe conditions imposed by temperatures above 500°F and high radiation. They are highly aromatic in structure, resulting in viscous fluids with high pour points and poor viscosity-temperature characteristics. However, they possess somewhat high thermal stability (600–700°F), fairly good oxidation resistance, and lubrication properties, being somewhat higher than petroleum oils in these respects. Polyphenyl ethers have high spontaneous ignition temperatures (1000°F) but rate only fair in fire-resistance.

An example of polyphenyl ether nomenclature is *m*-bis(*m* phenoxyphenoxy) benzene, which has a melting point of 109°F, a pour point of 40°F, a flash point of 540°F, and a viscosity range from 332 cSt at 100°F to 0.63 cSt at 700°F. This family of synthetic lubricants shows relatively better stability under radiation than other materials, such as the esters, being more susceptible to viscosity increases with higher molecular weight, but with a greatly reduced change in viscosity at higher temperatures, say 700°F. Some types show a slower rate of oxygen absorption under irradiated conditions than others.

Uses. Because of their high temperature and radiation characteristics, the polyphenyl ethers are recommended for use as nuclear reactor coolants and as high-temperature hydraulic fluids for nuclear-powered aircraft and space vehicles. They were originally developed for the lubrication of mach-speed turbo-jet engines. They are also suitable as dielectric and insulating liquids in electrical power equipment.

SILANES. Silanes are silicate compounds with high molecular weight and having an aromatic radical, an alkyl radical, or a mixture of the two as substituents. The radical or radical combinations govern the thermal and viscometric characteristics of the organosilanes. The tetraalkylsilanes were developed as high-temperature base stocks for high-temperature greases and hydraulic fluids.

A typical organosilane structure may be shown as

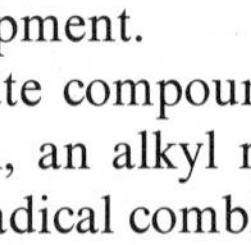

$$
\begin{array}{c} R \\ | \\ R\text{—}Si\text{—}R \\ | \\ R \end{array}
$$

in which R represents the radicals mentioned above. Since oxygen does not appear in the structure, the silanes are not polymers as are the silicones.

The silanes possess good thermal stability and fairly good viscosity-temperature characteristics, but they are not strongly resistant to oxidation. They are fairly soluble in other lubricants, so they do not present too much difficulty in their use as additives. Their flammability characteristics are not good, but they have fairly good lubrication properties and resistance to hydrolysis.

Uses. Like the polyphenyl esters, they are suitable as base stocks of high-temperature greases and hydraulic fluids.

SYNTHETIC HYDROCARBONS. The requirement for a lubricant of decomposing and volatilizing at elevated temperatures without leaving any residue is satisfied by certain polymers of petroleum derivatives. The major product of this type is the polybutenes, but other polymers such as polyethylene and polypropylene have similar properties and are classified as synthetic hydrocarbons.

Polybutenes are formed by the polymerization of butene, a product of petroleum cracking. Their typical structure has the formula

$$—(—CH_2CH_2CH_2—)_n$$

In contrast to petroleum oils, they are characterized by better low-temperature characteristics and lighter color commensurate with viscosity. They also possess better electrical properties. Their viscosity varies with molecular weight, which can be extended according to the structure shown above and ranges from about 5 to greater than 600 cSt at 210°F.

Uses. The important use of polybutenes is in the field of high-temperature lubrication, where regular petroleum oils tend to leave coking residues behind upon volatilizing. Bearings and conveyor links of ovens and furnaces are representative of the components to which synthetic hydrocarbons are applied.

Liquid Metal Lubricants

The requirement for liquid metals as lubricants is associated chiefly with closed-cycle space power systems. It is estimated that temperatures in such systems will vary from 400 to 1500°F, which is considered minimum because of the inconvenience of providing auxiliary cooling. Such temperatures preclude the use of the petroleum and synthetic lubricants discussed in this text from the standpoint (1) of lubricant degradation and (2) sealing difficulties.

The use of liquid metals poses a number of problems, as does the weightlessness associated with space environments. Briefly presented, they are as follows:

1. Liquid metals do not possess good lubricating qualities.
2. The bearing metals compatible with liquid metals are generally poor in friction and wear properties.

3. High lubricant temperatures tend to remove contaminant layers (oxides) from surfaces that are useful under boundary conditions required minimal for antifriction bearings.

4. Weightlessness tends to cause unbalance of rotating shafts in journal bearings to initiate whirl and disturb the development of hydrodynamic films.

5. The application of the lubricant by hydrostatic means requires special equipment.

6. Auxiliary cooling of the power system requires the additional weight of radiators on the space vehicle.

7. Liquid metals have low viscosities on the order of 10^{-7} reyn at 200°F. The viscosity of an SAE 30 motor oil is approximately 20 times that value at the same temperature.

8. Thermal growth within liquid metals due to lower specific heat properties and higher thermal conductivity requires special consideration in determining bearing clearances.

9. Low viscosity reduces the load-carrying capacity of bearings but also reduces power loss due to the decrease in fluid friction.

10. Cavitation tends to develop if a vapor-liquid phase state exists due to a drop in system pressure below the vapor pressure of the liquid metal and/or if the bearing temperature approaches too closely or reaches the boiling point of the lubricant.

11. The low viscosity of the liquid metal requires greater volume flow through the bearings to maintain hydrodynamic conditions. This condition encourages turbulent flow with subsequent power loss and increased friction (torque).

The potential liquid metal lubricants which also act as the cycle-working fluids for turbo-generators of space vehicles include those shown in Table 6-4.

Sodium-potassium is used as a nuclear reactor coolent and lubricant in the vicinity of atomic reactors where petroleum lubricants would be subject to chemical change due to radiation.

Table 6-4. BOILING POINTS AND VISCOSITIES OF POTENTIAL LIQUID METAL LUBRICANTS

Metal	*Boiling Point (°F)*	*Viscosity at 1200° F (reyns)*
Mercury	674	10×10^{-8} (13×10^{-8})
Rubidium	1295	2.2 (2.2)
Potassium	1395	2.0 (1.9)
Sodium	1630	2.9 (2.2)
Lithium	2430	2.7 (2.0)
Sodium-potassium	1456	2.2 (2.0)

Note: Values shown in parentheses are viscosities at the boiling point of the corresponding metal.

BEARING MATERIALS. The materials used for bearing pairs lubricated with liquid metals must be compatible with the latter. The combination of both rotor and stator materials is also an important consideration as are the temperatures involved and the type of liquid metal. Protection against friction and/or wear is dependent on a thick lubricant film developed either hydrodynamically or hydrostatically, the former under low-load operation.

Compatibility tests based on 10,000 hr of exposure of component materials, such as ferrous, nonferrous, and refractory metals and ceramics, to liquid metals have been made and published in various technical papers.

Water as a Lubricant

The poor lubricating properties of water render this substance an inadequate substitute for mineral oils or synthetic fluids. Its high specific heat of 1.0, however, promotes cooling and for this reason water is used to dissipate heat generated in a nuclear reactor. Because the reactor coolant is also used to lubricate its pressurizing equipment, the bearings of the latter must have high wear resistance.

The most promising materials for sliding surfaces within the reactor cooling circuit appear to be cobalt-base metals with a Brinell hardness of about 440 in the temperature range involved. Cobalt-base metals are used in conjunction with an opposing surface made up of either certain types of stainless steel (e.g., martensitic) or nickel-base metals.

Such combinations present the necessary hardness of opposing sliding surfaces to resist wear and galling and to resist the corroding effect of water.

PLASTIC LUBRICANTS

The plastic lubricants are grease and petrolatum, with grease having a much wider application. The greases discussed here are all plastic under operating conditions, but at room temperature they may range from liquid to firm plastic.

Plastic lubricants are indicated where design conditions will not permit the use of oil, usually because of leakage.

The following discussion of grease must be in general terms, because of the large number of chemical variations, processing methods, grease manufacturers, and available ingredients, and the quality range of the ingredients. Statements made regarding the characteristics of grease or any lubricant must represent a typical rather than an overall conclusion.

For example, it is the consensus of many lubrication engineers that some frictional requirements are best served by using a lime-lead grease in which the soap is lead naphthenate. To them, any beneficial characteristic recognized in this type of grease is credited to the use of lead naphthenate, all other qualities being equal, and they specify such a soap for their plant. Thus, when typical characteristics of a lime-lead soap base grease are reported, lead naphthenate is inferred and not lead oleate or any other kind of soap.

Another example makes an important point for interpreting statements of lubricant behavior. In the discussion of flexible couplings, the recommendation of aluminum soap base grease is made for some types. The recommendation is made on the basis that a typical aluminum-soap base grease is an effective lubricant, and as an important supplemental benefit, the oil will not readily separate from the soap under high centrifugal force. Another characteristic of aluminum soap greases is that they are thixotropic, i.e., they will regain consistency after being softened somewhat by working. To possess all the beneficial qualities, however, the grease must contain selected ingredients and be manufactured correctly. There are such greases and they are the ones of reference.

Lead is used as an additive for some aluminum base greases as a film strength improver.

Greases

Greases are made, generally, by saponifying a metal base with a fatty substance, such as tallow, and adding oil to the soap. The process, which includes agitation, is carried on under controlled temperature conditions, in either an open kettle or a pressure cooker. The bases used include calcium, sodium, aluminum, lithium, and lead. Barium and strontium are also used. Lubricants produced using various metal bases are identified as calcium- (lime-) soap, soda-soap, lithium-soap, aluminum-soap, etc., greases. To obtain the advantages of more than one soap, mixed-base greases are also available. The physical characteristics of a grease depend principally on the base used in their manufacture. The consistency of greases becomes more firm with greater soap content. The latter defines greases as soap-thickened oils. Oils of different viscosity are used in manufacturing grease, the choice of which depends on service considerations. Oil viscosities range from 60 to over 1000 Saybolt seconds at 100°F.

Greases made with mineral oils contain varying amounts of water that act as a bonding agent. Water content of greases varies from a trace to approximately 3.0% except for *cold-set* greases, made with a mixture of lime, rosin, and mineral oil, which contain over 8% water. These types of greases set or jell in their shipping containers in which they are poured after saponification is complete.

The property of water solubility varies with greases made with different bases. Lime-soap greases are water insoluble, while soda-soap greases are highly soluble.

Greases are treated with various additives for special purposes. Oxidation inhibitors extend the life of greases in the same manner with which they act on mineral oils. Solid materials such as graphite, molybdenum disulfide, talc, mica, and other fillers are added to some types of greases to assure protection to frictional surfaces and maintain a low coefficient of friction in the event of grease film rupture under rough service. Heavy pressure greases are developed by combining the film-strengthening characteristics of lead and zinc soaps (neither of which is a thickening agent to any great extent) with the thickening qualities of lime-soap. Such products are called *mixed-base* greases, which include the effective leaded greases used extensively for certain steel mill requirements. Lead-soap is also used as a film strength additive for heavy pressure oils.

Greases are classified according to the type of thickener used in their manufacture, but they may be classified by physical or chemical peculiarities, e.g., cold set, brick, etc., or synthetic, nonsoap, etc. Synthetic greases may further be divided into those containing synthetic fluids thickened with soap, such as a silicone fluid thickened with lithium soap, or the same fluid thickened with carbon black.

Classification of conventional greases is shown in Table 6-5, which also indicates their temperature ranges, water soluble characteristics and general service usage.

Table 6-5. GREASE CLASSIFICATION AND TYPICAL USES

Base (Soap)	*Temperature Limitations (°F)*	*Water Solubility*	*Typical Uses and Description*
Calcium (lime)	175	Insoluble	Cup grease. Plain bearings. Antifriction bearings of moderate speed and temperature. Water repellent (Pump glands). Buttery in texture NLGI Nos. 1 to 6
Sodium (soda)	200–350	Soluble	Antifriction bearings of high or moderate speeds. Will emulsify with water. Use under dry conditions generally. Will withstand higher operating temperatures than lime-soap greases. Do not require water as a stabilizer. Smooth, fibrous, or spongelike in structure. NLGI Nos. 1 to 6
Aluminum	180–190	Insoluble	Withstands shock loads without splashing. Cohesive and adhesive. Water repellent. Will not separate readily under centrifugal force. Auto chassis, flexible couplings, eccentrics, cams, small open gears. Smooth and stringy in structure
Lithium	Up to 300 Down to −100	Slightly soluble	Practically water repellent. With low-pour-point oils can be used at extremely low temperature. With high-viscosity oils can be used at higher temperatures. General purpose service. Buttery to stringy in structure

Table 6-5 (*Continued*)

Base (Soap)	*Temperature Limitations (°F)*	*Water Solubility*	*Typical Uses and Description*
Calcium complex	300+	Insoluble	Water repellent. Fairly high-temperature lubricant. Good pumpability. Usually has high film strength when fortified with proper additives. Good stability. General purpose lubricant
Mixed-base lime-lead	200	Insoluble	Water repellent. Heavy-pressure service and severe operating conditions. Good pumpability for central grease systems with long feed lines. Smooth texture
Mixed-base lime-lithium	250	Slightly soluble	Antifriction bearings subject to relatively high or low temperatures (depending on oil ingredient)
Graphite grease	175	Insoluble	Water repellent. Usually a lime-soap grease to which has been added up to 20% colloidal or powdered graphite. For wet service conditions. Rotating water screen chains, pump plungers, hydraulic rams (water). Not good for high temperatures or for antifriction bearings unless colloidal graphite is used. Wet or dry plain bearings
Cold-set (lime-rosin)	180	Insoluble	Buttery in consistency. Can be applied to open gears, axle bearings, rough machining in general. Water repellent
Block (sodium)	Up to 450	Soluble	Hard and brittle. High-temperature plain journal bearings where grease rides on rotating journal. Will form an emulsion with water which is sometimes desirable. Often wrapped in cloth before applying to journal for control of attrition. Large slow-turning journals. Some block greases contain other soaps in addition to sodium to impart high-temperature characteristics
Residuum	About 200	Will wash away unless specially treated	May or may not contain soaps. Heavy asphaltic type material of normally heavy body and high viscosity. Used principally for lubrication of large open gears. Often adulterated with a solvent to facilitate application. Solvent volatilizes upon application, leaving tough adhesive lubricating film. Lighter grades produced by blending with other petroleum stocks, or by distillation to desired consistency. Also called gear shields, flux, etc. Sometimes treated to improve film strength and water repellency
Nonsoap	250–500	Insoluble	High temperature range. Insoluble in both hot and cold water. Acid resistant to some extent. Inhibits rust

DESCRIPTION OF GREASES. Greases have color, texture, bloom, luster, and other identifying characteristics that are often used in their description. Their color may be amber, brown, black, yellow, orange, etc. They may have a fluorescence or bloom described as green, blue, or other predominating color. Greases usually have either a dull or bright luster. Their texture varies and can be described as being buttery, stringy, resilient, brittle, or fibrous. The types with long fibers are commonly referred to as sponge greases.

GREASE CONSISTENCY. The consistency of greases ranges from that of a semifluid to hard blocks, with intermediate states of softness and firmness. Semifluid and soft greases are made so for several reasons, an important one being that of pumpability. Firmness of greases is dictated by service conditions, of which leakage and handling are prime requisites. Hard greases are called block greases and do not necessarily require a reservoir, except as a receptacle for holding them in a desired location on the journal.

The consistency of greases is measured in terms of penetration. Penetration values are no criterion of grease quality or kind, but merely indicate relative hardness, or softness, as measured by a penetrometer under ASTM standard conditions. A penetrometer consists of a standard cone weighing 150 grams. The cone is connected by suitable linkage to a gauge designed to measure millimeter changes in cone elevation. The cone is held in suspension with its tip just touching the surface of the test sample of grease contained in a standard cup. At this point in the test, the gauge reads zero and the grease is at a standard temperature of 77°F. Penetration of the grease occurs when the cone is released. After a lapse of 5 sec, during which the cone exerts penetrating force, the gauge is read to determine the distance of penetration; i.e., the number of millimeters the cone dropped. The depth of penetration made by the cone in tenths of a millimeter is called the penetration number. If the grease sample has been undisturbed before testing, the results are expressed as *unworked penetration*. If the sample has been subject to extrusion prior to testing, by reciprocating a perforated piston 60 times (60 double strokes) through a cylinder containing the grease, the results are expressed as *worked penetration*. The softer the grease, the greater will be the depth of cone penetration. Consequently, the softer the grease, the higher will be its penetration number and vice versa.

To identify the relative consistency of greases, the National Lubricating Grease Institute has devised a system in which a number represents the penetration range for a given grease. The system is devised so that zero represents the softest grease (semifluid) and 6 the hardest. The numbers given the various consistencies, as shown in Table 6-6 are not necessarily those used by all grease manufacturers to identify their products. It does, however, act as a universal method for expressing equivalent values.

DROP POINT OF GREASES. The drop point of a grease is the temperature at which it liquefies when heated and drops from a cup through a bottom orifice (Table 6-7). The cup and the thermometer are placed in a test tube

Table 6-6. RELATIVE GREASE CONSISTENCY—
NLGI NUMBERS VS ASTM PENETRATION

NLGI Number	*Relative Consistency*	*ASTM Worked Penetration at 77°F*
000	Very fluid	445–475
00	Fluid	400–430
0	Semifluid	355–385
1	Very soft	310–340
2	Soft	265–295
3	Semifirm	220–250
4	Firm	175–205
5	Very firm	130–160
6	Hard	85–115

and the tube immersed in a heated bath of oil. When the consistency of the grease is reduced sufficiently, it will flow through the orifice in the bottom of the cup. The temperature at which the first drop occurs is the drop point. The drop point gives some indication of the consistency of a grease when exposed to a known operating temperature. It does not indicate the maximum temperature at which the grease will perform satisfactorily.

There is no standard rule for selecting a grease based on a relation of drop point to type of soap. Ingredients and manufacturing quality and processes control this property. This is not to infer, however, that certain type soaps do not tend to impart better characteristics at high operating temperature than others. When operating temperatures are of such high level or duration as to interfere with the water-stabilizing effect within a hydrous grease, a separation of oil and soap results. The oil runs away, leaving the soap to char and probably ruin the bearing.

CALCIUM- (LIME-) SOAP GREASES have good water resistant qualities and remain physically stable at temperatures within range of normal operation conservatively specified as being within the range of 40 to 150°F. They are easily applied with slight pressure but are not considered to possess favorable

Table 6-7. DROP POINT OF GREASES

Base	*Drop Point (°F)*
Calcium	200
Aluminum	200
Sodium	350–400
Lithium	350–400
Calcium EP complex	500+
Inorganic (nonsoap) silica and clays	500+
Organic base (nonsoap)	500+
Polyalkylene glycol (lithium-soap thickener)	370
Silicone (carbon black, or fine silica)	None
Fluorocarbons (fine silica thickener)	None
Diester (lithium-soap thickener)	380

slumpability even with consistencies of NLGI Nos. 0 and 1. When fed automatically and directly from the shipping drum, a follower plate is required to prevent pump starvation when the grease level falls to about one-third the drum height.

Calcium greases give excellent service on plain bearings operating under water wet conditions, as at the wet end of paper machines. They are used with good results on plunger packing where a No. 3 or 4 consistency should be used when applied by screwdown cups, depending on the temperature of the water being pumped. Calcium EP complex and lithium-soap greases have replaced the conventional lime-soap base-grease in many applications.

Like other greases, lime-soap greases change in their physical nature according to the type and viscosity of oil used in their preparation. A variation in the soap content produces changes in consistency from a fluid to a solid, i.e., the more soap, the firmer the grease.

SODIUM- (SODA-) SOAP-BASE GREASES are not water repellant as are lime-soap greases but remain more physically stable at higher temperatures. The emulsion formed serves as an effective lubricant for short periods in bearings on which the seals may be defective, but still capable of preventing rapid escape (washout) of the slurry formed. However, the seals of such bearings should be renewed as soon as possible. Soda-soap greases are recommended for rolling bearings and, when fortified with an antioxidant and effectively sealed within a bearing, they provide lubrication for the ordinary life of a bearing if temperatures are held below 300°F. To determine water repellency of a grease, place a small amount in the palm of one hand, wet with water, and rub vigorously with one or two fingers of the other hand. If the grease is of soda-base an emulsion will appear.

LITHIUM-SOAP GREASES have good water resistance, mechanical stability, and fairly good chemical stability. Their structure tends to retain the lubricating oil. Stability to shear may be imparted to lithium-soap greases by proper manufacturing processes. They show very little or no change in behavior after operating at a temperature above 225°F and cooled during operation. They have excellent low-temperature as well as high-temperature characteristics. They serve as aircraft and multipurpose lubricants. Lithium-soap forms the thickening agent in a number of synthetic greases.

In general, the properties of lithium-soap base greases are associated with the oil used in their formulation:

Low-temperature properties are improved with paraffin oils.

Oil separation is somewhat reduced with naphthene oil. Evaporation loss is lower with paraffin oils, although the flash point of the oil is a factor here. The paraffin oils show greater resistance to oxidation and a higher drop point.

Higher viscosity oils have higher drop points and less evaporation loss than lower viscosity oils, while better low temperature properties are obtained with the latter.

BARIUM-SOAP-BASE GREASES are smooth to fibrous in texture and will resist the washing effect of water. They are usually made by combining barium hydrate with animal fats or fatty acids to form a thickening agent for mineral oil. Water, up to 0.4%, is used as a stabilizer. However, several types are made possible because of the complexities of the soaps that can be chemically produced. The kind referred to here may be described as normal barium-soap greases.

Barium-soap-base greases are made in any of the standard NLGI consistencies, with the No. 2 consistency the most widely used.

The most general application of barium-base grease in multipurpose service is that it is applicable for water pumps, universal joints, chassis, and tractor crawlers. It is also recommended for bearings operating up to about 350°F. Having no recognized drop point below 400°F leakage at less elevated temperatures presents no problem. Barium-soap greases do not change appreciably through cycles of high and normal temperatures.

A combination of barium and soda soap dispersed in oil produces an effective block grease with a drop point of 400°F.

STRONTIUM-SOAP GREASES are made with strontium stearate or hydroxide, compounds that accept many kinds of fats and fatty acids, which when combined with oil produce a grease with some desirable properties, including resistance to water washing, protection against corrosive substances such as salt water, leaching by petroleum solvents, work stability, and good high-temperature characteristics (350°F).

BLOCK GREASES, or brick greases, are hard greases having a consistency from about 35 to 50 penetration (unworked, at 77°F). Block greases are usually made with a sodium-soap base, but a mixture of sodium and barium soap results in a grease of higher drop point (above 425°F). Block greases are applied by cutting the brick (usually 4 in.^{3}) to fit the open bearing cavity in which it rests in contact with the rotating shaft. The temperature of the contacting surfaces between the shaft and grease governs the melting point of the grease required in order to maintain a lubricating film. The melting point along with the texture of the grease controls the consumption, which should not be excessive. The texture, in turn, depends on all the ingredients, including the soap (usually sodium stearate), the oil, and water. The lubricating oil varies from those with low viscosity (about 200 SUS at 100°F) to cylinder stocks with high viscosity (about 100 SUS at 210°F). Free alkali is sometimes used in small amounts, as is barium soap. The difference between the basic types is a matter of the service to which they are to be applied. One characteristic that must be present in any block grease is the presence of tiny oil reservoirs throughout their mass that supplies the lubricant when applied. If the mass is too firm for the temperature to which the grease will be exposed, a starved bearing results. If too soft, consumption will be excessive.

Many of the open cap bearings have been replaced or modified by more positive methods of lubricant application, but some of them still remain in service. The forerunner of block grease was animal suet, usually heaped on a journal rotating in a half bearing, of which horse fat appeared to be superior. The drawback experienced with such lubricants was their ultimate putrefaction accompanied by myriads of maggots. This condition existed in the United States as recently as 1948 in equipment operating in dank and ill-lighted labyrinths located below ground of older paper mills.

Block greases find their widest use on heavy equipment of some paper mills, steel mills, cement plants, rag digesters, and steam locomotives. Made in squares, they are extruded in perforated presses to appear in the form of sticks, which are cut in suitable lengths and inserted in holes that afford application to rod pins of locomotives. Block grease is applied to the propellor shaft of older ocean-going vessels as tunnel grease. Blocks are sometimes wrapped in cloth before being placed in the bearing cavity in an effort to reduce consumption resulting from high temperature. Glazing of the grease at the surface of journal contact is often experienced with block greases that are either poorly chosen or made.

SEMIFLUID GREASES are used primarily when liquid oil would leak from the machinery reservoirs. Such greases will flow slowly at room temperature, because they contain only about 5% soap. This compares with 15 to 30% soap present in lime-soap greases having NLGI consistency numbers from 1 to 5. These greases become too firm and channel below the freezing temperature of water, while some are capable of maintaining sufficient body structure to perform satisfactorily at temperatures considerably above the boiling point (225 to 250°F).

Semifluid greases are also used when lubricant leakage would be detrimental to the product produced, such as textiles and food, or where leakage would be hazardous or a nuisance, as in machine shops and underground mines. They are applied by bath, as in gear cases, or distributed to various bearings from a central reservoir. Semifluid greases are made with either soda-soap or lime-soap as a thickener.

NONSOAP GREASES are thickened with gelling agents derived from natural clays, such as bentonite, fuller's earth, saponite, and attapulgite. Nonsoap greases are used principally to impart special qualities to greases, the most important of which include suitability for high temperature and resistance to water wash.

One of the commonly used clays is bentonite, from which the *bentones* are derived by the reaction of montmorillonite (the predominant mineral in bentonite) with amines containing a radical having at least 10 carbon atoms, such as ammonium salt. Montmorillonite is hydrous magnesium aluminum silicate. The consistency of bentone greases becomes greater as the amount of the modified clay is increased. For example, when no additives are used, an ASTM penetration of about 350 is obtained with 10% bentone by weight, whereas it is reduced to about 430 with 7% bentone. The latter values are greatly reduced (heavier consistency) when a dispersant is used in the mixing process.

High-viscosity oils are used to make bentone greases. With a drop point of 500°F+, they are suitable for high-temperature applications. They are water-resistant and evidence no melting point, but some of them will degel somewhat if held at about 300°F for several hours. Desired additives, such as oxidation inhibitors, are usually included in the bentone greases. The temperature at which bentone greases can be used for extended periods is limited by the stability of the lubricating oil used.

Another clay thickener used in nonsoap greases is attapulgite, a colloidal form of fuller's earth, which, like bentonite, is a hydrated magnesium aluminum silicate. However, the mixture of attapulgite and lubricating oil must be supplemented with a *surfactant* (surface active agent) suitable for the oil used, in order to produce a satisfactory lubricating grease. The choice of surfactants is based on low interfacial tension between water and oil, its resistance to oxidation and thermal decomposition, and water solubility. The literature lists Victamine C, Victamine D, and Ethomeen 18/15 as suitable surfactants for particularized oils in amounts ranging from 0.7 to 2.2% by weight.

Zeolites, synthetic clays used in water treatment, are also considered as gelling agents for nonsoap greases.

Carbon black of certain origins is a thickening agent for grease because of its ability to absorb a substantial volume of oil. It also offers properties that result in structural firmness. Silica gel has similar properties.

GREASE FILLERS are materials that introduce bulk as a factor in controlling consistency. The fillers used are inorganic compounds, some of which are

Graphite (colloidal, powdered and flaked)	Copper flake
	Talc
Asbestos	Magnesia
Mica	Red and white lead
Lead powder	Carbon black
Zinc dust	

Some fillers are actually lapping compounds that act to smooth out the asperities of bearing surfaces, while others act to form a protective film over such surfaces. One of the disadvantages of lapping agents is that they continue their polishing activity to the point of excessive wear. For this reason, their continued use on precision plain and antifriction bearings is to be avoided. Talc, mica, and asbestos, all of which are silicates, are among the polishing compounds. Graphite, on the other hand, is an adsorbant material and is often added to greases to increase load-carrying capacity, especially under thin film conditions. Under high vacuum of, say 10^{-6} torr, graphite loses its adsorption properties because of subsequent evaporation of the atom-covering moisture required by the material. Graphite has some thickening influence on grease when added in sufficient amounts, i.e., more than 10%.

Graphite is added to lime-soap greases (resistant to water wash) in the preparation of *curve grease,* used on railway curves to reduce wear of wheel flanges and rails and to prevent "squealing" due to dry solid friction. The product is usually applied by a trip applicator actuated by the flange, which depresses the pump lever in passing over it. The grease contains about 8 to 10% graphite and a consistency of NLGI Nos. 1 and 2, for winter and summer application, respectively.

Carbon black added to lime-soap grease enhances the metal drawing capability of the mixture.

Asbestos is used as a filler to compensate for large bearing clearances that occur either by wear or design. Such conditions are prevalent on older construction equipment. Asbestos greases usually consist of heavy-bodied cylinder or black oil mixed with asbestos, although soap-thickened greases may also act as a vehicle for the asbestos filler. Asbestos is an abrasive filler.

Mica is used as a filler for wagon axle greases.

Talc, the chief ingredient of soapstone, is a filler that enhances the metal-reducing ability of some drawing compounds because of its slipperiness.

Aluminum powder is added to a wire rope lubricant to impart a golden sheen, which identifies the manufacturer of the rope. Another manufacturer colors one of the strands of its wire rope purple for the same purpose.

Metal oxides, sulfides, sulfates, and carbonates are additional inorganic fillers used in greases for various purposes, including such metal-reducing processes as drawing, forming, upsetting, and cutting.

SYNTHETIC FLUIDS IN GREASES. Some of the synthetic fluids discussed previously are variously thickened to produce greases having special characteristics not as readily attainable with thickened mineral oils. The more important synthetic fluids used include—

Chlorofluorocarbon polymers thickened with inert materials, such as fine silica to produce a grease having properties similar to those of the base stock, having a service operating range of from −40°F to 350°F.

Dibasic acid esters thickened with lithium soap. This results in a grease (1) having a temperature range from −67°F to 250°F, (2) resistant to water wash, and (3) wear resistant, all of which add up to an excellent multipurpose lubricant. When fortified with an EP additive, ester type greases are important for the lubrication of precision aircraft frictional components.

Polyglycols containing graphite or molybdenum disulfide are excellent high-temperature lubricants.

Silicone fluids thickened with lithium soap produce greases of good thermal and mechanical stability and a high temperature range up to 400°F.

They are resistant to water wash. Higher service temperatures are attainable (600°F) with silicones thickened with carbon black.

Polyphenyl ethers with inert thickeners (e.g., fine silica) produce wide temperature range greases (from −65°F to 300°F) that excel the performance of mineral oil base greases. This is also reportedly true in a radiation environment.

Fine silica thickeners used with appropriate synthetic fluids such as silicones also produce greases having an excellent temperature range of from −40°F to 400°F.

The thickeners employed include lithium-soap, fine silica and carbon black. Lithium soap, because of its excellent service life at fairly high temperatures (300°F), water resistance, low starting torque characteristics at low temperatures, fair pumpability, and excellent operating stability, appears to be the most suitable for the special considerations given to synthetic greases.

SPECIAL USES FOR GREASE. *Water pump grease* is required to be resistant to water wash and heavy enough to provide effective sealing. It is generally composed of lime-soap and an oil of 300 to 750 SUS at 100°F. A consistency of No. 4 NLGI provides the necessary sealing properties. It can be applied either by hand packing, screw down cup, or hand-pressure grease gun. Where pump bearings are lubricated by circulating water, a small amount of soluble oil should be added to the water (about 1 part to 40 parts water).

Wheel bearing grease is required to lubricate automobile front wheel bearings of the tapered roller type, generally. Front wheel bearings are subject to extended periods of use at temperatures above 225°F, so that oxidation resistance is required. The grease must also continue to be firm enough in consistency (shear stability) to prevent drippage at maximum operating temperature. However, permanent channeling is a fault, since it is necessary that flecking of minute particles of grease from the mass should continually occur to feed the frictional surfaces of the bearing. An effective wheel bearing grease was found to be of a No. 5 NLGI consistency having a soda-soap base and an oil of 75–85 SUS at 210°F. Wheel bearings are usually hand packed.

COMPLEX GREASES. Some greases are described as "complex." The complexities of many of the conventional greases are due to the variation in characteristics with any change in basic ingredients. Their complexity increases when certain other materials are added that change both their structure, their characteristics, and often their capabilities. Such changes in a soap-base grease often produce a lubricant whose performance behavior is a decided transformation. When such a transformation is beneficial to the service intended, the new grease is usually adopted as a replacement for the conventional types.

The simple lime-soap or cup greases, as well as the simple soda-soap or fiber greases, still fulfill many requirements in which operating conditions of load, temperature, and speed are not extreme.

Complex greases are those whose basic ingredients have been fortified, modified, or treated so that they give exceptional performance in a given application. For example, a lime-lead grease manufactured under controlled procedures withstands higher temperatures, has greater water-washing resistance, and has a higher load capacity. The added advantages result in its adoption for multipurpose applications.

GREASES AND RADIATION. Both petroleum-soap greases and most synthetic greases are unstable when exposed to radiation, but the synthetics appear to be more resistant. Much work continues to be done in developing oils and greases that will perform satisfactorily under radiation for practical periods of operation. Lubricants that have been somewhat successfull in this respect are both highly sophisticated chemically and economically expensive. In some instances such factors are secondary to the process for which they are required.

Wool yarn saturated with a semifluid grease is used in journal boxes in which it is carefully packed to maintain contact with the rotating journal. The wool yarn is elastic enough to maintain a slight pressure against the journal while the fluidity of the grease promotes the flow of the lubricant by capillary action or gravity.

Petrolatum

Petrolatum, or petroleum jelly, is an amorphous wax derived from petroleum as the residue left by fractional distillation. Like other hydrocarbon compounds, it varies in chemical structure, ranging between $C_{17}H_{36}$ and $C_{21}H_{44}$. It has a microcrystalline structure and varies in melting point from about 115 to 130°F, and it has an ASTM cone penetration averaging about 130 at 77°F. Vaseline is a refined petroleum jelly with a proprietary name.

Petrolatums are used in the lubrication and coating of wire rope, for which high refinement is not always required. They are also used as electrical insulating substances which require a high dielectric strength (30 KV at 212°F). As lubricants, they should be refined by mechanical filtration at temperatures above their melting point.

SOLID LUBRICANTS

Solid lubricants include graphite, molybdenum disulfide, polytetrafluoroethylene (PTFE), polychlorotrifluoroethylene (PCFE), talc, mica, various sulfides of titanium and tungsten, tellurides, and selenides. Talc (soapstone) and mica are typical fillers for some heavy-duty greases such as axle grease. They are also used as lapping compounds for smoothing out frictional and decorative surfaces.

Typical applications for solid lubricants are in extreme conditions—of temperature, vacuum, load, and nuclear radiation—under which conventional oils and greases do not perform satisfactorily. (See the discussion of "Extreme Environments" in Chapter 21.) Some of the factors associated with lubricant limitations are—

High temperature: promotes oxidation and evaporation of volatile constituents.

Low temperature: causes congealing or solidification of fluid and plastic lubricants.

High vacuum: causes evaporation and outgassing of most substances, including some solid materials.

Nuclear radiation: affects lubricants in various ways, including ionization. One of the results of ionization of organic lubricants is less viscosity at first, followed by increasing viscosity to the point of solidification.

Solid lubricants such as graphite and molybdenum disulfide are sometimes used in their dry state, but more often they are used as an additive of oil and grease or are combined with a binder or carrier to place them on the target surface and keep them there. Organotins, such as tetraethyl lead (TEL), methylcyclopentadienyl manganese tricarbonyl (AK-33 X), and di-*n*-butyltin sulfide (Bu_2SnS) have been tested and found to form surface films on specific steels of lead, manganese, and tin.

Dry solid lubricants are usually applied on bearing surfaces subject to high temperatures, such as oven chains, or are located in dusty environments where the use of grease or oil would be impractical.

Molybdenum disulfide is added to grease in amounts up to 10% for use as a multipurpose lubricant on automotive equipment parts, such as chassis joints, steering joints, fifth wheels, universal joints, kingpins and shackle pins. It also tends to reduce fretting on close fits and performs excellently on slow-moving surfaces. Tests have shown higher mean Hertz load values with MoS_2 in diester oil base grease than with WS_2 and a synthetic graphite on a Shell four-ball extreme pressure tester at 1800 rpm and 77°F.*

Bonded coatings of molybdenum disulfide are used in aerospace applications and on metal-cutting tools, cams, sleeve bearings, small gears, universal joints, self-adjusting brake screws, and lubricated valves. Bondings include organic resins, inorganic compounds such as metals, ceramics, and sodium silicate. The selection of a binder is important, since it governs operating characteristics.

Solid lubricants are added to sintered metals and plastics to produce self-lubricated parts. Molybdenum disulfide serves to lubricate parts constructed of reinforced Teflon and phenolic parts, as well as nylon, epoxy and polycarbonate plastics.

*Reference *NLGI Spokesman,* January 1964.

Solid lubricants such as graphite and molybdenum disulfide are often carried to high-temperature environments, such as conveyor bearings in an oven, by a fluid that subsequently evaporates, leaving the solids behind to provide lubrication. They are also rubbed or sprayed on frictional surfaces and mixed with grease that holds them in place on surfaces having no other means of retention, such as hydraulic accumulator plungers.

Space vehicles, rocket engines, high-altitude aircraft, and equipment operating in cold climates require lubricants capable of combatting one or more of the limiting factors outlined above. This demand has led to the development of numerous material combinations that range from pure metals to sophisticated compounds containing such elements as tungsten, selenium, molybdenum, sulfur, fluorine, chlorine, silicon, gallium, lead, and iron, as well as the resins and oxides required to form surface films and bonds.

The solid lubricants may be classified according to their origin as being (1) organic (graphite and molybdenum disulfide), (2) inorganic (PTFE and PCFE), (3) metallic (lead, indium, gold, silver, barium), and (4) combination (molybdenum disulfide-graphite-sodium silicate, molybdenum disulfide-graphite-molybdenum/sodium silicate, etc.).

Solid lubricants may be in the form of powder (colloidal) dispersed in grease or fluids, surface films as oxides, or bonded by phenolics, epoxy, polyimide resins, fused ceramics, etc., or as retainers, such as the cages of ball bearings. Bonded solid lubricants may be applied by spraying, brushing, or dipping. Other methods include plasma spraying and electrophoretic deposition.

Surfaces to which the binder in the solid film must adhere are usually treated prior to deposition to increase wear life (Table 21-13).

Some examples of service compatibility of solid lubricants relative to the environmental factors listed above are as follows:

Graphite, molybdenum disulfide and PTFE successfully combat higher temperatures than are possible with other types of lubricants at atmospheric pressures and above. Molybdenum disulfide oxidizes at about 750°F, and PTFE is limited to light loads and temperatures to about 600°F.

Molybdenum disulfide satisfactorily performs in lubricating control rod mechanisms of nuclear reactors that are directly exposed to radiation.

A dual phenolic resin binder with molybdenum disulfide, graphite, and tin is an organic material that shows favorable outgassing properties in a high vacuum.

Thin, lead-coated (film thickness) balls and raceways of ball bearings, lightly loaded, proved reliable under space vacuum for up to 15,000 hr of operation.

A phenolic-resin-bonded molybdenum disulfide successfully allowed the extendible legs of the lunar module that landed on the moon in July 1969 to operate in a deep vacuum at −250°F.

At slow speeds and press fits, molybdenum disulfide showed negligible stick-slip and low kinetic friction force as compared with other solid materials, including tungsten diselenide.

The lowest kinetic coefficient of friction ever measured for a solid was 0.009 at 200,000 psi by the use of tantalum disulfide (TaS_2).*

Unlike graphite, which requires an adsorbed film of moisture to perform satisfactorily, molybdenum disulfide causes an increase in kinetic friction with the presence of moisture. This is also true to some extent with resin-bonded solid lubricants, as well as with rubbed-on films. Frictional heat buildup due to increased speed causes a decrease in coefficient of friction with molybdenum disulfide probably caused by moisture removal during the greater energy process.

To be classified as a true lubricant, solids of small particle size (powder, plates, etc.), must possess fluid-like qualities wherein adjacent molecular layers of a film readily slip over each other in the same manner described for liquids in Chapter 9. Such solids possess a greater bonding attraction between the atoms that *form* the layers (layer-lattice) than do the same kind of atoms that *make up* adjacent layers. For this reason, they are described as laminars that possess low shear strength.

It is often believed that, because solid lubricants cannot slump or flow back to a surface from which they have been displaced, film vacancies are left that result in metal-to-metal contact. This is untrue when the solid material is bonded to the frictional surface as has been described.

Layer-lattice solids are stable under high temperature, which makes them adaptable to frictional systems in ovens and similar conditions.

Graphite

Graphite is a crystalline form of the element carbon. It is a product of coke or hard coal. It is also mined. As a lubricant, it is in a powdered or colloidal form. It can be used alone or mixed with oil, grease, or other carriers. Graphite possesses an affinity for metal surfaces and imparts a high degree of lubricity to bearings. It has high chemical stability to withstand high temperatures and will not dissolve in water. Graphite in the form used as a lubricant will enter minute clearances. Oil compounded with colloidal graphite has had some measure of success as a metal-cutting fluid. It is mixed with lime-soap greases in varying quantities (2 to 20%) to produce graphite grease, an excellent lubricant especially suited for wet operation. Graphite depends on an adsorbed film of moisture (moisture condensed on its surface) for its lubricating action. If an adsorbed film of moisture is not present,

*P. M. Magie, "A Review of the Properties and Potentials of New Heavy Metal Derivative Solid Lubricants." Paper presented at the 21st Annual Meeting of ASLE, Pittsburgh, May 1966.

graphite acts as an abrasive. Such a condition may be brought about at high sliding velocities, during which a sufficient rise in temperature occurs to evaporate moisture when dry graphite is used. Because of its resistance to heavy pressures and high temperatures, graphite can be used under extremely severe operating conditions. Graphite has a low coefficient of friction. It has laminar structure.

Molybdenum Disulfide

Molybdenum disulfide is like graphite in that it has a laminar structure, being made up of alternate layers of molybdenum and sulfur atoms. It is produced by purifying the mineral molybdenite. Molybdenum disulfide is used (1) in powder form as a dry lubricant applied to clean metal surfaces or (2) mixed with oil, grease, or other carrier, such as silicone or polyalkylene glycol.

Molybdenum disulfide provides excellent lubrication under high or low temperature (−100 to 750°F). It will oxidize at temperatures above 750°F if not protected against the atmosphere. At ordinary temperatures, molybdenum is relatively inactive.

Some tests appear to show that an excellent lubricating film can be established with a mixture of molybdenum disulfide and corn oil. Because it resists heat, cold, water, and chemicals, molybdenum disulfide is finding wider use for extreme operating conditions. It is noncorrosive, possesses high film strength and low shear resistance, and has an affinity for metal surfaces. It is used as an additive in oils to provide protection when heavy pressures or wiping action destroy hydrodynamic films. It acts as an antiflux to prevent welding when the oil film becomes ruptured under heavy loading.

Molybdenum disulfide has a melting point of 2165°F, is blue-gray to black in color, has a molecular weight of 160.08 and 4.8 to 5.0 specific gravity. It has a coefficient of friction of about 0.032.

PTFE—Polytetrafluoroethylene

Polytetrafluoroethylene, a familiar form of which carries the trademark name Teflon, has the formula $(CF_2)_n$. It is made by breaking down the halogen compound chlorodifluoromethane by heating (pyrolysis) to obtain the gas tetrafluoroethylene. The gas molecules are then linked together by the process of polymerization, which yields PTFE.

The nonsticking property of PTFE and its resistance to high temperatures make it well known in the household. It is also chemically stable and resistant to wear and has a low coefficient of friction. It is therefore adaptable to bearings having no other means of lubrication.

PTFE is usually applied in thin films on the surface of supporting materials to provide the low coefficient of friction required for all lubricants.

In the form of thin films, it will withstand a pressure of 50,000 psi and higher, although the yield point of bulk PTFE is much lower. PTFE is well known for its heat-resistance, and when used as a self-lubricated surface coating, it can withstand temperatures up to 600°F. The coefficient of friction for PTFE ranks with some of the better lubricating substances, having a value of about $f = 0.04$.

PTFE is a thermoplastic, although it cannot be handled as one because of its high viscosity above the transition temperature of 620.6°F. It is formed by methods similar to those used in forming metal powders, i.e., by subjecting it to high pressure and sintering above its transition temperature.

To prepare PTFE for lasting wear it must be mixed with a porous material, woven with other fibers, or made into a fiber itself. When incorporated with lead and introduced into a bronze matrix, it produces a low-friction bearing with a coefficient of friction of $f = 0.02$ under substantial loads and low velocity and about $f = 0.16$ at higher speeds and lighter loads.

PTFE is sometimes used to impregnate porous metals to reduce friction. Other solid lubricants used to impregnate bearings to provide self-lubrication are nylon, graphite, and molybdenum disulfide.

When PTFE is drawn into fibers and woven with another fiber such as cotton, it produces a material which, when applied to a metal backing, results in a bearing that has low friction and is long wearing. The material is especially adaptable to slow speeds, even down to levels approaching one foot per hour.

PTFE is used with other materials as either a matrix or mixture to successfully meet the abnormal requirements of high temperature and high vacuum although some outgassing does occur when subject to both simultaneously.

Materials with which PTFE is used include thermosetting resins, thermoplastics, fiberglass, carbon, mica, graphite, molybdenum disufide,

Table 6-8. LOAD-VELOCITY CONSTANTS, $L \times V = K$*

	Mild Steel† etc. (hr)		*Hard Steel, Brinell 540 (hr)*	
	1,000	*10,000*	*1,000*	*10,000*
Thrust bearings	25,000	12,000	30,000	15,000
Slides	12,000	6,000	13,000	7,000
(a) With unidirectional loads (load fixed relative to bushing)	16,000	12,000	25,000	19,000
(b) With rotating loads (load rotating relative to bush)	25,000	19,000	30,000	24,000
(c) With fixed load and oscillating shaft	30,000	23,000	33,000	24,000

* From "Lead," Lead Industries Assn.

† These figures apply also to cast iron, austenitic stainless steel, anodised aluminum, and mild steel electroplated with lead, cadmium, or nickel.

asbestos, bronze, epoxy, and lead. When incorporated with lead and introduced to a bronze matrix, it produces a versatile PTFE-filled bearing that will support thrust, sliding, oscillating and rotating loads. The Lead Industries Assn. gives the load-velocity (load × velocity) constants shown in Table 6-8 for various types of motion in relation to the number of hours run for mild and hard steels used as backing materials. They have reportedly operated successfully at temperatures ranging from −450°F to +536°F. PTFE bearings backed with lead, bronze, and steel are employed in various services including helicopter rotor hubs, jet engine bearings, and a number of automotive, submarine, and aircraft applications. They are also applicable where dirty atmospheric conditions are prevalent, as in cement mills and coal mines.

Soap

Soap is a form of solid lubricant used primarily as a drawing compound, but it may be used to lubricate molds and other surfaces on which oil or grease would be deleterious, such as rubber and many plastics. There are different kinds of soaps, including the more commonly used stearates, such as those combined with zinc, barium, lead, aluminum, calcium, and magnesium. Ordinary laundry soap rubbed into screw threads facilitates driving screws into wood.

CLASSIFICATION BY VISCOSITY

One method of classifying mineral lubricating oils is according to viscosity. This method of classifying lubricating oils is according to the following outline:

Low viscosity class (60 to 220 at 100°F, SUS)
- Spindle oils
- Light machine oils
- Light turbine oils

Medium viscosity class (220 to 420 at 100°F, SUS)
- Medium-bodied machine oils
- Compressor oils
- Medium turbine oils

Heavy viscosity class (420 to 1500 at 100°F, SUS)
- Heavy machine oils
- Gear oils

Extra heavy viscosity class (1500 to 3000 at 100°F, SUS)
- Extra heavy machine oils
- Heavy gear oils
- Cylinder oils

This is an old method of classifying lubricating oils, and because it is based entirely on viscosity, it is not sound.

Various degrees of purity can be obtained, depending on the amount as well as the quality of refining to which an oil is subjected. For some purposes, as in refrigerating, turbine, and internal combustion engine systems where the lubricant may be affected by operating conditions and/or is reused for extended periods, a highly refined (chemically stable) oil is essential. Otherwise, chemical changes occur that result in the formation of gums, varnish-like materials, sludges, and carbon that deposit on the working surfaces of the lubricant system. When present on pistons, piston ring grooves, valves, cylinder heads, bearings, heat-transfer surfaces, and the interior walls of oil lines, such deposits interfere with mechanical processes, restrict the flow of oil, and eventually cause machine failure. The life expectancy of a mineral lubricating oil is reduced by one half (50%) for each 18°F rise in its temperature. It follows that only chemically stable oils having inherent longevity should be selected for lubricating systems requiring their continual reuse, or in which conditions of operation may cause rapid deterioration of an oil of inferior quality.

For systems in which the oil is lost after serving its purpose for a short period, such as reservoirless bearings and chains, any fairly good oil, having suitable characteristics otherwise, will suffice in many cases. In modern machinery, the latter type of lubrication system is diminishing, although some conditions of design will not permit its elimination entirely.

SERVICE CLASSIFICATION

Spindle oils. These are light-bodied fluids used to lubricate machine spindles rotating at moderate to very high speeds. They are highly refined mineral oils containing appropriate additives, with a viscosity range of from 30 to 100 SUS at 100°F.

Turbine oils. Turbine oils are slightly heavier in body than spindle oils to carry greater loads at the speeds encountered. Turbine oils are a basic type of high quality light-colored fluids usually containing a rust and oxidation inhibitor having an oxidation test value of 1000 to 4000 hr, depending on base oil quality. The viscosity range is from 130 to 400 SUS at 100°F.

Hydraulic oils. Hydraulic oils are in the same family as turbine oils but are given a separate classification in recognition of the commercial importance of hydraulic power. Some hydraulic oils contain antiwear agents to meet MS-sequence requirements for use in high-pressure, high-output pumps. Their viscosity ranges from 70 to 400 SUS at 100°F.

Circulation bearing oils. This family of oils includes the turbine and hydraulic oil types, but also those of higher viscosities to meet greater load requirements. They usually contain rust and oxidation inhibitors and also

Table 6-9. Classification of Lubricating Oils According to Service, Type, Viscosity, and Additives

Classification	*Service*	*Typical Viscosity (SUS at 100°F)*	*Comment and Commonly Used Additives*
I	Spindle oils		Highly refined mineral oils
	Spindle speeds >3600 rpm	35–100	FS, RI, MD
	Spindle speeds <3600 rpm	100–150	FS, RI, MD
	Spindle speeds >1800–<300	150–900	RI, OI
II	Turbine oils		
	Direct connected	150	RI, OI
	Geared	300	RI, OI
	Marine type, bleeder	400	RI, OI
III	Hydraulic oils		Highly refined mineral oils or phosphate esters
	Vane pumps	150–300	RI, OI
	Angle and radial piston pumps	150–900	RI, OI
	Axial piston	150–300	RI, OI
	High pressure, high output pumps	300	AW
	Gear pumps	150–600	RI, OI
IV	Circulating bearing oils		
	Light loads	150–400	Turbine type oil, RI, OI
	Intermediate loads	400–900	Turbine type oil, RI, OI
	Heavy loads	900–2500	Leaded compounds or turbine type, RI, OI
	Extremely heavy loads	1500–3500	Leaded compounds
V	Gear oils (enclosed gears) splash		
	Spur, herringbone, and bevel	600–1800	Turbine type oil or leaded compounds
	Heavy or shock loading		
	Worm gears	1500–2500	Leaded compound
		2500	Compounded steam cyl-oil or leaded compound
	Hypoid gears	2500	EP oil
VI	Engine oils (motor oils)		Bright stock blended with neutral oil
	Gasoline engines	SAE 30	OI, CI, Dt, High VI
	Diesel engines	SAE 30	OI, CI, Dt, High VI
	Gas engines	SAE 30, 40	OI, CI, Dt, High VI
	Propane engines (LPG)	SAE 30	OI, CI, Dt, High VI
	Dual trifuel engines	SAE 30, 40	OI, CI, Dt, High VI
VII	Way oils		
	Light loads	150–300	AW, A, Df, OA
	Intermediate loads	300–500	AW, A, Df, OA
	Heavy loads	900	AW, A, Df, OA
	Combustion way and hydraulic oil	150–300	AW, OI

Table 6-9 (*Continued*)

Classification	*Service*	*Typical Viscosity (SUS at 100°F)*	*Comment and Commonly Used Additives*
VIII	Cylinder oils (steam)		
	Wet steam	2500	Steam cylinder oil 3 to 10% FO
	Dry steam, up to 500°F superheat	3000–4000	Steam cylinder oil No compound to 6
	Dry steam, over 500°F superheat	4000–6000	Steam cylinder oil No compound
IX	Compressor cylinder oils (other than refrigeration)		
	Air and inert gases, up to 150 psi	300	OI
	Air and inert gases, 150 to 1000 psi	500–600	OI, AW, CI
	Air and inert gases, 1000 to 2000 psi	1500	OI, AW, CI
	Air and inert gases, 2000 to 4000 psi	1500–2500	OI, AW, CI
	Air and inert gases over 4000 psi	2500	CI, Df
	Hydrocarbon gases	1600–1700	OI, AW, CI
	Pneumatic tool cylinders	200–500	E
	(Oxygen)	(Water)	
	Expansion engines (cryogenic $< -300°F$)	Nonlubricated cylinder No lubrication	
X	Refrigerating oils any refrigerant (except SO_2)		
	Reciprocating and rotary compressors		
	Evap. temp. above $-50°F$	200–300	Straight mineral oil (preferred)
	Evap. temp. below $-50°F$	150	Low floc and pour-po
	Centrifugal compressors (bearings)	300	RI and OI permitted
	Hermetic reciprocating compressors		
	Evap. temp. minimum $-40°F$	300	Straight mineral oil Low floc and pour p
	Evap. temp. -40 to $-70°F$	150	Straight mineral oil Low floc and pour p
	(Sulfur dioxide)	(150–300)	(White oil)
XI	Machine oils		
	Light* loads, high speed	35–150	Spindle oil type
	Medium* loads, medium speeds	150–600	Way or turbine oil typ
	Heavy* loads, slow speeds	600–1800	Way or gear oil type
	Very heavy loads, slow speed	Up to 4000	Steam Cylinder type o leaded compounds
XII	Automotive transmission oils		
	Conventional transmission and differential	SAE 90–250	Cylinder type oil with Cylinder type oil with
	Hypoid gears	SAE 90–140	AW, Df
	Automotive fluid transmissions	200	
XIII	General purpose (intermittent application, generally)		
	Light duty service	150–300	FO
	Intermediate duty service	400–900	FO
	Heavy duty service	1500–2500	FO

Table 6-9 (*Continued*)

Classification	*Service*	*Typical Viscosity (SUS at 100°F)*	*Comment and Commonly Used Additives*
XIV	Air line Lubricants		
	Small pneumatic tools	200	E, AW, OA
	Pneumatic control cylinders	500	E, AW, OA
	Large air cylinders	800	E, AW, OA
XV	Vacuum pump oils		
	0.05 microns Hg	300–400	Low vapor-pressure oil
XVI	Open gear fluids		
	Hand, spray, or drop feed application		
	Without diluent	400–1500 at 210°F Diluent used for higher viscosities	Black residual
	Sump, dip application	2000–7500 at 100°F	Leaded compounds
XVII	Black oils		
	Circulation system	400–2000	A. Better quality oils are dark green
	Swab, pour, and dip application	2000–5000	

Key

A	Adhesive Agent	E	Emulsifier	OA	Oiliness Additive
AW	Antiwear Agent	EP	Extreme Pressure Additive	OI	Oxidation Inhibitor
CI	Corrosion Inhibitor	FO	Fatty Oil	RI	Rust Inhibitor
Df	Defoamant	MD	Metal Deactivator	VI	Viscosity Index
Dt	Detergent				

* May include adhesive compounds and AW

an antiwear agent. They are made from high quality oil stocks, and their viscosity ranges from 150 to 6000 SUS at 100°F.

Gear oils. These include both the higher viscosity turbine oils and oils compounded with lead soaps to impart antiwear and film strength properties. Their viscosity range is from 300 to 6000 SUS at 100°F.

Engine oils. Commonly referred to as motor oils, engine oils are designed primarily for lubrication of internal combustion engine cylinders, bearings, cams, valves, etc. They are ordinarily blends of bright stock (see Steam Cylinder Oils) and neutral oils (filtered oils obtained by distillation), but some of the high premium engine oils are made from select neutral stocks. Engine oils are compounded with appropriate additives to achieve oxidation stability, detergency (dispersancy), corrosion protection, antiwear properties, antiacidity, foam resistance, variously designed to meet ML, MM, MS, and DG standards, according to the API Classification System. They are produced to meet viscosities specified for SAE Nos. 5W to 50 and are suitable for the lubrication of diesel, gasoline, and LPG (low-pressure gas) engines. Their viscosity index ranges from 95 to 150.

Way oils. This family of oils is specifically made for the lubrication of machine ways and other parallel opposing surfaces having relative translatory motion between which boundary conditions tend to exist. Because of the absence of a hydrodynamic film, any lubricant applied to such surfaces squeezes out. This leaves only a very thin film conducive to metal-to-metal contact and the occurrence of intermittent sticking and slipping, or chatter under heavy loading and slow motion. To prevent stick-slip, reduce friction, and retard wear, way oils are compounded with oiliness, antiwear, and adhesive agents. Their viscosity ranges from 150 to 1000 SUS at 100°F.

Steam cylinder oils. These are composed of a heavy-bodied residue of paraffin base crude distillation called *cylinder stock,* which may or may not be filtered. When processed by dewaxing or clay filtering to improve color, they are called *bright stock* and are used to blend with neutral oils to form some motor oils and other lubricating oils. Cylinder stocks may be used with or without compounds and are used as lubricants for steam cylinders. Cylinder stocks intended for "wet" steam conditions are compounded with 3 to 10% animal oil, such as acidless tallow, to provide emulsion properties as well as oiliness required for the lubrication of worm gears. Dry (high superheat) steam conditions do not require such compounding. The viscosity ranges from 2300 to 6500 SUS at 100°F.

Compressor cylinder oils (Other than Refrigeration). These are high quality lubricants with a high viscosity index and are similar to turbine oils used on compressor cylinders and bearings handling dry air up to 150 psi discharge pressure. To cope with wet gas conditions, including nitrogen, CO_2, air, and other inert gases, an oil compounded with antiwear and anticorrosion agents, as well as an oxidation inhibitor, is required. High pressures in the magnitude of 2000 to 7000 psi require oils that resist deposit formation at the higher temperatures. Viscosity ranges from 300 to 2500 SUS at 100°F.

Refrigeration oils. These are special straight mineral oils of low floc and pour points. The floc point may measure as low as −70°F and lower. Some refrigerants have solvent effects on unsaturated hydrocarbons, which may be precipitated when subject to a sudden drop in temperature as occurs at the expansion valve of a refrigeration system. This results in a clogged throttling device. As a result, the use of high quality oils is mandatory in refrigeration systems in which the lubricant enters the low-pressure side along with the refrigerant. Additive type oils should not be used to lubricate refrigeration compressors, particularly those handling ammonia. However, a high quality rust and oxidation inhibited oil with a viscosity of 300 SUS at 100°F is preferred for lubricating the bearings of centrifugal compressors on which effective sealing maintains the system fairly free of in-leak oil.

Machine oils. This classification includes any oil from light spindle to heavy cylinder types whose quality may or may not be that required by turbine type oils. The definition as applied here refers to those oils subject to *all-loss* applications where the extra cost of premium grades may not be justified. Their methods of application may include bottle oilers, wick feed cups, drop feed devices, and by swab, brush, spray, pad, waste packing, and drip. They may also substitute for high quality hydraulic oils in systems subject to excessive leakage loss. Downgrading of lubricants should never be carried to the extent of jeopardizing the wear protection of equipment by permitting the formation of gum and other deposits in frictional equipment that may eventually cause it to cease operation, such as from the binding of a bearing.

Automatic fluid transmission oils. Here are included those oils generally made for use in automatic and semiautomatic fluid transmissions, torque converters, and power steering systems. They are predominantly used in passenger automobiles and light trucks and for some industrial purposes in variable torque drives. They are high quality oils with a high viscosity index, good chemical stability, low pour point, and antifriction and antiwear characteristics. Their viscosity ranges from 100 SUS at 100°F (for some bus hydraulic drives and transmissions) to about 200 SUS at 100°F (for conventional passenger car automatic fluid drives).

Automotive gear transmission oils. These are relatively heavy oils of the cylinder oil type employed for the lubrication of automotive gears other than hypoid, including those used in transmissions, differentials or axles, some final drives or power takeoffs. They are made to meet SAE 90 (900 SUS at 100°F), SAE 140 (2800 SUS at 100°F) and SAE 250 (5000 SUS at 100°F) viscosity standards. Automotive gear transmission oils also include the *extreme pressure* (EP) *lubricants* possessing antiweld characteristics (with sulfur, chlorine, and phosphorous combinations as additives). They are recommended for hypoid gears used in any type service, particularly automotive equipment and some paper mill drives. Viscosities range from SAE 90 to SAE 140, having approximately the same values as the non-EP oils above. A defoamant is usually added to all transmission oils. The SAE 90 extreme pressure oil is predominant in the lubrication of industrial hypoid gears.

General purpose oils. Lubricants referred to here fit the definition of machine oils, multi-purpose lubricants, and those meeting multiviscosity or multigrade specifications. Any nomenclature requiring modification is not a good one. For example, a turbine oil may be employed as a spindle oil, turbine oil, hydraulic oil, air line lubricant, circulating bearing oil and engine oil, and so becomes general in nature. A multiviscosity oil is usually an engine oil or automotive gear transmission oil that may be *generally* used both winter and summer. The nomenclature usually applies to lubricants used in all-loss applications and so fits the viscosities shown in the basic table for various load levels.

Air line lubricants. Oils that are especially compounded to lubricate pneumatic tools and machine power cylinders. They are similar to turbine type oils containing a small amount of emulsifying agent to combat the washing effect of condensate that precipitates from the compressed air upon expansion (decreased dew point temperature). Viscosity ranges from about 200 SUS at 100°F for small tools to about 800 SUS at 100°F for larger cylinder wall areas.

Vacuum pump oils. These oils possess low vapor tension to combat the evaporating effect of high vacuum environment. They serve to lubricate the cylinders of reciprocating vacuum pumps and to provide a seal between piston rings and cylinder wall to prevent blow-by.

Open gear fluids. For gears not provided with oil sumps, heavy residual type lubricants are used, applied by hand, spray, or drip. Heavy duty open gear lubricants are too heavy bodied to apply without heating (flux) if they are not diluted with a volatile solvent or diluent that evaporates from a spread surface. Types that are more fluid require no diluent but do not "set up" sufficiently to form a dry type film over the gear teeth. Leaded compounds that are heavy bodied (too high in viscosity to circulate) may also be used on fully open gears applied by spray or swab. Lighter bodied leaded compounds are used to lubricate open gears provided with a sump for dip application. The latter is a more satisfactory method of lubricating large open gears, but unfortunately some machine design situations lack the space to provide a sump, or their process develops contaminants in such quantities that the use of open bodies of lubricant is impossible. Steam cylinder oils may also be used for dip-fed open gears. So-called black oils, many of which are dark green, containing an adhesive agent, may also be used, but their lack of antiwear properties makes them a mediocre substitute for heavy duty service.

Black oils. These oils were familiar lubricants in past years, especially around mines, quarries, sand banks, steel mills, and iron works. Machines, including heavy machine tools, steam shovels, rolling mills, locomotives, and forging hammers, were rough and usually massive, and lubrication appeared to be a "necessary evil." Black oils were the machine lubricants of the day and differed only in their viscosity, generally classified as "winter" and "summer" black oils, which were respectively formulated by adding more or less distillate oils to still residuals. They are still used widely in some of the excavating, mining, and construction type industries for heavy duty plain bearings, wire rope, open gears, swivel joints, chains, roller tracks, waste packed car journal bearings, and crane and shovel axles. Today's substitutes are more refined than the old type black oils and are usually dark green in color. They are also made in a variety of viscosities ranging from about 400 to over 5000 SUS at 100°F, and some contain an adhesive agent. They are quite effective as lubricants because of their property of "staying put" on hot surfaces. The author prefers a black oil base for oils used on pipe thread chasers and other heavy metal shaping processes, such as upsetting, some forming operations, and other cold working processes. Unfiltered cylinder stocks have a similar property, and when blended with sperm oil and sulfur become a good utility heavy lubricant for cold working.

7

LUBRICANT CRITERIA AND TESTS

Specifications of lubricating oils are developed by standard test procedures. The conditions and methods of testing lubricants usually follow the standards prescribed by the American Society for Testing Materials (ASTM) in their publication "ASTM Standards on Petroleum Products and Lubricants."

While there are a number of tests required by the producer for purposes of control and development of lubricating oils, the user need be acquainted with only a few. Lubricant specifications are necessary in selecting a suitable product for a given requirement, but they should not be adopted as the sole criterion for predicting final results. Buying lubricants on the basis of specification alone is an extremely poor policy, especially where expensive machinery is concerned, or where oil deterioration will interfere with work processes. Quality and adaptability to a given lubrication requirement must also be considered. These factors have no numerical equivalency that can be adopted as a specification for the interpretation of service results. Quality lubricants cost more to produce than poor ones. The relative difference in unit price will usually result in lower maintenance cost, in fewer interruptions in machine operation, and in the long run reduced overall lubrication cost.

Specific Gravity

Gravity is associated with the weight per unit volume of oil. Specific gravity is the ratio of weight per unit volume of oil to the weight of an equal volume of water when both are at 60°F, or

$$\text{specific gravity} = \frac{\text{weight per unit volume of oil at } 60^\circ\text{F}}{\text{weight of equal volume of water at } 60^\circ\text{F}}$$

To facilitate gravity readings, the method proposed by the American Petroleum Institute (API) is used by the oil industry. Determined by an API hydrometer (having a scale of some part of the range from 0 to 100 degrees API), values are expressed in terms of API gravity. An equivalent value of API gravity may be determined if the specific gravity of an oil is known by the following equation, using 141.5 as a modulus:

$$\text{API gravity} = \frac{141.5}{\text{specific gravity } (60/60^\circ\text{F})} - 131.5$$

API gravity values correspond to specific gravity in the order of

$$\begin{aligned} 0 \text{ degree API} &= 1.076 \text{ sp gr} \\ 10 \text{ degrees API} &= 1.00 \text{ sp gr} \\ 100 \text{ degrees API} &= 0.6112 \text{ sp gr} \end{aligned}$$

EXAMPLE: Determine the API gravities for lubricating oils having specific gravities of 0.94, 0.91, and 0.88.

$$\text{1st case} \quad \text{API gravity} = \frac{141.5}{0.94} - 131.5 = 19.0$$

$$\text{2nd case} \quad \text{API gravity} = \frac{141.5}{0.91} - 131.5 = 24.0$$

$$\text{3rd case} \quad \text{API gravity} = \frac{141.5}{0.88} - 131.5 = 29.3$$

Note that API gravity values increase as specific gravity decreases. In other words, the API gravity becomes correspondingly less with heavier (in weight) oils. Also note that gravity differences are more discernible when expressed in terms of API gravity (Table 7-1).

Gravity is an indication of the type of crude from which an oil is produced. Oils from different crudes will have gravities approximating values shown in Table 7-2 for oils of similar viscosity.

Most petroleum oils are lighter than water and therefore have API gravities above 10, usually in the range of 20 to 30 for lubricating oils.

Flash Point

Flash point is the temperature at which an oil vaporizes sufficiently to sustain momentary ignition when exposed to a flame under atmospheric conditions and controlled test procedure. Placed in an open cup (Cleveland open cup) with a thermometer immersed, the oil is heated at the rate of 30°F/min temperature rise until it reaches 100°F below the probable flash

Table 7-1. Conversion of API Gravity to Specific Gravity and Corresponding Weight

API Gravity Degree	*Specific Gravity 60/60° F*	*Pounds per Gallon*	*API Gravity Degree*	*Specific Gravity 60/60° F*	*Pounds per Gallon*
1.36	1.065	8.870	24	0.910	7.578
1.99	1.060	8.829	25	0.904	7.529
2.62	1.055	8.787	26	0.898	7.481
3.26	1.050	8.745	27	0.893	7.434
3.91	1.045	8.704	28	0.887	7.387
4.56	1.040	8.662	29	0.882	7.341
5.21	1.035	8.620	30	0.876	7.296
5.38	1.030	8.578	31	0.871	7.251
6.55	1.025	8.537	32	0.865	7.206
7.23	1.020	8.495	33	0.860	7.163
7.91	1.015	8.453	34	0.855	7.119
8.60	1.010	8.412	35	0.850	7.076
9.30	1.005	8.370	36	0.845	7.034
10	1.000	8.328	37	0.840	6.933
11	0.993	8.270	38	0.835	6.951
12	0.9861	8.212	39	0.830	6.910
13	0.979	8.155	40	0.825	6.870
14	0.972	8.099	41	0.820	6.830
15	0.966	8.044	42	0.815	6.790
16	0.959	7.989	43	0.810	6.752
17	0.953	7.935	44	0.806	6.713
18	0.947	7.882	45	0.802	6.675
19	0.940	7.830	46	0.797	6.637
20	0.934	7.778	47	0.793	6.600
21	0.928	7.727	48	0.788	6.563
22	0.922	7.676	49	0.784	6.526
23	0.916	7.627	50	0.780	6.490

point. The rate of heating is then reduced gradually until flash occurs at about 10°F/min for the final 50°F. A flame is passed over the oil in the region immediately above the surface at intervals of 5°F temperature rise. The temperature at which a definite, self-extinguishing flash occurs on the surface of the oil is the flash point.

The flash point will vary with oils of similar viscosity made from different crudes. In this respect, paraffinic base oils have higher flash points than other types, with mixed, asphaltic, and naphthenic base oils following in that order. The flash point will increase with higher viscosity oils from

Table 7-2. API Gravities of Lubricating Oils From Various Type Crudes

Base of Lubricating Oil	*API Gravity of Lubricating Oil (approximate)*
Paraffinic base oils (Pennsylvania)	29.0–32.0
Mixed base oils (Mid-Continent)	27.5–29.5
Asphaltic base oils (California)	24.0–27.0
Naphthenic base oils (Gulf Coast)	19.0–23.0

the same crude. For example, light spindle oils may have a flash point of 350°F or less, while a cylinder oil from the same crude may not flash until temperatures above 600°F are reached. The flash point is a measure of lubricant volatility.

Fire Point

The fire point is the temperature at which an oil will sustain ignition continually when exposed to a flame under atmospheric conditions and controlled test procedure. The methods of determining both fire and flash points are similar, except that heating continues until the surface of the oil ignites and continues to burn for a period of at least 5 sec. The temperature at this point is the fire point. The difference between flash and fire point temperatures varies with different oils by an amount ranging from 10 to 100°F, depending chiefly on viscosity. For general purposes, interpolation of fire points can be made on the basis of the flash point–fire point correlation, shown in Table 7-3.

Pour Point

The lowest temperature at which an oil will flow when tested (chilled) under certain specified conditions is called the pour point. The oil to be tested is first placed in a testing jar tightly closed at the top by a cork, which also supports a thermometer immersed in the oil. The test involves, first, heating the oil in a bath to 115°F and then allowing it to cool in air (or in a water bath not over 77°F) to a temperature of 90°F. The jar of oil is then placed in a suitable cooling medium, such as crushed ice and calcium chloride crystals, and reduced in temperature. Beginning at a temperature of 20°F above the expected pour point, the jar is removed at each lower interval of 5°F and tilted to determine fluidity. The jar must be moved from one cooling medium to another having a lower temperature range at stipulated temperature levels above the pour point. The test continues under the above conditions until the oil will not show any movement when the jar is held in a horizontal position for 5 sec. The pour point of the oil tested will be 5°F above the temperature at which the sample solidified.

Table 7-3. FLASH POINT—FIRE POINT CORRELATION

Flash Point (°F)	*Fire Point (°F)*
200	210
300	330
400	450
500	570
600	690
650	750

The pour point is a guide in predicting the temperature at which oil will flow by gravity under otherwise static conditions. Operational factors tend to contradict test results in many instances, but discrepancies occur for reasons not always apparent. Pressures and previous conditions to which oils are exposed in service affect the pour point and often in a direction of advantage rather than disadvantage. Treated with intelligence, laboratory results will serve as a safe guide in governing the use of oils, not only as they concern pour point, but in many other physical tests as well.

Oils employed under low-temperature conditions or where they ultimately will find their way into cold operations must have low pour points. It is well to select oils having pour points substantially below (1) the minimum operating temperature of a system or (2) the minimum surrounding temperature to which the oil will be exposed. Refrigeration systems operating with low evaporator temperatures; worm gears of elevators having their drive mechanism located in unheated penthouses; fluid couplings, gears, pumps, chains, hydraulic systems, etc., operating out-of-doors during the winter; and many other systems operating under subnormal temperature conditions rely on the fluidity of the low pour point oils for maximum efficiency and wear protection.

Cloud Point

If the temperature of an oil is reduced sufficiently, wax will start to precipitate. The cloud point is the temperature at which wax precipitation starts. When precipitation occurs, the oil takes on a cloudy appearance, hence the name. Cloudiness is not always due entirely to wax crystallization, but may be caused partly by the presence of moisture. For this reason, it is well to use oils of low cloud point for the lubrication of refrigeration compressors, particularly in low-temperature systems.

Refrigerants that are miscible in oil, such as some of the fluorinated hydrocarbons, tend to reduce the temperature at which wax precipitation occurs. Therefore, the cloud point is not a true criterion of oil behavior in refrigeration systems employing miscible refrigerants. In the case of ammonia and other nonmiscible fluids, the cloud point must be considered if wax deposits in low-temperature evaporators are to be avoided. For miscible refrigerant systems, the floc point is of greater importance than the cloud point.

Floc Point

The floc test is somewhat similar to the cloud test, except that a mixture of 90% R-12 and 10% oil serves as the test specimen instead of oil alone. The floc point is the temperature at which flocculation starts to occur in the mixture when chilled. While a more practical test would employ a mixture containing the miscible refrigerant of immediate concern, R-12 is adopted as a standard for reporting floc points of oils. The relative quantity of oil used in determining floc point appears to be reasonable considering that the amount of oil entering a refrigeration system does not usually exceed 10% of the total fluid circulated. In most cases, the percentage is less than 5%, depending on the design and the condition of the compressor.

The floc point is important in selecting lubricants for low-temperature systems employing miscible refrigerants. It is of little value where immiscible refrigerants are used and need not be considered in any case where evaporator temperatures are maintained above the freezing point of water.

Dielectric Strength

Dielectric strength is pertinent to the study of lubrication and refrigeration because the insulating value of oils is directly associated with moisture content. Well-refined lubricating oils that are relatively free of moisture will resist the passage of an electric current through them. The less moisture they contain, the better insulators they become.

Where dehydrators can be used, small amounts of moisture may be removed from a system after it enters by whatever means. In cases of hermetically sealed units where every precaution is taken to eliminate moisture, by drying the unit or other means, and on which a dehydrator cannot be used after final assembly, the moisture content of oil becomes an important consideration. Therefore, it is necessary to use oils that have been dehydrated to a condition of high dielectric strength.

Lubricating oils can be dehydrated to a condition of extremely low moisture content, resulting in a dielectric strength of 27,500 V (volts) or more. For the majority of refrigerating installations where some moisture enters by various means and subsequently is removed by cartridge dryers, it is poor economics to demand oils that are especially dehydrated. The reason for this is that the refiner normally prepares refrigerating oils having a dielectric strength of at least 25,000 V, which indicates a very low moisture content.

Because mineral oils are hygroscopic (i.e., absorb moisture from the air), they should be kept covered and otherwise handled carefully to prevent undue exposure. This applies particularly to oils in which any significant amount of moisture may prove deleterious.

Emulsification

The subject of emulsification involves the old belief that "water and oil will not mix." The statement has some semblance of truth, but only under extraordinary circumstances. New highly refined straight mineral oils possess qualities of good water separability. Such qualities will remain to a high degree just so long as conditions of handling oils are proper. They will not continue to demulsify under poor operating conditions where contaminants not only are permitted to enter the oil but also are allowed to remain indefinitely. Carbon dioxide from the air forms carbonic acid when dissolved in water that may be present in an oil in service. Acids form sludges that have more or less emulsification properties. Add to the latter numerous other contaminants that find their way into an oil and the conglomerate results in a chemical complexity that leads not only to emulsions, but also to other deteriorating influences.

Except in special cases, such as soluble oils used in metal cutting processes, emulsification is undesirable. Emulsification permits oil to retain water particles that (1) ultimately lead to lubricant deterioration, (2) displace oil in frictional areas, causing mechanical failure, and (3) cause rust and corrosion. Reservoir foaming also results in some cases.

The problem of oil emulsification is of great importance in circulation or other reuse systems. It cannot be tolerated to any extent in lubrication systems serving high-speed journals, such as in steam turbines.

Laboratory tests dealing with the emulsification tendencies of lubricants include the steam emulsion test. This involves passing steam (212°F) through a test tube of oil (20 ml at room temperature) for a period of 4 to 6.5 min. The test tube containing the oil-condensate mixture is then moved to a separating bath (200 to 203°F). The time in seconds required for the oil to separate from the water is called the *ASTM steam emulsion number,* or simply the S.E. number. If 20 ml of oil does not separate in 20 min, the sample will be reported as having an S.E. number of 1200 plus.

Certain type additives used in turbine and other type oils tend to raise the S.E. number. Therefore, the result of a steam emulsion test is not a true criterion on which to base conclusions relative to pertinent chemical changes in the oil due to operation.

DEMULSIBILITY. Rapid separation of water from oil is an important characteristic related to circulating oils, especially those used for the lubrication of steam turbines and rolling mills. The ability of water to separate depends on the purity, viscosity, and temperature of the oil. Additives and certain contaminants slow the separating process so that older demulsibility tests are not practical for steam turbine oils. The tests are conducted by mixing oil and water in some definite proportion, depending on the test procedure (of which there are several) and agitating the mixture at a given rate and temperature, usually 130°F, for a given period of time. The rate at which separation occurs during a settling period following agitation is the relative measure of the separating ability of the test oil. The rate is usually measured in cubic centimeters of oil separation per hour. Under conditions of the Herschel demulsibility tests and a new, high grade straight mineral oil, a value of 1620 is considered desirable and is specified for circulating oils servicing rolling mill roll neck bearings.

Carbon Residue

The carbon residue test (Conradson) determines the percentage of carbon residue left by an oil upon evaporation under specified test conditions. The test sample is placed in a crucible within a crucible and is heated under atmospheric conditions until it is evaporated, that is, until its vapors no longer ignite. The residue is then cooled and weighed. The result is reported on the basis of weight ratio of residue to oil sample. The Ramsbottom test is another method of determining carbon residue by evaporation. Some discrepancy exists between the results obtained by the two methods.

The carbon residue test serves to permit some doubtful conclusions to be drawn relative to the carbonization properties of an oil. However, the nature of carbon may be deduced by residue inspection. Hard carbon is formed by paraffinic-base oils, while naphthenic-base oils produce soft, fluffy carbon. The amount of carbon residue of an oil varies with different crudes, refining methods, and viscosities. Predicting results based on carbon residue without consideration of all other factors that might influence operation is an unsound practice.

Viscosity

Viscosity reflects the fluidity of fluids. It is the measure of intrinsic friction during flow. The rate at which a lubricating oil will change in viscosity with any temperature change is indicated by its viscosity index. Viscosity is discussed in detail in Chapter 8.

Color

The color of oils varies from water clear to black, including different shades of amber, red, and green. Color alone is no indication of purity or service capacity. It may be associated with some type of services, such as the familiar ambers of spindle oils or the greens of cylinder stocks and heavier-bodied motor oils. A black oil may be either a new oil of dubious quality or a highly refined motor oil containing a detergent that has been in automotive service for a very short time. While color is of some importance to the oil refiner as a processing control indicator, variation in the color of a new oil is no guide for most machine operators.

A change in color of an oil in service is quite another consideration. Rapid darkening of turbine oils may indicate some change in the system that should be investigated. A gradual change in turbine oil coloring is normal. A change from some shade of amber to apricot means water in the turbine system, and if accompanied by some degree of emulsion, it becomes a matter of concern. But to judge the expected performance of an oil on the basis of its being an ASTM color number 2.5 or 5 for any service has no logical basis.

With most oils, the higher the color number (standard test procedures), the darker it appears. This is not true with what are commonly called "white" oils, which more correctly corresponds to the appearance of pure water (colorless). In the case of white oils, the higher the number, starting with −16, the closer they approach water white, rated 30+ on the Saybolt chromometer. White oils in the 30+ category are usually odorless and tasteless, and most of them pass the United States Pharmacopeia standards relative to food purity. Meat hooks, for example, on which butchers hang meat carcasses to be passed through an abattoir for processing are coated with a white oil which must pass U.S.P. standards, which permit no odor or taste.

Color is specified variously according to the method used, the more familiar one being the ASTM color standard (D-1500). This procedure determines the color expressed by a number starting with 0.0 on the color scale and graduating upward to 8.0 ASTM. It is determined on the ASTM colorimeter and represents a color comparison with the color of standard glass discs by transmitted light. Colors darker than 8.0 are determined by diluting the test oil with kerosene having a water white color of +21 Saybolt or above, using a dilution of 85% kerosene and 15% oil. Color specifications of oils so determined include the designation "Dil" after the number. Other color designations identify the procedures used, such as the older ASTM D-155-45T, determined on the Union colorimeter, TAG Robinson, Union Petroleum Company standard, and National Petroleum Association numbers. Corresponding values of those mentioned are given in Table 7-4.

Neutralization Number

The neutralization number is the number of milligrams of potassium hydroxide (KOH) required to produce a pink color (by titration) in a mixture of test oil, distilled water, alcohol, and phenolphthalein. The result of the

Table 7-4. CORRESPONDING OIL COLOR DESIGNATIONS

ASTM D-1500	*TAG Robinson*	*Union Petroleum Company*	*National Petroleum Assn.*	*ASTM 155-45T*
0.5	21	G	1	1
1.0	17.5	H	1.5	1.5
1.5	12.5	I	2	2
2.5	10	J	2.5	2.5
3.0	9.75	K	3	3
3.5	9	L	3.5	3.5
4.0	8.25	M	4	4
5.0	5	N	4.5	4.5
5.5	3.5	O	5	5
6.5	2.25	P	6	6
7.5	2	Q	7	7
8.0	1	R	8	8

test for neutralization number must not be construed as a measure of absolute acidic or basic properties of an oil. It is intended to determine the pressure of acidic or basic constituents in an oil. The American Society for Testing Materials lists as constituents of new or used oils considered to have acidic characteristics, the following: organic and inorganic acids, esters, phenolic compounds, resins, salts of heavy metals, inhibitors, detergents, and others. Basic constituents include: organic and inorganic bases, amino compounds, salts of weak acids (soaps), salts of heavy metals, inhibitors, detergents, and others. Some of the constituents are the result of changes that may occur in an oil due to the deteriorating influence of oxidizing conditions. Such constituents may promote further deterioration either alone or in conjunction with some types of construction metals that act as catalysts.

The criterion for changing used oil in a lubrication system should not always be based on the level of the neutralization number as much as on the rate at which it reaches that level, assuming that the lubricant condition warrants such consideration. A decided rise in neutralization number accompanied by an increase in viscosity and other pertinent changes is an indication that a serious condition has been reached, or is being approached. Operations involving steam turbines, large internal combustion engines, hydraulic systems, and large refrigeration compressors are usually closely attended. The operational history of their oil charges should be followed and recorded. A correlation of an immediate oil condition with its record furnishes the soundest basis on which to decide what to do with an oil charge.

To more strongly express neutralization number of an oil, additional tests are required to find the values that better represent the influence of the separate constituents causing chemical changes. The neutralization number indicates the number of milligrams of potassium hydroxide needed to titrate all acidic constituents present in 1 gram of oil (a volumetric analysis). The value may be better expressed as the *total acid number* (TAN). Such a number represents all the acidic constituents, including those from certain additives, combustion blow-by, and oxy-materials. TAN is equivalent to a quantitative titration required to raise the pH of the sample to 11. New, highly refined straight mineral oils have a neutralization number of 0.10 or less.

The strong acids in the sample may be reported by titrating such acids present in 1 gram of oil. The number of milligrams of KOH representing the *strong acid number* (SAN) is that required to bring the pH to four.

The weak acids and the effect of some additives are indicated by the difference between TAN and SAN.

The *strong base number* is the quantity of acid, expressed in terms of the equivalent number of milligrams of KOH, required to titrate all material reactive to an acidic titrant to a pH of four, assuming that the initial pH is greater than four.

The pH number is not of sufficient significance for the lubrication engineer to make a decision based on acid content as to change the oil in a system. It does, however, help the laboratory to evaluate the acid or alkaline levels of an oil based on total acid number, strong acid number, and strong base number (Appendix A-11).

LABORATORY ANALYSIS. A typical laboratory analysis of used lubricating oil from, say, a gas engine driving a compressor would include the following data:

Operating data:
- Laboratory number
- Identification of oil (brand) and new oil viscosity
- Make, type, and size of engine
- Machine number
- Location
- Date sample received by laboratory

Analysis of sample
- Gravity, °API
- Color, ASTM
- Viscosity at 100°F
- Viscosity at 210°F
- Neutralization number
- Percentage water
- Percentage insoluble in pentane
- Percentage ash

Ash, spectrographic analysis
- Major Iron (for example)
- Minor Lead, silicon, barium (for example)

Remarks
- Including products found by microexamination, such as magnetic iron, carbonaceous materials, abrasive particles

Possible causes
- Including suspicion of, say, water leaks, possible low-temperature conditions due to intermittent operation, etc.

Suggested action to be taken
- Determine source of water (condensation, leakage from cooling systems, etc.), raise operating temperature, install crankcase ventilation systems, etc.

CREEP BARRIER FILMS

One of the properties of some liquids, particularly oils of various kinds and their acid derivatives, is the ability to spread out over the surfaces on which they are deposited. This property of "creep" is also called oiliness and is a function of surface tension.

Creep is measured by the equilibrium angle of contact that a drop of liquid makes with the surface on which it is deposited. The angle of contact is a function of two factors:

(1) Cohesive force of the liquid molecules
(2) Adhesive force between liquid and surface material on which the liquid rests.

If both liquid and surface are of such nature as to promote a greater adhesive force between them than the cohesive force of the liquid, creeping occurs. The action is called surface wetting. Wetting occurs when the surface tension of the liquid is less than the critical surface tension of the surface material. Surface tension is expressed in terms of force and distance, usually in dynes per centimeter.

Low surface tension of liquids is an advantage for numerous requirements:

(a) For boundary lubrication where it is desirable to maintain wetted bearing surfaces.

(b) For adhesion to opposing bearing surfaces under hydrodynamic lubrication conditions to develop the separating wedge.

(c) To allow a liquid to creep (penetrate) into minute crevises to separate surfaces stuck together, as in loosening nuts that bind on their bolts and freeing keys wedged in their ways.

The disadvantage of liquid creep is apparent in lubricants (1) that are preapplied to bearings of equipment that is stored for extended periods of time, or (2) whose initial lubricant application may be its only one, as practiced for so called "packed-for-life" bearings.

While both situations, represented by (1) and (2), involve grease, subsequent bleeding accompanied by creeping of oil from the grease leaves a somewhat solid soap residue. Because such residues are incapable of providing adequate lubrication bearings of the latter categories fail prematurely in operation. In some cases bearings fail after serving only about 25% of their life expectancy.

The loss of oil from a bearing due to creep can be reduced (1) by using oils having natural or imposed high surface tension and (2) by providing a creep barrier around the frictional area of the bearing to prevent escape of the oil.

The use of oils having high surface tension, either as a liquid or an ingredient of grease, is self-defeating in many cases especially where oil spread is essential within the bearing.

This leaves the second alternative, which may be accomplished by surrounding the bearing area with a thin surface coating with minimum wettability.

Fluorinated polymers having critical surface tension of wetting of approximately 10.5 dyn/cm have been found to provide a nonwettable barrier to creep by any of the lubricating oils, either natural or synthetic, in use at the present time.

An example of the method used to employ creep barriers is the application of a thin coating of a suitable polymer, such as fluorinated methacrylate, to the outer ring faces of ball bearings. The polymer held in a solution of 98% solvent and 2% polymer forms a tenacious film about 1 micron in thickness over the ring face surface upon evaporation of the diluent.

Successful results have been obtained with the use of fluorinated methacrylate in retaining such liquid lubricants as mineral oil, silicone, and diesters within the confines of barriers formed by the fluorinated material.

In any industrial program barrier films should be used with oils of low volitility to prevent their loss by evaporation; otherwise the advantage of barrier films would be nullified. Care must also be taken to keep the frictional surfaces of bearings free of barrier film material that would prevent the establishment of a protective lubricant film in the areas so contaminated.

INTERFACIAL TENSION

Circulating lubricating oils should possess properties that facilitate their separation from water that may enter the system. This is particularly true of steam turbine lubricating oils. For uninhibited oils the separability of water and oil emulsions can be evaluated by the *steam emulsion number.* The SE number is not applicable to the same kind of oils when they have been inhibited against rust and oxidation. Inhibited oils are tested by a method that determines the presence of any water-attracting (hydrophilic) compounds. Another test determines the *interfacial tension* (IFT) of an oil against water by the ring method described in detail under ASTM designation D971-48T.

Briefly, the IFT test is conducted by means of a tensiometer to which is attached a torsion wire that lifts a fine platinum ring up from the surface of distilled water into an upper thin layer of oil, both of which are contained in a beaker. Interfacial tension is the measure of force (in dynes per centimeter) required to detach such a ring from the surface of a liquid of higher surface tension (distilled water) upward from a water-oil interface.

High interfacial tension indicates a resistance to oil and water separation, so that the force values are reliable indicators of the presence of hydrophilic compounds. While new inhibited oils may show an interfacial tension of about 25 dyn/cm, the force diminishes to a low of 10 dyn/cm after a period of service.

SURFACE TENSION. The resistance of a liquid to wet out over a surface is called *surface tension.* It is due to the cohesive action of the molecules that make up the liquid. The cohesive forces act in all directions below the liquid surfaces which are not counteracted by any force above the surface. The unbalanced condition develops a tenseness, or to borrow an expression from J. R. Battle, "a drumhead-like effect" at the surface. The surface tension of water is relatively high at 72 dyn/cm.

FRICTION AND WEAR TESTS

Appropriate test mechanisms are employed to determine the capacity of lubricants to combat friction, wear, and seizure. The test principle selected depends on the end result desired and the type of lubricant involved. The results of such tests are reported variously in terms that relate to friction, seizure, stick-slip, wear, temperature, and load capacities under constant or different conditions. All lubricants are tested, including straight mineral oils, hydraulic fluids, extreme-pressure lubricants, leaded compounds, metal-cutting oils and compounds, greases, solid lubricants, and bonded materials.

Some of the more widely used wear tests include the Timken, the Falex, and the Shell Four-Ball methods. There are a number of others, including the Almen-Weiland and the hydraulic pump tests.

TIMKEN LUBRICANT TEST. The Timken test determines the load at which a film of lubricant will fail and cause scoring. The test consists of a revolving test cup (race) that rubs on a stationary block. Speeds are infinitely variable, but usually run at a race speed of 800 rpm. Load is applied through a lever arrangement, at the end of which the necessary weights are gradually added automatically at a rate of 2.3 lb/sec. Refined Timken test machines, capable of infinitely variable loading from 0 to 1000 lb, measure friction, wear, and frictional temperature. Failure of lubricant is indicated by scoring of the

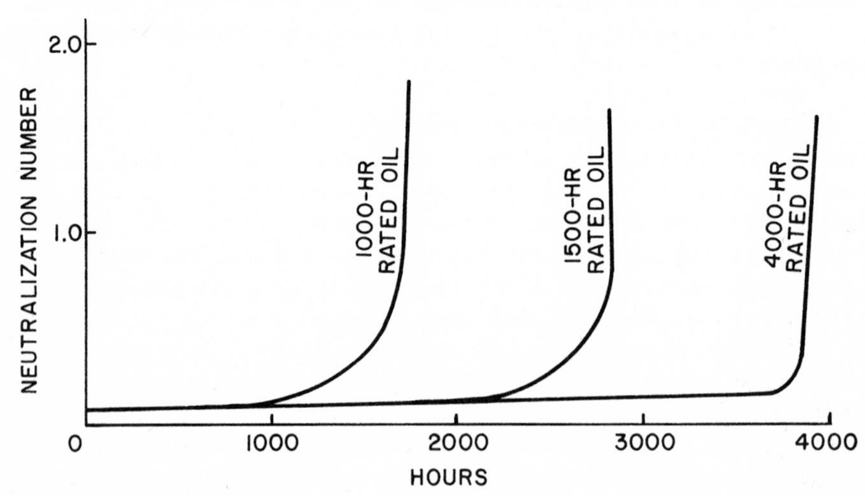

Fig. 7-1. Neutralization number trend due to heat exposure of oils of various chemical stabilities for given periods.

test block. Tests of oils, greases, compounds, and solid lubricants can be made.

Typical acceptable load values for leaded gear compounds obtained just prior to scoring range between 30 and 55, with an average of 40 expressed as the Timken OK load.

FALEX LUBRICANT TEST. The purpose of this test is to measure bearing loads and wear at constant speed and temperature. It consists of a 1/4-in.-diam pin rotating at (usually) 290 rpm between opposing V-blocks (jaws) on which the load is applied. The load is variable up to 4500 lb. Falex tests can also be made at 390 and 750 rpm. Wear is measured by a spring gauge micrometer through a ratchet type loader which compensates for wear at the rate of 0.0000556 in. per tooth movement. The pin and V-blocks are submerged in the sample of oil during the test.

PRECISION—SHELL FOUR-BALL WEAR TEST. This test determines the relation of friction to wear when a load is applied to a four-ball arrangement of an equilateral tetrahedron. The steel balls are placed in a pot with 10 ml of test oil, and the upper ball is rotated at 1800 rpm. Operating temperature is maintained (say at 130°F), and loads ranging up to 1200 kg (late models) are applied. The test can be made to measure torque by a special mechanical arrangement and measuring device. The apparatus is capable of determining seizure and weld loads, and also *Hertz load,* by calculation from measured wear at various loads from resulting scar (indentation) diameters.

Mineral oils show a typical coefficient of friction (steel on steel) equal to 0.053 and a scar diameter of about 0.5 mm under a 5-kg load. The same data for an extreme pressure lubricant show typical readings of $f = 0.04+$ and a scar diameter of 0.30 mm.

The Shell apparatus will test oils, greases, pastes, and dispersions such as graphite plus carrier.

ALMEN-WEILAND FAILURE LOAD TEST. This test will measure any lubricant, including bonded solids. The apparatus consists of a rotating pin (1/4 in. diam) and conforming bearing shells that wrap nearly around the pin to which the load is applied. The apparatus will determine the failure load of a lubricant or measure the friction and wear at loads up to 4400 lb. With metal-cutting oil containing appreciable amounts of dissolved sulphur, the results are usually reported as "pin blackened at Almen pin reading of 30." The average for straight mineral oils is from 6 to 8.

ABRASION TEST. The wear by abrasion caused by various dispersions, or powders (such as graphite and molybdenum disulfide), can be determined on an abrasion tester. The test is conducted by circulating a mixture of solid lubricant powder (15%) dispersed in a standard mineral oil through a double row ball bearing (1.26 in. o.d.) and shaft, the shaft rotating at 1750 rpm. The test is run for about two hours and the wear is determined by the resulting weight loss of the bearing.

HYDRAULIC PUMP TEST. This test was devised to determine the oxidation stability and wear characteristics of hydraulic oils. It consists of a system through which a given volume of oil is circulated by a Vickers vane type pump under fixed conditions of pressure, temperature, and time. Periodic and final analysis of the oil is made to determine changes in viscosity, neutralization number, and any other pertinent condition of the test oil. The difference between the original and final weights of pump parts is also measured during the test run.

OXIDATION STABILITY TESTS

OXIDATION-INHIBITED OILS. There is a test for determining the oxidation stability of oils, particularly those used for the lubrication of steam turbines, usually containing rust inhibitors. The procedure of the test consists of passing oxygen through a sample of oil (300 ml) in contact with a catalyst, maintained at a temperature of about 200°F, and observing the rate at which the neutralization number rises (Fig. 7-2). The pattern for this number is a gradual increase followed by a rapid rise. The neutralization number is the determining factor in the test, the limits of which may be established between 0.25 and 2.0 mg of KOH per gram of oil. The number of hours between the start of the test and the point at which the acid number rises rapidly indicates the relative oxidation stability of the test oil. Specifications that include the oxidation factor usually report the value in terms of 1000, 1200, 1500 or 4000 hr, the higher being the more stable (Fig. 7-1).

The data resulting from oxidation stability tests are important in selecting fluids for steam turbines, hydraulic circuits, and all equipment employing reuse oil systems in which service life is a factor.

GREASE OXIDATION TEST. This test is conducted for the same purpose as the oil oxidation test, i.e., to determine the resistance of a grease to oxidation under certain conditions. Because of the relatively high pressure involved (110 psi), a strong, well-sealed bomb is required for the test. Five dishes of test grease, each holding 4 grams of grease, are placed on shelves one above the other in the bomb. Then the bomb is purged of air by the repeated introduction and release of oxygen (four times at 100°F). The fifth filling of oxygen is made at 100 psi, and the bomb is allowed to stand 24 hr at room temperature. If no leaks occur, the bomb is placed in an oil bath at 210°F, which will cause the pressure to rise. By appropriate release of oxygen, the pressure is finally stabilized and maintained at 110 psi for 2 hr. If there are no leaks, any drop in oxygen pressure after final stabilization indicates oxidation activity within the grease structure. The process is allowed to continue for 100 hr, and then the drop in pressure is noted. Any pressure drop less than 10 psi is considered proportionately excellent to good.

TYPE OF MACHINE	TEST PRINCIPLE	TEST CONDITIONS
Falex Tester	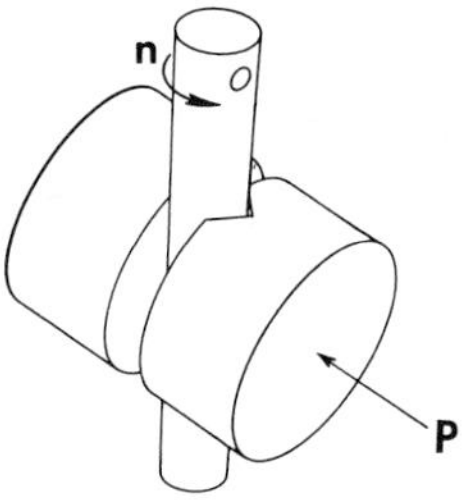 n, P	Test Specimens: ¼″ diameter pin and ½″ diameter V-blocks. Speed: 290 r.p.m. (can be readily changed to 390 and 750 r.p.m.) Load: Variable up to 4500 lbs. Procedure: a) Increase load on pin to failure. b) Friction and wear on specimens at constant load. Machine will test: Oils, Dispersions, Greases, Bonded Solid Lubricants.
Almen-Wieland	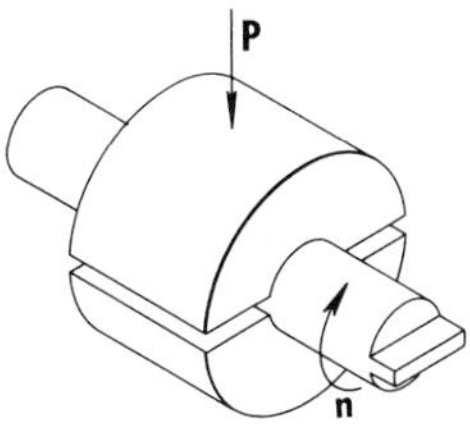 P, n	Test Specimens: ¼″ diameter pin and conforming bearing shells. Speed: 200 r.p.m. Load: Variable up to 4400 lbs. Procedure: a) Increase load on pin to failure. b) Friction and wear of specimens at constant load. Machine will test: Oils, Dispersions, Greases, Pastes, Bonded Solid Lubricants.
Reichert Wear Tester	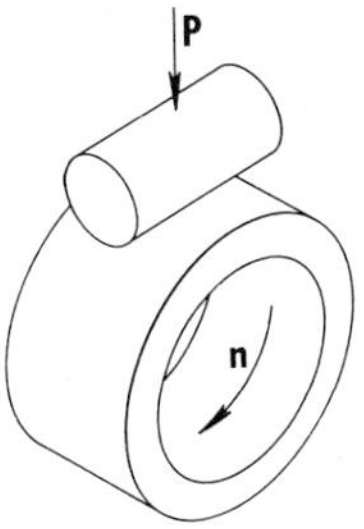 P, n	Test Specimens: Crossed cylinders Speed: Fixed 900 r.p.m. Load: Step loading up to 130 lbs. Procedure: Measure wear after a predetermined length of sliding. Machine will test: Oils, Dispersions, Greases, Pastes, Bonded Solid Lubricants.
Shell 4 Ball EP Tester	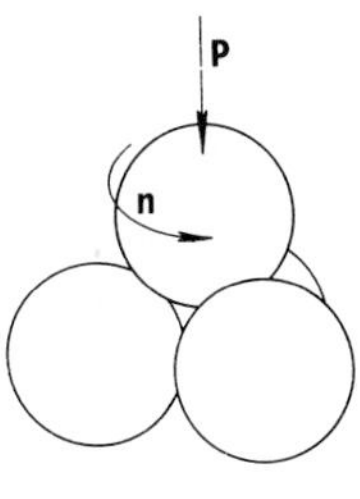 P, n	Test Specimens: ½″ diameter balls Speed: 1500 r.p.m. with 50 cycle current; 1800 r.p.m. with 60 cycle current. Load: Variable up to 1200 kilograms Procedure: a) Seizure load: One minute test at increasing load to seizure. b) Weld Load: One minute test at increasing load to welding. c) Hertz Load: Calculation from measured wear at various loads. Machine will test: Oils, Dispersions, Greases, Pastes.

Fig. 7-2. Film strength tests. *Courtesy: Dow Corning.*

TYPE OF MACHINE	TEST PRINCIPLE	TEST CONDITIONS
Bartel Lubri-meter	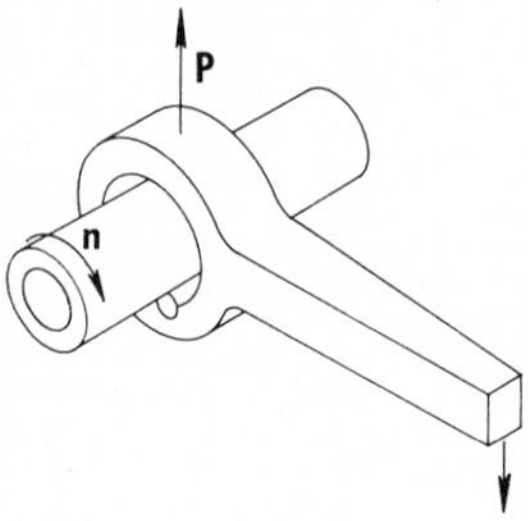 P n	Test Specimens: Hollowed journal and two clamped bearing cylinders. Speed: Variable .005 to 3000 r.p.m. Load: Variable up to bearing pressures 230,000 p.s.i. Procedure: Friction, wear, and friction temperature measured under a wide variety of speeds, loads, and ambient temperatures (up to 1832°F). Machine will test: Oils, Dispersions, Greases, Pastes, Bonded Solid Lubricants.
Press Fit Test	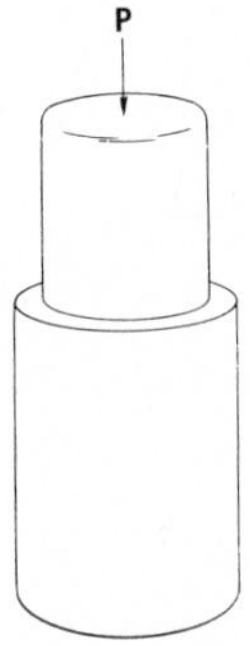 P	Test Specimens: Pin and bushing Speed: Normally 0.6″ per minute, variable up to 2″ per minute. Load: Variable up to yield point of specimen material. Various press fit interferences available. Procedure: Measure static and kinetic coefficient of friction; determine seizure and stick-slip behavior. Machine will test: Powders, Oils, Greases, Pastes, Dispersions, Bonded Solid Lubricants.
Alpha Lubricant Tester Model LFW-1	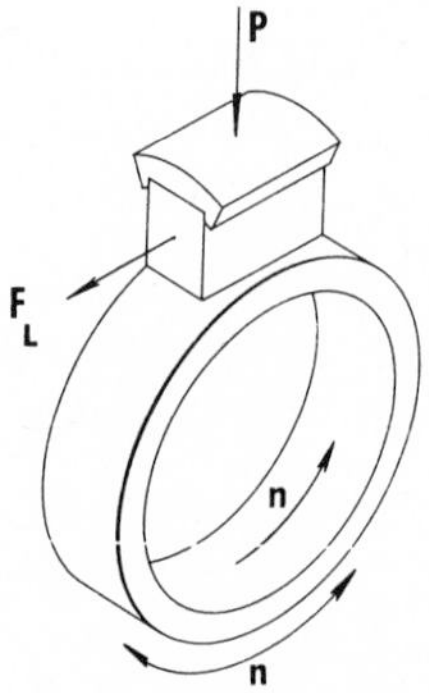 P F_L n n	Test Specimens: Test block and 1⅜″ diameter Timken bearing outer race. Speed: 12.5, 35, 72, 120, and 197 r.p.m. Load: 30 to 630 lbs. Procedure: a) Step load to 630 lbs.; measure friction and wear at constant speed. b) Same as a) with oscillating motion. Machine will test: Oils, Dispersions, Greases, Bonded Solid Lubricants.

TYPE OF MACHINE	TEST PRINCIPLE	TEST CONDITIONS
Alpha Lubricant Tester Model LFW-2 (Refined Timken)	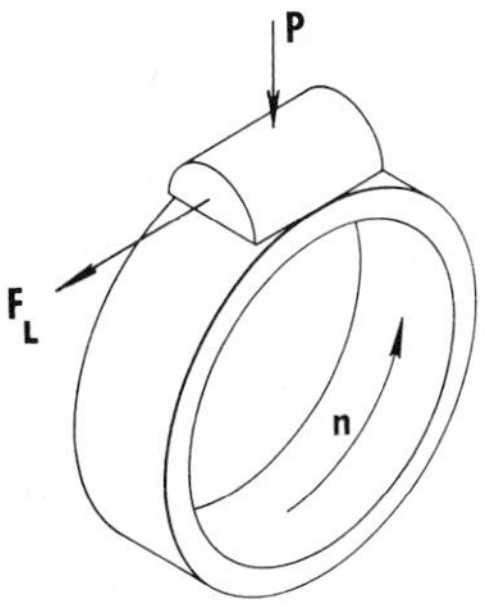 P F_L n	Test Specimens: Test block and 1.94" diameter Timken bearing outer race. Speed: Infinitely variable 0 to 1100 r.p.m. Load: Infinitely variable 0 to 1000 lbs. Procedure: Measure friction, wear, and frictional temperature under widely varying speed and load conditions. Data are fully recorded. Machine will test: Oils, Dispersions, Greases, Pastes, Bonded Solid Lubricants.
Alpha Lubricant Tester Model LFW-3 (Environmental Research)	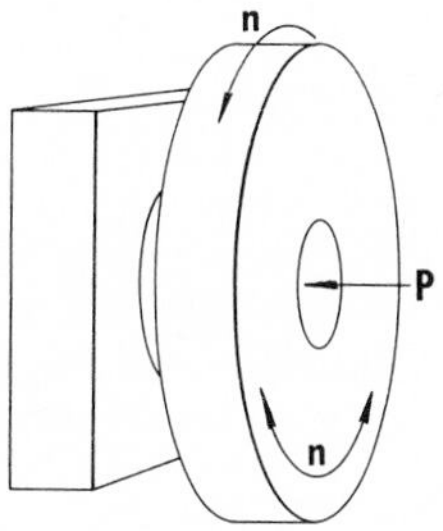 n P n	Test Specimens: Annular ring contacting flat block. Speed: Uni-directional; variable from 9 to 325 r.p.m. Oscillatory: 0° to 120°, variable from 6 to 120 c.p.m.; 0° to 60°, variable from 6 to 227 c.p.m. Load: Infinitely variable 0 to 5000 lbs. Procedure: Measure friction and wear in air and other environments at temperatures to 1200°F. Machine will test: Pastes, Bonded Solid Lubricants.
ASTM Abrasion Tester for Solids	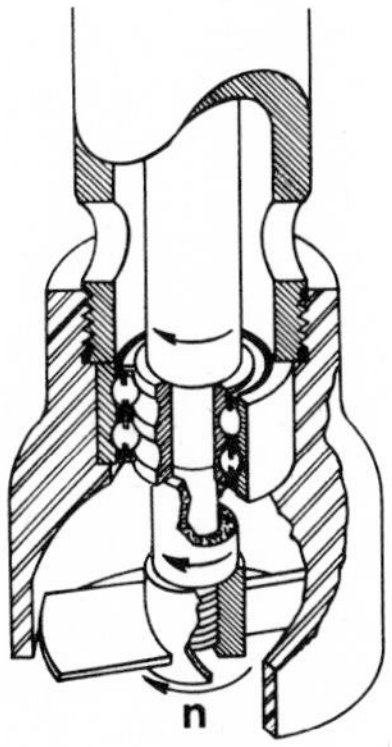 n	Test Specimens: Standardized double row ball bearing 1.26" O.D. Speed: 1750 r.p.m. Load: 0 Procedure: Measure wear by weight loss of the test bearing after two hours operation with a slurry of 15% by weight solids in a standardized petroleum oil. Machine will test: Dispersions, Powders.

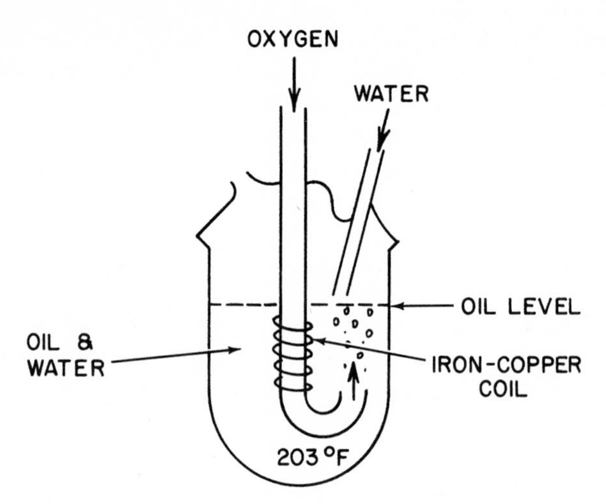

Fig. 7-3. Oil oxidation test (schematic). Purpose is to measure the time in hours for oil to attain a neutralization number of 2.0. The time ranges from 75 hr for straight mineral oil to 1500–4000 hr for inhibited oil.

TESTS FOR USED OILS

Most marketers of lubricants and fuels operate field service laboratories, which along with other functions cooperate in determining the physical and chemical conditions of used oils. Such products are taken from the users' equipment for various reasons. One reason may be to determine the conditions of an oil in which a decision may be made as to the practicality of its continued service or its immediate rejection. Another reason may be to furnish physical and chemical data pursuant to some real or imagined

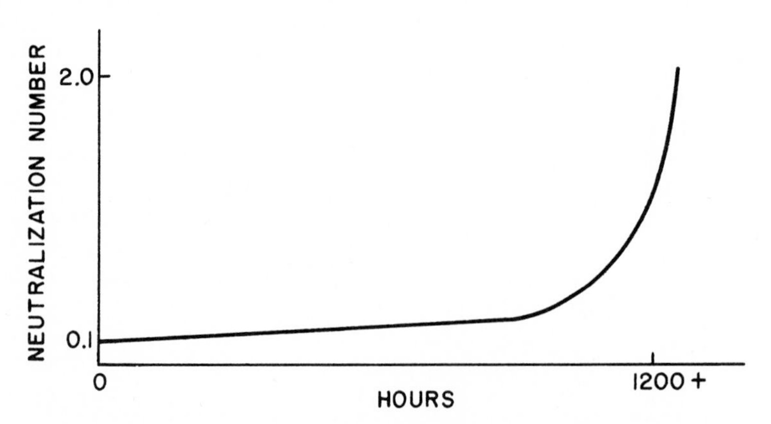

Fig. 7-4. Oxidation development as determined by rise in neutralization number.

difficulty allegedly caused by the lubricant, or to help determine other intrinsic causes for the trouble. A third reason may be only in the interest of research and development of pertinent products.

Many field investigations are concluded by simple observations not requiring the costly service of a laboratory. For example, the presence of water in a circulation system is often obvious, requiring only the determination of where it originates and how to eliminate the leak. Foaming, sediment, emulsions, and gradual discoloration of lubricating oil are examples of unusual operating conditions that more often than not can be corrected at the site of the equipment involved. The same applies to such mechanical problems as electrolysis, rapid wear, abnormal temperatures, clutch release failures, coupling grease separation, water-washing of grease, excessive lubricant consumption or deterioration, diesel fuel oil dilution, and many others.

One such problem was a rapid rise in neutralization number and subsequent oil darkening of repeatedly changed batches of new oil used in the circulating systems of a large geared steam turbine. It was found to be caused by the absence of a vent in the gear case. A pyrometer applied to the component immediately gave rise to the suspicion that the cause originated at that point.

TOOLS FOR TROUBLESHOOTING. The tools of the troubleshooter should include a thermometer (clad), hand type viscometer (such as described in Chapter 8), a magnet pyrometer, magnifying glass, tachometer, pressure indicator (or compressor gauge), calipers (internal and external), micrometer, rulers (both 6 in and 6 ft with depth extension, or winding tape), stop watch, litmus paper, water detector,* thief, emulsion beaker (graduated stem), test lamp (with low-wattage bulb), small directional compass, flashlight, screw drivers (including Phillips), pliers, small file, adjustable wrench, feeler gauges, small level, scriber, blotters, spatula, pocket knife, awl, 1-oz. jars (for grease), and several pint cans for carrying samples of oil.

ROUTINE OIL TESTS. To determine the condition of an oil after a period of service routine tests are usually run. This service is applied only to large circulating oil systems where the expense of periodic testing is warranted. Unless extenuating circumstances are involved, sufficient information can be gained to judge the condition of an oil by conducting or having conducted the basic tests:

1. Viscosity
2. Neutralization number
3. Water and sediment

*Water detectors are proprietary materials that will change color when submerged in water. Water may be detected at the bottom of a tank filled with oil if the material is applied to the end of a stick or rod used as a probe.

If the viscosity of the oil has changed appreciably from that of the original batch, further investigation may be necessary. Such a change accompanied by an insignificant increase or no increase in neutralization number may indicate that the wrong oil was added to the system.

Any decided visible darkening of the oil without an increase in viscosity or neutralization number may indicate copper contamination and will only occur above about 10 to 15 ppm. The effect on the physical condition of the oil is not serious, and the copper, if present, will eventually be filtered out. If the condition continues to worsen, an investigation of parts containing copper should be made as soon as practicable. If the condition stabilizes, it can await scheduled periodic overhaul or inspection.

Any significant increase in neutralization number should be followed by an investigation, particularly of high local temperatures. Providing the difficulty is found and rectified, and the neutralization number is below 0.35, routine makeup will usually return the oil condition to some safe level of acidity.

Amber-colored oils, such as those used for steam turbines, will turn to an apricot hue when sufficient water is present. The first step is to drain the oil reservoir until water-free oil appears. Repeat the process every half hour. If the quantity of water diminishes with each drain, continue operation with close attention. However, if the quantity of water increases during the first two or three drains, the system should be shut down and the inleak found and rectified.

Traces of water can be removed either from the bottom of the system reservoir (settling), in absorbent type filters, or by centrifuging.

Soot can be detected and the relative magnitude estimated by observing the shade of darkness of spots made on conventional blotting paper. The spots should set for about an hour to allow for oil migration. The presence of sediment requires an inspection of oil-filtering devices and procedures. Use only clean, lint-free cheesecloth when adding new oil to system reservoirs.

Laboratory technicians or chemists will usually call attention to any suspicious conditions detected during routine tests. For example, any hard abrasive particles present in the sediment may lead to other tests to determine their nature. This may require a simple magnetic test, microscopic examination, or the more complex spectrographic analysis.

INFRARED ANALYSIS

Lubricant analysis by the infrared spectrophotometer is not limited in scope to used oils and greases. It is included here because of its advantages in determining certain chemical changes that may provide solutions to current problems or predict service life limitations of reuse oils, especially of engine lubricants. Infrared analysis not only matches the performance of neutralization number tests, but is more articulate in its findings, less difficult, and faster.

The presence of new type additives in modern lubricating oils poses analysis problems not generally prevalent in the past. This, coupled with the trend to incorporate "cure-all" inhibitors that are often more harmful than helpful, has given impetus to the adoption of infrared analysis, a relatively new laboratory technique.

The practical use of infrared analysis as a preventive maintenance tool has been established in one way, for example, by its ability to detect excess nitrogen products, that develop within engine oils. Such products lead to rapid deterioration of the oil, and the advantages gained by their early detection are quite obvious. Other capabilities of this type of engine oil analysis include the detection of antifreeze solutions (glycol), fuels, oxidation products, foreign contaminants, blow-by products, and additive depletion. Any chronological pattern on the part of internal combustion engines to produce these deteriorating conditions (1) allows for early rectification of those caused by mechanical defects (leakage in the case of glycol, etc.) or (2) leads to the establishment of safe oil change periods.

The infrared analysis of materials is based on the tendency of all organic materials, including mineral oils and some of the constituents of greases, to absorb certain electromagnetic radiations to which they are exposed. The radiations involved here are those that emanate from the electromagnetic spectrum having wavelengths from 0.7 to 200 microns (Fig. 7-5), which are the infrared or heat rays. The width of the infrared band is further divided into three separate ranges, called the near infrared (0.7 to 2.5 microns), mid-infrared (2.5 to 15 microns), and far infrared (15 to 200 microns). It is the range between 15 and 50 microns that is used in most infrared analysis.

Unlike light rays, infrared rays are too long to be visible.

APPARATUS. Techniques of infrared analysis all use a spectrophotometer and involve these four steps (Fig. 7-6):

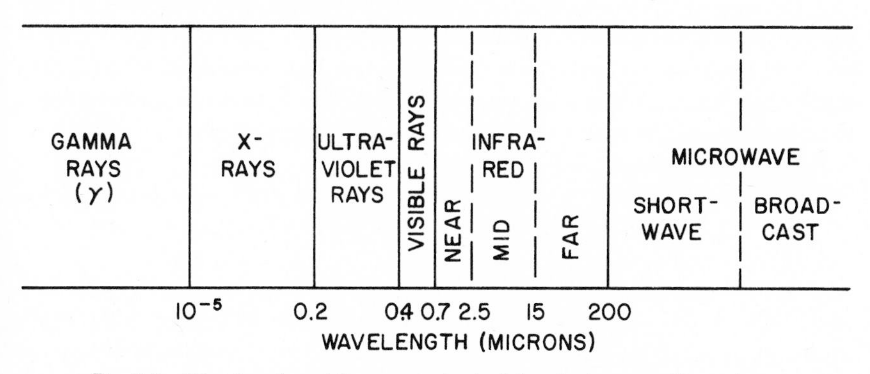

Fig. 7-5. Wavelength and frequency chart of the electromagnetic spectrum.

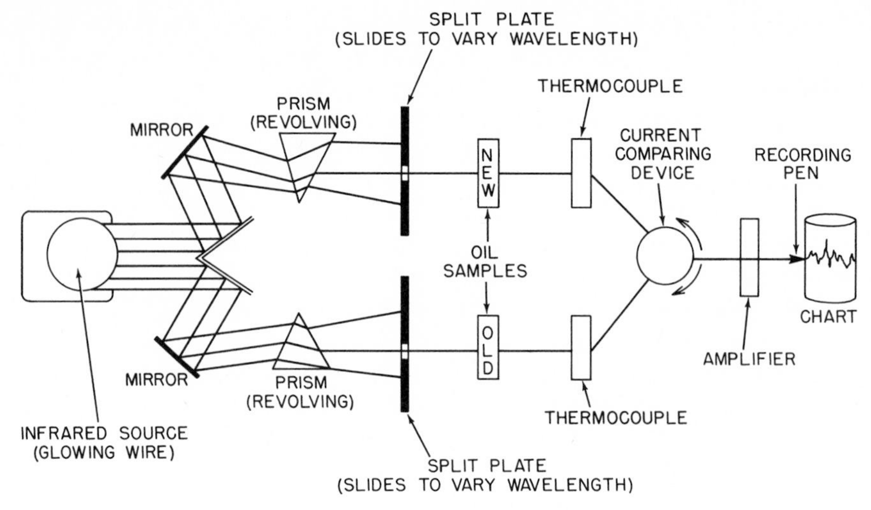

Fig. 7-6. Diagrammatic plan of infrared principle.

1. Production of an infrared spectrum band of a given wavelength range. Thus a source of infrared radiation is required.

2. Separation of the continuous infrared radiation into narrow wavelength intervals. This is done by passing the infrared rays through a prism, or grating (narrow slits that isolate the spectrum colors according to wavelength). This is accomplished in a part of the spectrometer called the *monochrometer.*

3. Detecting and amplifying the radiant energy that is separated by the monochrometer in the detector-amplifier system of the apparatus. The magnitude of the energy is measured by an electrical instrument of fine sensitivity called a *bolometer,* or by *thermocouples.*

4. Recording the amplified signal from the bolometer or thermocouples. This is usually done by a servo-mechanism, driven by the amplified radiations that have been converted into electrical energy. The signal is recorded as percentage absorbance.

The result is a graph showing the characteristic wavelength (beam) of each of the organic compounds present, and the percentage absorbance is interpreted as the quantity of such compounds. The result is compared with a second beam of identical infrared radiation spectrum that has not passed through any substance before reaching the bolometer or thermocouples.

PRINCIPLES. The principle of infrared analysis for detecting substances in a mass is based on the phenomenon that infrared radiation of a molecule of a given substance causes the atoms of the molecule to increase their activity only at a radiation frequency characteristic of the substance. This leads to identification of the substance.

By passing a beam containing the entire infrared spectrum through a sample substance, part of the energy will be absorbed and the remainder will pass through. The beam then continues through a dispersion grating or prism to be divided into narrow bands having different wavelengths. These bands are focused on a detector which portrays such changes in wavelength as the pattern of absorption versus free flow of energy through the sample.

Thus each substance has a characteristic spectrum. No two materials have the same spectrum.

By comparing the energy in the beam passed through the sample for each wavelength band with the energy of an identical reference beam which has not passed through the sample, the difference in energy can be detected. Unknown substances can be identified by their absorption magnitude (percentage of radiation transmitted) and by their wavelength.

One example of the value of the spectroscopic analysis is in the determination of concentrations of oxidation and nitration products that are undesirable, particularly in internal combustion engine oils. This is made by differential infrared (DIR) analysis, by which the presence and amount of such products are determined. Because nitration products are associated with certain types of engine oil deterioration (the presence of peroxides that act as oxidizers), this kind of analysis in conjunction with other tests has become quite important in establishing practical change periods. In DIR analysis, the pattern of the used oil is compared with the pattern of an unused sample of the same oil as shown with the reference beam when both are run simultaneously.

The DIR analysis is used to determine dilution of used gasoline engine oils. Diesel engine oils usually contain so much soot that it is difficult to make a spectrographic pattern of nitration products unless they are clarified.

A recent proprietary innovation in the recording of spectrum patterns is the division of products detected. Readings *above* a base line indicate the properties that were originally present but were depleted or exhausted during use of, say, the engine oil sample, and the readings *below* the base line indicate the products of chemical change, or the presence of contaminants, thus providing a simple display of the oil condition.

Depleted or exhausted materials shown by this type of graph are usually the additives introduced into the oil when new.

The results of infrared analysis may be summarized by the data shown on the spectrum of a given organic material, namely, (Fig. 7-7).

(a) Each constituent will be indicated by a dip in the graph at its individual wavelength band.

(b) The amount of the constituent in a sample is indicated by the percentage of transmittance or vertical elongation of the dip in the graph.

An analysis of a diesel or gasoline engine lubricant would include

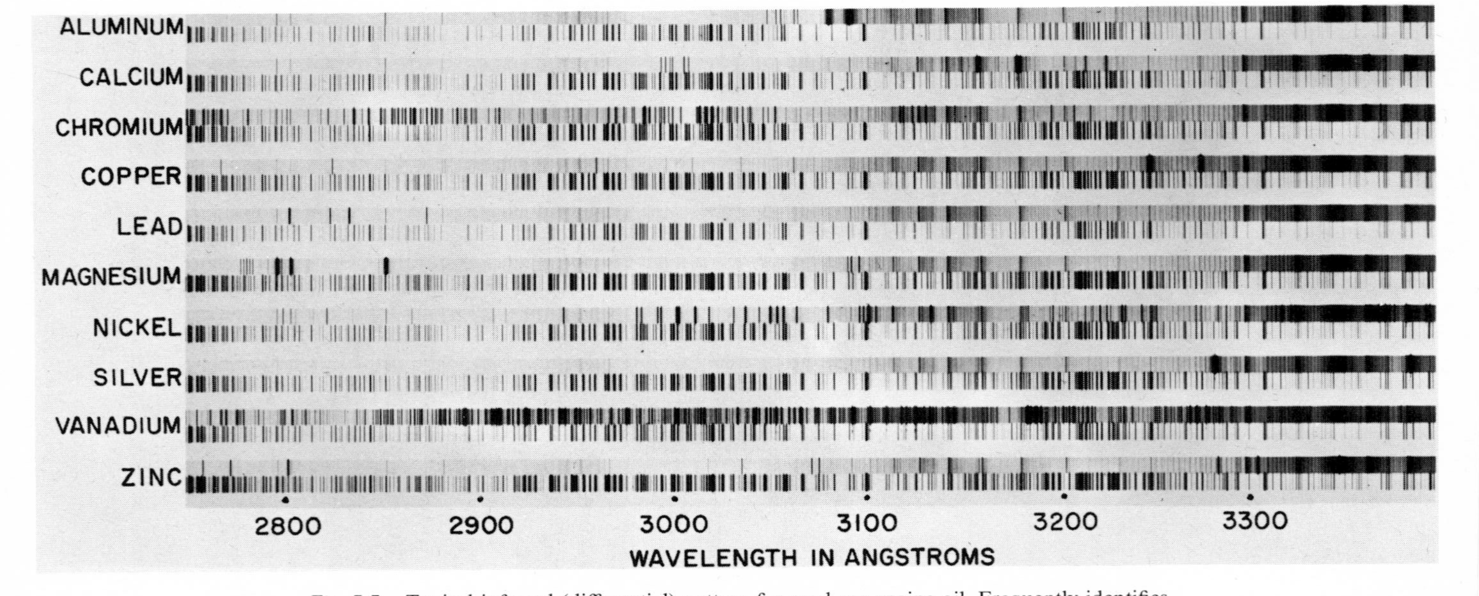

Fig. 7-7. Typical infrared (differential) pattern for used gas engine oil. Frequently identifies compounds (water at 2.9, oxidation products at 5.8). Absorbance indicates amount of contamination. New oil is straight line. Blow-by would appear at 6.4.

percentage dilution and magnitude of nitrogen fixation, as determined by infrared analysis. The products identified by differential infrared are located on the absorbance curve as follows:

2.9 μ	Water (or glycol)
5.8 μ	Oxidation products
6.2 μ	Organic nitrates*
6.5 μ	Nitrogen compounds
7.9, 11.8 μ	Confirms nitration at lower values

* Adverse engine operating conditions may be deduced when absorbance value at this wavelength is high.

EMISSION SPECTROGRAPHY

The presence of mineral elements in used lubricating oil can be determined by emission spectrography. The metals found by this method of analysis often provide clues for determining the trouble areas within a frictional system. This method can detect wear of bearings, gears, and cylinders; leaks in cooling systems; fuel contamination; and contamination from external sources. Spectrographic analysis can be made of either the oil or its ash, the latter providing the more accurate estimates of the presence and concentration of the elements present.

The atoms of all mineral elements emit radiation when heated to incandescence (Fig. 7-8). Each element has its own characteristic radiation wavelength, which is the basis of its identity. In emission spectrography the

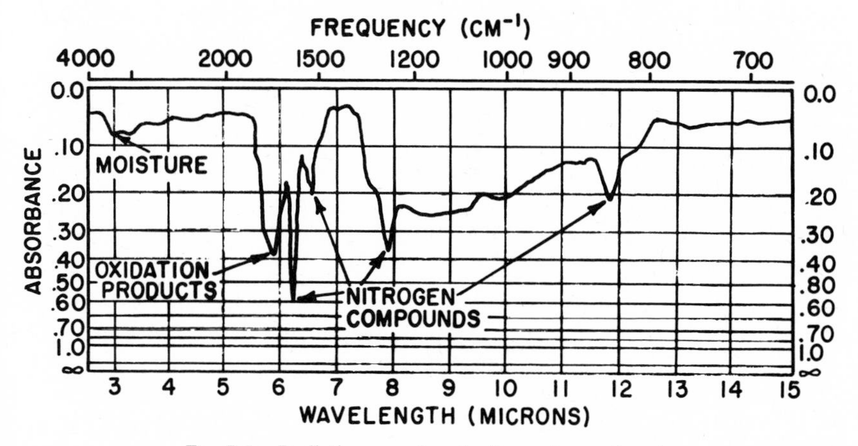

Fig. 7-8. Radiation wavelengths for various minerals.

sample is brought to incandescence by electric arcing. The radiation produced is passed through a prism or grating, which sorts the various wavelengths just as a prism separates the various basic colors or spectra from a light source. The spectra of the oil are photographed to form a spectrogram. A reference spectrum of pure iron is taken simultaneously and placed beside the sample to help analyze the spectra of the minerals contained in the sample.

The standard unit of measurement of short wavelengths is the angstrom, equal to one-hundred-millionth of a centimeter.

By applying additional spectrographic techniques, trace amounts of metals can be detected so that a report of major and minor elements present can be provided. Such information helps determine the source of suspected system problems.

SOLUBLES AND INSOLUBLES

Selected solvents are used to determine certain constituents of used oils and sludges taken from a reuse mineral oil lubrication system. It is the technique of the elimination process whereby the constituents are identified and their quantity estimated by their solvency or insolvency when diluted with certain solvents and centrifuged to separate the precipitate.

The solvents used are benzene (petroleum ether) and normal pentane ($H_{12}C_5$).

Dust, metallic substances, and soot are insoluble in benzene, and their presence indicates generally the need for a more effective system for filtering the oil circulated.

Solids and resinous materials are insoluble in pentane. Solids most generally can be removed by filtration.

The difference between the amount of pentane and benzene insolubles represents the insoluble resins. Such materials are the oxidation products of either the fuel or the oil or both. The resinous materials form the lacquer and varnish on parts subject to high temperature, such as ring zone areas, combustion chambers, and valves of internal combustion engines.

When the difference between the pentane and benzene insolubles approaches or exceeds 0.75%, both oil and filter should be changed. The presence of solids such as metal particles, dust, and dirt indicates that the filter should be changed.

ANILINE POINT

The aniline point is the lowest temperature at which equal volumes of aniline and an oil or other petroleum product are either soluble each with the other or are completely miscible. A value called the mixed aniline point is defined similarly, except that the minimum equilibrium solution temperature is prescribed for a mixture of two parts aniline, one part oil sample, and one part *n*-heptane of specified purity.

The aniline point of an oil is used as a criterion of the suitability of a material, particularly rubber and rubber products, to be used as an oil seal.

Having the ring structure of aminobenzene ($C_6H_5NH_2$), aniline is more soluble in aromatic and naphthenic oils than paraffinic base oils at similar temperatures. The aniline number, which indicates the temperature of mutual solubility, must therefore be higher for the latter type chain structural products. Because most oils are a mixture of several type structures, a high aniline point indicates a greater constituency of *chain* type materials, whereas a lower number indicates a greater constituency of *ring* type materials.

It is generally true that the higher the aniline point of an oil, the less effect the oil has on the swelling properties of rubber, Buna S, and Neoprene. This means that oils of substantially greater paraffinicity will have less swelling effect on rubber and some synthetic rubbers. Thus by controlling the aniline point by blending certain proportions of aromatic oils with paraffinic oils, a method can be developed for controlling oil leakage from a reuse system by causing limited swelling of seals. The method works only as long as the system oil temperature remains within the range for which the blending proportion is established. Figure 7-9 shows the volume increase due to swelling caused by lubricating oils according to increasing aniline point.

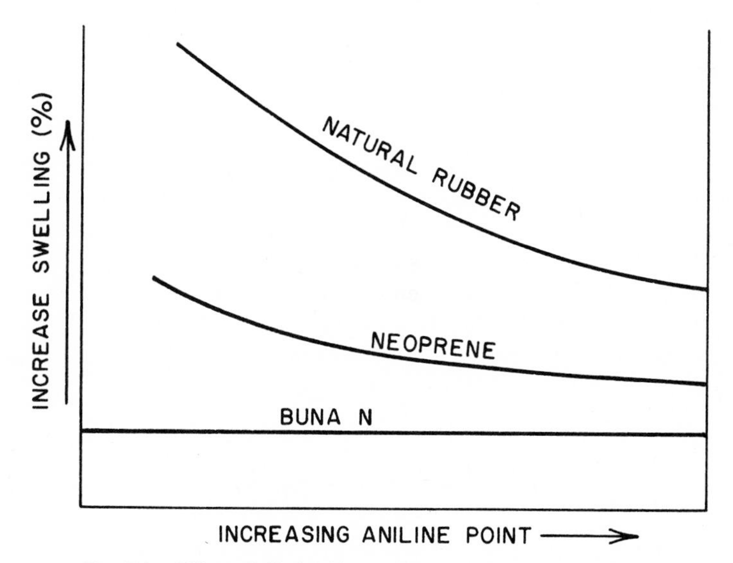

Fig. 7-9. Effect of diminishing aniline point on seal materials.

GREASE CRITERIA

CHANNELING, in lubrication, is the formation of grooves in the lubricant left by a moving object, such as a gear tooth, ring, chain, race, or wheel. Permanent channeling occurs when the walls formed by the action fail to close in to fill the space after the moving object passes. The tendency to fill the space increases as the slumpability of the grease becomes greater.

SLUMPABILITY is the tendency of a mass to lose shape by gravitation. The thinner the grease, the greater its slumpability. A certain degree of slumpability is desired in grease that is pumped from a central reservoir to prevent cavitation or pump-starving. With insufficient slumpability, a *follower plate* is required to maintain pressure automatically on the top of the grease and assure its presence at the pump suction continuously. A follower plate is a weight having about the same area and shape as the cross section of the container in which it is installed. Some clearance must be allowed between the plate and container walls to enable the follower plate to move downward as the grease mass diminishes.

Slumpability is a function of a lubricant's texture and *yield value,* the latter being the pressure at which a grease begins to flow. The *yield point* is the temperature of a grease when it begins to flow, usually determined in a pressure viscometer.

PUMPABILITY is a measure of grease mobility or ability to flow under certain conditions of pressure, temperature and geometry of the conduit. It is an important property, especially in central grease systems involving feed lines of substantial length, and where ambient temperatures may vary to considerable extent. While the problem of pumpability appears to be one of consistency selection, i.e., a soft grease being relatively more yielding, which is true under simple application conditions, the hypothesis does not always hold true when based on the premise of consistency alone. Other considerations include the oil separating and coking characteristics of the grease. Any separation of oil from the soap results in a residue that compresses to a hard substance, which abetted by elevated temperatures is reduced to an effective plug that blocks the feed lines. Under low ambient temperature grease should maintain sufficient mobility to reach its ultimate destination. Separation under pressure should not be construed to be a criterion of the susceptibility of a grease to *bleed.* Tests for pumpability are usually run at 0°F and 150 psi, during which the flow in grams per second is measured.

BLEEDING is the oil separation tendency of a grease under fixed *static* conditions. It is determined by measuring the weight of oil collected after subjecting the grease to a temperature of 212°F for 50 hours.

VISCOSITY

Viscosity is a fundamental aspect of the lubrication art. It is the property of a fluid that resists the force tending to cause the fluid to flow. It is also a measurement of the rate at which a fluid is capable of flowing under specified conditions. Fluidity decreases with higher viscosity, i.e., flow rate varies as viscosity.

Viscosity is as important to the lubrication engineer as strength of materials is to the architect, or as temperature is to the metallurgist. Viscosity of a lubricant is a major factor in developing film thickness in a hydrodynamic wedge; it determines load-carrying capacity, the ability of an oil to flow, operating temperature levels, and often wear rate.

To simplify discussion, the word fluid in this chapter will be applied only to liquids, although some of the laws discussed also apply generally to gases used as lubricants.

Some liquids flow more easily or readily than others. The same liquid will flow more easily when heated. It will flow less easily if it is subjected to a higher pressure. Therefore, any comparison of liquid viscosities must be made under specified conditions. Accordingly, the discussion below assumes a temperature of 60°F (standard density level) and a pressure of 29.92 in. Hg (standard atmospheric pressure, or 14.69 psi at sea level).

The relationship of flowability of two liquids may be expressed as one being more or less viscous than the other, but such expressions are as ambiguous as describing the temperatures of two substances by saying that one is hotter or colder than the other. To eliminate this ambiguity, reduce description to numerical values based on an appropriate, recognized standard.

Water flows very readily because it is light-bodied, i.e., has a low relative viscosity. Glycerine on the other hand flows slowly because it is heavy-bodied, i.e., has a high relative viscosity. A motor oil such as used in the engine of a passenger automobile flows less freely than water but more readily than glycerine because it is medium-bodied, i.e., has an intermediate relative viscosity. Describing liquids according to body refers to their viscosity and not to their color or weight.

The rate of flow of any liquid is a function of force applied to a given mass to cause motion. Viscosity is the measure of the rate of flow when a liquid (1) seeks its own level, i.e., is subject to the pull of gravity, (2) is displaced by an external mechanical force, such as a pump, or (3) is sheared by drag, as between the surfaces of two bodies in relative motion as in a journal-bearing arrangement. If the force causing flow is gravity, the rate of flow or viscosity can be expressed as a function of time when all other parameters—volume, temperature, pressure, and flow area (orifice)—are equal. Viscosity can also be expressed in terms of the *ratio of the flow rate* of one liquid to that of a standard liquid, such as water, both moved by gravity and subject to the same rules as described above.

Viscosity can also be expressed in terms associated with the function of the force necessary to cause flow at a given temperature. In this case, the force is externally applied to a given mass having a specified area and thickness of film to attain motion at a given velocity. This system of determining the flow rate, more accurately called the *shear rate* (the rate at which the layers of molecules slide over each other in laminar flow), was first developed by Sir Isaac Newton, the English physicist.

Viscosity is the property of lubricating fluids that reflects the friction developed between adjacent layers of a given substance during flow. The friction developed is known as *fluid friction,* which governs the rate of shear that occurs between the molecular layers of a given substance. When reduced to numerical values, this property serves as the major criterion for determining the physical limitations, i.e., temperature, pressure, etc., to which a given fluid lubricant can be exposed.

The numerical values of viscosity are based on the test methods employed in their determination. While such values differ by reason of the methods used, they are all mutually convertible, providing the conversion is related to common temperatures, i.e., the same temperature in each case.

Determining Relative Viscosity

The simplest method of determining whether one noncontaminated liquid is of higher viscosity than another is to shake both of them vigorously in a bottle containing some air. The one in which the air bubbles rise most rapidly is the liquid of lowest viscosity. The liquids must be allowed to stand for a sufficient length of time to assure both being at room temperature during the test.

Another simple method of determining the relative viscosity of two liquids, called the falling ball method, is to drop a ball bearing in each and observe their descent from the surface to the bottom of the container, preferably a glass beaker or flask. Providing both liquids are at the same level and temperature, the one in which the ball reaches the bottom first is the one lowest in viscosity.

A more precise method of determining relative viscosity using the above principle is to use a glass test tube having an inside diameter much greater than the diameter of the ball, and calibrated at varying heights relative to the bottom of the tube. The oil-filled test tube is immersed in a temperature-controlled water bath and the descent time of the ball is measured in seconds. The time required for the ball to fall through the distance measured from a selected point of calibration to the bottom of the tube indicates the absolute viscosity of the liquid tested. This method has been suggested as a method of determining viscosity values of liquids of sufficiently high viscosities to permit accurate time readings, and calibrated according to some standard liquid.

A field instrument used by the author for years to obtain approximate viscosity levels consists of two parallel glass tubes of the same size mounted on a metal bracket. One tube is completely filled with a standard fluid or oil and tightly sealed at both ends. The other tube is empty when not in use and is equipped with a plunger at one end for drawing a sample of liquid to be tested in through the opposite end. Both tubes contain a steel ball that moves freely from one end to the other. The mounting bracket is calibrated along the surface separating the two tubes. The test oil is drawn into the empty tube until it is completely filled, and the instrument is allowed to stand until the fluid in both tubes reaches ambient temperature. The instrument is then tilted until both balls are located at the same end of each tube. The test begins when the instrument is tilted in the opposite direction (about 45°) when the balls begin their downward rolling movement. Close attention is given both balls until one reaches the extreme length of its tube. The instrument is immediately maneuvered to a level position and the location of the remaining ball observed. The viscosity of the test oil is then determined from the calibration at which the ball in the test tube reached when the instrument was brought to a level position. Fairly accurate results can be obtained with such an instrument if the temperature of both the standard and test liquids can be brought to 100°F.

This instrument is based on the same principle as other types of rolling ball viscometers designed to produce results of high precision using standard liquids of known viscosity. One such viscometer measures viscosity at elevated pressures.

The following method of determining the dilution of diesel engine lubricating oil by fuel oil by measuring relative viscosities is somewhat crude, but it can be used in the absence of more refined methods. It is conducted by depositing, from the same height, a drop each (by an eye-dropper) of the used lubricating oil and the same but unused lubricating oil onto the surface of an inclined plane consisting of a flat metal plate. Note their relative rapidity in flowing toward the lower edge of the plate. If the used oil runs down substantially faster than the new oil, fuel dilution is the probable cause, since diesel fuel is generally lower in viscosity than engine oil and tends to "thin out" the latter. Making up several batches of new oil-fuel mixtures in which the amount of fuel oil is varied from 1% up to, say, 5%, or until the runoff time of one of the batches matches that of the used lubricating oil, a somewhat accurate estimate of fuel oil dilution can be made. Consideration of all operating conditions, hours of operation since last oil change, etc., assists in deciding the recommendation to be made concerning future operations and corrective measures.

The two methods for estimating relative viscosity described above are strictly field tests and should be used only as expediencies in lieu of laboratory service. Their value in solving operating problems involving liquid lubricants lies in the care with which they are performed.

Accurate lubrication analysis is based on sophisticated data relative to pertinent parameters, including the viscosity of liquid lubricants. Therefore, each individual oil is subjected to viscosity tests during which all factors are meticulously controlled and possible errors mathematically corrected. The nomenclature employed in reporting viscosities of the various liquids is based on the test method employed. As previously stated, such nomenclature may be in terms of force, time functions, or, in some cases, the ratio of values of test liquids to those of a standard liquid, each determined by the same method. Because any given liquid retains its individual intrinsic fluid friction magnitude under the same physical conditions regardless of the test method, their variously reported values are convertible from any nomenclature to any other.

VISCOSITY MEASUREMENT

The Rankine and Kelvin temperature scales are based on the state of matter wherein energy ceases to exist, *absolute zero*. Man has closely approached that point, but never reached it. The elusive level of zero energy, therefore, is accepted as the true base of all thermal activity. Man determined his own starting point thermodynamically by selecting a datum of *zero enthalpy* and *zero entropy* at the coincidence of temperature scales, i.e., where $-40\,^{\circ}\mathrm{F} = -40\,^{\circ}\mathrm{C}$. Weight is based on the force of the earth's gravity, and velocity is based on the speed of sound under specific conditions related to a vacuum. Electrical capabilities are dependent on the ampere, which is the quantity of electrical energy in transit per unit of time and is based on the amount of current that will deposit a given weight of silver per second under specified conditions. The question now arises—on what datum is the internal friction of fluids or *absolute viscosity* based? Newton furnished the most plausible answer to this question when he published his work on the subject in 1668.

FUNDAMENTALS OF VISCOSITY

Just as one solid body resists the action of sliding when in physical contact with another, so does each molecular layer of a fluid mass resist sliding over its neighboring layers during the motion called *flow*. The action of displacing parallel layers of a substance in the general direction of their line of contact is called *shear*. Flow of a fluid is a complexity of shearing actions between the multiplicity of molecular layers that form the mass. The force that resists shear in fluids is, as is true between sliding solid bodies, due to friction identified as internal fluid friction. Viscosity is the measure of the force required to overcome internal friction within a fluid mass.

A change in thermal energy by the addition or subtraction of heat varies molecular activity within any substance. Because temperature is indicative of such molecular activity, viscosity must always be related to the temperature at which it was determined. Lubricating oils decrease in viscosity as their temperatures increase, and vice versa.

Viscosity as it applies to fluid lubricants must be identified with certain physical conditions of the substance. In the discussion of hydrodynamic lubrication, it is shown that a liquid film present between two parallel plane surfaces, one of which moves relative to the other, divides itself into a series of layers which move parallel to the plane and to each other. It is also shown that each succeeding layer from the stationary plane to the moving plane moves at increasing velocity. The film next to the stationary plane adheres to the surface so does not move at all, while the one next to the moving plane moves at the velocity of that surface induced by an appropriate force. The layers between the two extremes will have a velocity that is directly proportional to their distance from the stationary surface. The *force per unit area* of the opposing planes necessary to produce motion between the layers is known as the *shear stress*. The relative motion of one layer to another is identified as the *shear strain rate*. The ratio of velocity of a given layer to its distance from the stationary plane is called the *rate of shear*, or *velocity gradient*, represented by v/h, where v is the velocity of a given layer and h is the film thickness measured perpendicularly from that layer to the stationary plane. The velocity gradient is constant for each of the several layers, and *laminar flow* results. The parameters mentioned above are shown in their mathematical relationship in the discussion of Newton's hypothesis, following. The important point here is that the viscosity of a given liquid remains the same at any selected temperature and pressure. It is also independent of rate of shear. The shear velocity will therefore correspond to the force necessary to produce relative motion of the layers in respect to each other. Any fluid that conforms to the laws of viscosity as defined by Newton and described above so that the shear stress is directly proportional to the rate-of-shear is called a *Newtonian fluid.* Development of the hydrodynamic film is dependent on the laws that define Newtonian liquids. Plastic materials such as grease and petrolatum that have properties of yield and whose viscosities are changed by the effects of shearing are described as *non-Newtonian substances.* Flow properties of such substances are described as *apparent viscosity,* discussed later in this chapter.

NEWTON'S HYPOTHESIS, ABSOLUTE VISCOSITY. On the premise that an internal frictional force existed in flowing fluids similar to solid friction, Newton developed the basic concept on which absolute viscosity is still recognized as the nearest approach to the true viscosity datum. He used two concentric cylinders, one placed within the other, as shown in Fig. 8-1. The outside diameter of the inner cylinder was slightly less than the inside diameter of the outer one, which provided space for the still water in which they were immersed. By the use of weights and pulleys, one cylinder was made to rotate at an ever-increasing rate until its velocity caused the free cylinder to also rotate at some relative velocity in respect to the driven cylinder. The force necessary to bring the cylinders into relative rotating motion he accepted as being the measure of internal fluid friction of the fluid tested. With the same apparatus and under similar conditions, the force required to bring about relative motion between the two cylinders varies with different fluids or with the same fluid at different temperatures. Because all factors related to the apparatus remained constant, the difference in force

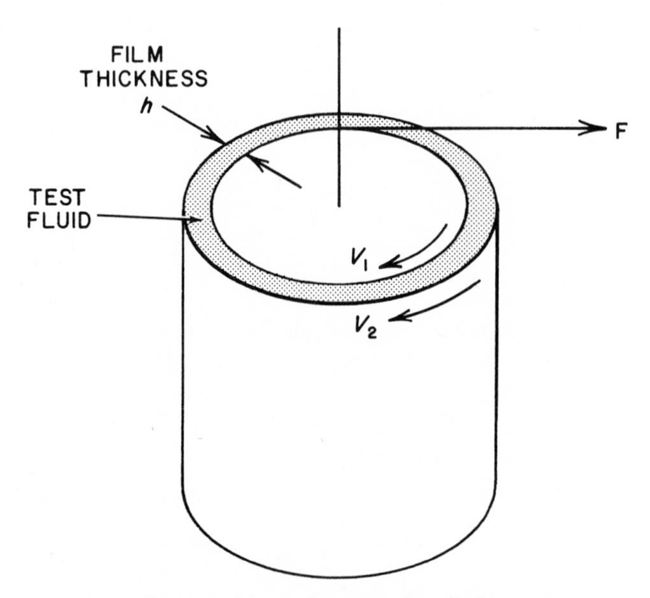

Fig. 8-1. Newton's concentric cylinders.

represented the difference in internal fluid friction for various liquids, or for any given liquid at different temperatures. In consideration of any similar apparatus in which the swept cylinder area and/or clearance space may or may not vary, the relationship of force to all other factors was found to be as follows:

Force is proportional to area.
Force is proportional to rotating speed.
Force is inversely proportional to film thickness.

Mathematically, the force magnitude is modified by the resistance to shear, represented by a factor or coefficient that is constant for a given fluid at some specific temperature. This constant is the *absolute viscosity* of the fluid tested for the temperature at which the test is conducted. The relationship of the above factors may be expressed as follows:

$$F = \mu A(v/h)$$

where

F = force
μ = absolute viscosity
A = area of effective working surfaces
v = velocity
h = distance between cylinders, or film thickness
v/h = velocity gradient or rate of shear (see hydrodynamic film)

The absolute viscosity (also called the coefficient of viscosity) is shown by solving for mu:

$$\mu = \frac{F/A}{v/h} = \frac{\text{shear stress}}{\text{rate of shear}} \qquad \text{(proportionality constant)}$$

To evaluate absolute viscosity, the parameters must have some standard measurable values. Both the English and metric systems were used in order to give numerical value to the viscosity coefficient so that the fluidity of one fluid could be compared with that of another. By the English system the force (μ) required to cause relative motion of the cylinders is registered in pounds. To be consistent, the area must be registered in square inches, film thickness in inches, and velocity in inches per second. By the metric system the force (z) is in dynes, and the area, film thickness, and velocity are registered in square centimeters, centimeters, and centimeters per second, respectively. To determine the value of v, then, first transpose the above equation $F = \mu A(v/h)$

$$\mu = \frac{Fh}{Av}$$

or

$$\mu = \frac{Fh}{A(d/t)} = \frac{FAt}{Ad}$$

where

$$t = \text{time in seconds}$$
$$A = \text{square inches}$$
$$d = \text{distance}$$
$$d/t = \text{velocity } v, \text{ in./sec}$$

Therefore

$$\mu = \frac{\text{lb} \times \text{in.} \times \text{sec}}{\text{in.}^3}$$

$$\mu = \text{reyn (after Reynolds)}$$

or

$$\mu = \frac{\text{dyn} \times \text{cm} \times \text{sec}}{\text{cm}^3}$$

$$\mu = \text{poise (after Poiseuille)}$$

One *dyne* (abbreviated dyn) is the force acting on a mass of 1 gram to produce a velocity of 1 cm/sec in 1 sec. The poise and reyn, therefore, are the names that identify absolute viscosity according to the measurement system used to determine their values. To bring their values to more convenient numerical values, the reyn is reduced to the microreyn and the poise to the centipoise. The centipoise is more universally employed when corrected according to the density of the fluid to which it applies.

The latter may be explained by describing the nature of two fluids having the same absolute viscosity but different in density, i.e., weight per unit volume. The density of various fluids is represented by their specific gravity or the ratio of the weight of a given volume of any fluid at a given temperature to the weight of a similar volume of water at the same temperature. The rate of gravitational flow of a liquid through an orifice or capillary tube is dependent upon its density. The one of higher density will flow faster, especially when the force is solely a function of *head* (pressure developed by height or depth) of fluid. The absolute viscosity does not reflect the influence of density differences.

Consider a liquid flowing through a capillary tube having a fixed internal diameter from an overhead reservoir into a flask below, instead of the mass resting on a plane surface and having area and thickness, as in Newton's apparatus. The volume (v) of liquid that would flow in unit time is proportional to the fourth power of the tube radius (R) and to the pressure (P) on the liquid and inversely proportional to the length of the tube (L) and to the absolute viscosity (μ) of the liquid, or

$$\mu = \frac{P\pi R^4}{8vL} \qquad \text{(Poiseuille's equation)}$$

If pressure depends on gravity, then

$$P = gh$$

where

$$g = 32.2 \text{ ft/sec/sec}$$
$$h = \text{head}$$

so that

$$\mu = \frac{gh\pi R^4}{8vL}$$

In consideration of fluid density (ρ) as a factor governing the flow rate, the head becomes a function of ρgh in Poiseuille's equation, so that

$$\frac{\mu}{\rho} = \frac{gh\pi R^4}{8vL}$$

KINEMATIC VISCOSITY. The ratio of absolute viscosity μ of a liquid to its density ρ is called the *kinematic viscosity,* and the unit of measurement is the *stoke* (after Stokes), represented by ν, or

$$\nu = \frac{\mu}{\rho}$$

and

$$1 \text{ stoke} = \frac{1 \text{ poise}}{\text{density}}$$

$$1 \text{ centistoke} = \frac{1}{100} \text{ stoke}$$

The *centistoke* (cSt) is more convenient to use and is therefore the value on which most calculations and standard conversion tables are based.

The specific gravity of water at 68°F is given an arbitrary value of 1.0. The specific gravity of all other liquids, including water at temperatures other than the standard 68°F, is based on the weight of 68°F water. Therefore, ρ is equal to the specific gravity of the test liquid. For 68°F water, $\mu = \nu$.

The kinematic viscosity of 68°F distilled water is 1.007 cSt (at 68.36°F, $z = 1.00$). At 100°F and 130°F, it is 0.689 and 0.158 cSt, respectively.

To convert absolute viscosity μ to kinematic viscosity ν, or vice versa, then,

$$\nu = \mu/\rho = \text{centistokes}$$
$$\mu = \nu\rho = \text{centipoises}$$

Kinematic viscosity is thus not a method of determining viscosity, but a system of expressing viscosity.

Table 8-1. VISCOSITY AND FLUIDITY OF WATER

Temperature (°C)	*Viscosity* (cP)	*Fluidity* (l/P)	*Specific Viscosity* (0° C)
0	1.792	55.8	1.000
5	1.519	65.8	0.847
10	1.307	76.5	0.730
15	1.140	87.7	0.636
20	1.005	99.5	0.561
20.4	1.000	100.0	0.558
25	0.894	111.9	0.499
30	0.801	124.9	0.447
35	0.723	138.4	0.403
40	0.656	152.5	0.366
45	0.599	167.0	0.334
50	0.549	182.0	0.307
60	0.469	213.3	0.262
70	0.406	246.3	0.227
80	0.357	280.5	0.199
90	0.317	315.9	0.177
100	0.284	352.3	0.158

SPECIFIC VISCOSITY is the ratio of the viscosity of any substance to that of water at some specified temperature, usually 0°C, or 32°F (Table 8-1). For example, the viscosity of water at 32°F is 1.792 cP; at 20°C it is 1.005 cP. The specific viscosity of 20°C water, then, is 1.005/1.792 = 0.5608. The specific viscosity of 32°F water is 1.000.

Fluidity is the reciprocal of absolute viscosity in poises, or 1/P. The fluidity of 68°F water having an absolute viscosity of 1.005 cP to poises is expressed as

$$\text{fluidity} = \frac{1}{\text{poises}} = \frac{1}{\text{centipoises}/100} = \frac{100}{\text{centipoises}} = \frac{100}{1.005} = 99.5$$

or

$$\text{fluidity} = \frac{1}{\text{poises}} = \frac{1}{\text{centipoises}/100} = \frac{1}{0.01005} = 99.5$$

The *micropoises* (μP) is the unit in which the absolute viscosity of gases is usually reported (Table 8-2). For example, air at 32°F has an absolute viscosity of 170.8 μP.

Viscosity of solids is usually reported in poises, e.g., the absolute viscosity of glass at 1067°F is 11×10^{12} = (110,000,000,000)P.

To summarize the terms associated with viscosity, note the following:

(a) Absolute viscosity is expressed in terms of (1) poises or centipoises, or (2) reyns or microreyns. They represent (force × time)/distance.

One poise (*P*) is equal to the viscosity of a fluid requiring the force of 1 dyn to move a mass of 1 cm^3 at a velocity of 1 cm/sec when the thickness is the distance between a moving and a stationary surface. In the metric system of measurement, the poise is expressed in dyn × sec/cm^2.

Table 8-2. VISCOSITY OF GASES

Gas	*Temperature (°C)*	*Viscosity (μP)*
Air	0	171
	70	180
	74	210
Argon	0	210
	100	270
Helium	0	186
	100	228
Hydrogen	0	84
	130	110
Methane	0	103
	100	133
Neon	0	297
	100	365
Nitrogen	0	161
	130	222
Oxygen	0	189
	130	258
Propane	20	80
	100	100
Sulfur dioxide	0	116
	100	161

One reyn is the viscosity of a fluid requiring the force of 1 lb to move a mass of 1 in.^{3} at a velocity of 1 in./sec when the thickness is the distance between a moving and a stationary surface. (The English system of measurement is expressed in lb $\times$ sec/in.^{2})

$$1 \text{ reyn} = 68{,}944\ P = 6.894 \times 10^6 \text{ cP}$$

$$1\ P = 68{,}944 \times \text{reyns}$$

$$1 \text{ reyn} = \frac{1}{68{,}944}\ P$$

$$= 1.45 \times 10^{-5}\ P$$

$$1 \text{ microreyn} = 1.45 \times 10^{-7} \times 1{,}000{,}000 \text{ cP}$$

$$1 \text{ reyn} = 1.45 \times 10^{-5}\ P$$

μ represents absolute viscosity in reyns. or any viscosity providing it relates to the applicable system of parameters.
Z represents absolute viscosity in centipoises.
ν represents kinematic viscosity in centistokes.

Centipoises and reyns can be compared by showing their values for air and water:

	Absolute Viscosity	
	Centipoises (z)	*Reyns (μ)*
Air	0.018	0.0026×10^{-6}
Water	1.005	0.145×10^{-6}

Table 8-3. VISCOSITY OF SOME LIQUIDS

Liquid	*Temperature* (°*C*)	*Viscosity* (*cP*)
Air, liquid	−192	0.172
Ammonia	−50	0.317
	−28	0.214
Carbon tetrachloride	0	1.33
	100	0.384
Alcohol, ethyl	20	1.2
Glycol, ethylene	20	20.0
	100	2.0
Glycerin	0	12,110
	20	1,490
	70	500
Castor oil	10	2,420
	20	986
	100	17
Olive oil	10	138
	40	36
	70	100
Rapeseed oil	10	385
	20	163
Sperm oil	15	42
	100	5
Turpentine	0	2.25
	20	1.49
Honey	70	1,500.00
Mercury	70	1.5

(b) Kinematic viscosity is the ratio of absolute viscosity to the density of a liquid at the temperature of the viscosity test. Therefore, kinematic viscosity is a scale, not a measurement.

Kinematic viscosity is equal to the absolute viscosity divided by the density of a liquid at 60°F.

Absolute viscosity is equal to the kinematic viscosity multiplied by the density of a liquid at 60°F.

Kinematic viscosity is represented by ν, density by ρ. Therefore,

$$\mu = \nu \times \rho$$

or

$$\nu = \mu/\rho$$

One poise is equal to 1 stoke when gravity is 1.00. Along with Saybolt Universal Seconds (SUS), kinematic viscosity is most widely used in the United States.

To convert kinematic viscosity to or from Saybolt Universal Seconds, the following equations may be used with a fair degree of accuracy:

(a) To determine Saybolt Universal Seconds from centistokes:

SUS at 100°F = kinematic centistokes at 100°F $\times$ 4.62
SUS at 130°F = kinematic centistokes at 130°F $\times$ 4.629
SUS at 210°F = kinematic centistokes at 210°F $\times$ 4.6524

(b) To determine centistokes from Saybolt Universal Seconds:

$$\text{Kinematic centistokes at } 100°\text{F} = \text{SUS at } 100°\text{F} \times \frac{1}{4.62}$$
(or 0.2165)

$$\text{Kinematic centistokes at } 130°\text{F} = \text{SUS at } 130°\text{F} \times \frac{1}{4.629}$$
(or 0.216)

$$\text{Kinematic centistokes at } 210°\text{F} = \text{SUS at } 210°\text{F} \times \frac{1}{4.6524}$$
(or 0.215)

VISCOMETERS

Various techniques are used to determine the viscosity of fluids, most of which have been employed in the field service and research laboratories of lubricating oil companies. The techniques use both force and flow time, but regardless of the technique, the apparatus used is called a viscometer, or viscosimeter.

The use of force demonstrated in the discussion of Newton's apparatus involving concentric cylinders is actually a measurement of torque, the force of rotation or twist. While Newton employed cylinders in his experiment, disks, cones and paddles can also be rotated within a liquid, either *at constant speed to determine torque,* or *at constant torque to determine rotational speed.* The purpose of all viscometers is to show absolute viscosity of liquids. The following forms of torque rotating apparatus are other types.

MacMichael Viscometer. This is basically a disk suspended by a wire in a cup containing the test sample of liquid. The cup is rotated at constant speed, while the drag of the oil tends to turn the disk and twist the suspension wire. An indicator located at the tip end of the wire measures the arc through which twisting occurs. The arc magnitude is related to the absolute viscosity of the test oil. As with all laboratory viscometers, the rotating cup holding the liquid is immersed in a controlled temperature bath. This viscometer is particularly adaptable to highly viscous liquids.

Stormer Viscometer. This viscometer consists of a fixed cup that holds the test liquid in which an inverted cylinder is suspended and rotated by a pulley and weight arrangement. The viscosity of the liquid is measured according to the time it takes for the cylinder to make 100 revolutions. Like the MacMichael apparatus, the Stormer viscometer is adaptable to highly viscous liquids.

Doolittle Viscometer. This is also a torsion apparatus, much like the MacMichael viscometer. The basic difference lies in the reverse method of measuring torque. In the Doolittle viscometer, the suspension wire is manually twisted (wound up) 360° or one full revolution either to the right or left. Upon release of the spring, the cylinder attached to the lower end of the wire and immersed in the test oil tends to return through its original position and beyond. If the test oil cup is empty, the momentum developed by the release of twisting stress within the wire would continue to rotate the cylinder an additional 360°, or two revolutions. However, the resistance or drag of the test oil causes the cylinder to stop rotating short of that point. The resistance displayed by the oil as evidenced by the point at which the cylinder ceases to rotate short of the additional 360° is the measure of viscosity of the test oil. The test is continued by repeating the rotation of the cylinder in the opposite direction of the first test. The mean of both runs is acceptable as the true reading. This type of viscometer is not in general use, but is included here to describe an apparatus that employs the principle of *torque reaction* in determining viscosity.

TIME-MEASURED VISCOSITIES. In the foregoing discussion, viscosity was described as being the measure of torque required to overcome the resistance of a liquid to produce rotational motion of a body immersed in the fluid mass. The action is one of drag and infers the occurrence of shear involving force. Viscometers that measure viscosity in time units employ short tubes, or capillaries through which a liquid is passed either by gravity or pressure. The action is one of flow. The viscosity of a liquid so measured is reported in terms of *efflux time,* or the number of seconds (usually) for a given volume to flow out of a reservoir through an orifice of specified area and length at a fixed temperature. While the short tube viscometer is strictly a time-reporting apparatus, the capillary type reports viscosity in terms of kinematic viscosity (centistokes), because density of the liquid is intrinsically involved.

Capillary Viscometer. Capillaries are relatively long tubes with very small bores. Capillarity is the action that causes the surface of a liquid to rise, or be depressed, when in contact with the solid wall of the tube. Thus the action may be one of attraction or repulsion, depending on the wetting characteristics of the liquid, or the relative surface energy (tension) of the solid in respect to the liquid.

The use of the capillary tube in determining viscosity is based on certain laws derived simultaneously by Poiseuille and Hagen. These laws as related to capillary tubes are:

(a) Steady viscous flow (laminar flow) must exist, shown by Reynolds that the relationship of all tube parameters meet the limitations $rvd/\mu < 2000$.

Where r = bore radius, d = liquid density, v = mean velocity of the fluid, and μ = fluid viscosity.

(b) Rate of liquid flow is proportional to the pressure, or density.

(c) Rate of liquid flow is inversely proportional to the length of tube.

(d) Rate of liquid flow is proportional to the fourth power of the cylindrical bore radius.

The absolute viscosity is given by the following relationship of the above laws, called Poiseuille's equation:

$$Q = \frac{P_d \pi r^4}{8 \mu \ell}$$

where

Q = flow rate (volume per unit of time)
P_d = pressure difference between the two ends of the capillary tube
r = radius of tube bore
ℓ = length of capillary
μ = viscosity of liquid

or, transposing,

$$\mu = \frac{P_d \pi r^4}{8 Q \ell}$$

This equation measures frictional resistance only. Substituting density ρ, gravity g and mean head h for pressure difference P_d, then

$$\mu = \frac{\rho g h \pi r^4}{8 Q \ell}$$

This equation shows absolute viscosity proportional to liquid density. Transposing,

$$\frac{\mu}{\rho} = \frac{g h \pi r^4}{8 Q \ell}$$

but

$$\frac{\mu}{\rho} = \text{kinematic viscosity}$$

so that

$$V = \frac{g h \pi r^4}{8 Q \ell} = \text{stokes or centistokes}$$

The basic capillary viscometer is shown in Fig. 8-2, representing the suspended-level instrument of the *Ubbelohde* single capillary type. There

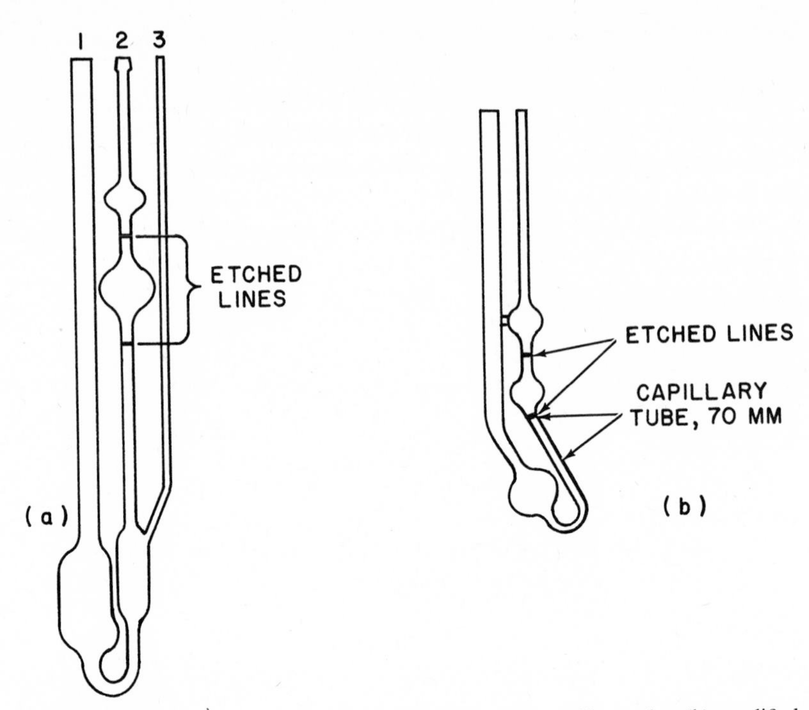

Fig. 8-2. Kinematic viscometers. (a) Ubbelohde single capillary tube, (b) modified Otswald viscometer.

are a number of different kinds of capillary viscometers, varying in size and design according to their purpose and kind of fluid tested. All are made of glass to allow visibility of liquid flow. The complete test apparatus includes a timer, thermometer, and a constant temperature bath in which the viscometer is immersed during the test. Various bath fluids are used according to the test temperature, which is either 100°F or 210°F for lubricating oils. They include water (100°F) and white oil, or ethylene glycol (210°F).

The test procedure is as follows:

1. Introduce the oil to be tested (about 10 ml) into the instrument through tube 1 after filtering it.

2. Immerse the instrument into the bath to about 5 cm from the top.

3. Allow sufficient time for the test oil to reach bath temperature.

4. By suction, draw up the test oil into tube 2 to some point above the upper etched line, while stopping tube 3 by placing a finger over its opening.

5. Release the suction and unstopper tube 3, which permits the oil to flow out of the capillary tube.

6. The time required for the meniscus to drop from the upper to the lower etched line is measured in seconds. Two test runs are made, and the

average time of both represents the factor for calculating kinematic viscosity in centistokes.

For accurate determination of kinematic viscosity, certain modifications must be made relative to (1) the physical characteristics of the instrument and (2) its design. The modifications are constants determined by calibration, using two liquids of known but different kinematic viscosities in the instrument to be calibrated. If A represents the constant for step 1 above, then

$$A = \frac{v_2 t_2 - v_1 t_1}{t_2^2 - t_1^2}$$

If B represents the constant for step 2 above, then

$$B = \frac{t_1 t_2}{t_2^2 - t_1^2}(v_2 t_1 - v_1 t_2)$$

where

v_1 and v_2 = kinematic viscosities of the two known liquids
t_1 and t_2 = respective flow (efflux) time of the two known liquids

Knowing the value of constants A and B, we can determine the kinematic viscosity of the test oil:

$$\nu = At - (B/t)$$

If distilled water is used as the liquid of known ν and t, the kinematic viscosity of the water is accepted as being 1.007 cSt at 68°F, 0.689 at 100°F, and 0.518 at 130°F. If efflux time is relatively long, B/t can be eliminated without affecting the practical accuracy of test results.

Short-Tube Viscometers. In contrast to a tube length of 10 to 12 cm of the capillary viscometer, the outlet tube of the short-tube (Saybolt Universal) viscometer is 1.225 cm long. The principle of this type of viscometer is based on the time required for a given volume of liquid to flow through the outlet tube from an upper reservoir into a receiving flask below. Like the capillary instrument, the oil liquid reservoir (oil tube) is immersed in a constant-temperature bath. The short-tube viscometers may also be classified as kinematic instruments, as well as the Ubbelohde and Otswald viscometers, but their rapid flow rate is inaccurate in the lower viscosity range.

There are several short-tube viscometers, the more important being the *Saybolt, Redwood No. 1,* and *Engler* instruments. The Saybolt viscometer is the standard instrument in the United States, and the Redwood and Engler viscometers hold similar recognition in England and Germany, respectively.

The operation of the short-tube apparatus may be understood by referring to Fig. 8-3, showing a Saybolt universal viscometer. Oil is intro-

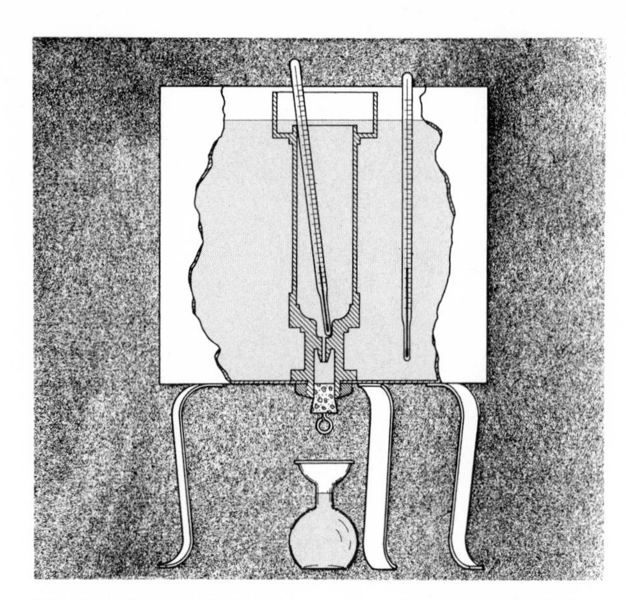

Fig. 8-3. Saybolt Universal viscometer. *Courtesy, Gulf Oil Corp.*

duced into the oil tube (reservoir) after proper filtering. It is allowed to stand until the test oil reaches the selected temperature maintained in the bath. The excess test oil is drained by a pipette from the gallery surrounding the upper rim of the oil tube. The cork is then pulled from the outlet end of the Saybolt outlet tube. The time, in seconds, required for 60 cm^3 to flow into the receiving flask is reported as the Saybolt Universal Seconds (SUS) viscosity of the test oil at the bath temperature. For the Saybolt apparatus, the standard temperatures are 100°F, 130°F, and 210°F. Oils with an SUS of less than 32 sec are never tested with the Saybolt viscometer.

Another type of Saybolt instrument is the Saybolt Furol viscometer, used to measure the flow time of high-viscosity fluids, such as heavy fuels and road oils, from which the name "Furol" is derived. The outlet tube of the Furol viscometer being almost twice that of the Universal apparatus, the outflow time is approximately one-tenth that of the latter. The results of this instrument are reported as Saybolt Furol Viscosity, in seconds, which may be converted to an approximate Universal value at the same temperature by multiplying by 10, or,

$$\text{Saybolt Universal Seconds} = 10 \times \text{Saybolt Furol seconds}$$

The Saybolt Furol viscosity tests are reported for 122°F usually, but may be shown for 77°F, 100°F, and 210°F also.

While the *Redwood No. 1* viscometer varies in the volume of oil allowed to flow (50 cm^3) in determining efflux time, also in seconds, the method of testing is essentially the same as that of the Saybolt instrument. Standard test temperatures for Redwood 1 viscosity are 70°F, 100°F, 140°F, and 200°F.

The Engler viscometer is based on the efflux time of 200 cm^3 of oil and an outlet tube length of 2.0 cm. However, the method of reporting is not in seconds but represents a ratio of the efflux time of the test oil to that of water at 68°F, and it is called the *Engler degree.* The latter may be accepted as being 51 sec (average). For example, for an oil having an Engler efflux time of, say, 220 sec, the Engler number would be 220/51 = 4.3 Engler°.

VISCOSITY CONVERSION

It is often necessary to convert the viscosity value expressed in terms of one system to that of another. This is particularly true in mathematical solutions pertaining to lubrication problems in which absolute viscosity values are required. Field lubrication engineers often find it necessary to convert viscosity recommendations made by foreign machinery manufacturers into terms understood in the country to which the equipment is shipped.

The two important factors that must be considered in making such conversions are (1) temperature and (2) density as shown below:

Density Correction. To correct for specific gravity taken at any other temperature than 60°F, apply the following equation:

$$\text{Density at } t\,^\circ\text{F} = \text{sp gr } 60/60 - (0.0004)(t - 60)$$

EXAMPLE: Determine the specific gravity of an oil at 210°F whose specific gravity at 60°F is 0.950.

$$\begin{aligned}\text{sp gr at } 210^\circ\text{F} &= \text{sp gr at } 60^\circ\text{F} - (0.0004)(t - 60)\\ &= 0.950 - (0.0004)(210 - 60)\\ &= 0.950 - 0.06\\ &= 0.89\end{aligned}$$

KINEMATIC VISCOSITY TO ABSOLUTE VISCOSITY

EXAMPLE: Determine the absolute viscosity in centipoises at 210°F of an oil having a kinematic viscosity 31.9 cSt at 210°F and a specific gravity of 0.925 at 60°F.

$$\begin{aligned}\text{sp gr at } 210^\circ\text{F} &= 0.925 - 0.06\\ &= 0.865\\ \text{Absolute viscosity at } 210^\circ\text{F} &= \text{kinematic viscosity at } 210^\circ\text{F}\\ &\quad \times \text{density at } 210^\circ\text{F}\\ &= 31.90 \times 0.865\\ &= 27.59 \text{ cP}\end{aligned}$$

CONVERSION OF KINEMATIC VISCOSITY TO SAYBOLT UNIVERSAL SECONDS (SUS). To convert kinematic viscosity (in centistokes) to Saybolt Universal Seconds at 100°F or 210°F, it is more convenient to use the appropriate table. It can also be done by referring to the curve that includes the specific value within range of viscosities covered. Conversion can also be accomplished by calculation, using the empirical equation developed for the purpose.

Table 8-4 converts kinematic, Saybolt Universal, Engler, and Redwood No. 1 viscosities each to the other at any given temperature.

Table 8-4. CONVERSION FACTORS, KINEMATIC VISCOSITY TO SAYBOLT UNIVERSAL VISCOSITY

	Conversion Factors	
Temperature (°F)	*Factor A, 75 cSt and Under*	*Factor B, Over 75 cSt*
0	0.994	4.604
10	0.995	4.607
20	0.995	4.610
30	0.996	4.613
40	0.996	4.615
50	0.997	4.618
60	0.998	4.621
70	0.998	4.624
80	0.999	4.627
90	0.999	4.630
100	1.000	4.632
110	1.001	4.635
120	1.001	4.638
130	1.002	4.641
140	1.002	4.644
150	1.003	4.647
160	1.004	4.649
170	1.004	4.652
180	1.005	4.655
190	1.005	4.658
200	1.006	4.661
210	1.007	4.664
220	1.007	4.666
230	1.008	4.669
240	1.009	4.672
250	1.009	4.675
260	1.010	4.678
270	1.010	4.680
280	1.011	4.683
290	1.012	4.686
300	1.012	4.689
310	1.013	4.692
320	1.013	4.695
330	1.014	4.697
340	1.015	4.700
350	1.015	4.703

Courtesy American Society of Testing Materials, ASTM (D2161)

Kinematic to Saybolt Universal Seconds (SUS)

$$\text{SUS} = \frac{(4.605 + 0.000297t)\nu}{1 - 10^{-0.07445\nu^{0.9538}}}$$

where

ν = centistokes at temperature t

or where kinematic viscosity is in excess of 70 cSt

$\text{SUS} = \nu \times 4.635$ for SUS at 100°F (approx.)

$\text{SUS} = \nu \times 4.667$ for SUS at 210°F (approx.)

Kinematic to Redwood No. 1

$$\text{Redwood No. 1 seconds} = \frac{(4.037 + 0.000344t)\nu}{1 - 10^{-0.06668\nu^{0.9704}}}$$

or, where kinematic viscosity is in excess of 70 cSt,

Redwood No. 1 seconds at 70°F = $\nu \times 4.06$
Redwood No. 1 seconds at 100°F = $\nu \times 4.08$
Redwood No. 1 seconds at 140°F = $\nu \times 4.10$
Redwood No. 1 seconds at 200°F = $\nu \times 4.18$

Engler Degrees to Kinematic ν (E° from 1.35 to 3.2)

$$\nu = 8.0\ \text{E}^\circ - \frac{8.64}{\text{E}^\circ}$$

Engler Degrees to Kinematic ν (E° above 3.2)

$$\nu = 7.6\ \text{E}^\circ - \frac{4.0}{\text{E}^\circ}$$

Redwood No. 1 to Kinematic ν

$$\nu = 0.260 \times \text{Redwood No. 1} - \frac{179}{\text{Redwood No. 1}}$$

where Redwood No. 1 seconds are between 34 and 100

and

$$\nu = 0.247 \times \text{Redwood No. 1} - \frac{50}{\text{Redwood No. 1}}$$

where Redwood No. 1 seconds are above 100.

Redwood No. 2 (Admiralty) to Kinematic ν

Redwood No. 2 between 32 and 90 sec

$$\nu = 2.46 \times \text{Redwood No. 2 seconds} - \frac{100}{\text{Redwood No. 2 seconds}}$$

Redwood No. 2 above 90 sec

$$\nu = 2.45 \times \text{Redwood No. 2 seconds}$$

Kinematic to Engler Degrees (above 70 cSt)

At any temperature — $\text{E}^\circ = 0.132\nu$

Saybolt Furol to Kinematic ν

Saybolt Furol between 25 and 40 sec:

$$\nu = 2.24 \text{ Saybolt Furol} - \frac{184}{\text{Saybolt Furol}}$$

Saybolt Furol above 40 sec:

$$\nu = 2.16 \text{ Saybolt Furol} - \frac{60}{\text{Saybolt Furol}}$$

Another means of converting kinematic viscosity to Saybolt Universal Seconds is by the use of conversion factors given in Table 8-4. Note that the factors change for viscosities under and over 75 cSt for the various temperatures limited.

The use of the table is demonstrated by the following examples:

EXAMPLE 1: Determine the SUS viscosity for an oil having a kinematic viscosity of 64 cSt at 150°F.

Refer to Table 8-4, factor *A* opposite 150°F, to find 1.003. Multiply 64 cSt $\times$ 1.003 = 64.19 sec SUS at 150°F.

EXAMPLE 2: Determine the SUS for an oil having a kinematic viscosity of 254 cSt at 210°F.

Refer to Table 8-4, factor *B* opposite 210°F, to find 4.664. Multiply 254 cSt $\times$ 4.664 = 1184.6 sec SUS at 210°F.

REASONS FOR VISCOSITY CHANGES IN LUBRICATING OILS

The viscosity of lubricating oils varies with changes in temperature and pressure according to the following laws:

(a) *Viscosity decreases as the temperature of the oil increases and increases as the temperature decreases.* The magnitude of change varies with different types of oil and is designated by a parameter known as the viscosity index (VI). It is based on the incremental change in viscosity per degree temperature increase or decrease. As the magnitude of viscosity change becomes less per degree change in temperature, the VI value becomes higher, and vice versa. Thus, lubricating oils differ in their viscosity reaction to temperature changes, as evidenced by their individual viscosity index values, which may range between zero (low) to above 150 (high).

As indicated previously, a Newtonian fluid is one in which the shear stress is directly proportional to the rate of shear and whose viscosity is constant at a given temperature and pressure, but independent of the rate of shear. Fluids that do not obey the Newtonian concept, i.e., are affected by the effects of shearing so that their viscosity depends on the rate of shear, are classified as *non-Newtonian fluids* whose viscosity decreases as their rate of shear is increased. Lubricating greases are non-Newtonian, of the plastic type, while lubricating oils above their pour point temperature are Newtonian. Because the viscosity of greases changes with shear rate, their multiviscosity values are described as *apparent viscosity,* a subject discussed later in this chapter.

(b) *Viscosity increases as pressure on the oil increases and decreases as pressure decreases.* Again, as with temperature, the change in viscosity of a lubricant (including synthetics) per unit change in applied pressure varies with different type (base stock) oils. Table 8-5 shows the pressure viscosity relationship of lubricating oils. With pressures from 175,000 psi to over 200,000 psi encountered in antifriction bearings and hypoid gears, the viscosity-pressure relationship in lubricants is very important. The effect of pressure is greater (1) at higher magnitudes than at lower magnitudes for similar increases, (2) for higher viscosity oils than for low viscosity oils, and (3) for naphthenic oils than for paraffinic oils.

It is often a question if unit bearing pressures compensate for any viscosity reduction due to the increase in operating temperature that is induced by higher pressures. The author believes that the increase in viscosity due to pressure in combination with microhydrodynamic films prevents severe metal-to-metal contact in a number of heavily loaded bearings. Under temporary extremely high pressures, as with shock loading or during intermittent heavy loads, it is suspected that an oil will increase in viscosity to the point of becoming an elastic solid. The occurrence is described as one of *viscoelasticity,* which takes place in an environment of rapidly changing stresses and exists only for short periods of time, after which the oil will return to its normal viscosity commensurate with the pressure and temperature involved.

Table 8-5. EFFECT OF PRESSURE ON VISCOSITY OF TURBINE OIL (100°F)

Pressure (psi)	*Saybolt Universal Seconds*	*Kinematic Viscosity (cSt)*	*Absolute Viscosity (cP)*
Atmospheric	154	32.8	28.18
1,000	176	37.7	32.39
2,000	203	43.6	37.50
3,000	234	50.5	43.38
4,000	271	58.4	50.17
5,000	316	68.2	58.59
6,000	365	78.7	67.61
7,000	422	91.2	78.35
8,000	491	105.9	90.98
9,000	567	122.3	105.07
10,000	658	142	122
15,000	1,282	277	238
20,000	2,417	522	448
25,000	4,561	984	845
30,000	8,621	1,860	1,598

The addition of certain *polar compounds,* such as fatty oils, will decrease the viscosity of a mineral oil. In some instances, as in the case of steam cylinder stocks, the decrease will amount to as much as 15% with the addition of about 5% acidless tallow oil. Part of the reason for this is the lower viscosity of the compound in some cases, while the reduction in intrinsic molecular friction, a function of the polar material, represents a substantial influence in reducing the resistance of the fluid to flow.

The viscosity of hydrocarbon oils increases when exposed to high *nuclear radiation.* Oils will solidify when subjected to long periods of high dosage. Paraffinic oils appear more susceptible to viscosity change due to radiation than naphthenic oils, while some synthetic lubricants are more susceptible than either.

Oils containing additives for improving their viscosity index (viscosity-temperature relationship) become less viscous under low radiation dosage, but when the radiation continues and accumulates, the viscosity increases, sometimes to the extent of several times the initial level.

Because density varies with oil temperature, it follows that *viscosity varies directly with density.*

A study of the *dimensionless equation* ZN/P, in which Z represents the viscosity of the lubricating oil, shows that a closer approach to a hydrodynamic condition, or even the establishment of such a wedge, is made with higher Z values. Speed N, a critical parameter in establishing a hydrodynamic wedge, becomes somewhat nullified at lower magnitudes by an increase in viscosity. There remains a limit of course to which compensation between Z and N can be considered (see Chapter 9).

The Reynolds number (Chapter 9) is greatly influenced by the viscosity of the flowing fluid. This is shown in the equation $R = DV/\nu$ (D = diameter of conduit in feet, V = velocity of flow in ft/sec, and ν = kinematic viscosity in in.^{2}/sec), in which the Reynolds number R becomes less, to approach or establish laminar flow ($R < 2000$) with fluids of greater viscosity values.

The viscosity of an oil affects the temperature of the system in which it is used. Assuming that the viscosity of an oil is sufficiently high to develop a safe ZN/P value, light-bodied oils tend to reduce the operating temperature. The reason for this lies in greater pumpability and lower fluid (internal) friction of lower-viscosity oils as compared with heavier-bodied oils.

For example, without any other changes in a 20,000 kW geared steam turbine, except to replace a circulating oil of 410 SUS at 100°F with a similar type of 300 SUS at 100°F, the following temperature reductions resulted:

Turbine

No. 1 bearing (thrust)— from 175° to 163°F
No. 2 bearing— from 164°F to 145°F

Gear set

No. 3 bearing, pinion, turbine side— from 140°F to 128°F
No. 4 bearing, pinion, generator side— from 137°F to 124°F

No. 5 bearing, gear, turbine side— from 159°F to 143°F
No. 6 bearing, gear, generator side— from 156°F to 140°F
Reduction gears (ratio 2 : 1)— from 188°F to 170°F

Generator

No. 7 bearing, generator, turbine end— from 119°F, no change
No. 8 bearing, generator, far end— from 106°F to 111°F, increase

Cooler

Oil, in— from 140°F to 129°F
Oil, out— from 104°F to 100°F

Oil from all bearings flowed into the main reservoir (sump) through the gear case below the gear. A greater reduction in gear temperature was attained after all oil, except the spray oil, was directed onto the gear at the opposite side of the spray header.

There is evidence that automotive crankcase temperature decreases with lower-viscosity oil. The average reduction is 15°F for each reduction in SAE viscosity number, i.e., from about 200°F with an SAE No. 30 oil to about 185°F with an SAE No. 20 oil.

Lighter-bodied straight mineral oils have greater water separation than heavier-bodied oils. This ability diminishes when an oil contains foreign materials that cause emulsification. Included with such materials are detergents and acid sludges. With water-soluble oils, such as those used for metal cutting, the addition of sulfuric acid to the emulsion (oil plus water) will break the emulsion and cause separation of the oil and water.

Low-viscosity oils are preferred for automotive engines, both gasoline and diesel, especially those with a high viscosity index, to provide better low-temperature starting. Light-bodied oils show better contaminant filtering capability and provide higher fuel economy because of their lower internal friction. Heavier contaminant particles tend to settle out of a low-viscosity oil more readily than from higher-viscosity oils.

Viscosity Index

Mineral lubricating oils decrease in viscosity as their temperature is raised. For example, a cylinder oil having a viscosity of, say, 3000 sec at 100°F (SUS) can be reduced to 1075 sec at 130°F and again to 175 sec at 210°F. The reduction from 3000 to 175 sec with a temperature increase of 100°F to 210°F, respectively, is said to be the rate of viscosity change for that oil. Another cylinder oil of 3000 sec at 100°F (SUS) may have a lower viscosity at 210°F of, say, 160 sec (SUS). The second oil then would be considered to have a higher rate of viscosity change per degree temperature rise.

The rate at which an oil changes viscosity with a rise or drop in temperature is designated by a comparative number called *viscosity index* (VI).

A VI of 100 indicates that an oil with this value "thins out" less rapidly than oils having a VI of zero. Values in between indicate intermediate viscosity changes according to the comparative VI number (Fig. 8-4).

The VI for any oil having a viscosity change between the two extremes can be determined by the Dean and Davis formula (Table 8-6):

$$\text{VI} = \frac{L - U}{L - H} \times 100$$

where

L = viscosity SUS of low VI (0) oil at 100°F
H = viscosity SUS of high VI (100) oil at 100°F
U = viscosity SUS of unknown VI oil at 100°F

The viscosity at 210°F of oils represented by H and U must be the same.

EXAMPLE: The viscosity of a paraffinic base oil has a viscosity at 100°F of 340 SUS and of 55 SUS at 210°F, and is rated at 100 VI. A naphthenic base oil having viscosities of 600 SUS at 100°F and 55 SUS at 210°F is rated at 0 VI. The VI of another oil having the same viscosity at 210, but between 340 and 600 sec at 100, say 440 sec, would have a VI as follows:

$$\text{VI} = \frac{L - U}{L - H} \times 100$$

$$\text{VI} = \frac{600 - 440}{600 - 340} \times 100$$

$$\text{VI} = 61.5 \text{ or } 62 \text{ (approx.)}$$

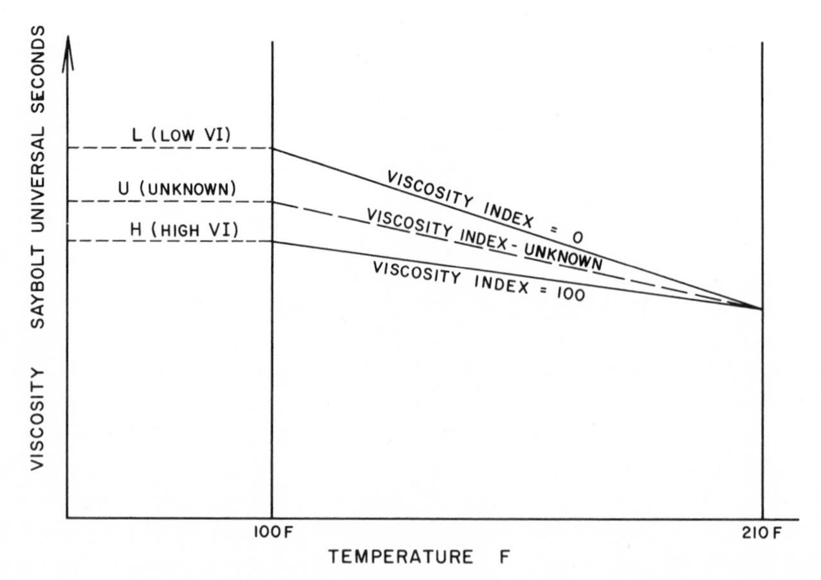

Fig. 8-4. Graphic demonstration of viscosity index determination.

Table 8-6. VALUES OF H AND L IN DEAN AND DAVIS FORMULA FOR DETERMINING VISCOSITY INDEX

Saybolt Universal Viscosity at 210°F	*H*	*L*	*(L − H)*
60	425	780	355
65	514	976	462
70	604	1,182	578
75	697	1,399	702
80	791	1,627	836
85	888	1,865	977
90	986	2,115	1,129
95	1,087	2,375	1,288
100	1,189	2,646	1,457
105	1,294	2,928	1,634
110	1,401	3,220	1,819
115	1,510	3,523	2,013
120	1,620	3,838	2,218
125	1,733	4,163	2,430
130	1,848	4,498	2,650
135	1,965	4,845	2,880
140	2,084	5,202	3,118
145	2,205	5,570	3,365
150	2,328	5,949	3,621
160	2,580	6,740	4,160
170	2,840	7,573	4,733
180	3,109	8,450	5,341
190	3,385	9,370	5,985
200	3,670	10,333	6,663
225	4,418	12,930	8,512
250	5,217	15,796	10,579
300	6,967	22,340	15,373

Courtesy American Society for Testing Materials, ASTM D567: "Standard Method for Calculating Viscosity Index."

EXAMPLE: Using Table 8-6, determine the VI of an oil having a viscosity of 70 sec at 210°F and 640 sec at 100°F.

$$\mathrm{VI} = \frac{L - U}{L - H}$$

$$= \frac{1182 - 640}{578} \times 100$$

$$= 94$$

The viscosity index is extensively used in connection with lubricants for the internal combustion engine. The VI increases as the oil pressure increases. It also increases with the addition of VI improvers (Chapter 10).

The Dean and Davis formula is accurate for oils having viscosity indices below 100–105. But as the value of VI becomes increasingly higher than 100, the formula becomes less reliable, which made it necessary to devise a different method of determination. The old method, or Method A (Dean and Davis) is identified as ASTM Method D567. The revised method, Method B, is identified as ASTM Method D2270, and is accurate for all viscosity indices and is referred to as VI_E. Summarizing,

VI or VI_E — accurate for values 100 or below
VI_E — the only accurate reference for values above 105.

VI_E is determined as follows:

$$\nu_{210}{}^{N} = \frac{H}{U}$$

$$VI_E = \frac{(\text{antilog}, N) - 1}{0.0075} + 100$$

where

$$N = \frac{\log H - \log U}{\log \nu_{210}}$$

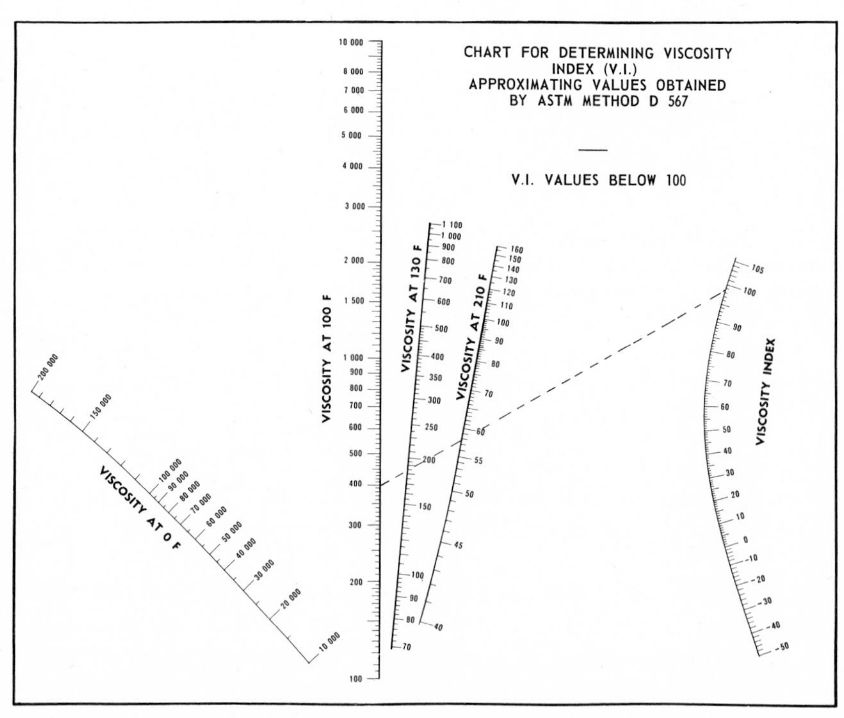

Fig. 8-5. Nomograph for determining viscosity index when Saybolt universal viscosity at two standard temperatures is known. In the example, an oil has a viscosity of 400 SUS at 100°F and 58 SUS at 210°F. A straight line through these points on the appropriate scales and extended to the viscosity index scale indicates a VI of approximately 98.

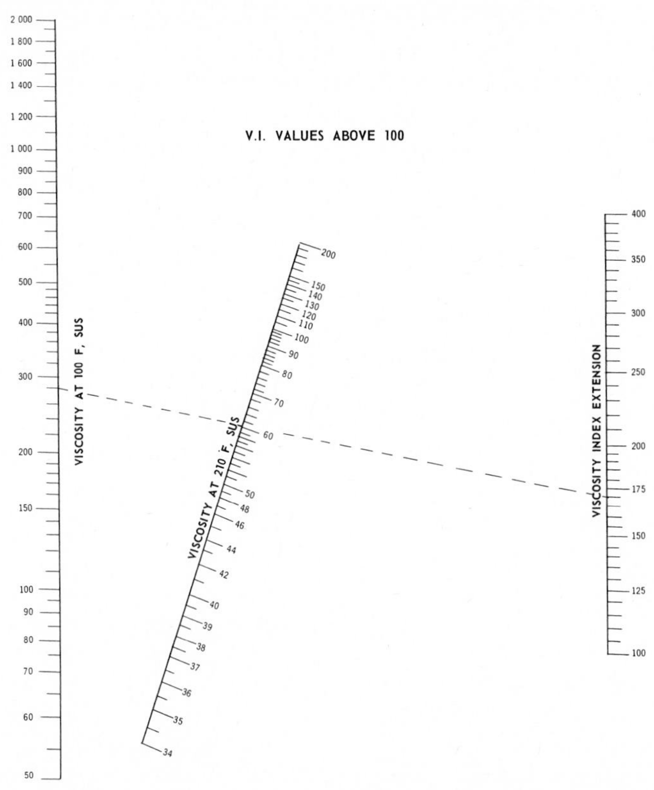

Fig. 8-6. Nomograph for determining viscosity index when Saybolt universal viscosity at two standard temperatures is known. In the example shown, an oil has a viscosity of 280 SUS at 100°F and 60 SUS at 210°F. A straight line through these points on the appropriate scales and extended to the viscosity index scale indicates a VI_E of approximately 170.

H = kinematic viscosity at 100°F of the 100 VI oil H referred to in the Dean and Davis formula, i.e., having the same viscosity at 210°F as the unknown (test) oil.

U = kinematic viscosity at 100°F of the unknown (test) oil

ν_{210} = kinematic viscosity at 210°F of the unknown (test) oil

The kinematic values of H can be converted from the Saybolt value given in Table 8-4.

A convenient method of determining the VI for any oil for which the SUS viscosities at 100°F, 130°F, or 210°F are known is by the use of the charts in Figs. 8-5 and 8-6.

Paraffinic oils lose their viscosity less rapidly with increased temperature than naphthenic oils and subsequently have a higher VI generally. However, with newer methods of refining (solvent treating) and the use of certain type

Table 8-7. SAE NUMBERS OF CRANKCASE OILS

SAE Viscosity Number	*Viscosity Range, Saybolt Universal Seconds*			
	0° F, min.	*0° F, max.*	*210° F, min.*	*210° F, max.*
5W		4,000	—	—
10W	6,000*	Less than 12,000	—	—
20W	12,000†	48,000	—	—
20	—	—	45	Less than 58
30	—	—	58	Less than 70
40	—	—	70	Less than 85
50	—	—	85	110

* Minimum viscosity at 0°F can be waived provided that viscosity is not below 40 SUS at 210°F.

† Minimum viscosity at 0°F can be waived provided that viscosity is not below 45 SUS at 210°F.

additives, the viscosity-temperature relationship of naphthenic oils has improved. The viscosity index of an oil is a valuable specification in predicting the lubricating-film-forming characteristics of an oil when exposed to elevated temperatures.

Viscosity Relation to SAE Numbers

A general designation of lubricating oil viscosities was devised by the Society of Automotive Engineers to simplify recommendations for automobile frictional components. SAE numbers classify lubricating oils on the basis of viscosity only and do not represent any other qualities or characteristics. Each number represents a range of viscosity at a given temperature. For the lower viscosity oils, the range represents Saybolt Universal Seconds at 0°F. For higher viscosity oils, 210°F readings are designated. Tables 8-7 and 8-8 show the classification of lubricating oils according to SAE numbers. Engine oils are represented by the SAE numbers 5W to 50, inclusive.

Table 8-8. SAE NUMBERS OF TRANSMISSION AND AXLE LUBRICANTS

SAE Viscosity Number	*Viscosity Range, Saybolt Universal Seconds*			
	0° F, min.	*0° F, max.*	*210° F, min.*	*210° F, max.*
75	—	15,000	—	—
80	15,000*	100,000	—	—
90	—	—	75	120†
140	—	—	120	200
250	—	—	200	—

* Minimum viscosity at 0°F can be waived provided viscosity is not below 48 SUS at 210°F.

† Maximum viscosity at 210°F can be waived provided viscosity is not greater than 750,000 SUS at 0°F (extrapolated).

The viscosity of gear lubricants in this classification for use in transmissions and drive axles should not be less than 40 SUS at 210°F.

APPARENT VISCOSITY

The term viscosity, as previously discussed, applies to liquids that obey the Newtonian definition, i.e., a fluid whose shear stress is directly proportional to the rate of shear. This means that a Newtonian fluid, such as a mineral oil whose temperature is above its cloud point, will start flowing simultaneously with any force applied to its mass, including gravity. The viscosity of Newtonian fluids, therefore, is independent of the rate of shear at any one temperature. This would not be true if the same oil was thickened sufficiently by a polymer or by soap (Fig. 8-7). An excellent example of a product that does not obey the Newtonian principle is ketchup that requires vigorous agitation, or shock, before it will flow from a bottle. Because the ketchup requires an external force to cause it to flow, it is *non-Newtonian* in character.

NON-NEWTONIAN SUBSTANCES. Rheology is the science involving the flow of substances. It deals with shear rates, shear stresses, and yield stresses of fluids. It also treats of the ultimate reaction of some materials during and after their subjection to shear, or "working." Such reactions are particularly evident in non-Newtonian substances of different kinds, each having their individual behavior patterns in respect to shear. Non-Newtonian substances may be divided into types according to their behavior under conditions of shear and its aftereffects. The three types closely associated with lubricating greases containing mineral oil of some kind are of interest here, namely, *plastic, thixotropic,* and *rheopectic substances.*

Plastic substances include most lubricating greases. Such materials are reduced in viscosity with higher shear rates. When at rest, plastic non-Newtonian substances will remain so until some measure of shear stress is

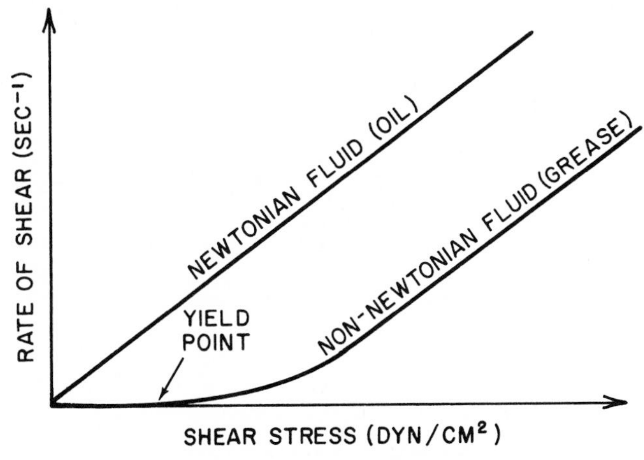

Fig. 8-7. Rate of shear-shear stress relationship of Newtonian and non-Newtonian fluids showing flow properties of oil vs grease.

applied to its mass to cause flow. The stress magnitude required to cause flow or shear is identified as the *yield stress,* or *yield point,* of a given grease. This shear is accomplished only upon an accumulation of force or a build-up of potential energy whose magnitude depends on the structural characteristics of the plastic material. The relationship of viscosity, shear stress, and rate of shear of plastic non-Newtonian substances is discussed in the next section, "Apparent Viscosity Determinations."

Thixotropic substances, like plastic substances, also include lubricating greases of the mineral oil type. Greases that are soap-thickened mineral oil soften, i.e., they are reduced in *consistency* due to structural breakdown when subject to a shearing action. For this reason, the numerical designation representing grease consistency is usually reported for *worked penetration,* in contrast to *unworked penetration*. This reduction in consistency continues toward lower levels with prolonged shear time at a constant rate. However, the decrease in apparent viscosity occurs immediately upon stress application and continues downward as the rate of shear increases, which distinguishes the difference between consistency and apparent viscosity of plastic substances.

Thixotropy is characterized by the tendency of some non-Newtonian substances, including grease, to return to their original structural form and revert to their initial state of viscosity and consistency upon the discontinuance of the shearing process. Again the difference between consistency and viscosity changes becomes manifest. The recovery of hardness in grease is a function of time, whereas its viscosity increase is immediate upon cessation of shear. The response in both cases, however, varies with different greases, both as to hardening time and degree of structural rejuvenation. If the substance returns to its original viscosity upon cessation of shear, the property is said to be *reversible thixotropy*. If reversal is less than complete, the property is one of *irreversible thixotropy* (Fig. 8-8).

The property of thixotropy becomes immediately evident when considered in light of the advantages of grease-lubricated bearings:

1. Because of the immediate increase in apparent viscosity, leakage is prevented or retarded when motion of the journal ceases.
2. The increase in consistency that occurs during the bearing rest period provides a state of hardness approaching unworked conditions when the journal is again put into motion.
3. That portion of the grease forced out of the races of an antifriction bearing tends to harden and provide a seal, while the remaining portion continues to lubricate properly at a reduced viscosity.

Rheopectic substances retain a higher consistency even after a period of shearing and resting. Unlike thixotropic substances, however, they increase in viscosity in time to some maximum value upon shear. While greases

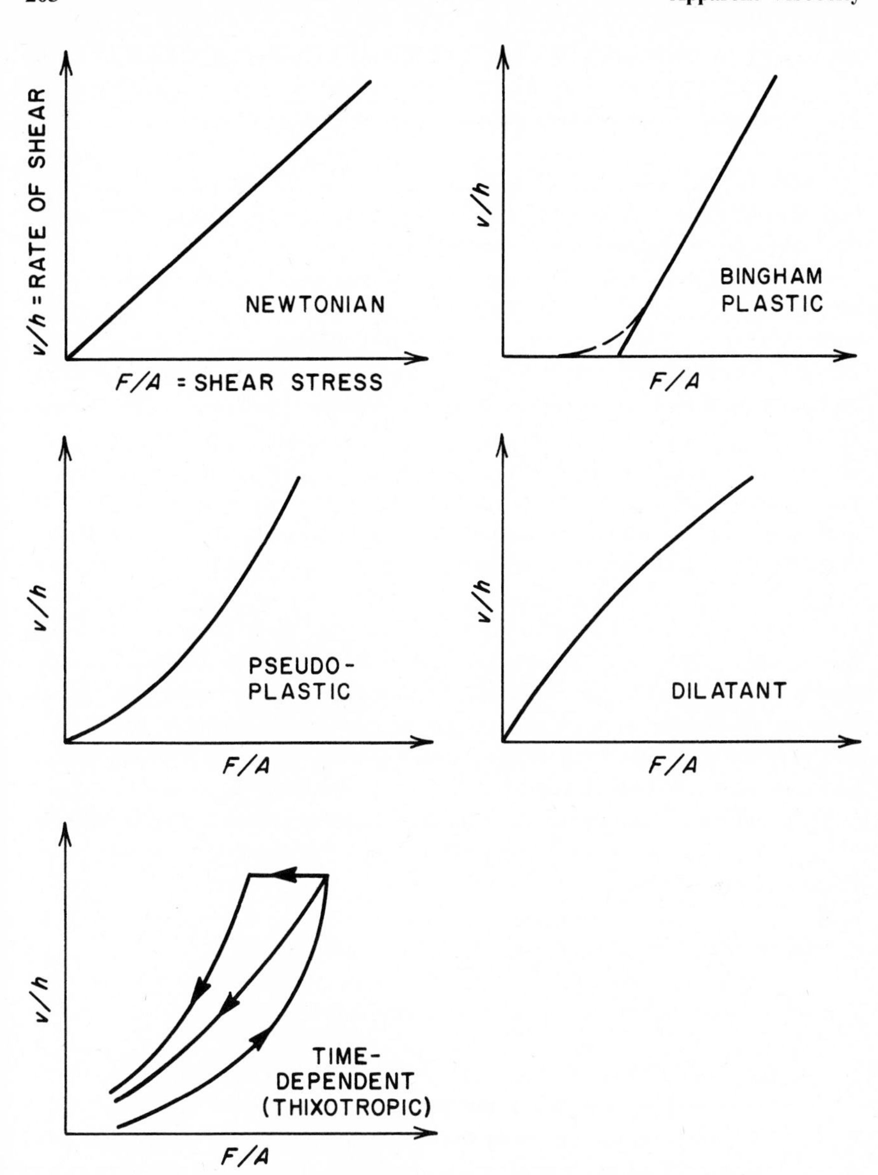

Fig. 8-8. Flow curves of fluids, showing rheological behavior of Newtonian and various types of non-Newtonian substances.

never attain full rheopectic properties, their behavior as described above provides certain advantages:

1. Application at a lower viscosity facilitates flow, especially through extended feed lines leading to bearings.

2. Rheopectic substances retain their place in a bearing after an increase in apparent viscosity above that of application due to the shearing effect of the journal.

Other non-Newtonian substances associated with lubricants in addition to grease are polymer-thickened oils, solidified oils by reason of pressure or low temperature, and oils whose temperature is below their cloud point.

Unlike Newtonian fluids, then, the viscosity of non-Newtonian fluids is dependent on the rate of shear. The flow property of non-Newtonian substances, such as lubricating grease, is called *apparent viscosity,* which is the ratio of shear stress to shear rate. Apparent viscosity serves to characterize the flow capacity of lubricating greases.

Apparent Viscosity Determinations

Apparent viscosity is determined according to Poiseuille's formula and the unit of measurement is the *poise* (P).

Because the viscosity of grease depends on the rate of shear acting on it, it follows that any given grease possesses a multiplicity of viscosities, each of which corresponds to the shear rate at any point in a series of shear rate measurements at constant temperature. *Therefore, the apparent viscosity of a grease has meaning only when expressed in absolute units as a measure of flow resistance at a given rate of shear.* Apparent viscosity becomes less as the rate of shear increases.

The differences between the viscosity of a Newtonian fluid (oil) and the apparent viscosity of a non-Newtonian substance (grease) according to their definitions given above are—

(a) The viscosity of an oil remains constant at any rate at which it is forced to flow.
(b) The apparent viscosity of a grease varies according to the rate at which it is forced to flow, i.e., the resistance of a grease to flow diminishes as the pressure causing flow increases.
(c) Oil viscosity is measured in terms of efflux time in seconds, which can be converted to stokes or centistokes.
(d) Apparent viscosity is the measure of shear stress (pressure) in poises at a certain shear rate given in reciprocal seconds, or more simply, in fractions of seconds (Table 8-9).

A similarity in viscosity and apparent viscosity determinations is the specification of temperature at which their tests are conducted. The temperatures mostly associated with apparent viscosity are −100°F, −65°F, −40°F, 0°F, 77°F, 100°F, 130°F, and 210°F, or at any temperature at which a given grease is to be employed. Both oils and greases become more viscous with any reduction in temperature.

Table 8-9. EFFECT OF RATE OF SHEAR ON APPARENT VISCOSITY OF GREASES

Rate of Shear (sec^{-1})	*Apparent Viscosity in Poises* (*at 77° F*)	*Apparent Viscosity, Approx.* (*SUS at 77° F*)
0.1	10,000	5,000,000
10.0	500	250,000
1,000.0	12	6,000
10,000.0	6	3,000
100,000.0	4.5	2,250

Reference: M. H. Arveson, "Flow of Petroleum Lubricating Grease—1," *Industrial and Engineering Chemistry,* Vol. 24, No. 1, January 1932.

APPARATUS. A pressure viscometer is used to determine the apparent viscosity of grease. The basic components of the instrument are—

1. A power system made up of an electric motor, speed reducer (200:1 reduction), and a variable-speed unit consisting of a 42-tooth gear that can be mated with either a 40-tooth gear or 64-tooth gear.
2. Hydraulic gear consisting of a fixed-displacement (constant-volume) zenith gear pump, driven by the variable-speed gear unit, piping, and attached alternate pressure gauges with ranges of 0–600 psi and 0–4000 psi, safety head, and valve for return of the oil to the pump reservoir. Oil with a viscosity of 2000 cSt at the test temperature is used in the hydraulic system.
3. A grease system composed of a piston-cylinder arrangement into which the hydraulic oil enters through the top above the floating piston, and into which the grease sample to be tested (at least ½ lb) is installed below the piston. Exit of the grease is arranged through a capillary tube located in the bottom bulkhead of the cylinder.
4. A controlled-temperature bath (optional) into which the hydraulic and grease systems are immersed. In place of a bath, the apparatus may be installed in a cabinet or a test room in which the appropriate temperature level can be maintained.

PROCEDURE OF TEST. The cylinder is filled with the test grease, assuring absence of air. The hydraulic system is filled with oil, including the gauges, gauge piping, and the upper end of the cylinder. Air from the hydraulic system is removed by circulating the oil via return valve. The return valve is closed until equilibrium pressure is reached, using the 40-tooth gear. Under equilibrium pressure, which is recorded, the piston in the cylinder starts to move, forcing grease out through the No. 1 capillary at a constant flow rate, during which the temperature drop and pressure drop across the capillary are measured. The system is then switched to the 64-tooth gear to change the flow, and the new equilibrium pressure is observed and recorded. The pressure is relieved, the No. 1 capillary is replaced by the No. 2 capillary, and the operation is repeated. By varying the gear ratios that drive the pump and the dimensions of the capillary, apparent viscosities can be obtained at shear rates from 10 to 15,000 reciprocal seconds. The complete test comprises eight different capillaries subjected to the flow rates of both the 40- and 64-tooth gears, totaling 16 pressure recordings. The entire procedural details are described in the American Society for Testing Materials (ASTM), Designation: D1092-62 (reapproved 1968), "Standard Method of Test for Apparent Viscosity of Lubricating Greases."

CALCULATION. The grease is forced out through the capillary tube at a constant flow rate. The back pressure or equilibrium pressure in the cylinder represents the viscosity of the grease. The pressure in pounds per square inch must be converted from dynes per square centimeter. The flow rate is the displacement of the fixed displacement pump in cubic centimeters per second. Substituting the above parameters, including flow rate and pressure drop, and inserting other known values, such as capillary radius and length, into the Poiseuille equation, the apparent viscosity of the test grease can be calculated.

By definition, apparent viscosity (η) is

$$\eta = \frac{\text{shear stress in dyn/cm}^2}{\text{shear rate in reciprocal seconds}} = \frac{F}{S} \qquad \text{(poises, P)}$$

Tests are conducted with gauges calibrated in pounds per square inch, while viscosity is reported in poises, so that Poiseuille's equation must be modified accordingly by using the conversion factor 68,944, which is

$$\frac{\text{444,688 dyn (equivalent to 1 lb force)}}{6.45\ \text{cm}^2/\text{in.}^2}$$

or, to expand the scope of this factor,

$$\text{number of poises} = 68{,}944 \times \text{number of reyns}$$

$$\begin{aligned}\text{number of reyns} &= \frac{1}{68{,}944} \times \text{number of poises} \\ &= 1.45 \times 10^{-5} \times \text{number of poises} \\ &= 1.45 \times 10^{-7} \times \text{number of centipoises}\end{aligned}$$

Applying the conversion factor to the Poiseuille equation, we get

$$\eta = \frac{F}{S}$$

where

$$F = \frac{P\pi r^2}{2\pi r \ell} = p\pi r^4$$

$$S = \frac{4(v/t)}{\pi r^2} = 8\ell(v/t)$$

therefore

$$\frac{F}{S} = \frac{P\pi r^4}{8\ell(v/t)}$$

or

$$\eta = \frac{F}{S} = \frac{P\pi r^4}{8\ell(v/t)} = \frac{P\pi r^4}{8\ell(v/t)} = 68{,}944$$

where

η = apparent viscosity (P)
F = shear stress (force in dynes)
S = rate of shear (reciprocal seconds, 1/sec)
p = pressure observed (dyn/cm^2)
P = pressure observed (psi)
ℓ = length of capillary
r = radius of capillary
v/t = flow rate (cm^3/sec)

The period of test during which the pressure rises from zero to equilibrium pressure represents a stage of flow resistance of the grease sample. At the moment that equilibrium pressure occurs, i.e., when pressure ceases to rise, the grease starts to flow through the capillary, the level of which is known as the *yield stress* of the grease tested.

Practical Value of Apparent Viscosity

By determining apparent viscosity at different rates of shear, a logarithmic graph, such as shown in Fig. 8-9, can be constructed. From such

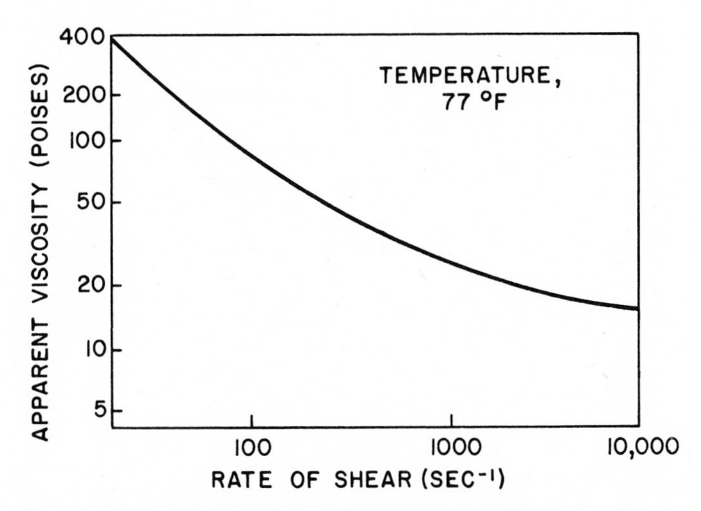

Fig. 8-9. Apparent viscosity vs rate of shear (hypothetical).

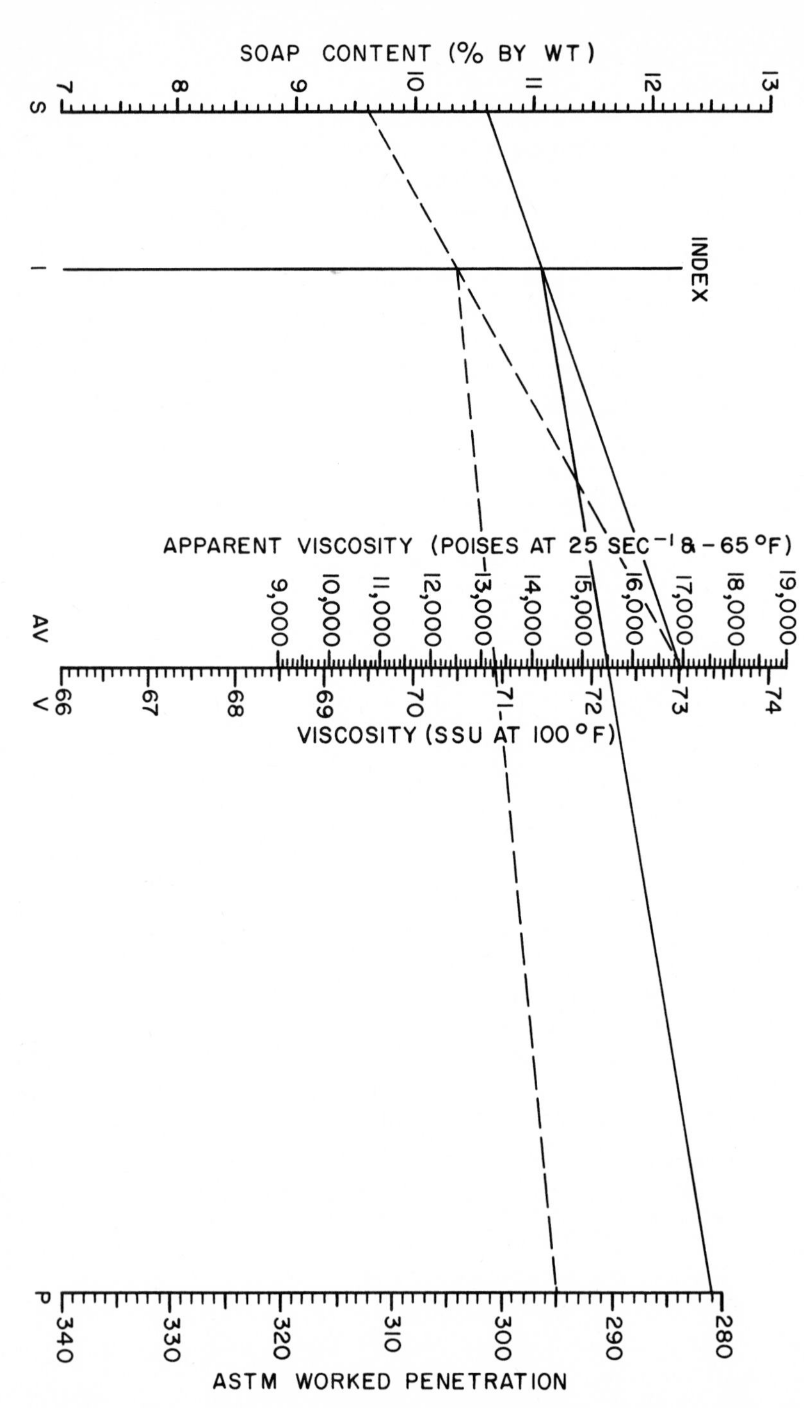

Fig. 8-10. Nomograph for calculation of apparent viscosity. Directions: (1) Connect points on S and V. (2) Connect intersection on 1 with point on P. (3) Read apparent viscosity on AV. *Courtesy, Institute Spokesman.*

a graph the apparent viscosity of a given grease can be determined at any rate of shear.

The apparent viscosity of a grease serves as an indicator of *pumpability*. Greases having lower apparent viscosity are caused to flow with less applied force than greases of high apparent viscosity. Apparent viscosity determinations can also be used to predict the lowest temperature at which a lubricating grease can be pumped and dispensed. It will also indicate the tendency of a grease to leak or remain in a bearing at a given temperature.

Because apparent viscosity is affected by the mineral oil specifications and by the amount and nature of the soap used in the manufacture of grease, it becomes a criterion in the selection of such ingredients in obtaining products possessing desired properties. The combinations of such ingredients are numerous, making it impossible to demonstrate a typical set of apparent viscosity values, representing individual grease categories.

The dependency of apparent viscosity of greases on the soap content, viscosity of the mineral oil, and worked penetration has been developed statistically. One such relationship developed is demonstrated by the following equation:

$$\text{apparent viscosity} = \text{AV}^* = 6200 - 85\text{P} + 1020S + 305V$$

where

AV = apparent viscosity (P, at 25 sec^{-1} and $-65°\text{F}$)
P = ASTM penetration
S = soap content (percentage by weight)
V = viscosity of mineral oil (SUS at 100°F)

A nomograph based on the equation above is shown in Fig. 8-10, by which the apparent viscosity of a grease can be conveniently determined. The confidence limits of correlation by the use of the equation range from within 930 P of predicted values for 50% of the test determinations to 2710 *P* for 95% of determinations.

The nomograph (Fig. 8-10) can be used to determine the relative magnitude of the three variables required to meet a given apparent viscosity limitation (within the scope of this particular nomograph). Such limitations may be demanded by the capability of a pump to handle a grease only when the apparent viscosity is below a given value. A grease having an apparent viscosity above the pump limitations can be modified by the addition of more oil to the mixture combination in an amount determined from the nomograph. While the equation and nomograph are based on certain ingredients (were established with a lithium-calcium base grease) and variable combinations, the system is suggested as an approach to the possibility of predicting apparent viscosity on a quantitative basis as compared with the qualitative methods so widely used.

* *NLGI Spokesman*, January 1955, etc.
Note: In above discussion, P = ASTM penetration; *P* = poises

9

THEORY OF LUBRICATION

Lubrication is the science of reducing friction between two solid bodies in relative motion by interposing a lubricant between their rubbing surfaces. It is not an exact science, in that a precise solution cannot be determined for any given practical problem involving friction, except those possessing the more simple parameter combinations. The reason for this lies in the many variables and the complexities of many of those variables associated with the design, operation, and ambient operating conditions of practical friction systems. Design factors often remain constant, but operation conditions usually vary, and ambient conditions are the least constant of all factors. In fact, the latter may be subject to change from one moment to the next.

Through scientific experiments, mathematical analysis, and experience, and even by accident, the lubrication art is permeated with a myriad of conclusions based on theory, hypothesis, assumption, empiricism, contradiction, and ambiguity. Out of all this some semblance of rationale has emerged to furnish fairly excellent guides in solving practical frictional problems. Bearing designers and machinery builders are often responsible for the mechanics of frictional systems, especially as they relate to proprietary equipment. The necessary speeds, loads, and other dimensional requirements are governed by the ultimate use of the frictional components. Position, location, and other geometrical considerations often limit the methods of lubricant application, as well as the type of lubricant that can be applied. Frictional systems are all too often designed and installed without regard for the ambient conditions to which they will be exposed, which include high temperature, moisture, and dust. The deleterious results of such conditions become pronounced and in many cases disastrous, when critical bearings, chains, and gears are installed in remote locations where their accessibility is difficult, and impossible during operation of their host machine.

Despite much improvement in recent years, it is still occasionally left to the ingenuity of the field or plant lubrication engineer to devise means of maintaining ill-applied and ill-designed frictional systems in serviceable condition. Where all other requirements are substantially satisfied as to design, location, lubricant application, and protection, it becomes necessary for those responsible for system operation only to select and properly apply the correct lubricant and arrange for its care and handling.

Mindful of the foregoing discussion, then, the practical lubrication engineer should consider with caution any solution derived theoretically as demonstrated in the following text and resort to such compromises as experience and logic directs. Caution should also accompany any attempt to build into lubrication solutions safety factors of such magnitude as to prove disappointing. The old adage, "When in doubt, use 600W," does not always hold true.

LUBRICANT PHENOMENON

The purpose of practical lubrication is to combat or avoid solid sliding friction. Where the art of lubrication is successfully administered, the following primary results are obtained:

1. Wear is retarded.
2. Temperature rise is minimized.
3. Power requirement is decreased.

Secondary results include those leading to economic benefits:

(a) Extended useful life of frictional components.
(b) Reduction in unscheduled machine downtime, resulting in increased production.
(c) Lower production cost as a result of (1) continuity of machine operation, (2) lower maintenance labor cost, and (3) reduced cost of replacement parts.
(d) Decreased power consumption.

Protection of rubbing surfaces against wear depends on the type of lubricating film applied to the areas in contact with each other. Surface conditions in this regard may be classified according to the four basic kinds described below:

1. *Perfectly clean surface condition.* Surfaces free of any lubricant or

contaminant are referred to as clean, or dry. This condition develops the greatest frictional resistance to motion.

2. *Boundary condition.* Surfaces are slightly contaminated, such as by an adsorbed film (oxides, vapors, etc.) or are covered with a very thin film of lubricant. The film thickness is considerably less than the dimensions of surface asperities and is often referred to as a thin film condition. This condition approaches or borders on a clean, dry surface effect. The most effective boundary condition, used to combat heavy loading, is described as *extreme pressure lubrication.*

3. *Mixed-film condition.* This is intermediate between boundary and thick-film conditions, the latter being capable of completely separating the opposing bodies intermittently or simultaneously over different areas of the sliding surfaces. The mixed-film condition is also called quasi-hydrodynamic.

4. *Thick-film condition.* When a film of lubricant is thick enough to maintain complete separation of the opposing surfaces, the conditions of dry surface, boundary, and mixed film are eliminated. When the opposing surfaces are inclined relative to each other, or nonparallel, the thick-film condition develops a hydrodynamic converging wedge that forces the surfaces apart during relative motion. Taking its name from this phenomenon, this condition is identified as *hydrodynamic lubrication,* the most important of all frictional systems.

Another form of lubrication associated mainly with rolling friction is called *elastohydrodynamic lubrication.* Elastohydrodynamic films involve the combined effects of elastic deformation of metal surfaces and the pressure-viscosity relationship of lubricants. Experiments have developed an elastohydrodynamic lubricant film parameter for rating the effectiveness of lubrication in ball and roller bearing applications, as demonstrated in the text on rolling friction.

HYDRODYNAMIC LUBRICATION

The prinicple of hydrodynamic lubrication is based on the formation of a thick fluid film of characteristic geometrical profile that develops automatically between opposing nonparallel solid surfaces having relative motion to each other. The characteristic geometrical profile resembles a converging wedge having maximum thickness at its leading edge and minimum thickness at its trailing edge. Between the two edges exists the pressure zone of the two surfaces as a result of external loading.

The leading edge of the hydrodynamic or converging wedge represents the point of lowest pressure within the film, while the trailing edge represents the point of maximum pressure. In between, the pressure increases from the leading edge to the trailing edge, the pattern of which describes the *pressure gradient.* The magnitude of the pressure gradient varies according to the amount of force exerted by the load, which tends to press the opposing surfaces together in metal-to-metal contact.

As with all wedges, the fluid, or hydrodynamic wedge, usually an oil, acts to cause separation of the opposing bodies between which it is forced by the pumping action of the moving component. It is the velocity of the moving component, such as a journal, that governs the length of the wedge along the x-axis (Fig. 9-1), or in the case of full journal bearings, the magnitude of the pressure arc. As velocity increases the wedge buildup, it causes a reduction in eccentricity, i.e., the journal and bearing approach more nearly a concentric relationship.

The bodies referred to above include those components that transmit or carry the machine load while in relative motion. They include journal bearings, rider and ways, mating gear teeth, cams and followers, and piston-cylinder arrangements.

The formation of the hydrodynamic wedge depends on the geometrical relationship of the opposing surfaces. Some inclination relative to each other must exist in the case of translating motion on flat surfaces, such as thrust bearings. Rotating motion as in journal bearings requires a slightly larger bearing diameter than that of its journal to provide the necessary eccentricity to form the hydrodynamic wedge. While methods differ in achieving proper geometry, they all tend to develop a converging film profile whose thickness gradient conforms to that of the wedge (Fig. 9-1).

The magnitude of the minimum thickness of film which governs the load-carrying capacity of the hydrodynamic arc is a function of the three important parameters mentioned above, which are (a) relative velocity of opposing surfaces, (b) load per unit area, and (c) viscosity of the lubricant at the operating temperature. It will be demonstrated later in the text how the load-carrying characteristic of any bearing is a function of the viscosity-velocity-load relationship.

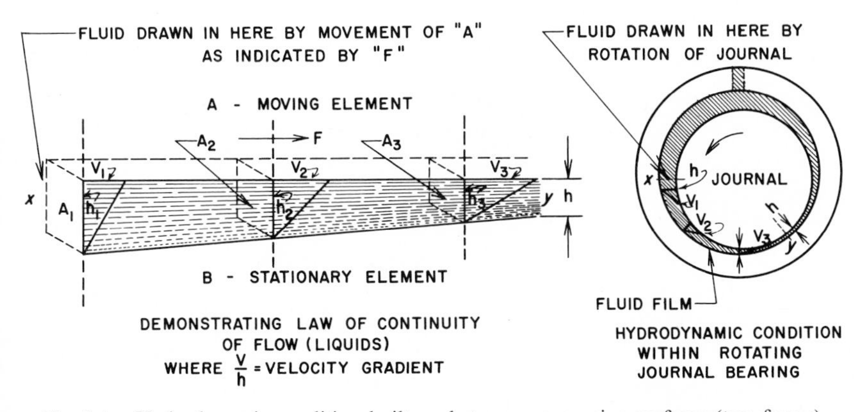

Fig. 9-1. Hydrodynamic condition built up between converging surfaces (two forms).

Another important factor associated with the development of the hydrodynamic wedge is a continuous and adequate supply of lubricant to the critical bearing area during motion.

Hydrodynamic theory evolved as a result of a series of physical experiments and classic mathematical analyses chronologically performed by a number of individuals. The more important individuals involved and their major contribution to the science of lubrication are as follows:

Sir Isaac Newton. Viscosity of fluids (1687).

Beauchamp Tower. Existence of pressure, and pressure patterns of a lubricating oil film in journal bearings (1883).

Arnold Sommerfeld. Extended Reynolds' theory for application to journal bearings of infinite length and all values of eccentricity (clearance) as expressed by Sommerfeld's variable, $(\mu N/P)(r/C_r)$.

F. W. Ocvirk. Extended Sommerfeld's variable to allow end-leakage.

Osborne Reynolds. Developed physics associated with behavior of fluids during flow.

HYDRODYNAMIC PARAMETERS

The following discussion involves the functions of various parameters as they relate to hydrodynamic lubrication. The parameters and their identifying symbols are listed below:

D = diameter of bearing (in.)
R = radius of bearing (in.)
r = radius of journal (in.)
B = breadth of bearing (in.)
L = length of bearing (in.)
A = projected area of bearing, $L \times D$ (in.^{2})
C_d = diametral clearance (in.)
C_r = radial clearance, $C_d/2$ (in.)
L/D = ratio, length to diameter
e = eccentricity, distance 0 to 0_e (center of bearing to center of shaft during rotation)
ε = eccentricity ratio = e/C_r
W = total load (lb). Also power (hp or ft-lb)
P = unit load (psi)
P_L = unit load per inch of bearing length
N = journal speed (rpm)
n = journal speed (surface ft/min)
U or n_s = journal speed $\left(\text{sfps} = \dfrac{2\pi RN}{60}\right)$
T = temperature (°F)

dT = incremental change in T (°F)
H = thermal energy in transition or kinetic thermal energy (Btu)
K_h = H (Btu) per square foot per hour per degree Fahrenheit
μ = absolute viscosity (reyns or other appropriate unit)
Z = absolute viscosity (cP)
ν = kinematic viscosity (cSt)
f = coefficient of friction
ρ = density of fluid
h = thickness of film (in.)
h_{min} = minimum thickness of film (in.)
h_{max} = maximum thickness of film (in.)
π = 3.1416, or circumference of journal per unit diameter
0 = center of bearing
0_e = center of journal not concentric
m = clearance modulus = r/C_r, or D/C_d
C = dimensionless number = ZN/P, or bearing characteristic number
Re = Reynolds number
S = Summerfeld number, $\left(\frac{\mu N}{P}\right)\left(\frac{r}{C_r}\right)^2$
K = constant = 1.31 × degrees of journal circumference
k = constant based on L/D = 0.0025 for $L/D \leqq 3$
ϕ = attitude angle of moving journal center as related to bearing center
θ = angle, leading edge of hydrodynamic film in direction of rotation
α = angle, trailing edge in direction of rotation

Sir Isaac Newton in 1687 reported his discovery that *the shearing stress of a fluid is directly proportional to the rate of shear at* any fixed temperature. Lubricants that flow according to Newton's Law are described as Newtonian fluids. Those that do not (grease generally) are considered non-Newtonian.

Newton's conclusion as stated above defines the fluid property of *viscosity*. Generally defined, viscosity expresses the frictional resistance that opposes the flow of molecules relative to each other comprising the fluid mass. Viscosity is the measure of flow rate of any given fluid. It is the most important property of lubricating fluids and one of the fundamental parameters underlying hydrodynamic phenomena. Newton's development of the laws of viscosity is discussed in more detail in Chapter 8.

Non-Newtonian substances include the following types:

1. *Plastics.* Capable of being molded, such as putty and grease.
2. *Dilatant.* Increasing in *apparent viscosity* with prolonged agitation, i.e., as rate of shear increases. Pigmented materials, such as paints, and some starches are examples of dilatant substances.

3. *Thixotropic.* Decreasing in apparent viscosity with prolonged constant shear. Quicksand, and a special mud-like clay called rotary fluid used in deep oil well drilling operations, are thixotropic. Some thixotropic materials tend to return to their original state when agitation ceases (reversible thixotropy). Some greases are thixotropic; others, such as a mineral oil whose temperature is below the congealing level, will return to an apparent viscosity less than its original value upon cessation of agitation (irreversible thixotropy).

4. *Rheopectic.* Some greases in their preagitated state (fresh and never before used) can be pumped with little difficulty. Subject to shearing stresses, however, they become more *consistent,* i.e., their apparent viscosity increases to the point at which they act as an effective bearing seal. This is particularly important in antifriction bearings operating in wet or dusty environments, to augment any mechanical seal provided. Rheopectic qualities enhance the *limited cavitation* properties of some greases, such as those employed for automobile front-wheel bearing lubrication. In the latter, some particles of grease must be continually entering the frictional area at a controlled rate, an action known as *flecking*.

Newtonian substances are fluids whose viscosity remains constant at any given temperature and remains independent of shear rate. They obey Newton's Law. Petroleum oils at temperatures above their cloud point are Newtonian liquids.

Beauchamp Tower in 1883 accidentally discovered that a pressure developed within a lubricating oil of a journal bearing during the rotation of the journal. His experiments involved a partial bearing (about 157°) located on the upper part of the journal as employed in railroad car journal-bearing arrangements. An oil bath method of application was used during the experiment at the time of his discovery.

By the use of pressure gauges appropriately placed, he also discovered two important pressure patterns (Fig. 9-2):

1. Maximum pressure occurred at a point in the middle of the bearing which decreased *longitudinally* to zero at the ends.

2. The same pressure pattern was duplicated when measured *radially*, during which maximum pressure occurred *near* the center, tapering off to zero at the sides.

Other interesting and important observations were made during these experiments:

1. When a load having a unit-bearing pressure of 330 psi was applied, a gauge inserted in a hole at the center of the bearing registered a pressure of 625 psi, or about 1.58 times the unit load.

2. The gauge registered an increase or decrease, respectively, with any increase or decrease of load.

3. The unit pressure exerted by the load was equal to the average pressure of all pressures measured at the various locations between the center and bearing extremities.

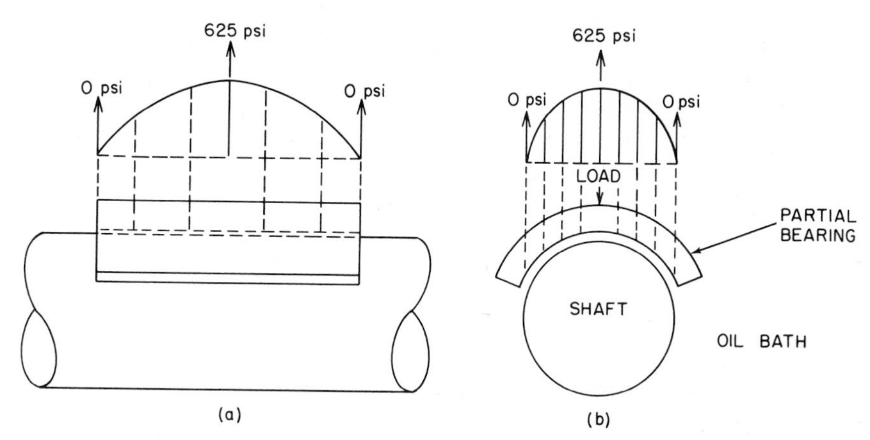

Fig. 9-2. Distribution of pressure in a partial bearing according to Tower's experiment.

4. The total pressure exerted upward by the oil film was apparently equal to the total pressure exerted downward by the load according to 3 above and by actual measurement.

In 1886, Osborne Reynolds demonstrated mathematically that the pressure pattern developed in Tower's journal-bearing experiments was the result of hydrodynamic action. His analysis established the capability of an adequately thick oil film of an incompressible nature to maintain separation of inclined flat surfaces having relative motion under load. While Tower established the presence of the phenomenon, Reynolds analyzed it and thus founded the principles underlying hydrodynamic lubrication.

The conditions conducive to hydrodynamic action can be described as those of a pump continuously primed by a copious supply of oil fed in an appropriate manner, location, and direction between opposing surfaces. The lubricating fluid is drawn into a converging stream under relatively low pressure between mechanically elastic surfaces by the pumping action of the moving member (e.g., rotating journal), raised to a higher pressure at a point of minimum thickness (h_{min}), and discharged into the low-pressure region in the direction of motion. Under ideal conditions, the fluid maintains nonturbulent or *laminar* flow throughout the process.

Three basic directions are involved with the hydrodynamic phenomenon as related to practical bearing operation:

1. Direction of flow between and parallel to the opposing surfaces. For purposes of coordination, this is indicated as the x-axis.

2. Direction of force generally perpendicular to the axis of flow, indicated as the y-axis.

3. Direction of constant fluid film profile as measured along the axis of rotation (in the case of a journal bearing), indicated as the z-axis.

The critical factors in establishing and maintaining a hydrodynamic film are dependent on (a) the load-carrying ability of the fluid based on thickness of film at the point of maximum pressure, (b) maintaining continuous laminar flow through the pressure area in the direction of flow, and (c) effects of end-leakage of oil, i.e., out the ends of the bearing.

Reynolds' equation is based on either two or three dimensions and is complex, to say the least. However, fairly accurate interpretation representing general solutions for practical bearings is demonstrated in the following section.

HYDRODYNAMIC THEORY

Hydrodynamic lubrication is variously called fluid-film, thick-film, or flooded lubrication. A hydrodynamic condition is obtained by interposing a sufficiently thick film of lubricant between opposing surfaces of solid bodies in relative motion. A sufficiently thick film is one that will separate the solid surfaces and prevent contact between their asperities (roughness projections). When one body moves over the other under hydrodynamic conditions, fluid friction is substituted for sliding friction. Because fluid friction develops a lower coefficient of friction than does sliding friction, a general decrease in friction between the two solid bodies occurs. The amount of decrease occurring depends on the rate at which the molecules of the lubricant are capable of sliding over each other. The disposition of the molecular layers of a fluid to slide over each other may be described as its *rate of shear*. Fluidity and inherent lubricity (slipperiness) of lubricants govern their relative rate of shear.

The hydrodynamic theory may be demonstrated by describing the action taking place in Fig. 9-1 (*left*). Body A having a weight W moves by a force F over the stationary body B. The space filled with broken lines between and running parallel with the body represents a lubricant film having some definite thickness. The film separates the opposing surfaces. Each broken line represents a molecular layer of fluid. Differences in length of the dash marks indicate the velocity of each layer relative to its companion layer. In the illustration the lines grow shorter with diminishing velocity. The velocity of the fluid layer contiguous to the moving surface of A is equal to the velocity of that surface. The velocity of each succeeding layer diminishes until a zero velocity is reached in that layer contiguous to the stationary surface of B. Thus, fluid layer velocity ranges from maximum, at the speed of the moving or fastest moving member, to minimum, corresponding to the speed of the slower moving member. In the case of one stationary member, as it is here, minimum velocity equals zero.

Because relative velocity exists between companion layers of fluid as shown, a multiplicity of sliding or shearing actions occur throughout the fluid mass. The rate of shear between each layer is a function of the speed of A, assuming that all other factors remain the same. The speed of A as a *free body** depends on the magnitude of fluid friction within the film mass. Stated differently, the speed of A as a free body is governed by the disposition of companion fluid layers to slide over each other. As a *captive* body, the speed of A will depend on the amount of power applied continuously parallel to motion. With fluids of high internal friction, a greater amount of power would be required by a captive body than would be necessary using fluids of low internal friction.

By interposing a fluid film between the surfaces of a bearing and its rotating journal, as shown in Fig. 9-1 (*right*), the same hydrodynamic condition may be established as described above.

Flat Surfaces, Parallel

A lubricant film of h thickness separating a moving body A having a force F across a stationary body B is shown in Fig. 9-3 (*left*). The surfaces of bodies A and B are parallel. Assume that both surfaces are of indefinite width (perpendicular to the page) so that no side-leakage of lubricant occurs. Under such conditions, (a) all the lubricant entering the space h at the leading end x will leave at the trailing end y during motion of A; (b) the velocity of the lubricant layer contiguous to the stationary surface is zero, and the layer contiguous to the moving member has a velocity equal to the latter (assuming no slippage). With the same quantity of oil flowing at all cross sections of h throughout the length of the surfaces A and B (x to y), and a constant relative velocity existing between each fluid layer at every point along xy, no increase in pressure is allowed to develop at any one

*A free body is one that continues to move unaided after having received an initial push by some external force.

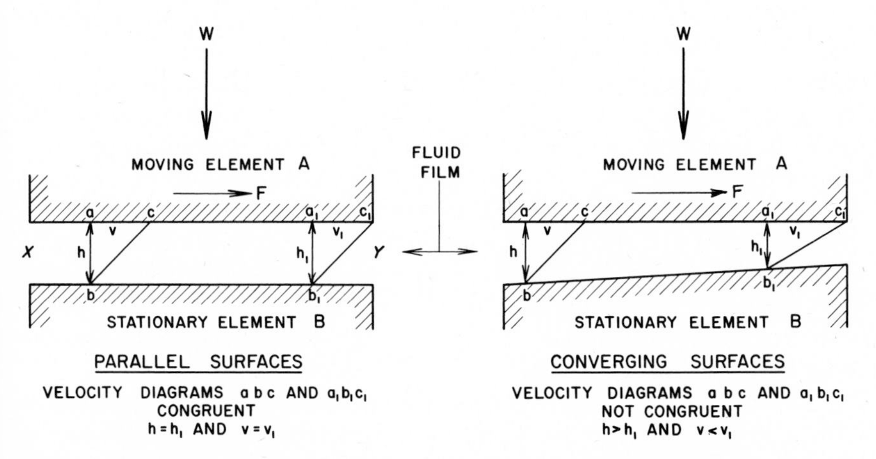

Fig. 9-3. Demonstrating the instability of a fluid film between parallel surfaces vs fluid wedge formation with converging surfaces.

point above that at any other point. Thickness h then becomes a function of load W. When W is great enough, h shrinks sufficiently in magnitude to allow the surfaces of A and B to make solid contact.

Thus, parallel surfaces do not sustain a hydrodynamic film where motion is accompanied by any appreciable loading. Guide bearings, crossheads, guides, and ways are frictional components whose surfaces are parallel to the opposing surfaces of their riding elements.

Flat Surfaces, Converging

When two opposing flat surfaces are not parallel, any lubricant interposed will develop a film having a wedge-shaped cross section, as shown in Fig. 9-3 (*right*). Assuming that the direction of the moving body A is toward the point of smallest (separation) fluid film cross section, a fluid lubricant will be drawn between the bodies at the leading end x of the wedge. The same amount of lubricant must leave at the trailing end y as enters at x. Because the area at the outlet is smaller than the inlet area, an ever-increasing velocity occurs from, and in the direction of, x to y. Thus, the relative velocity between fluid layers changes at each increment of length between x and y—i.e., with diminishing film thickness. The change in velocity is called the *velocity gradient* and is expressed by the ratio V/h, where V represents velocity and h is the film thickness at any point along xy. The load W attempts to force A more closely in contact with B and further restrict the outlet area at y. This tends to crowd the fluid, resulting in gradual pressure rise along the route xy. It is this rise in pressure that counteracts the load and maintains a fluid film thick enough to keep the bodies separated. Points x and y in Fig. 9-3 (*right*) are in the same relative positions as shown in Fig. 9-3 (*left*).

From the above it may be stated that for constant-speed, nonparallel surfaces employing the same fluid,

(a) Fluid film at the point of minimum clearance decreases in thickness as the load increases;
(b) Pressure within the fluid mass (hydrodynamic force) increases as the film thickness decreases due to load;
(c) Pressure within the fluid mass is greatest at some point approaching minimum clearance and lowest at the point of maximum clearance;
(d) Fluidity decreases, i.e., fluid becomes more viscous, as applied pressure increases;
(e) Thickness of film increases at the point of minimum clearance as load decreases.

It is also true that

(f) Any increase in relative speed increases film thickness at the point of minimum clearance, and where all other factors remain the same,

(g) Film thickness at the point of minimum clearance increases with the use of more viscous fluids.

Thrust bearings having tapered or beveled lands, pivoted thrust shoes, and tilting pads are all nonparallel, flat surface bearings. The surfaces mentioned may represent part of the stationary member, although the wedge developing component may also be the moving member. Kingsbury, Mitchell, and Gibbs thrust bearings are some of the types designed to produce an oil wedge and develop hydrodynamic lubrication.

Curved Surfaces—Full Circular Bearing

The clearance allowed between a journal and its bearing creates an *eccentric* relation between the two. Eccentricity is the distance between the center of the journal and the center of the bearing. The clearance space assumes the shape of a crescent, due to eccentricity, when the journal lies at rest, as shown (exaggerated) in Fig. 9-4 (*left*). At rest, the journal exerts its weight on the bottom surface of the bearing. If the journal is rotated, it would tend to climb up the side of the bearing much in the same manner as a wheel rolls uphill. At the moment rotation is initiated, a position would be assumed, as shown in Fig. 9-4 (*center*). Due to the weight imposed on it, however, the journal would continually slip back toward its original position.

If sufficient oil were poured continuously into the bearing through the hole in the top to keep the clearance space filled while the journal was being rotated, a condition similar to that described for nonparallel flat surfaces would develop. The crescent-shaped clearance space would act to provide the oil wedge.

Upon initiating rotation, the journal attempts to climb the side of the bearing. As it starts to climb, the journal enters upon the thin oil film at the end of the crescent-shaped oil mass. In doing so, two actions occur. First, the journal loses traction rapidly by reason of the fluid film and slips back to its original position immediately. Second, oil adheres to the surface of the journal and is pulled between it and the bearing. As speed increases, the rotating journal acts as a rotary pump, pulling more and more oil beneath it, causing its separation with the bearing surface. The journal now rides on a fluid film. At this point, a similar condition exists between journal and bearing as described for nonparallel flat surfaces. The oil, entering between the journal and the bearing, continues to build up, during which it develops hydrodynamic pressure along the side of the bearing opposite the direction of rotation. The greatest pressure occurs at the tip of the crescent (converging wedge), which is the zone of thinnest oil film. During the initial stage of rotation, this zone is located slightly to the left of the

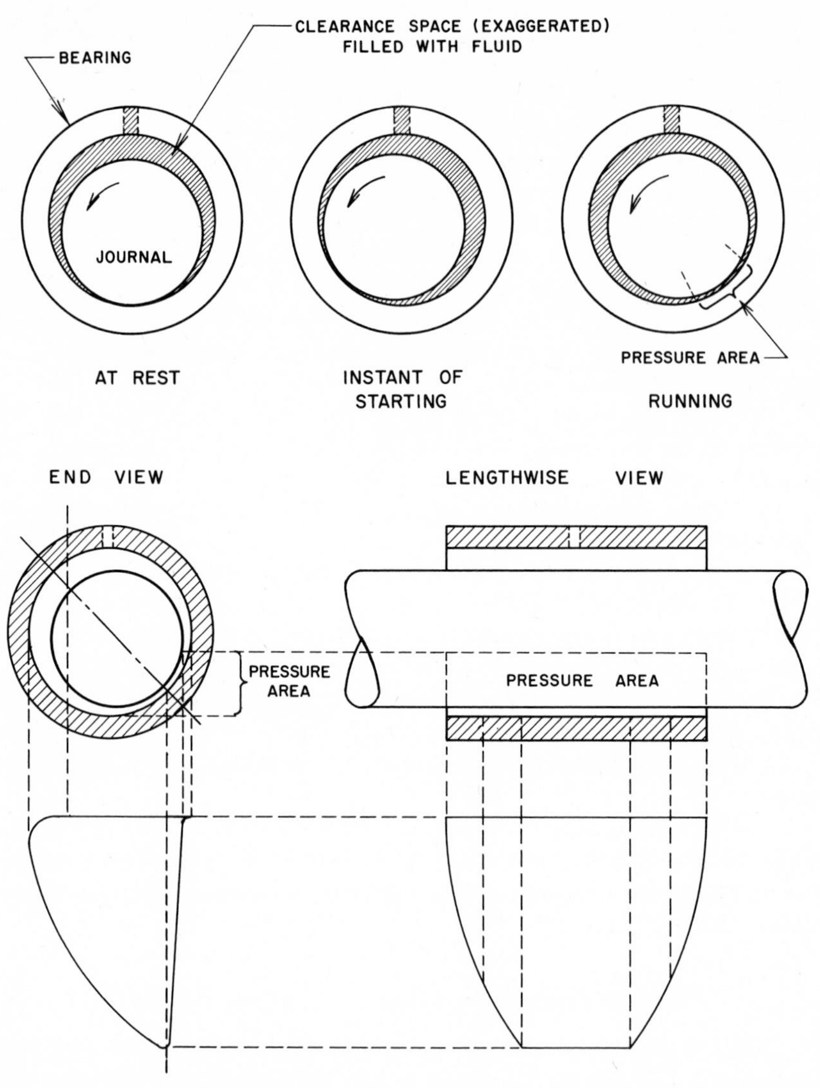

Fig. 9-4. (Top) Hydrodynamic film developed within a full journal bearing. (Bottom) Pressure distribution in a full journal bearing.

extreme bottom of the bearing, assuming that the journal is turning counterclockwise and load is vertical.

Upon development of hydrodynamic pressure along the side opposite rotation, in the same manner it developed from x to y (Fig. 9-1), the journal is forced over to the other side of the bearing. The thinnest film is now located slightly to the right of the extreme bottom of the bearing, and the crescent has shifted to its permanent running position, as shown in Fig. 9-4 (*right*). In this position, the journal continues rotation on a fluid film subject only to fluid friction, the amount of which depends on the rate of shear of the lubricant used. The actual relation of the rotating journal to its bearing, i.e., the position of the crescent, depends on load and speed. Higher speeds tend to shift the journal farther to the right (or to the left, depending on direction of rotation), while heavier loads attempt to keep it at or near the extreme bottom of the bearing. It will finally settle in a position of equilibrium governed by the resultant force of both speed and load. The zone of thinnest fluid film is the pressure area of a given journal bearing.

The dimension of the thinnest film when all other factors remain constant depends on the fluidity of the lubricant. Its relative thickness is based on the following rules:

(a) With any load the thickness of fluid film in the pressure area will increase as the viscosity of the fluid introduced into the bearing becomes greater.

(b) With a given fluid, the thickness of the film will decrease as the load becomes heavier.

(c) With a given load and fluid, the thickness of the film will increase as speed is increased.

(d) Fluid friction increases as the viscosity of the lubricant becomes greater. Conversely, fluids of high fluidity develop low fluid friction.

From the above rules, the following may be concluded for plain journal bearings.

1. For light loads and high speeds, use a lubricant of high relative fluidity.
2. For intermediate loads and reasonable speeds (within the normal range for the work system), use a lubricant of medium relative fluidity.
3. For heavy loads and slow to medium speeds, use a lubricant of low relative fluidity.

Pressure Distribution in Full Journal Bearings

The hydrodynamic forces developed between a journal and a bearing, as shown in Fig. 9-4 (*right*), follow nearly the same pattern as the forces developed between nonparallel flat surfaces. As the journal rotates, it tends to crowd oil between the two opposing surfaces in the direction of rotation. The clearance space being crescent-shaped, i.e., diminishing in a cross-sectional area, requires a higher flow velocity to develop in order to allow the same quantity of fluid to pass through it as is forced to enter it. A pressure increase develops coincident with velocity increase and reaches its highest level at the point of minimum film thickness. This rise in pressure tends to lift the journal and increase the thickness of this fluid film throughout the pressure zone or pressure area. The lubricant passes through the pressure area and escapes into a region beyond the point of minimum film thickness where the clearance space starts diverging. The pressure drops rapidly in the diverging region.

FLUID FLOW

Reynolds developed much of the physics associated with the behavior of fluids in flow. An important contribution involved the degree of flow stability of a fluid as it is forced through a conduit at different velocities. It was found that at some critical velocity any given fluid changes from streamline or *laminar flow* to *turbulent flow,* or vice versa, depending on whether the velocity is decreased or increased, respectively.

Reynolds determined criteria for predicting the critical velocity through relatively smooth circular conduits by expressing fluid behavior regarding laminar or turbulent flow as a function of fluid velocity, viscosity, and diameter of conduit. His criteria in the form of a dimensionless number, called the *Reynolds number* (Re), is determined by the relationship

$$\mathrm{Re} = \frac{\text{velocity} \times \text{diameter}}{\text{kinematic viscosity}}$$

or

$$\mathrm{Re} = \frac{V \times D}{\nu}$$

Kinematic viscosity ν being equal to the absolute viscosity μ divided by the density ρ of the fluid, the equation may be expressed by appropriate substitution, or

$$\mathrm{Re} = \frac{V \times D \times \rho}{\mu}$$

Reynolds numbers as shown below act as the criteria for the flow behavior of fluids through circular pipes:

Re less than 2000 predicts laminar flow.

Re of 2000 or greater and less than 4000 is the zone of transition from laminar to turbulent flow, or vice versa, and no accurate prediction is possible.

Re greater than 4000 predicts turbulent flow.

Using the value 2000 for Re in the equation above to determine critical velocity for a given set of conditions leads to the equation:

$$V_{\text{critical}} = \frac{2000\mu}{D\rho}$$

Applying the equation to practical bearings, the radial clearance dimension ($\frac{1}{2}$ diametral clearance) may be substituted for the diameter of the conduit D,

$$\text{critical velocity} = \frac{2000\mu}{h\rho}$$

where

$$h = \text{radial clearance (film thickness)}$$

The Reynolds number has become an important factor in fluid mechanics.

The transition from laminar flow to turbulent flow, or critical velocity, is determined by passing a given liquid through a glass capillary tube in which a dye is fed through a separate smaller tube into the center of the stream. Under circumstances of laminar flow, the dye will flow with threadlike continuity along the fluid stream. By gradually increasing the velocity of the stream, a point is reached when the dye thread commences to waver, and eventually permeates the entire area of stream cross section. The velocity at which wavering starts is the critical velocity for the particular combination of factors involved.

Laminar flow is variously known as streamline, viscous, and nonturbulent flow.

The Reynolds number can be established similarly when any quantity measurement is given, provided that the correct conversion factor and viscosity measurement are employed.

BEARING CHARACTERISTIC NUMBER

In the foregoing discussion of hydrodynamic action and laminar flow, several parameters were involved. Included were fluid velocity, load, and fluid viscosity, which represent the major factors that govern frictional behavior in bearings. It is well to consider the influence that each exerts by stating several appropriate rules.

Fluid velocity depends on the velocity of the journal, or rider. It is expressed in surface feet per second, or for a journal of given diameter, in revolutions per minute.

The higher the speed for a constant load, the more work is performed, resulting in a greater transformation of mechanical to thermal energy.

Any increase in relative velocity tends toward a decrease in eccentricity of the journal-bearing centers accompanied by greater *minimum film thickness*.

Load (*unit pressure*), as it increases, tends to decrease minimum film thickness h_{min}. This, in turn, increases the pressure within the film mass to provide a counteracting force. Because the counteracting force is exerted in all directions, it also tends to squeeze the oil out the ends of the bearing. Any increase or decrease in unit pressure affects fluid viscosity, as described below.

Viscosity is the most important property of a fluid subject to hydrodynamic action. A lubricating oil becomes more viscous as unit pressure increases. This is offset somewhat by the resultant rise in temperature due to the increased work load. Where other factors remain the same, the minimum film thickness becomes greater with lubricants of higher viscosity. The converse is also true.

Analyzing the effects of each of these three parameters on frictional behavior of journal bearings as governed by minimum film thickness, the following may be adduced:

1. An increase in velocity up to the critical level or region of turbulence has a favorable or positive influence on minimum film thickness, or $h_{\min} \propto V$.

2. An increase in viscosity of the lubricant also affects a favorable or positive influence on minimum film thickness, or $h_{\min} \propto Z$, where Z represents viscosity in centipoises.

3. An increase in load (unit pressure) has an unfavorable or negative influence in the minimum film thickness, $h_{\min} \propto 1/P$.

These rules are based on the assumption that the remaining parameters are maintained at nearly constant magnitude with a change in any one of the three.

Because the minimum film thickness remains the criterion for the frictional state of any bearing, it becomes necessary that the velocity, viscosity, and unit load be contained within such appropriate limits as to assure a hydrodynamic state, if so desired. While the mathematical relationship of the three parameters cannot predict the film thickness dimension without resorting to experimental or empirical modification, it can be used to determine the *frictional characteristics of bearings* operating under specific combinations of velocity, viscosity, and unit load.

By combining the three parameters in an expression according to their influence on minimum film thickness, as shown above, the following equation results in a dimensionless number:

$$\frac{\text{viscosity} \times \text{velocity}}{\text{unit load}} = \text{a dimentionless number}$$

or

$$ZN/P = C$$

where C represents the bearing characteristic number.

By experience, the value of C has been found to indicate with some accuracy the frictional state of a given bearing having known or selective values of Z, N, and P, whether the lubrication is hydrodynamic, mixed film, or boundary in nature. The critical value of C is the point of transition from a hydrodynamic to a lower frictional condition, and it may vary according to bearing surface conditions, but a dimensionless number of 35–40 is acceptable for most practical purposes. However, a value as low as 5 is minimum under some surface conditions.

For example, where the viscosity of the oil film is, say, 19 cP and the journal carries a load exerting a pressure of 350 psi (projected area) at 1750 rpm, then

$$C = \frac{ZN}{P} = \frac{19 \times 1750}{350} = 95.0$$

A value well above the safe dimensionless number of 35 characterizing the establishment of hydrodynamic conditions.

If the viscosity of the oil were reduced to say, 12 cP, then

$$\frac{ZN}{P} = \frac{12 \times 1750}{350} = 60.0$$

Reducing speed to 900 rpm and using the original oil, then

$$C = \frac{ZN}{P} = \frac{19 \times 900}{350} = 48.85$$

Increasing the load to 500 psi and maintaining original oil and speed,

$$C = \frac{ZN}{P} = \frac{19 \times 1750}{500} = 66.5$$

If the speed were reduced to 50 rpm in the original example, then

$$C = \frac{ZN}{P} = \frac{19 \times 50}{350} = 2.71$$

These examples show the effect of changes in any of the three factors on the ZN/P number. A significant change is that of speed. Using an oil of higher viscosity in the last example, a higher ZN/P number will result, and if increased sufficiently, the characteristic of the bearing may be changed to support a hydrodynamic film.

Selecting an oil having a viscosity of 330 cP,

$$C = \frac{ZN}{P} = \frac{330 \times 50}{350} = 47.1$$

Thus by substituting an oil of 1800 SUS at 100°F for one of about 100 SUS at 110°F, a hydrodynamic condition develops within the bearing to eliminate the extreme boundary condition of lubrication offered by the less viscous oil.

The coefficient of friction of a given bearing may range from about 0.002 for an ideal hydrodynamic condition to about 0.250 for an extreme boundary condition, with a value of $f = 0.02$ to 0.10 for a mixed-film condition. This relationship is portrayed by the abrupt upward swing of the curve to the left of the transition zone in Fig. 9-5.

There is also a tendency for the coefficient of friction to increase gradually as the hydrodynamic film thickness increases in some direct proportion to the value of ZN/P. This is portrayed by the upward grade of the curve to the right of the transition zone, shown in the figure.

Because a definite relation exists between ZN/P and the fluid film thickness, it follows that the coefficient of friction is affected also by any single or multiple change in viscosity, speed, and load. The coefficient of friction developed between a journal and a bearing is the ratio of tangential force required to turn the journal to the total load on the bearing. If a cord were wound around a loaded journal and pulled with sufficient force, the journal would rotate. By pulling the cord through a spring balance, the amount of force required to turn the journal could be measured. Under dry (nonlubricated) conditions, it might be established that the tangential

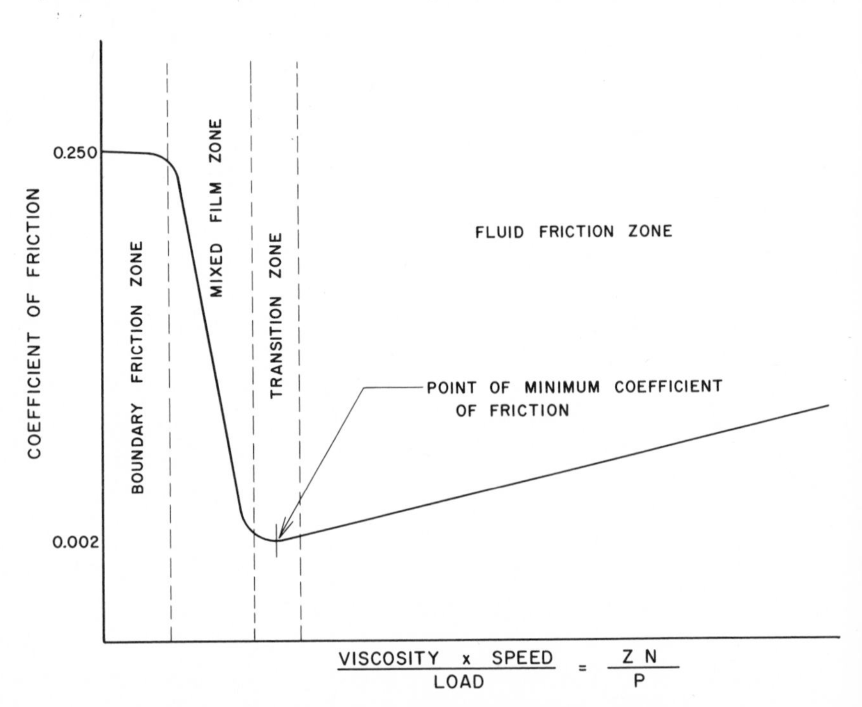

Fig. 9-5. Relation of coefficient of friction to ZN/P.

pull would be 0.25 to 0.40 of the load, determined by the ratio (tangential force in pounds): (load in pounds). Therefore, the value 0.25 to 0.40 would be the coefficient of friction f for that bearing by definition.

When a small amount of lubricant is applied to the bearing to wet the surfaces slightly (boundary condition), and when force is applied to the cord, the coefficient of friction is approximately 0.1 to 0.15. Adding more lubricant to obtain a partial (mixed-film) fluid film would decrease f to about 0.02 to 0.10. Under hydrodynamic conditions (fluid film), f would be below 0.002. The amount that f falls below 0.002 depends on the hydrodynamic efficiency of the bearing system under the load, the viscosity of the fluid film, and the speed at which the journal is turned.

With the journal operated at constant speed and load with oils having different viscosities, a series of ZN/P numbers could be established for that particular bearing. If the power required to turn the journal were measured during each change of lubricant, a relationship between lubricant viscosity and f would be established. By the same processes, ZN/P and f could be determined for any variable combination of load, speed, and viscosity.

Using values of f as the ordinate and ZN/P as the abscissa, a pattern of frictional behavior within the bearing could be plotted, having the general form shown in Fig. 9-5. Note that for low ZN/P numbers boundary conditions exist. As ZN/P values increase, a transition occurs to establish a partial or mixed-film condition conducive to a reduction in f. The lowest value of f is established at a point representing the initial phase of a fluid film. It is at this point where metal-to-metal contact ceases to exist and fluid friction is at minimum magnitude due to the thinness of film (minimum shear). It may be said that perfect lubrication exists slightly beyond or to the right of the low point in the curve and should be selected as the ideal running condition. However, any adverse change in Z, N, or P (shock, temperature rise, etc.) would cause the thin film to rupture and initiate boundary conditions. For this reason, the safest condition lies in the zone of higher ZN/P to the right of the low point, where f is slightly higher due to increased fluid friction (thicker film).

The following rules may be appropriately given here apropos of ZN/P versus frictional behavior:

1. Continually increasing the unit load decreases the ZN/P value, assuming speed and viscosity remain constant. The same results are obtained by reducing either the speed or viscosity, or both, providing the unit load remains constant or is increased.

2. Fluid friction varies with viscosity, i.e., any increase in viscosity increases fluid friction.

3. Fluid friction is proportional to velocity at lower speeds, but varies inversely to velocity at higher speeds.

In many cases, bearings are designed to meet given load and speed requirements. Assuming that the correct method of lubricant application is also provided, the only variable remaining to be considered by the field lubrication engineer to obtain a desired ZN/P value is the proper viscosity of the oil at the inlet temperature.

Factors not considered in the value of ZN/P number are the difference between bearing and journal diameter, or diametral clearance C_d and the ratio of bearing length to diameter L/D. These will be discussed later in the text.

Before leaving the discussion of Reynolds' contribution to the science of lubrication, it is important to outline both his conclusions as generally interpreted and his assumptions relative to the hydrodynamic fluid wedge:

REYNOLDS' CONCLUSIONS

1. Pressures within an oil film are not uniform.
2. A pattern of pressure variations exists within the oil film.
3. The exact location of the points of maximum and minimum pressures along the line of oil flow can be demonstrated and therefore predicted.
4. The wedge-shaped oil film is essential in developing the pressure gradient along the direction of oil film flow, or x-axis.
5. The wedge-shaped oil film possesses a load-carrying capacity commensurate with its magnitude, both as to thickness and length. This applies to any bearing with nonparallel surfaces.
6. Opposing surfaces must be slightly inclined relative to each other (nonparallel).
7. Load-carrying capacity is directly proportional to the viscosity of the oil and also to the velocity of the journal.
8. Load-carrying capacity is inversely proportional to the thickness of the oil film.

REYNOLDS' ASSUMPTIONS

1. Viscosity remains constant at all points within the oil wedge.
2. No significant end-leakage of oil occurs.
3. Lubricity (oiliness characteristics) of the lubricant may be neglected.
4. An adequate supply of oil is continuously applied in the proper location relative to the pressure area as governed by load direction and magnitude.
5. There is no deformation of surfaces, either by waviness or distortion of either member.
6. The oil is highly adhesive to surfaces.
7. There is uniform film pressure between longitudinal extremities of the bearing, which in turn are exposed to an external atmospheric pressure.
8. The flow is laminar with Newtonian fluid.
9. The fluid is noncompressible.

As will be shown in the following discussion, some of these assumptions cannot be applied to practical bearing problems, particularly where ambient temperature is subject to substantial change, bearings have low L/D ratios, surface deflection occurs due to load inbalance, and oil aeration permits some compressibility.

Sommerfeld in 1904 applied Reynolds' equation for plane surfaces to journal bearings of finite length and having various diametral clearances. In doing so, he developed an equation expressing the influence of journal eccentricity (relative to the bearing center) on friction as associated with rotating members. The equation relates to bearings of infinite length and so does not allow for end-leakage of lubricant from the pressure area. *This omission is of little importance for bearings whose lengths are in excess of four times their respective diameters, in which case, end-leakage may be neglected.*

Sommerfeld succeeded in equating the values of velocity, viscosity, and unit load modified by the influence of eccentricity to determine more favorable solutions concerning frictional magnitudes in journal bearings of various clearances. The recognition of clearance variation in the Sommerfeld equation is made apparent by the use of a factor known as the *clearance modulus* m, which is the ratio of journal radius r to radial clearance C_r, or if preferred, the ratio of journal diameter D to diametral clearance C_d.

The result is another dimensionless number known as the Sommerfeld number S, expressed by the equation

$$S = \left(\frac{ZN}{P}\right)\left(\frac{r}{C_r}\right)^2$$

To bring the value of S to a more convenient number, it should be multiplied by 10^6. Because of the modification of ZN/P by incorporating the influence of clearance factor, the Sommerfeld variable becomes a preferable criterion for interpreting bearing frictional conditions. But because S values do not consider end-leakage, a function of L/D, it is not a preferred criterion by bearing designers.

Another way of expressing the Sommerfeld variable is by denoting the clearance modulus in terms of the clearance ratio r/C_r represented by m, or $m = C_r/r$. Substituting m in the Sommerfeld equation, it may be expressed as

$$S = \left(\frac{ZN}{P}\right)\left(\frac{1}{m}\right)^2$$

where m is the clearance modulus of the bearing in inches of clearance per inch of radius, or $r_c = mr$. For industrial type bearings $m = (0.001$ in. $\times D) + 0.002$ in., or $m = (0.0005$ in. $\times r) + 0.001$ in., whichever is preferred, since they are numerically equal.

One of the advantages of the Sommerfeld number is that, when plotted as the abscissa against a modified coefficient of friction—$f(r/C_r)$, or $f(1/\text{m})$—as the ordinate, a single curve results (Fig. 9-6).

ECCENTRICITY. Eccentricity e is the distance from the center of the journal to the center of the bearing during rotation. When the journal is in contact with the bearing as it is when the journal is at rest or when the load is infinite, the distance between centers is equal to the radial clearance,

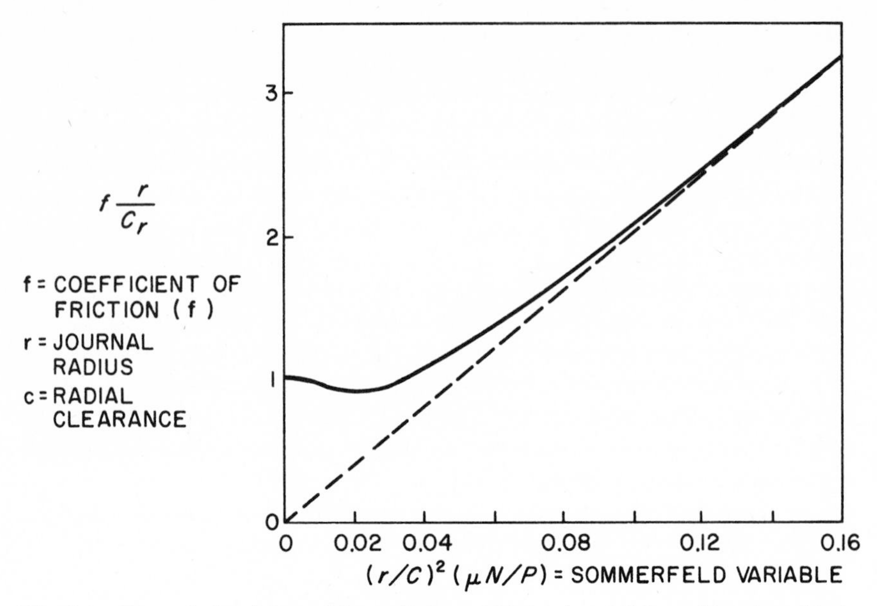

Fig. 9-6. Theoretical variation of the coefficient of friction in a full journal bearing with the Sommerfeld variable. *Courtesy, Gulf Oil Corp.*

which represents maximum eccentricity. If the journal rotates under conditions during which the radial clearance is constant around the entire periphery, the centers bear on a mutual axis and eccentricity is zero. If the two extremes are designated as 1.00 for maximum eccentricity and 0.0 for zero eccentricity, then any incremental distance between the two as a result of hydrodynamic balance (ZN/P) is reported by an appropriate decimal called the *eccentric ratio* ε:

$$\varepsilon = \frac{\text{eccentricity}}{\text{radial clearance}} = \frac{e}{C_r}$$

There is one other important variable related to the path along which the center point of eccentricity moves with any change in hydrodynamic conditions, or ZN/P. While the distance e may represent the minimum film thickness, the other variable referred to represents the angular location of the center line of eccentricity relative to the common line of centers, i.e., the load line and bearing center line that are congruent when the journal is at rest. The resulting angle ϕ is known as the *attitude* of the journal.

The load tends to displace the center point of the journal along its own path, or in the direction of the line of force. However, due to the vector forces developed by the pressure generated in a hydrodynamic film and the effect of end-leakage, the center point swerves to scribe an arc connecting the bearing center point with the center point of the journal. A line drawn radially through the two center points and extended to the hydrodynamic region between the opposing surfaces indicates the location of *minimum film thickness* and the point of *maximum film pressure*. If the load is continually increased to sufficient magnitude, the hydrodynamic film will eventually break down, whereupon the arc (locus) will swing around in a semicircle and return to the load line accompanied by an increase of eccentricity ratio to 1.0 (Fig. 9-7). For practical bearings, the point of minimum film thickness will be located less than $\phi = 90°$ relative to the load line in the direction of rotation. Sommerfeld predicted the locus to be a straight line along the center line 90° relative to the load line, probably due to the assumption of infinite length bearings (no end-leakage). It bears out Sommerfeld's prediction for very light loads only.

The diameter of the circle in Fig. 9-7 is equal to the diametral clearance (exaggerated) of a hypothetical journal bearing, and it therefore sets the limit of eccentricity as scribed by the center point of the journal relative to the center point of the bearing.

CLEARANCE. The discussion of eccentricity implies the importance of clearance magnitude, or the difference between the internal diameter of the

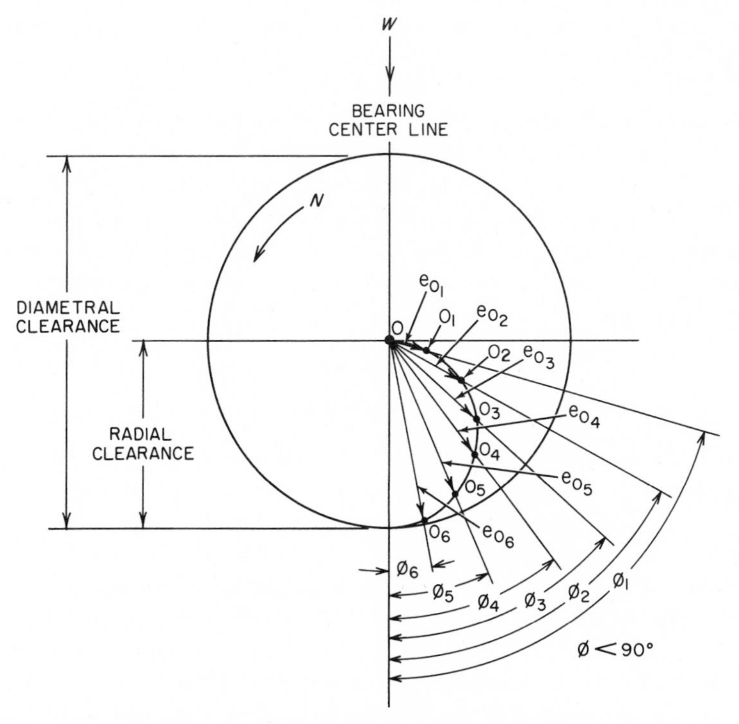

Fig. 9-7. Arc scribed by center point (axis) of journal of 180° journal bearing.
0 = axis of bearing and journal in concentric position.
0_1, 0_2, 0_3, etc. = axis of journal in various running positions.
W = load. N = direction of rotation.
0–0_1, 0–0_2, 0–0_3, etc. = eccentricity for respective attitudes ϕ_1, ϕ_2, ϕ_3, etc.

bearing and the maximum diameter of the journal. It is essential that the ratio of clearance to the journal diameter be accurately calculated to fit the design of the link per se. Character includes all dimensions, all materials, all operating factors, intended service, relative position, and design.

Mechanical considerations such as vibration, whirl, and deflection limit the magnitude of clearance enlargement, and seizure due to the effect of temperature on dimensions, abrasive factors, and fluid flow restrictions limit any approach toward zero clearance.

The importance of clearance is demonstrated in the following example.

EXAMPLE: An exciter located at the end of a direct-connected 5000-kw, 3600-rpm turbo-generator linkage caused a shutdown three times in as many months. Called in to investigate, the author found that a plain bearing clearance of 0.006 in. was allowed for its 2.25-in. shaft. A clearance of 0.0004 was recommended and adopted, ending the problem. An oil with a viscosity of 150 SUS at 100°F was continued in use in the circulation system.

Actually in any bearing in which it is essential to maintain a hydrodynamic film, it is necessary that a sufficient minimum thickness of film be constantly maintained at the point of maximum pressure exerted within the film.

Sommerfeld succeeded, making a series of assumptions (substitutions), in solving for pressure magnitudes at increments of angularity within the range of ϕ. His solution is expressed in the equation

$$p = \frac{\eta U r}{C_r^2}\left[\frac{6n(2 + n\cos\theta)\sin\theta}{(2 + n^2)(1 + n\cos\theta)^2}\right] + p_0$$

where

p = pressure in the oil film
θ = angle from line of centers to line of any pressure p, not necessarily attitude
p_0 = pressure of film at $\theta = 0$, which is oil pressure at entry to the bearing
U = velocity
r = radius of journal
C_r = radial clearance
π = circumference of journal per unit diameter
η = viscosity (reyns)

Maximum pressure occurs in the half of the bearing under load. Cavitation is evident in areas approaching the loaded region of the bearing.

The load-carrying capacity of any bearing depends on the thickness of lubricating film at any moment. Therefore, the magnitude of the minimum film thickness of a hydrodynamic wedge is the important parameter in the establishment of safe lubrication conditions regarding friction and wear.

Practical applications of bearings are made according to predetermined loads and speeds so that their frictional characteristics are intrinsic (built in) for any individual bearing. Size and service dictate the clearance magnitude. The remaining variable for frictional control is the viscosity of the fluid employed, assuming, of course, that the method of application provides a sufficient quantity of lubricant. Selection of the lubricant viscosity, then, is based on the maximum load, average speed, and operating temperature. The latter, while mentioned here for the first time, is of major importance.

OCVIRK'S EQUATION FOR FINITE BEARINGS. Summerfeld's solution for modifying the ZN/P value by employing the square of the clearance modulus m = journal radius/radial clearance applies to bearings of infinite length, or for all practical purposes, to bearings longer than four times their diameter. For bearings of finite length, or those having a length-to-diameter ratio less than four, a further modification by extension of pertinent functions must be made to allow for end-leakage or axial flow which imparts a substantial influence on the load capacity of the journal bearings.

The omission of short bearing approximation for full journal bearings approximation was pointed out by A. G. M. Michell in 1929. The solution was developed by F. W. Ocvirk in 1950 by extending the Sommerfeld variable. He multiplied it by the square of the L/D ratio of the bearing involved. The result is an equation that determines the *capacity number*, C_n, for short bearings as follows:

$$C_n = \left(\frac{r}{C_r}\right)^2 \left(\frac{ZN}{p}\right)\left(\frac{L}{d}\right)^2$$

or

$$C_n = S\left(\frac{L}{d}\right)^2$$

The reciprocal of C_n is the *load number*, and it is also dimensionless, used for convenience in plotting curves.

The load number may be expressed as follows:

$$\frac{1}{C_n} = \frac{P}{\eta N^1}\left(\frac{C_d}{d}\right)^2\left(\frac{d}{L}\right)^2$$

where

$P = \dfrac{\text{load}}{dL}$, unit load (psi)

N^1 = journal speed (rps)

η = viscosity (reyns)

C_d = diametral clearance (in.)

d = diameter (in.)

L = length of bearing (in.)

The load number increases as the eccentricity ratio increases rapidly up to about $\varepsilon = 0.50$, and then tapers off above $\varepsilon = 0.70$ so that $\varepsilon = 0.6$ corresponds to $1/C_n = 50$, indicating the relationship of ε to $1/C_n$. This relationship develops the pattern of a decreasing minimum film thickness (h_{min}) as the load (number) increases.

If ε is known, h_{min} can be determined by the equation

$$h_{min} = C_r(1 - \varepsilon)$$

(If C_r is in inches, h_{min} is also.)

The film thickness is a function of the load and oil viscosity.

PRESSURE-VELOCITY RELATIONSHIP. The pressure exerted by the film is a function of velocity (speed of the journal). The reason for this is that a greater quantity of oil will be pumped into the critical point (h_{min}) with increased velocity. This forces the journal further away from the bearing, thus generating a greater amount of pressure. Load, then, imparts a force that tends to cause the critical pressure point to locate at the junction of the load line and the point of the load line's intersection with the bearing surface, i.e., a straight line congruent with the center line of the bearing in the direction of the load. This is the condition when the journal is at rest and the oil film is of negligible thickness. When the journal rotates, the oil is pulled between it and the bearing where it develops a pressure of its own, which if high enough lifts the journal away from the bearing. The pressure developed by the resulting wedge is caused by the pumping action of the rotating journal, the magnitude of such pressure depending on the relative speed of the opposing surfaces. Therefore, the velocity of the journal is one of the major functions of critical pressure (p_0).

It is this critical pressure that opposes the pressure of the load in proportion to the angularity (vector force) of its location within the clearance space relative to the load line. It is this angularity and the thickness of film at the point of critical pressure that governs the location of the center of eccentricity which also becomes the rotational center of the journal relative to the fixed center of the bearing. The minimum film thickness must be of sufficient magnitude to keep the surface asperities entirely separated and possess adequate supporting pressure to either sustain such thickness or increase it.

Under *similar loading,* film thickness will increase with more viscous oils, providing that the consequent thermal development does not increase sufficiently to offset the advantage.

Any increase in velocity, as stated above, tends to pump more oil into the converging area to increase the minimum thickness, consequently, increasing the pressure opposing the load (hydrodynamic action) and decreasing the eccentricity. Experiments show the relationship of supporting pressure to velocity by the formula

$$p = 7.5\sqrt{V} \text{ (approx.)}$$

where

$$p = \text{unit pressure (psi)}$$
$$V = \text{velocity (ft/min)}$$

The coefficient of friction of a journal bearing varies approximately with the square root of velocity for speeds above 100 ft/min.

With increased temperature, the coefficient of friction varies inversely with temperature, i.e., it decreases with temperature increase. It follows that the coefficient of friction of a hydrodynamic system decreases as viscosity is reduced.

TEMPERATURE. The temperature of bearing surfaces is a function of transmitted power and efficiency and is determined by Joule's equivalent for thermal energy in relation to horsepower. This energy must be constantly dissipated in the form of heat (thermal kinetic energy or thermal energy in transition), either by natural conduction, convection, or radiation to an external source, or by forced cooling as by air or water flow. Depending on surrounding conditions, the temperature of the oil should be from about 30°F to 60°F above ambient. The temperature is measured at the bearing outlet of a circulating system or within the oil mass of a bath or splash oil reservoir.

The dissipated heat from a bearing housing can be estimated for a noncirculating oil system by the equation

$$H = K_h \times A \times (t_2 - t_1)$$

where

H = heat dissipated from bearing housing (Btu/hr)
K_h = 2 Btu/ft²/hr/°F for still air, or 6 Btu/ft²/hr/°F for forced air flow having a minimum velocity of 500 ft/min
A = Exposed area of bearing based on 20 × diameter × length (ft²)
$(t_2 - t_1)$ = temperature difference between oil film and ambient

Much of the heat generated in a bearing is carried away by the oil supplied under circulated or recirculated applications.

FORMULAS FOR JOURNAL BEARING VALUE DETERMINATIONS

Coefficient of friction f: (McKee and McKee, *ASME Trans.*, 51, 1929)

$$f = K\left(\frac{ZN}{P}\right)\left(\frac{r}{C_r}\right)10^{-10} + k$$

for a 360° bearing with a length-to-diameter ratio of from 0.75 to 4.0.

$$f = 1.31 \times 360° \left(\frac{ZN}{P}\right)\left(\frac{r}{C_r}\right) 10^{-10} + 0.0025$$

Dimensionless product of ratio $\dfrac{\text{diameter}}{\text{diametral clearance}}$, or $\dfrac{D}{C_d}$ and coefficient of friction f is a function of the eccentricity ratio ε expressed as $(D/C_d)f$.

The ratio of $(D/C_d)f$ to the Sommerfeld variable S is based on the factors related to S, or

$$\left(\frac{D}{C_d}\right) f \sim \frac{\mu N}{P}\left(\frac{r}{C_r}\right)^2$$

for a bearing having a given L/D. The value of S provides a method of establishing a single curve for the relationship of variations in S and $(D/C_d)f$ for a given L/D.

To determine the above relationship, each combination of factors must be calculated separately. The curve shown in Fig. 9-8 shows the solution of such calculations for various L/D ratios.

Minimum oil film thickness, h_{min} is important: (1) It must be of greater magnitude than the surface asperities to maintain complete separation of opposing surfaces. (2) It must provide a permanent cushion between opposing surfaces to prevent intermittent metal-to-metal contact during surges in load, sudden shock, and vibration. It is a function of all factors related

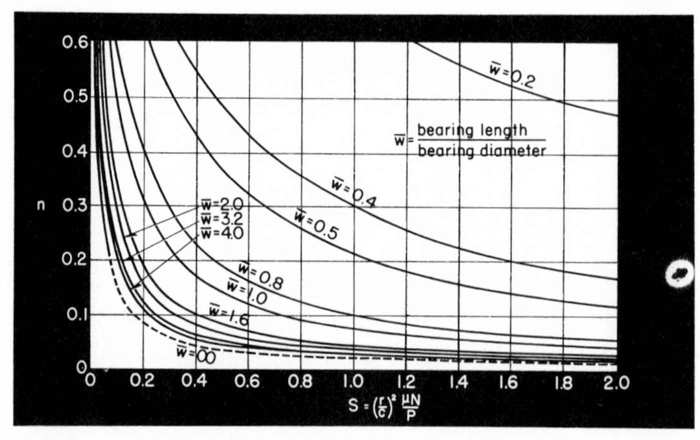

Fig. 9-8. Variation of the journal eccentricity n with the Sommerfeld variable for flood-lubricated journal bearings of finite length. *Courtesy, Gulf Oil Corp.*

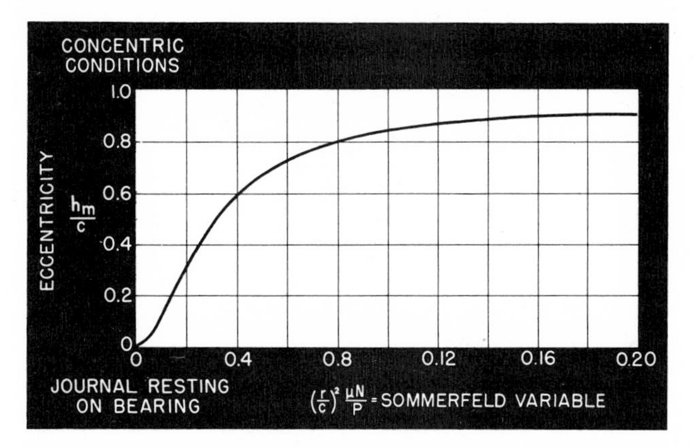

Fig. 9-9. Theoretical variation of the minimum film thickness in a full journal bearing with the Sommerfeld variable. *Courtesy, Gulf Oil Corp.*

to the Sommerfeld variable and eccentricity ratios, or

$$\varepsilon \sim \frac{ZN}{P}\left(\frac{r}{C_r}\right)^2$$

where

$$\varepsilon = e/C_r$$
$$h_{\min} = C_r - \varepsilon C_r = C_r(1 - \varepsilon)$$

The eccentricity is determined from Fig. 9-9, a curve plotted for various $(ZN/P)(r/C_r)^2$ values against eccentricity e.

The bearing area required depends on the total load to be supported by the bearing. The area is determined by the total load divided by the permissible unit load, which is governed by bearing design and materials. The quotient derived by the ratio of total load to unit load is known as the projected area. It represents the area determined by bearing length times bearing diameter, or

$$\text{projected area} = \frac{\text{total load (lb)}}{\text{unit load (psi)}}$$

$$A = W/P$$

or

$$A = LD$$

so that

$$P = W/LD$$

Hydrodynamic solutions make use of the ratio L/D, which is a function of end-leakage. The following equations apply to bearing size as related to projected area, which is governed by the permissible unit load W, as shown above.

$$W = (L/D)D$$

where

$$D = \sqrt{\frac{\text{projected area}}{L/D}}$$

Power loss due to friction can be determined theoretically if the values f, h_{min}, ε, and S are known.

$$\text{horsepower} = \frac{2\pi_r NWf}{33{,}000 \times 12}$$

The denominator 12 converts r to feet.

EXAMPLE: Space allows ample bearing dimensions for a total load of 6000 lb. Design specifies a unit load limit of 300 psi at 1800 rpm. Consider an operating temperature of 130°F and the use of an oil having a viscosity of 300 SUS at 100°F. Solve for projected area, L/D, coefficient of friction, eccentricity ratio, minimum film thickness, and loss of power due to friction for a full (360°) journal bearing. Assume an oil with a specific gravity (density ρ) of 0.880.

1. Projected area

$$A = \frac{W}{p} = \frac{6000}{300} = 20 \text{ in.}^2$$

2. To reduce the effect of end-leakage to a minimum, select a length-to-diameter ratio of 4.

3. Diameter $= \sqrt{\frac{A}{L/D}} = \sqrt{\frac{20}{4}} = 2.236$ in.

   say, 2.25 in. $\left(L = \frac{L}{D} \times D = 4 \times 2.25 = 9 \text{ in.}\right)$

4. Diametral clearance $= (2.25 \times 0.001 \text{ in.}) + 0.001 \text{ in.} = 0.00325$ in.

5. Journal radius $r = \frac{2.25 - 0.00325}{2} = 1.1233$

6. Absolute viscosity Z is equal to the kinematic viscosity of the oil at the operating temperature times its specific gravity (density ρ). From Table 8-4, it is found that an oil having a viscosity of 300 SUS at 100°F and a viscosity index of, say, 95 has a kinematic viscosity of 64.8 cSt at 100°F. The same oil is reduced to 148 SUS at 130°F operating temperature. From the same chart, find the kinematic viscosity at 130°F of 31.2, or 31.2 × 0.880 (specific gravity) = 27.45 absolute viscosity (Z).
7. Coefficient of friction f:

$$f = K\left(\frac{ZN}{P}\right)\left(\frac{r}{C_r}\right)10^{-10} + k$$

$$= 1.31 \times 360° \times \left(\frac{27.45 \times 1800}{300}\right)\left(\frac{1.1233}{0.001625}\right)10^{-10} + 0.0027$$

$$= 0.00808, \text{ or } 0.0081$$

8. Minimum thickness of film h_{min}:

$$h_{min} = C_r - \varepsilon C_r = C_r(1 - \varepsilon)$$

It is first necessary to determine the value of the eccentricity ratio, which is a function of $(ZN/P)(r/C_r)^2 \times 10^{-6}$.

$$\frac{ZN}{P}\left(\frac{r}{C_r}\right)^2 = \frac{27.45 \times 1800}{300} \times \left(\frac{1.1233}{0.001625}\right)^2 \times 10^{-6}$$

$$= 165 \times 692^2 \times 10^{-6}$$

$$= 79.01 \text{ dimensionless}$$

From Fig. 9-9, using 79.01 as abscissa, find ε (ordinate).

$$\varepsilon = 0.545 \textit{ eccentricity ratio}$$

$$h_{min} = 0.001625\ (1 - .545)$$

$$= 0.00074 \text{ in. } \textit{minimum film thickness}$$

9. Power loss in horsepower:

$$\text{power loss} = \frac{2\pi rNWf}{33{,}000 \times 12}$$

$$= \frac{2 \times 3.1416 \times 1.233 \times 1800 \times 6000 \times 0.0081}{33{,}000 \times 12}$$

$$= 1.71 \text{ hp}$$

ELASTOHYDRODYNAMIC LUBRICATION

When a rounded steel object, such as a ball, roller, or rocker, is pressed against a surface of another steel object, a deformation occurs in the form of a depression on both surfaces. If the force exerted remains within the elastic limits of both materials, their surfaces will return to their original geometry upon pressure release.

Let the ball and roller represent the rolling members of antifriction bearings while the opposing surfaces are their races. Also, let the rocker represent the surfaces of mating gear teeth.

The thickness of a hydrodynamic film in a sliding bearing is a function of viscosity, speed, and load modified by eccentricity as represented by the Sommerfeld variable and further modified by end-leakage, as determined by Ocvirk's equation. In a journal bearing, the maximum thickness of film is limited by the radial clearance in the direction of the y-axis, assuming that the eccentricity equals zero.

Consider now a journal bearing in which a journal rotates, separated from the bearing by a series of rollers placed in the periphery that normally would be the clearance space. The energy lost in such a bearing would be due to rolling friction rather than fluid friction if the clearance space were filled with oil rather than rollers. And the surfaces of all opposing members would be subject to deformation upon load application from the journal. The clearance space represented by the diameter of the rollers at no-load would then become equal to the roller diameter, plus some possible increase due to surface depression when load is applied.

Consider again that the space within the periphery not taken up by the rollers is filled with oil of any given viscosity and maintained so, say by immersion. The journal is then rotated at a given speed under a given load.

Because the oil is fluid, it will flow, assisted by the pumping action of the rollers, into the depressions formed by the related geometry of the bearing and rollers. In the continuous attempt by the rollers to climb out of such depressions, the roller action becomes a combination of (1) rolling and (2) sliding. The sliding action of the roller pulls a portion of the oil between it and the surfaces of the journal and bearing (guiding members) to develop a hydrodynamic film similar to that formed between a journal and plain bearing. The minimum thickness of film thus formed depends on the magnitude of surface deformation, which is a function of surface elasticity.

Hertz stresses are often used in connection with wear tests to help the lubrication engineer to evaluate the probable wear pattern of bearing materials under various load characteristics (Appendix A-10).

One other parameter entering the process at this point is the increase in viscosity of the oil relative to the pressure of the film, a function of the load and various radii.

The solution of elastohydrodynamic parameters is complex. Manufacturers of antifriction bearings and gears have developed such parameters for their own products. They are analogous to the ZN/P factor as related to journal bearings.

Where the geometry and modulus of elasticity are inherent within the test results of a given component, such as a ball bearing of a given series, a factor of behavior is determined that considers both parameters. The factor, then, becomes a constant K for that series of components, which when used in computations eliminates the need to further consider either the design or material. The remaining parameters are lubricant properties *during operation,* speed and load. Combined, the following equation is representative of a dimensionless number indicating the lubrication characteristic of the component:

$$K(\mu_0, \alpha)^n P_0 = \Lambda$$

where

Λ = elastohydrodynamic film parameter
K = constant for a given component
$(\mu_0, \alpha)^n$ = lubricant properties
P_0 = equivalent load

The lubricant properties are dependent on the origin of the fluid, either paraffinic, naphthenic, or synthetic, where μ_0 is dynamic viscosity (lb-sec/in.^{2}) and α is the pressure coefficient of viscosity 1/psi.

If Λ is found to be of sufficient magnitude, say 2.5 or above, the bearing characteristics are considered "healthy" and conditions for increased operating life exist. Below that value, the conditions gradually deteriorate as the value approaches a critical surface distress region of about 1.5 or below.

The friction torque of a rolling bearing is a function of design, load, viscosity, oil volume, and speed. While the description above assumed an immersed oil condition, it was made only to assure the reader that an adequate quantity of lubricant is necessary to develop hydrodynamic film. Actually, with more than an adequate amount of oil to develop the film, the friction torque will increase with an increase in speed.

Figure 13-16 shows the bearing characteristic number Λ (or elastohydrodynamic film parameter) plotted against film magnitude reported as percentage film. For various types of rolling bearings, the minimum viscosity of oils at their *operating temperature,* as shown in Table 9-1, will generally result in a lubrication characteristic within the region of increased life, providing that an adequate quantity of lubricant is constantly available.

The grease recommendations for rolling bearings are shown in Table

Table 9-1. MINIMUM OIL VISCOSITY AT OPERATING TEMPERATURE

Type of Bearing	*Minimum Viscosity, SUS, at Operating Temperature*
Ball bearings	75
Cylindrical roller bearings	75
Spherical roller bearings	100
Spherical roller thrust bearings	150

13-7, which conforms with those of the Anti-Friction Bearing Manufacturers Association, Inc.

It is the oil in such grease that develops the hydrodynamic film under elastohydrodynamic conditions.

Microhydrodynamic Lubrication

Hydrodynamic lubrication is usually associated with fluid films of such thickness as to separate opposing surfaces having some relative smoothness and geometric relationship. Waviness and asperities, the latter being made up of high and subsequent low points, are present on all surfaces. When conditions of load, speed, and oil viscosity are such as to force opposing high points into close proximity during sliding motion, a hydrodynamic action tends to develop between them. The situation is one in which a pattern composed of a myriad of microscopic hydrodynamic actions occur on the opposing surfaces. It is also a situation in which a similar pattern of interspersed boundary conditions exist. It may be said, then, that the mean coefficient of friction that develops under such conditions depends on the ratio of the number of microhydrodynamic points to the number of boundary points. The major point here is that numerous microhydrodynamic pressure systems develop having film thicknesses of less than 1 micron.

Both friction and thickness of film under microhydrodynamic conditions depend on certain basic parameters of which they are a function:

(a) Surface Material. (1) Elasticity of each of the opposing metals, as determined by Young's modulus, E. (2) *Deformation,* or the ratio of lateral unit deformation to lineal unit deformation within the elastic limit, Poisson's ratio, λ. (3) The *coefficient of elasticity* of both (opposing) materials, E'/E. (4) The *coefficient of deformation* of both materials, λ'/λ.

(b) Lubricant. (1) *Viscosity* at inlet to pressure area Z. (2) Temperature coefficient of viscosity (VI). (3) Pressure coefficient of viscosity, (Z'_p/Z_p). (4) Compressibility of the fluid, V'/V.

(c) Relative velocity between rotor and stator.

(d) Thermal Factors. (1) Heat conduction from both material and lubricant, k. (2) Thermal energy capacity of the lubricant (sp ht).

(e) Geometry of frictional system.

Microhydrodynamic lubrication prevalent over a major area of the opposing surfaces tends toward the zone of transition, but sinks to a boundary situation during periods of vibration, or shock loading. See the f versus ZN/P curve.

BEARING WEAR. All bearing materials are subject to wear, the rate of which depends on the material, condition of the service, load and velocity. The load (P) and velocity (V) are combined to provide a value related to the physical capabilities of such materials. This PV value is determined by multiplying the load in pounds per square inch of projected area by the velocity in surface feet per minute. Each type of material has a *limiting PV* which defines the maximum load and velocity of a given material without regarding wear.

Bearing designers are concerned with the limiting PV as well as the wear rate of materials available for use. Each kind of material has a wear factor associated with nonlubricated rubbing contact determined by tests. The wear factor is a proportionality constant (K), so that the amount of radial wear in a given time (R) may be related to load (P), velocity (V) and time (T) as follows:

$$R = KPVT$$

When the wear factor K of a given bearing material and the allowable wear of a particular application is known, the allowable PV value can be determined. Unit wear can also be predicted for a required PV value.

10

ADDITIVES

Petroleum oils, synthetic lubricants, and fuels often require modification or fortification to meet certain demands imposed on them by modern mechanical equipment. Such modifiers and fortifiers are called additives, or inhibitors, depending on their function. While the use of additives is not new, some of the more complex compounds have recently been found that impart desirable properties to lubricants, both oil and grease. Fixed oils have been added to mineral oils for many years to combat the washing effect of condensate in steam cylinders. Since shortly after World War I, sulfur has been used in metal-cutting oils to prevent welding of the chip to the tool and provide better cooling and finish. Oleic acid was known to provide greater lubricity to mineral oils since the latter part of the nineteenth century; up to World War II, a small amount (10%) of used steam turbine oil was mixed with new oil upon changing the batch, to resist rusting and prevent foaming. A sad commentary on past practices may be made of the addition of sawdust or cork to the oil in automobile differentials by used car dealers to quiet the worn gears temporarily.

ADDITIVE NOMENCLATURE. Additives are selected substances that are combined with oils and greases, both mineral and synthetic, for the purpose of imparting one or more beneficial properties that do not exist in sufficient magnitude in the straight product. Additives can have one or more beneficial effects on lubricants:

1. Increase the capability of a lubricant to perform more effectively under extreme operating conditions.

2. Retard the deterioration of the lubricant by chemical forces that develop within it as a result of changes incurred by its environment or activity.

3. Protect the lubricated surface against contaminants in the lubricant.
4. Depress pour point, break foam, and improve the viscosity index.

Some additives fortify film strength, improve pour point temperature, reduce wear rate, and impart stringiness, or tackiness. Others act to passivate (render inert) or deactivate (neutralize) metallic catalysts, and others retard the chemical forces that tend to destroy or incapacitate both the lubricant itself or the surfaces it contacts. Most metals are to some degree catalytic.

Additives that (1) fortify or substitute for the lubricant in successfully providing a protective film strong enough to prevent solid surface contact and (2) act to disperse solid materials, suppress foam, or form an emulsion are referred to as *agents*. This category includes film strength agents, wear prevention agents, cleaning agents (detergents), dispersants, and emulsifiers.

Additives that retard the chemical forces that result in the development of deleterious substances that form either gradually or by chain reaction are known as *inhibitors*. These include oxidation inhibitors, or antioxidants, which inhibit rust and corrosion.

Agents that lower pour point are usually identified as *depressants,* and antifoam agents may be referred to as *suppressants*.

Proper blending is accomplished only if the material and the oil are compatible. This means that selective base stocks are used, the kind depending on the "marriageable" properties of the additive in respect to the oil. Also, the amount of additive used is a matter of *useful saturation* in some cases, and magnitude of *effect desired* in others. Oleic acid serves as a good example of useful saturation because it imparts its maximum effect of frictional reduction to mineral oils when the volume added reaches 1.5 to 3.0%. Above this amount no further improvement is obtained. On the other hand, the percentage of detergent may vary from less than 1% to about 30%, depending on the degree of cleaning desired or required.

Additives may be further divided into two groups, (1) unifunctional and (2) multifunctional. The first group imparts the primary function desired only, such as increasing film strength or improving the viscosity index. The second imparts more than one function, (a) all of which may be desirable, or (b) only one or two of which are desirable, while the others may have side effects that may or may not be advantageous. In the second group are found additives that act as detergents and also as corrosion inhibitors, or as both antioxidants and corrosion inhibitors. On the other hand, some detergent type additives were found to be oxidation accelerators.

The complete oxidation reaction, briefly, is the formation of hydroperoxides, unstable oxy-materials, when oxygen and hydrocarbon combine chemically. The reaction is regenerating in that additional materials are developed that become progressively more susceptible to oxidation in a pattern resembling a chain reaction.

Deterioration of Lubricating Oils

The most widely used and the more economical lubricants are the mineral oils produced from petroleum. Except where a process is accompanied by an inherent fire hazard, such as molten metal, welding sparks, or high temperature, most lubrication systems contain petroleum oils. This text enumerates many advantages gained through the use of petroleum oils as lubricants. This chapter deals with some of the disadvantages of such oils when not modified or treated to meet certain operating conditions conducive (1) to their own deterioration and (2) to the destruction of engineering materials, including seals and conduits. Many of the disadvantages of using straight mineral oils are interrelated and are influenced by the progressive changes due to technological advances made in machine performance and efficiency. Such advances have brought about greater temperature extremes, higher speeds, increased unit loading, and in some instances, longer periods of continuous operation under the aggravated conditions mentioned.

Petroleum lubricating oils are susceptible to deterioration due to oxidation. Oxygen atoms present in such environmental substances as water and air have an affinity for other kinds of atoms that are not locked together with still other kinds of atoms to form a tight bond. Or they will react with other type atoms whose bonds are weakened by surrounding physical or chemical conditions.

For example, if an unsatisfied (loose) atom of carbon is exposed to one or two atoms of oxygen, it will form carbon monoxide (CO), or because carbon has a valence of 4 and oxygen has a valence of 2, it will more readily form carbon dioxide (CO_2). Carbon dioxide combines with water to form carbonic acid. Considering that the elements hydrogen, carbon, nitrogen, sulfur, and others, either inherent or foreign, are present in operating environments involving petroleum oil, the possibilities of creating various kinds of acids are great when the molecular chains that compose lubricating oils begin to become dissociated.

The rate of oxidation depends on a number of factors:

1. *Oil stability* (chemical), which is governed by the degree of *refining*. Modern solvent treatment of lubricating oils has produced highly stable products that resist oxidation during service.

2. *Temperature.* Oxidation of oil increases with higher temperatures.

3. *Catalytic action.* The presence of such metals as iron, lead, and copper acts to initiate and continue the oxidation process.

4. *Aeration.* Air, of course, is about 21% oxygen, so that its presence is conducive to oxidation. Its influence becomes greater where turbulence (splash caused by gears, crankshafts, etc.) is prevalent, especially if the oil is dispersed into mist or small droplets.

5. *Water,* generally due to condensation, is usually present in the cooler

parts of oil systems, such as reservoirs of steam turbine systems, automobile crankcases, and gear cases. A number of acids are formed by the reaction of active oxides (SO_3, N_2O_3, etc.) with water, e.g., sulfur trioxide plus water yields sulfuric acid. The formation of sulfuric acid is more prevalent in diesel engines using fuels containing sulfur in amounts ranging from 2.5 to 5%. In an oil body where any one or a number of acids are either probable or possible, it is difficult to isolate any one or more in particular. By certain tests, including neutralization number, nitrogen determination, or spectographic analysis, the presence of acid is proven and the elements can be found which may be responsible for the chemical condition of the oil.

DETERGENTS

Lubricating oils used in internal combustion engines of both the Otto and diesel types, commonly referred to as motor oils, are exposed to high-temperature operating conditions. Such conditions are conducive to chemical changes of the oil, particularly oxidation. Oxidation products originally appear as fine particles that are impervious to the parent oil, both as to solvency and washing effect. The particles of oxidation ultimately agglomerate or become deposited on various internal parts of the engine. The agglomerate settles to the bottom of the crankcase in the form of a tar-like or slimy sludge, or in a granular form resembling coffee grounds. That which is deposited on the piston skirt develops into a hard, tenacious material resembling lacquer or varnish, as a result of baking at high temperature. It also forms a hard carbonaceous deposit in the piston ring grooves and on other parts of the piston, including the underside and top, as well as on the internal surface of the engine head, valve stems, and guides. Included in the agglomerates and deposits are some of the products of combustion of the fuel (gasoline, diesel fuel, propane, various gases), consisting of soot and resins. Resins also act as a binder to hold the particles together. Oxidation products convert a clean, efficient engine into a dirty, inefficient one.

Just as pure water is unable to wash away grease, straight mineral oil cannot remove oxidation residue from engine surfaces. However, by providing a suitable detergent agent in either case, the contaminants can be removed, held in suspension, or both.

Basically, detergents are metal soaps of high molecular weight, soluble in oil, and sometimes described as metal derivatives of organic compounds. They consist of any one of a number of metals and an organic component which imparts, or is alkylated to impart, oil solubility. The metals used, generally, are barium and calcium, although aluminum, magnesium, tin, and zinc are employed occasionally. The organic components include phosphates, phenolates, sulfonates, and alcoholates. To *impart* oil solubility when necessary, an alkyl radical is introduced to the organic compound, either by substitution or addition, a process known as *alkylation*. The principal classes of detergent compounds capable of removing gums, lacquers, and other deposits from engine surfaces include

1. Metal sulfonates
2. Metal phosphonates, or thiophosphonates
3. Metal phenates, or phenate sulfides, the phenol or phenol sulfide being alkylated to attain oil solubility of the metal.

An *ashless* type of detergent is also employed. Being metal-free, it leaves no metallic residue and serves to good advantage in maintaining cleanliness in the ring zone. Ashless detergents are polymerized olefins, or iso-olefins having high molecular weight. They include such groups as the polyglycols, amides (acrylated amine), pyrrolidones, and carboxylic acids.

Detergent Action. The activity carried on by detergent additives in oil is not entirely understood. They contain a metal atom, usually a water-repellent group, oil-solubilizing groups, and a free hydroxyl. The detergents also are multifunctional, acting as they do as an antioxidant, bearing corrosion inhibitor, as well as a cleansing agent. However complex, they succeed in dispersing and maintaining dispersion of the fine particles of oxidized oil products so that they remain suspended in the oil mass. This allows such materials to be drained from the engine oil system during periodic oil changes. Otherwise, as previously stated, they tend to become deposited on hot surfaces or gravitate to the bottom of the crankcase. Detergent oils become progressively darker as they continually accumulate suspended oxidized particles. Early discoloration of such oils should not be a matter of concern, except that the cause does not add to the lubricating capability of the oil.

The action of detergents is presumed to affect two basic functions: (1) to thin the gummy binder that would otherwise hold the fine particles together, and (2) to surround the particles and render them incapable of agglomerating by mutual attraction. In the combustion chamber, detergents which are metallic join with carbonaceous residues as oxides, thus creating a fluffy material that is easily blown out of the engine with the exhaust gases.

Detergents, then, do not correct the cause of oxy-product development, but act only to prevent its buildup in critical areas. Such critical areas include ring grooves, piston skirt, underside of piston, valve guides, valve stems, crankcase, and filter.

Oxy-materials in the form of lacquer and varnish, as well as hard carbonaceous deposits, cause ring sticking, piston binding, and filter plugging. They also interfere with valve motion, heat conduction, and spark plug firing. Accumulation on engine head and piston top surfaces causes an undesirable increase in compression ratio.

Automotive vehicles that are subject to "cold" operation, as with local delivery trucks requiring constant stops and starts, require an oil having a cleansing additive effective at the lower temperatures. Such additives, called *dispersants,* are used in conjunction with detergents effective only at high temperatures, thus meeting all conditions of operation.

Care must be taken with detergent motor oils in their introduction to engines formerly operated with nondetergent oils for any length of time. In such engines, it is possible that deposits form behind the oil rings and, in progress with such formations, force the ring outward to keep some semblance of snugness with the cylinder wall as the ring wears. An oil of high detergency will tend to loosen such deposits and leave the ring to float free in the groove. In doing so, the ring loses its oil-wiping ability, and this results in high oil consumption. The material so rapidly loosened from the affected areas also tends to clog the oil feed lines and filter. This was the experience of a large public utilities fleet, which, against the author's advice, changed over all units, both new and old, to high-detergency oil upon a notification of the car builder, which had referred only to trucks with low mileage.

Some authorities consider the term detergent a misnomer, because such additives used in lubricating oils do not have the properties of true, water-soluble detergents. Other authorities advocate identifying these products as detergent-dispersants, because they perform both functions at appropriate temperatures.

Metal-free polymers are used to decrease valve deposits and eliminate oxidant-producing agents.

DISPERSANTS

If the residues removed from hot engine surfaces by the cleansing action of detergent oils were allowed to take their own course, they would accumulate in the various parts of the engine. When enough of the solid particles accumulated to form a substantial mass, the oil lines, including the filter, would become restricted and eventually become completely blocked. There are at least two methods of preventing such extensive accumulations. One is to continually drain off such residues, which means changing the oil more often than is necessary commensurate with the good chemical condition of the oil. The second method is to isolate the individual particles from each other and keep them separated. This is accomplished by dispersing the fine particles and keeping them suspended in the oil mass as they are carried with it throughout each cycle of circulation through the engine.

Dispersion is accomplished by a group of compounds that prevents the finely divided oxidized particles from becoming attached either to each other or to any other substance, such as engine metal surfaces. They are usually small enough to preclude their separation from the parent oil, even by

Table 10-1. ADDITIVE TYPES FOR CRANKCASE LUBRICANTS

Dispersants		*Oxidation and Bearing Corrosion Inhibitors*	*Antiwear Additives*	*Viscosity Index Improvers*
Metallic	*Ashless*			
Salicylate ester salts	Copolymers of methacrylates or acrylates with monomers containing polar groups such as amines, amides, imines, imides, hydroxyl, ether, etc.	Organic phosphites (DF)*	Organic phosphites	Polyisobutenes
Sulfonates-neutral Basic	N-Substd long-chain-alkenyl succinimides	Metal dithiocarbamates (DF)*	Sulfurized olefins	Polymethacrylates
Phosphonates and thiophosphonates	High MW esters and polyesters	Sulfurized olefins (DF)*	Zinc dithiophosphates	Vinyl acetate-fumaric acid ester copolymers
Neutral Basic	Amine salts of high MW organic acids	Zinc dithiophosphates (DF)*	Alkaline additives	Polyacrylates
Phenates and phenol sulfide salts	Vinyl acetate-fumaric acid ester copolymers	Phenolic cmpds (AO)†		Copolymers of acrylates and methacrylates with monomers containing polar groups
Alkyl substd salicylates		Selenides (AO)†		*Pour Point Depressants*
		Amines (AO)†		Alkylated wax
		Phospho-sulfurized terpenes (AC)‡		Naphthalenes
				Polymethacrylates
				Alkylated wax phenol

* Dual function. † Antioxidant. ‡ Anticorrosion.
Courtesy, American Society of Lubrication Engineers.

microfiltration. Filters will, however, pick up those larger agglomerates that may have escaped the influence of the dispersing agent present in the oil because of their association with dirt from external sources that entered through the carburetor or breather.

Detergents help to disperse sludge particles, but they are more effective in keeping engine surfaces clean, especially in the regions of high temperature. Effective dispersion is accomplished only with dispersants, which act under relatively lower temperatures than the detergents. Engines dependent on detergent action alone find that, under certain cold operating conditions, a formation known as *cold water sludge* forms in the crankcase. It is more correctly called *low-temperature sludge,* but it does contain a substantial percentage of water.

DISPERSANT MATERIALS. The dispersed particles of sludge that are afloat in the oil mass and circulate with it throughout the engine system do not enhance the lubricating qualities of the oil. However, they are preferred in the latter location than as a residue deposited on critical engine parts.

The difference between a detergent and a dispersant was defined earlier in this chapter. To control the activity of sludge particles under both high and low temperature operation, it is the practice to combine both types of additives in the same oil. The dispersant keeps the solid particles that form the tar-like residues uprooted by the detergent in suspension within the oil body. The dispersing action is better described as *peptization,* during which the particle size is held within micro-dimensions. The dispersant provides a characteristic film around the particle by polar forces, which repel other particles to prevent agglomeration by an *anionic* electrical charge. Sludges can be removed by petroleum solvents, while lacquers and varnishes cannot.

Dispersant additives are polymers in which a nitrogen-containing monomer is present, having a surface action conducive to the attraction of polar constituents of the sludge-forming products. Included in the latter group are the esters of various compounds, e.g., methacrylates, acrylates, alcoholates and polycarboxylic acids.

A dual roll played by the dispersant polymer is that of a viscosity index improver.

OXIDATION INHIBITORS

A chemical reaction occurs when petroleum lubricating oils are exposed to oxygen-bearing materials such as air and water. Oxidation is accelerated by an increase in temperature, aeration, and the presence of catalytic metals. Oxidation is a process which increases the proportion of oxygen or acid-forming element or radical in a compound. *Hydroperoxides* are produced when oxygen and hydrocarbons combine. The hydroperoxides, or oxy-materials, decompose to form other materials that are more susceptible to oxidation. The process is one of regeneration, a form of chain reaction. Oxidation of lubricating oils used in automobile engines is most prevalent, embracing as they do all the factors leading to rapid deterioration of the lubricant, i.e., (a) high combustion temperatures, (b) water due to combustion and condensation, (c) aeration of large exposed areas of the liquid as it is broken up into droplets by the splashing effect of the crank mechanism, and (d) in some cases, the presence of copper (a highly active catalyst) in the bearings. It is believed that peroxides present also assist in the oxidation process.

Products of oxidation include sludge, gum, varnish, lacquer, carbon deposits, and acidic compounds.

SLUDGE. The type of sludge depends on the conditions (1) under which it is created and (2) to which it is afterward exposed. Its composition includes water, carbon, engine oil, organic residue, and dirt that finds its way into an engine through the breather and carburetor. Sludge can be removed by petroleum solvents. As found in the crankcase, it is for the most part a black tar-like substance that can be broken into fine oily particles when squeezed by hand, at which time abrasive grit may be felt. This grit may be sand (foreign material), or hard carbonaceous particles formed from the oil in high-temperature regions of the engine. In crankcases not properly ventilated, such as those located in the rear of buses, the sludge often has the appearance of coffee grounds. Sludge usually settles to the bottom of the crankcase. Lighter particles, however, circulate with the oil, and some become separated in the filter. The major factor in the creation of engine sludge is temperature. Water, where conditions tolerate its presence, acts in its development. Metals such as copper act as catalysts. Acids in such forms as sulfur and carbon compounds accelerate its formation. The discussion on chemical refining in the appendix describes the reaction of sulfuric acid on oil to cause an accumulation called acid sludge. The test for *neutralization number* is one of the accepted methods of determining the acidity of lubricating oils.

Dividing sludges into two extreme classifications results in (a) high-temperature sludge and (b) low-temperature sludge. All other kinds represent some combination of these two extremes. This means that there are as many types of sludge as there are engine temperatures involved.

ENGINE GUM. A combination of phenolic compounds formed by selected hydrocarbons and formaldehyde. It is generated in the crankcase, generally, under high temperature and other environmental conditions prevailing there. Gum also forms in the combustion area and acts as a binder causing the residue to adhere to piston rings, ring grooves, valves, etc., and also causing agglomeration.

VARNISH. A product of oxidation involving both the lubricating oil and fuel, particularly gasoline. Other fuels are also susceptible to gumming, which is the basic cause of varnish, but not to the extent of gasoline. Petroleum gum when exposed to the temperatures prevailing on piston and bearing surfaces bakes to a moderately soft oil-insoluble substance and/or is ironed out to form a tenaceous coating by the motion and pressure of such components.

LACQUER. A thin layer of converted varnish that has been baked and ironed out under extended periods of high temperature and pressure. Like varnish, it is also insoluble in oil.

CARBON DEPOSITS. A combination of soot (from excessive fuel), tar-like carbon residues from oxidized lubricating oil, and foreign materials drawn into the engine and held by the gasoline gum. It bakes to a black, hard substance found in piston ring grooves and on the internal surfaces of the engine head, top of pistons, valves, etc.

ACIDIC COMPOUNDS. The deterioration of lubricating oils is accompanied by the formation of certain oil-soluble acids that not only cause metal corrosion but also react with elements contained in the oil to hasten the decomposing process, which, in turn, initiates a continuing cycle leading to a condition approaching complete oil destruction.

An oxidation inhibitor, or antioxidant, is introduced into a number of kinds of lubricating oils to prevent, retard, or modify the oxidation process. This additive is a true inhibitor by definition and retards the development of oxidation materials, which in straight mineral oils may double for every 18°F increase in temperature. Oxidation inhibitors have the dual function of inhibiting the corrosion of bearing metals.

Antioxidants perform their task in different ways, and the processes involved are usually more on the order of retardation than of prevention.

PREFERENTIAL OXIDATION. The types of additives in this category include those that are more susceptable to the attack by environmental oxygen than the oil. Thus the oxygen is swept up by the antioxidants, producing inert compounds leaving the oil molecules free of any chemical change. It might be said that such additives are more decoys than barrier compounds. Because they are expendable, the parent oil will eventually become prey to oxidation and revert to a straight mineral oil whose life will depend on its basic quality.

It is economically wise to use oils with a highly refined base stock and good chemical stability. In addition to the resistance to oxidation that is an inherent quality of the latter, it has become a chemical axiom that well-refined oils have better additive response.

The preferential oxidation additives are classified as *chain stoppers,* or chain breakers, and they have the function of destroying the oxidant molecule. In doing so, a chain reaction occurs when the energy so dissipated by the chemical action in destroying the oxidizing molecule is transferred from one molecule to the next.

Another chain-stopper reaction involves the decomposition of *peroxides* formed during deceleration of internal combustion engines. Peroxide is an oxidizing agent and must therefore be controlled.

METAL DEACTIVATORS. Oxidation cannot be prevented entirely so that every means must be taken to (1) neutralize the catalytic action of pertinent metal either by smothering such particles present in the oil, or (2) by providing a barrier between the metal parts and the main oil body. The additives responsible for bringing about such action are called metal deactivators, or anticatalysts.

Metal deactivators have two functions:

1. To react with the soluble metal compounds to form other harmless compounds, thereby having a stabilizing effect on the oil.

2. To form a protective coating over the metal surfaces of the machine parts exposed to the oil. The action of nullifying the catalytic effect of metals is called *catalytic poisoning*.

The additives used for oxidation control may be said to have either an *inhibiting* action or a *retarding* action. The inhibitor acts to stop the oxidant from oxidizing numerous oil molecules. The retardant is a compound that becomes an inhibitor as a result of the oxidation process. Both the inhibitor and retarder are associated with their own characteristic time factor. The inhibitor prevents oxidation for an extended period of time, which finally ends with a rapid reaction. The retardant is involved in a rapid reaction to begin with, but its effect gradually tapers off to some constant lower level. The latter is better described as an *autoretardant* because of the negative effect of the oxidation compounds resulting from the initial reaction.

ANTIOXIDANTS. A summary of the foregoing discussion relating to the action of antioxidants shows their dual roll as an inhibitor and a metal deactivator or passifier. This means that the materials used as antioxidants may also act to prevent corrosion of machine metal parts. For example, one additive commonly used is zinc dithiophosphate, which is an excellent antioxidant and corrosion inhibitor. Actually, this additive is highly versatile and has several functions: (a) decomposes peroxides, (b) acts as a metal deactivator or passifier, (c) inhibits oxidation of oil by preferential oxidation, (d) inhibits corrosion, and (e) reduces wear.

Other oxidation inhibitors are the principal ones listed below according to their classification:

1. *Sulfur compounds.* These consist of sulfurized esters, terpenes, olefins, phenols, and aromatic sulfides. Sulfurized sperm oil, a fatty ester, was one of the first additives used.

The terpenes in the form of sulfurized dipentane are effective inhibitors. Phenol sulfide in the form of diamyl phenol sulfide and dibutylphenol sulfide are more complex additives.

2. *Phosphorous compounds.* These consist of the dithiophosphate

mentioned earlier, triphenyl phosphite, tributyl phosphite, and calcium cityl phosphate. Because both sulfur and phosphorous in their elemental form are corrosive to nonferrous metals, they are always combined, as shown by the compounds mentioned above to provide oil-soluble organic additives. Tests show that both an oxidizing agent and an acid are necessary for attack on automotive type bearings. Zinc dithiophosphate is capable of combating both by (1) peroxide decomposition (oxidizing agent) and (2) protective film formation.

3. *Sulfur-phosphorus compounds.* Zinc dithiophosphate is included here again because it is prepared by a reaction involving phosphorous pentasulfide.

4. *Amines.* The amines are represented by phenyl alpha naphthylamine.

5. *Phenols.* The major types of phenols are 2, 6-diteriary-butyl-methyl phenol (paracresol) and metal salts, such as calcium salts of alkyl phenol formaldehyde.

Additives that are introduced to lubricating oils for other basic reasons, such as inhibiting rust and reducing wear, also act to passivate catalytic metal surfaces.

CORROSION INHIBITORS

Corrosion is taken here to mean the etching of engine parts, particularly bearings, and it is chiefly caused by the acids formed during the process of oil oxidation. The solution to the corrosion problem lies basically in providing a protective film over the metal surface with a compound that is catalytically inactive and noncorrosive and that has a strong affinity for the metals it is intended to protect. The affinity requirement prescribes a film that adheres tightly to the metal and resists displacement by the washing effect of the oil or other compounds.

Of the additives listed as oxidation inhibitors, the phosphorus compounds appear to be the more effective metal passivators, particularly zinc dithiophosphate, although sulfurized compounds, such as sulfurized olefins, have been used successfully.

RUST INHIBITORS

Temperatures above 32°F and below 212°F tolerate the presence of water and an atmosphere containing moisture. Gear cases, turbine oil systems, hydraulic systems, etc., are usually within that range. Water enters these systems chiefly by condensation of atmospheric moisture, especially during periods of high relative humidity. Oil, being hygroscopic, readily absorbs moisture. Cold-running and intermittently operated internal combustion engines are susceptible to water-ladened crankcase oil. In running cold, engines are not continuously hot enough to evaporate water so that it can be exhausted by proper ventilation. With intermittent operation, condensation progresses with great rapidity as the engine continues to cool after each shutdown.

While the presence of water is a major factor in the formation of cold sludges, as discussed previously, it is also responsible for rusting of ferrous metals in all types of mechanical parts made of iron or nonstainless steels. Those parts of equipment not in continuous contact with the lubricating oil are the major targets of rust.

To prohibit or retard the formation of rust on susceptible metals, there are three approaches:

1. Construct an entirely enclosed air-tight system equipped with an expansion chamber. With the exception of the expansion chamber, the system is otherwise filled to capacity with a highly refined, dehydrated (high dielectric strength) oil. Heat transfer systems are usually of this type.

2. Construct a system that is partially closed, having an open, but protected, oil reservoir which is appropriately ventilated and maintained under a slight vacuum (0.5 in. water). The amount of oil in such a system is held to a level to assure each frictional part an adequate supply and is always maintained above the entrance to the pump suction. A centrifuge is employed to remove any water entering the oil from either the reservoir or from leaky steam seals or oil cooler joints.

This is the system generally used for steam turbine lubrication, in which specially refined oils are employed. While every effort is made to eliminate water by the methods described, some moisture does enter the upper surfaces of oil conduits and bearing housings.

3. Use an oil containing an additive that inhibits rust. When this approach is combined with the system of approach 2 above, a rust-free system results.

Prior to 1939 rust was somewhat combated by the addition of about 10% used turbine oil to a new oil at the time of batch change. This had the effect of reducing the interfacial tension relationship of oil to water. This procedure obviously down-graded the quality of the new oil, resulting in a shorter service life. This method of incorporating rust inhibiting characteristics to reuse lubricating oils has been replaced by the addition of more sophisticated chemical compounds to the new oil prior to its introduction to a system. Such compounds are called *rust preventives* and some of them have the capability of combating oxidation and acidic corrosion.

RUST PREVENTIVE ACTION. The prevention of rust on ferrous metals of mechanical equipment, such as turbines, gears, hydraulic systems, and compressors, depends on a thin film of oil remaining on their surfaces, both during operation and through periods of shutdown. The presence of such a film is particularly important on uppermost surfaces left exposed to the atmosphere when the batch oil does not entirely fill the conduits and housings. Straight mineral oil that coats these surfaces during periods of turbulence, either by splash or vaporization, will eventually drain off during less turbulent flow or during shutdown, leaving them unprotected from rust.

By incorporating into the oil a compound with a high polar attraction to metal surfaces, a continuous, tenaceous film will provide a barrier against moisture. The action of the rust inhibiting compound, then, is *preferential adsorption* initiated by polar type surface-active substances having an attraction for metals. The process is called *plating*.

The plating process tends toward depletion of the additive, an occurrence known as *plating out*. Because of this, it is often necessary to replenish the additive after the oil has been in service for a period of time, especially if the system is new.

RUST-PREVENTIVE MATERIALS. The materials referred to here include only those employed as additives introduced into lubricating and hydraulic oils and should not be confused with those usually described as slushing compounds. A slushing compound is an oil containing a substance having preferential wetting properties that displaces water deposited on metal surfaces or protects them from water.

Ferrous surfaces submerged in a good quality straight mineral oil, free of water, are protected against rust because they are not exposed to atmospheric oxygen. If the oil does contain some water, however, rusting will occur. By introducing a suitable preferential wetting agent into the oil to provide a thin coating over the ferrous surfaces, rusting will not occur in either case.

Rust preventive additives are compounds that will displace moisture that may be already present on metal surfaces or provide a film capable of preventing any contact between the surfaces and surrounding moisture. This preferential wetting characteristic means that they are more *surface active* than water. The metal surface activity displayed by such components is due to *polar characteristics* whereby their adsorption behavior resembles a magnetic field in which the molecules become lines of force whose direction attracts the molecules of the metal to which they are exposed. Rust inhibitors, therefore, are compounds having sufficient polar action to adhere to metal surfaces, but which will not in any way exert further chemical action that may prove injurious. Depending on the type of antirust compound used, they are added to lubricating oils in quantities ranging from 1 to 2.5% by volume. Excessive quantities interfere with the lubricating quality of host oils, which may be of serious consequence to gear teeth, for example.

Like some of the other lubricating oil additives, rust inhibitors represent a number of different compounds. Some of the compounds effectively used are metal sulfonates—the radical of which (sulfonate) imparts solubility to the metal—amines, fatty acids, phosphates, metallic soaps of fatty acids, halogenated derivatives of some fatty acids, and oxidized wax acids. Rust inhibitors are not completely soluble in oil, nor would that be desirable, since it would preclude their concentration on metal surfaces.

Some additives have been known to remove the paint from the interior surface of reservoirs. Investigation has shown that paint removal is caused by oil containing amines as an additive. Amines are compounds formed by the replacement of one or more atoms of ammonia with a base-radical. The harm is not in the removal of the paint so much as in the subsequent fouling of machine parts and oil fuel lines of the machine, particularly if a hydraulic system is involved.

There is little need to paint interior surfaces of oil reservoirs. If the practice is deemed necessary, it should be made certain that the paint and oil are compatible.

ANTIFOAM AGENTS (SUPPRESSANTS)

Foam or froth is caused in one way by the entrapment of air within an oil mass in motion. Air enters oil in several ways:

1. Through any pump intake not fully submerged in oil
2. By allowing oil to discharge into the reservoir or sump at a point above the oil surface
3. In a splash system, where oil is broken up in small droplets and falls back into the main body
4. With the forceful ejection of oil from small orifices
5. During the filling of an oil reservoir through its only opening, the fill hole, by allowing the funnel to rest on the perimeter, during which the oil pouring through the spout provides an effective seal to entrap the air within the reservoir.

When air enters oil, it tends to rise to the surface in bubbles, a condition called *aeration*, the cause of foaming. Bubbles vary in size. The larger ones break rapidly when they surface; the smaller ones do not. The rate of breaking depends also on the surface tension of the oil. Some additives, especially detergents, and contaminants increase surface tension and present a serious foaming problem. With additives that plate out, foaming diminishes with time.

The presence of water encourages foaming, as do low temperature and high speed. A higher operating temperature and greater reservoir surface area will often reduce foaming to a safe level. Foaming often occurs more rapidly and to a higher degree with a new oil than with a used oil. The bubbles are more difficult to break in heavy weight oils than in lighter oils.

Foaming is a nuisance at any time, but when it causes an overflow through the breathers and fill-pipes or is responsible for a false oil level, as indicated in a bull's eye, it becomes harmful to the machinery as well as a safety hazard. Excess foaming in a bank of gear sets, for example, can cause the oil to spew out of the vents all over the sets and the surrounding floor.

Pure oil possesses negligible compressibility. When it contains air, the mass becomes compressible to an extent corresponding to the amount of air held. In a hydraulic system, any compressibility of the working fluid results in an interruption in both power and speed. In a lubrication system, foam interferes with the formation of hydrodynamic films, causing mechanical failures. Therefore, any abnormal foaming in an oil circulation system is reason enough to shut down the equipment to determine the cause and rectify it. The first effort in this regard is to locate any air leakage into the oil. Except when the inherent properties of the oil are the chief reason, foaming can be controlled by proper system design, which should include the following measures:

1. Vent the reservoirs properly (see 6 and 7 below).
2. Return oil through pipes extending into the reservoir oil body.
3. Eliminate oil splashing wherever possible, as by baffling.
4. Assure the exclusion of air from the pump suction opening by submerging it in the oil body or by repairing any damage to the suction pipe.
5. Maintain a constant but slightly elevated temperature level, especially in the reservoir. Temperatures maintained should be appropriate to the system and its work process.
6. Exclude condensate water by reservoir ventilation. Never cross ventilate.
7. Maintain a slight vacuum above the reservoir oil surface. This can be done simply by installing an exhaust fan at what should be the *only* vent or vent pipe leading out of the reservoir.
8. Provide an additional opening in the top of the reservoir for exhausting displaced air during the filling operation. This opening should be tightly capped after filling is completed.
9. Maintain an operating oil level capable of completely flooding all feed lines.
10. Furnish a pump large enough to maintain essential oil pressure throughout the system, using bypass control where there are pressure reductions to mechanical parts, such as bearings.
11. Reduce air intake to journal bearings that create low-pressure areas within the housing during operation.

ANTIFOAM AGENTS AND THEIR MECHANISM. It has become necessary in modern mechanical systems to depend on additives to reduce the influence of natural forces that lead to lubrication problems. This is true with foaming, as well as with oxidation and rusting. As insurance against foaming, appropriate chemicals are added to oils during their processing and, in some cases, after they are in service.

Foam is a group of air bubbles existing in a fluid mass, in this case, a combination of oil and atmosphere. They continue to exist in the atmosphere because they continue to be completely enveloped in a thin skin of oil which is tough enough and of such elasticity that they remain intact for a period of time. Heavy foaming occurs when bubbles are created faster than they are dissipated. If for some reason inherent in the oil or its contaminants, such as soaps or additives, the bubbles cannot be dissipated as fast as they are created, some step must be taken either to extend the size of the bubble and stretch the oil skin beyond its elastic limits, or reduce the elastic limit of the oil film so that it will rupture even under slight stretching, as with small bubbles. The elastic limit of fluids is more correctly called surface tension. Surface tension is measured in dynes per centimeter, or as ergs per square centimeter for straight surfaces. For bubbles, the question of internal pressure is involved, where pressure P in dynes per square centimeter is due to surface tension on a bubble having a radius of r cm for a liquid whose surface tension is T dyn/cm. Pressure is expressed as

$$P = \frac{2T}{r}$$

or, where r is equal to the mean radius of a bubble,

$$P = \frac{4T}{r}$$

The term surface tension is preferred to interfacial tension here, since the latter relates to the surface tension at the interface between two liquids, such as oil and water, whereas the discussion here involves oil and air.

To increase the rate of destruction of bubbles, it is necessary either to reduce the surface tension T of the oil, or increase the bubble size r as it appears on the surface of the oil mass. The expression "rate of destruction" is used here because all oils foam, but it is the rate at which they dissipate that is important. The purpose, therefore, is to prevent the formation of a stable foam.

Silicones are the more commonly used antifoam agents. In the form of methyl silicone polymers having viscosities ranging from about 1400 to above 4000 SUS at 100°F, they are only slightly soluble in mineral oils. Being of low surface tension, it is presumed that they impart sufficient impact on the oil skin that surrounds the air to break the bubble, or to allow small bubbles to combine to form larger ones. If the surface of the oil is maintained somewhat free of small bubbles, larger bubbles are encouraged to form, resulting in more rapid breakage.

The old saying that "if a little is good more is better" does not hold true with foam breakers. Because effective lubrication depends on the persistence of an oil film, any tampering with its surface tension is self-defeating. Therefore, only small doses of a foam inhibitor, on the order of <5 ppm, should be added, which is safe and in many instances all that is necessary for good results.

Antifoam agents are usually added to oils used in turbines, electric motors, spindles, and hydraulic and other lubricant-circulating systems. They are also added to internal combustion engine oils as well as to nonaqueous fire-resistant fluids.

VISCOSITY INDEX IMPROVERS

In the discussion of hydrodynamic oil films, the importance of viscosity at the operating temperature was emphasized. If the viscosity of lubricating oils remained constant throughout the working range of temperature change in any given system, the problems related to some systems would be simplified. Because lubricating oils are subject to a decrease in viscosity with an increase in temperature, it poses the problem of obtaining adequate frictional protection at the extremes of the range, particularly when the difference between minimum and maximum temperatures is relatively great. A typical example is the automobile engine, in which the operating range may be as high as 600°F, as measured in the winter from a cold start to normal operating temperature in the ring zone.

It has been determined experimentally that most engine part wear occurs during warm-up periods and not during the much longer periods of driving. The wear problem becomes more acute if the engine is started while the lubricating oil is congealed, or nearly so, due to low ambient temperature. There is also the problem of engine starting in cold environments due to higher torque.

There are many mechanical systems in which the normal change in viscosity is of little significance because their operating temperature range is not particularly wide. This is especially true for equipment that is in continuous operation for long periods, such as steam turbines, gear sets, and hydraulic systems. The foregoing statements, however, do not preclude other advantages gained by the use of oils having less viscosity-temperature response than others in a number of such systems.

The basis of this discussion can be summarized thus:

1. Lubricating oils vary in viscosity according to temperature changes.
2. Viscosity decreases or increases as the temperature rises or falls, respectively.
3. Some types of lubricating oils are subject to a greater viscosity change in response to temperature change than others.
4. The phenomenon of viscosity change with temperature may be

described as the viscosity-temperature susceptibility characteristic of an oil and is determined by the magnitude of viscosity change per degree of temperature change.

5. The viscosity-temperature susceptibility characteristics of an oil are signified by a value known as *viscosity index*. (The method of determining the viscosity index (VI) is described in Chapter 7.)
6. The value of the viscosity index becomes higher as the viscosity-temperature susceptibility characteristics of an oil become less.

The viscosity index may be improved in several ways:

(a) Removing the predominantly aromatic compounds from the oil during the refining process.
(b) Blending oils of light and heavy viscosity where in some instances the viscosity index of the blend is greater than that of the oils blended.
(c) By the use of polymers with a high molecular weight (10 to 20,000), known as viscosity index improvers, to sustain a low viscosity at low temperatures and a high viscosity at elevated temperatures.

The purpose of VI improvers is two-fold, (1) facilitate oil flow to assure adequate circulation at low temperatures and (2) provide maximum viscosity at high operating temperatures.

VI IMPROVERS AND THEIR MECHANISM. VI improvers are capable of raising the viscosity of an oil by a greater proportion at higher temperatures than at lower temperatures. They are therefore less susceptible to changes in temperature than the oil to which they are added. The reason for this is a matter of solubility and/or dispersion (colloidal) associated with substances having both a high molecular weight (800 to 20,000 is typical) and a high viscosity (3000 SUS at 210°F, typical).

Viscosity changes occur in the improvers themselves in an amount that depends on the shear stresses within the mechanical system and on the molecular structure of the compound. Short-chain polymers exert greater resistance to viscosity change than the longer-chain types. Any change in viscosity of the improver due to high mechanical shear stress is accompanied by a VI drop of the host oil. In some instances, a combination of different improvers are introduced into a base oil.

The more important VI improvers include the following polymers:

Polyacrylates (longer chain)
Polymethacrylate esters
Polyisobutylenes (shorter chain)

The shear stability of these improvers is greater in the order of their descending listing in contrast to the ascending order of VI effectiveness.

Because of this variation, VI improvers must be selected with meticulous care in order to meet different operating conditions. Other polymers include alkylated styrene polymers, polybutenes, polymerized olefins, and various selected copolymers.

VI improvers are used in motor oils (internal combustion engines) and in automatic fluid transmission oils.

Another capability of the VI improver is to modify the viscosity of motor oils so that the range between their low- and high-temperature limits spans two or more viscosity requirements specified by the Society of Automotive Engineers, according to SAE viscosity numbers. For example, one oil, called a multigrade oil, will meet the minimum viscosity range standard stated for an SAE 10W oil at 0°F and also the maximum viscosity range standard stated for an SAE 30 oil at 210°F. With this property the same oil can fulfill all the operating demands that would ordinarily require three oils, i.e., SAE 10W, 20W and 30, through the range of corresponding temperature levels. Multigrade oils are produced in several viscosity categories, including gear oils, SAE 80-90.

POUR-POINT DEPRESSANTS

Mineral lubricating oils congeal at some reduced temperature, depending on the amount of wax they contain. A congealed oil is no longer a Newtonian fluid. The reason for this is that the wax crystallizes to form a porous solid matrix (crystal growth) within an otherwise fluid mass when the temperature of the oil is reduced below its pour point. All petroleum oils are subject to gelling action, but those made from paraffinic base crudes have higher pour point temperatures and those of highly naphthenic crudes have much lower pour points. Being thixotropic, the gelled substance can be broken up by mechanical agitation to promote flow, after which it will return to its original state if the low temperature is maintained. Actually, the wax becomes less soluble in oil as the temperature is reduced.

A low pour point is important in oils exposed to low-temperature operation. In such cases, the pour point of an oil should be lower than the ambient temperature. Low-temperature refrigeration systems and automotive engines in winter operation require lubricants with low pour points. In refrigerators, congealed oil on the surfaces of the evaporators would interfere with heat transfer capacity, and in automotive engines, a thick oil would add to the resistance of engine starting and effective lubricant circulation.

The pour point of oils can be reduced by

1. Dewaxing the lubricating oil stocks during the refining process as described in Chapter 6.

Table 10-2. GEAR OIL ADDITIVES

Type of Additive	*Function in Lubricant*
POUR-POINT DEPRESSANT High-mol.-wt. polymer or wax naphthalene condensation product	Reduce the pour point or channel point of the lubricant so that it will flow freely at reduced temperatures.
VISCOSITY INDEX IMPROVER High-mol.-wt. resin	Improve the viscosity-temperature relationship of the lubricant to minimize viscosity spread over service temperature range.
FOAM DEPRESSANT High-mol.-wt. silicone polymers	Prevent foam or accelerate foam collapse.

EXTREME PRESSURE AND POLAR AGENTS

	Function	*Type of Gear*	*Normal Operation*	*Heavy Duty Extreme Pressure Lub. Required*
Type 1 No additive	General lubrication	Spur	yes	
		Bevel	yes	
		Spiral bevel	yes	
		Annular or internal	yes	
		Helical or herringbone	yes	
		Worm	yes	
		Rack and pinion	yes	
Type 2 Noncorrosive type sulfurized fat	Prevent welding by forming an easily sheared film: iron sulfide	Spur	yes	
		Bevel	yes	
		Spiral bevel	yes (and break in)	

Table 10-2, Continued

Type 3 Noncorrosive type Lead soap Sulfur Phosphorus	Prevent welding by (a) forming an easily sheared film: iron sulfide, lead sulfide, etc. (b) forming a low melting point phosphorus alloy	Spur Bevel Spiral bevel Annular or internal Helical or herringbone Worm		yes yes yes yes yes yes
Polar agent (optional)	Reduce friction: Metal-to-metal contact or between easily sheared films	Rack and pinion		yes
Type 4 Corrosive type Lead soap Active sulfur	Prevent welding by forming an easily sheared film; iron sulfide, lead sulfide, etc.	Hypoid	yes High speed—low torque with shock loading	
Type 5 Noncorrosive type Chlorine Sulfur Phosphorus	Prevent welding by (a) forming an easily sheared film: iron chloride. This action is catalyzed or speeded up by the presence of sulfur and its compounds (b) forming a low melting point phosphorus alloy	Hypoid	yes High speed—low torque	yes Low speed—high torque
Type 6 Noncorrosive type Lead soap Sulfur Chlorine	Prevent welding by forming an easily sheared film: iron sulfide and iron chloride. The formation of iron chloride is catalyzed by the presence of sulfur and its compounds.	Hypoid	yes High speed—low torque	yes Low speed—high torque
Polar compound (optional)	Reduce friction: Metal-to-metal contact or between easily sheared films			

Courtesy, American Society of Lubrication Engineers.

2. Adding selected compounds called pour point depressants.
3. Applying both processes 1 and 2. It is more economical to combine the processes than it is to obtain an ultra-low pour point by refrigeration means alone.

POUR-POINT DEPRESSANTS AND THEIR MECHANISM. The mechanism of pour-point depressants is based on their ability at low temperatures to coat wax crystals so as to reduce or prevent their growth, to modify the wax crystal form, and also to inhibit their joining together (interlacing) to form the gelatin-like mass or matrix.

The materials used for depressing the pour point of lubricating oils include:

polymethacrylate of high molecular weight
wax alkylated or polyalkyl-naphthalenes
polyalkylphenol esters
methacrylate polymers
other polymers with high molecular weight

Pour-point depressants do not affect the cloud point or viscosity of the free oil within the congealed mass. They are able to depress the pour point of low-viscosity oils up to 50°F with 1% depressant, although the usual levels are within the range of 0.01 to 0.3% wherein the pour point reduction of about 30°F is obtained with the addition of 0.15% by weight.

ANTIFRICTION COMPOUNDS

The basic purpose of the antifriction compounds, also known as *friction modifiers,* is to reduce the coefficient of friction between the surfaces of two rubbing solid bodies under boundary conditions. Such compounds possess high persistance of film, but they do not necessarily serve in the same capacity as *film strength additives,* or *extreme pressure additives,* because the mechanisms differ in each case.

Antifriction compounds have the property of *lubricity,* associated readily with fixed oils such as castor, sperm, and lard oils. Any of these three oils has a greater degree of oiliness than, say, an acid refined mineral oil and can raise the level of oiliness of the mineral oil with proper blending. The oiliness property alone does not necessarily preclude wear or welding of opposing surfaces, any more than would a straight mineral oil, under boundary lubrication conditions. Therefore, the compounds as they are considered here are strictly friction-reducing agents. Most substances associated with oiliness have pronounced wetting characteristics, a term often used as a synonym for oiliness. Oiliness is also related to the phenomenon of creep, which, in turn, is a function of metal surface energy, better known as surface tension, at the fluid-metal interface. It is the ability to creep that qualifies some fluids as good penetrating compounds.

Oiliness is not an intrinsic property of an oil, but is manifest only when it becomes a working element between the opposing surfaces of two rubbing solid surfaces. Mineral oils that are solvent treated to a high degree lose their natural oiliness, because the process extracts much of the constituents that are responsible for the property of lubricity.

The oiliness factor is related to the size of molecules as measured by molecular weight, increasing as the size of molecules increases. The proportionality of increase, however, is a function limited to those substances that also possess adequate polar characteristics, i.e., the affinity (electrostatic) of the fluid relative to metal surfaces. This affinity is the measure of the capacity of the compound to form a monomolecular surface film essential for effective boundary lubrication. While the monomolecular film is not sufficient to protect a surface against metal-to-metal contact under pressure, it does provide a base upon which additional layers (stratification) may exist.

The theory holds that an unbalanced electrostatic charge upsets the normal neutral (positive charge = negative charge) state of molecules by an unequal distribution within its mass. This causes the molecules to act as having an unbalanced charge and so become reactive with any field of electrostatic force with which it comes in contact. This field is found in close proximity to metal surfaces having some degree of activity (surface energy), depending on the kind of metal involved. An orderly reorientation of the molecules due to attraction of their unbalanced ends to the metal occurs, causing them to adhere to the surface as an *absorbing layer* of *polar chain molecules*. It is this layer that forms the stationary phase of both the boundary and hydrodynamic films described in Chapter 9. If the field of electrostatic forces induced by the metal molecules and the mutual attraction of the fluid molecules is strong enough, a film composed of multimolecular layers may be formed. Beyond this field, perpendicular to the surfaces, the fluid molecules are free to become displaced by an external force imposed upon them. However, the greater the ability of an oil to develop a monomolecular layer, the better it will behave as an antifriction compound, because such behavior is an indication of greater oiliness. Because a boundary lubrication condition makes impossible the development of a hydrodynamic film, it must depend entirely on the oiliness of the lubricant. Antifriction compounds are often added to lubricating oils used in systems capable of developing hydrodynamic films to protect the frictional parts during a boundary condition prevalent during the starting-up process.

The oil that shows the greatest reduction in coefficient of friction between the same two bodies, when all other conditions remain the same, is the one having the greatest oiliness. Because of the persistence of film resulting from oiliness, both the static and dynamic friction are reduced, leading to a decrease in starting torque and running power requirements. The question of wear is not considered in relation to oiliness unless accompanied by another property associated with the strength of the film or the resistance of the film to be ruptured under heavy or shock loads. The extreme pressures that cause seizure and welding are also not considered under the present subject. The latter two conditions represent the variations in behavior mechanisms mentioned in the introductory paragraph of this discussion. It is discreet to use compounds that possess all the qualities appropriate to the operating circumstances. For this reason, it is usual for lubricating oils to be compounded with additives having both oiliness and film strength properties identified by the general name of *film strength improvers.*

Oiliness varies with the nature of the surface materials and also with the method used in the preparation of those surfaces. Superfinishes tend to reduce the effect of oiliness to reduce friction, although such surface conditions increase load-carrying capacity when considered in terms of area in contact.

Compounds employed as oiliness additives include lard oil, sperm oil, organic fatty acids and their esters, amines and other oiliness carriers that are more or less acidic and contain oxygen, sulfur, halogens and similar elements in molecular bond. Oleic acid is an effective oiliness agent and reaches its maximum friction-reducing effect when blended with mineral oils in quantities of about 3% by volume. It is corrosive to zinc, which should not be used in dispensing equipment, such as galvanized reservoirs.

Moisture has a decided effect on the friction-reducing ability of some fatty acids. When fatty (saponifiable) oils are compounded with mineral oils, the effective ingredient in reducing friction is the free fatty acid formed by hydrolysis during lubrication. The moisture that either develops through chemical action or is introduced from the surrounding atmosphere (hydroscopically) is necessary to increase the affinity of fatty compounds for metals. Oiliness is imparted to mineral oils by the introduction of free fatty acids. Formerly fixed oils were used for this purpose.

ANTIWEAR AGENTS

In the discussion of oiliness additives, it was brought out that their primary purpose is to reduce friction. While friction is one of the causes of wear, it follows that any reduction in friction results in a corresponding decrease in wear rate. This is true under some circumstances related to operational behavior and conditions, e.g.: accurate alignment, light loads, absence of shock, continuous running and somewhat constant speed. Where conditions are contrary to those mentioned, the property of pure oiliness does not normally provide a strong enough film to deter rupture and prevent metal-to-metal contact. The reason is principally one of higher temperature in the latter case.

The three basic categories of operating pressure-temperature levels for boundary lubricated frictional components are normal, severe, or heavy duty, and extreme pressure.

Normal Temperature-Pressure Levels

Under mild boundary or mixed film lubricating conditions, this category is usually satisfied with an oil containing an oiliness additive having sufficient persistence of film to maintain a low coefficient of friction at the film temperature involved. If continually exposed to temperatures above decomposition levels of the fatty materials used, the monomolecular film becomes displaced and metal-to-metal contact occurs. An increase in friction accompanied by rapid wear follows such displacement. It is also necessary to consider the melting point of the additive in selecting compounded wear or friction reducing oils for low-temperature operation. This is true for all additives whose chemical reaction to bearing surfaces does not occur rapidly enough below their melting point to form a protecting film of sufficient magnitude.

The operating conditions that limit the use of oiliness additives alone were previously enumerated. If for any reason, the limitations were to be violated, as they would be with load surges, misalignment, vibration, etc., oiliness additives should be supplemented with a wear agent whose chemical reaction to bearing surfaces occurs at or near the decomposition temperature of the former (Fig. 10-1). It is customary to so treat a number of lubricating oils for the reason described and also to give greater versatility to individual

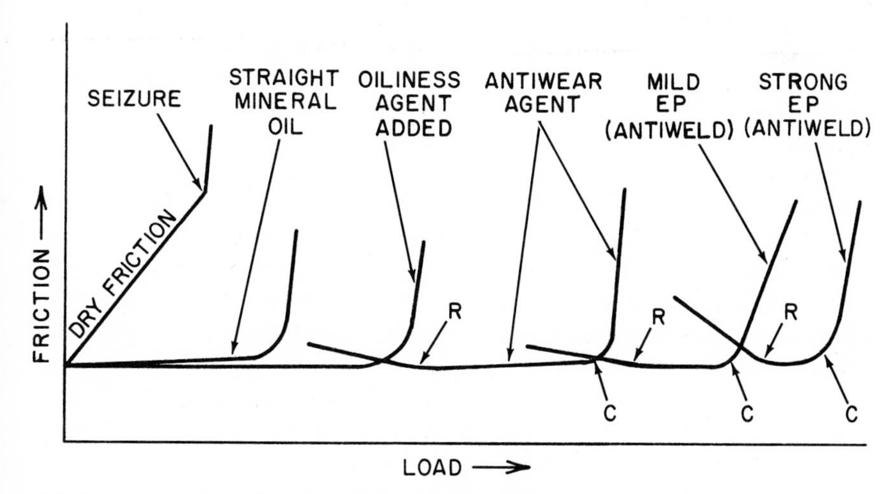

Fig. 10-1. Frictional behaviour of lubricants with increasing loads, and higher temperatures due to greater unit pressures. R, Point of reaction of additives with metal surfaces to form coatings that will sustain load and prevent welding. C, point at which reaction ceases.

lubricants. Where operating temperature conditions warrant, the additive possessing pure oiliness may be omitted from the blend, depending solely on the antiwear additive to supply the required film.

Normal conditions involves those operations for which hydrodynamic lubrication to combat sliding friction is not possible. Journal bearings to which oil is applied periodically (all-loss) are reduced to *sparce* lubrication sometime between applications, open-throated worm gears, guides (nonstep), and lead screws are some of the boundary lubricated components that require at least an oiliness additive.

Heavy Duty

This category represents those operating conditions where loads are of such magnitude as to sustain or threaten to sustain boundary lubrication. The high pressures involved accompanied by a rise in film temperature preclude the use of a straight mineral oil, or such oils compounded with a pure oiliness additive as effective lubricants. What is needed is an additive having a high chemical reaction to metal surfaces in order to develop a strong persistence of film having a low shear strength. It is within such films that the rubbing action occurs and not at the opposing metal surfaces.

The surface energy displayed by the additive compounds required in this category is a function of the thermal conditions existing at any given time. It was noted previously that oiliness compounds required certain temperature levels to become effective. The same holds true with antiwear agents used successfully to meet the severe operating conditions represented in this category. Again, as with anti-friction compounds, antiwear agents have their temperature range limits. At the low end is the temperature at which they become effective metal surface reactors. At the high end is the temperature of their decomposition or deterioration. Beyond those limitations in either direction, an increase in friction occurs. Except during periods of initiating motion (startup), deficiency of the additive is not too serious at the low end because the temperature is not sufficiently high to displace the film formed by the oil or additive carrier. Beyond the temperature of decomposition, however, the film capable of preventing metal-to-metal contact does not exist, resulting in excessive wear or galling.

The reaction of antiwear additives with metals produces films of sulfides, chlorides and phosphide, depending on the constituents of the agent employed.

Wear occurs when the protective lubricating film breaks down to permit metal-to-metal contact. Films break down when interfacial temperatures reach the critical level of the film material. Interfacial temperatures vary according to the load magnitude, rising as the pressure between rubbing surfaces increases. Breakdown temperature differs for different materials. Therefore, antiwear agents must be selected to meet the pressure-temperature characteristic of the frictional system to which they are to be applied.

One other consideration involves the surface metals and their susceptibility to corrosion by individual compounds used as antiwear agents. Some metals are less susceptible than others. Highly active compounds produce films of greater effectiveness, but they are also more corrosive, how much so depending upon the kind of metals used. This means that greater liberties may be taken with some types of steel on steel than with steel on copper, for example. It is significant that steels are capable of withstanding higher loads than any other engineering material. For softer metals the problem is not significant, because they are unable to withstand the crushing pressures of load approaching extreme levels.

Some antiwear agents, such as the metallic soaps, provide a "slippery" film to afford effective lubrication even though the interfacial temperature remains below their softening point. The reason for this is that such films have a lower shear strength than the base metal. Such additives replace the pure oiliness compounds in the areas of lower pressure-temperature values.

Where the P-T condition is relatively mild, the phosphorous compounds are highly effective antiwear agents. They include the older reliable compound *tricresyl phosphate* and the versatile *zinc dithiophosphate*. The versatility of the latter is evident by its added ability to decompose peroxides and thus act as an antioxidant in automotive engine lubricating oils, and also as a metal deactivator. Zinc dithiophosphate is particularly effective in hydraulic oils, as well as in motor oils. Its use has resulted in a much greater service life of hydraulic pumps than any antiwear agent previously used.

Other antiwear agents are the lead soaps with a somewhat higher P-T range. The more effective leaded lubricants are those compounded with lead naphthenate, although lead oleate is sometimes used. Lubricants containing lead soaps form films that are strong and highly persistent, which enable them to combat high unit pressures and shock loads. They are ideal lubricants for heavily loaded gears and are gradually replacing the oiliness compounds, such as compounded cylinder oils for worm gears. Leaded oils of high viscosity are used for the lubrication of large open gears and screw-down threads of rolling mills employed in heavy industry.

Organic sulfur and chlorine compounds, as well as sulfurized fats and certain amines, are also used. Such compounds form low-shear-strength films, such as iron chloride and iron sulfide.

Extreme Pressure

This category includes those antiwear agents capable of combating the extreme pressures developed in hypoid gears. The P-T conditions in this type of transmission reach levels that are conducive to localized welding of high spots on opposing steel surfaces during metal-to-metal contact. To prevent this, an antiwelding material is necessary, such as the element sulfur. In addition to the antiwear qualities required by heavy duty, extreme pressure additives must possess high surface reaction accompanied by antiweld characteristics.

The most effective hypoid gear lubricant appears to be an oil containing an additive made up of a combination of sulfur and chlorine compounds. The sulfur phase provides the antiweld action, and the chlorine phase also supplies some measure of cooling to the process, probably due to its antifriction properties.

To distinguish antiwear agents from extreme pressure additives consider two factors:

(a) Wear prevention agents act to reduce local temperatures by a chemical polishing action that causes a somewhat uniform distribution of the load over a large surface area.

(b) Extreme pressure additives react with the metal under high pressure to form a coating that will either sustain the load, or failing to do so, prevent the contacting metal surfaces from welding together.

The term "mild extreme pressure lubricants" is used too loosely. It applies to the form of sulfur that may be present in the additive. The form of sulfur may be *active* or *inactive* so-called, which applies to the staining effect of either type on copper, brass, or bronze at reduced temperatures, say, below 300°F. Active sulfur will stain such materials even at low temperatures, whereas the inactive type will not when the film is maintained below 300°F. The term "mild extreme" is *not* objectionable when used according to the above definition. It is objectionable, however, when it is used to describe the capabilities of lubricants containing antiwear properties, but not endowed with the feature of antiwelding. So used in the market place, it becomes a subtle tool for the unscrupulous or uninformed vendor of lubricants.

ADHESIVE COMPOUNDS

It is often desirable to use an oil or grease that will neither splash or throw readily as a result of centrifugal force, impact, or turbulence. The reason for this is that lubricants often cause spoilage of certain manufactured products by contamination or soil. Paints, foods, pharmaceuticals, and textiles are products in this category. Because mineral oils will splash and/or throw with ease corresponding to their viscosity, it is necessary to modify their adhesiveness and cohesiveness to reduce this tendency. Adhesive modifiers are used to impart *tackiness* in an effort to prevent particle separation, either from the lubricant body or from surfaces to which the lubricant is applied. Adhesive type oils are also able to reduce drip, due to gravity or squeeze action, from frictional parts. Thus, such lubricants provide a sustained protective film between opposing surfaces of all-loss bearings that are lubricated at extended intervals.

The adhesive property of lubricants so modified should not be confused with the film developed by antiwear additives through chemical reaction. The action is one of stickiness which causes some substances to "string out" with any attempt to separate two bodies when such substances are sandwiched between them. For this reason, adhesive compounds are often called *stringiness additives*.

Rubber in the form of latex is one of the earliest adhesive additives employed, both for oils and greases. One of the disadvantages of rubber is that it loses its stringiness characteristics after a short time, probably due to oxidation, especially at slightly elevated temperatures. The additive in common use takes the form of a linear polymer of isobutylene having a molecular weight of from 30,000 to 100,000. Both of the above materials impart a high degree of adhesiveness and cohesiveness to new oils. They are used in aluminum base lubricating greases, a product ideal for flexible couplings, because when such greases are so treated, a high resistance to oil separation due to centrifugal force is imparted.

Oils and greases containing adhesive additives are excellent for use in all-loss systems for the purpose mentioned previously. The success of stringiness oils is due to the large molecules of adhesive additives. When, for some reason, the size of the molecules is reduced, the oil carrier will tend to revert to its original state but without tackiness. The shearing effect of working components, such as hydraulic pumps, will cause such a molecular size reduction. Claims that adhesive type oils will reduce leakage from circulation systems is true for short periods. However, repeated cycling diminishes this possibility because of the reasons stated. The problem is best solved by good seal maintenance or through the use of oils containing seal expanders described next.

SEAL EXPANDERS

Circulating oil systems are subject to leakage due to seal shrinkage or deterioration and loose-fitting mechanical joints. The rate of oil leakage increases with the pressure maintained within the system. The common cause of seal deficiency is due to shrinkage during service. Shrinkage does not indicate a complete loss in resiliency, although if allowed to continue it leads to deterioration. Loss of oil as referred to here does not include that due to incorrect or careless seal installation. That can be corrected only by manual corrective action or replacement.

New systems, particularly hydraulic operations or systems subject to recent overhaul, will usually have tight seals. Leakage, however, increases gradually with time and use by reason of seal wear or shrinkage. Seals so affected do not lose their sealing potential if one of two actions is taken. The first would be to reduce the size of the orifice to meet the amount of seal shrinkage, an impractical, if not an impossible task. The second is to expand the seal sufficiently to accomplish the same task in reverse, a practical and economical solution.

Seals are usually made of synthetic materials having some of the required properties of natural rubber. Oil, however, acts as a rubber solvent, softening the seal and leading to its final deterioration. There are a number of synthetic rubber materials used, such as neoprene and Buna-S. These materials, in general, shrink to some extent when subjected to frictional heat. They also wear when in snug contact with a moving surface.

Seals are made in many forms and shapes, usually identified with a letter in the alphabet which it may resemble, such as: O-ring, U-ring, and M-ring.

Retardation or elimination of leakage through the static and dynamic joints of an oil circulation systems can be accomplished by exposing synthetic rubber materials to compounds added to the base lubricant that cause the seals to swell. Since this expansion must be limited, the swelling compounds must be carefully selected and measured.

Compounds employed as seal expanders include *aromatics* (cycle hydrocarbons having relatively high specific gravity and solvent properties), hologenated hydrocarbons, and organic phosphates.

EMULSIFIERS

Pure oil and pure water do not mix with each other, that is, they are mutually insoluble. They will form a temporary emulsion when shaken but will immediately separate when turbulence ceases, with the lighter weight oil rising to the top.

By adding a third ingredient, such as an acid sludge or fatty compound, to the oil-water mixture, a more permanent emulsion will occur. An emulsion usually appears as a froth similar to that formed as suds in soapy water. The third ingredient is an emulsifier, defined as a substance that will bring about an emulsion of an oil by causing intimate contact between it and the water. The emulsifier acts as a wetting agent to reduce interfacial relationships. This results in one of the substances becoming microscopic droplets within the other. Whether the oil or water becomes the droplet depends on the type of emulsifier.

There are two basic types of emulsion, *water-in-oil* and *oil-in-water*. For example, an emulsifier made up of sulfurized sperm oil will form a water-in-oil solution, known as an *invert emulsion*. Because sulfurized sperm oil is an effective antiwear agent, it produces emulsions capable of protecting

Table 10-3. ADDITIVES—PURPOSES, MECHANISMS, AND INGREDIENTS

Additive	*Purpose*	*Mechanism*	*Ingredients*
Detergents Typical use: engine lubricating oils	Cleansing agent. Acts on surfaces exposed to lubricating oil	Causes oxidation products to be soluble in oil. Chemical actions	Metallic soaps Sulfonates Phosphonates Thiophosphonates Phenates or phenate sulfides Ashless types Polymerized olefins Acrylated amines Pyrrolidones Carboxylic acids Polyglycols
Dispersants (Detergent) Engine lubricating oils	Suspension agents	Isolates oxidized particles and keeps them in suspension within the oil mass. Peptization	Esters Methacrylates Acrylates Alcoholates Polycarboxylic acid
Oxidation inhibitors (Antioxidant) Typical use: engine lubricating oils, circulating oils (turbines, gears, etc.), fluid clutches. Greases. Reuse systems	Retard chemical decomposition of oils due to oxidation	Preferential oxidation and metal deactivation by coating (passifiers)	Sulfur compounds Sulfurized esters, terpenes, olefins, phenols, and aromatic sulfides Phosphorus compounds Dithiophosphates Triphenyl phosphite Tributyl phosphite Amines Phenyl alpha naphthylamine Phenols 2,6-ditertiary-butyl-methyl phenol Metal salts

Table 10-3 (*Continued*)

Additive	*Purpose*	*Mechanism*	*Ingredients*
Corrosion inhibitors Typical use: Engine lubricating oils, fluid clutches	Prevent corrosion of metal parts by acids in oil contaminants	To neutralize the acids formed by oxidized oil materials and provide a protective film on metal surfaces (passivators)	Zinc dithiophosphate Sulfurized olefins Metallic phenolates
Rust inhibitors Typical uses: circulating lubricating oils, gear oils	Retard or prevent rust	Coating of surfaces exposed to moisture in oil or atmosphere, or during cold operation (preferential adsorption)	Metallic sulfonates Amines Fatty acids Phosphates Oxidized wax acids
Antifoam agents (Suppressants) Typical uses: Fluid clutches, gear oils	To break bubbles that cause foam and to prevent the occurrence of a stable foam	Reduce surface tension	Silicone polymers
Viscosity index improvers Typical use: Engine lubricating oils, fluid clutches, aircraft hydraulic systems	Reduce the change in viscosity with any change in oil temperature	VI improvers are less susceptible to viscosity change than the oil. A function of solubility and/or dispersion of substances having high molecular weight and high viscosity	Polymers Polyacrylate (longer chain) Polymethacrylate Polyisobutylenes (shorter chain)
Pour-Point depressant Typical use: gear oils, refrigerating oils, low-temperature systems	Reduce pour points of lubricants to achieve flow at low temperature	To coat wax crystals and prevent their growth, modify crystal form, and prevent their coagulation to form a matrix	High-molecular-weight polymers Polymethacrylate of high molecular weight Wax alkylated—or polyalkyl—naphthalenes Methacrylate polymers, etc.
Antifriction compounds Typical use: boundary lubrication systems, metal cutting fluids	Reduce coefficient of friction between rubbing surfaces. Reduce temperature. Reduce static coefficient of friction (startup)	Slipperiness, oiliness, lubricity. Absorbing layer of polar chain molecules versus greater freedom of flow of multimolecular layers imposed	Fatty oils Sperm Lard Organic fatty acids and their esters Oleic acid Amines

Antiwear agents Typical use: heavily loaded bearings and gears	Reduce friction and wear. Prevent scoring, seizure, and reduce temperature of rubbing	Chemical reaction on metal surfaces forms protective film having lower shear stress than metal on which it is formed. Prevents bonding of opposing surfaces in the case of fluid film rupture	Zinc dithiophosphate Tricreasyl phosphate Lead soaps Organic sulfur and chlorine compounds. Sulfurized fats (e.g., sulfurized sperm oil)
Extreme pressure agents Typical use: hypoid gears	Acts similarly to antiwear agents, except that their antiwelding characteristics are much more pronounced. Actually the sulfur-chlorinated compounds listed for antiwear agents are antiweld compounds used for hypoid gears, the most vulnerable frictional system in this regard.		
Adhesive compounds Typical use: lubricants in all loss systems, eccentrics, cams, open gears, impact surfaces, centrifugal motion systems	To impart greater adhesiveness and cohesiveness (stringiness) to lubricants to prevent throw, splash, etc., by centrifugal force, impact or turbulence (splash)	Substances used impart tackiness to the lubricant mass. Additive of high molecular weight, takes up space between adjacent molecules to resist their ease of separation	Latex Linear polymer Isobutylene
Emulsifiers Typical use: steam cylinders, air tools, stern tubes, metal-cutting coolants.	To secure an emulsion of oil and water	Acts as a wetting agent to reduce interfacial tension. Forms oil-in-water or water-in-oil emulsion	Blown rapeseed oil Acid sludge Acidless tallow oil Lard oil
Seal expanders or swellers Typical use: hydraulic fluids	To increase tightness of seals by swelling in order to reduce leakage	Elastomers are chemically modified by certain aromatic materials to cause limited growth	Aromatics Halogenated hydrocarbons Organic phosphates

the rubbing surfaces of pneumatic cylinders, as in air tools. Moisture forms in pneumatic cylinders when the compressed air is reduced in volume sufficiently to raise the humidity level above the saturation point. Other invert emulsifiers in addition to the sulphonates include fatty esters, phenolic derivatives, and carboxylates. They are recommended for the lubrication of air compressor cylinders in which pressures are raised to about 150 psi or higher. Invert emulsifiers are also used in cylinder oils used to lubricate steam cylinders, particularly in high-pressure superheat systems. They also serve as oiliness agents in lubricants for worm gears and in emulsion type fire-resistant hydraulic fluids in which they are further fortified with other agents, such as metal dithiophosphates, to improve antiwear characteristics.

Lard oil and blown rapeseed oil are two fatty oils widely used to produce an oil-in-water emulsion. Cylinder oils compounded with 5 to 10% emulsifier are used to lubricate steam cylinders under low superheat or saturated conditions where moisture is continually produced due to the conversion of thermal energy to mechanical energy. The emulsion film that forms on rubbing surfaces withstands the washing effect of water, whereas a straight mineral oil film would soon be washed off. Cylinder oils compounded with oil-in-water emulsifiers are also used to good advantage in worm gears.

Blown rapeseed oil is used in marine industries to combat the washing effect of seawater, such as that which leaks into the stern tube bearings of ocean-going vessels. Many of the older ships are equipped with lignum vitae lined stern tubes that are dove-tailed into the casting. A bronze sleeve seated onto the propeller shaft, to prevent external corrosion, completes the bearing pair, except for the necessary packings and stuffing box. Bearings of this kind are lubricated successfully with only seawater because of the low coefficient of friction developed between wet lignum vitae and bronze. However, lower cost materials replace the bronze on vessels sailing in water containing abrasives (sand, mud, etc.). Babbitt and cast iron frictional surfaces require an oil compounded with blown rapeseed oil for their lubrication and protection against water-wash. Stern tube lubricants are usually of high viscosity, on the order of SAE 50.

Emulsifiers such as acid sludges are used to produce some metal-cutting oils of the soluble type. The latter are also oil-in-water emulsions, sometimes fortified with sulphur compounds to attain some exteme pressure characteristics.

Emulsions are to be avoided in most lubricating oil systems. Avoidance is difficult however in some circulating systems that are subject to the entrance of water along with acidic foreign materials that promote emulsions.

11

LUBRICANT APPLICATION AND PURIFICATION

Effective lubrication is a function based on the use of a properly selected lubricant for a given frictional requirement, and the assurance that the lubricant is correctly applied. The following general rules apply in all cases:

Type of lubricant. Select the lubricant that meets the physical and chemical characteristics required as determined by frictional conditions of load, speed, temperature, and environment according to the frictional laws involved.

Quantity of lubricant. A sufficient amount of lubricant must be continuously supplied to sustain an adequate minimum film thickness and maintain safe temperature levels.

The rate of oil application varies according to the method of application. The difference is shown by comparing the oil feed rate of a wick-feed oiling system with a mist oil system serving the same bearings, using a 600 SUS at 100°F oil in both cases:

Wick-feed oiling system	
Oil consumption per 24 hr, 1 bearing	3 oz
Oil consumption per 24 hr, 100 bearings	300 oz
Mist oil system	
Oil consumption per 24 hr, 1 bearing	0.6 oz
Oil consumption per 24 hr, 100 bearings	60 oz

Thus in this example the mist oil system requires 240 oz less oil in 24 hr for each 100 bearings.

Interval of lubricant application. Lubricant must be made available at frictional interfaces with regularity and thus prevent both film deficiencies and excesses. Intervals are specified at the end of the chapter.

Method of lubricant application. The application of a lubricant should be made in a manner that assures complete distribution between the frictional surfaces.

The method of application depends on five major factors: (1) design of the frictional system, (2) type of lubricant required, (3) potential frictional heat, (4) functions of the lubricant, and (5) required frequency of application. Some variables associated with each of the five major factors are:

(a) Bearings, gears, or cylinders may be involved.
(b) The lubricant may be an oil, grease, or other type.
(c) The lubricant may function also as a cooling agent.
(d) The lubricant may function also as a hydraulic medium. Turbine oils, for example, serve to lubricate bearings (and gears) and also to actuate the governor mechanism.
(e) Frequency of lubricant application depends on one or more of the following:

Temperature	Grooving techniques
Speed	Type of frictional part
Load	Type of lubricant used
Cleanliness	Method of lubricant application
Leakage	
Effect of seals	
Clearance	

Methods of applying lubricants are numerous, ranging from simple hand oiling to automatic central distributing systems serving many frictional points from a common reservoir. The methods used are described below.

OIL APPLICATION

Manual Oiling

Common oil can operated by thumb pressure. Intermittent application permits boundary lubrication only, resulting in a "feast or famine" type of lubricant supply.

The oil holes through which oil is fed to the bearings by oil cans should be protected against entrance of contaminants by hole covers (Fig. 11-1).

Semiautomatic Oil Appliances

Bottle oiler, mounted on bearing cap of plain bearing, applies oil directly to journal (Fig. 11-2). Actuated by pumping action caused by vibration

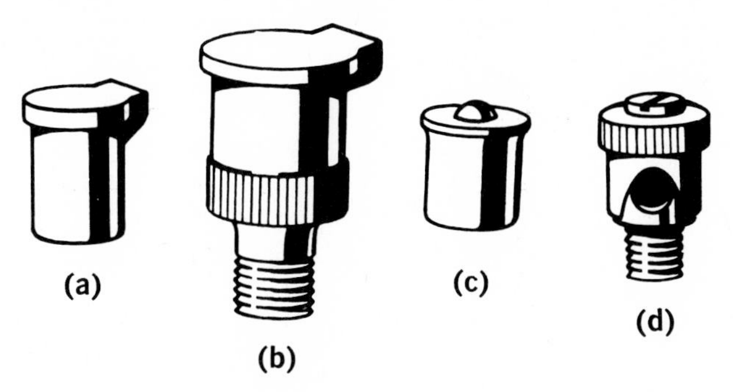

Fig. 11-1. Oil cups and hole covers. (a) Plain drive, spring hinged cover, (b) threaded type with spring-hinged cover, (c) ball valve, (d) revolving sleeve. *Courtesy, Gits Brothers Mfg. Co.*

transmitted through stem or plunger from journal (on which stem rides) to oil in the bottle. The quantity of oil fed depends on journal speed and bearing temperature. One bottle per 6 in. of bearing length is the usual requirement. The bottle oiler reduces consumption on an all-loss system and ensures a steady supply. It does not feed when journal is at rest.

Constant level oiler maintains oil level in main reservoir of ring-oiled bearings and similar oil application methods. It is actuated by a drop in reservoir oil level, which causes air entering the oiler to force more oil into the reservoir until the constant level covers the opening to the oiler.

The *drop-feed oiler* is actuated by gravity, and flow is turned on or off by a shutoff valve (Fig. 11-3). A needle adjustment controls the rate of oil supply. Gravity-actuated drop-feed oilers are mounted on the bearing cap, gear case, etc., or they may be located remotely with feed lines and sight feed valves to direct flow (Fig. 11-4) into one or more feed lines. Drop feed is an efficient method of lubricating bearings, slow-speed gears, and short ways and guides. When oil is fed by gravity from a drop-feed dispenser to one or more frictional points through sight feed valves, a solenoid can be used to stop the oil flow automatically (Fig. 11-5).

To provide sufficient pressure to enable the oil to be sprayed or raised above the dispenser, air pressure is used (Fig. 11-6). It is a central oil reservoir that supplies oil to remote or elevated bearings where gravity feeding is not feasible. The type shown includes filter, pressure regulator with gauge, relief valve, and filter cap. The air line pressure of 80 to 100 psi is reduced to below 30 psi by the regulator, although some dispensers are equipped for air pressures up to 125 psi.

Wick-feed oiler is actuated by capillary action through wool or cotton wicking (Fig. 11-7). It is discussed later in a separate article.

Pad oiler consists of a felt pad held against a journal by springs and fed by wicks or drop-feed oiler (Fig. 11-8). Oil usually drops into a bottom reservoir from the pad and is reused continuously. It is used on partial sleeve

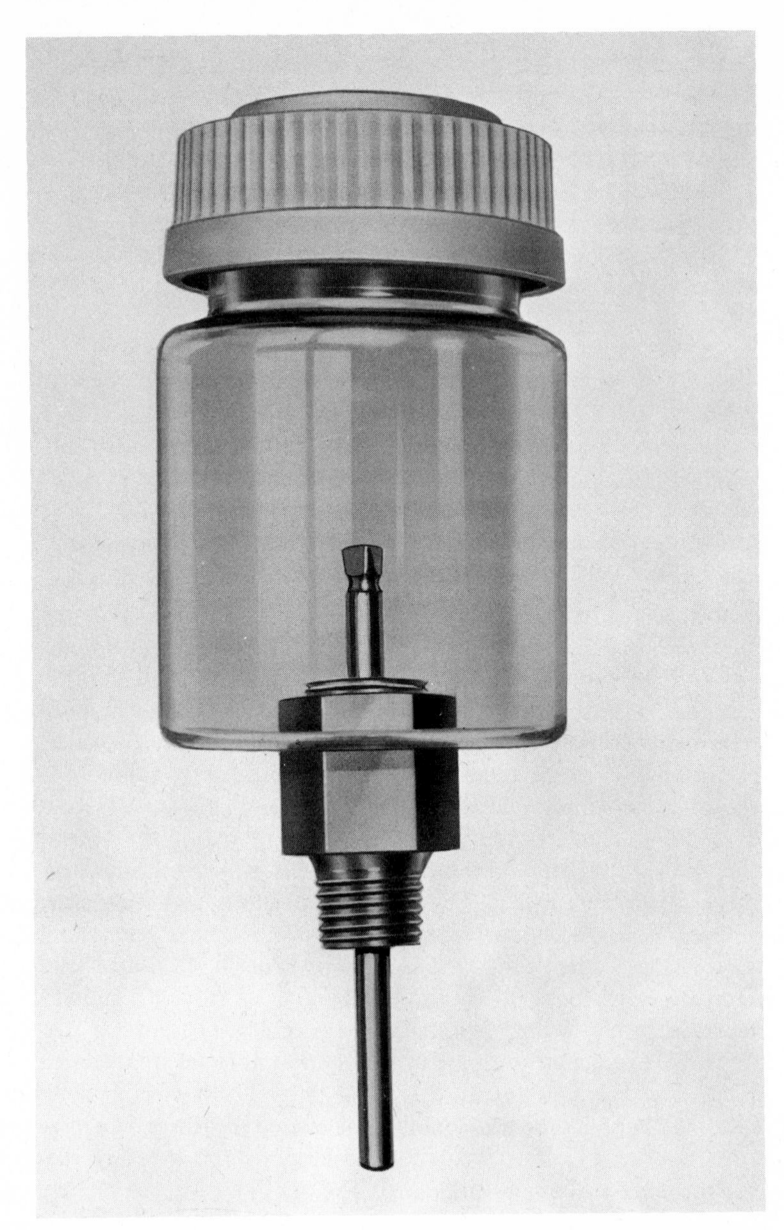

Fig. 11-2. Vibrating rod oiler (bottle oiler). *Courtesy, Oil-Rite Corp.*

bearings where the load may or may not be transmitted to the journal through the bearing.

Waste-packed oiler contains waste (usually wool yarn) held against the journal by its own resiliency after proper packing. The oil feeds through the waste by capillary action from its own reservoir (Fig. 11-9). Waste is

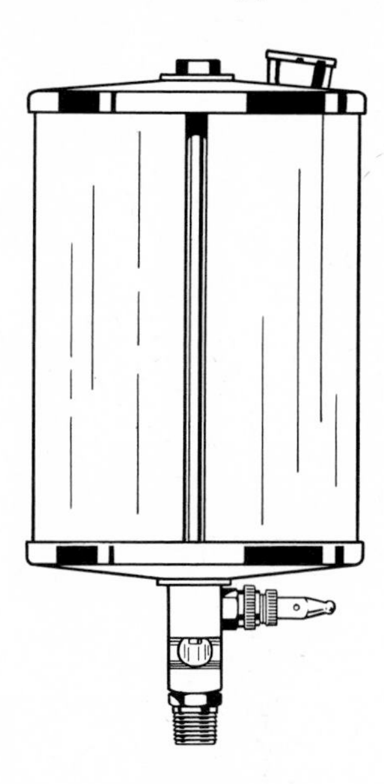

Fig. 11-3. Drop-feed oiler, large type with toggle shut-off. Capacities range up to 1 gallon. *Courtesy, Oil-Rite Corp.*

soaked in oil and drained before installing. Horsehair mixed with waste imparts excellent resiliency. The oil in the reservoir below is reused continuously. This system is used in electric motors and railway car journals. Glazing of either waste or pads should be avoided.

Central oiling system (*manual*) consists of a hand pump, oil reservoir, feed lines, and measuring valves (Fig. 11-10). Measuring valves control oil feed to any number of frictional parts. The rate of feed is controlled by pump adjustment and the frequency of pump actuation by hand.

The Farval DD measuring valves (Fig. 11-11) operate in the following *double outlet* sequence:

1. Pressurized lubricant, entering valve through line 1, forces pilot piston down, allowing pressure to be applied to top of main piston. Main piston begins to move down.
2. Main piston, moving down under pressure, forces lubricant from its chamber, past the lower land of the pilot piston and out port "B" to the bearings.
3. During the next operating period, pressure is applied to line 2 and the above operating sequence is repeated with the pistons moving in the opposite direction. This time lube moves out port "A."
4. Indicator stem shows when valve is operating. Valve discharge can be varied by adjusting screw in packing gland to alter main piston travel.

Single outlet sequence of Farval DD valves: When a Farval Dualine DD lubricating valve is cross-ported (converted to single discharge) the

Fig. 11-4. Sight feed valves. (a) General purpose, (b) with toggle shutoff, (c) multiple arrangement. *Courtesy, Oil-Rite Corp.*

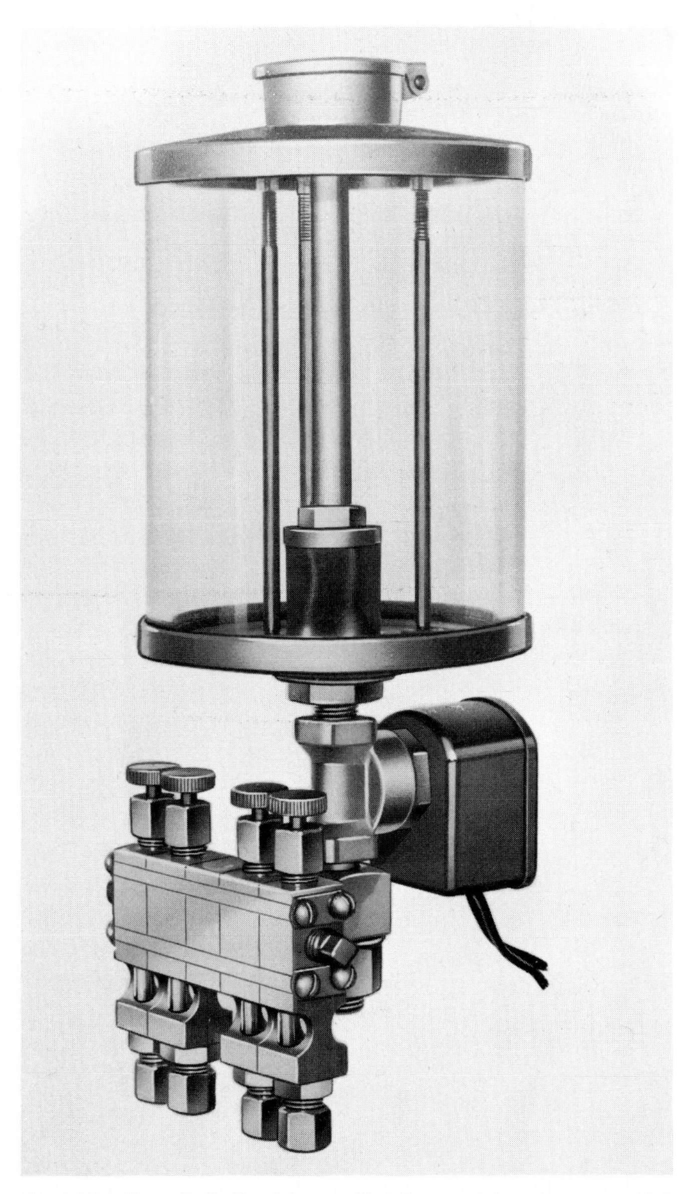

Fig. 11-5. Drop-feed oiler (electro oiler) for supplying oil at controlled rate to remote or widely separated bearings. It acts as an automatic central oiling system by gravity through a solenoid valve to single or multiple gauge mounted sight-feed valves for individual drop-feed control. It has a built-in filter. *Courtesy, Oil-Rite Corp.*

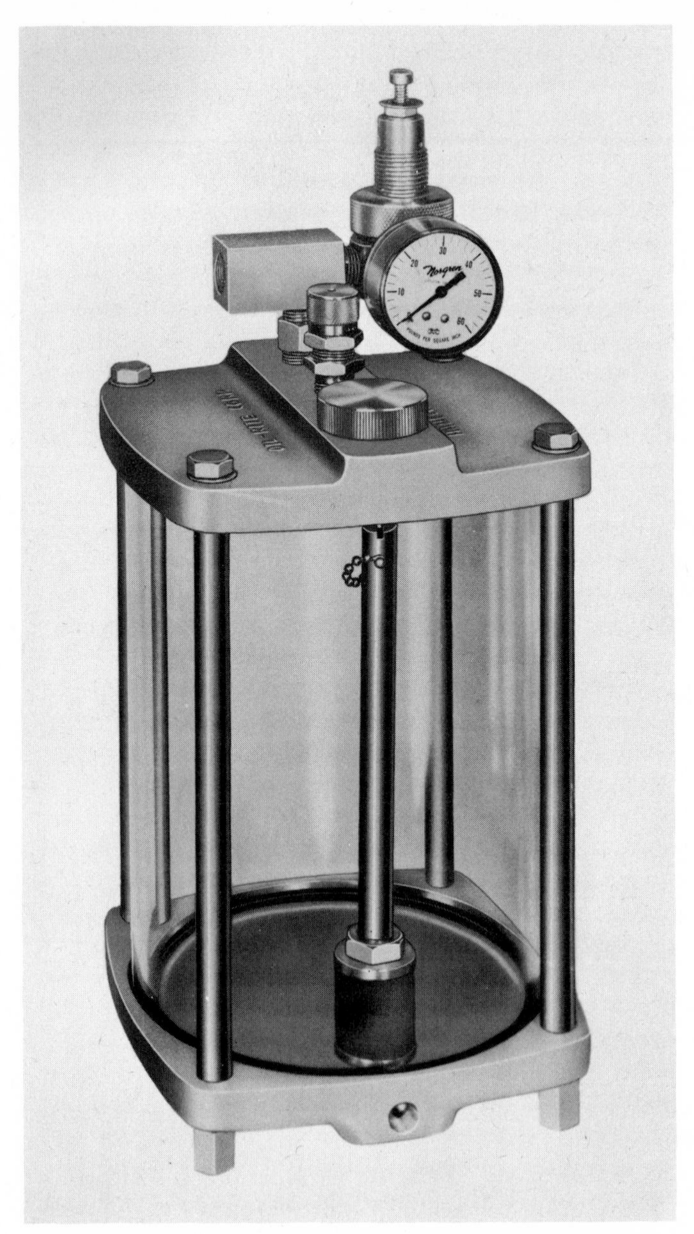

Fig. 11-6. Air-operated oil dispenser. *Courtesy, Oil-Rite Corp.*

operational sequence of the valve itself does not change. As shown above, supply lines are pressurized alternately and piston movement remains the same. The only difference is that the total discharge of each valve operation is directed out one outlet to a single bearing. Note: One discharge outlet

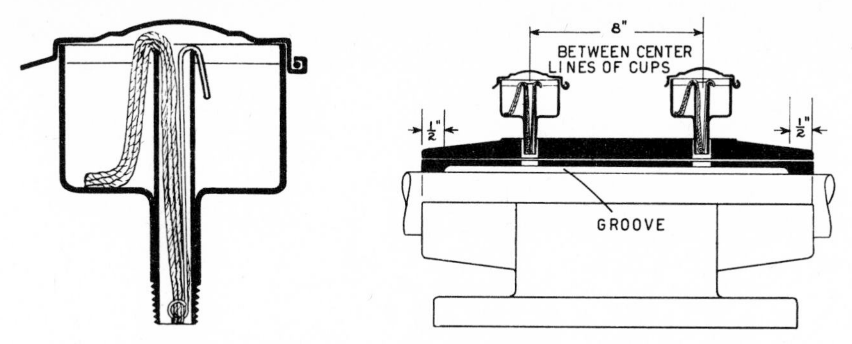

Fig. 11-7. Wick-feed oiler. Note arrangement of installation on long bearings. *Courtesy, Mobil Oil Corp.*

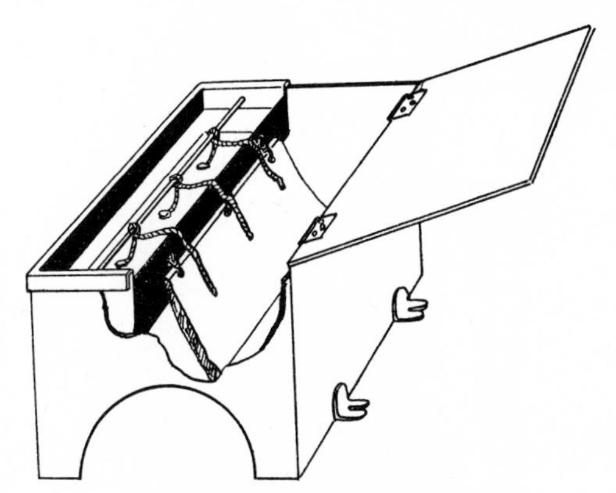

Fig. 11-8. Pad oiler (felt). *Courtesy, Mobil Oil Corp.*

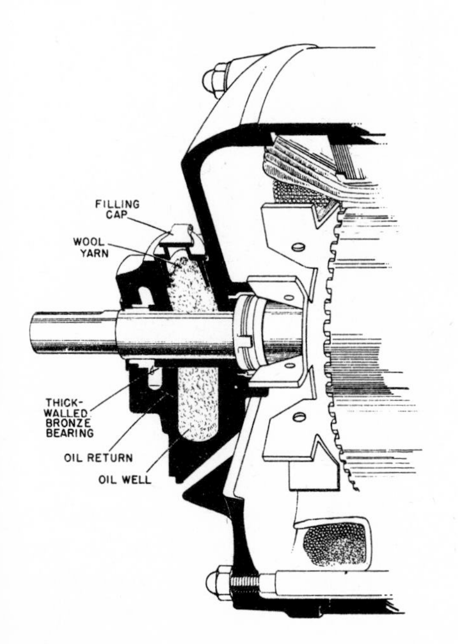

Fig. 11-9. Waste-packed oiler for single bearing. *Courtesy, Mobil Oil Corp.*

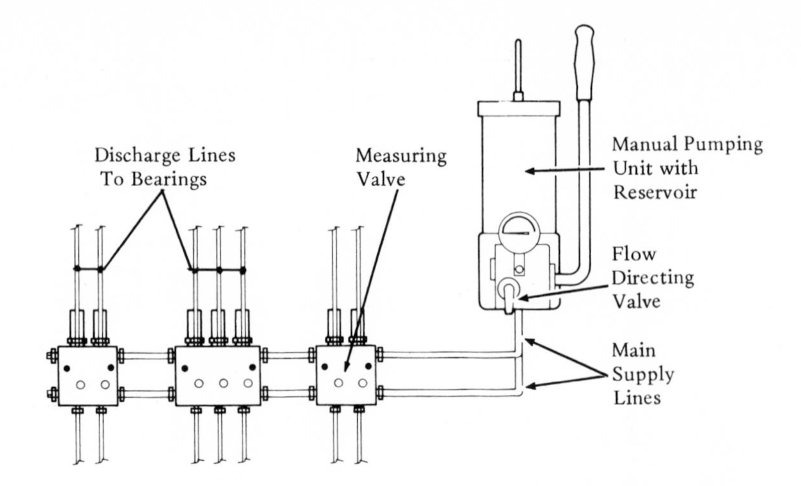

Fig. 11-10. (Above) Farval Dualine manual system in which the lubricant is directed alternately into first one and then the other of the two supply lines by means of a push-pull flow-directing valve (DD) located in the base of the pump. (Right) Manually operated pump. *Courtesy, Farval Division, Eaton Yale & Towne, Inc.*

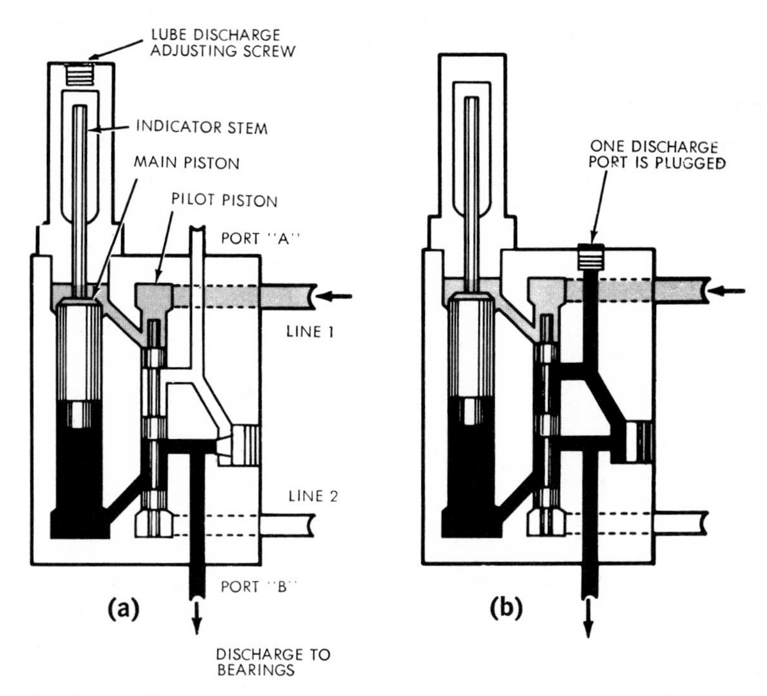

Fig. 11-11. Farval DD measuring valves. (a) Double outlet sequence, (b) single outlet sequence. *Courtesy, Farval Division, Eaton Yale & Towne, Inc.*

must be plugged or valve will discharge only random amounts of lubrication through both outlets.

Automatic Oil Appliances

OIL MIST LUBRICATOR. This device is fed air at 10–15 psi, which vaporizes the oil and sends it to the bearing surfaces at a pressure slightly above atmospheric (Figs. 11-12 and 11-13).

The Oil Mist Lubricator, a product of the Alemite Division, Stewart-Warner Corp., handles any good quality lubricating oil up to a medium heavy-bodied oil with a viscosity of 1000 SUS at 75°F, including extreme pressure oils. As oil viscosity increases, the quantity of oil circulated within the unit decreases and the mist output is reduced. Maximum viscosities at minimum recommended lubricator temperatures are shown in Table 11-1.

Ultra-high-speed spindles require special consideration before they are included in an oil mist system.

FARVAL VORTEX MIST LUBRICATOR. The operating principle of the Farval Vortex Mist Lubricator is described below with references to Fig. 11-14.

1. Entrance of air from pressure regulator (not shown). Bypass air

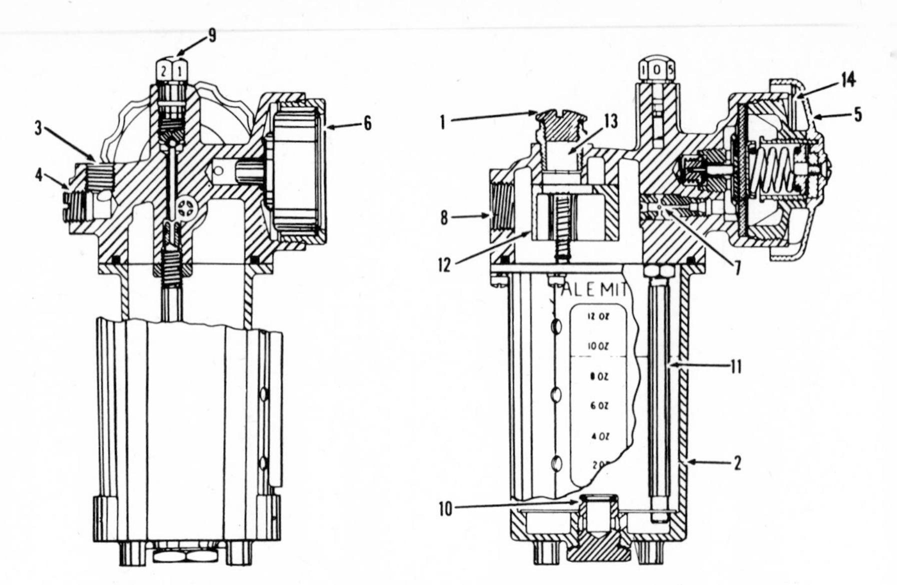

Fig. 11-12. Oil mist lubricator. *Courtesy, Alemite Division, Stewart-Warner Corp.*

1 Oil inlet	7 Venturi
2 Reservoir	8 Main supply line
3 Air feed	9 Oil flow regulator (needle valve)
4 Air feed	10 Screen
5 Air pressure regulator	11 Venturi tube
6 Air pressure gauge	12 Baffle

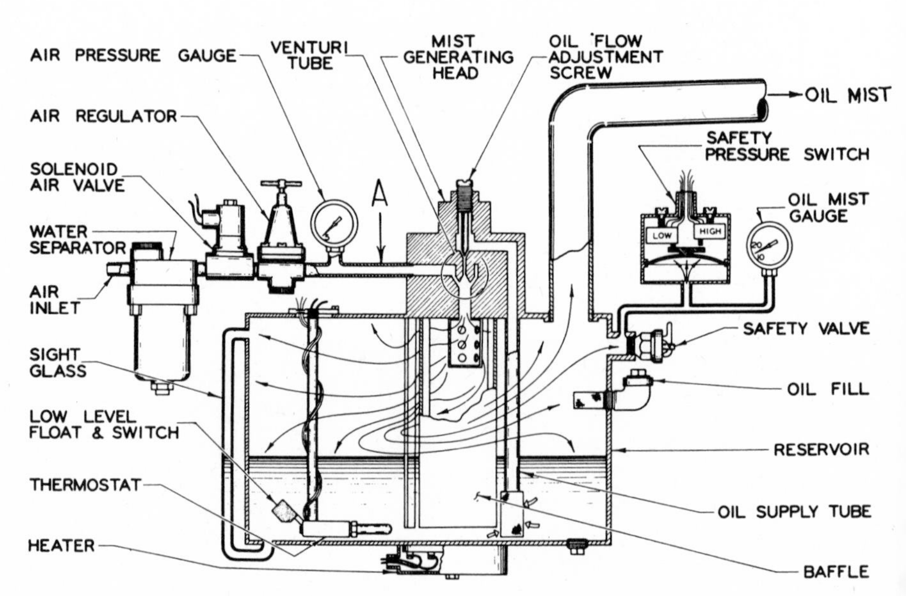

Fig. 11-13. Operational diagram of oil mist lubricator with heater and water separator. *Courtesy, Alemite Division, Stewart-Warner Corp.*

Table 11-1. MINIMUM TEMPERATURE AT OIL MIST LUBRICATOR

Oil Viscosity (SUS *at 100°F*)	*Minimum Recommended Temperature at Lubricator*
1000	75°F
500	55°F
300	45°F
200	32°F
100	8°F

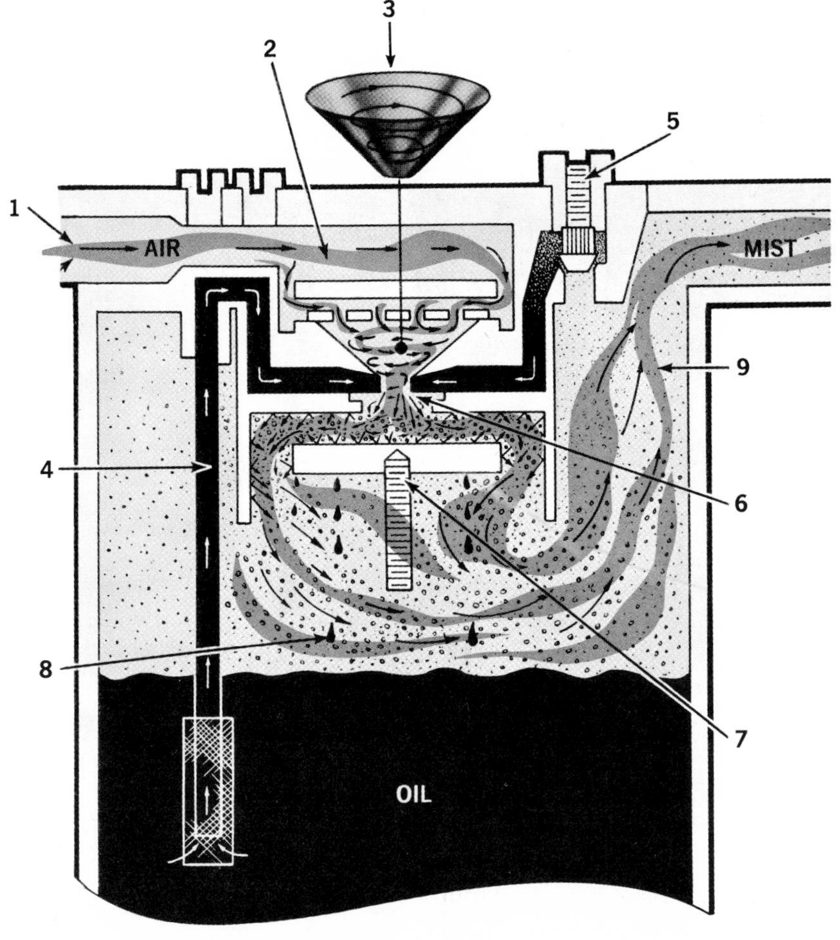

Fig. 11-14. Farval Vortex oil mist lubricator. *Courtesy, Farval Division, Eaton Yale & Towne, Inc.*

control (not shown) permits increase in pressure (and flow) through mist manifold distribution system.

2. Air enters inlet port and flows downward into air ring and continues to vortex chamber.

3. Air is forced inward with increasing velocity toward exit orifice. Conical jet of spiraling air leaves exit at sonic velocity and cyclonically swirls downward.

4. Oil is drawn up suction tube into vortex chamber. Velocity oil jet has low-pressure (vacuum) boundary layer about its outside and inside surfaces, causing high suction effect in oil gap. It is then drawn toward high-velocity jet and into mixing zone.

5. Oil flow regulator permits reduction in oil flow. When needle valve is opened pressurized oil mist is allowed in the oil stream to interfere with the suction and reducing oil flow to vortex nozzle.

6. Oil is torn apart in the oil gap by jet and expands and swirls downward, being highly turbulent and oscillating in a three-directional pattern with frequencies in the sonic and ultrasonic range. This tends to reduce droplets to optimum mist particle size.

7. Jet is deflected by a particle size regulator disk horizontally outward, where it strikes baffle ring.

8. Oil mist particles too large to be transported into manifold "wet out" against baffle ring and drop back to reservoir.

9. Oil mist is forced toward reservoir outlet and into distributing system.

Density control is provided by moving the regulator (adjusting screw) for dryer mist or downward for wetter mist.

A mist density meter is available (Farval) with solid state control features. Placed in the distributing pipe, it allows visual meter readout of relative mist density and permits vortex mist head tuning to optimum output. The meter can be adapted for location up to 250 ft from the control panel.

A mist lubricator can be used in a system that requires application in the form of spray and liquid as well as mist. At the bearing area, the outlet is provided with a fitting, or reclassifier, that delivers the oil in the required form (Fig. 11-15).

Most lubrication reclassifier fittings are of three types. The *mist fitting* is designed to develop minimum air turbulence through the orifice. It is applied to high-speed gears, chains, and rolling bearings. The *spray fitting* is designed to cause limited turbulence by passing the oil and air through a long orifice. It is applied to medium-speed components. The *condensing fitting* increases turbulence with baffles in the oil-air stream, causing droplets to form at the end of the reclassifier and drop onto the frictional surface. Applied to slow-speed plain bearings and ways.

The amount of oil mist delivered increases with greater feed air pressure

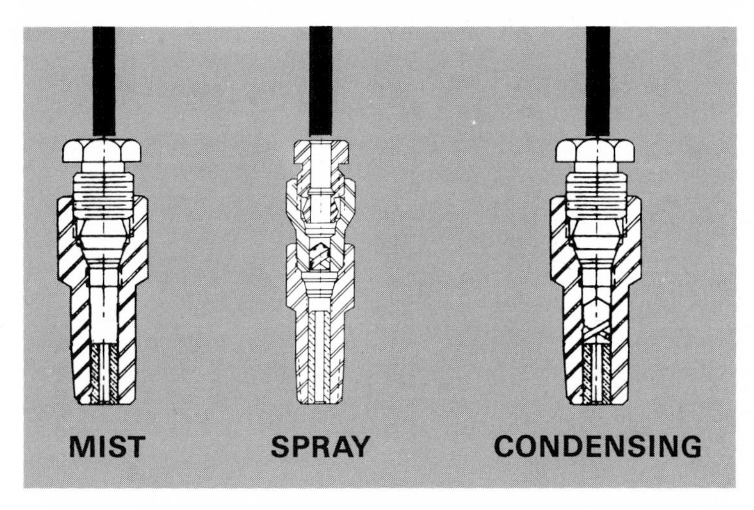

Fig. 11-15. Oil reclassifiers. *Courtesy, Farval Division, Eaton Yale & Towne, Inc.*

up to the point of critical turbulence that causes oil to drop out of the oil-air stream within the manifold. It also increases with lower-viscosity oils, varying inversely as the viscosity decreases with straight mineral oils. Greater output can be obtained by heating the oil.

Mist flow should be directed *through* the bearing. It is not recommended for electric motor bearings unless the exit stream can be diverted away from the windings.

The maximum number of straight mist fittings that can be served by a single lubricator is 60. The Oil-Mist Lubricator (Alemite) shown in Fig. 11-13 has a rated capacity of 30 *bearing in.* for shafts up to 5 in. in diameter. A bearing inch is equal to one 1-in. shaft diameter or two ½-in. shafts, etc.

For shafts larger than 5 in. in diameter, the following rule applies:

6- to 8-in. shaft diameter—four bearings maximum per lubricator.
9- to 12-in. shaft diameter—two bearings maximum per lubricator.

Larger shafts require special considerations best made by the manufacturer.

Air line lubricator, or oil fog lubricator, acts on the principle of a common atomizer, either by syphon or capillary action. It is installed in the compressed air line leading to pneumatic tools and machines and is usually coupled with an air filter. The oil feed is adjustable (Fig. 11-16).

OPERATION OF THE NORGREN LUBRO-CONTROL UNIT. As compressed air enters the filter, air flows through the louvers (A) forcing the air into a whirling pattern (Fig. 11-17). Liquid particles and heavy solids are thrown against the inside wall of the bowl by centrifugal force and run to the bottom. The baffle (B) maintains the "quiet zone" in the lower bowl and prevents

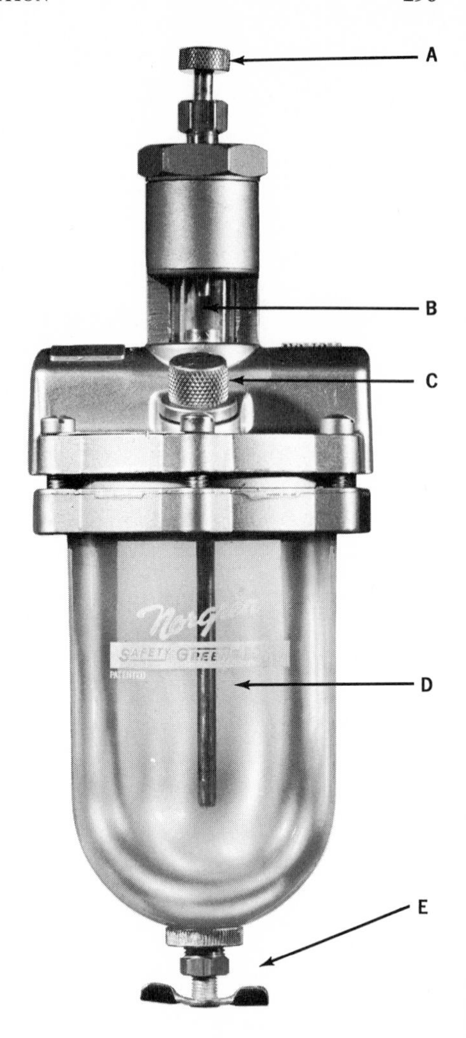

Fig. 11-16. Oil fog lubricator. A: thumb screw for precise oil feed. B: sight feed glass for visible oil feed. C: oil filler plug. D: bowl, or reservoir. E: drain cock. This model can be filled under pressure, and air flow is bidirectional. *Courtesy, C. A. Norgren Co.*

air turbulence from recirculating liquid into the air stream. Air leaving the bowl passes through the filter element (C), removing the solid impurities.

From the filter the air goes to the pressure regulator. When no load is applied to the adjusting spring (D), the regulator valve (E) is closed. As the adjusting screw (F) is turned in, it applies a load to the spring which is transmitted to the valve through the flexible diaphragm (G), opening the valve. The air pressure side of the diaphragm is connected to the outlet port so that regulated pressure is exerted against the diaphragm. As the regulated pressure increases, the pressure against the diaphragm increases forcing the diaphragm to compress the adjusting spring until the load exerted by the spring is equal to the load exerted by the regulated pressure.

Air from the filter-regulator then enters the lubricator (H) via the

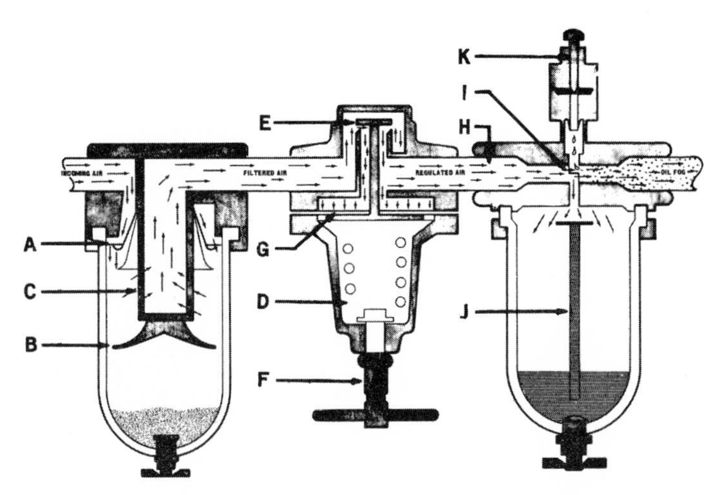

Fig. 11-17. Air line or oil fog lubricator with filter and pressure regulator combined to form a Lubro-Control Unit. *Courtesy, C. A. Norgren Co.*

venturi tube (I) pressurizing the oil bowl. The flow of air through the lubricator creates a reduced pressure area as it passes through the venturi section which causes the oil to flow up the siphon tube (J) to the chamber above the drip gland. The needle valve (K) controls the desired rate of feed from the drip gland through the sight-feed chamber and into the air line. As the oil enters the air stream, it is atomized into an airborne oil fog which is carried to the pneumatic device.

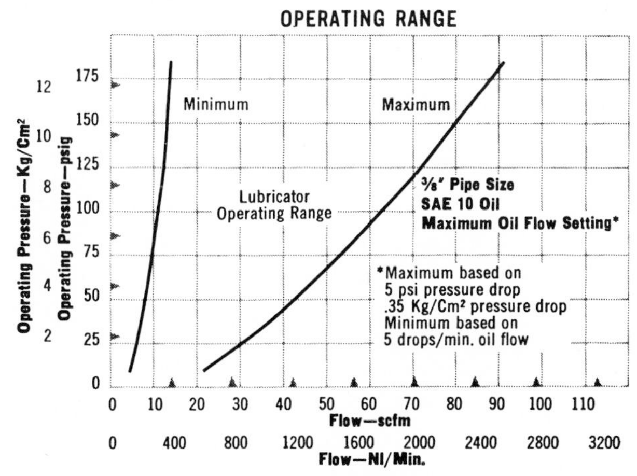

Fig. 11-18. Operating range for oil fog lubricator shown in Fig. 11-17 for 70°F and 29.92 in. Hg and for operating pressures of 0 to 175 psi using an oil of 150 to 220 SUS at 100°F. *Courtesy, C. A. Norgren Co.*

Air line oilers of the pendulum and expanded bulb type are designed for the lubrication of rock drills (Figs. 11-19a and 11-19b). It is recommended that they be placed in the air line not more than 12 ft from the drill. The pendulum design operates on the weighted oil principle which assures oil flow in any horizontal position.

The pendulum type lubricator, Fig. 11-19a, shows these parts: (1) Air inlet—can be connected at either end. (2) Air port—entrance to oil chamber. (3) Pendulum design—weighs oil intake and assures oil flow in any horizontal position. (4) Particles of oil to air level. (5) Oil flow adjustment.

The operation of the expanded bulb type (Fig. 11-19b) is described as follows: As the oil is used trom the reservoir (R), the rubber bulb (E) fills with air. The air pressure forces the oil past the adjustable metering pin (M), thru passage (P), and into the air flow stream (F). When the oil reservoir (R) becomes empty, the forward section of the expanded bulb depresses the shutoff valve (S) to shut off the air supply. Check valve (V) keeps air from the reservoir opening when the filler plug (O) is removed for refilling the reservoir. The simple, trouble-free adjustment metering pin on the outside of the filler plug allows immediate regulation and adjustment of the oil flow.

MECHANICAL FORCE-FEED LUBRICATOR is an adjustable oil feed pump (Fig. 11-20). Usually, but not always, it is actuated by the machine on which it is used through suitable linkage. It may have one or several pump plungers, all of which are driven from a common shaft through cams. Serving

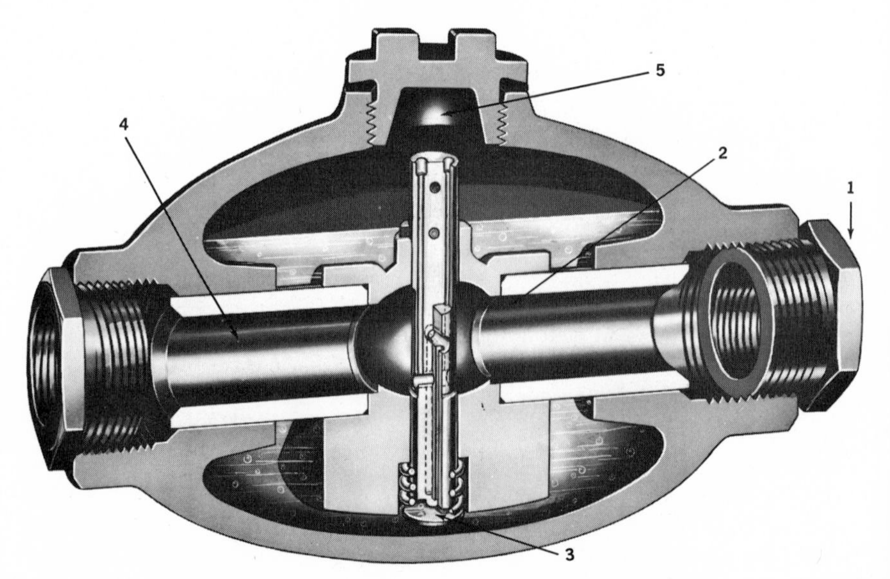

Fig. 11-19a. Air line lubricator, pendulum type, for pneumatic tools, especially rock drills (Gardner Denver model LO 8). *Courtesy, Henry Pohs, Gardner-Denver Co.*

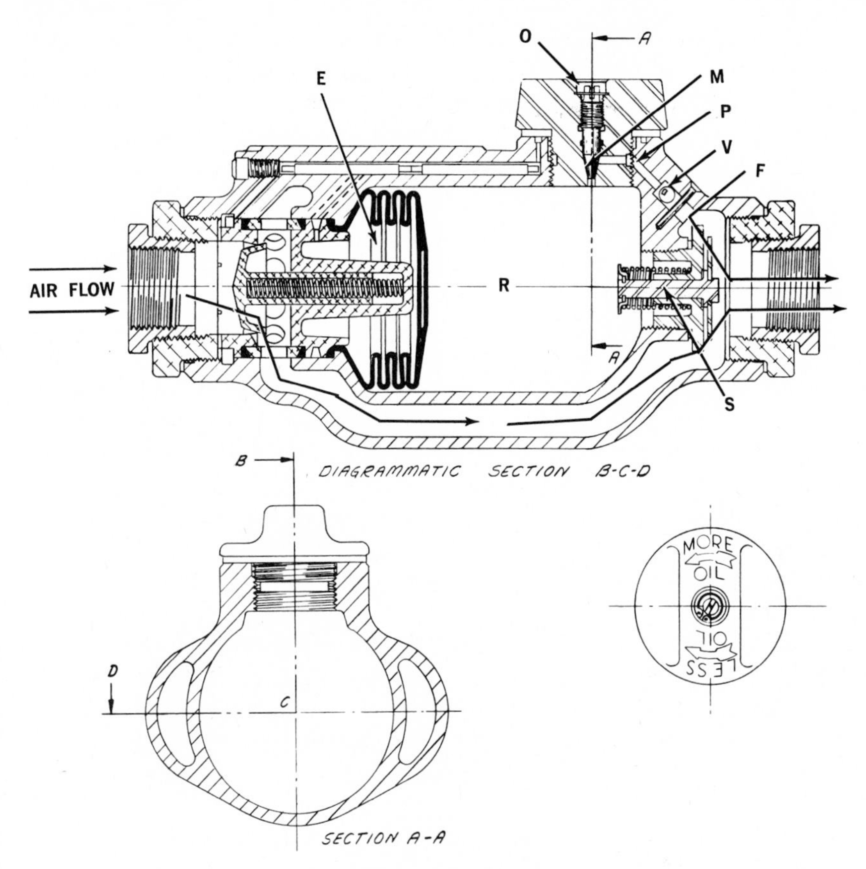

Fig. 11-19b. Air line lubricator with expanded bulb for large rock drills (Gardner Denver model LO 12A). *Courtesy, Henry Pohs, Gardner-Denver Co.*

one or more oil feed lines, it is used to lubricate all types of compressor and engine cylinders, as well as bearings, gears, etc. The rate of feed is determined by means of a sight glass in the oil discharge line and is adjustable, usually in drops per minute. Mechanical force-feed lubricators can have multiple feeds operated automatically or by hand (Fig. 11-21).

HYDROSTATIC LUBRICATOR works on the principle of a head differential acting on a fluid column. It is used chiefly for steam cylinder lubrication by injecting oil into the high-pressure steam line leading to the steam chest. The lubricator (Fig. 11-22) is located at the bottom of a bypass line, one end of which leads into the steam pipe an appropriate distance above the other end, which also opens into the same steam line. Steam enters the bypass through the top connection (A) where it condenses and forms a column of water (B). The water column bears on the oil in the lubricator reservoir (C). The steam pressure bearing on the top of the water column,

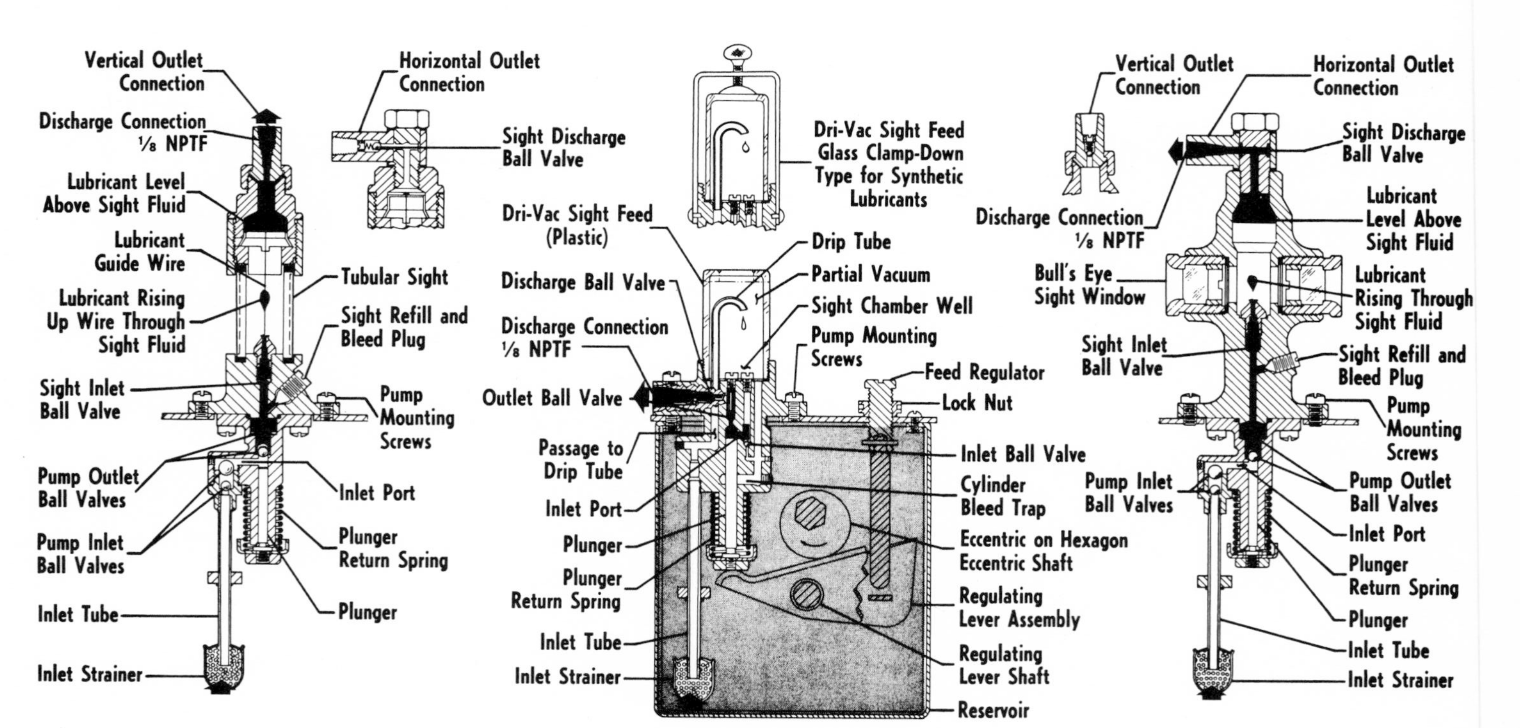

Fig. 11-20. Mechanical force-feed lubricator. The principal method of applying oil to cylinders of various types of engines and compressors. *Courtesy, Manzel, Houdaille Industries, Inc.*

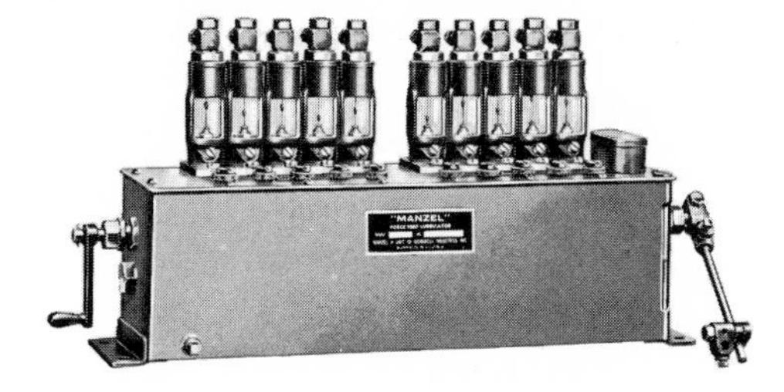

Fig. 11-21. Mechanical force-feed lubricator, ratchet drive, with 10 feeds. *Courtesy, Manzel, Houdaille Industries, Inc.*

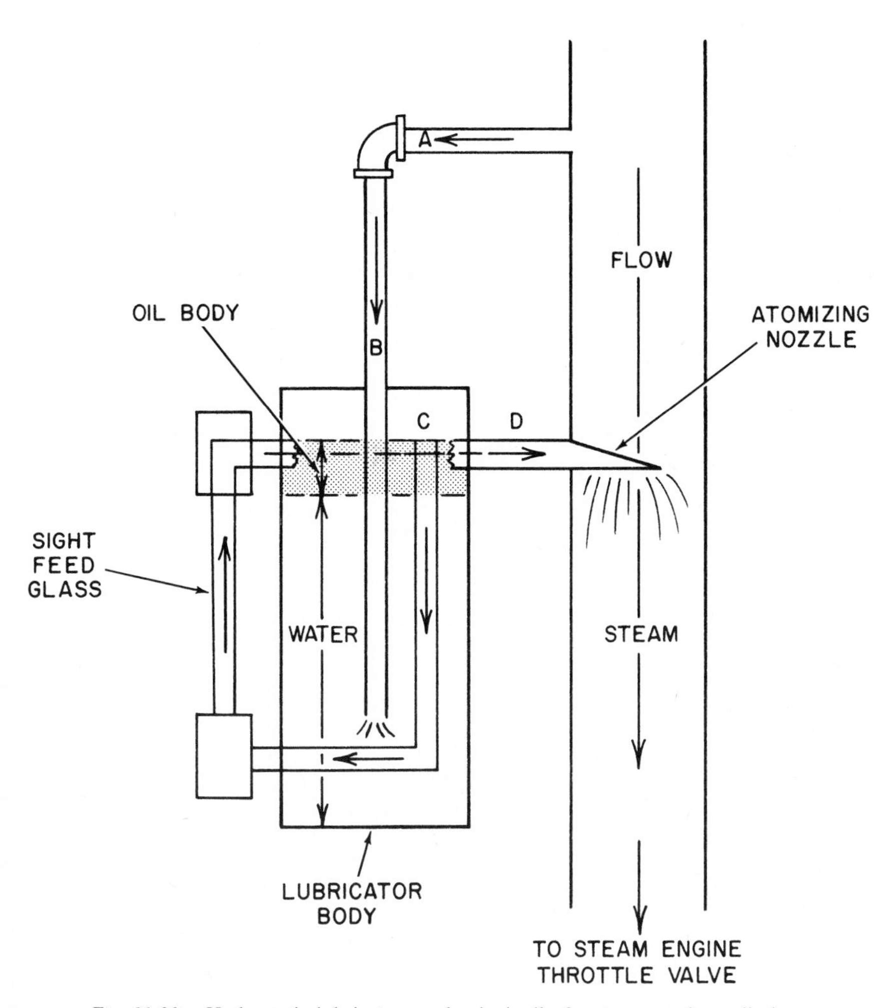

Fig. 11-22. Hydrostatic lubricator used principally for steam engine cylinders.

plus the weight of the water forming on the column, produces a pressure slightly greater than the steam pressure at the entrance of the bottom bypass connection. This pressure differential forces the oil into the steam pipe through the bottom connection (D). As the oil enters the high-velocity downward flowing steam, it breaks up into fine particles (atomizes) and is carried into the cylinder (Fig. 11-23). It then deposits on the valves and the cylinder walls to provide lubrication. The hydrostatic lubricator should be installed at least 6 ft from the steam chest, or any intermediate fittings, in order to obtain complete atomization.

CENTRAL OILING SYSTEM (automatic) is designed like its manual counterpart and operates similarly. Instead of a hand pump, automatic types are motorized and feed oil at intervals governed by a timing device (Fig. 11-24).

LIQUID-FILLED SIGHT FEED. To assure oil flow and determine the rate of oil flow, usually in drops per minute delivered from a mechanical force-feed lubricator or hydrostatic lubricator, a transparent sight-feed indicator is used. It consists primarily of a vertical glass tube filled with liquid and connected at the bottom to a restriction nozzle fed by the lubricant plunger. The oil from the plunger collects at the nozzle, and when the mass becomes great enough, it leaves periodically in the form of drops. The drops are

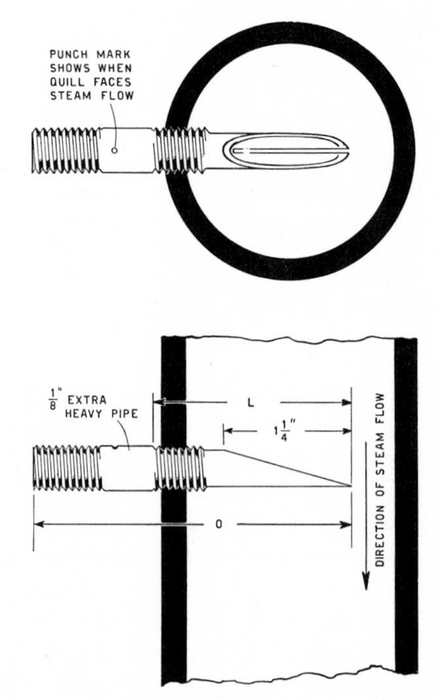

Fig. 11-23. Nozzle for atomizing oil in a steam line from a mechanical force-feed or hydrostatic lubricator.

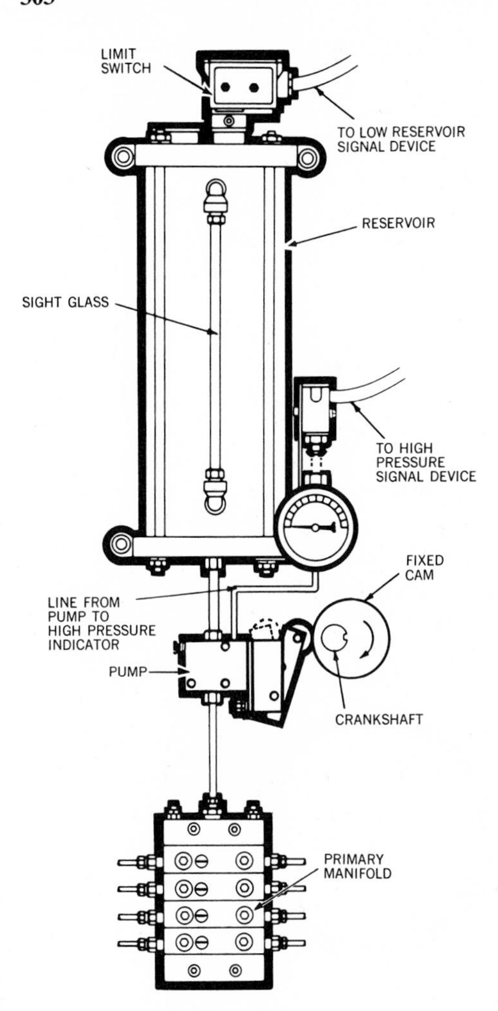

Fig. 11-24. Central lubricant pumping station with pump-operated takeoff. *Courtesy, Farval Division, Eaton Yale & Towne, Inc.*

guided up the center of the tube by a wire, thus keeping the oil from smearing the tube walls. The tube is connected at the top to a vertical oil feed line leading to the point of lubrication. A modified ball type relief valve at the top outlet allows entrance of the oil into the feed line.

The liquid is usually a solution of about 30 to 40% distilled water and glycerine. The glycerine, having a higher specific gravity than water, stabilizes the solution and prevents it from rising into the discharge line. Glycerine also lowers the freezing temperature of the water, which is important where cold weather conditions prevail. A liquid-filled sight flow indicator is shown atop the mechanical force feed lubricator in Fig. 11-20.

Where additive type oils tend to cloud the water and glycerine solution, a mixture of 50% water and 50% calcium nitrate is usually found satisfactory.

Continuous (Reuse) Oiling Methods

Ring oiler consists of a free metal ring that rides on the journal and carries oil from a reservoir located below the bearing (Fig. 11-25). The bearing is slotted to allow the ring to ride on the journal. There is a speed limit above which oil will be thrown from ring by centrifugal force. Below a certain speed, oil will drain from the ring or the ring will not revolve, resulting in an insufficient supply of oil to the journal. The bearing is designed with grooves so that oil is distributed along the top of the journal. The reservoir should be equipped with constant level oilers to assure safe oil supply.

Chain oiler works on the same principle as the ring oiler, except that a chain is substituted for the ring. A chain has greater oil-carrying capacity than a ring and can be used in the slow-speed range within which a ring cannot function effectively. If driven too fast, a chain will whip against the sides of the reservoir, lose contact with the journal, and churn the oil (foam).

Fixed-collar oiler works on the same principle as the ring and chain oiler, except that the collar is fixed to the journal. A fixed collar is capable of delivering heavy oil to slow-speed journals, which ordinarily would interfere with the rotation of a chain or a ring.

Bath lubrication provides lubricant to frictional parts by partially or wholly submerging them in a reservoir of oil (or grease, in some cases).

Splash lubrication is similar to that of the bath system, except that it includes the action of oil throw to other frictional parts within the same or connected housing.

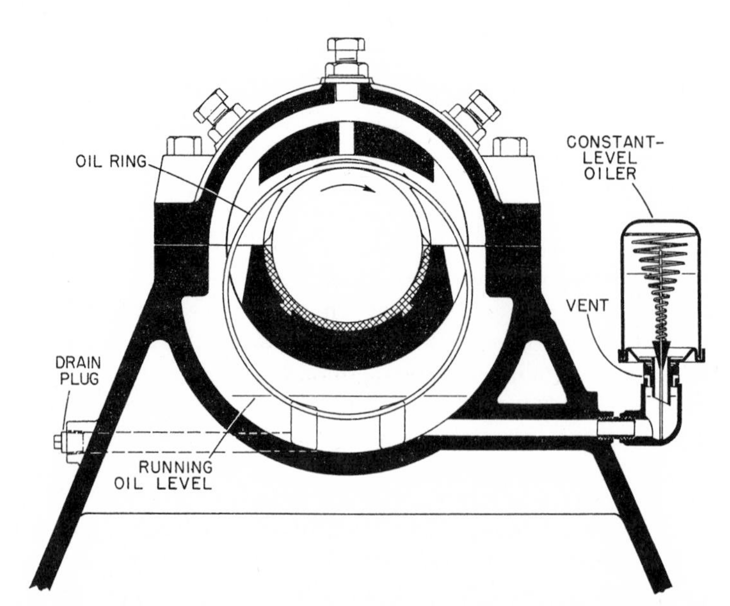

Fig. 11-25. Ring-oiled bearings with constant level oiler. *Courtesy, Mobil Oil Corp.*

SPECIAL APPLICATION METHODS

Splash and Bath Oil Systems

To apply a lubricant to a frictional part by immersion or partial immersion as it passes through a body of oil is called application by *bath*. If the part to be lubricated passes through an oil body with such velocity as to cause some of the lubricant to leave the body and become deposited on other frictional parts not accessible to the bath, the system is one of *splash* application. Splash is sometimes developed by slingers, ram troughs, discs, scoops and other devices. In both methods of application, the oil is returned to the oil reservoir by drippage. Some systems that appear to be strictly splash or bath types also include a circulating phase by which oil from the reservoir is pumped under pressure to various frictional parts, either through individual tubing or through passages drilled in the various components, such as shafts and connecting rods. Another bath-circulating system is one in which the oil is caused to flow within the system by pumping action of the frictional part, such as a rotating vertical spindle.

The bath application method is best demonstrated by a horizontal worm gear partially submerged in its lubricating oil.

Splash systems are exemplified by many small or medium size gear sets and air compressors, wherein the cylinders and wrist-pin are flooded by splash caused by the revolving crankshaft. Slingers, discs, etc., distribute oil to the remote frictional parts of enclosed chain drives and large enclosed gears.

In the automotive engine the oil is distributed by a combination of bath, splash, and circulation.

One of the critical parts of ring-spinning frames used in the textile industry are the spindles, some of which operate at speeds up to 11,000 rpm. The spindle on which a bobbin is held is supported by and revolves within a bolster held in place by a spindle rail. The spindle includes a pulley, or whorl, which maintains frictional contact with a belt that keeps it turning. The bottom part of the spindle below the whorl is inserted in the bolster and rests on a step bearing at the thrust end. A guide bearing, which is either plain or rolling, is located at the whorl. The points of greatest friction are the step and guide bearings. Spindles are lubricated with an oil of low viscosity, which is introduced through an oil inlet to fill the clearance space between the spindle body and bolster. At the speeds involved, the oil is pumped from the step bearing up through the clearance space, on up to the guide bearing, and thence it tends to drop back to the step bearing in continuous circulation, acting both as lubricant and coolant.

The oil used in ring-spindle lubrication is of light body, having a viscosity of about 60 SUS at 100°F to 100 SUS at 100°F, depending on the severity of the service from light to heavier duty, respectively. Such an oil should contain a film strength agent, oxidation inhibitor, and rust inhibitor and should be changed at least once per year. During the change the reservoir should be thoroughly cleaned, including the removal of lint that may have entered the system.

Conveyor Lubricator

The trolley wheels of overhead rail conveyors in many plants are grease lubricated, the grease being applied by a hand pressure gun. The task of servicing all wheels is time consuming and can only be accomplished during shutdown periods. A conveyor lubricator (Fig. 11-26) is available that accomplishes the same task automatically during operation. The conveyor lubricator is a self-contained unit mounted on the rail and actuated by the moving conveyor. Air pressure is required to move the grease from a

Fig. 11-26. Automatic conveyer lubricator designed to inject either oil or grease in measured quantity (adjustable) into conveyor wheel bearings. *Courtesy, J. N. Fauver Co., Inc.*

Fig. 11-27. Automatic conveyor lubricator. Photograph shows method of lubricating chain only. Can be equipped with unit to lubricate wheels also, or can be converted to lubricate wheels only. *Courtesy, J. N. Fauver Co., Inc.*

reservoir to the nozzle through which it is applied to the wheel grease fitting.

As a trolley wheel approaches the lubricator, the hub engages the nozzle of the pumping unit, which is automatically brought in contact with the grease fitting. The wheel carries the nozzle with it, and when properly aligned with the fitting, a measured quantity of grease is discharged under pressure into the wheel bearing. The nozzle is carried slightly beyond this center position and then released from the trolley to return to its pickup position. The amount of grease discharged can be adjusted within the range of 0 to 0.1 cubic inch.

The apparatus is made up of a grease reservoir equipped with a follower plate actuated by external air pressure. An air control panel is also part of the equipment. The follower is subjected to just enough air pressure to keep a steady, constant supply of grease flowing to the pumping unit. The air pressure may be varied according to the pumpability of the lubricant.

The pumping assembly can be taken out of service by merely swinging it around to a lockout position.

A conveyor chain lubricator is shown in Fig. 11-27.

Spray Lubricators

One of the more effective methods of distributing lubricant uniformly over a frictional surface is by the use of pressurized spray nozzles. Unlike the mist or fog device, the spray nozzle releases a stream of atomized lubricant particles in liquid form rather than in a vaporous state. The size of such particles and the spray pattern depend upon the pressure head, the size and type of nozzle, the viscosity of the lubricant at spray temperature, and the distance between nozzle discharge outlet and target surface.

The application of spray may be done manually or automatically, and intermittently or continuously. Intermittent automatic application is controlled by timing devices (Fig. 11-28). Narrow frictional surfaces may be serviced by a single nozzle, whereas a bank of any number of nozzles may be required for wider surfaces. Nozzles are designed variously to handle lubricants of different viscosities, which may range from a light-bodied penetrating oil to a heavy-bodied residual type lubricant for open gears. The latter is usually diluted with a volatile liquid, however, to facilitate flow.

A number of nozzles mounted at different locations on the same machine, or several adjacent machines, may be supplied with lubricant from a central source. One of the important points to remember in the successful operation of any spray system, particularly a centralized system servicing a number of outlets, is that the lubricant must be pumpable. If pumpability can be accomplished by heating the lubricant (such as a heavy residual type) to reduce viscosity, then this should be done under controlled conditions.

The requirements of a spray system include a source of air at a pressure of from 50 to 100 psi, a lubricant reservoir from which the lubricant is pumped to the spray nozzles, spray control valves (metering device), with indicator stem, nozzles, and appropriate piping. On automatic intermittent systems, a time clock is also used to control application frequency and interval of actual discharge.

Nozzles are classified according to their spray pattern, which may be round, wide angle, or flat. They may be of the internal or external atomizing type. Open-gear lubricants are usually applied by the nozzles designed with external atomization, having either a round or flat pattern. The most effective distance between nozzle and target is 8 to 10 in.

When used to lubricate gears, the nozzles should be arranged so that their spray is directed at an angle of about 30° in respect to the centerline of the gear, causing impingement on the leading tooth face. A distance of 6 to 8 in. between the nozzle and pitch diameter of the gear is recommended.

Spray application has been extended to serve the lubrication requirements of gears, chains, wire rope, dies, punches, shears, and plain bearings where clearances allow. Spray application is usually limited to gear speeds of about 1000 ft/min and chain speeds of 600 ft/min, but they can be successfully used at higher speeds in both cases if discharge factors are modified to soften the shock of target impingement. This can be done by

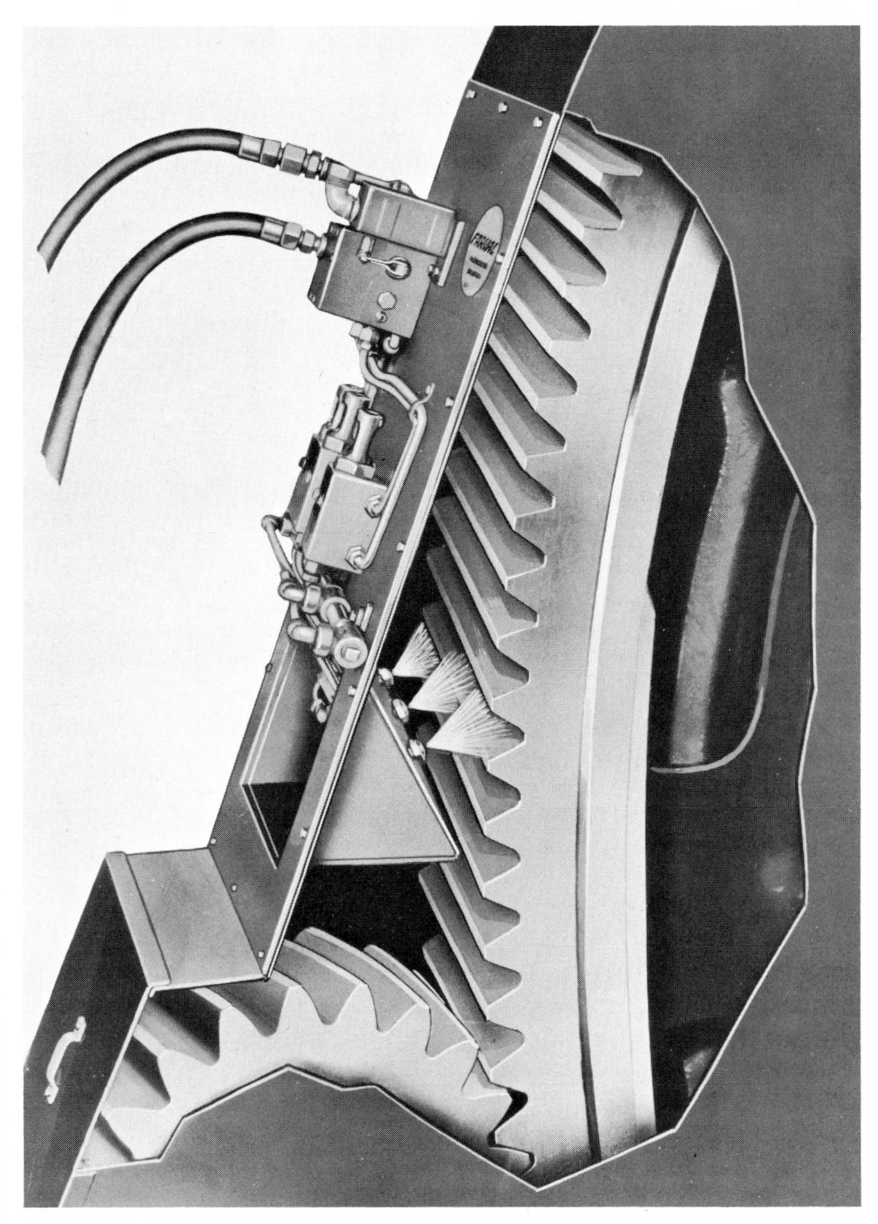

Fig. 11-28. Lubricant spray arrangement as applied to a large gear and pinion. *Courtesy, Farval Division, Eaton Yale & Towne, Inc.*

lowering head pressure, reducing volume per shot, reducing spray frequency and increasing viscosity commensurate with reduced head pressure and service requirements.

On bull gears and pinions, care should be taken not to apply excessive residual type lubricant. Unless there is some means of periodically removing the drippage, it will accumulate in the bottom of the casing and interfere with the operation of the gear. On such gears, the period of each discharge should equal the time it takes for the bull gear to make one revolution. The amount of oil applied depends on the operating viscosity and the wiping resistance of the lubricant.

Lubrication by spray has eliminated hand application by swab, brush, drip, and similar methods. The result has been savings in labor costs and a reduction of as much as 75% in lubricant consumption. One of the hazards of spray systems is that the nozzles may be knocked off or damaged if vulnerably located in respect to either moving machine components or maintenance procedures.

Gunther Lubrication System

High-speed gears and high-temperature bearings can be effectively lubricated by an appropriate liquid refrigerant in conjunction with a lubricant selected for the operating conditions. It was originally designed to protect the teeth of gears operating at speeds of 100,000 rpm and higher. At such speeds, high fluid friction and rapid attrition occur when lubrication is attempted by conventional systems.

The system consists of injecting the proper lubricant into the gear housing with a high-pressure liquid refrigerant, say R-12, through an adjustable three-way valve. This develops an oil mist that deposits on the moving surfaces to supply adequate lubrication. Simultaneously the liquid refrigerant, entering a lower-pressure environment, flashes into a vapor, causing a refrigerating effect. This counteracts the frictional heat generated by the meshing gears.

At high speeds, the direction of the injected fluids should be in the same direction of rotation as the gear to which it is aimed. This will reduce the impinging effect of the fluid stream with the gear teeth.

The used refrigerant gas can be allowed to escape into the atmosphere through vents or reclaimed by conventional refrigeration methods after it is properly filtered.

The amount of refrigerant required will depend on the power transmitted through the gears and the heat conduction capability of the gear housing.

Wick-Feed Oilers

If the wick in a lamp were extended and turned down so that it hung outside the lamp to a point lower than the lamp reservoir, the fuel would eventually all drip from the lower end of the wick. This action is the result of the capillary siphoning phenomenon, and it has specific applications in lubrication.

In the wick-feed oil cup of Fig. 11-7, the supply leg of the wick is submerged in the oil while the delivery leg is draped over the lip of a standpipe or delivery tube. For an oil with an SUS of 600 at 100°F, the outer leg should be from $3/8$ in. to 2 in. below the bottom of the reservoir.

The farther a wick hangs outside its reservoir the greater is the rate of flow. Also, as the level of the liquid in the reservoir lowers, the flow to the outside decreases, since the liquid has farther to climb before it starts to fall.

There are certain important rules associated with the use of wick-feed oilers:

1. White wool yarn or worsted used in knitting and embroidery has good absorption and lifting properties. It should be free of dye, mildew inhibitors, and waterproofing chemicals.

2. The rate of oil feed from a reservoir (cup, pan, etc.) will increase—(a) with the greater number of wicks used, (b) with a lesser distance maintained between the oil level in the reservoir and the top of the delivery tube, (c) as the length of the delivery leg is extended further downward, and (d) with any decrease in oil viscosity.

3. A wide, shallow reservoir delivers a more uniform feed than a deep, narrow one.

4. As inferred in (b) (2) above, the rate of feed is greatest when the reservoir is full, but gradually diminishes as the level continues to drop.

5. Wicks also act as filters and should be cleaned often (about every three months), by stripping any debris from the strands in the reverse direction to fluid flow. They should be changed at least once a year.

6. Do not use galvanized cups with oil containing oiliness additives of the fatty acid variety.

7. Wick-feed oilers should be lifted from the delivery tube during periods of shutdown since they will otherwise continue to feed and waste oil. A wire shaped somewhat like a Christmas tree ball hanger can be used by wrapping one end around the strands and draping the other end over the lip of the delivery tube.

8. A single reservoir may be equipped with a number of delivery tubes and wicks to distribute oil to a group of friction parts. Such an arrangement should be equipped with pet-cocks at the end of the delivery tube.

9. A four-ounce oil cup using wool wicks will provide lubrication from 4 to 10 hr between fillings, depending on (a) the ratio of oil surface area to depth of oil when full, (b) the number of strands, (c) the length of the delivery leg, and (d) the viscosity of the oil at ambient temperature.

10. On bearings having an upward load, the oil from the wick-feed oiler should be led into the side of the bearing *ahead* of the pressure area.

11. Bearings whose length exceeds, say, 8 in., should be served by two

or more wick feed cups. Their entrance holes should be joined by grooving extending to within about $\frac{1}{4}$ to $\frac{1}{2}$ in. of each end. See "Grooving of Plain Bearings," Chapter 12.

12. Maintaining wick-feed oil cups at a given level by a constant level oiler assures a uniform rate of feed at all times.

UNDERFEED WICK OILER. The underfeed wick oiler is upright, like a lamp wick, or at a slight angle, with its upper end touching and transferring oil to the journal or other part (Fig. 11-29). The top of the wick should never be more than 2 in. from the surface of the reservoir oil at its lowest level. It has the advantage over the wick-feed oiler that it does not have to be moved when operation ceases; it stops lubricating automatically.

Wool worsted has the property of not glazing easily, which makes it especially suited for underfeed wicks.

Constant Level Oiler

The principle of the constant level oiler is similar to that used for watering chickens. Inverting a bottle nearly filled with oil and stopping the mouth causes the oil to fall as far as the stopper (Fig. 11-30). The space left above the liquid is free of air, causing a partial vacuum. By supporting the inverted bottle as to allow the mouth to rest at a given height above the bottom of the pan or reservoir and removing the stopper, the oil will flow by gravity into the reservoir until the surface of the reservoir rises to the level of the mouth, thereby preventing further entry of air into the bottle and sealing it. The air in the bottle exerts some pressure on the liquid column, but not sufficient to counteract the atmospheric pressure exerted over the surface of the reservoir. As soon as the mouth is uncovered because of a drop in reservoir level, more air will enter the bottle and enough oil will flow out to maintain the constant level.

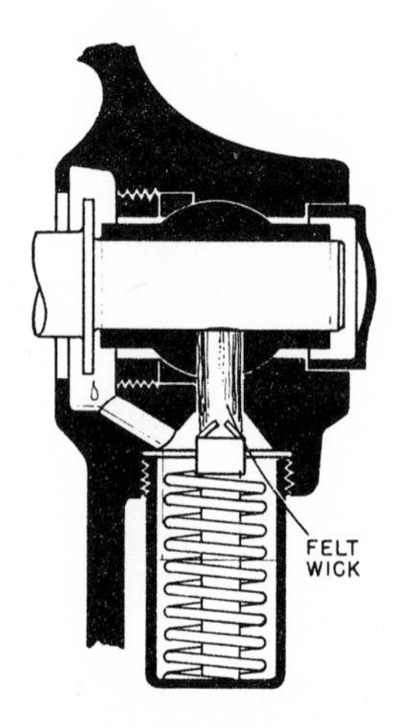

Fig. 11-29. Inverted wick-feed oiler.

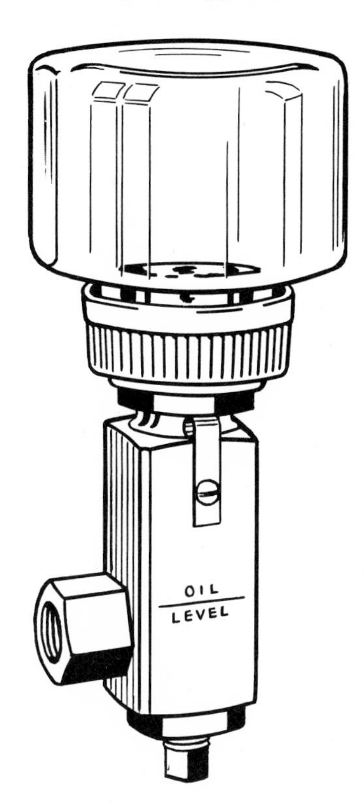

Fig. 11-30. Constant level lubricator. *Courtesy, Oil-Rite Corp.*

In the case of the ring oil bearing where the reservoir is tightly covered, the constant level oiler must be vented in some way to allow air to enter the bottle or container. Otherwise, a strong vacuum will form and prevent flow.

Absorbent Oilers

On some plain bearings, lubrication is supplied directly to the revolving shaft from an oil-absorbent pad by a wiping action. The pad, usually felt, is placed in a cavity cut in a nonpressure area of the bearing so that it is in constant contact with the shaft assisted by its own resiliency. The pad is kept oil-wet from either an external source or, if the load is upward, from its own reservoir as for some printing presses.

Another absorbent oiler is the *felt-roll oiler* used on open bearings. The roller is rotated by contact with the journal, thus providing oil for the bearing. A wick feed cup or a similar arrangement keeps the felt roll oil-wet. The rolling action of the felt roll reduces the danger of glazing or charring the surface, to which stationary pads are susceptible.

Some plain bearings, such as those used on paper mill dryer rolls, are built with a housing located above the roll neck for holding block grease that rests on the rotating roll neck. The wiping action of the neck on the block grease lubricates the bearing. If the softening point of the block grease is too high for the operating temperature, or if surface glazing occurs, it is necessary to supplement the grease with an oil.

Block grease is no longer used on most paper machines and steel rolling mill equipment; instead, an absorbent felt pad is hung from an oil tray placed in the grease housing to make contact with the neck. To control oil feed, the pad is attached just below the top of the tray but outside it and it is fed by the capillary action of wicks leading from the oil placed in the tray over the edge to the pad. This method of application can be used on many existing open bearings (half bearings) where space for the oil tray is available.

SLIPPER OR SPINDLE COUPLINGS

To obtain an offset drive between a drive shaft and the roll shafts of rolling mills, a slipper joint located at one or both ends of a spindle is used. The geometry of such joints is described in Chapter 15. The lubrication of these heavy-duty couplings has been a problem for years, being manually attended during shutdown periods until recent times. The methods used to lubricate them automatically while in operation are numerous, and some have been quite successful. The lubricant is applied from an external or internal reservoir. External application may be actuated by a pump or air supply, while internal application depends upon either the centrifugal or inertial forces developed by the rotating coupling to force entrance of the lubricant to the slippers.

External methods often require a lubricant passageway to be drilled through the roll shaft (Fig. 11-31). The coupling is connected to a feed line from a lubricant metering valve through a swivel or revolving joint located at the blind end of the roll shaft. The drilled shaft method has proven the most satisfactory of the external application methods.

An oil spray system is another method of external application to a spindle coupling. A divided stationary casing surrounds the coupling, the upper half of which houses an oil distributor head and connected nozzle. The lower half acts as an oil catch pan that receives the sprayed oil after it drips from the coupling to which it was directed by the nozzles. The oil may be fed to the nozzles from the same circulation system serving the mill drive gears and bearings and to which it will be returned from the catch pan, or the pan may be equipped with its own pump to form a self-contained lubricator. In such a case, some means of cooling the oil is required.

A reversing mill type of slipper lubricator externally fed is the *slip collar* type (Fig. 11-32). This consists of two stationary collars wrapped side by

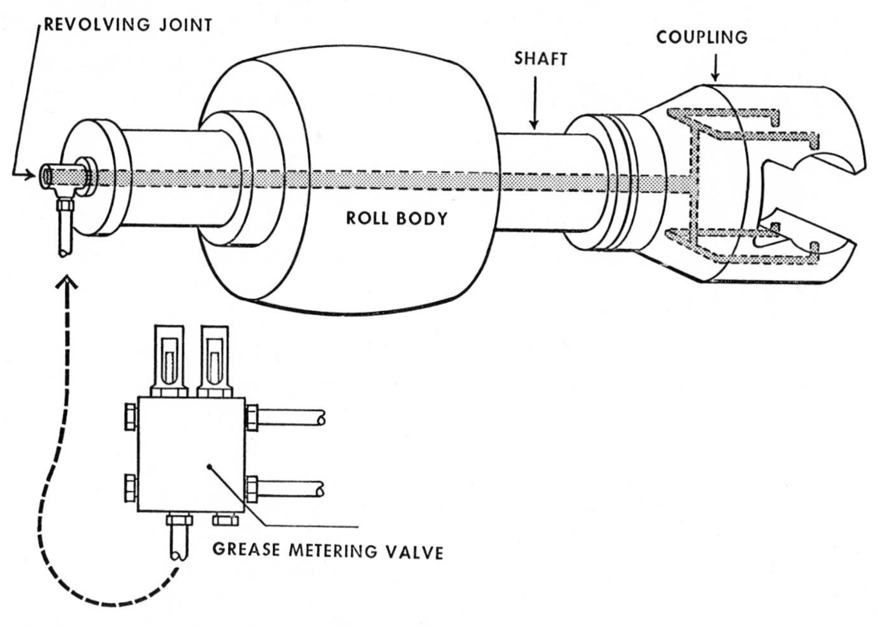

Fig. 11-31. Steel mill spindle coupling lubrication system. Drilled shaft method of application. *Courtesy, American Society of Lubrication Engineers.*

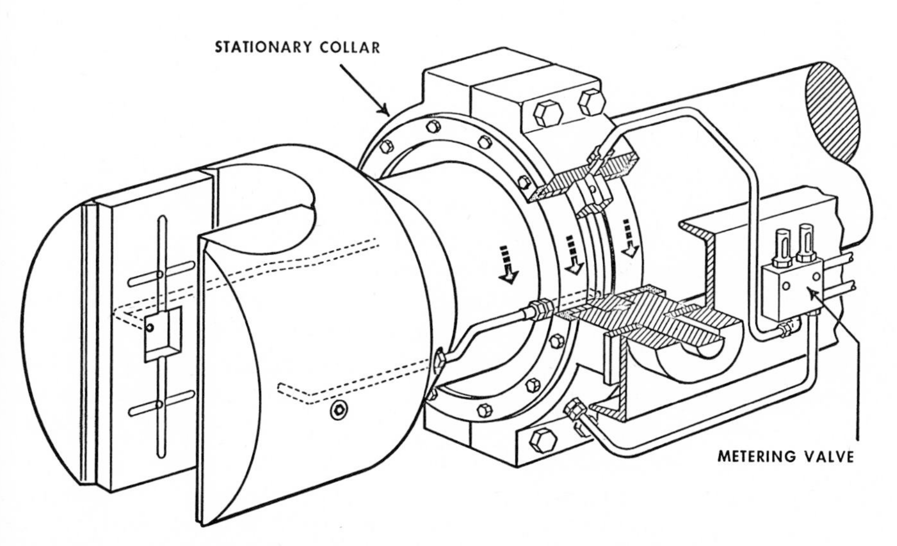

Fig. 11-32. Steel mill spindle coupling lubrication. System using slip collar. No drilling of spindle body. U-method. *Courtesy, American Society of Lubrication Engineers.*

side around the coupling head and sealed to prevent leakage of lubricant between the inner surface of the collars and the outer surface of the head. Holes in the underside of the collars are intermittently aligned with passages drilled in the coupling head as it rotates. The collars are fed externally with grease under pressure. One collar feeds the slipper area under pressure, when rotation is in one direction, and the other feeds the slipper area under pressure in the reverse direction.

Self-contained or internal lubricators differ in the method by which the lubricant is forced onto the frictional surfaces of the coupling. Several use a pump for the purpose; others use the centrifugal forces of the coupling.

On reversing mills, a pump is actuated by an inertial weight during mill reversals. The assembly, called a *unitized spindle lubricator* (Fig. 11-33), consists of a divided cylindrical casing fixed around the spindle near the coupling and rotated with it. One-half of the casing houses the pump and feed valves, and the other half acts as the oil reservoir. Each time the mill reverses direction, the inertia of the weight strokes the pump to feed lubricant (leaded compound of about 1800 SUS at 100°F), through feeder lines drilled in the coupling, to the jaws of the coupling.

On either reversing or unidirectional mills, the pump may be actuated by a cam and follower, the cam being attached to the spindle bearing housing. This type, known as the *palm end lubricator* (Fig. 11-34), as on the unitized lubricator, causes the whole assembly, including the lubricant reservoir, to rotate with the spindle. It is designed to handle grease.

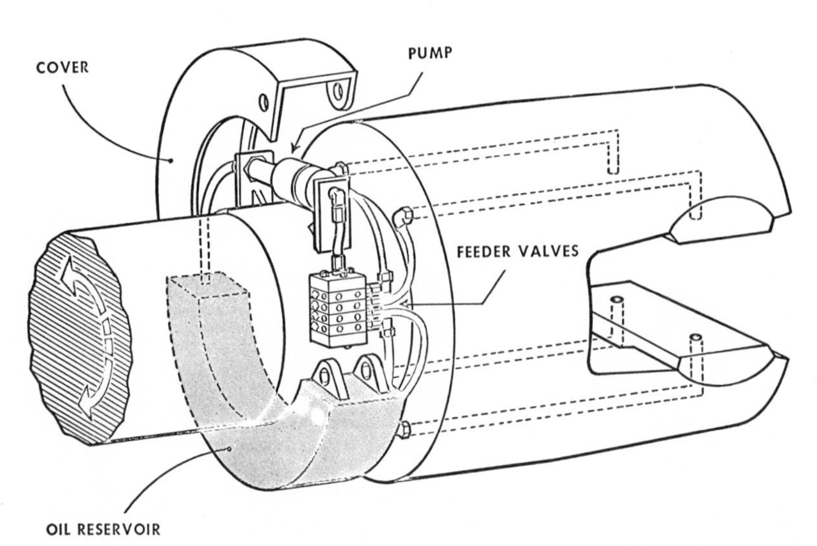

Fig. 11-33. Steel mill spindle coupling lubrication system. Unitized lubricator, for reversing mill only. *Courtesy, American Society of Lubrication Engineers.*

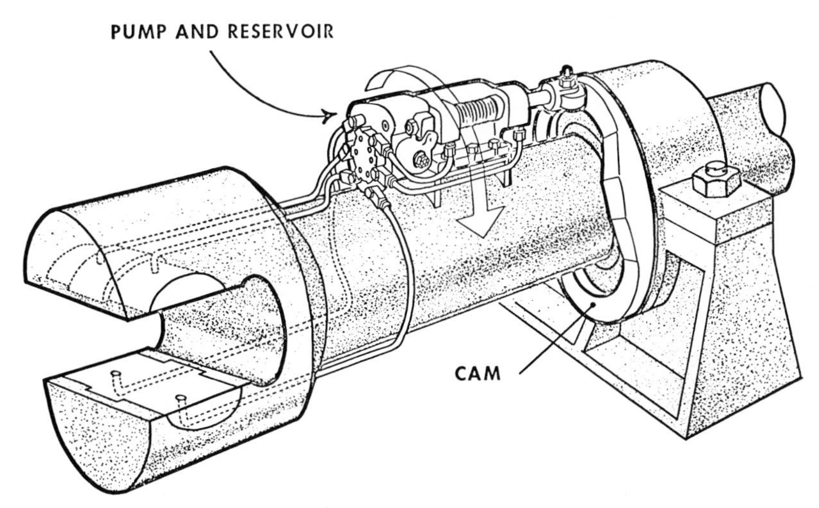

Fig. 11-34. Steel mill spindle coupling lubrication system. Palm end lubricator. *Courtesy, American Society of Lubrication Engineers.*

The slip collar method of applying lubricant to the slipper joint may be modified for use of a heavy oil under low air pressure. The oil is injected intermittently into a constant airstream by which it is carried to the slipper surfaces. The air helps to purge the bearing system and keep it free of abrasives. A mist oil system is also employed in connection with slip collars.

Lubricant may also be applied through the spindle bearing. The spindle is drilled radially to carry the lubricant from the carrier bearing to the center of the spindle, from where it splits and is carried by laterally drilled holes to the slippers on each end of the spindle.

An innovation of the fixed oil reservoir in order to have a self-contained oil applicator is to erect an enclosed housing about the tapered portion of the slipper head, which becomes the reservoir. Holes are drilled laterally through each jaw, and these are intersected by holes drilled radially from the slipper bearings. An adjusting screw placed at the intersection controls the flow of oil (Fig. 11-35). To maintain atmospheric pressure within the reservoir, a hole is drilled in the fill plug, which is automatically covered with a ball check when the plug passes through the lower arc of rotation. The same effect is obtained by attaching an extended open tube to the fill plug long enough to reach nearly to the spindle body when the plug is tightly screwed into its hole.

Grease must be pressurized to reach the slipper bearings through the passages drilled into the spindle or the slipper head. The grease should possess adequate pumpability at the operating temperature. Where inertia, centrifugal force, or gravity supplies the activating energy, oil should be used.

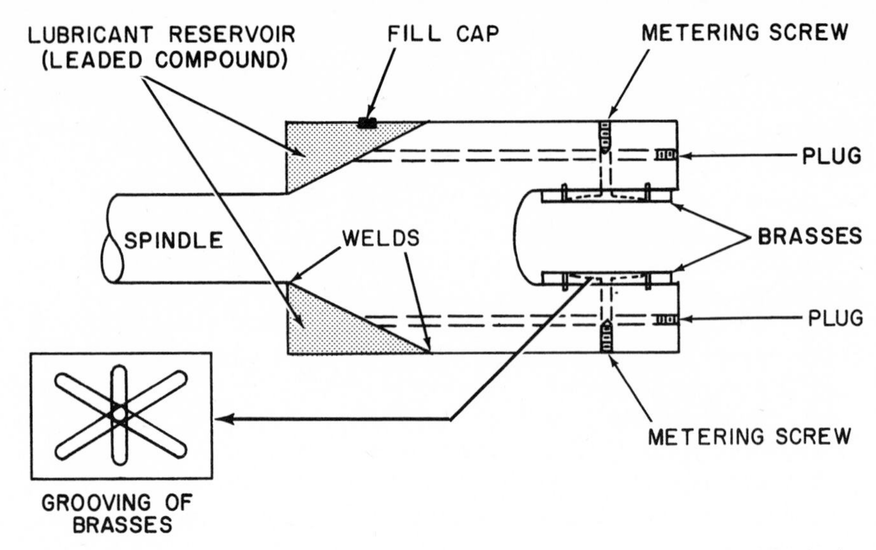

Fig. 11-35. Steel mill spindle coupling (slipper bearing), automatic self-contained lubrication system.

OIL CIRCULATION SYSTEMS

An oil circulation system is one in which the oil is returned to a central reservoir for reuse. Sizes of such systems range from small ones involving less than a pint of oil to large turbine installations handling several thousand gallons.

The system may be pressurized, in which case the reservoir is located below the frictional system and the oil is pumped to the bearings under any required pressure and returned by gravity. Or the frictional system may be fed by gravity from an oil reservoir located above, and the oil is returned to it by a pump.

The circulation system is ideal for certain purposes, because oil temperatures, cleanliness, and volume can all be controlled. The cycle of oil use can be shortened or extended, depending on the flow rate (pump capacity and size of conduits) and the quantity of oil contained in the system (size of reservoir).

A pressurized circulating system (Fig. 11-36) usually has the following components:

A reservoir with a sloping bottom, the lowest end of which collects any water that gets into the system. There is a drainpipe and valve arrangement by which the water is removed. The slope should be at least $^3/_4$ in. per foot of length.

A pump large enough (1) to force oil to all the frictional parts and (2) to carry away frictional heat in the system and maintain safe operating temperatures.

Feed piping large enough to carry the required volume of oil from the pump to the frictional parts at a safe unit pressure. Feeder lines are entirely flooded.

Drain piping large enough to return all the oil to the reservoir without the possibility of feedback. The drains of open systems are usually made so that they are about half full of oil being returned to the reservoir under gravitational force.

An oil cooler designed with sufficient heat transfer surface to reduce oil temperature to 100°F or lower when cooling water reaches its maximum seasonal (summer) temperature. Maximum seasonal temperature of river water is about 90°F, and for well water 70–75°F.

Filters are essential in removing foreign materials from the circulated oil. They are usually of the full-flow, self-cleaning type, although magnetic and cartridge (renewable) filters are available.

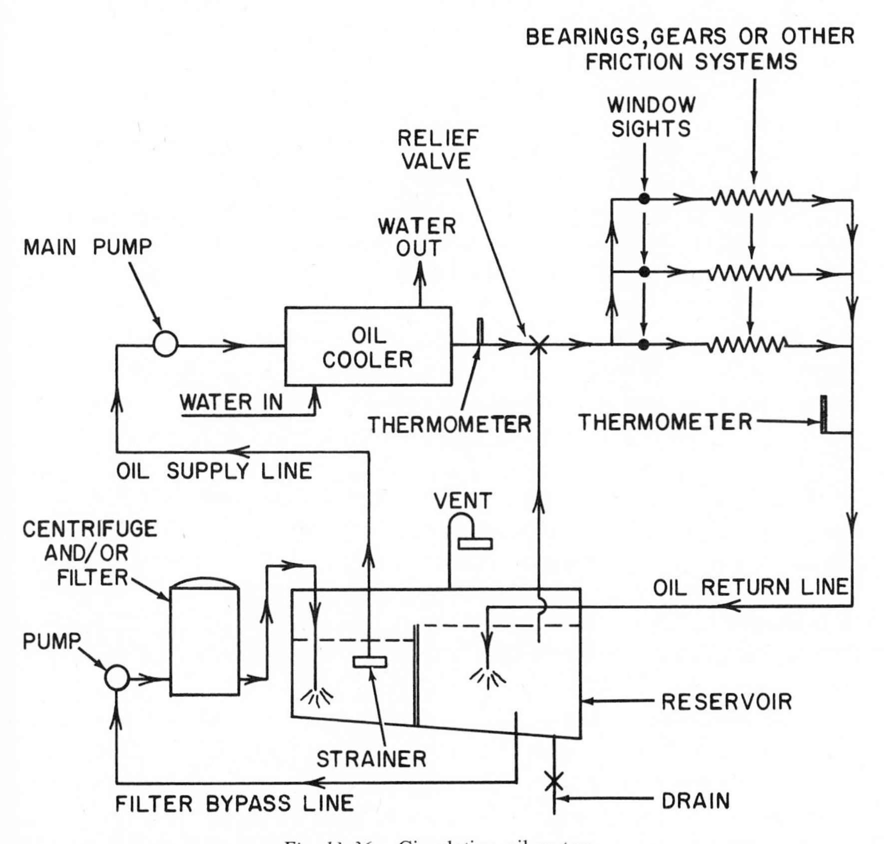

Fig. 11-36. Circulation oil system.

A strainer of lintless cheesecloth or fine-mesh wire is placed over the reservoir fill opening when new oil is being introduced either as makeup or during a change in the entire batch. It is essential when used oil is returned to the reservoir after repairs.

A centrifuge, either full-flow or bypass, is required for systems exposed to water, such as steam turbines (condensate) and some rolling mills. Where leakage is slight, the centrifuge may be operated periodically.

Heaters are required on systems exposed to low temperatures during winter months to maintain a desirable oil viscosity level and assure flow to the most remote frictional part.

Thermometers, pressure gauges, sight gauges, and liquid level gauges are installed at appropriate locations throughout the system to determine the temperature, flow pattern, oil pressure, and reservoir level at any time during operation.

Bypass lines are installed around filters. If a filter becomes clogged, the bypass line opens automatically.

Alarms are located at critical points within the system to communicate any oil pressure failure that occurs, either by sound or light.

Evacuators or vapor extractors are used to continuously remove moisture from the space above the oil in reservoirs and maintain a slight vacuum in that region. This may be a simple motor-driven exhaust fan placed over the breather or an exhaust line leading from the breather to the tailpipe (water discharge pipe of the condenser) of a steam turbine system. The action of the latter is similar to that of an aspirator. Where it is desirable to eliminate such moisture and accompanying oil fog from the vicinity of the reservoir, the exhaust fan may be located in a wall opening or window and connected to the breather by a duct. A dehumidifier is sometimes used to draw the moist air out of the reservoir, dry it, and return it, having been reduced to the latter temperature by a water cooled condenser in the dehumidifier. The resulting condensate then drains from the dehumidifier through a ball float arrangement.

Oil fog can be removed from the exhaust air by placing a cylindrical screen at the exit of the breather opening, which is supported in a well from which the oil drains back to the reservoir.

Gravity Feed System

Gravity feed systems consist of an oil reservoir located above the equipment to be served, feed lines, an oil sump, and a pump to return the used oil to the overhead reservoir. A filter is usually installed in the return line.

Two methods are used to feed the oil to the frictional parts. The first method has individual lines serving each bearing or frictional part. The second, the *cascade method,* has the oil from the reservoir entering the uppermost bearing and passing through to the next lower bearing, and thence to all succeeding bearings in turn until it drops into the sump.

The cascade gravity method has been widely used for paper mill calender roll bearings (Fig. 21-21) for many years, with a highly refined 900 SUS oil at 100°F.

OIL PURIFICATION METHODS

In order for a lubricating oil to remain in a satisfactory chemical and physical condition and to perform its intended service without harmful effects on the engineering materials to which it is exposed, it must be continually protected against catalytic influence, excessively high temperature, and harmful contaminants.

CATALYTIC INFLUENCE. A catalyst is a substance that initiates and/or accelerates a chemical action between two or more other substances without itself combining with either or undergoing change. Certain metals act as catalysts to promote chemical changes in mineral oils. Constant exposure of oils to such metals leads to their ultimate deterioration both chemically and physically. Chemical change involves the formation of acids during oxidizing processes that cause both the destruction of the oil and the corrosion of structural materials within the system. Chemical change also creates sludge and other carbonaceous materials that tend to accumulate and become deposited on various areas of the circulation system to hinder its normal processes.

The most important physical change in the oil is its increased viscosity. The most apparent physical change is a darkening of the oil.

Highly refined lubricating oils are fortified with additives to counteract the forces that deteriorate both oil and structural materials, but additives must never be relied on as an excuse to avoid taking the measures necessary to keep both the oil and system clean and safe.

Brass, iron, and aluminum are compatible with lubricating oils, but copper and zinc act as catalysts and should be eliminated from the system wherever possible. The effect is aggravated when such materials are present in the oil as fine particles. The initial products of oxidation are probably certain peroxides which act as catalysts to continue the process.

TEMPERATURE. A straight mineral oil is subject to some degree of deterioration when its temperature is increased, and the rate of deterioration accelerates as the temperature rises. The useful life of an oil is reduced 50% for about every 18°F rise in continuous bulk temperature. An initial oil life expectancy is substantially longer with a chemically stable or highly refined oil than with secondary or commodity oils.

Local oil temperatures usually vary throughout a pressure circulating oil system, but the bulk temperature should be kept to a maximum of 140°F and preferably cooler at the inlet of the oil cooler. This can be accomplished by (1) circulating sufficient oil and (2) cooling it enough to bring the system into a thermodynamic balance.

CONTAMINANTS. Any circulating substance other than the oil and its additives is a contaminant. Contaminants may be gaseous, liquid, or solid. The gases carbon dioxide and sulfur dioxide, for example, may be absorbed from the atmosphere as is the liquid contaminant, water. Air itself is a contaminant, causing foaming due to excessive oil turbulence or splashing. Solid contaminants may be particles from damaged bearings or gears or from some external source, e.g., fly-ash, coal dust, metal filings. Or the solids may be carbonaceous material developed by oxidation of the oil itself.

Some gaseous contaminants combine with water to form acids. Acids cause corrosion and also initiate a deteriorating chain reaction in the oil, evidenced by the appearance of sludge. Water also combines with metallic soaps that may be in the oil, resulting in an emulsion. Solids are usually abrasive and cause wear, jeopardizing the life of frictional components.

All three contaminant types must be continually removed from the circulating oil. Each must be treated individually unless a system has the facilities to remove them all in one process.

Prevention of Contamination

Oil can be purified in a number of ways and by various combinations of processes, but the basic principles involved remain the same. The principles, which will be discussed in turn, are straining, filtering, adsorption, magnetic separation, centrifugal separation, and settling.

First consideration, however, should be given to removing the cause of contamination or reducing the amount. This applies to both accumulated contaminants, such as sludge from oxidation, and to those that are introduced in each flow cycle, such as water.

For example, oil tends to vaporize and form a coke in a wet-sump bearing operating at elevated temperatures, such as paper mill dryer roll bearings heated by steam at 70 to 120 psi. The coke not only builds up in the bearing to interfere with its proper functioning, but also carries over into the oil system, where it will eventually clog the piping and filters. Selective additives can retard deterioration, but the logical step would be to change the wet sump to a dry sump and reduce the time that oil is in contact with the hot bearing. A similar condition exists when heating elements are incorrectly installed.

Water can get into the oil from one or more of the processes carried on within the system, such as condensate from a steam turbine. Maintaining the oil pressure in coolers above that of the cooling water forces oil leakage out rather than allowing entrance of water into the system.

A new system should be flushed with the oil that will be used in it in order to clear the passages of debris left during installation and construction. All new piping should be inspected to assure freedom from rust-preventive films which must be removed if present. A new batch or makeup oil should be strained into the reservoir.

After all reasonable measures have been taken to keep the oil in good condition, a program of purification-maintenance should be planned, based on regular analysis of the oil.

SAFETY INDICATORS. There are various methods of protecting frictional equipment against damage due to low oil levels, line blockage, and other failures that prevent lubricants from reaching their point of application. Reservoir oil level and feed line pressure are two clear indexes of how a system is functioning.

The simplest device for assuring adequate oil feed to a frictional component is the constant level lubricator that automatically maintains a favorable condition within the bearing or gear sump to which it is connected. It is effective, of course, only as long as the oil level in the constant level device itself is maintained above the point of discharge. A periodic visual inspection is necessary.

Oil levels within sumps, tanks, and reservoirs should be visible through window sights and liquid level gauges, which also must be periodically observed. Window sights also allow visual inspection of oil flow in feed lines of circulation systems.

Neglecting visual inspection, either through carelessness or difficulty due to remoteness of oil reservoirs, may result in mechanical failure.

Grease systems are best checked by strategically located pressure gauges. Another method to check grease flow is with blow-out indicators at various outlets. Any blockage upstream of the indicator will cause a pressure rise to hydraulically force an indicator pin outward. The protrusion of the pin not only gives evidence of blockage, but helps locate the blocked region.

Lubricant flow failure, whether by blockage or breakage, usually requires immediate attention. The associated pressure changes should be utilized either to shut down the operation or give warning.

AUTOMATIC WARNING DEVICES. A simple device to assure adequate oil level in a level device or in a manually filled oil reservoir is a guided ball float containing a magnet whose energy field acts to close a sensitive reed switch located at the critically low level of the reservoir. When the switch closes, a red light flashes or a horn blows. The low level switch may also be interlocked with a pump or solenoid value installed in a makeup oil level to automatically refill the reservoir. Upon reaching a given level, the magnet within the ball acts on a high level switch to shut off the flow of makeup oil.

Ball floats are used in various ways to control the high and low levels of liquids within tanks and reservoirs, including trip mechanisms, float valves, and electrical make-and-break switches. On large storage tanks, a radioactive tracer system is used to indicate liquid levels.

A FULL-FLOW PURIFICATION SYSTEM (Fig. 11-37) is one in which all the oil circulated passes through the purifying process during every cycle. Such systems should be equipped with an alarm system to warn of partial or complete interruption in oil flow. Otherwise they should also include an automatic bypass arrangement (Fig. 11-38) to assure an adequate oil flow during blockage in the oil purifier. A combination full-flow and bypass filter is shown in Fig. 11-39.

Five filters are shown in Fig. 11-38—three bypass (B) filters and two full-flow (F) types. In any particular hydraulic system, usually only one filter is used in addition to a suction screen. Location B-1 is most common for a bypass filter, but B-2 and B-3 are sometimes convenient. With B-2, a low-pressure relief type valve, set to maintain about 50 psi across the filter, must be placed in the high-pressure relief valve drain line. With B-3, the low-pressure relief type valve must be placed as shown in the main return line. Bypass filters take from 5 to 15% of the hydraulic fluid pumped, and they are not usually built to withstand full system pressure. Pressure and flow are limited by means of orifices and valves, which may be built into the filter. Full-flow filters (F-1 or F-2) take 100% of the flow and must withstand the full line pressure. A bypass around the filter must be provided. Otherwise, a high pressure drop across a clogged filter may cause a machine to be shut down. Sometimes, the bypass is built in. The line pressure at

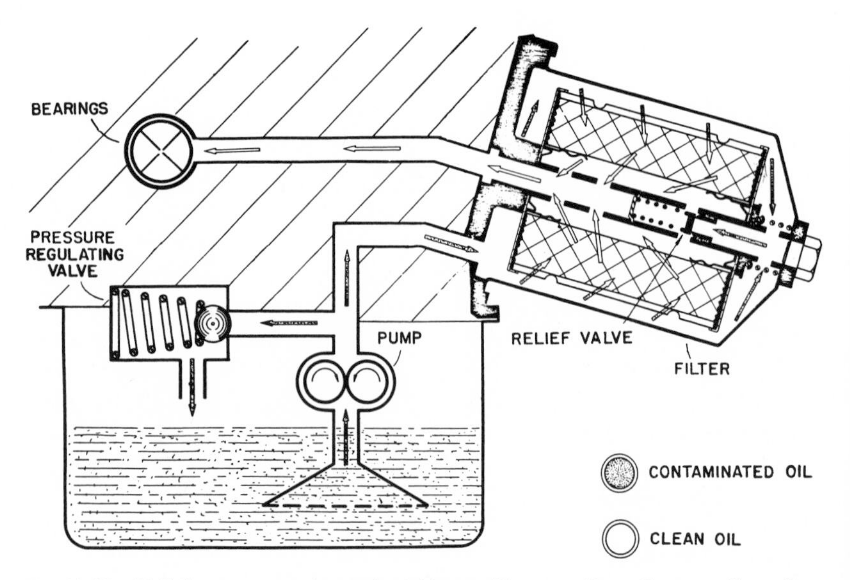

Fig. 11-37. Full-flow system using replaceable cartridge type filter. *Courtesy, Fram Corp.*

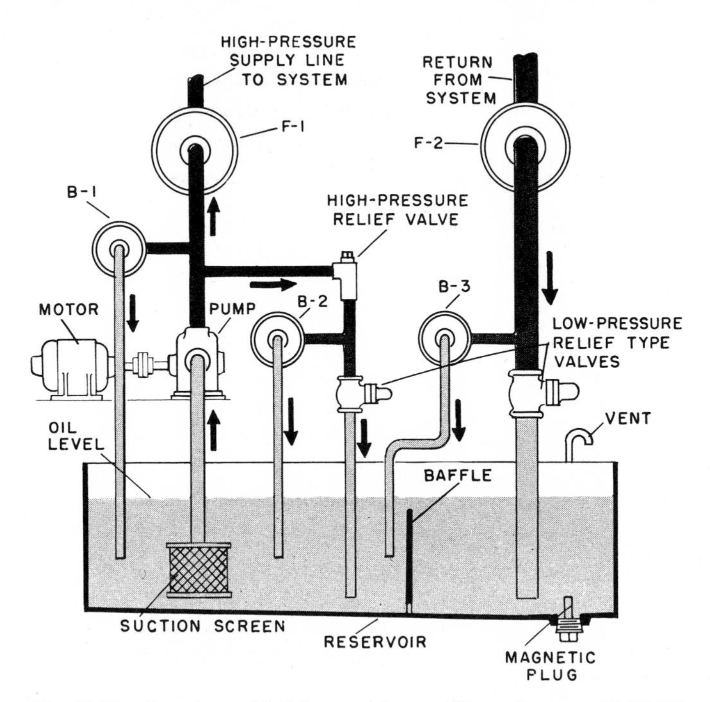

Fig. 11-38. Locations of full-flow and bypass filters. *Courtesy, Mobil Oil Corp.*

location F-2 is, of course, low. In addition to the locations shown, the *independent* system is common. In this, a separate pump takes oil from the dirty oil side of the reservoir (to the right of the baffle), puts it through a filter, and returns it to the clean oil side.

The normal procedure is to purify continually a fractional part of the entire charge or batch. This is accomplished by shunting or bypassing about 5 to 15% of the oil circulated through the purifying units, after which it is returned to the main circuit. Such systems are respectively known as *shunt* and *bypass purification systems* (Figs. 11-40 and 11-41).

Systems such as those serving hydraulic, turbine, engine, and compressor installations employ a separate system, or *independent purification system*. This involves the continuous or periodic withdrawal of oil from the main reservoir to be purified and then returned to the reservoir. Continuous independent purification systems are usually installed as part of the host machine, while periodic purification is accomplished, usually, with a portable pump-purifier arrangement. This method is ideal for servicing hydraulic systems of injection molding machines which may be continued in operation during the purification period.

Another purification procedure uses a *batch settling method*. This usually

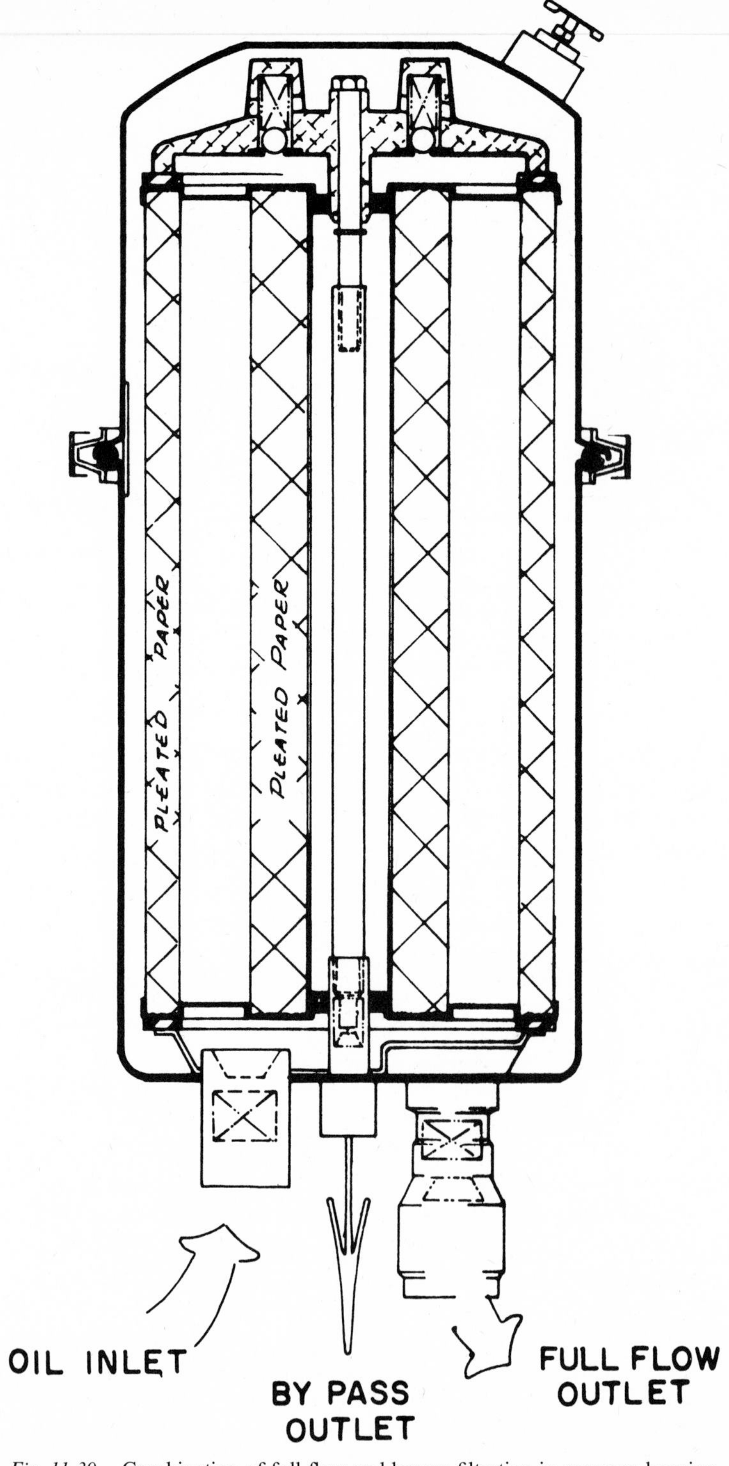

Fig. 11-39. Combination of full-flow and bypass filtration in common housing. In operation, most of the oil flows through the full-flow cartridge. *Courtesy of Fram Corp.*

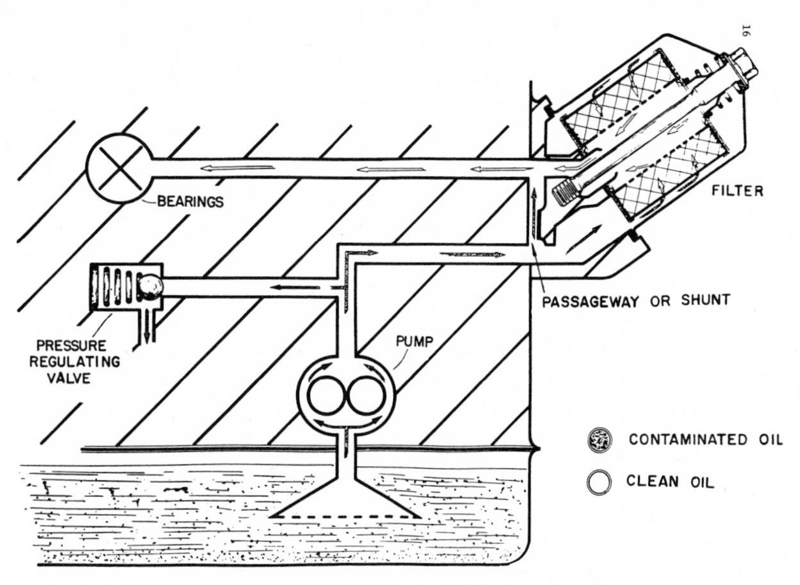

Fig. 11-40. Shunt system using replaceable cartridge type filter. *Courtesy of Fram Corp.*

requires operation shutdown, but a purified second batch may be ready so that operation can resume as soon as the system is filled. This method, discussed later in this chapter under Settling, may be supplemented with other purification techniques, such as filtering and centrifuging.

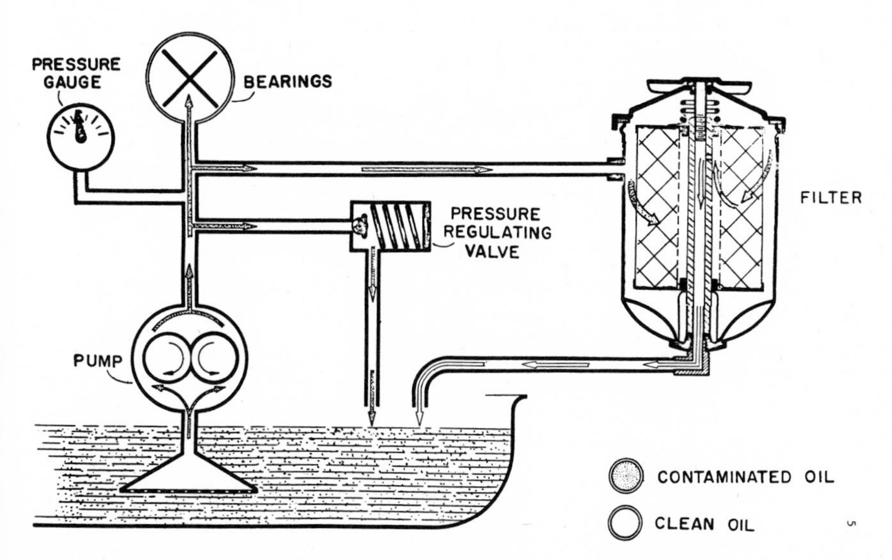

Fig. 11-41. Bypass filter system using replaceable cartridge type filter. *Courtesy of Fram Corp.*

Strainers

A permanent strainer consists of fine mesh wire placed across an opening to catch solid contaminants. The line may be a pump suction pipe, a breather pipe, a fill opening, a standpipe, a vent, ventilator, and other openings leading to a reservoir, crankcase, storage tank, and other containers.

Strainers have limited filtering capabilities, but their effectiveness is proportional to the fineness of mesh. Mesh values are reported according to the number of wires per inch of wire cloth and range from about 35 mesh (35 per inch) to over 200 per inch (for air). Good fabric used for filtering air is capable of separating dust particles as small as 0.000020 (20 millionths) inch, which corresponds to one-half micron and 99% efficiency. (1 micron = 0.000039 in.) A 270-mesh wire cloth will stop the passage of particles larger than about 50 microns. Wire cloth may be made to remove particles as small as 3 microns.

The separating capability of any type of filtering medium must be correlated with the initial pressure drop corresponding to the required rate of fluid flow. To remain within reasonable limits, some compromise is often necessary for which there are four variables:

Pressure differential. Increasing the delivery pressure on a filter increases the pressure drop through it and increases flow velocity.

Viscosity. Decreasing viscosity increases the flow. This can be done by raising the temperature of a given oil.

Mesh value. Efficiency is conversely proportional to flow rate. Any increase in mesh value allows larger particles to pass through.

Area of mesh. Mesh area can be varied by forming the strainer material into various shapes, such as bags, cylinders, balls, and accordion folds, which may be installed outside or within the delivery space.

Varying mesh area is the most practical method of controlling straining capacity and rate of flow, and this applies to both strainers and filters.

Size Filtering

Size filters, absorption filters, and magnetic filters are used to purify oils. Size filters use the principle of the strainer but can trap finer particles. Strainers may consist of one layer of mesh, whereas size filters usually have depth. An exception is the surface filter, which is described later. Size filters are encased in a housing designed so that the separated debris can be flushed out or so that it can be discarded and replaced with a new one. The types of size filters available include strainer, depth filters, edge filters, and surface filters (Fig. 11-42).

REPLACEABLE FILTER CARTRIDGES. Figure 11-42 shows four types of replaceable filter cartridges. The strainer type is used primarily to protect the pump and circuit from large particles. It is most often used on the suction

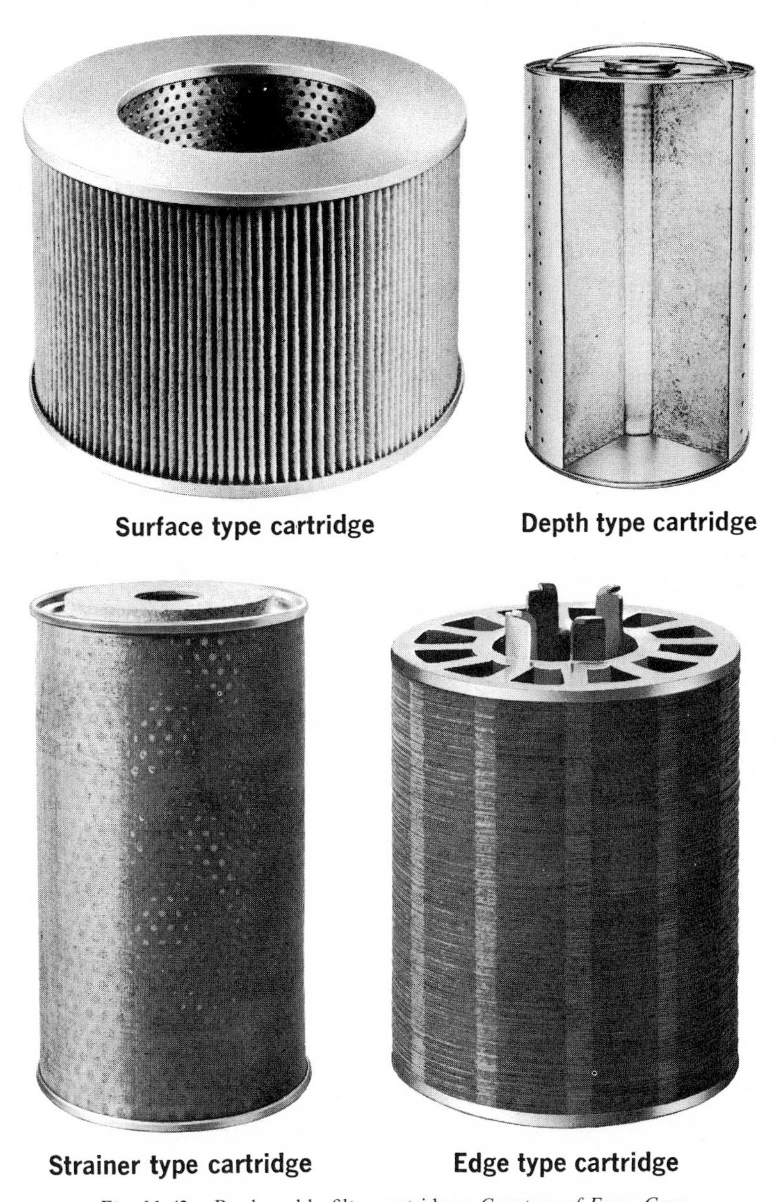

Fig. 11-42. Replaceable filter cartridges. *Courtesy of Fram Corp.*

side of the pump. It is cleanable. It offers coarse filtration, low restriction, and high flow. The depth type cartridge (bulk cotton, felt, cellulose fibers, etc.), 10 to 50 microns (not cleanable), offers a good degree of filtration combined with the ability to absorb quantities of water and resins. The edge type cartridge (resin-treated ribbon or metal disks), 2 microns for paper disks (not cleanable), 40 to 70 microns for paper ribbon (cleanable), offers finer filtration than the strainer with a reasonable low restriction and high flow. The surface type cartridge (resin-treated paper or fabric) is in the 2 to 40 micron range. The limited porosity of this type filter requires a large area; thus most cartridges are pleated paper. It offers fine filtration, low restriction, and high flow.

DEPTH TYPE FILTERS. The element of a depth type filter is usually a hollow cylinder whose walls are thick layers of cellulose, felt, or waste (Fig. 11-43). The oil enters the housing and flows through the element to the hollow space within the cylinder and is then discharged. The separating capabilities of depth filters vary with radial thickness and density of material. Flow-through volume capacity is a function of surface area, which varies according to the length and circumference of the cylinder. Separating capabilities of depth type filters range from 10 to 50 microns.

Depth-type filter elements separate larger solid particles at the inlet surface, while the smaller particles are trapped within the material mass. Fibrous cylinders tend to become graded, i.e., they gradually become more dense toward the internal surface as a result of compression of their fibers during the wrapping process. This accounts for internal trapping and a longer service life.

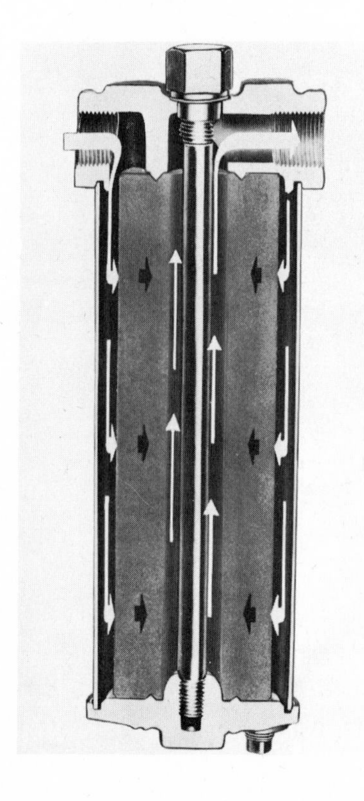

Fig. 11-43. Depth type cartridge filter. Capacity to filter particles sizes of 10 to 50 microns. Not cleanable. *Courtesy of Cuno Division, A.M.F.*

The absorbent properties of some filter materials will remove water from the oil, but saturation will ultimately cause clogging. This principle is often used to remove water in the final stages of oil dehydration processes but is not recommended for large circulating systems into which an appreciable amount of water finds its way.

Materials such as cotton waste are not structurally self-supporting and must be packed into a cylinder of wire or perforated metal. Machine-packed waste elements can be purchased for replacement and have proved to be more effective than hand-packed cylinders fabricated in the plant. The reason for this lies in the uniform density of the machine-packed mass which prevents channeling so prevalent when fabrication is accomplished by piecemeal buildup of tangled waste. The correct packing process is particularly important with bulk type filters in which a loose mass of waste is supported by the retaining housing of the filter unit. On circulating systems serving large internal combustion engines, packing is done periodically by plant personnel.

Another depth type filter consists of stacked cells drawn tightly together by a wing nut and through-bolt. The cells are made up of asbestos and cellulose disks sealed at the outer edges and connected to spacers at the inner edges. The materials are water absorbent, which may be removed by drying.

EDGE TYPE FILTERS. The edge filter differs from the depth type in that it consists of disks stacked on each other to form a laminated pile (Fig. 11-44). The spaces between the adjacent disks form radial passages through which the oil flows from the outer periphery to the inner well formed by the center of the disks. In some types the pile is held in place by a vertical shaft or spindle running from a keeper attached to the bottom end of the shaft up through the cap of the housing. The upper end of the shaft is fitted with a handle to facilitate turning it occasionally in either direction for cleaning. Cleaning is accomplished by a pile of stationary blades having concave edges that coincide with the curvature of the disks. The shaft has a square cross section to prevent rotary motion of the blades, which are oriented with the gaps between the wheels so that they scrape the trapped debris from the filter elements when the latter are turned by the shaft handle.

The separating capabilities of edge filters depend on the gap permitted between the adjacent disks. This gap is governed by spacers inserted between disks and varies according to service demands to sizes below 40 microns, or about 0.0015 in.

Cellulose disks, or ribbons, wound to form a continuous helix (flat surface on flat surface) are also used in place of the wheels described above, to accomplish edge-type filtration. The filtering elements are usually resin treated so that they can be reused after periodic solvent cleaning. Filter capabilities of disc types range from 0.5 to 10 microns, while about 40 microns is the minimum particle size that may be separated by ribbon types.

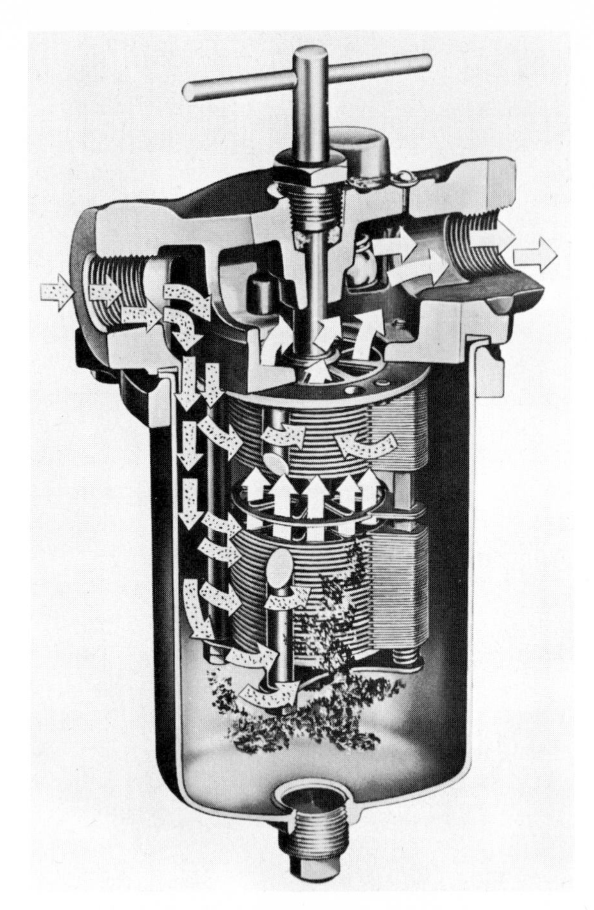

Fig. 11-44. Edge type filter. Mechanically cleanable. *Courtesy, Cuno Division, A.M.F.*

SURFACE FILTERS. This type of filter employs the strainer principle to cause size separation of solid particles, depending on a closely woven cloth or pleated resin treated paper to separate the solid particles from the oil passing through it. Maximum flow-through is accomplished by the geometry of the element, effective forms being the accordian-pleat and corrugated types. Except for the depth principle, the flow pattern of the surface filter is the same as that of exploiting material thickness.

The element may be of paper, which is replaced when it becomes unserviceable, or cloth that may be washed and reused several times before replacement is required. Paper elements are made that can remove particles 1 micron in diameter, whereas cloth cannot remove particles smaller than 50 microns.

A combination depth and surface filter is shown in Fig. 11-45.

Table 11-2. SEPARATING CAPACITIES OF VARIOUS FILTERS WHEN CLEAN. (DIRTY FILTERS MAY HOLD BACK PARTICLES OF SMALLER SIZES THAN SHOWN.)

Type of Filter Element	*Approximate Range of Particle Sizes*
Depth type	1 to 75 microns
Surface type, cylindrical pleated treated paper	1 to 10 microns
Edge type, metal disks	100 microns or less, depending on space between disks
Edge type, cellulose disks	0.5 to 40 microns

Microns	*Inches*	*Microns*	*Inches*
0.5	0.000019	50.0	0.001950
1.0	0.000039	100.0	0.00390
10.0	0.000390		
25.0	0.000970		

$$\text{Inches} = 0.000039 \times \text{no. of microns}$$

$$\text{Microns} = \frac{\text{No. of inches}}{0.000039}$$

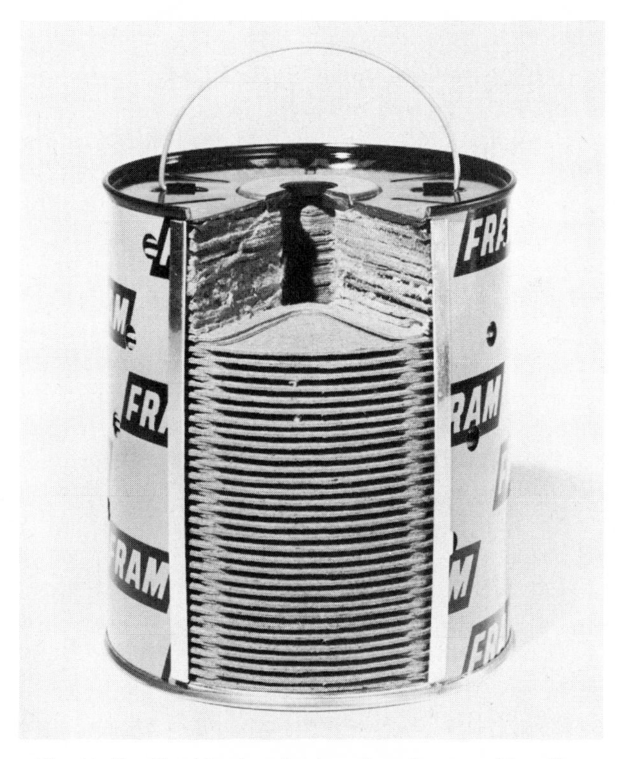

Fig. 11-45. Combination depth and surface cartridge filter. Edges of paper act as surface filter while combined bulk of disks provides depth. *Courtesy, Fram Corp.*

Adsorption

Adsorption is the adhesion of a very thin molecular layer of gases, dissolved substances, or liquids to the surface of solids, called adsorbents, with which they come in contact. Purification by adsorption is not recommended for oils containing additives (steam turbine lubricants) because the adsorbent will remove an additive as readily as it will remove a contaminant. Typical additive-free oils that are purified by adsorption are those used for transformers and refrigeration systems.

The most frequently used adsorbents are Fuller's earth, bentonite, activated alumina, bauxite, charcoal, and chemically treated cellulose.

Fuller's earth is a soft natural clay used for decoloring and clarifying mineral oils, fats, greases, and vegetable oils. It can remove certain oil-soluble oxidation products. It is the most widely used adsorbent, serving as either a contact or percolation material. Fullers' earth will remove color bodies, solids, and some liquid compounds.

Bentonite is a colloidal clay with a hydrophilic or water-attracting property that enables it to absorb five times its own weight of water.

Activated alumina, which consists of beads of aluminum trihydrate that have been partly dehydrated, has a strong affinity for moisture and gases and is employed as a dessicant. It can be repeatedly regenerated. Excessive water causes activated alumina to swell and burst.

Bauxite is an important aluminum ore used as filtering medium and catalyst in petroleum refining. It is an effective filtering material. One of the advantages of activated bauxite in oil filtering is that, like activated alumina, it can be regenerated indefinitely.

Charcoal is valuable for filtering gases and colored substances from solutions. It is more widely used as an air purifier and water sweetener than for filtering oil.

It is best to precede an adsorbent purifier with a centrifuge or some other method of removing water which acts to turn the clay filtering material to mud. Because in some types of purifiers, the adsorbent forms a bed through which the oil percolates, such units also act as effective depth filters. Another method of using adsorbents is to mix it with the oil (about 30 lb per 100 gal of oil) to be purified and then passing the slurry through a size filter or filter press to remove the adsorbent. The filter press consists of a series of cloth filters, canvas, wire mesh, and perforated plates, in that order.

Heating the oil to about 150°F increases the adsorption activity of the clay which results in more rapid flow through the purifier.

Transformer and additive-free refrigeration oils are typical fluids purified by adsorbent separators. Passing oil through a bed of fuller's earth is called a *percolation* process.

Magnetic Separators

To separate magnetic fines from lubricating oil, magnets are placed in strategic locations within a circulation system or reservoir. Either electromagnets or permanent magnets may be used. In their simplest form, they are made a part of reservoir drain plugs which are removed periodically and cleaned. One type of magnetic drain plug is designed so that the drain will remain closed while the plug is taken out for cleaning.

Magnetic separators are installed in circulating oil lines in the same manner as depth filters. One type consists of a grid arrangement energized by permanent magnets, all of which are installed in a supporting housing. The oil enters the housing at the top, drops to the bottom between the housing wall and grid cylinder from where it is forced upward through the grids and discharged at the top opposite the inlet. The grid consists of soft iron washers stacked on top of each other and held together by a through-bolt and wing nut. When the unit is disassembled for cleaning, the grid becomes deenergized and the washers can be cleaned easily.

A self-cleaning magnetic separator consists of a revolving drum to which a series of permanent magnets are attached. A second cylinder encircling the magnets and turning with the drum attracts any magnetic fines from the oil passing under and counter to it. A scraper removes the debris from the outer surface of the cylinder and discharges it from the unit.

Magnetic separators are installed on oil systems related to metal-cutting operations and metal-surface running-in processes. Machine tools require magnetic separators. Planers have them installed in the way oil line, and grinders have them located in the metal-cutting fluid system. Central oil systems servicing a bank of new internal combustion engines during the running-in period require magnetic separators in conjunction with size filters to eliminate featheredging of piston rings and abrasion of other frictional parts.

Magnetic plugs may be installed in series of two or more somewhere prior to the pump section in hydraulic systems, way lubricating oil lines, and reservoirs. Periodic cleaning is necessary.

Centrifugal Purifiers

The method of separating contaminants from a fluid by centrifugal force is based on the principle that water, being heavier than oil, will accelerate outward at a greater rate than the oil and will tend to leave the oil body if the mass is whirled about in a bowl.

The major purpose of a centrifuge (Fig. 11-46) in any circulating oil system is to remove water, which is the most damaging and prevalent contaminant, and water-soluble materials. Some of the heavier solid particles

Fig. 11-46. Centrifugal purifier (centrifuge). *Courtesy, DeLaval Separator Co.*

may be removed with the water, but the machine should not be considered as a substitute for filters in systems capable of removing appreciable amounts of oil-soluble acids and other oil-soluble materials.

The oil enters a receiver at the top of the revolving centrifuge (Fig. 11-47) and falls through a hollow shaft to the bottom. It is then forced outward to become distributed between disks. The disks have holes that permit the clean oil to flow along their upper surfaces toward the axis, while the water flows outward from the axis along the lower surfaces. The clean oil is continually forced inward and upward along the hollow shaft to be discharged through the outlet spout. The water is forced outward and upward along the periphery of the bowl and discharged through a spout located below the clean oil spout. An overflow spout (not shown) is located above the other two spouts to carry away the excess influent from the top of the hollow shaft just below the receiver. The purpose of the overflow is to prevent the introduction of more oil than the centrifuge can handle.

Where the centrifuge is used to separate solid particles as with some

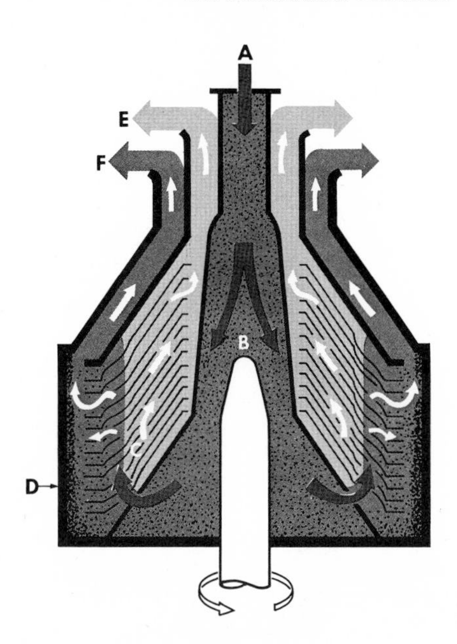

Fig. 11-47. Centrifuge. A, used oil entrance; B, used oil flow; C, separation of oil and water; D, wall (solids adhere to the inside); E, clean oil exit; F, water exit. *Courtesy, DeLaval Separator Co.*

diesel fuel and lubricant systems, the through-put should be reduced to about one-half its rated capacity.

Since centrifuging is more efficient when the oil is warm, both in releasing fines and in separating water from emulsion, this process should immediately follow when the oil leaves the lubrication cycle.

Adding water to the oil as it enters the centrifuge and immediately removing it during the centrifuging process acts to purge the oil of various contaminants, including acids and alkalis. Water washing is quite effective with straight mineral oils, but should not be practiced with inhibited oils since some of the additives will be removed with the water.

The function and soluble characteristics of turbine oil additives are listed in Table 11-4. The oxidation inhibitors are not affected by size filtration or centrifugal purification but can be removed by low vacuum or clay filtration. Rust inhibitors can be removed only to the degree that they bond to water when subject to purification devices. Silicones are water insoluble

TABLE 11-3. TYPICAL CAPACITIES OF CENTRIFUGAL SEPARATORS, DELAVAL SERIES MAB 200 (GALLONS PER MINUTE)

Model	*Light Turbine Oil, 150 SUS at 150°F*	*Diesel Fuel, 70 SUS at 100°F*	*Diesel Lube*	*Heavy Fuel, No. 6*
MAB-205	500–800	850–1050	500–550	450–600
MAB-206	900–1100	1200–1500	750–900	550–900
MAB-207	1500–1800	2000–2250	1000–1500	750–1200
MAB-209	2400–2800	3000–3200	2000–2400	1800–1850

Courtesy, DeLaval Separator Co.

Table 11-4. ADDITIVES IN TURBINE OIL

Additive	*Function*	*Solubility*
Debutyl paracresol	Antioxidant	Oil soluble
Alkylated succinic acid	Rust inhibitor	Slightly water soluble
Silicones*	Antifoam	Insoluble in oil

* Some loss of silicones occurs when turbine oil is passed through a one-micron filter.

so will not be removed by centrifuging. Dispersion of silicone may result in droplet sizes ranging from less than one micron to five microns, of which up to 80% are between 1 to 5 microns. Loss of valuable silicone additive can result if purifiers are specified to remove a particle size one micron or less.

The centrifuge is a much more rapid method of removing water from oil than the batch settling method described in the following section. The difference between the two methods lies in the gravity factor, which is constant for the settling procedure but may be varied according to speed (centrifugal force) for any given fluid in a centrifuge.

The nomograph of Fig. 11-48 is prepared for sizing purifier requirements when approximate gallons in system and flow back are known. Put the edge of a ruler on line 2, intersect line 3 at the desired percentage, and read the capacity on line 4. If reservoir capacity is known, use line 2 as a starting point. Results will be relatively accurate. Line 1 can be used for estimates of reservoir capacity if only the megawatt output is known. The

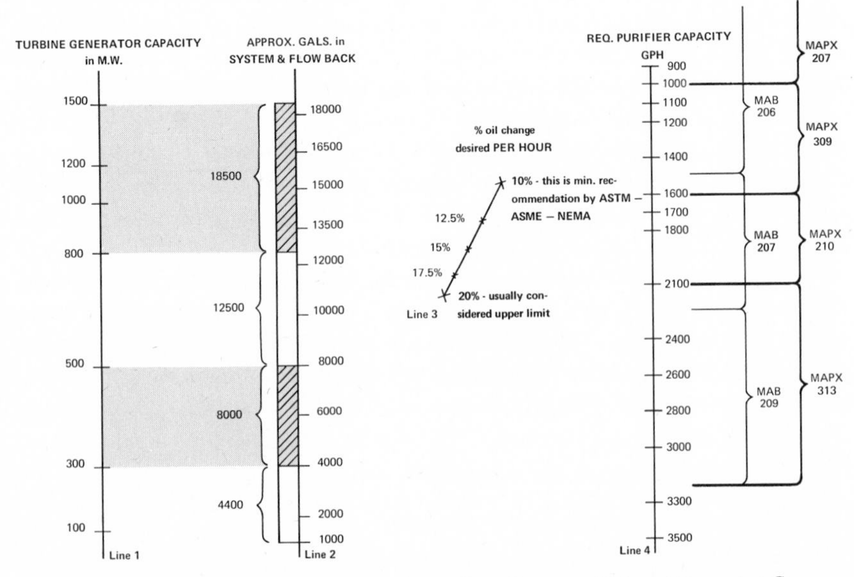

Fig. 11-48. Nomograph for sizing oil purifier. *Courtesy, DeLaval Separator Co.*

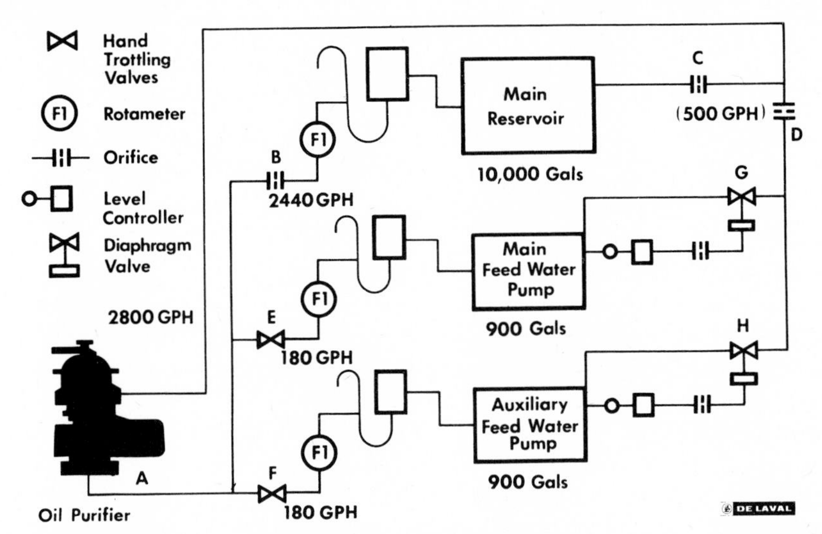

Fig. 11-49. Simultaneous purification of three oil reservoirs. *Courtesy, DeLaval Separator Co.*

block correlation between lines 1 and 2 is such that, for example, generators with a capacity between 800 and 1500 MW would use an 18,500-gal system.

A typical turbine oil system and a flow sketch for the simultaneous purification of three oil reservoirs are shown in Fig. 18-11 and Fig. 11-49, respectively.

SIMULTANEOUS PURIFICATION OF OIL RESERVOIRS. The system shown in the flow sketch (Fig. 11-49) offers definite cost savings over the installation of an individual purification device on each reservoir. For example, a centrifugal oil purifier sized to protect the main turbine of a 500 MW unit would cost x dollars. A unit to protect each of the two boiler feed pump auxiliaries would cost $\frac{1}{3}x$ dollars or a total investment of approximately $1.6x$ dollars. A larger centrifuge, capable of handling the flow rate of all three reservoirs, would cost $1.2x$ dollars, or approximately 30% reduction in the overall cost of purification protection.

The system is designed to operate as follows: The flow to and from each reservoir is metered. The flow returning to the main reservoir is maintained at a constant rate by an orifice shown at C. The remaining flow passes through a limiting orifice D, where level controllers regulate the flow into the smaller reservoirs. Vented sight overflow devices keep the operating level from falling below safe standards. The design is such that a failure will not interrupt turbine service because of low oil level. Separate pumps are not required, as the centrifugal oil purifier has built-in pumps, both feed and discharge.

The differences in centrifuge types are shown in Fig. 11-50 and described as follows: Original batch-type centrifuge (A) receives feed at 1. Feed

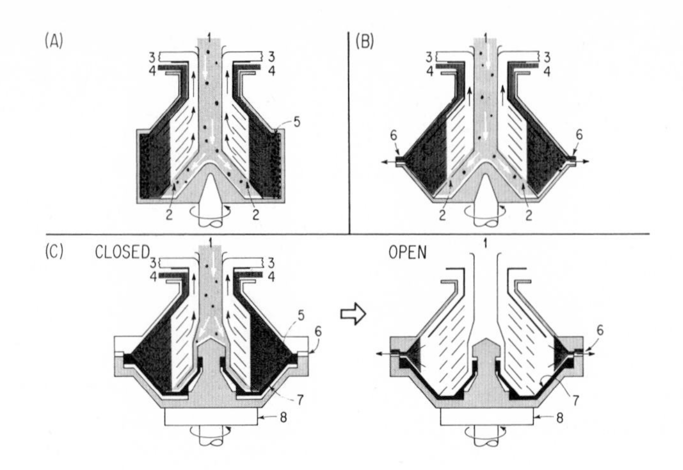

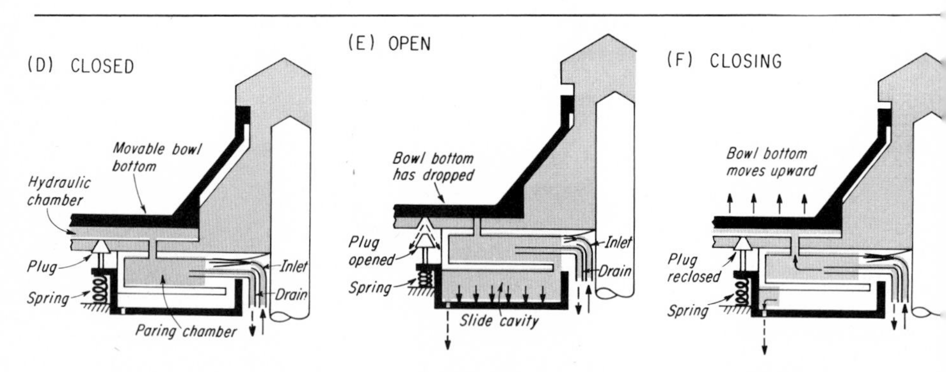

Fig. 11-50. Difference in centrifuge types. *Courtesy, DeLaval Separator Co.*

flows to disk stack at 2 and separates into oil, which leaves at 3, and water, which leaves at 4. Sludge and solids collect at 5. Periodic cleaning prevents sludge from re-entering water and even oil flow.

Two designs that largely eliminate frequent manual cleaning of the bowl are continuous type with nozzles (B), and the split-bowl type (C).

In the continuous type with nozzles 6, the bowl wall tapers outward to give an easier path for sludge leaving the bowl.

The split-bowl type, shown closed and open, is essentially a nozzle type with large, intermittently opened nozzles 6. Moving bowl bottom 7 and actuating mechanism 8 raise cost considerably but make clogging less likely.

The movable bottom referred to above is illustrated in Fig. 11-50 (D) closed, (E) open, and (F) while closing:

(D) Closed. The bowl bottom is raised in the closed position for regular operation. A water trickle through inlet keeps paring chamber partly filled. Excess water is pared off and pours into drain. Paring chamber water, via small orifice, fills the hydraulic chamber under pressure developed by centrifugal force, thus keeping bowl bottom raised in closed position. All the elements shown in sketch, except inlet and drain, are spinning at the centrifuge's normal speed.

(E) Open. To open the bowl, programmed control sends a sudden extra flow of water through the inlet. This flow exceeds removal rate of drain, and water floods into slide cavity, pushing it down against spring force and opening plug of hydraulic chamber. Water in hydraulic chamber drains immediately, and bowl bottom drops to its open position. Some slide-cavity water drains out but is continuously replaced by water from the paring chamber.

(F) Closing. To close the bowl, control cuts off extra flow of water through inlet. Water in paring chamber recedes to its normal limit and no longer supplies slide cavity. Slide-cavity water drains out, eliminating force that originally moved the slide cavity down. Spring then forces the plug upward to seat and seal the hydraulic chamber, which then fills with pressure water through the orifice. The pressure raises the bowl bottom, closing it again.

Settling

Purification of oil by batch settling is accomplished by keeping oil out of circulation in a separate container and letting it sit for several days to several months. During the static period, heavier-than-oil substances such as water and solid particles will settle to the bottom of the storage container, from where they can be drained off (Fig. 11-51).

As in the centrifuge, heating the oil to a maximum of 180°F in the *initial* stage of settling causes the viscosity to decrease, which tends to increase the rate of settling and helps to break any oil-water emulsions.

Settling begins only when the heat is turned off and the batch starts to cool, because the application of heat generates convection currents that keep the contaminants in suspension. Settling continues for about 10 days or until all the contaminants that can be separated reach the bottom. The debris at the bottom should be drawn off first. Clean oil leaving the drain indicates that most of the waste product has been disposed of, after which the purified oil may be carefully drawn off at some point higher than the drain and the container cleaned in preparation for a new batch.

Batch settling is done either during periodic inspection, cleaning, or repair of the main unit or by alternating two batches of oil, one in operation and the other settling.

Another method of removing contaminants from a pressure circulation system is by augmenting a continuous settling process with strainers and depth type filters. The purifier, known as a dry type oil conditioner, consists of trays up through which the influent flows to remove water followed by

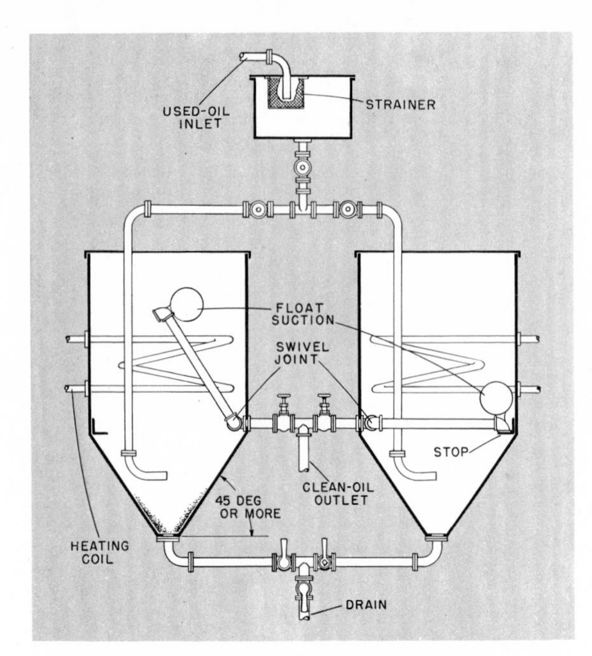

Fig. 11-51. Settling tank for use in batch purification of circulating oil. *Courtesy, Mobil Oil Corp.*

a bank of bag strainers and finally a cellulose depth type filter. The water from the trays settles to the bottom of the conditioner cabinet and is removed by an automatic overflow. As the water-free oil enters the submerged bag strainers from their outside surface to their inside, the solid particles are removed. The bags are supported on wire screens. The purified oil is then forced back into the main stream through depth type filters, which remove any remaining moisture and smaller fines. Vapor extractors located above the oil in all three process compartments reduce the humidity throughout the cabinet.

Oil Washer and Strainer

In the plant-built oil purifier shown in Fig. 11-52 the oil is introduced through an aluminum downspout to the bottom of an oil drum in which a water level of about 18 in. is maintained. As the oil enters the water, it is washed and rises through a waste strainer, supported on a perforated plate, to remove any remaining solid particles and break any emulsions that may form. The clean oil finally rises to an overflow located at some level above the strainer. Only straight mineral oils should be purified in this washer, since any additives would be removed in the washing process. This type of oil purifier is used effectively for the reclamation of refrigerating oils.

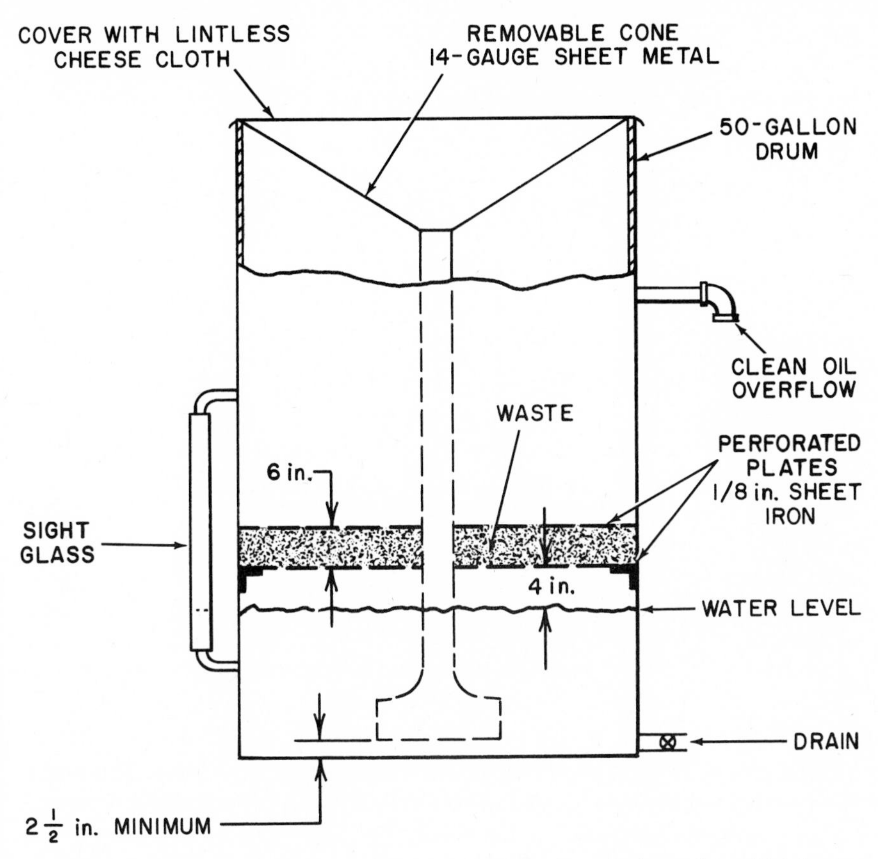

Fig. 11-52. Oil filter made with an oil drum that may be used for filtering reclaimed lubricating oil for further service in all-loss systems. Its capacity is about 25 gallons per day.

Wood shavings are sometimes used in place of the waste, but they present the danger of the oil absorbing resins and pitch which are carried back to the refrigeration system to foul the compressor and other parts. Their use is not recommended for this reason.

Oil Reclaimer

Oil trapped from various low points of ammonia refrigeration systems can be reclaimed and returned to service. One effective reclaiming process first strains the oil, then settles it starting at 180°F and cooling, and finally filters it again. Any ammonia is driven off and vented during the heating process. If this method is used to reclaim transformer oil, a dehydrator must follow the settling and filtering processes. Effective dehydration to the dielectric strength level required is accomplished by heating an oil under vacuum, or by an electrical process in which the water is precipitated after becoming charged between opposing poles of an alternating current of high voltage. The rapidly changing polarity develops a high frequency vibration that breaks the water-in-oil emulsion and causes the water to drop out.

GREASE APPLICATION

Manual Greasing

Hand application of grease is accomplished by either *swabbing or packing*. Methods are crude, wasteful, and in many cases, as with packing, not very effective.

Pressure grease guns, hand operated (Fig. 11-53), are of the screw plunger, lever, or push types. Each bearing must be equipped with a fitting that fits the nozzle of the gun (Fig. 11-54). They do not require an external source of power. Frequent filling of the gun cylinder is required where numerous bearings are serviced regularly. The push type is dangerous around moving machinery.

Pressure grease gun, power operated, requires external pressure for actuation, such as compressed air or motor-driven pump. There are several types, the most convenient being one that is connected by a flexible hose with the grease container mounted on a conveyance (for mobility), or on the floor or platform where mobility is not a factor. A follower plate is installed in the container (in which the grease is shipped) that acts as a piston to force grease through the line leading to the pressure gun.

Screw-down (compression) grease cups are fitted with a screw cap or other

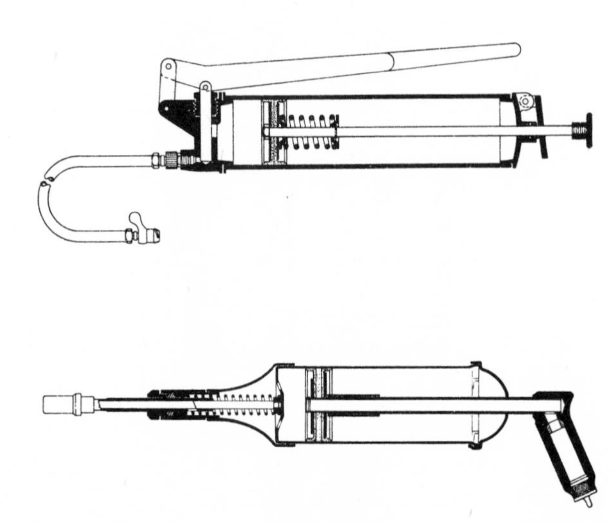

Fig. 11-53. Grease guns. (Top) Hand operated, (bottom) air operated.

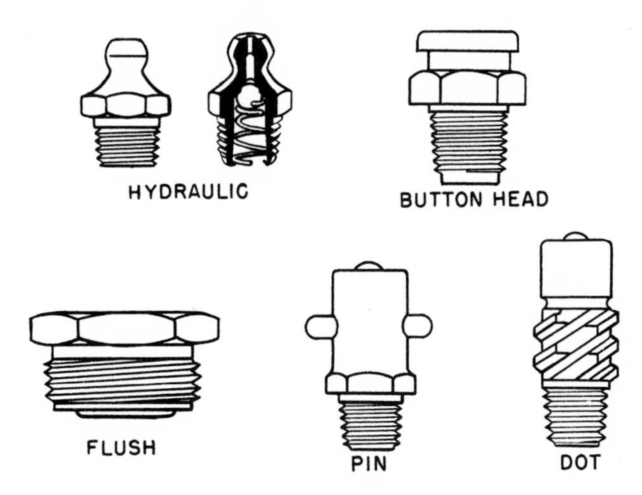

Fig. 11-54. Pressure grease gun fittings.

arrangement which, when turned on its thread, moves inward to force grease through an orifice in the floor of the cup (Fig. 11-55). The cap must be turned slightly at periodic intervals until the grease supply is exhausted, at which time the cup is refilled. It is more positive than hand packing.

Semiautomatic Grease Appliances

Spring-loaded (*compression*) (Fig. 11-56) *grease cups* act on the same principle as hand screw-down cups, except that a spring located in the cap

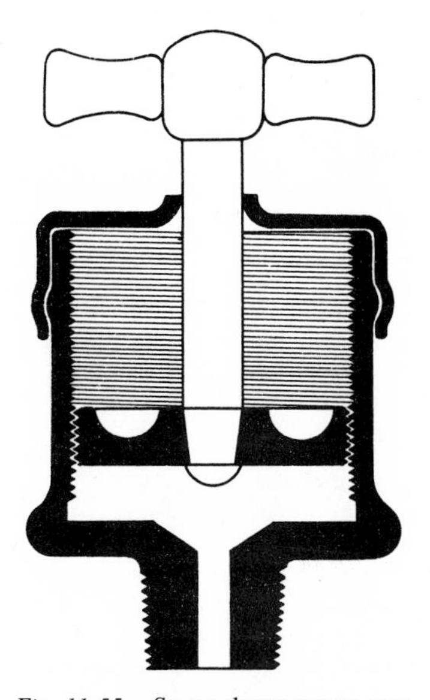

Fig. 11-55. Screw-down grease cup.

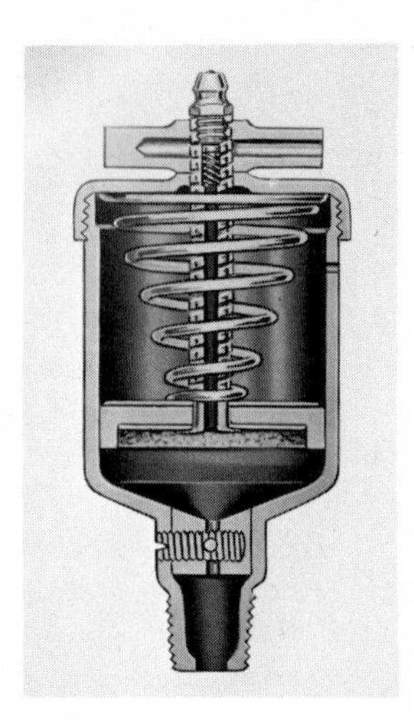

Fig. 11-56. Automatic spring compression grease cup. *Courtesy, Oil-Rite Corp.*

presses a follower plate or plunger to maintain a continuous pressure on the grease and initiate flow as needed. This type of cup is filled by a pressure grease gun through a fitting located at the bottom. As grease is forced into the cup, it compresses the spring between the top of the cup and the follower plate or piston. For best results, the cup should be kept filled with grease of NLGI No. 2 consistency. Sometimes this type of cup is fitted with a grease level indicating stem.

Air-loaded (*compression*) *grease cups* (Fig. 11-57) act on the same principle as spring-loaded cups. However, grease is forced into the bearing by air trapped in the cup when it expands during a bearing temperature rise. Positive application diminishes as air reaches maximum expansion relative

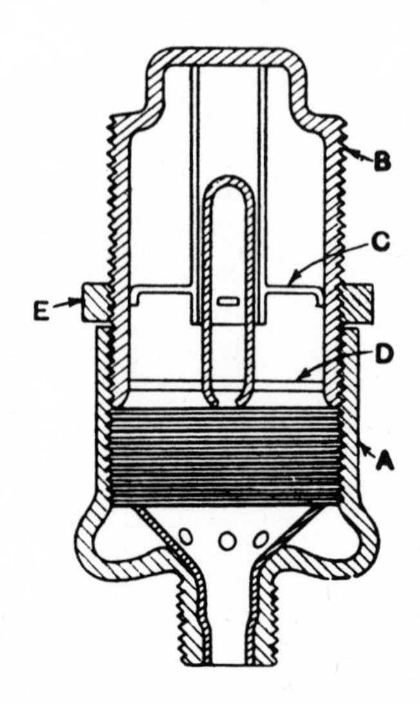

Fig. 11-57. Compressed air grease cup. A, bottom grease retainer. B, screw cap. C, fixed disk. D, movable disk. E, locknut. Air is trapped between C and D when B is turned down.

to bearing temperature. Some types are equipped with a compensating spiral to obtain more positive action as the effectiveness of either the spring or air diminish.

Automatic Central Grease System

Central grease systems are similar to central oil systems. Higher pressures are required for greases than for oil. Also a follower plate in the grease container is often necessary to assure movement of the grease to the pump suction. Central grease systems may be operated manually or automatically by a hand or power operated pump, respectively. They may also be fed from a central station by pressure grease gun (Fig. 11-58). Timing devices are employed to control feeds of automatic central systems, whereas manually operated systems depend on periodic hand operation, say, once or twice per shift.

Grease consistency is an important consideration in all systems having long feed lines. An NLGI consistency grease of No. 0 is used for winter

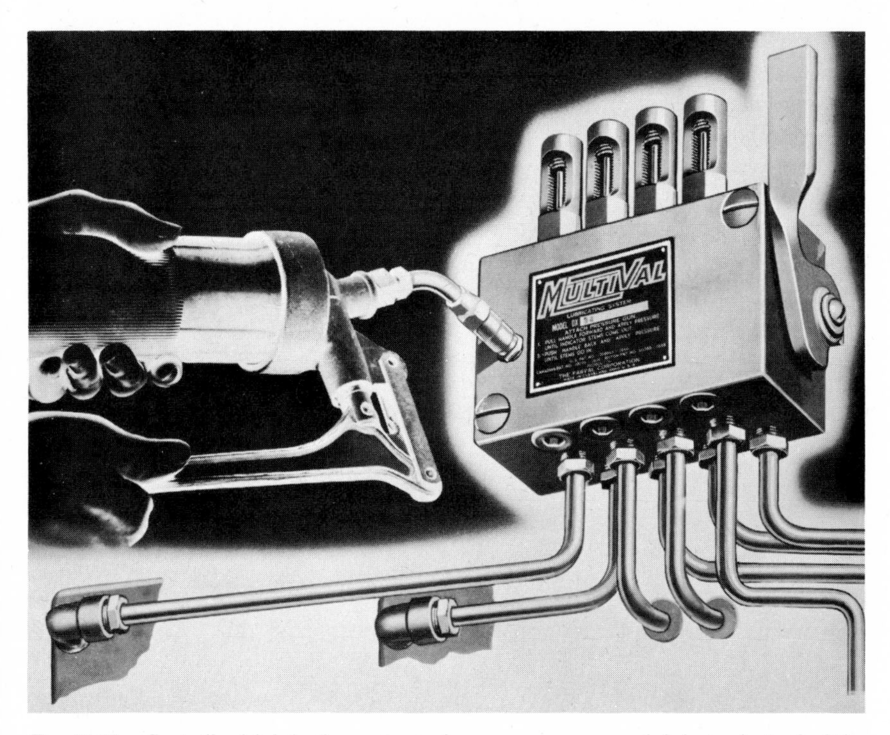

Fig. 11-58. Centralized lubricating system using a pressure gun as lubricator instead of the manual and power-operated pumps. A 90-degree movement of the rotary valve handle directs the flow of lubricant alternately to opposite ends of the measuring chamber. Measured charges of lubricant are delivered. *Courtesy, Farval Division, Yale & Towne, Inc.*

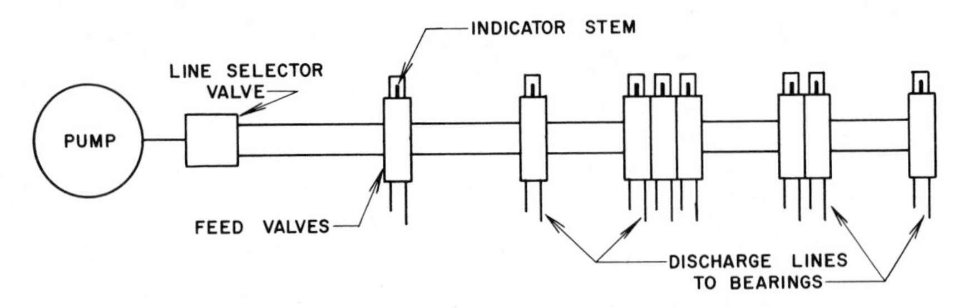

Fig. 11-59. Central grease system—two-line, nonprogressive, end-of-line type.

operation, No. 1 for summer. For Type I Accumatic System, the Alemite Division, Stewart-Warner Corporation, recommends the system tube dimensions listed in Table 11-5 for any fluid up to 5000 SUS at 100°F or soft grease (maximum NLGI No. 1).

Frequency of application may be intermittent or continuous. More frequent feeding of small amounts of grease is more advantageous then less frequent application of large amounts. Overfeeding should be avoided for both the safety of the bearing and cleanliness. Systems vary in their means of grease distribution. All are equipped with pumps, necessary feed lines, and feeder valves that control flow from main lines to branch or discharge lines. A control valve (line-selector) is used on two-line systems, while a reversing valve is found on single-line loop systems of the progressive type.

The various types of central grease systems include the following:

Two-line, nonprogressive, end-of-line system (Fig. 11-59)
Two-line, nonprogressive, loop system (Fig. 11-60)
Single-line, progressive, loop system (Fig. 11-61)
Single-line, nonprogressive, end-of-line system (Fig. 11-62)
Single-line, progressive, end-of-line system (Fig. 11-63)

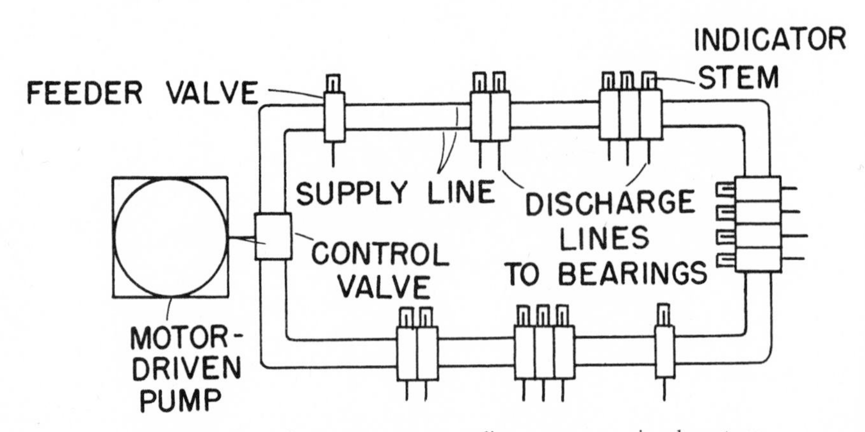

Fig. 11-60. Central grease system—two-line, nonprogressive, loop type.

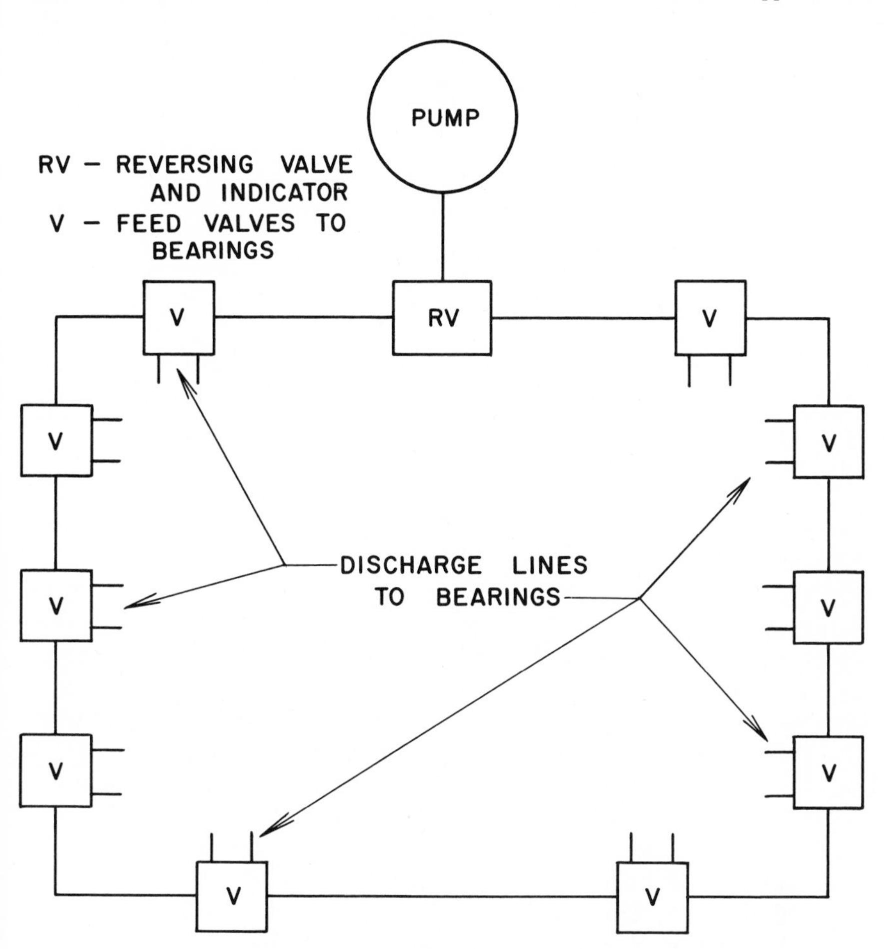

Fig. 11-61. Central grease system—single-line, progressive, loop type.

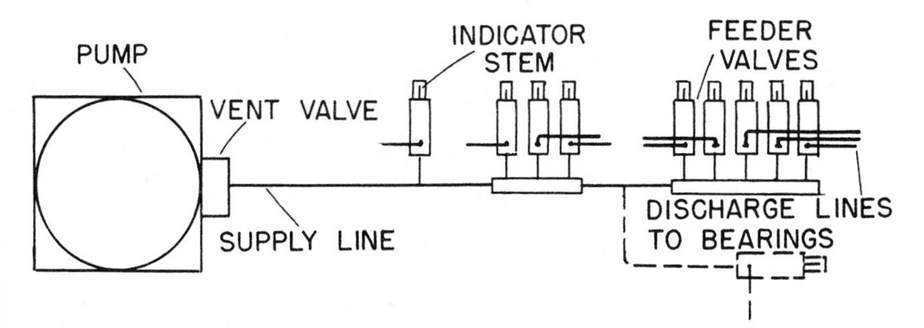

Fig. 11-62. Central grease system—single-line, nonprogressive, end-of-line type.

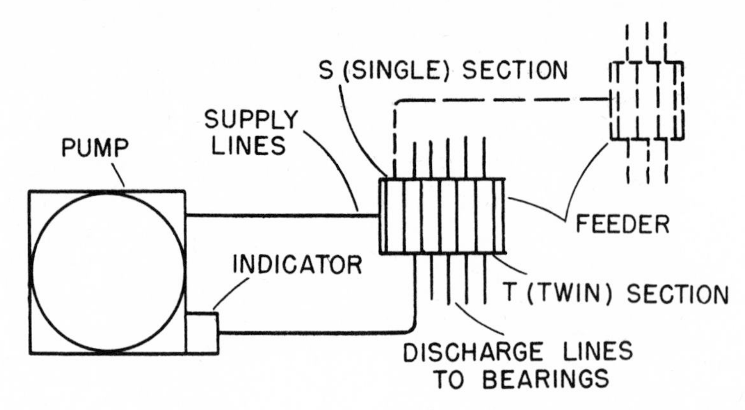

Fig. 11-63. Central grease system—single-line, progressive, end-of-line type.

Central grease systems should be designed and operated in consideration of the following:

(*a*) Eliminate dead-ends where grease may remain permanently static to finally harden in the line. Keep grease moving at frequent intervals.
(*b*) Bends should be designed with long sweeps, not sharp elbows.
(*c*) Avoid excessive length of run where possible.
(*d*) Insulate or isolate feed lines from high-temperature areas.
(*e*) Use large enough tubing having sufficient strength to withstand maximum pressures.
(*f*) Avoid crimping or depressing the internal area of tubing.
(*g*) Install all parts in accessible locations for servicing.
(*h*) Use grease having good pumpability commensurate with size of tubing.
(*i*) Avoid air pockets. Introduction of grease should immediately follow a charge of oil pumped through the lines to remove air.
(*j*) Keep contaminants (dirt, chips, etc.) out of the system.

Components of an Automatic Central Grease System

Reservoirs and Pumps. Automatic central grease systems are variously designed to lubricate a relatively few frictional parts of an individual machine or numerous bearings located throughout an industrial plant, such as a steel mill. The grease reservoir can be (a) a specially built type for mounting on a machine or nearby wall (Fig. 11-24), (b) the container in which the grease is received from the supplier from which it is pumped by an air-operated barrel pump (Fig. 11-64), or (c) a large bulk grease tank requiring a heavy-duty motor-driven pump (Fig. 11-65).

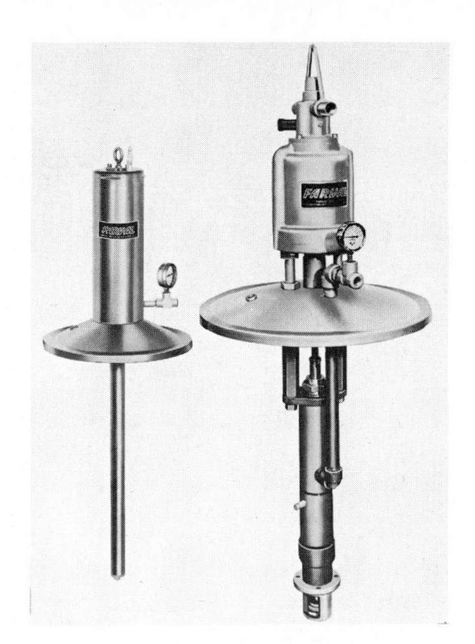

Fig. 11-64. Air-operated barrel pump. Used when oil or grease drum is the central lubricant reservoir. Designed for medium- to heavy-duty service and when rapid buildup of system pressure is desired. *Courtesy, Farval Division, Eaton Yale & Towne, Inc.*

Fig. 11-65. Automatic pumping station for central grease system. *Courtesy, Farval Division, Eaton Yale & Towne, Inc.*

Valves and Manifolds. Valves and manifolds are required to direct and meter the flow of lubricant from the main feed line to either branch lines or directly to the frictional part. Some types of valves are equipped with indicators to signify any stoppage of lubricant flow. Valves are also made to serve either a single-line or two-line system.

A single-line system (Fig. 11-66) is the type used for large bulk grease systems serving a number of machines located in several plant areas.

Fig. 11-67 shows a two-line central lubrication system in which one complete operating cycle operates as follows:

1. Timer starts pump motor. Pump builds up pressure in one main loop until all measuring valves have operated. The other main loop is relieved to the reservoir.

2. Pressure continues to build up until it is high enough to operate the reversing valve, which directs the flow of lubricant to the other main line and trips an attached double throw electric switch. The switch stops the motor, ending the cycle.

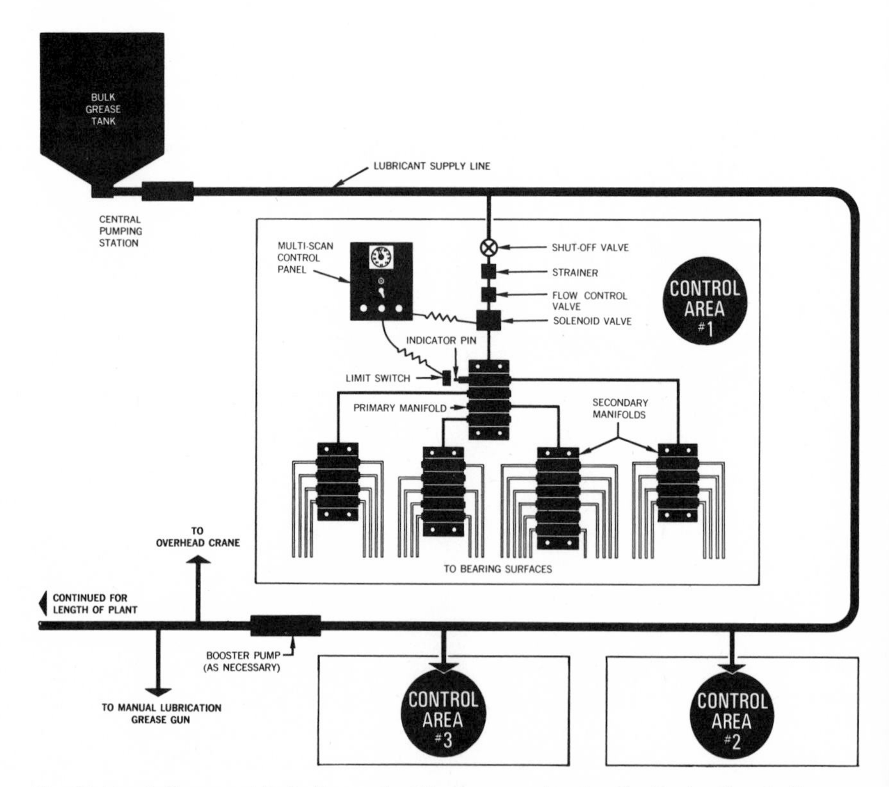

Fig. 11-66. Bulk grease (single-line, end-of-line) system, showing distribution from bulk grease tank to any number of take-off points of application with multiscan control panel. Note primary and secondary manifolds. *Courtesy, Farval Division, Eaton Yale & Towne, Inc.*

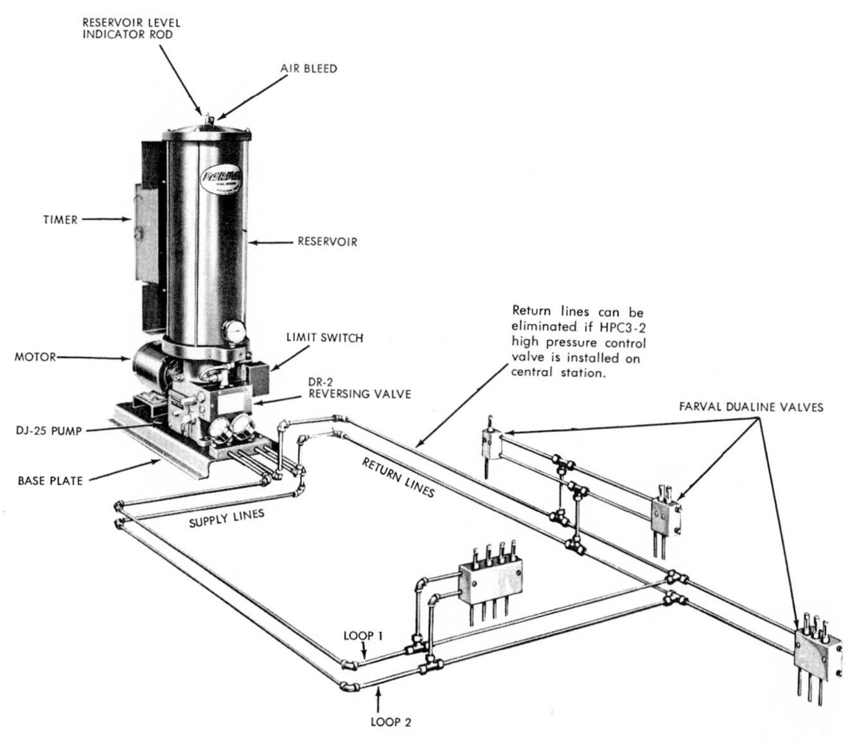

Fig. 11-67. Farval two-line system. *Courtesy, Farval Division, Eaton Yale & Towne, Inc.*

3. When the timer next starts the pump motor, the above sequence is repeated in the other main loop.

The measuring valves in this two-line system are of the type shown in Fig. 11-68, and their operating principle is described in Fig. 11-69 for double outlet operational sequence.

Another two-line central grease system is the Accumatic II system (Fig. 11-70). Note the alternative use of a hand pump and a power-driven pump with automatic control. The valves used in this system are shown in Fig. 11-71, and they operate as follows:

The Type II Accumatic Valve is served by two lubricant supply lines which are alternately pressurized.

When lubricant is pumped into supply line No. 1, the lubricant enters the valve through Inlet Passage No. 1 (Fig. 11-71a), moving the Admitting Valve No. 1 forward, which closes Outlet Passage No. 2. Continued line pressure causes the lubricant to flow, bypassing the head of the Admitting Valve No. 1, through the Valve Port No. 1, into the cylinder behind the piston. This incoming lubricant causes the piston to traverse the cylinder, forcing the lubricant through Valve Port No. 2. In doing so, the Admitting Valve No. 2 is moved backward, sealing off the Inlet Passage No. 2 and opening Outlet Passage No. 1 to the bearing.

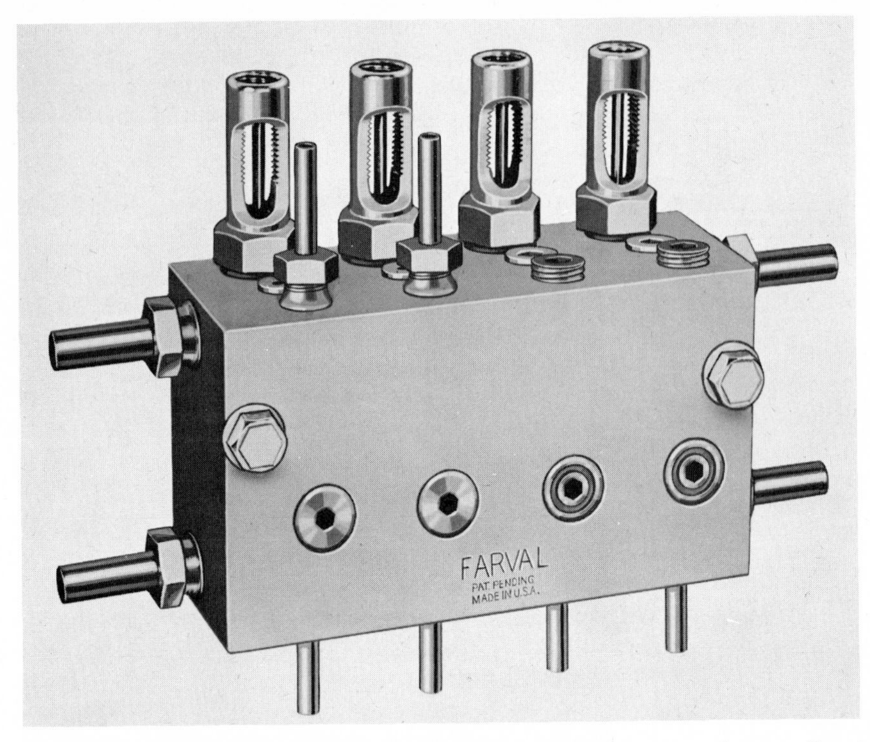

Fig. 11-68. Measuring valves used for central Dualine lubrication system. *Courtesy, Farval Division, Eaton Yale & Towne, Inc.*

When pressure in supply line No. 1 is relieved (vented) and applied to supply line No. 2, the action of the valve is reversed. Lubricant from supply line No. 2 enters Inlet Passage No. 2 (Fig. 11-71b), Outlet Passage No. 1 is closed, and the piston is driven to the opposite end of the cylinder, discharging lubricant from the cylinder ahead of the piston through Outlet Passage No. 2. Pressure in supply line No. 2 is then relieved in readiness for the next lubrication cycle.

Recommended Oiling and Greasing Periods

The following lubricant application and oil change periods are to be considered as guides, to be modified according to operating conditions, particularly those tending to deteriorate, contaminate or remove the lubricant.

Grease-Lubricated Plain Bearings

Intermittent operation	20 operating hours
Continuous operation	8–10 operating hours

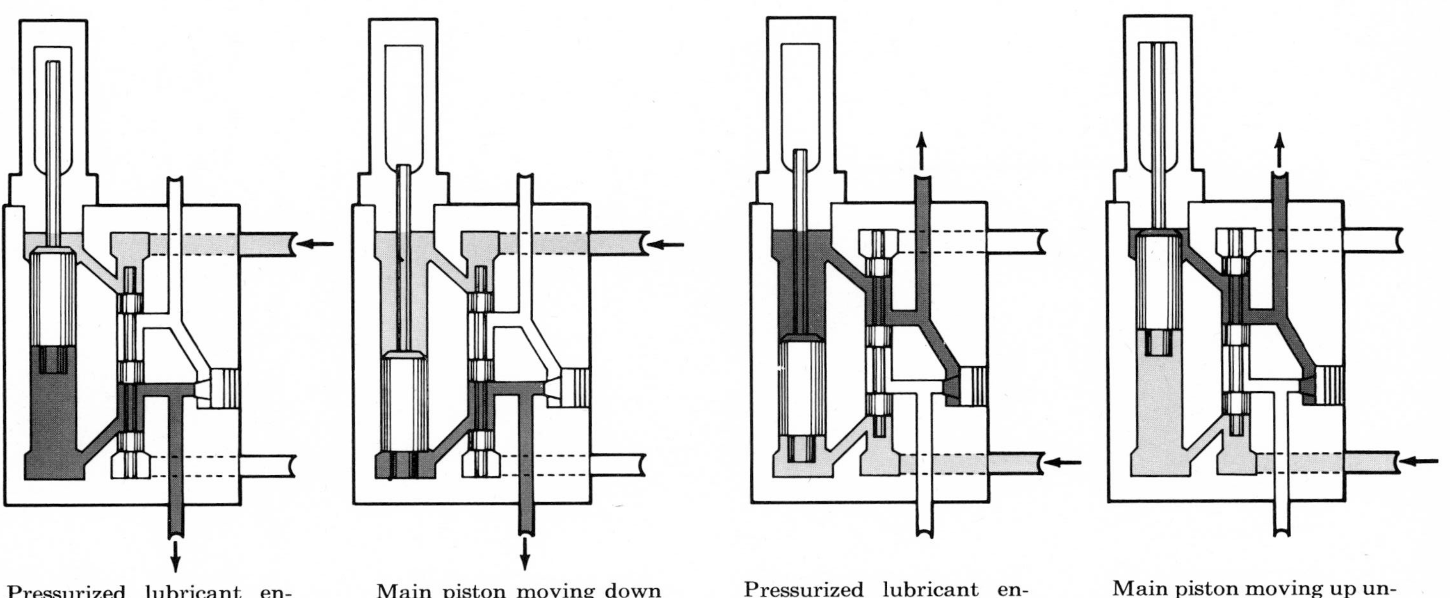

Fig. 11-69. Double outlet operational sequence of Dualine measuring valves. *Courtesy, Farval Division, Eaton Yale & Towne, Inc.*

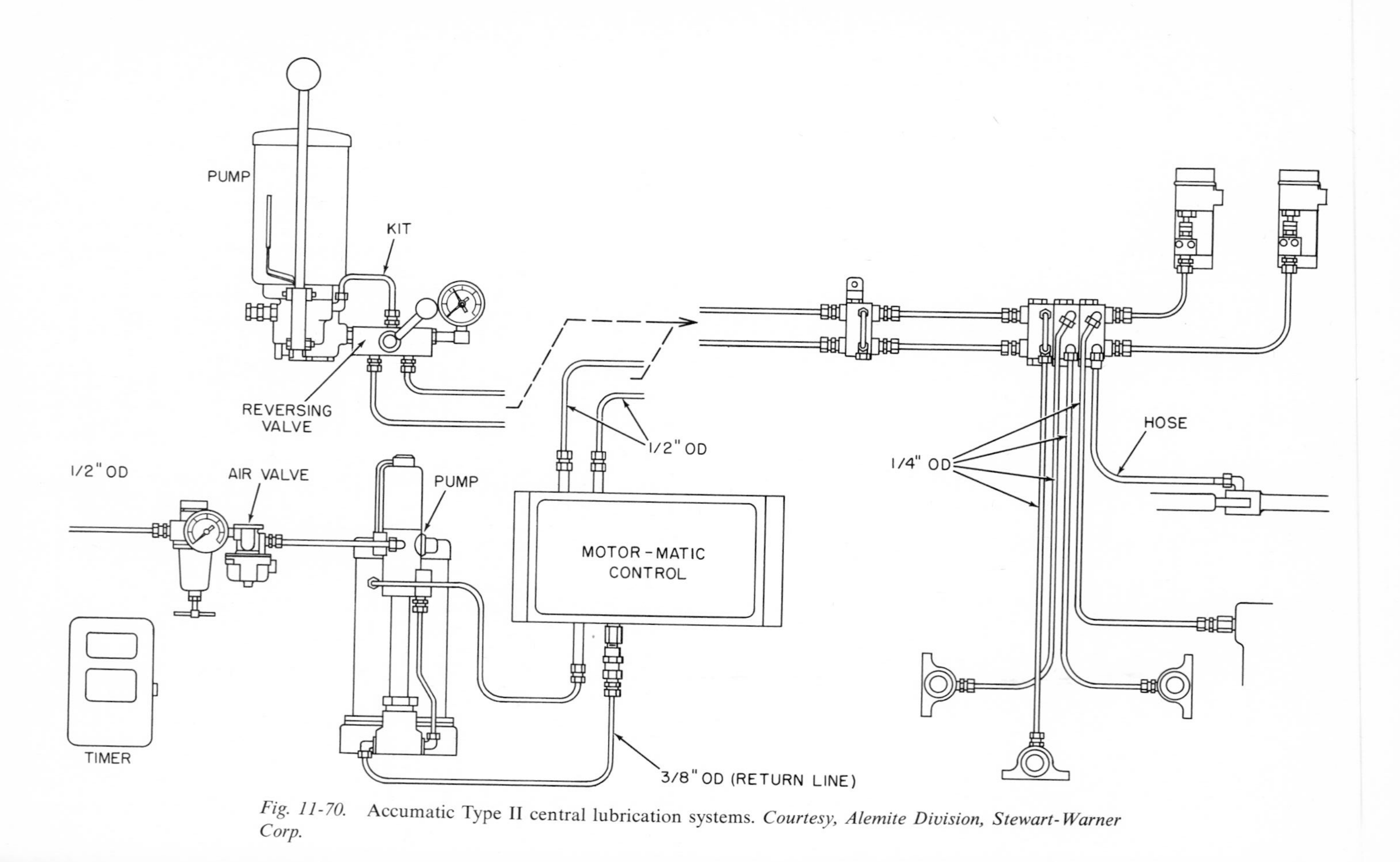

Fig. 11-70. Accumatic Type II central lubrication systems. *Courtesy, Alemite Division, Stewart-Warner Corp.*

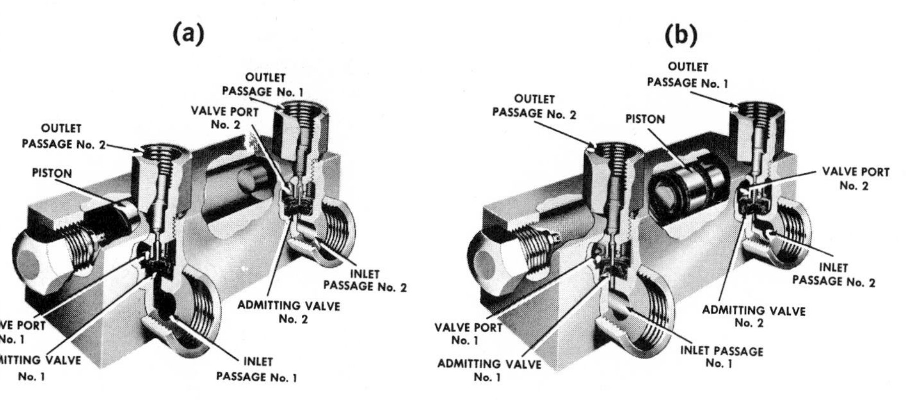

Fig. 11-71. Accumatic Type II system valves. *Courtesy, Alemite Division, Stewart-Warner Corp.*

Grease-Lubricated Rolling Bearings

Operating temperature (relatively clean conditions)	
Up to 140°F	480 operating hours
140 to 180°F	160 operating hours
Over 180°F	20 operating hours (Flush old grease and replace with new grease)

Table 11-5. LUBRICANT SYSTEM TUBE DATA FOR ACCUMATIC SYSTEM FOR FLUIDS UP TO 20,000 SUS AT 100°F OR SOFT GREASE UP TO NLGI NO. 2

Tube O.D. (*in.*)	*Wall* (*in.*)	*Maximum Length* (*ft*)
¼	0.028	10 (Branch line feeders and outlet lines from valve only)
⅜	0.042	100 (Main lines and outlet lines over 10 ft)
½	0.065	200 (Main lines)
⅝	0.083	500 (Main lines, max. for hand-operated system)
¾	0.095	750 (Main lines)
⅞ (or ¾-in. extra heavy pipe)	0.109	1000 (Main lines)
1	0.120	1500 (Main lines)
(1-in. extra heavy pipe)		2000 (Main lines)

Use Model 382200-A1 Motor-Matic Control for automatic systems over 500 ft. Temperature: 50°F minimum. (0°F, reduce line length by 50%.) Courtesy, Alemite Division, Stewart Warner Corp.

For dirty conditions, reduce hours ½ to ⅓, according to extent of contamination. For heavy moisture or water conditions, apply grease at least once every 40 hours. Force all water out of bearing cavity during grease application.

Oil-Lubricated Plain Bearings

Intermittent application (Hand-oiled or one-shot pump)	4 operating hours
Continuous application (Bottle, wick-feed, or drop-feed oiler)	Keep reservoir filled and adjust flow to meet load, speed, and temperature conditions.
All lubricator reservoirs (oil cups, MFFL reservoirs, etc.)	Clean annually

Reuse Oil Systems—Oil Change Periods

Oil reservoirs (enclosed circulation splash systems)	Annually
Hydraulic systems	Semiannually or according to condition of oil found by analysis
Worm gears	Annually

All oil reservoirs located in dusty environments (lint, cement, rock dust, etc.), should be cleaned more often than indicated above.

12

JOURNAL, THRUST, AND SLIDER BEARINGS

The primary function of a bearing is to provide support and constraint to a moving link of a kinematic chain or mechanism. A secondary function is to allow motion to occur smoothly and with a minimum of effort.

As a connector within a mechanism, a bearing may form the joint between (1) a moving link and the stationary or fixed link, usually the frame, or (2) between two moving or floating links, such as a connecting rod-crankshaft linkage.

There are three basic types of bearings:

1. *Journal bearings* support and constrain rotating motion subject to radial loading, i.e., the principal force acts perpendicular to the axis of the shaft. The journal is the section of the shaft that rests on the bearing. This type of bearing is also referred to as a radial plain bearing.

2. *Thrust bearings* support and constrain rotating motion subject to axial loading, i.e., the principal force acts parallel to the axis of the shaft. Some thrust bearings prevent the shaft from drifting horizontally; others support the vector forces exerted by the weight of the rotating assembly and any additional force acting to move the shaft axially in a vertical direction or tending toward a vertical direction.

3. *Slider bearings* support and constrain motion of translation wherein the principal force may be substantially divided into vector forces as a result of both parallel and perpendicular loading relative to the plane of sliding. Ways, gibs, crossheads and guide bearings are substantially the same, differing only in the force exerted on them by the sliding member.

Rolling bearings (roller and ball) serve purposes similar to those described above for journal and thrust bearings and are rapidly replacing the latter where appropriate, and they are fully discussed in Chapter 13.

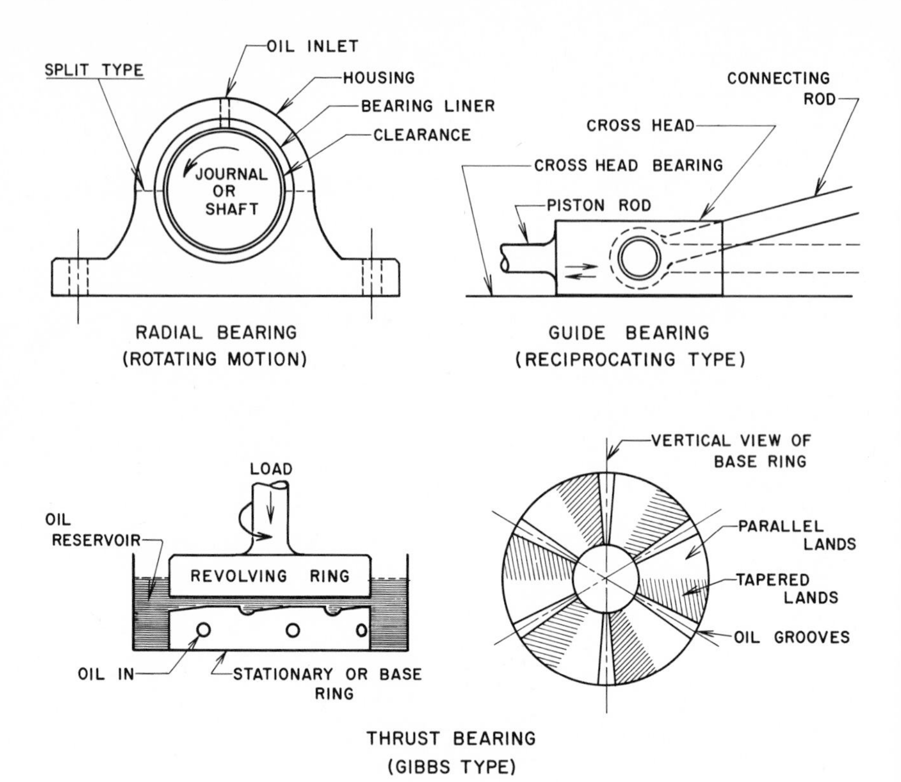

Fig. 12-1. Types of bearings.

JOURNAL BEARINGS

Radial, plain, or journal bearings represent the majority of frictional components in general service. They may be described as a tube made to encircle or partially encircle a round shaft. Their radius of curvature is always slightly greater than that of the shaft encircled. The difference in radii is called *radial clearance.*

Journal bearings include the following types:

(a) Sleeve, or bushing—full circle bearings that completely encircle the shaft without *parting*. Bushings are usually made from an alloy selected to meet a particular requirement. Their use is mandatory when loads are subject to directional changes. Bushings may be cast, rolled, or sintered. Rolled types are bronze strips having their ends butted together.

(b) Split—consists of two halves to form full circle. Used to facilitate installation of long shafts by removing the top half.

(c) Half bearing—semicircular, covering only half the journal circumference. Used only when the direction of load is constant.

(d) Segmented—having a radial arc of less than 180 degrees.
(e) Multipart—composed of more than two segments that together form a full circle.

BEARING MATERIALS

Journal bearings may consist of one or more materials:

1. One material, homogeneous or heterogeneous, composing both the basic structure and frictional surface.
2. Two materials (bimetal), one of which is the basic structure and the other the lining (overlay) that acts as the bearing to meet the frictional forces. The structural component may be the link itself, formed to support the lining, or it may be a separate backing that not only supports the lining but also permits some flexibility.
3. Three materials (trimetal), made up of three parts laminated to (a) provide structure, (b) impart cushioning, and (c) combat friction in the order named from inside out. Grid bearings consist of a soft metal bearing surface held in place by a matrix of a harder metal or alloy.

Plain journal bearing surfaces are constructed of both hard and soft materials. *Hard bearing metals* include cast iron, alloys with a base of copper, cadmium, leaded bronze, and tin bronze. *Soft bearing metals* include tin base and lead base babbitts. The latter are subject to further softening at temperatures above 250°F, although tin-base babbitt may withstand slightly higher thermal limits. The copper-base metals, such as copper-lead and copper-bronze, possess good load-carrying capacity and resistance to elevated temperatures. Grid bearings may be formed with lead having a copper matrix.

New bearing materials are being developed or investigated for use in highly radioactive environments. These are intended to be lubricated with the liquid metals employed as nuclear reactor coolants, such as mercury, lithium, sodium, sodium-potassium alloy, and rubidium. The new bearing materials include tungsten, molybdenum, chromium, beryllium, cobalt, graphite, aluminum and beryllium oxides, pure iron, and several ferrous metals. In addition to having good structural properties, these materials must also show high resistance to corrosion under the temperatures involved and be compatible with the specific reactor coolant (lubricant) to which they are exposed.

Another type of bearing material is *sintered metal,* which is made by compressing powdered metals in a reducing atmosphere at high temperature, which fuses the powder. Bronze and iron are two of the metals used, which upon fusing result in a compact porous mass which is impregnated with an oil. Their purpose is to serve for extended periods without relubrication, hence their common description, "oil-less bearings." However, these bearings must be relubricated at regular intervals.

Metal Bearings	*Nonmetal Bearings*
Bronzes	Hard maple (wood)
Copper-lead-tin	Lignum vitae (wood)
Silver	Rubber
Aluminum alloys	Plastics, including Teflon and Bakelite
Copper-lead	Nylon
Cadmium alloys	Glass
Cadmium-silver-copper alloy	Ceramics
Tin-base babbitt	Carbon (bushings)
Lead-base babbitt	
Copper-lead with overlay	
Silver with overlay	
Porous metals (sintered)	
Aluminum-tin alloy	
Aluminum alloy with overlay	
Gridded copper with lead or babbitt	
Graphited alloys	
Cast iron	
Special refractory metals (radioactive environments)	

SELECTION OF JOURNAL BEARING MATERIAL. The selection of bearing material should be based on economic and operating considerations. Economically, the following factors are important:

1. Bearing life
2. Power consumption per bearing
3. Degree of precision required
4. Cost and facility of installation
5. Cost and facility of maintenance
6. Safety factors

Operating factors to be considered are:

1. Journal speed
2. Maximum load
3. Shock load
4. Diameter-length ratio
5. Clearance allowance
6. Temperature, operating and ambient
7. Abrasiveness of possible contaminants
8. Lubrication factors—type of lubricant and possible methods of application

CHARACTERISTICS OF AN IDEAL PLAIN BEARING MATERIAL. The ideal material for plain bearings would possess the following qualities:

1. *Compressive strength*—Sufficient strength to support maximum loads without collapsing (crushing), crumbling, or rupturing.

2. *High fatigue strength*—The ability to withstand continuous variable, and/or cyclic stresses imposed by the load. An example of bearings subject to extreme cyclic stresses are wrist pin bushings of internal combustion engines. Steam turbine bearings are representative of the type subject to continuous steady loads. Pitman bearings of mechanical presses are exposed to variable loads.

3. *Conformity*—Allowance for some deformity due to shaft deflection during which the bearing conforms to moderate alignment changes. Readjustment to distortion of shaft or bearing is often the result of convenient wear (without damage otherwise to either bearing or shaft), or wiping. Such action usually occurs at the ends of the bearing (bell shape), where the greatest stresses exist during periods of misalignment or deflection. Bearings are often made self-aligning to meet this problem.

4. *Embeddability*—Soft enough to allow any abrasive particles that may enter the bearing to become embedded, i.e., they are pressed into the bearing surface. Otherwise, they remain between the bearing and journal and cause scratching or gouging of the opposing surfaces.

5. *Low coefficient of friction*—Reduction in friction between journal and bearing results in less power consumption, both in startup (static friction) and during operation (dynamic friction).

6. *Low thermal expansion*—Size remains nearly constant during periods of operation subject to temperature change.

7. *High thermal conductivity*—The ability to dissipate quickly heat due to friction.

8. *Wettability*—An affinity for lubricants so that they adhere and spread to form a protective film over the entire bearing surface.

9. *Corrosion resistance*—Resistance to attack by acids formed in lubricating oils subject to chemical deterioration due to heat and subsequent oxidation.

10. *Antiweld characteristic*—The ability not to bond with the journal. Welding of a bearing causes seizure and results in shaft scoring and failure of the bearing. Seizure usually occurs when the bearing surface temperature reaches a level conducive to plastic flow of the bearing metal. The antiweld property is also described as the antiseizure characteristic, or score resistance.

11. *Relative hardness*—The bearing should be softer than the journal to prevent shaft wear but hard enough to resist adhesive and abrasive wear of its own surface.

12. *Elasticity*—Elastic enough to allow the bearing to return to original shape upon relief of stresses that may cause temporary distortion, such as misalignment and overloading.

13. *Low cost*—The lowest cost consistent with proper function and long life.

14. *Availability*—The material should be readily available, not only for initial installation, but also to facilitate replacement in the event of bearing failure.

Metals Used in Plain Bearing Alloys

No one metal possesses all the desirable qualities of an ideal bearing, but a sufficient number of those desirable properties can be obtained by selecting two or more metals (or other elements) to produce a special alloy. Some of the more important metals used in bearings are described below.

Aluminum. Very light in weight, with a specific gravity of 2.6 and a melting point of 1216°F, aluminum is a lustrous white metal with a bluish cast. It is ductile and malleable and a good conductor of heat. When exposed to air, an oxide forms on the surface of aluminum protecting it against corrosion. It is strong and durable and alloys readily with copper, an element that further strengthens and hardens it. The alloys of aluminum and nickel have a low coefficient of expansion and are used for pistons and cylinder heads of internal combustion engines. Aluminum alloys in general have a low coefficient of expansion and a surface conducive to an effective oil film.

Antimony. Antimony is a bluish-white metal. It is brittle and has a crystalline scale-like structure. The physical properties of antimony are those of a metal, but its chemical properties are similar to those of a nonmetal. It has a melting point of 824°F, a specific gravity of 6.62, and a Brinell hardness of about 6.0. It powders easily. Antimony is used in soft-metal alloys, such as tin-base and lead-base babbitt, to increase hardness and improve surface smoothness.

Bismuth. Bismuth is a grayish-white metal. Like antimony, it is brittle and powders easily. It has a melting point of 520°F, a Brinell hardness of 75 and a specific gravity of 9.75. Bismuth has a low coefficient of expansion and is a poor conductor of heat. It reduces shrinkage of bearing metals upon cooling and increases hardness of lead and tin alloys by refining the grain structure. It also lowers the melting point of lead and tin alloys, so that it is not chosen for bearings that become hot.

Cadmium. A very ductile, silvery white crystalline metal, cadmium melts at 608°F and has a specific gravity of 8.6. It is used as a base for bearings that must withstand higher temperatures than those with a lead or tin base. Cadmium increases bearing strength and imparts properties that tend to reduce fatigue failures. It has a hardening effect on copper and lowers the coefficient of friction of bearings at higher operating temperatures. Cadmium alloys possess greater load-carrying capacity than babbitt. Cadmium is subject to corrosion of oxided hydrocarbon materials but this can be counteracted by the addition of indium.

Copper. Copper is a tough, ductile, and malleable metal, yellow-red in color, having a melting point of 1980°F and a specific gravity of 8.91. It is combined with zinc to make brass and with tin to produce bronze. Copper-base bearings possess a hard surface and a high load-carrying capacity. When added to babbitt in relatively small amounts, copper imparts hardness and toughness.

Indium. A silvery-white metallic element, indium is soft (having a Brinell hardness of 1), ductile, and malleable. It has a melting point of 311°F and a specific gravity of 7.30. Indium will harden and increase the tensile strength of silver, copper, and lead alloys. It increases corrosion resistance of lead alloys and is used to resist corrosion when plated on cadmium-base bearings.

Iron. A gray metal, iron melts at 2777°F and has a specific gravity of 7.85. It is used in small quantities to harden bearing alloys containing zinc, with which it combines to form $FeZn_7$ and Fe_3Zn_{10}. However, iron reduces malleability in such bearings even when added in small quantities. Any more than 1% iron in an alloy makes a brittle bearing. Brittleness diminishes in bearings containing iron when the copper content is increased.

Lead. Bluish-gray in color, lead is soft and very malleable and has a melting point of 621°F. It is a heavy material, having a specific gravity of 11.4. It alloys readily with tin and other metals and forms the base for lead-base babbitt. It is also used in varying amounts in other bearing materials, including tin-base babbitt, lead bronze, phosphorous bronze, copper bronze, aluminum-base and copper-base alloys. Lead has good antifriction qualities, but there appears to be no advantage in using more than 15% lead in bronze bearings to gain self-lubrication properties. It is a very stable metal, although its surface oxidizes readily, turning it to a color approaching black. It is this oxidation that causes lead to become harder and more brittle following each remelt. For this reason, bearings should be made with lead not already subjected to repeated melting. The tensile strength of lead is low. To overcome this disadvantage and still exploit its antifriction qualities, good embeddability and score resistance, grid type bearings have been developed. Grid bearings are made with a copper matrix, to provide strength, the recesses of which are filled with lead or lead-base babbitt to which the journal is exposed. Lead-indium, or lead alone, is used to coat silver bearings to lend better score resistance and antiseizure properties.

Nickel. Highly resistant to corrosion, nickel is a silvery-white metal with a melting point of 2646°F and a specific gravity of 8.84. Nickel has high tensile strength and is used in small quantities to lend this property in some types of bearings, including cadmium-base and aluminum-base alloys. It also increases the melting point of bearing materials with which it is alloyed, decreasing somewhat their embeddability. Nickel improves the finish of aluminum alloys. It imparts anticorrosive properties to stainless steel.

Silver. Silver is a lustrous, white metal, harder than gold but softer than copper. It is ductile and malleable and has a melting point of 1760°F and a specific gravity of 10.5. It is an excellent conductor of heat. Being very soft, silver is never used in its pure state for bearings, except when overplated on other metals, but it is alloyed with other materials, such as cadmiun. Cadmium and copper have a hardening effect on silver. Silver possesses high load-carrying capacity and good fatigue resistance. Silver alloys, unlike those of aluminum, have surfaces on which it is difficult to establish an oil film due to their low wettability. Silver added to copper-lead alloys acts to prevent local concentration of the lead within the mass and promotes a dense microstructure.

Tin. Tin is a soft and malleable metal, silvery-white in color, having a melting point of 464°F and a specific gravity of 7.30. It resists corrosion and is used to coat other metals to protect them against oxidizing. Tin has some lubrication value and therefore, improves the antifrictional qualities of bearings in which it is used. It forms the base for tin-base babbitt, a heavy-duty material compared with the compressive strength of lead-base babbitt. Being soft and malleable, it produces bearings having high embeddability when used in sufficient amounts.

Zinc. A lustrous bluish-white metal, zinc is rather brittle at room temperature, but softens and becomes ductile at 180 to 300°F. It has a melting point of 788°F and a specific gravity of 7.10. It is highly resistant to corrosion, and for this reason it is used as a protective coating for iron. The process of dipping iron in molten zinc is called galvanizing. Zinc is an ingredient of brass, bronze and German silver. Zinc-base alloys are used to produce bearings of high compressive strength. However, the presence of zinc reduces antifriction qualities and in excessive amounts will cause adhesiveness. It also prolongs the wearing-in period of bronze bearings. Zinc combines with iron to form $FeZn_7$ and Fe_3Zn_{10}. Malleability of zinc is reduced when even small amounts of iron are present. Zinc-base bearings may contain 80 to 86% zinc, 10% copper, and from 2.5 to 4% aluminum, while tin and lead are also found in some types. Zinc gives greater hardness, more resistance to deformation under shock, reduced friction, and better wearing qualities in copper-tin-lead alloys.

The Bronzes

Basically, bronze is a combination of copper and tin, a relatively hard, strong alloy. However, to obtain or improve certain properties, other elements are selectively used. These include lead, zinc, nickel, iron, aluminum, antimony, phosphorous, and silicon. Lead, for example, lends softness to bronze. High-lead low-tin bronzes possess greater embedability and antifriction characteristics, and are less apt to score than the harder, high-tin low- or no-lead bronzes. However, because of their relative softness, leaded bronzes do not have the strength and therefore the load-carrying capacity, of the harder materials.

A typical analysis for a hard, strong, high-tin low-lead bronze is approximately 88% copper, 10% tin, 0.3% lead, 2% zinc, 1.0% nickel, and small percentages of iron (0.10–0.15%) and aluminum (0.005%).

A typical analysis for a softer, high-lead low-tin bronze is approximately 72% copper, 3% tin, 23% lead, 3% zinc, and a small percentage of other elements, including iron (0.30%).

The two formulas given above represent somewhat the two extremes of hardness and softness in bronzes, but there are numerous modifications that produce any degree of compromise, or even changes, in physical properties of this type alloy.

Bronze bearings are, in general, hard, tough alloys with excellent fatigue strength, good load-carrying capacity, and a high melting point. Their relative antifriction property, as stated above, is contingent chiefly on lead content. Bronze bearings have low conformability and embeddability and, therefore, score readily, especially when subject to overheating. Their range of service may be defined as moderate relative to loads, speeds, and temperature.

Typical compositions of some bearing metals are shown in Tables 12-1 and 12-2.

AUTOMOTIVE ENGINE BEARINGS. Bearings of internal combustion engines are represented by lead-base and tin-base babbitts, cadmium-base alloys, and copper-lead alloys. The softer materials have been replaced by harder ones in modern automotive engines. The babbitts and bronze alloys, along with cast iron, represent a large percentage of industrial bearing materials. Cadmium-base alloys have received little attention for any purpose in recent years.

The combination of metals employed to create a suitable bearing, of course, rests with the characteristic factors outlined above, as required by the operating conditions to which a particular bearing is exposed. For example, any acid condition that develops within the bearing environment requires a material highly resistant to corrosion. One step that can be taken to combat the corrosive effect of an acid environment is to plate the surface with indium. In the case of internal combustion engines, however, corrosion resistance is given to motor oils by the addition of the corrosion inhibitors (Chapter 10). This method of inhibiting corrosion drastically curtailed the preferential removal of lead by chemical decomposition and also the copper at low temperature in the presence of moisture, from copper-lead bearings used in internal combustion engines. Tin is added to copper-lead bearings in order to reduce cracking and to improve strength. Tin-base babbitt, in general, has a better load-carrying capacity (about 1800 psi) than lead-base babbitt (about 1500 psi). The order may be reversed by the addition of arsenic or some other strength-improving element to the lead-base material.

Table 12-1. TYPICAL ANALYSIS OF SOME BEARING METALS

Bearing Metals (Embeddability)‖	*Brinell Hardness (Av)*	% *Tin*	% *Lead*	% *Copper*	% *Anti-mony*	% *Arsenic*	% *Silver*	% *Zinc*	% *Nickel*	% *Cad-mium*	% *Phos-phorus*
SOFT METALS:											
*Tin base babbitt (a)	25	75–93	0–18	2–6	5–15	0.1					
† Lead base babbitt (a)	17	1–11	74–89	0.5–0.6	9–16	0.1–1.10					
HARD METALS:											
Copper base alloys (copper-lead) (b)	25	0.25	24–40	60–75			0–1.0				
‡ Cadmium base alloys (a)	35			0.4–0.75			0.5–1.0		1.0–1.5	98	
Bronze base alloys											
Tin bronze (Admiralty) (c)	70	10	0.5–4.0	88				2			
Lead bronze (c)	55	5	25	70							
Phosphorus bronze (c)	90+	10	10	80							0.3–0.4
§ Aluminum base alloys (c)	50	6.5	1.0–2.0	(90% pure aluminum)					1.0		
Silver (overplated) (c)	25	(overplated on steel or copper-lead)									
Sintered bearings (c)		10	(2–4)¶	88	(1½% graphite)						
Grid bearings (c)	(lead in a copper matrix)										

* Heavy-duty material.
† Light-duty material.
‡ Indium plated for corrosion resistance.
§ One type contains 2% copper, 1% nickel, and about 1% magnesium.
‖ Embeddability: (a) good, (b) fair, (c) hard surface.
¶ 2 to 4% lead added to some types.

Table 12-2. Composition of Some Bearings Used in Steel Mills

Metal	Cu (%)	Zn (%)	Sn (%)	Sb (%)	Pb (%)
Red brass	85	15			
Yellow brass	65	35			
Bronze No. 1	85		15		
Bronze No. 2	82	15	3		
White metals			10–15	12–20	65–80
Babbitts			59–91	4–12	0.35–26

LINING THICKNESS. The thickness of bearing lining or insert material must be increased with the diameter of the journal. Linings vary from about 0.02 in. for journal diameters up to 2.5 in. to about 0.06 in. for diameters up to 12 in. (Fig. 12-4). For other than unidirectional loads, the lining thickness may be reduced slightly as with some internal combustion engine bearings.

The average load-carrying ability of some plain bearing lining materials is shown in Table 12-3.

SHAFT PLATING. To prevent scoring and extend the service life of diesel

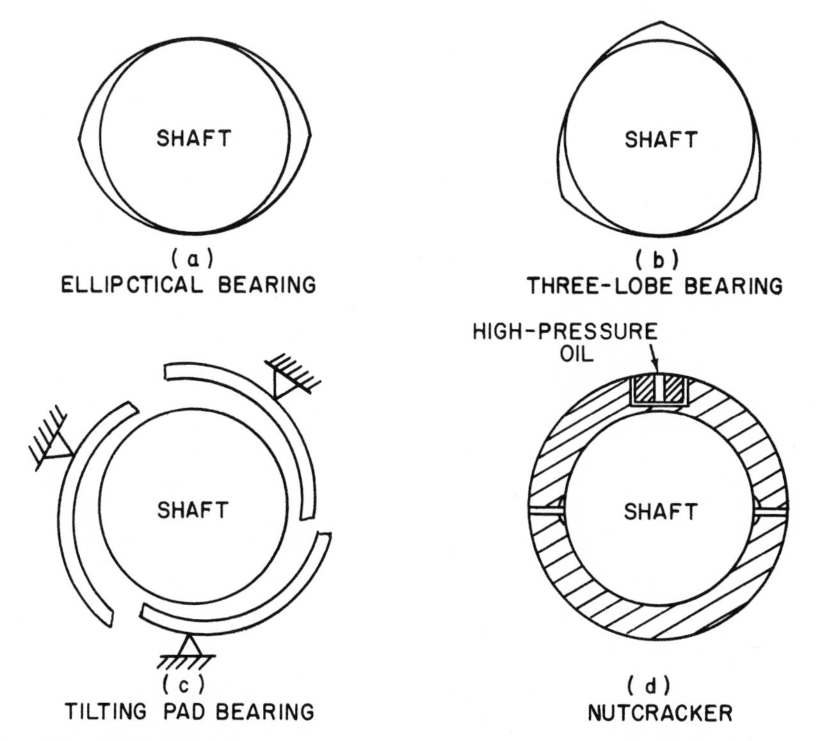

Fig. 12-2. Methods of inducing internal artificial loading to eliminate whirl in a journal bearing.

Table 12-3. MAXIMUM PERMISSIBLE UNIT PRESSURE FOR PLAIN BEARING LININGS

Bearing Alloy	*Maximum Permissible Unit Pressure (psi)*
Babbitt, high lead (84%)	1500
Babbitt, high lead (84%) (1% arsenic)	1800
Babbitt, high lead (96%)	2400
Babbitt, high tin (89 to 91%)	1800
Copper-lead	3200, requires hard shaft*
Aluminum-tin	4000, requires hard shaft*
Cadmium-silver-copper	2600, requires hard shaft*

* Brinell hardness of 300 minimum.

engine camshafts, they are plated with various metals. Plating with silver, copper, and lead, as well as phosphating did not prove entirely satisfactory, especially with higher-prices cars using chilled cast iron camshafts that are induction hardened. A hard chromium plate seemed the best answer until it was found that a substantial coating of hard chromium overcoated with a layer of tin of 3 to 4 microns thickness proved to be the best solution. The process is patented by Daimler-Benz.

Anodizing is the formation of an oxide coating on a metal surface through a chemical action between the metal and active oxygen, making the metal anodic in an electrolyte. In the latter case, the nascent oxygen liberated by the passage of a dc current (or ac with magnesium) reacts chemically with the metal to form a coating. The coating formed by sulfuric or chromic acid is porous and harder than the metal. Tests on a Shell Four-Ball extreme pressure tester on an aluminum test block with a dry lubricant showed improved wear resistance, lower temperature, and longer surface life. Thickness of coatings vary up to 0.001 in., but practical results are obtained with coatings of 0.10 micron or less. Loads calculated on the basis of wear scar measurements and test cycles approached 50,000 psi.

Journal Bearing Parameters

Operating factors: loads, speeds, temperature, type and fluidity of the lubricant used and the method of application.

Design factors: projected area, ratio of bearing length to journal and bearing surface (bearing fits), journal hardness, lining thickness, and lubricant-distributing grooves and chamfers.

LOADING. The criterion for establishing permissible loads is not a matter of total weight supported by a bearing; it is the mean unit pressure that must be considered and expressed in terms of pounds per square inch. Mean unit-bearing pressures vary from light to extremely heavy, depending on the service. Based on projected area (described below), load magnitudes may be described as light, moderately heavy, heavy, and extremely heavy.

The demarcation value in psi between each of the load categories is not fixed, but the divisions are made on the basis of various types of service (Table 12-4).

Light loading. Line shaft bearings support the shaft between the points of power source and takeoff. The torque forces within the shaft may be high, but the pressures on the bearings are between 40 and 100 psi of projected area. The length-to-diameter ratio is 1 : 2.

Moderate loading. Steam turbine bearings not only carry the weight of the rotor assembly but also are subject to some vector forces that may be developed by any imbalance of steam admission. These combined forces give bearing pressures of 150 to 400 psi of projected area. The length-to-diameter ratio is 1 : 2.5. Many other examples of moderate loading include the plain bearings of machine tools, reduction gear sets, and electric motors having length-to-diameter ratios between 0.75 and 2.0.

Heavy loading. Crank-pin bearings and some wrist pin bushings of internal combustion engines receive pressures that reach as high as 2700 psi of projected area, with length-to-diameter ratios of 0.5 to 1.5. Heavy loading is found also on main journal bearings of some mechanical punch presses and metal shears.

Extremely heavy loading. Pressures over 4000 psi of projected area do not in any way represent maximum pressures developed between mating surfaces, since local pressures of 100,000 psi occur with some types of gears. Pressures of 4000 to 5000 psi of projected area exist for wrist pin bushings of some gasoline and diesel engines when their pistons are in firing attitude. Length-to-diameter ratios range from about 1 to 1.5. Wrist pins are usually supported by bushings (continuous, full circular bearing), which are applicable for pressures as high as 5000 psi under slow speeds or oscillating motion.

The important influence of journal load magnitudes (p) is its inverse effect on the thickness of hydrodynamic lubricating films relative to speed of the journal (N) and the operating viscosity of the lubricant (Z). The effect is expressed as a dimensionless number determined by the relationship of ZN/P.

Speeds. While the speed in revolutions per minute is an indication of the relative motion between a journal and its bearing, a more significant

Table 12-4. Mean Unit Bearing Pressures, psi of Projected Area

Service Load Magnitude	*Unit Pressure (psi)*
Light loading	40 to 100
Moderate loading	150 to 400
Heavy loading	1000 to 2500
Extremely heavy loading	2500 to 5000
Hypoid (gear) loading	100,000

value is obtained by converting to velocity in feet per minute. For example, two journals with diameters of 3 in. and 6 in. both rotate at 1800 rpm, but their velocities are 1414.5 and 2829 ft/min, respectively. Errors in judgment may often be prevented by realizing the difference in potential between the two values. This is demonstrated by determining the centrifugal force per pound of rim weight developed by two gears rotating at the same speed, say 150 rpm, but having pitch diameters of 2 and 3 ft, respectively.

$$\text{Gear 1. Centrifugal force} = \frac{Wv^2}{gr} = \frac{1 \times 15.7 \times 15.7}{32.2 \times 1} = 7.65 \text{ lb}$$

$$\text{Gear 2. Centrifugal force} = \frac{Wv^2}{gr} = \frac{1 \times 23.56 \times 23.56}{32.2 \times 1.5} = 9.25 \text{ lb}$$

where

$$r = \text{radius, ft}$$

$$v = \text{velocity, ft/sec} = \frac{\text{diameter} \times \pi \times \text{rpm}}{60 \text{ sec/min}}$$

$$W = \text{rim weight} = 1 \text{ lb}$$
$$g = 32.2 \text{ ft/sec/sec} = \text{gravitational force}$$

The variation shown by this conversion can be used in considering lubricant throw from open frictional components having different velocity-diameter relationships. In the latter connection, velocity values may dictate the fluid qualities of a lubricant and/or the method used to distribute lubricant to frictional surfaces, such as those used in large gear cases.

As with loads, speeds can be described in terms of high, intermediate, and slow. Ultra-high speed refers to velocities above 100 ft/sec.

A journal's speed and load govern the viscosity of the lubricant. Bearings of higher-speed journals are more adequately served by oils of lower viscosity. This is based on the assumption that high speed and heavy loading do not generally occur together. This assumption can be expanded to include all speed-load categories expressed as follows:

Heavily loaded bearings supporting slow-speed journals are more adequately served by high-viscosity lubricating oils.

Lightly loaded bearings supporting high-speed journals are more adequately served by low-viscosity lubricating oils.

It follows that intermediate load-speed combinations call for oil viscosities that are commensurate with any given level between the two extremes.

Journals subject to lighter loads and higher speeds tend to stabilize their axis at some point approaching the center of their bearings. The higher the speed, the closer becomes the congruency of both axes, resulting in more uniform clearance configuration and greater minimum film thickness. The action is one in which the attitude of the journal approaches an eccentricity ratio of zero, i.e., it approaches concentricity with the bearing, which never occurs. Thus the squeeze action between the journal and bearing surface diminishes with higher speed, allowing progressively lower viscosity oils to be used. Pumping capacity of the pair increases as viscosity becomes lower so that temperatures are reduced by greater oil throughput.

Speed is also considered in the "state of the film" factor ZN/P, in which it is represented by N. It is shown in the formula that film thickness is increased in direct proportion to the speed, whereas load P has an inverse effect. As shown in Chapter 9, higher ZN/P values indicate proportionally greater film thickness.

The positive effect of speed holds true up to the point at which turbulence commences. The point at which turbulence is determinable by the Reynolds equation is based on the relationships of fluid velocity, conduit diameter, and fluid viscosity.

Critical speed. When a shaft is at rest, it is assumed to be in a position of equilibrium, which may be considered as its normal or original shape. If an external force is momentarily applied to some point along its length, some deflection occurs, at a magnitude depending on the rigidity of the shaft and its supporting system. If the external force applied is within the elastic limit of the shaft material, the tendency is for the shaft to return to its original shape or state of equilibrium. In doing so, inertial forces abetted by elastic forces cause it to pass through its position of equilibrium. Lacking additional external disturbing forces, the shaft will deflect back and forth through its original shape with diminishing displacement until it finally returns to a state of equilibrium. The action is known as *vibration,* but in the case of the stationary shaft, it is more accurately categorized as the *free* or *natural vibration* of that system. The number of displacements or vibrations completed in one second is called the frequency of natural vibration.

By causing the same shaft to rotate, the disturbing factor of centrifugal force develops to some degree and acts on the flexure of the shaft. Everything else considered, the measure and effect of centrifugal force depend on the speed of rotation. The speed at which centrifugal force becomes effective varies with each shaft flexibility, load distribution, supporting structure, and any imbalance within the rotating mass. All the factors mentioned cause some deflection regardless of the speed. However, when that deflection (*forced vibration*) reaches the displacement frequency of natural vibration, the system is said to be at *critical speed.* The critical speed of any shaft system, may be described as the speed at which the rotating frequency and natural vibrating frequency approach synchronization, i.e., become about equal to each other. When this occurs, maximum displacement may develop, which means that the amplitude of the system becomes great enough to bring about a state of *resonance.* If continued in this state, the system will destroy itself.

The speed at which any shaft system becomes critical can be reduced or increased, but it should never be allowed to exist within the range of operating speeds. This is exemplified in an automobile engine that reaches a distinct vibration level at a speed close to the maximum allowed for turnpike driving. In this event, it becomes necessary to increase speed or reduce it, either of which may prove hazardous or inconvenient.

With shaft at rest, the energy transferred to it by a momentary external force is gradually reduced to zero in combatting the internal elastic forces, and the body comes to rest. This may be described as a self-damping process. In a rotating shaft, the external forces are continuous, so that any vibrational damping must be built into the system.

This is accomplished (1) by increasing the diameter of the shaft to promote greater stiffness and force the critical speed to occur above the operating speed, or (2) by increasing shaft flexibility and causing the critical speed to drop below operating speed. With long span shafts operating at high speed, the critical speed needs to be more than one-half the operating speed to avoid *half-frequency whirl.*

Whirl and Whip. Eccentricity is the distance between the axes of the journal and bearing, and the length and direction of the eccentricity are functions of the sum of all vector forces involved, including load, speed, and lubricant viscosity. The reason for the eccentricity is deflection of the shaft under load. During rotation of a deflected shaft, two circular motions evolve. One is *rotation* of the shaft about its own axes and the other is a *revolving* motion, which would not exist if the shaft remained concentric with its bearings as it tends to do at unusually high speed. A revolving motion is one in which the circular path is based, not on a fixed axis, but on the path scribed by the eccentric configuration of the axis of the shaft during its rotation. If not sufficiently restrained, the revolving becomes aggravated and vibrates at an amplitude dictated by the revolving frequency relative to the rotating frequency or revolutions per second. Such a motion is not rotation, but a whirl, or whip, which may be *conical* or *translatory.* With conical whirl, the axes of the shaft and bearing approach angularity. With translatory whirl, the path of the journal axis remains parallel relative to the bearing axis.

If the vibration frequency occurs at speeds below twice the critical speed of the journal, the axis of the shaft tends to become congruent with the bearing axis. The resultant decrease in hydrodynamic effect reduces the pumping capacity of the journal, and the subsequent loss in oil volume causes the wedge to travel around the bearing at its *average* velocity or one-half the surface velocity of the journal. This phenomenon is known as *half-frequency whirl.*

At above twice critical speed, or in the region somewhat below three times the critical speed, the displacement grows in magnitude until it tends to reach maximum (amplitude) at speeds above that level, and the whirl converts to a whip called *oil-film whipping.*

The difficulties of whirl and whip are experienced generally on high-speed, lightly loaded shafting, the axis of which may be either horizontal or vertical, although it occurs more readily with a vertical axis. Speeds of 100 ft/sec and above are considered high speed. This is equivalent to 22,320 rpm of a 1-in.-diam shaft.

Whirl is due to deficiency in loading relative to centrifugal forces developed by the speed, so that it can be suppressed by modifying either factor. If the speed is fixed at some desired level, then load becomes the manipulated factor. There are several means of accomplishing this: (a) by reducing the load capacity of the bearing and (b) by inducing artificial loading, either within or externally to the bearing.

The load capacity of the bearing can be reduced by decreasing the length of the bearing, thus increasing unit loading or by providing three or four longitudinal grooves equally distributed around the bearing to form the equivalent of partial bearings of 120° and 90°, respectively.

Induced internal artificial loading (Fig. 12-2) can be achieved by providing multi-oil wedges in the form of elliptical-shaped bearings, three-lobe bearings, and tilting or pivoted-pad bearings.

Induced external artificial loading can also be attained by using the nutcracker principle. This employs a piston-cylinder arrangement in the upper part of the bearing housing. By hydraulic means, the piston is forced against the top half of the bearing that is free to move vertically, to exert in conjunction with the stationary lower half a squeeze action on the rotating shaft (Fig. 12-2).

TEMPERATURE. Bearing temperatures should be held within the limits beyond which their appropriate hardness is affected. Any softening results in wiping and loss of the bearing. High temperatures also present a deteriorating effect on the lubricating oil as described in Chapter 10.

Thermal energy required to elevate temperatures is equivalent to the mechanical energy lost by friction developed within the bearing during relative motion of its journal. Under boundary conditions, the thermal

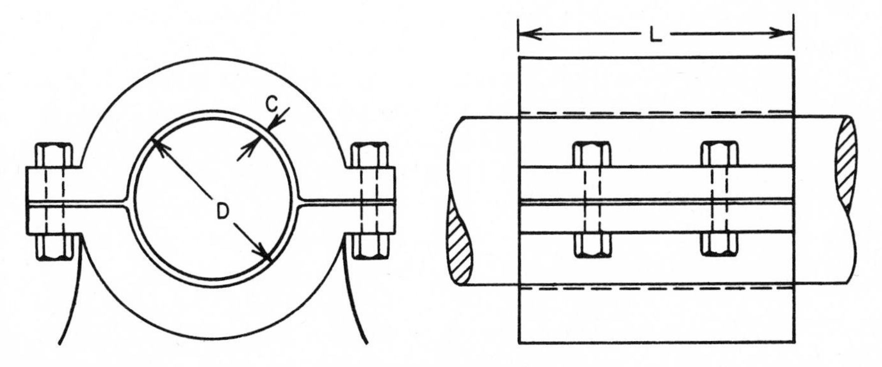

Fig. 12-3. Bearing parameters.

energy generated can be significant, depending on the lubricity of the lubricant, the nature of the opposing materials, and the quality of their sliding surfaces.

Under flooded lubrication conditions, the tendency to generate thermal energy is lessened by the substitution of fluid friction for boundary friction. Temperature can be maintained at a safe level by appropriate grooving, suitable ventilation or water cooling, and a copious supply of oil of selected viscosity.

Another source of heat is that reaching the bearings from surrounding processes, the best example being the internal combustion engine. Even here in such confined environment, with the circulation of oil and its subsequent cooling in ventilated crankcases, bearing temperatures may be maintained within safe limits.

For bearings that are otherwise in good physical condition, abnormal temperatures may be due to higher speeds, concentrated loading, incorrect lubrication and/or insufficient cooling. Where speeds cannot be changed for mechanical reasons, recourse lies in expanding bearing area to reduce unit pressure (or additional bearings), providing a sufficient amount of correct viscosity oil, and providing proper clearances and effective cooling. An example of effective cooling occurs in electric motors in which bearing cooling is effected by the same fan that cools the rotor. It is important, however, that any convection cooling applied must be so devised as to avoid condensation within the bearing system. A cold air blast on a turbo-fan bearing results in an appreciable water accumulation in the oil sump.

PROJECTED AREA. Permissible bearing pressures are usually based on projected area (Fig. 12-3). Projected area is the product of the length of the bearing multiplied by its diameter, both of which are given in inches. Thus a bearing 8 in. long and 4 in. in diameter would have a projected area of $8 \times 4 = 32\text{in.}^2$ If a total load of 48,000 lb is exerted on a bearing of that size, the unit load would equal 48,000/32 = 1500 psi of projected area.

The diameter of a circle is about ⅔ of its circumference divided by 2, which is equivalent to a partial circle of 120°. Considering thickness of oil film and clearance provided between bearing and journal, 120° represents an arc over which a load, normal to it, is distributed. Therefore, d in the equation $\ell \times d =$ projected area is equivalent to about 65% of the arc forming the bearing half carrying the load. Therefore, the projected area required to meet a given load is determined by the equation

$$\text{projected area (in.)}^2 = \frac{\text{total load}}{\text{load capacity of bearing material (psi)}}$$

LENGTH-TO-DIAMETER RATIO. Major considerations for length-to-diameter ratio (ℓ/d) values for bearings include end leakage of lubricant,

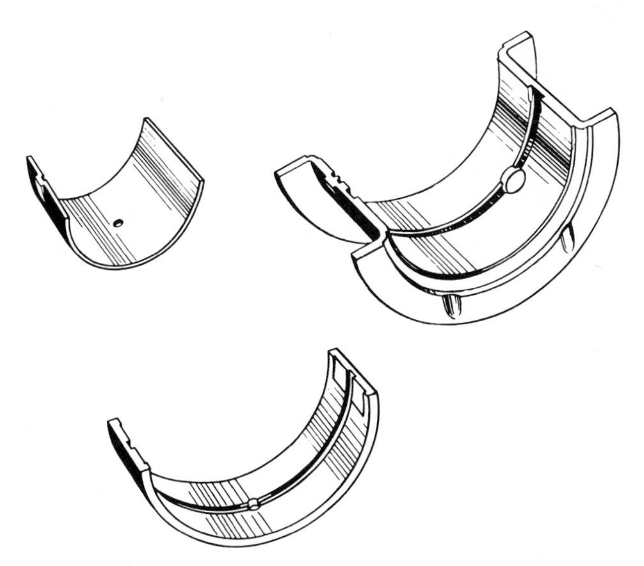

Fig. 12-4. Precision bearing inserts. *Courtesy, Mobil Oil Corp.*

vibration and accuracy of alignment. With copious oil supply and balanced systems, the misalignment factor becomes predominantly important because of greater initial end wear (bell mouth), followed by the loss of uniformity along the entire length of the journal-bearing surface. Ratios vary from $\ell/d = <1$ to 4+. Steam turbines rotating at relatively high speeds (3600 rpm) are supported successfully with an abundance of lubricant having a length-to-diameter ratio of 1.0 to 1.5. In contrast, line shaft bearings placed relatively far apart and subject to slow speeds (300 rpm or less) are designed with ratio $\ell/d = 3$ to 4. The trend has been to reduce the ratio on bearings of internal combustion engines to as low as 0.5 to 0.75, particularly crank-pin and wrist-pin bushings. Such reductions, however, are accompanied by other modifications, such as increasing load-carrying capacity by eliminating oil grooves and reducing the mass of bearing metal to provide more rapid cooling. Other factors leading toward smaller ℓ/d ratio in such engines include greater fluidity of multi-viscosity oils and the incorporation of selected additives to such lubricants.

Where it is necessary to reduce the length of a bearing for reasons of space, a required projected area may be attained by increasing shaft diameter.

Clearance. The difference between journal diameter and internal diameter of the bearing is called clearance. Its purpose is to allow for expansion and provide lubricant space. The preciseness of clearance fits depends on the precision requirements of the bearing system.

For general purposes, clearance based on an allowance of 0.001 in./in. journal diameter, plus 0.001 in., is sufficient. Kingsbury recommended the addition of 0.002 in. to the value of $d \times 0.001$. For heated bearings (steam joints), greater clearances should be allowed, depending on temperature, but should not exceed a clearance-to-diameter ratio of 0.002.

Bearings for precision machines operate with reduced clearance in an amount commensurate with the effectiveness of the lubricant and its supply method. Bearing and journal surfaces may be superfinished to below 10 microinches so that small clearances may be provided, in which case, low-viscosity oils containing antiwear agents are employed, with viscosities of 60 to 100 SUS for some machine tool spindles.

Clearances also vary according to bearing materials, but the ratio should be kept within the limits of 0.005 to 0.0018, with an average for most purposes of 0.0013. If clearances are too large, the loose fit may cause vibration; insufficient clearance may result in rapid wear and seizure.

Journal Hardness. Some bearing materials are not sensitive to the hardness of the journals; others are most responsive. It is not important for the softer bearings, such as the tin-base and lead-base babbitts. For cadmium-silver-copper alloys, the Brinell hardness should not be less than about 260. For copper-lead and aluminum-tin alloys, a harder steel is required, i.e., a Brinell hardness of about 300 or higher.

ZN/P Values. The value of $ZN/P = 40$ usually signifies that viscosity, speed, and load are so great relative to each other as to represent a hydrodynamic condition. It is not the minimum condition conducive to effective lubrication protection, however. This minimum ZN/P value varies according to the bearing material and its permissible unit loads. Values vary from about $ZN/P = 10$ to 15 (minimum) for tin-base and lead-base babbitts to $ZN/P = 3$ for the harder bearings, such as copper-lead and aluminum-tin alloys, which values are based on maximum permissible unit pressures with all materials.

THRUST BEARINGS

Thrust bearings support the end thrust of both vertical and horizontal shafts, i.e., loads acting parallel with the axis of the rotating member (Fig. 12-5). The principal function of the thrust bearing is to prevent the shaft from drifting, or moving parallel with its axis.

The pressure exerted on the thrust bearing of a horizontal shaft is usually limited to the axial force exerted by the load imposed on, or developed within, the rotating assembly. In addition to the axial force imposed by the load, thrust bearings located at the lower end of vertical shafts must also support the weight of the shaft, plus the weight of any machine parts attached to it. The moving surface exerted against a thrust bearing may be the entire end area of the shaft or part of it, or it may be the surface of a collar or ring fixed at any point to the shaft. The moving surface may

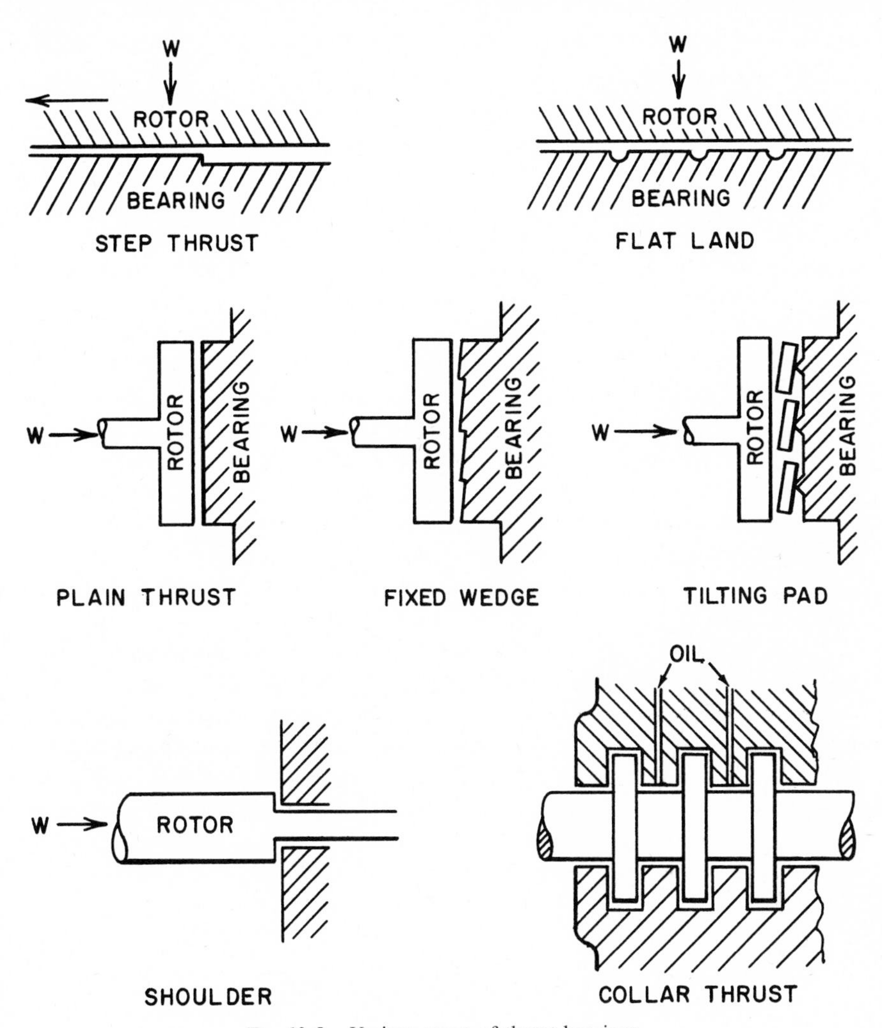

Fig. 12-5. Various types of thrust bearings.

also consist of a step (or steps) cut in the shaft, the shoulders of which bear against a stationary thrust collar or washer.

The basic types of thrust bearings are as follows:

1. *Plain thrust.* This is the simplest form, consisting of a stationary flat bearing surface against which the flat end of a rotating shaft is permitted to bear.

2. *Step bearing.* This consists of a raised (stepped) bearing surface upon which the lower end of a vertical shaft or spindle rotates, the entire assembly being completely submerged in lubricant. Some step bearings are designed to be lubricated hydrostatically, whereby oil is forced between rubbing surfaces by external pressure.

3. *Hydraulic thrust.* A collar fixed on a rotating shaft runs in an annular pocket cut into the machine housing. One side of the annular pocket provides a bearing against which the collar rubs during rotation of the shaft. A reverse thrust may be provided by attaching an additional collar further along the shaft and locating the bearing on the opposite side of its annular pocket. The entire assembly is submerged in lubricating oil (under pressure) which maintains a separating film between the collar and bearing to prevent metal-to-metal contact. The oil acts as a hydraulic medium between the rubbing surfaces, providing a protective cushion.

4. *Fixed pad, or Bibbs thrust bearing.* This consists of a ring of stationary sector-shaped (axis view), tapered (profile view) pads or lands separated by radial grooves, mounted on a stationary or base ring, upon which a rotating ring (fixed to the shaft) bears. Oil enters the center of the bearing through holes drilled radially in the stationary ring to continuously replace that which is forced centrifugally outward by the revolving ring.

5. *Tilting pad, or Mitchel thrust bearing.* This is a ring of sector-shaped pads having a flat upper surface and a bottom surface which is partly cut away to provide two levels. The edge formed by the two levels allows the pad to rock or tilt on the top of a stationary surface. A rotating collar attached to the shaft rides on the flat surfaces of the pads, of which there may be from two to ten (or more). The assembly is submerged in an oil bath. As the shaft revolves, the rotating ring carries oil into the space between it and the flat surfaces of the pads. This causes the pads to tilt backward (relative to the direction of rotation), allowing an oil wedge to be formed between the opposing surfaces. To initiate flow between the two opposing flat surfaces at the instant of starting, the leading edge of the upper pad surface is rounded. Thus the tilting pad acts similarly to the fixed pad, once rotation has commenced.

6. *Pivoted shoe, or Kingsbury thrust bearing.* This is similar to the tilting pad type except that the shoes (or pads) are mounted on pivots attached to the stationary ring, which provides the tilting or rocking action to form the oil wedge with the rotating thrust plate. The leading edges of the shoes are rounded to allow entry of oil between opposing surfaces upon starting.

7. *Horseshoe type thrust bearing.* This type of thrust bearing is made up of a number of adjustable bearing rings in the shape of horseshoes. Used in marine service, they are water cooled and act to counteract the thrust of the propeller. Thrust collars are provided by grooving the shaft. Because the bearing faces of this type of bearing are parallel, they do not provide the geometry to develop hydrodynamic films resulting in a high coefficient of friction. They have been improved by the use of thrust collars fitted with tapered lands.

Some difficulty is encountered with multiple collar thrust bearings unless adjustments are made to attain uniform distribution of the load on all opposing surfaces.

Collars of plain thrust bearings (having parallel surfaces) are usually of hardened steel, while their bearing rings are of bronze or babbitt, the latter of which possesses a lower load-carrying capacity.

SLIDER BEARINGS

Slider bearings, also called ways, or guides, support either rectilinear or curvilinear motion (Fig. 12-6). They are frequently associated with reciprocating motion. In cross section they may be flat, convex, concave, or V-shaped. They may be vertical, like elevator guides, horizontal, as are many machine tool ways, or in any position between. The riding member moves parallel with the bearing surface with a sliding action. Contact pressure due to gravity is maximum with horizontal ways and diminishes as they approach a vertical position.

Guide bearings are made from a number of the materials described in this chapter, but steel ways are more prevalent, as are those of cast iron and steel. Plastic inserts are also used with steel and iron for machine tool ways. The use of materials having a high coefficient of friction and improper lubricant must be avoided in the case of ways and guide bearings because of the consequent chattering of the rider, especially at low speeds. Good surface finish and an affinity of the material for lubricants (wettability) are attributes of efficient guide bearings.

Ways consist of a rider, called a *table way* because it carries the piece on which work is done, and a *bed way,* on which the table way slides in a constant plane. Ways are usually made of cast iron carefully machined and hand spotted and scraped to assure maximum smoothness. Flame-hardened to resist wear, they are often reinforced to gain greater surface toughness by the insertion of hardened steel strips, or to reduce friction by inserting plastic strips into their surfaces.

Ways are used to support the motion of worktables, carriages, saddles, toolholders and various type heads for all kinds of production machines, expecially of the machine tool variety.

Way Lubrication

To determine the correct lubricant and its method of application for the protection of ways against both friction and wear, the conditions related to their operation must be fully understood. The major conditions are as follows:

1. Flat surfaces that slide over each other are in area contact, which tends to increase adhesive wear and molecular attraction.

2. After a period of wear-in, the opposing surfaces approach a common plane to form a close fit and squeeze out and wipe away any lubricant mass placed in the path of motion.

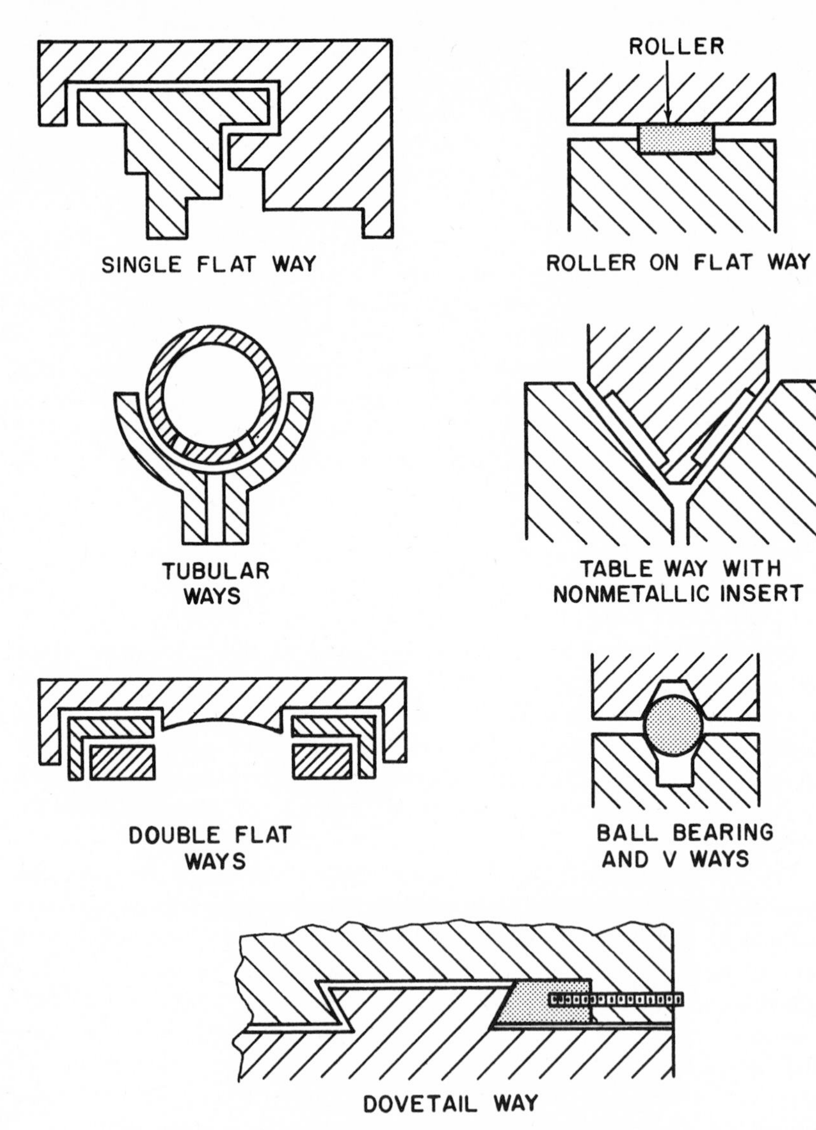

Fig. 12-6. Cross-sectional view of various types of ways.

3. Because of the squeezing and wiping action, only a thin-film of lubricant remains on the surface so that boundary conditions exist.

4. Boundary lubrication of ways is not a serious problem, because any thick film or hydrodynamic wedge formed between way surfaces lifts or floats the moving member, causing dimensional changes in the workpiece being produced. Therefore, any interference with plane motion causes loss of mechanical precision.

5. What oil film remains after wiping by the moving member must be

distributed over the entire way area. For ways having extended distances of travel, the lubricant must be supplied at intervals along the route to accomplish effective distribution and prevent excessive wear.

6. Scratching, gouging and abrasion often occur on ways as a result of metallic fines or chips that appear on the surfaces, either from a circulating lubricant or from the atmosphere.

7. In the absence of sufficient lubricant or the use of a lubricant lacking the essential qualities for way service, the force of friction may rise momentarily to magnitudes above the mechanical force required to overcome moving friction of the table ways. This causes chattering, or stick-slip.

8. Way lubricants are variously applied. Some machines are equipped with oil circulating systems in which the lubricant is constantly reused. Others depend on application devices to supply oil periodically to ways and slides, whereby the lubricant is used once and subsequently lost to the (all-loss) system.

Stick-Slip

When high-precision machine tools, such as jig borers and other metal-removing tools, were first introduced, lubrication of the slider bearings was a serious problem because the exact dimension of production parts limited the thickness of lubricating films between rider and way. This has been overcome but is still an area of extreme boundary condition. Not only is the shape of the workpiece important, but its surface must often be perfectly smooth.

The situation is critical because of the tendency of boundary lubricated opposing surfaces, especially when operating at very low relative velocity, to adhere to each other. *Adhesion occurs when the force of static friction becomes equal to or greater than the force applied to the moving surface components to overcome it.* Because of the machine power capabilities, adhesion is momentary, followed by a subsequent momentary sliding. The action of rapid alternations of adhesion and sliding (or stops and starts) is referred to as stick-slip. When occurring at high frequency, stick-slip causes chattering of the workpiece or tool, leaving a series of disfiguring marks on the finished surface of the workpiece.

The forces that develop stick-slip not only cause damage of the part being machined but also promote wear of the machine rider and way due to the nearly pure sliding friction that results.

Stick-slip is not overcome by hydrodynamic lubrication or by using an oil of greater viscosity. These measures would develop variations in lubricant film thickness whose erratic dimensions would be relayed to the plane of contact between tool and workpiece and cause (1) waviness of the finished surface and (2) loss of size control of the machined part. Thus any interference with the translating path of the worktable must be avoided.

Stick-slip is prevented by a combination of proper machine design and careful selection of lubricant, with attention given to its properties:

1. Viscosity. The viscosity depends on the load, which is determined by the weight of the table and workpiece plus the vector force exerted by the tool. The viscosity at operating temperature should assure a thin film developed by a combined squeeze action and *creep* distribution. Recommended viscosities range from 150 SUS at 100°F for light loads, such as is prevalent on tool and roll grinders, to 900 SUS at 100°F for heavy loads of some milling machine and large planer operations. Where a compromise is desired to minimize lubricant inventory and handling, a 500 SUS at 100°F can be used satisfactorily.

2. Wettability. The oil should possess less surface tension than the critical surface tension of the surface material. An oil having a surface tension of, say, 20 dyn/cm tends to creep more readily over surfaces having a critical surface tension of wetting above that value, the rate depending on the difference.

3. Oiliness. Oiliness, an inexact property, is associated with lubricants capable of producing a relatively low coefficient of friction. It is usually related to oils possessing good wettability (adsorption).

4. Film strength. The lubricant film should be tough enough to prevent mass penetration of surface asperities and excessive metal-to-metal contact.

5. Adhesiveness. The lubricant should adhere to the rubbing surfaces and not run off due to gravity or be easily washed off by soluble metal-cutting fluids that may be used.

No straight mineral oil has all these characteristics sufficiently to meet all way requirements, but an oil can be formulated so that stick-slip is eliminated.

The lubricants suited to machine tool ways are also applicable to other bearings geometrically similar but with less frictional surfaces. They include various slides, gibs, and crosshead guides (horizontal compressors and engines). Worm gears, feed screws, and columns are also potential applicants for oils having the above characteristics.

BLEEDER FOR WAYS. Some types of hydraulically operated machine tools use a bleeder arrangement for the lubrication of their ways. This entails the "bleeding" of oil from the hydraulic system. This is convenient, but it is not sound from the standpoint of chemical stability of the hydraulic medium if all the characteristics of the way oil are provided. Some of the agents used in way lubricants are subject to chemical change during extended periods of hydraulic service. To sacrifice such agents in the interest of hydraulic oil stability diminishes their capability as way lubricants.

Cylinders

There are two types of cylinders widely used in mechanical equipment. One is the reciprocating type through which a piston passes, lengthwise, back and forth to act as the power element of a steam or an internal combustion engine, or the pressure element of a compressor. The second type involves rotary motion in which a bladed rotor revolves within an inclosed cylinder, an example of which is the rotary compressor. Both types develop sliding friction between the moving members and the cylinder wall.

Reciprocating type cylinders assume either a horizontal, a vertical, or angular position. Frictional loads within a horizontal cylinder whose piston is connected directly with the crankshaft include the weight of the piston, angular thrust, and the pressure of gases acting on the piston rings to force them radially outward against the cylinder wall.

The weight of the piston exerts its greatest influence on horizontal cylinders, whereas in vertical cylinders the pressure on the cylinder wall due to this factor is nil. In radial type compressors the weight of the piston exerts a frictional force commensurate with the angularity of the cylinder, diminishing as the latter approaches a vertical position.

Angular thrust is always present in machines whose pistons are driven directly from the crankshaft through a connecting rod. The thrust occurs between the piston and the cylinder wall in the area adjacent to the piston pin bushings and on the right side of the cylinder if rotation is counterclockwise and on the left if clockwise. During operation, the force of connecting-rod thrust increases from zero to maximum as the crank moves from bottom-dead-center through 90 degrees of its revolution to maximum angularity at half-stroke. It then diminishes again to zero as it moves from half-stroke position to top-dead-center.

Where pistons are driven by a piston rod from a crosshead as in a double-acting compressor, the total angular thrust is taken by the crosshead guide bearing and therefore is not a factor in cylinders of this type machine.

In addition to the weight of the piston and angular thrust where either or both are present, there is a third force that creates considerable friction within a cylinder. This third force is the result of compression, combustion, or expansion, depending on whether a compressor, an internal combustion engine, or a steam engine is involved. The high-pressure gases fill the space within the piston grooves not taken up by the rings, causing the latter to expand outwardly and press against the cylinder wall. A downward pressure exerted by the same gas also causes the bottom of the ring to press against the groove surface (skirt side of groove) on which it bears. This force develops a high frictional load during the outward expanding motion of the ring.

The cylinder of a rotary vane type compressor is subject to the force imposed by the pressure of blades or vanes that are thrust radially outward from their rotor grooves. The force is developed by centrifugal action, becoming greater with increased shaft speed. In smaller compressors, centrifugal force is augmented by the tension of springs placed behind the blades to maintain a tight seal between blade and cylinder wall.

PISTON RINGS. Piston rings are placed in annular grooves cut in the piston for the purpose. The number and the type of rings used in any one piston vary according to the size of the piston and the service of the machine under consideration. The primary purpose of a piston ring is to form a seal between the piston and the cylinder wall. Because of the overall clearance necessary to prevent binding, a piston cannot be made sufficiently snug to provide an effective seal with its cylinder, especially under high compression ratios. In addition to the mechanical difficulties involved in obtaining a gas-tight fit, the temperature difference between the top and the bottom sections of the piston, and between the piston and the cylinder wall, make it necessary for the piston to be tapered. This means that the diameter at the head of the piston is less than at the skirt or open end. Tapering is necessary to allow for greater expansion of the piston area directly exposed to hot gases, which is at the top or crown. Subject only to conducted heat from the crown, the skirt of the piston has a lesser degree of expansion. It is also necessary to provide sufficient additional clearance to allow for the expansion variation between the piston and the cylinder wall because the latter is water-jacketed (cooled) and the former is not.

Piston rings are concentric, elastic, split metal rings capable of expanding and contracting in such a manner as to contact the wall of a cylinder as they ride with their piston along a reciprocating path. In doing so, they maintain a permanent seal between piston and cylinder. To maintain an effective seal, contact pressure must be uniform at all points around the circumference of the cylinder. The split makes expansion and contraction possible as the piston moves along the cylinder from the zone of minimum temperature to the zone of maximum temperature, or vice versa. When the ring reaches a condition of maximum expansion, which it does when the piston is at top-dead-center, the gap prevents crowding of the ring within the cylinder. The usual allowance for gap clearance is on the order of 0.004 in. per inch of cylinder bore, measured when the ring is fully compressed in its groove.

The two basic types of piston rings are (1) the compression ring and (2) the oil ring (Figs. 12-7, 12-8, and 12-9). The primary function of the compression ring is to provide a tight seal between the piston and the cylinder to prevent leakage of high-pressure gases from the compression chamber. A tight seal is also necessary in obtaining a vacuum within the compression chamber during the suction stroke.

The compression ring is designed with various cross-sectional shapes, including the rectangular and wedge or keystone types. Some are designed to allow full face (flat) contact with the cylinder wall. Others are tapered to provide minimum face contact. To decrease frictional resistance between ring face and cylinder wall, grooves are cut in the face of some type rings and filled with a material possessing a low coefficient of friction, such as tin or bronze. Rings plated with tin, chrome, cadmium, or zinc are also available and prevent scuffing during break-in periods. To allow oil to reach

Fig. 12-7. Piston rings, compression type. (Top to bottom) Butt or straight, angle or miter cut, and step cut rings. *Courtesy, Double Seal Ring Co.*

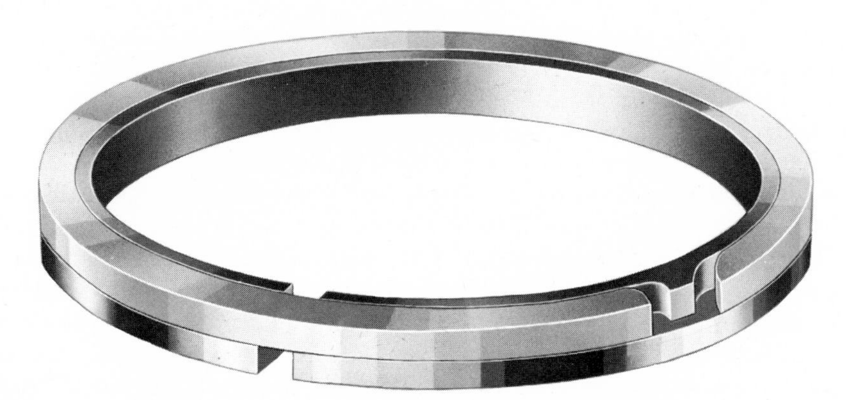

Fig. 12-8. Two-piece piston ring. *Courtesy, Double Seal Ring Co.*

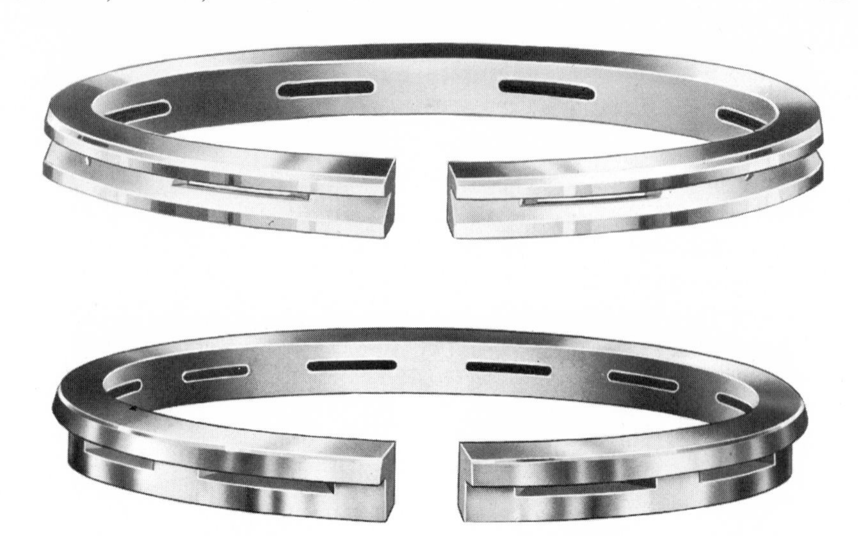

Fig. 12-9. Piston rings, oil type. (Top) Plain ventilated, (bottom) single scraper ventilated. *Courtesy, Double Seal Ring Co.*

the frictional surfaces between ring face and cylinder wall, the bottom edge is beveled on some types of rings. Beveled edges also allow free passage of rings across suction ports present in certain types of cylinders.

Oil rings usually are installed in the bottom grooves. Their function is to distribute lubricating oil evenly over the cylinder wall surface during the compression stroke, and scrape excess oil from the surface on the suction stroke. The cross sections of oil rings are also variously designed, but the principle of riding over the oil on the upstroke and scraping it toward the crankcase on the downstroke prevails in nearly all types. To function, it is necessary that the upper face edge be beveled and the lower face edge be relatively sharp. Double-facing by cutting an annular groove in the face completely around the ring and providing a bevel in the top of the upper face edge makes an efficient oil scraper ring. Usually such rings are ventilated by cutting slots (ventilators) in the bottom of the annular groove (Fig. 12-10). The oil is scraped from the cylinder wall by the sharp bottom edge of the upper face, enters the ring groove from which it is forced through the ventilators to the back clearance space of the piston groove. The oil then enters the crankcase from the back clearance space through holes drilled through the piston skirt. A modification of the above is a single-faced ring made by cutting away the bottom land to leave an ell-shaped ring having a small width of face in contact with the cylinder wall. The upper edge of the face is also beveled in this type of ring.

The split end of rings may be cut straight (butt), mitered, or in the form of a step. It is through the gap of an ill-fitted ring that appreciable leakage or blow-by occurs.

Where holes in the piston are not provided, excess oil drains down the cylinder wall into the crankcase. Annular grooves in oil rings, whether ventilated or not, permit excess oil to flow within them around the piston as they act in the role of moving dams during the downstroke. This action distributes oil completely around the cylinder, thus eliminating dry spots.

Segmental rings consist of two, three, or more segments mounted on a flat expanding hoop. Having stepped joints, they provide a seal for various type compressors, including double-acting types in which they combat pressure in two directions.

Because frictional load increases with greater rubbing area, piston rings should be designed to provide minimum face contact with the cylinder wall. However, face dimensions should be commensurate with efficient sealing and economical ring life.

Sufficient side clearance must be provided to allow rings to move freely in their grooves. To prevent excessive crowding of the ring which results in binding and ultimate damage to either the ring, the piston, or the cylinder (or all three), sufficient end clearance (gap) must be provided. The fixing

Fig. 12-10. Piston rings. (Top) Double scraper flexible ventilated oil ring, with expander, and (bottom) three-piece segmental rings. *Courtesy, Double Seal Ring Co.*

of piston rings at one end to prevent their turning to line up the gaps is questionable procedure and not recommended.

Cast iron rings of uniform cross section have proved to be highly successful in all types of reciprocating machines. Cast iron rings are often treated chemically to eliminate scuffing tendencies due to the presence of the soft pure iron (ferrite) contained in this metal. Chemical treatment converts ferrite to a depth of about 0.001 in. into iron salts, such as iron sulfide or iron phosphate. Steel rings are also used. Nonferrous metal rings employing bronze are satisfactory where operating conditions permit their use. Carbon rings have a low coefficient of friction and are employed in handling some types of gases where, because of fire hazard, the use of petroleum lubricating oil is prohibitive. Bakelite and other synthetic materials are employed for piston rings handling mild acid and alkaline gases.

To provide uniform pressure between the piston face and the cylinder wall around the entire circumference of the bore, stresses throughout the ring must also be uniform. Otherwise openings would occur in some areas to permit blow-by. Because of their design and their flexibility, rings have a tendency to become straight—i.e., to spread outward at the gap. This means that the greatest radial force would be exerted at the ends and at a point halfway around the ring from the gap. To balance the internal stresses and allow the ring to exert uniform radial pressure along its entire length of face, ring peening or hammering is practiced. This is done by peening the inside of the ring (as with a ball-peen hammer), increasing the peening force gradually as the action proceeds from the gap to a point halfway around the ring. From the halfway point the peening force diminishes in the same degree as it increased as the action continues toward the end opposite that of starting.

To increase the pressure between face and cylinder wall of oil rings, sometimes expanders are used. Expanders are abutment type springs designed to exert radial force. Placed at the back (around the inside circumference) of the ring, the radial force of the expander is transmitted to the ring, causing it to expand outwardly beyond its normal static limits. Compression type coils or ribbons made of spring steel make up the expander whose normal circumferential length is longer than the inside circumference of the ring when the latter is fully compressed to cylinder size.

GROOVING

To assure complete lubricant distribution continuously to all internal surfaces, particularly those that carry the load, strategically located narrow grooves, or chamfer cuts, are designed for bearing surfaces. The grooves are usually connected to the lubricant inlet.

Grooving of Journal Bearings

Grooves for horizontal bearings are cut axially and extend to within $\frac{1}{8}$ to $\frac{3}{4}$ in. of each end of the bearing, depending on its length (Fig. 12-11). Channel grooves are relatively narrow and rounded slightly at the trailing edge to eliminate wiping of the lubricant from the journal as it rotates. Chamfer grooves are also cut axially, but their trailing edge may be beveled, with a gradual slope extending upward as much as 45° before leveling off with the normal bearing surface.

PURPOSE OF GROOVING. Grooves cut in the surface of plain bearings serve the following purposes:

1. Act as lubricant reservoirs to store oil or grease, an important function particularly in bearings subject to boundary or near boundary lubrication conditions.
2. Assist in distributing the lubricant evenly through the bearing, especially to those surfaces that carry loads and are subject to the highest pressures.
3. Induce the formation of the hydrodynamic wedge essential in attaining fluid friction rather than boundary friction.

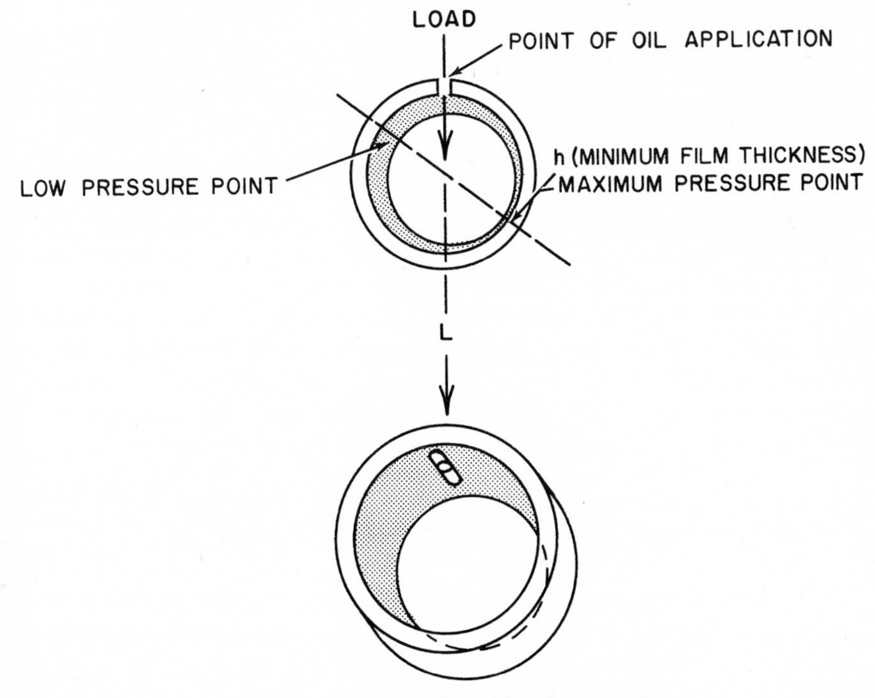

Fig. 12-11. Axial grooving of full sleeve bearing.

4. Allow sufficient oil flow through the bearing to obtain a greater cooling effect.
5. Provide basins for the collection of abrasives and other debris and help prevent their entrance into high-pressure areas.
6. Reduce premature side leakage via clearance space.
7. Collect oil for return to sump.

Channel grooves are primarily lubricant distributors and smaller reservoirs. Chamfers are larger reservoirs that not only act as distributors, but increase the lubricant holding capacity of a bearing to affect cooling and reduce end leakage (Figs. 12-12, 12-13, and 12-14). Both type grooves may be designed in such a manner as to guide the flow of some of the lubricant from the ends back toward the center of the bearing.

LOCATION OF GROOVES. With the exception of some hydrostatically lubricated (flow-through) bearings, the load-carrying surfaces of plain bearings should never be marred by grooves or other recesses for two major reasons. (1) The development of the hydrodynamic film is interfered with by disturbing the surface on which it depends for support. (2) Any removal of load-carrying bearing surface increases the unit pressure on the remaining area. This reduces the bearing load capacity in an amount corresponding to the area removed.

Hydrostatic bearings may be grooved if it is necessary to apply the externally pressurized fluid at one or more points within the area opposing load direction in order to separate the rotor and stator members of the pair. Grooves might also be in order for some internal combustion engine bearings, such as main and connecting rod bearings, where one bearing serves as a passageway for oil flowing to another bearing, or where application of oil cannot be attained otherwise. In the latter case, circumferential grooving is provided.

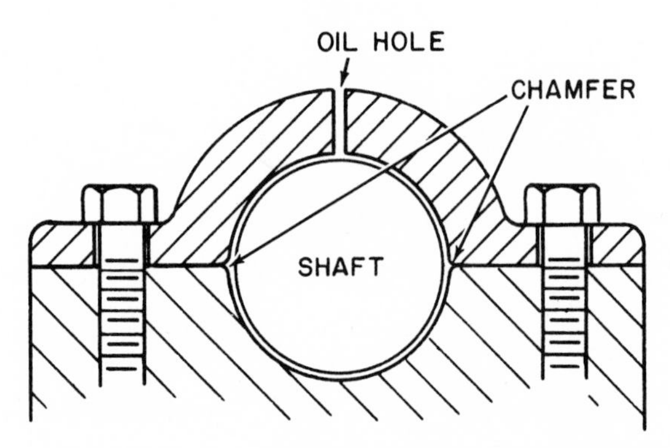

Fig. 12-12. Groove formed by slightly chamfering the parting edges of a split bearing within ½ in. of each end.

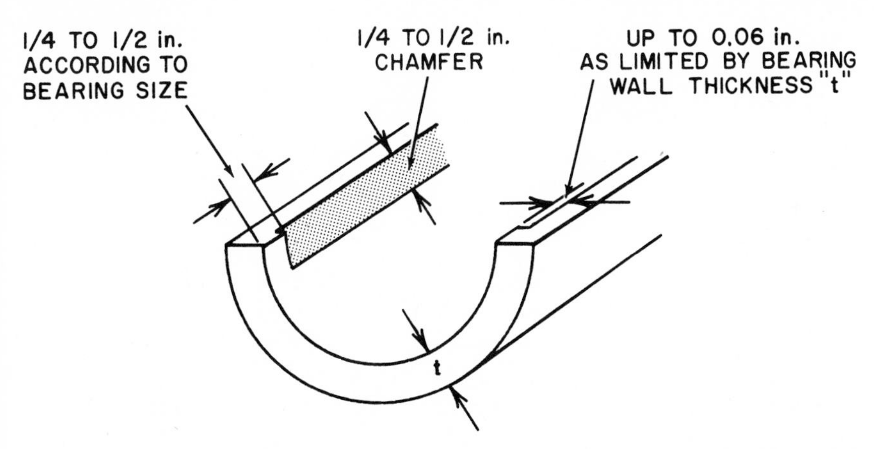

Fig. 12-13. Bottom half of split bearing with chamfer that provides axial distribution and reservoir for oil.

A groove whose primary purpose is to store and evenly distribute oil for the lubrication of a plain bearing should be located and oriented as to promote the development of a hydrodynamic or microhydrodynamic film. A journal bearing is subjected to a progression of incremental pressure changes alternating from minimum to maximum to minimum during one revolution under an applied load. With a vertical downward load, the minimum pressure exists in the upper half of a full circular bearing, while maximum pressure is developed at some angle beyond the load line in the lower half, the exact point depending on the eccentricity ratio, which is a

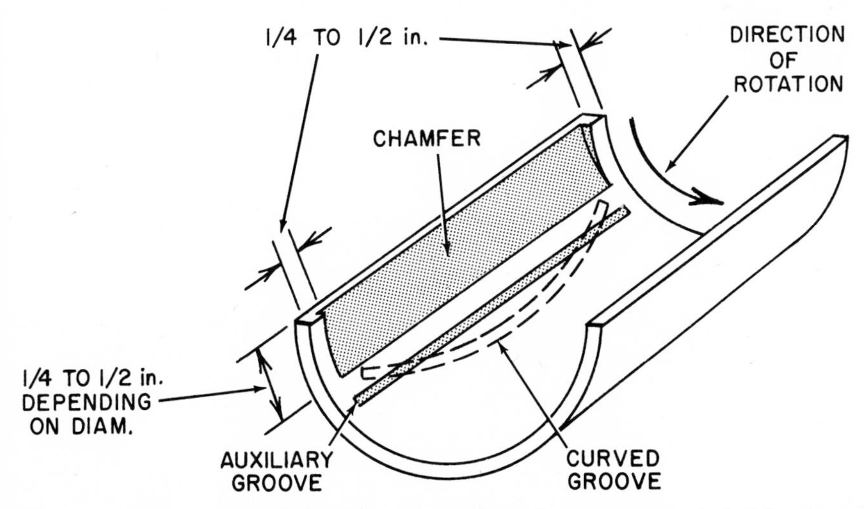

Fig. 12-14. Chamfer for split bearing, with auxiliary oil storage groove included only on bearings subject to intermittent oil application. Top half (not shown) grooved as in Fig. 12-11.

function of clearance, load, speed, and oil viscosity. If the values of all these factors are known and are constant, the most strategic location of an axial groove can be accurately calculated, or determined by vectors. Because such precision is not attainable for most bearing operations, a rule-of-thumb method must be devised to determine the safe location of the groove or the trailing extremity of a chamfer. The important rules to follow in locating and planning a groove are as follows:

1. Grooves should be located in the low-pressure area, preferably on a line *approaching* the beginning of the high-pressure zone.

2. The exact location of the trailing edge of an axial groove may be calculated according to the operating factors and clearance, as mentioned above.

3. If rule 2 is not feasible, locate the groove no closer than approximately 60° *before* the point of maximum oil film thickness. The same rule applies to grooves located for some reason beyond the point of maximum pressure. The various figures depicting the grooving of bearings show their typical location and orientation. Note that they are usually located somewhat beyond the limitation of 60° given above.

4. Grooves should not extend the full length of the bearing, since this would permit free flow of lubricant and cause premature drainage through the resultant opening. Rather, the grooves must be cut short of each end by a distance dictated by diameter of the journal which governs clearance magnitude. The distance between the end of the groove and bearing extremities is calculated on the basis of $\frac{1}{4}$ in. per unit of diameter, with limitations of about $\frac{1}{16}$ in. for very small bearings up to $\frac{1}{2}$ in. for large ones. The progressively larger clearance space with greater journal diameter allows sufficient overflow to lubricate the outer surfaces beyond the grooving.

5. There are no fast rules governing the width and depth of grooves, except that they should be able to hold and carry enough oil to feed the bearing. Where oil is applied directly into a channel type axial groove, its width should diminish slightly from the point of application to its terminals. This does not apply to other auxiliary grooves or chamfers.

Groove Design and Orientation

RADIAL, FULL CIRCLE BEARINGS. Locate groove as shown in Fig. 12-11. Parting lines of two- or four-part bearings should be slightly chamfered. The groove should always be located in the low-pressure area, i.e., in the top half of the bearing for downward loads and in the lower half for upward loads, etc.

Variable loading. A circumferential groove located in the center of the bearing, while splitting the surface into two load-carrying areas, resulting in a reduced capacity, is an effective method of distributing oil regardless of load angularity. Axial chamfering at the parting lines of the split bearing acts to spread oil and increase flow. Bearings of this type are used in internal combustion engines and compressors as crank-pin and wrist-pin bearings. They are fed from an oil pump via passages drilled in the crankshaft to the crank-pins through which the lubricant passes into similar passages drilled in the connecting rods to the wrist-pins. The grooves in the crank-pin bearings act as pathways for the oil to the wrist-pins, as well as distributors for the crank-pin bearings. Circumferential grooves should be clean-cut at the edges, but never chamfered, which would act only to further reduce the load capacity of the bearing.

RING-OILED BEARINGS. Chamfering at the parting lines is usually sufficient for oil distribution in ring-oiled bearings subject to downward loads only. Where variations in direction of load and reverse shaft motion are encountered, the continuous groove helps to spread the oil (Fig. 12-15). Used effectively in electric motors.

Infrequent oil application. To provide additional oil storage space for intermittently fed bearings, an auxiliary groove (Fig. 12-14) may be added. It should be kept clear of the pressure area.

VERTICAL BEARINGS. To reduce leakage, vertical bearings are usually grease lubricated. Where oil is used, a circumferential groove cut near the upper end of the bearing promotes effective distribution (Fig. 12-16). The oil is applied directly to the groove either by hand or a device. Leakage of oil is a major factor with vertical bearings. Many such bearings act only as guides for vertical shafts and therefore support light loads. A spiral groove progressing upward against rotational direction of the shaft reduces leakage by the pumping action engendered by such a design.

The author found that sparing amounts of an intermediate viscosity (600–700 SUS) oil containing an adhesive agent provided adequate protection with minimum leakage when applied occasionally to vertical bearings.

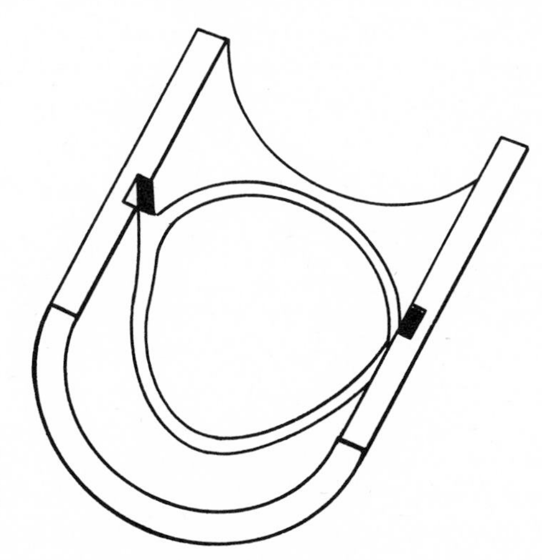

Fig. 12-15. Continuous groove in bottom half of ring-oiled bearing to meet changes in load direction and reverse shaft motion.

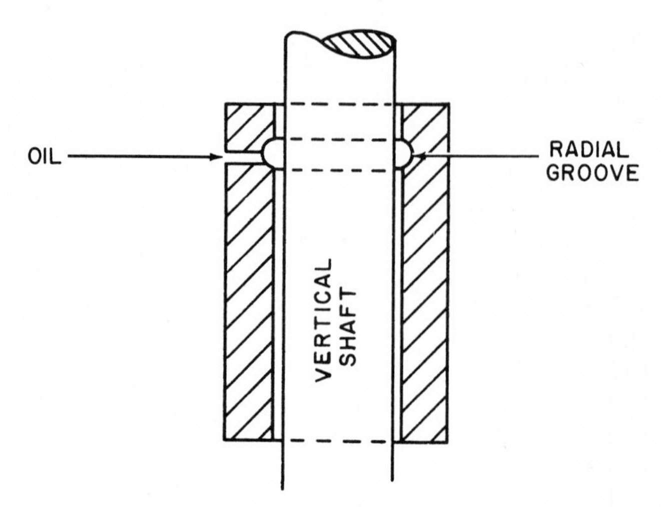

Fig. 12-16. Radial groove for vertical plain bearing.

THRUST COLLARS. An axial bearing is sometimes designed to resist axial motion due to either load or drift. Such bearings are usually located at the end(s) of the shaft or the shaft is provided with a suitable collar or shoulder against which the bearing reacts to support the thrust and maintain axial position. The ends of radially loaded bearings subject to moderate thrust loads are chamfered to allow the oil leaking from the bearing to flow onto the collar and become outwardly distributed by centrifugal force (Fig. 12-17). With multiple collars, the oil is fed directly to the bearing surface, reaching the thrust areas as a result of beveling and centrifugal force, as shown.

Thrust Bearings. Grooving of a tapered-land thrust bearing is shown in Fig. 12-1. The revolving ring is flat and not grooved. The stationary ring is grooved radially with a tapered-land leading to the load-carrying land in the direction of rotation. Step bearings are grooved.

Fig. 12-17. Grooving of thrust collar bearing.

COOLING CHAMFERS. Large, high-speed machines such as steam turbines require rapid dissipation of heat from their bearings. The flow of oil may be increased by providing a wide chamfer in the top half of the bearing. The chamfer should extend from the top of the bearing circulating to the parting line, where it converges with the normal surface of the bottom half in the direction of rotation. Lengthwise it should approach but not extend to the ends of the bearing. Circumferential grooves cut near each end of the bearing provide for return of the oil to the sump. The grooves cut at each end of this type bearing through which oil passes to return to the sump are known as *oil-collecting grooves.*

Curved Grooves. The dotted curved groove superimposed on the bearing shown in Fig. 12-14 returns the oil flowing toward the ends to the center of the bearing. Joining the curved groove and the parting-line chamfer at the ends to form a D is an effective means of meeting boundary lubricating conditions for slow-speed, heavily loaded bearings. Relieving the entire surface of the bearing within the D by proper chamfering (in the direction of rotation) has the same effect as the D groove and substantially reduces end leakage.

Upward load, horizontal bearings. The groove shown in the bottom of the bearing in Fig. 12-11 should be located similarly in the top half of the bearing when the load is upward. Also the auxiliary grooves in upward loaded bearings are located in comparable locations in the upper half on the side before the direction of rotation.

GREASE GROOVES. Grooves for grease should be wider and deeper than those cut for oil in order to facilitate the flow of the heavier-bodied material at reasonable pressure. The same principles of groove orientation and design as described for oil may be applied for grease. As in oil bearings, circumferential grooves cut near the ends reduce grease leakage and provide access

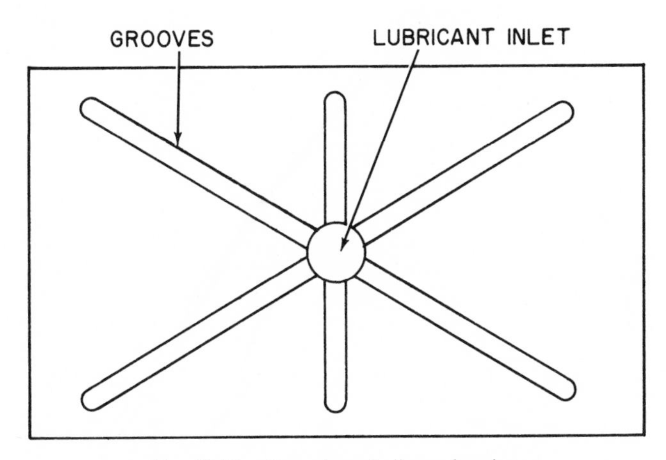

Fig. 12-18. Grooving of slipper bearing.

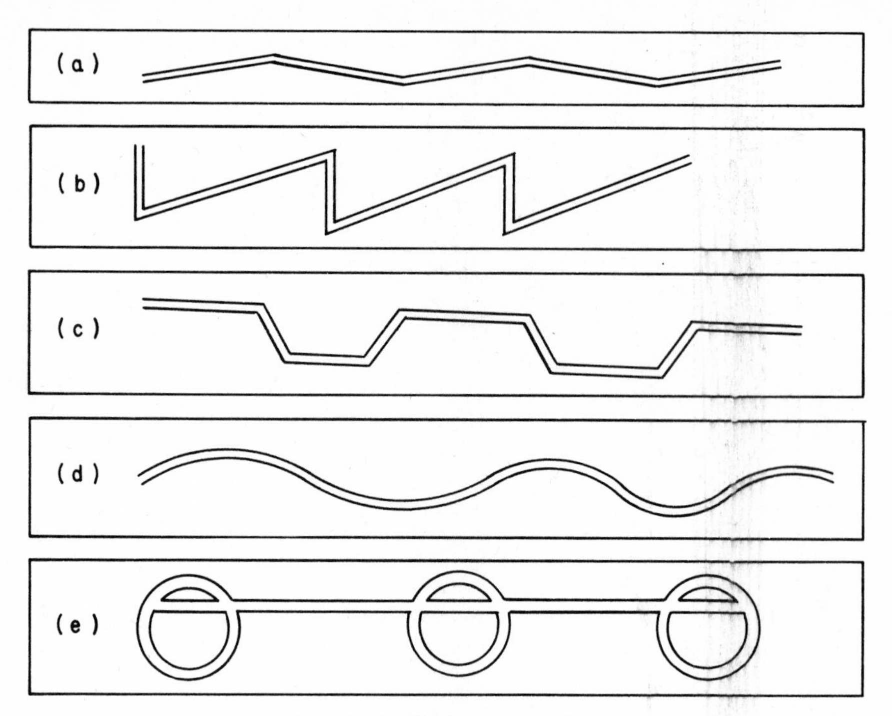

Fig. 12-19. Table way grooving.

of the lubricant to the load-carrying side of the bearing, which eventually is forced into the low-pressure areas to be carried to the high-pressure zone by the motion of the journal.

SLIPPER BEARINGS. Grooving shown in Fig. 12-18 is typical for steel mill slipper bearings, described in Chapter 11.

Grooving of Ways

Ways are also grooved to distribute oil over their entire surface. Some designs for way grooves are shown in Fig. 12-19.

13

ROLLING BEARINGS

Ball and roller bearings are not frictionless, but their coefficient of friction is exceptionally low. They generally consist of four parts:

1. *Inner ring,* or inner race, is grooved or channeled around its outer surface and guides the rolling elements as they travel 360° about a common axis. Its internal diameter, or bore, is usually reported in millimeters. The inner race is also referred to as the *cone.*

2. *Outer ring,* or outer race, is also grooved or channeled but on its inner surface, where it faces the groove of the inner race. The opposing grooves provide the annuler space that holds and guides the rolling elements. The outer race also acts to enclose and hold the bearing assembly and support it within the machine housing in which it is fitted. It is also called the *cup.*

3. *Rolling elements,* either balls or rollers, occupy the annular space formed by the grooves in the inner and outer races. (Fig. 13-1). Rollers are placed in the assembly either longitudinally or radially relative to the shaft, depending on whether the bearing is designed to meet radial or axial loads, or a combination of both. Rollers may be cylindrical, tapered, concave, or spherical. Relatively long, slender, cylindrical rolling elements are called *needle rollers,* and their bearings are *needle bearings.*

4. *Separator,* or *cage,* is located between the inner and outer race to form an endless series of pockets that hold and locate the rolling elements. Its primary function is to keep the rolling elements separated and uniformly spaced as they roll around the stationary race. It is also known as a retainer.

The separator prevents rollers from departing axial alignment, or *skewing.* There is a strong tendency on the part of rolling elements to spread apart when approaching or entering the area of greatest pressure. The separator prevents such spreading. When a separator is not used, the space

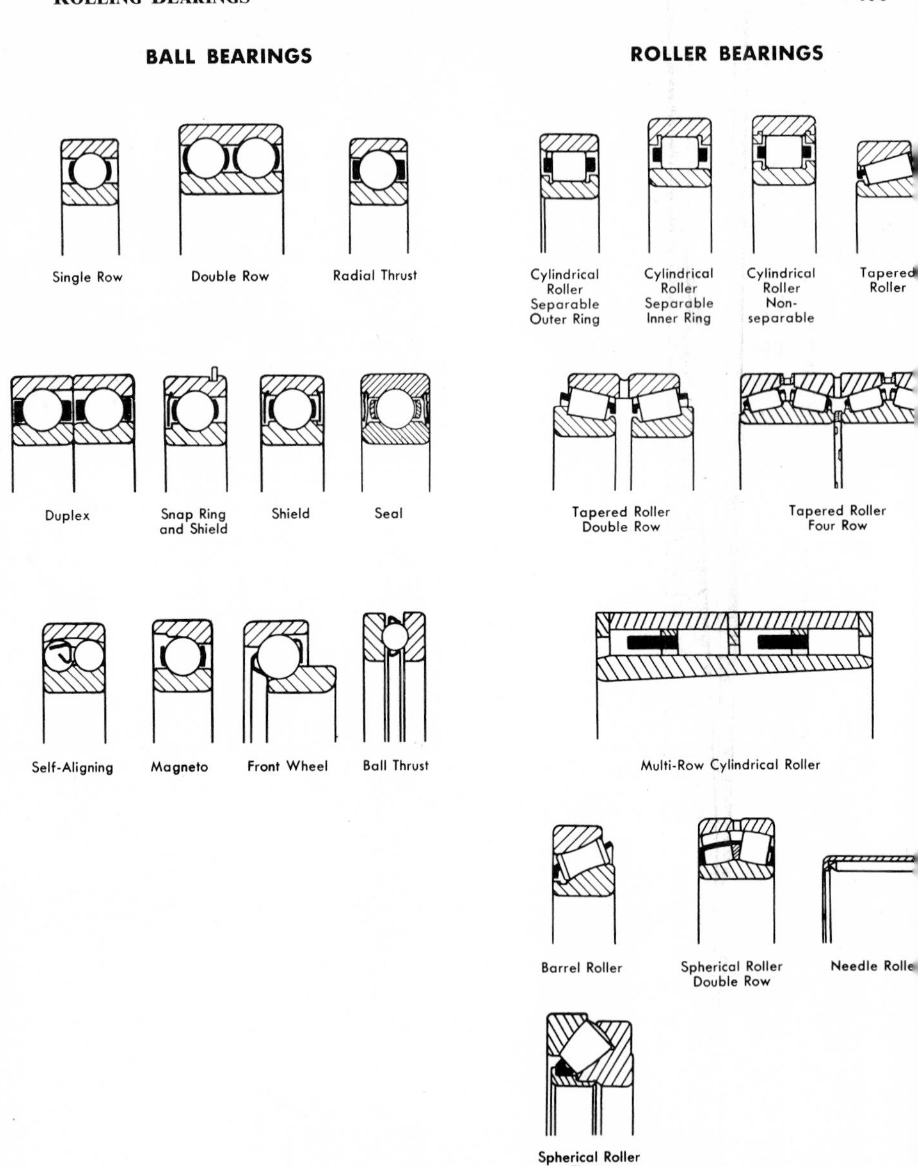

Fig. 13-1. Rolling bearing types. *Courtesy, SKF Industries, Inc.*

between the races must be completely filled with rolling elements to avoid their separation. Known as *full-complement bearings,* they have higher load-carrying capacity because they are equipped with more rolling elements than those with separators. Separators prevent the rolling elements from rubbing at high speeds, the rubbing being aggravated because such contacting surfaces are moving in opposite directions. They are made of pressed steel, machined bronze, or molded phenolic composition.

Seals and *shields* are often included in the rolling bearing assembly to keep dirt out and retain the lubricant. Seals are particularly required where free water is present, as at the wet end of paper machines.

Rolling bearing nomenclature is illustrated in Fig. 13-2.

BALL BEARINGS

Ball bearings are very smooth balls of hardened steel (Fig. 13-3). They are variously designed to meet one or more service requirements such as (a) high radial load, (b) high axial (thrust) load, (c) combined radial and

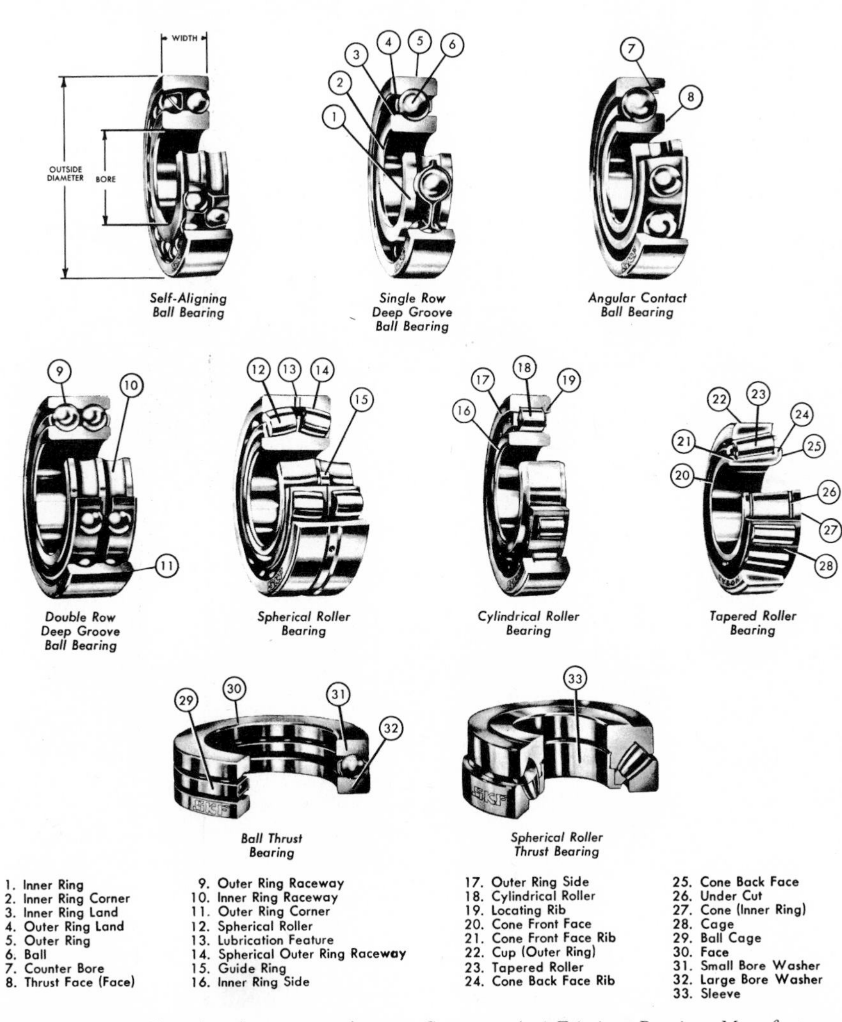

Fig. 13-2. Rolling bearing nomenclature. *Courtesy Anti-Friction Bearing Manufacturers Association, Inc.*

axial loads, and (d) misalignment and distortions of mounting. Variations in design to meet specific service requirements involve the curvature of the outer surface of the inner race and the inner and outer surfaces of the outer race. Breaking it down further, consider further service requirements: (e) high radial load and moderate thrust, (f) high thrust and moderate radial load, (g) pure thrust, and (h) misalignment, moderate radial load, and moderate thrust.

All bearings must possess some thrust capability to meet any slight axial shift of the supported shaft. To support an axial load, some means of retaining the balls in their path must be furnished. This can be accomplished by providing a barricade, either a deep groove or angular contact shoulders on the side of the race receiving the thrust. However, before providing such

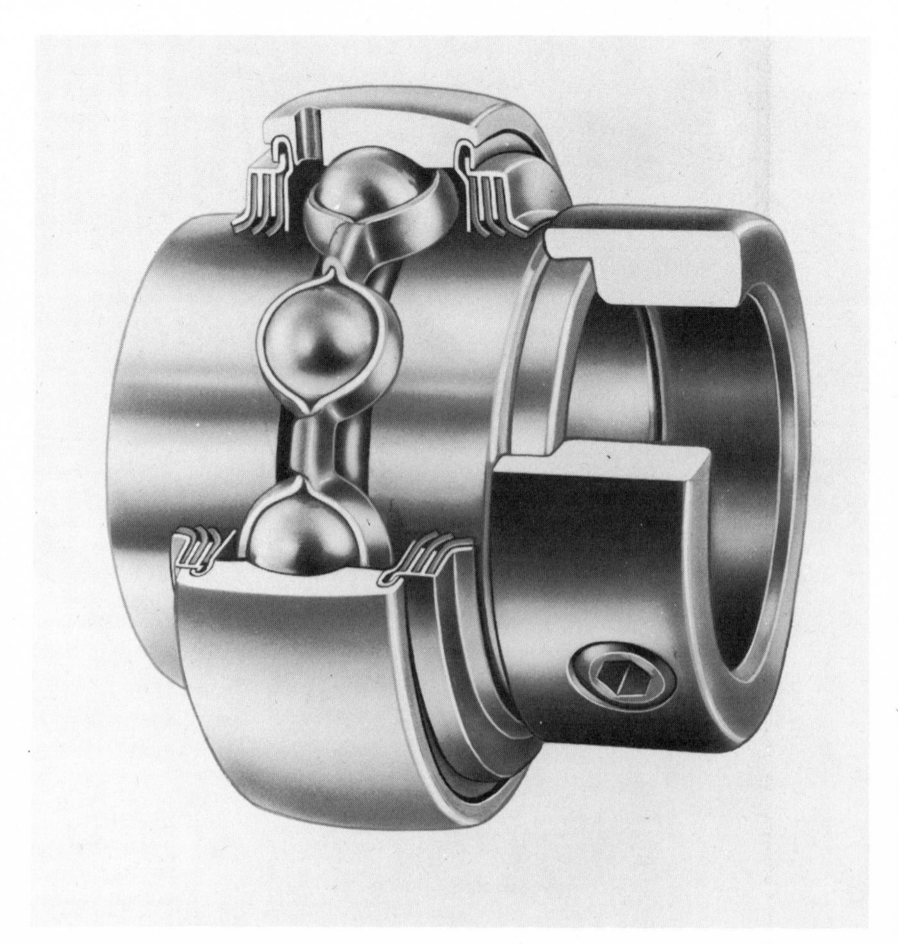

Fig. 13-3. Ball bearing with effective seals (Tri-ply). *Courtesy, The Fafnir Bearing Co.*

barricades, it is well to realize the disadvantages that accompany such conveniences.

1. The radial load capacity of a series of balls is greatest when the force is perpendicular to the axis of rotation and the balls travel in a perfect plane of that force.

2. Any force that interferes with the tendency of the balls toward perfect rolling action reduces their radial load-carrying capacity. Perfect rolling action requires a plane surface, hard enough to resist deformation (an impractical assumption).

3. The load-carrying capacity of ball bearings of similar overall dimension is increased as the number of rolling elements occupying the area of maximum pressure is increased.

4. Ball bearings designed for multipurpose service, e.g., radial and thrust loading, necessarily entail additional friction, which reduces load capacity, induces higher operating temperature, and causes rapid wear.

5. The physics that applies to radial bearings also applies to bearings whose major, or sole requirement is to support thrust loads.

Thus the grooves and shoulders along which rolling elements travel have a disadvantage, although they are necessary. Their disadvantage is sliding friction.

The convenience of overcoming slight irregularities in mounting and shaft deflection is provided by features that allow the bearing to automatically align itself. The principle of self-alignment lies in the ability of the critical parts to adjust their spherical attitude relative to the housing according to the magnitude of the irregularities. To accomplish changes in attitude the mating surfaces normal to the load are spherically ground to a radius at least equal to that of the distance between the surfaces and the bearing axis or center line. Designs vary according to the spherically ground surfaces to provide self-alignment:

Inner surface of outer race which is in contact with the balls.
Outer surface of outer race in contact with congruent spherical surface of housing cavity in which bearing is fitted.

Service Classification

Ball bearings are classified in relation to their service features, the most common type of which are

Single row, deep groove
Double row, deep groove
Angular contact
Self-aligning
Thrust (ball)

Single row, deep groove ball bearings have high radial load capacity and substantial thrust resistance in either direction. Close contact between balls and groove exists. Careful alignment is essential. The load line is perpendicular to the axis of rotation.

Double row, deep groove ball bearings are similar to the single row type except that the balls are usually smaller for bearings of the same overall dimensions, and their grooves are shaped so that load lines through the balls have angularity that permits their converging at some inward or outward point. Having two rows of balls gives this bearing greater radial load capacity.

Angular contact ball bearings have a high shoulder on one side of the inner ring and another on the opposite side of the outer ring. The steep contact angle of these shoulders gives this bearing a high thrust capacity in one direction. It also provides mechanical rigidity to support axial force. The shoulders may be mounted in pairs either to increase thrust resistance or to meet greater combined radial and axial loads. For greater unidirectional thrust, two angular contact bearings are mounted in tandem (Fig. 13-4). For greater combined radial and axial load capacity, two flush-ground angular contact bearings are mounted either face-to-face or back-to-back.

Self-aligning ball bearings may have either one or two rows of balls. Their spherical surfaces are ground either in the inner or outer surfaces of the outer ring while the balls are in close contact with a groove or grooves in the outer surface of the inner ring. The spherical feature precludes any shaft bending tendency on the part of the bearing.

Thrust ball bearings may be likened to two flat washers separated by balls (Fig. 13-5). The balls maintain their circular paths within either grooves or flat seats ground in the opposing inner faces of each washer. Such bearings

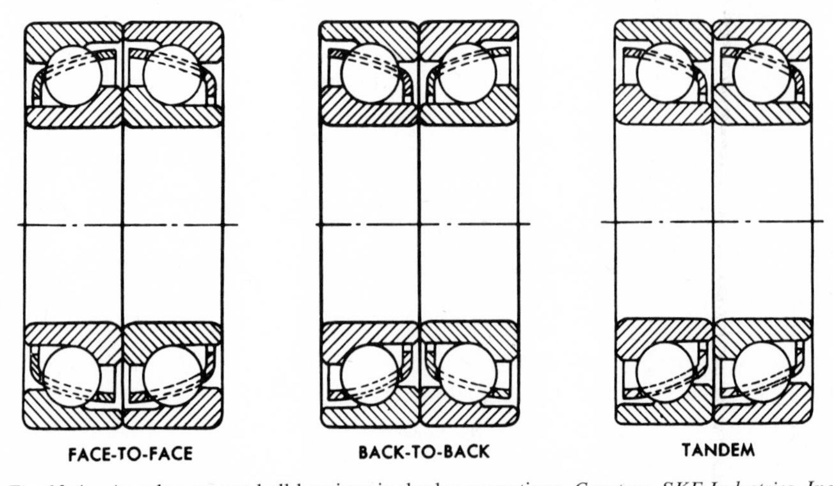

Fig. 13-4. Angular contact ball bearings in duplex mountings. *Courtesy, SKF Industries, Inc.*

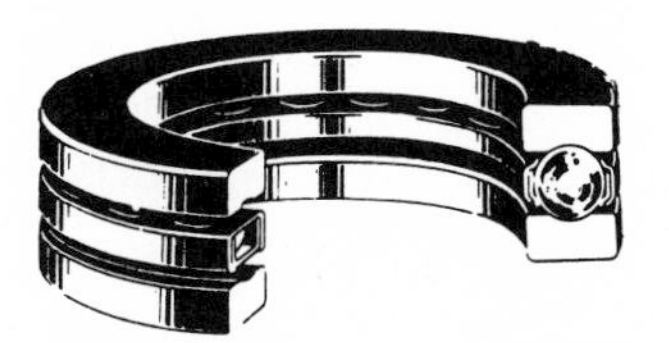

Fig. 13-5. Ball thrust bearing. *Courtesy, SKF Industries, Inc.*

are applicable to axial loads only, since the thrust is applied perpendicular to the face of the washer or parallel to the axis of the shaft. The washers are actually raceways in the same sense as those used for radial ball bearings. By using three raceways separated by two rows of balls to form five layers, a thrust in either direction is resisted. The middle raceway is made with smaller bore diameter to receive the ends of two shafts extending in opposite directions.

Magneto bearings are so called because they were originally made for use in magnetos. They are of the separable ball type having a shallow groove and angular contact with a shoulder on one side of the outer ring only. The magneto bearing is equipped with a separator and the outer ring is removable. They are small and are easy to replace. Since they are subject to slight end play, they are usually mounted in opposed pairs to maintain ball and shoulder contact by a slight thrust.

Filling slot or filling notch ball bearings are used to carry relatively heavy radial loads. They are made with a slot cut in the shoulder of the groove to a point above the raceway proper or at the bottom curvature of the groove. The purpose of the slot is to allow additional balls to be inserted and to increase radial load capacity. Because of the slot a slight axial load is necessary to maintain contact between ball and the shoulder at the opposite side of the outer ring.

ROLLER BEARINGS

Roller bearings are classified according to their longitudinal profile of which there are three major variations: cylindrical, tapered, and spherical. Roller bearings have wide versatility relative to radial loads, axial loads, and alignment (Fig. 13-6).

Cylindrical roller bearings have constant diameter rollers along their entire length. They vary in length and diameter according to the required service. Because cylindrical roller bearings develop low friction, they can be used for high speeds. They are guided by flanges located on both sides of the inner, or outer, races. The resultant channels take the place of the grooves used for ball bearings. In some instances the flange is provided on only one race which may be either the inner or outer race. The unflanged race is slightly crowned to prevent loading the roller edges if the bearing

Fig. 13-6. Two-row tapered roller bearing. *Courtesy, The Timken Co.*

is subject to slight shaft misalignment. While the flanges are substantial enough to resist moderate axial motion, they are not recommended for high continuous thrust.

Cylindrical roller bearings with separators are *cage-type* bearings. Those having no separator are called *loose rollers* or *full complement bearings.* The space left by eliminating the separator makes room for additional rollers to give the bearings a greater radial load capacity. Cylindrical roller bearings are practical for use on precision machinery where deformation of both roller and raceway surface is held to a minimum which is more characteristic of rollers than of balls.

Tapered roller bearings have rollers that gradually taper from one end to the other. Like spherical rollers the large end is in contact with the flange on the inner race.

The angles of the race cones and roller taper intersect at the bearing axis. The steepness of the slope varies according to whether the predominant load is radial or axial. With more pronounced thrust load the angle of slope becomes greater and vice versa. To relieve end loading of the roller, the raceway is slightly relieved at each side. Again, as with spherical rollers, the end surface of the tapered rollers and the guide flange are spherically mated. They are made with single or multiple rows (Fig. 13-7). The double-row type may have one inner race with two outer races, or one outer race with two inner races.

Tapered roller bearings are capable of carrying combined radial and unidirectional axial loads simultaneously. They are best employed in con-

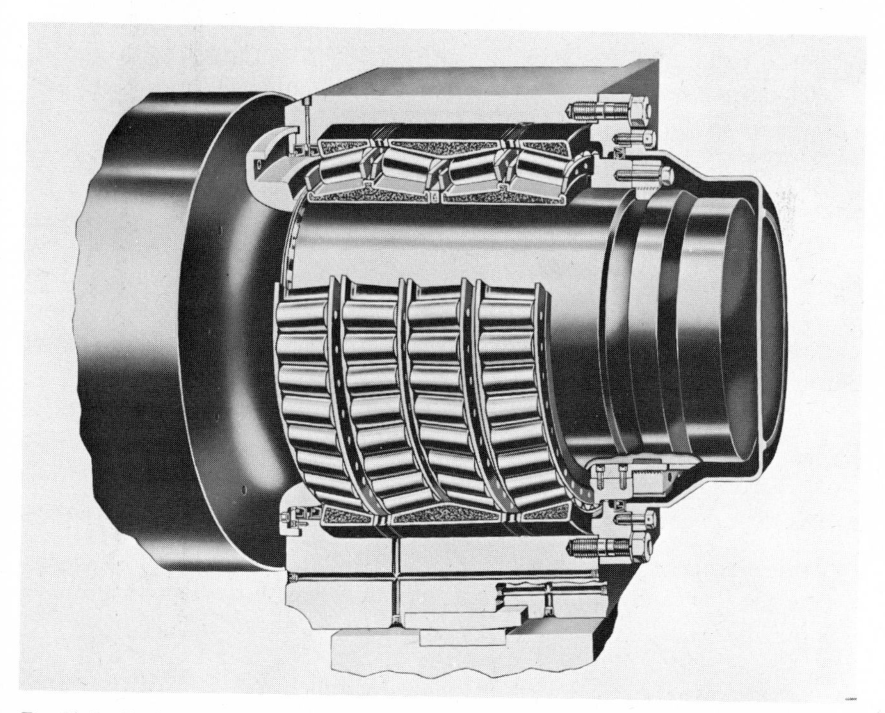

Fig. 13-7. Back-up roller bearing for rolling mill. Four-row tapered roller bearing. *Courtesy, The Timken Co.*

junction with another mating tapered bearing mounted in an opposed position on the same shaft to balance the thrust in both directions. The tapered roller bearing can carry heavy radial loads. Where both radial and axial loads are exerted, the angle of roller contact averages 15 degrees, and for heavy thrust loads the angle is usually about 28 degrees.

Spherical roller bearings have rollers that are convex in longitudinal profile and their diameter is larger at one end than the other. Thus they differ from the *barrel roller* which is the same diameter at both ends. Spherical rollers roll on a mating spherical inner surface on the outer unflanged race to permit self-alignment. The spherical principle is carried over onto the larger end of the roller which is in contact with the center guide flange on the inner race, which is also made spherical to mate with the roller ends. Because the roller is larger at one end, it tends to hug the center flange of the inner race while rotating. It is made in both single-row and double-row types. It has a high load capacity at relatively low speeds and combines simultaneous axial and radial load capabilities, especially when thrust is the predominating load.

Roller Thrust Bearings

Roller thrust bearings are designed to support axial loads only. Such bearings are similar to the thrust ball bearings. Roller thrust bearings cannot support a radial load, but they can take a heavy unidirectional thrust load.

The *cylindrical roller thrust bearing* has short rollers to minimize twist with its accompanying additional sliding action. To obtain high thrust load-carrying capacity, a double row of short cylindrical rollers are used. Only low speeds are permissible with this type of thrust bearing.

The *spherical roller thrust bearing* is similar to the ball thrust bearing except that the roller is at high angularity with its races. Load-carrying capability is high with this type of bearing. The roller and surface and its guide flange surface are both made spherical as with the radial type. However, the radii differ in order to enable a hydrodynamic lubricity film to develop.

The *tapered roller thrust bearing* is similar to the ball type thrust bearing (Fig. 13-8).

Needle Bearings

Needle bearings are a type of cylindrical roller bearing. The needle rollers are long and slender, with a minimum length-to-diameter ratio of 4 (Fig. 13-9). They are conducive to higher frictional resistance than other types of roller bearings because of the difficulty in keeping them parallel to the supported shaft. Needle bearings are suited for slow speeds, or for oscillating and intermittent motion which permits the rollers to return to

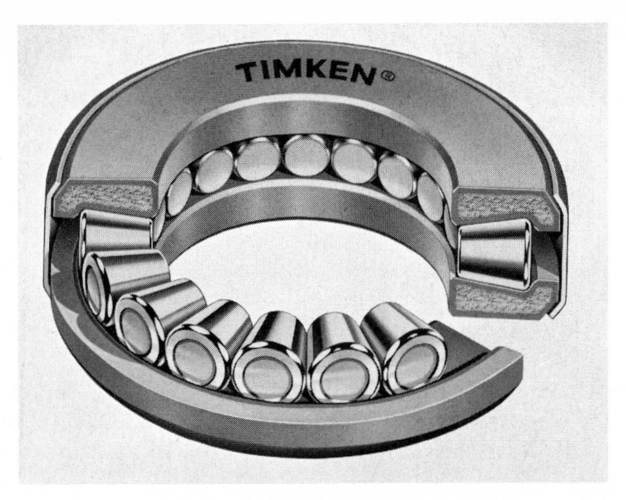

Fig. 13-8a. Tapered roller thrust bearing. *Courtesy, The Timken Co.*

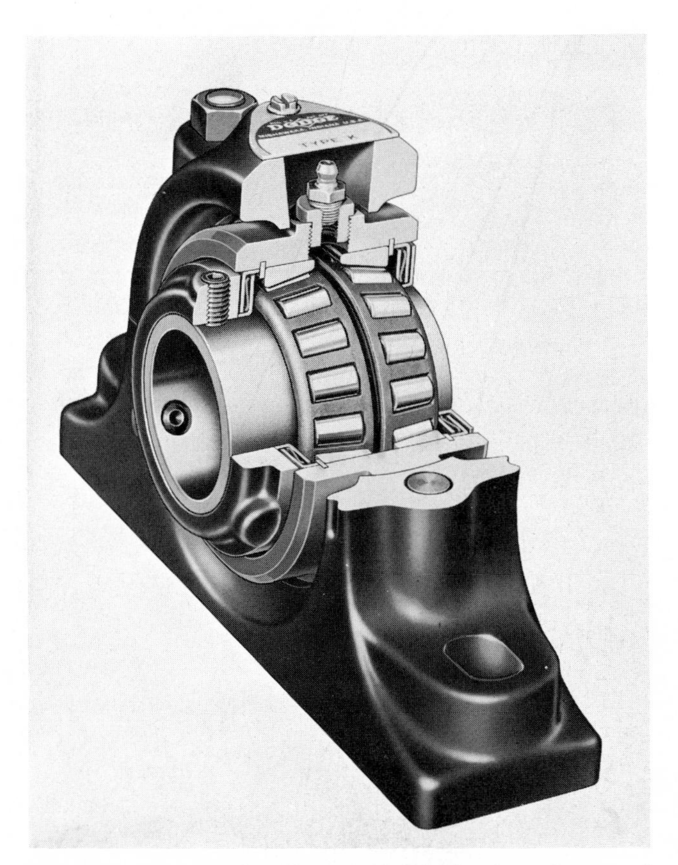

Fig. 13-8b. Pillow block with tapered roller bearings. *Courtesy of Dodge Manufacturing Co., Division of Reliance Electric Co.*

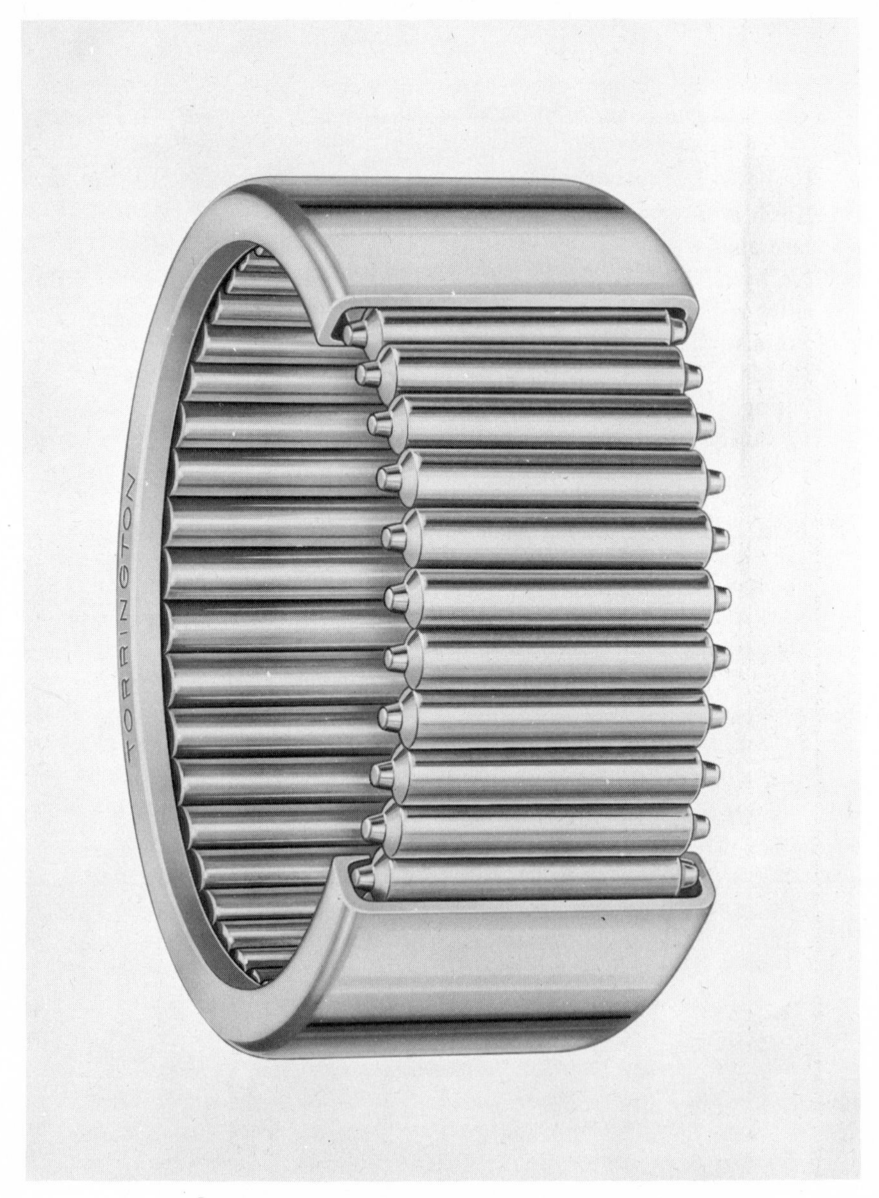

Fig. 13-9. Needle bearing. *Courtesy, The Torrington Co.*

their required position upon load relief. They are often designed with outer race and needle rollers only, in which case their hardened steel shaft acts as the inner race. The radial type needle bearing lacks thrust load capability, but needle thrust bearings are available (Fig. 13-10).

Seals and Shields

Seals and shields keep the lubricant in and contaminants out of rolling bearings (Fig. 13-11). Seals are either permanent or removable. Permanent seals are made with felt, rubber or leather rings on their inner circumference and are pressed into grooving located in the outer ring, or housing end plates. They form a lip over the outer edge of the inner ring on which they rub. They are also made to snug fit about the shaft. Removable seals are held in place by snap rings. Seals are usually of either the lip or rotary type. The latter rotate with the inner ring and act as an oil slinger. Labyrinth seals are also used.

Shields are usually attached to the stationary outer ring and enclose the bearing cavity to within two or three thousandths of an inch of the rotating inner ring. They are normally used when grease is the lubricant.

Specific Load Capacity

The specific load capacity of a rolling bearing is its relative ability to sustain an allowable load for a given number of revolutions (life). Capacity depends on several factors:

1. *Number of rolling elements.* Load capacity is proportional to the number of rolling elements because more balls or rollers give better load distribution and greater dynamical structional strength.

2. *Size of rolling elements.* The load-carrying capacity is proportional to the square of the diameter of the rolling elements as demonstrated by

Fig. 13-10. Needle thrust bearing. *Courtesy, The Torrington Co.*

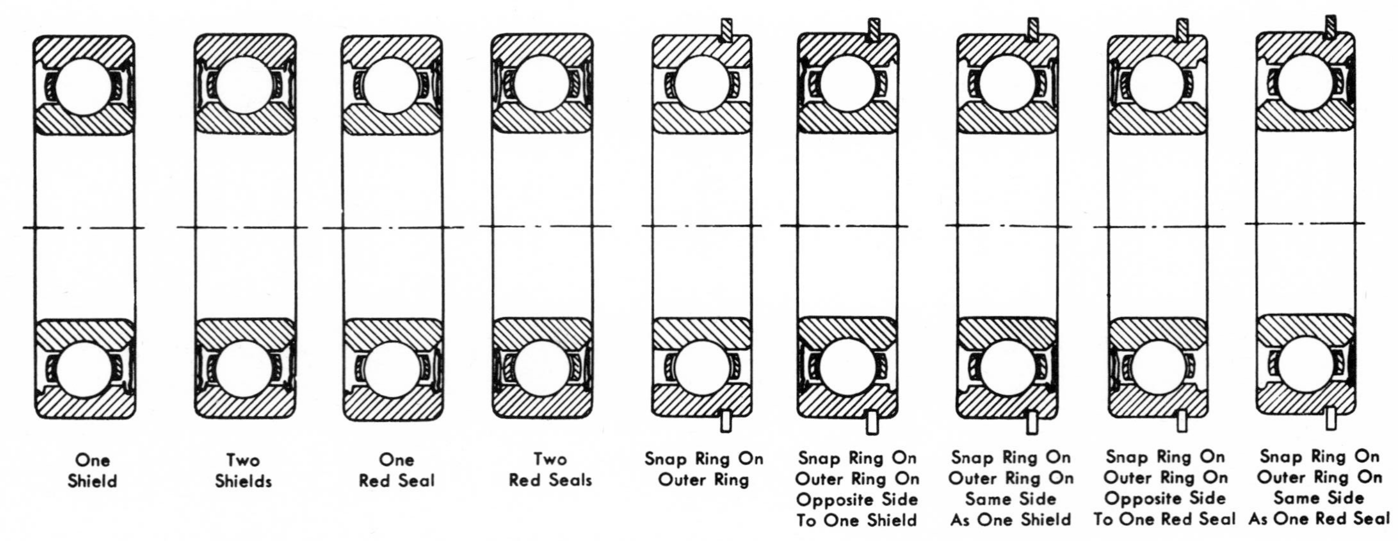

Fig. 13-11. Seals, shields, and snap rings. *Courtesy, SKF Industries, Inc.*

the Hertz theory. According to A. Palmgren,* however, small ball bearings have a relatively higher load-carrying capacity than large balls.

3. *Number of rows of rolling elements.* The load capacity is proportional to the number of rows of elements. For self-aligning bearings, the number is limited to two. In all other types of bearings, the value is predicated on the ability to obtain perfectly matched elements so that the race rests on all adjacent balls or rollers simultaneously.

4. *Bearing material.* The elastic property of the metal used governs the surface deformation and its effect as described in Chapter 4. It involves stress frequency imposed on the race surfaces and its effect on the dynamic service life of rolling bearings discussed next.

5. *Design of bearing.* Design involves the frictional relationship between the rolling elements and adjacent parts such as grooves, flanges, separators, and deformed race surfaces. It also relates to the length of rollers and the kind of load, i.e., radial, thrust, or combined. The physical condition of a bearing is obviously a factor. For theoretical purposes, it must be assumed to be good or better which results in a parameter value of 1, so may be disregarded.

Of these factors, the number and size of rolling elements and the number of rows have definite values. Because the material used for both rolling elements and races are constructed of hardened steel (usually SAE 52100) those components are expected to react similarly in bearings of the same size and design, when products of high quality manufacture only are considered.

A given bearing design carries an inherent frictional characteristic with a close average consistency when individual bearings of the same design are compared. Changes in design, however, are subject to variations in frictional characteristics, so that the average consistency of one design may differ from that of another.

The effect of factors 4 and 5 would be difficult to calculate with any degree of accuracy because of the intricacies posed by the combination of surface properties, closeness of contact between sliding surfaces, skewing of rollers, variation in rolling element size, and other physical causes. One way to evaluate the effect of the intangibles is by thoroughly testing a number of each type of bearing. Such tests determine constants that may be applied to all bearings of similar design, regardless of size. They also can determine the specific load capacity for a given type. Therefore knowing the number of rolling elements, their size, and the number of rows, and then finding the specific capacity of the various types of bearings, the constant can be numerically determined. The relationship of specific capacity is therefore proportional to the combined effect of all intangibles as represented by the value of the empirical constant.

*SKF Industries, "Ball and Roller Bearings Engineering."

Combining all the above variables results in the following equations for ball bearings (single row).

$$C_s = kn^{2/3}D^2 \cos\alpha \qquad \text{(Hertzian theory)}$$

for roller bearings (single row)

$$C_s = kn^{2/3}D^2L \cos\alpha$$

where

C_s = specific load capacity (kg)
n = number of rolling elements
D = diameter of rolling element (mm)
L = length of rollers (mm) when $L/D \lesseqqgtr 1.4$
($L/D > 1.4$ subject to skewing, therefore unpredictable)
α = contact angle (Fig. 13-14).
k = constant for various type bearings (Table 13-1)
i = number of rows of rolling elements

It has been found that small balls have a greater load capacity than balls of larger diameter. To correct the discrepancy based on Hertz's theory, a correction factor has been devised by Palmgren to modify the above equation for ball bearings as follows:

$$C_s = k\frac{n^{2/3}D^2 \cos\alpha}{1 + 0.02D}$$

For bearings of more than one row, multiply the numerator by i.

Table 13-1. VALUE OF k FOR VARIOUS TYPES OF ROLLING BEARINGS BASED ON DIMENSIONS IN MILLIMETERS
(applies to material of Rockwell hardness 61 to 65 at <100°C)

Bearing Types	k
Double row self-aligning ball bearings	2.25
Single row deep groove ball bearings without filling slot	4.5
Single row angular contact bearings	4.5
Double row angular contact bearings	4
Radial roller bearings with short rolls ($L < 1.4D$)	5
Single row thrust ball bearings	6
Spherical roller thrust bearings	12.5

(A. Palmgren, SKF Industries)

For thrust ball bearings, single row,

$$C_s = k\frac{D^2n^{2/3}}{1 + 0.02D}$$

and for thrust roller bearings, single row,

$$C_s = kDLn^{2/3} \sin\alpha$$

Misalignment of a supported shaft is usually caused by faulty installation, but also by distortion of machine structures due to settling of foundations, excessive tightening of hold-down bolts, temperature changes, and deflection due to loading. The temperature variation can be countered by using metals with the required coefficient of expansion. Shaft deflection is usually minimized by (1) making them strong enough, (2) eliminating overloading, (3) spacing bearings so as to balance the load, and (4) avoiding overhang wherever possible.

Axial shaft motion due to temperature changes is present in all mechanical systems. To avoid damage to rolling bearings that support a shaft, the practice is to fix one bearing (called a held bearing) in permanent position leaving the rest free. A *free bearing* is one whose housing is sized to provide sufficient axial clearance to allow the shaft to expand and contract axially and therefore prevent binding. The choice of the *held bearing* lies with the one supporting the greatest thrust load.

ROLLING BEARING LIFE

The life of rolling bearings is based on probabilities. It is based on the probable life of a given number of identical bearings operating under the same load. Life is measured in terms of number of revolutions or hours. Bearing life may be expressed as follows:

Rating life is the number of revolutions that 90% of the group can complete or exceed under a given load.

Minimum life is the number of hours that 90% of the group can operate at a given speed under a given load.

Average life is the average number of hours that a group of bearings can operate at a given speed under a given load.

For ball bearings of the same group, the average life may be five times minimum life, and for roller bearings about three times.

The probable life of rolling bearings is a function of the *basic load rating,* or rated load for a given group. The basic load rating is the radial load that 90% of the group can be expected to endure during 1,000,000 revolutions.

The life ratings are developed by actual tests during which the bearings are operated without deleterious influences such as abrasives, water, abnormally high temperatures, and shock that tend toward premature failure. Under such influences, the life of a bearing is not predictable. The limit of endurance for any bearing is the result of some inherent failure that occurs due to the mechanical relationship of the individual parts to each other under dynamic load. The components reach the point of metal fatigue, which normally occurs gradually.

Metal Fatigue. The rolling elements and raceways, being subject to deformation under load, develop alternating stresses. Assuming the load is kept within the elastic limits of the contacting materials, no permanent deformation occurs, but if the stress alternations are repeated often enough, the internal structure of the material gradually weakens, resulting in incipient surface cracks followed by flaking of surface material. Carried far enough, the removal of a flake develops spall (breaking of the surface into many small thin fragments or flakes). Spalling leaves a concentrated rough surface that will spread and eventually destroy the rolling element and raceway. Spalling is the result of metal fatigue.

The life expectancy of a rolling bearing is based (1) on the magnitude of the load and (2) the frequency with which the stresses are imposed by the load, or the speed in revolutions per minute at which the bearing operates.

If during a series of tests in which a group of identical bearings are used, it were found that the average life was 3800 hr at 1000 rpm under a load of 2200 lb, these values would form the basis of rating *all* bearings of the same kind.

The general relation of life expectancy to load at constant speed is

$$\text{life} \propto \frac{1}{\text{load}^3}$$

The general relation of life to speed at constant load is

$$\text{life} \propto \frac{1}{\text{speed}}$$

The relationship of life expectancy, load, and speed of rolling bearings is expressed more definitely by the equation

$$C_r = \frac{C_s}{F_s}$$

where

C_r = rated capacity (lb) at a speed other than test, or manufacturer's listed speed
C_s = capacity (lb) based on test speed as listed by manufacturer
F_s = speed factor

or

$$F_s = \sqrt[n]{\frac{\text{selected speed, rpm}}{\text{test or listed speed, rpm}}}$$

or

F_s = reciprocal of the nth root of the speed ratio where

$$n = 3 \text{ to } 4, \text{ depending on the manufacturer}$$

For example, the rated capacity of a ball bearing having an average life of 3800 hr at 1000 rpm and 2200-lb load when operated at 5000 rpm under the same load would be, assuming n to have a value of 3.0,

$$C_r = \frac{C_s}{F_s}$$

$$C_r = \frac{2200}{\sqrt[3]{\frac{5000}{1000}}} = \frac{2200}{\sqrt[3]{5}} = \frac{2200}{1.71}$$

$$C_r = 1286 \text{ lb}$$

If allowed to operate under a load of 1600 lb at 5000 rpm, the expected life of the same bearing would be determined as follows:

$$L_a = \left(\frac{C_r}{C_a}\right)^n \times L_r$$

where

L_a = expected or actual life of the bearing in hours at operated speed and load

C_a = actual load

L_r = average life based on manufacturer's rating at standard speed and load

$$L_a = \left(\frac{1286}{1600}\right)^3 \times 3800 = 0.8 \times 3800$$

$$L_a = 3040$$

In determining loads, the effect of both radial and thrust forces must be considered. However, both forces combined are expressed as *equivalent radial load.* Combined loads produced by the simultaneous action of radial and thrust forces may be expressed as the square root of the sum of the square of the forces, or

$$\text{combined load} = \sqrt{\text{radial load}^2 + \text{thrust load}^2}$$

$$L_c = \sqrt{L_r^2 + L_t^2}$$

Loads include the weight of structural members and the forces set up due to power transmission, torque, centrifugal action, etc. Loads are calculated on the basis of all factors involved in the operation of equipment whose forces impose either a radial or a thrust load. Bevel and hypoid gears, worm gears, propellers, and the weight of equipment associated with vertical or angular shafting are some of the components of thrust loads. Their influence can be determined by proper formulas. Radial forces rely on the location of the bearings in relation to the points of applied load.

For pure thrust (axial load), the combined load is equal to the thrust load, or $L_c = L_t$. For pure radial loads (L_r), $L_c = L_r$. For combined loads, however, the equivalent radial load is equal to radial load plus the thrust load, one or both of which may be modified, by appropriate coefficients, according to the type of bearing, magnitude of shock, rotating race, i.e., whether the inner, the outer, or both races turn, and other considerations. The relationship may be expressed as

$$L_c = L_r + L_t \text{ for some bearings}$$

and

$$L_c = xL_r + yL_t \text{ for others}$$

where

$$x = \text{coefficient of radial loads}$$
$$y = \text{coefficient of thrust loads}$$

The values of x and y differ for different types of bearings and different manufacturers. For example, the equivalent radial load for a single-row radial ball bearing may be

$$L_c = L_r + L_t \text{ at low speeds}$$

and

$$L_c = L_r + 1.5\ L_t \text{ at high speeds}$$

and for a single-row, steep-angle taper roller bearing the equivalent radial load may be

$$L_c = 0.66\ L_r + 0.75\ L_t$$

etc.

ROTATING OUTER RACE. When the inner race remains stationary while the outer race rotates, the effect is to direct the load toward the smaller area of the inner race. This combined with the higher rotating speed of the rolling members as compared with their speed if the inner race traveled at a comparable rate, increases the internal stresses and the stress frequency. Thus, by rotating, the outer race reduces capacity 10 to 30%, depending on the type of bearing and its manufacturer. This means that bearings will have a carrying capacity of 70 to 90%, of their rated capacity when standard speeds apply to the outer race.

For example, a single-row light series ball bearing rated at 4500 lb at 1000 rpm of its inner race will have its carrying capacity reduced, assuming

Table 13-2. CAPACITY MODIFYING FACTORS FOR ROLLING BEARINGS DUE TO SHOCK LOADS

Smooth, steady load	1.0
Light shock load	1.5
Moderate shock load	2.0
Heavy shock load	2.5
Extreme shock load	3.0

a rotating outer race coefficient of 0.87, to 0.87 × 4500 = 3915 lb. Because the coefficient varies with duty series classification, the probable value for a medium series bearing of the above type would be 0.84, and for the heavy series, 0.82.

SHOCK LOADS. Rolling bearing data are based on smooth, steady load. Therefore, some compensation must be made for shock and to a degree corresponding to shock magnitude. Shock factors range from 1.0 for smooth, steady load to 3.0 for extreme shock loads, as shown in Table 13-2. To meet a normal 4500-lb load having moderate shocks, the bearing selected would be required to have a carrying capacity of 2 × 4500 = 9000 lb. If a bearing having a normal rated capacity of 4500 lb were selected for the foregoing requirement, its life would be reduced to one third its rated average life.

LIFE FACTOR. The ratio of rated capacity at a given speed to actual load is called the life factor of a rolling bearing. The rated capacity at the desired speed should be modified to allow for all deteriorating influences. Table 13-3 lists the various influences affecting bearing life, including those already discussed, along with their probable range of magnitude (f).

The life factor for a given bearing may be expressed by the equation

$$\text{life factor} = \frac{\text{rated capacity at operating speed} \times f}{\text{actual load}}$$

$$= \frac{C_r \times f}{C_a}$$

Table 13-3. ROLLING BEARING LIFE FACTORS

Condition	*Range of F*
Shock loads:	
Extreme	0.3–0.5
Heavy	0.4–0.6
Moderate	0.5–0.7
Light shock (or vibration)	0.85
Temperature of operation:	
High (above 250°F)	0.6–0.75
Moderate (180 to 250°F)	0.7–0.9
Normal (below 180°F)	0.9–1.0
Rotating outer race	0.7–0.9
Rotating load	0.7–0.9
Oscillating (high degree)	0.75 (av.)

and the number of hours of expected life under conditions expressed by the life factor may be determined by the equation

$$\text{Service hours} = (\text{life factor})\, n \times K$$

where

n = 3.0 to 4.0 (according to the manufacturer)
K = number of hours on which manufacturer's ratings are based

ROLLING BEARING STEELS

Steels selected for rolling bearings must be hard and resist wear. Alloying materials include chromium, nickel, manganese, molybdenum, silicon, sulfur, and phosphorus. Tungsten and vanadium are also used in some bearing alloys. Carbon content ranges from 0.10 to 0.25% in low carbon alloys and from 0.85 to 1.10% in high carbon steels.

Chromium is beneficial in giving deep and uniform hardness penetration as well as imparting good wear and fatigue resistance to steel. The effect of chromium is to obtain *through hardness,* i.e., penetration of hardness from surface to core center. The alternative is to obtain high surface hardness with tough centers by carburizing (case hardening). While it generally follows that the carrying capacity of a bearing increases with greater hardness, fineness of grain and depth of hardness are important qualities in preventing surface failures. Sulfur and phosphorus are present in very small amounts ranging from 0.020 to 0.05% maximum for nearly all bearing steels.

The popular alloy known as SAE 52100 possesses high antifriction bearing qualifications. The specifications of this alloy are in the order as shown below:

SAE 52100

	Percentage
Carbon	0.95–1.10
Chromium	1.30–1.60
Iron	87+
Manganese	0.25–0.45
Silicon	0.20–0.35
Phosphorus	0.025 (max)
Sulfur	0.025 (max)

The latter steel and other alloys having slight modifications in contents as given above can all be heat-treated to adequate hardness to produce antifriction bearings of high load-carrying capacity.

To provide corrosion-resistant bearings, stainless steels containing 13 and 17% chromium are used. Carbon content ranges from 0.25 to 0.35%

and 0.95 to 1.05%, respectively. However, unless molybdenum is added in amounts approximating 0.50%, a hardness equal to that of properly heat-treated SAE 52100 (about Rockwell C-62) cannot be approached.

Hardness of Metals

The method of measuring metal hardness involves the use of an indentor in the form of a sphere or a cone forced into the surface of a test specimen under some arbitrary load. The hardness of the material tested is based on the depth of penetration or the diameter of indentation, depending on the test machine used.

Indentation tests include the Brinell, Rockwell B, Rockwell C, Vickers, and Shore Scleroscope methods, of which the Brinell and the Rockwell machines are the more widely used. For nonferrous metals and soft steels, the Rockwell B method is used. The hardness of nonferrous metals is expressed also in terms of Brinell number.

A conical diamond called a "Brale" is used as the indentor in the Rockwell C method of hardness testing. An initial penetration is gained by the cone tip upon application of a minor load of 10 kg. At this point in the test the depth-measuring scale is adjusted to an arbitrary value of 100. A major load of 150 kg is then applied under which greater penetration of the diamond cone into the test material is gained. The hardness number is read directly from the depth-measuring scale when the major load is removed and the force is returned to the original 10 kg. The scale is designed to register low penetration numbers for soft metals and progressively higher numbers with increased hardness. Rockwell C scale ranges from 15 to 70, the latter value representing the hardest material.

The Rockwell B method is similar to Rockwell C except that in the former a $\frac{1}{16}$-in. hardened steel ball is used under a major load of 100 kg. The minor load of 10 kg remains the same, but the depth-measuring scale is adjusted to an arbitrary value of 130 under Rockwell B procedure. Rockwell B numbers range from 0 to 100.

Brinell hardness is also determined by indentation. A hardened steel ball of 10 mm diameter is used. When steels are tested, a steady load of 3000 kg is applied to the ball. For softer metals, a steady load of 500 kg is applied, and 100 kg for very soft materials. The Brinell hardness number is determined by measuring the diameter of indentation and applying the equation

$$\text{Brinell number} = \frac{2P}{\pi D}\left(d - \sqrt{D^2 - d^2}\right)$$

where

P = load (kg)
D = diameter of ball (minimum)
d = diameter of indentation (minimum)

In the case of steels, the 3000-kg load should be held steady for at least 10 sec. For softer materials, the 500-kg load should be held for a longer period of 30 sec or more. Brinell numbers range up to 700. Above 500, however, an especially hardened steel ball should be used.

The relation of Brinell number (3000 kg) to Rockwell C hardness is about 10:1 up to about Rockwell C—60. Above this value the ratio increases from about 10.5:1 to about 11:1. The increase is due to the flattening tendency of the Brinell ball by harder test steels. Therefore a material having a Brinell number of 500 would have a value comparable to Rockwell C—50, approximately.

Steel Identification. In order to identify the various steels, a numerical index has been adopted. Classified according to numbers, adopted by the Society of Automotive Engineers, each number in the index represents a different type steel, as shown in Table 13-4.

Table 13-4. ALLOY NUMBERS FOR IDENTIFYING (SAE) STEELS

Carbon steel	1
Nickel steel	2
Nickel-chromium steel	3
Molybdenum steel	4
Chromium steel	5
Chromium-vanadium steel	6
Tungsten steel	7
Silico-manganese steel	9

The numbers shown in Table 13-4 are the first figures of any SAE steel classification and represent the basic alloying element. The second number indicates the approximate *amount* of the basic alloying element. The third, fourth, and fifth (if any) numbers indicate the approximate amount of carbon in the steel.

EXAMPLES: SAE 1020 steel is a carbon steel, as indicated by the first numeral 1. It contains 0.15 to 0.25% carbon, as indicated by the third and fourth numerals 2 and 0.

SAE 2115 is a nickel steel, as indicated by the first numeral 2. It contains 1.25 to 1.75% nickel (i.e., over 1.0%), as indicated by the second numeral 1. The third and fourth numerals 1 and 5 indicate a carbon content of 0.10 to 0.20%.

SAE 52100 is a chromium steel, according to the first numeral 5, of which there is approximately 1.5% (actually 1.3 to 1.6%) indicated by the second numeral 2. It contains about 1.0% carbon, denoted by the third and fourth numerals 1 and 0.

SAE 6140 is a chrome-vanadium steel containing about 0.35 to 1.10% chromium and 0.35 to 0.40% carbon. The minimum vanadium present in chrome-vanadium steel is 0.15%.

LOADING

RADIAL LOADS. Radial loads are the forces acting on a bearing in a direction perpendicular to the axis of the journal. The weight of machine parts which a bearing supports plus the external work forces acting through the various parts represent the total bearing load. Usually a journal is supported by two or more bearings. The amount of the total radial load carried by each bearing depends on their relative position in respect to the load.

If the load is exerted at a point halfway or is equally distributed on the journal between two bearings, the weight will be divided equally and each bearing will carry the same or one half the load. Figure 13-12 shows a shaft carrying loads that are not equally distributed or that are placed at any location relative to the bearings.

In Fig. 13-12 (*top*), the load on each of the bearings is determined by the equations

$$\text{load, bearing } A = \frac{Wy}{x + y}$$

$$\text{load, bearing } B = \frac{Wx}{x + y}$$

In Fig. 13-12 (*bottom*),

$$\text{load, bearing } C = \frac{Wy}{x}$$

$$\text{load, bearing } D = \frac{W(x + y)}{x}$$

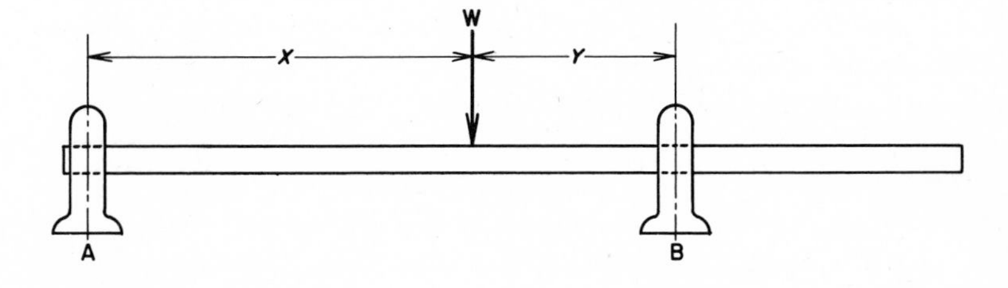

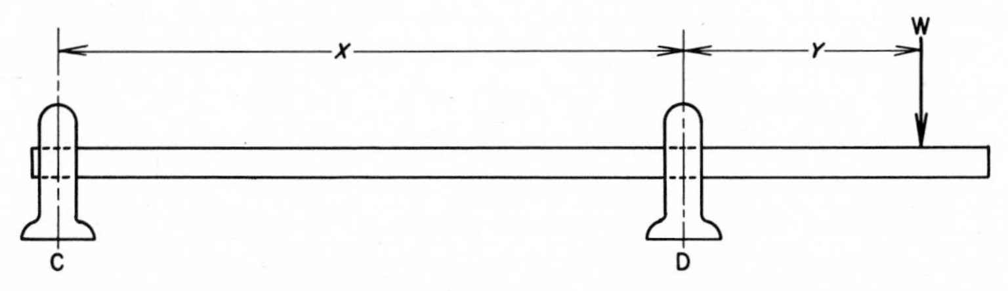

Fig. 13-12. Effect of bearing location on individual bearing load.

EXAMPLE:

1. If W equals 1000 lb, then the loads carried by A and B, respectively, when $x = 3$ ft and $y = 2$ ft, are

$$\text{load on } A = \frac{1000 \times 2}{3 + 2} = \frac{2000}{5} = 400 \text{ lb}$$

$$\text{load on } B = \frac{1000 \times 3}{3 + 2} = \frac{3000}{5} = 600 \text{ lb}$$

2. If W equals 1000 lb, then the loads carried by C and D, respectively, when $x = 4$ ft and $y = 1\frac{1}{2}$ ft are

$$\text{load on } C = \frac{1000 \times 1.5}{4} = 375 \text{ lb (load upward)}$$

$$\text{load on } D = \frac{1000\ (4 + 1.5)}{4} = 1375 \text{ lb (load downward)}$$

LOAD CAPACITY VS NUMBER OF ROWS OF ROLLING ELEMENTS. The load capacity of a rolling bearing increases with the number of rows of balls or rollers ie., a double row bearing has twice the load capacity as a single row bearing. Because of (1) the difficulty in obtaining perfectly matched rolling elements (i.e., diametrical precision) and (2) the deflection of the shaft tending toward load concentration alternately on one row or the other, the above rule applies only to double row self-aligning bearings.

THRUST LOADS. Axial loads may be supported as a combined load of a basically radial type bearing or by bearings designed to support thrust only. The latter type are represented by the ball thrust and by the cylindrical, tapered, and spherical roller thrust bearings. An angular contact ball bearing, will support up to 40% thrust of a combined load depending on angularity (thrust capacity increasing as the contact angle increases). Deep grooved ball bearings will also carry combined loads. Tapered roller bearings are made with roller contact angles up to 30 degrees and made to support heavy thrust loads although those of lesser angles (of about 16 degrees) will carry moderate thrust in one direction as well as radial loads. Single row bearings (hour glass) are suited only for very low axial loads. Double row spherical roller bearings are capable of supporting high thrust because of their angular attitude relative to axial loads. The specific capacity of ball and spherical thrust bearings can be determined by the equation given previously using the appropriate constant given in Table 13-2.

LOAD DISTRIBUTION in rolling bearings is portrayed for radial loads in Fig. 13-13. With a load in the direction shown by the line W_2 in the figure, the rolling elements are identified according to ascending or descending pressures exerted on each ball. The lower rolling elements designated in order from P_1 to P_5 would be subject to pressures distributed in some relative magnitude according to their angularity α with the load line. As the pressure

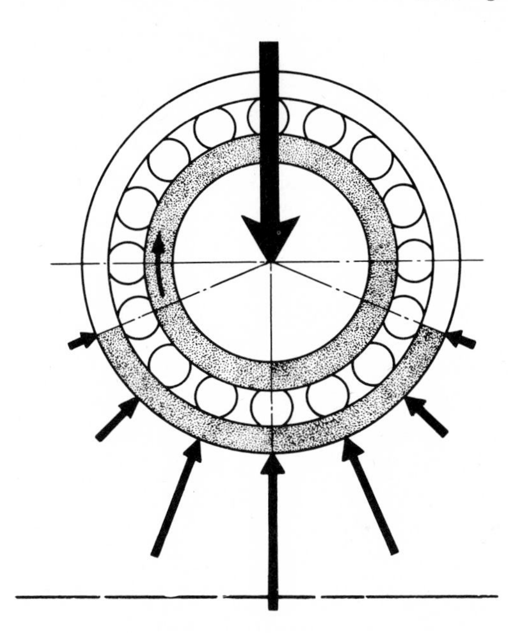

Fig. 13-13. Load distribution within a rolling bearing. *Courtesy, SKF Industries, Inc.*

on each increases as their angularity α approaches zero, the load proportionality is determined by the Strebeck formula

$$Q_{\max} = 5\frac{W_r}{n}$$

where

$$W_r = \text{radial load}$$
$$n = \text{number of rolling elements}$$
$$Q_{\max} = \text{maximum load per ball}$$

Tests show that, for the rolling elements under stress in the pressure zone, the proportionality of load on rolling elements at a location other than $\alpha = 0$ varies directly with the constant in the formula

$$\frac{nQ_{\max}}{W_r} = 1 \text{ at 55 degrees} \quad \text{(approx.)}$$

$$\frac{nQ_{\max}}{W_r} = 2 \text{ at } 47\tfrac{1}{2} \text{ degrees} \quad \text{(approx.)}$$

$$\frac{nQ_{\max}}{W_r} = 3 \text{ at } 42\tfrac{1}{2} \text{ degrees} \quad \text{(approx.)}$$

$$\frac{nQ_{\max}}{W_r} = 4 \text{ at 30 degrees}$$

$$\frac{nQ_{\max}}{W_r} = 5 \text{ at 0 degrees}$$

Angularity greater than 62½ degrees is shown to be $nQ_{max}/W_r = 0$ and maximum when $nQ_{max}/W_r = 5$. The pressure pattern for a rolling bearing under W_r load is shown in Fig. 13-14.

PRELOADING. Applying a force to certain parts of a rolling bearing prior to its introduction to a work load is called preloading. Its purpose is to eliminate or reduce the amount of play or movement of internal parts that are normally free to move some limited distance in one or more directions, such as balls and rollers. The amount of preloading may vary in stress magnitudes as to affect any range of clearance reduction between adjacent components up to maximum tightness of fit. The preloading effect is usually directed toward obtaining some contact pressure, prior to work loading, between the rolling elements and their raceways and/or shoulders. The contact pressure may be exerted radially, axially, or both simultaneously. Such an arrangement of parts would prevent any deflection or displacement of, say, a supported spindle of a machine tool designed for precision performance.

Preloading is applied to the bearing either during production or in place at the time of installation. The method of preloading depends on the type of bearing and the results desired. Unless accomplished by an experienced mechanic, irreparable damage to the bearing may occur. Excessive crowding of internal parts leaves no allowance for expansion resulting from operational heating and must be considered with preloaded bearings. It is only necessary to understand the factors that govern the life of any rolling bearing to realize that any magnitude of preloading leads to a corresponding reduc-

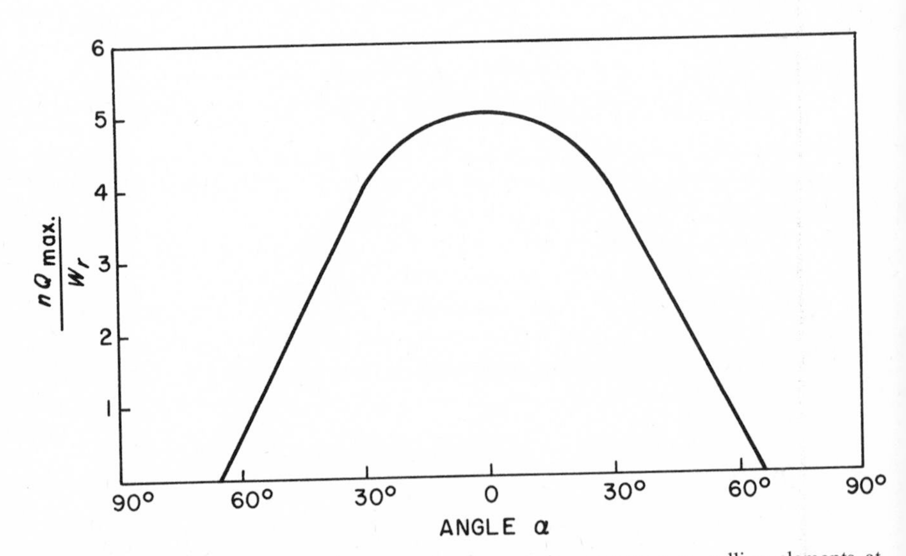

Fig. 13-14. Load proportionality curve showing relative pressures on rolling elements at various angles ∝. From A. Palmgren, *Ball and Roller Bearing Engineering,* SKF Industries, Inc.

tion in the time-service factor as computed according to the relationship

$$L_a = \left(\frac{C_r}{C_a}\right)^n \times L_r$$

where

L_a = actual life
C_r = rated capacity
$C_a = C_p$ (preload) + C_w (workload)
L_r = rating life

Preloaded bearings are produced by inserting balls having a larger diameter than the concentric space between the inner and outer raceways measured radially. This method furnishes a radial preloading which is also obtained by using a tapered bore that becomes stretched to increase radial diameter when pressed on a corresponding tapered shaft sleeve. Thrust preloading can be obtained by forcing duplex ball bearings, appropriately ground, against or away from each other by appropriate sleeve or spring arrangements. The tightening of end covers is one method of developing preload stresses.

Preloaded bearings tend to generate more thermal energy than normal clearance bearings. Thus it is essential not only to furnish a suitable lubricant to counteract the greater frictional load involved, but to design application systems that permit an adequate amount of oil to maintain maximum temperature reduction.

Moderate preloads on ball bearings may prove beneficial because of the reduction in deflection of the contacting surfaces accompanied by a diminished hysteresis reaction.

LUBRICATION OF ROLLING BEARINGS

As in all frictional systems, the lubricant of a rolling bearing serves more than one purpose, depending on the characteristics of the system: (a) reduce friction, (b) transfer heat, (c) prevent contamination (sealing), (d) flush contaminants from the bearing, and (e) resist corrosion.

TYPE OF LUBRICANT. The question is often asked "what is the best lubricant, an oil or a grease?" The choice lies with the circumstances surrounding the operation of the ball or roller bearing. The conditions to be considered are:

Speed	Ambient conditions
Size	Dust, dirt, etc.
Type of bearing	Moisture
Load	Acid, alkali, or neutral
Ambient temperature	Convenience of application
Induced heat	Torque

Dn VALUE. The shear rate of a lubricant is the governing factor in selecting a lubricant in consideration of speed. Speed in the case of rolling bearings is based on the velocity of the ball or roller as it travels on its race around the axis of the bearing. A criterion for velocity of any rolling bearing is the relationship of its diameter D to the number of revolutions n made by the supported shaft. This relationship is expressed by the formula

$$D \times n$$

The number becomes the criterion for determining the limits of a grease in meeting various speed ranges according to bearing size and type. The ranges overlap at their extremities, but they have been generally adopted for use as practical guides as to whether to use an oil or grease in rolling bearings.

For ball and cylindrical roller bearings—

Up to 50 mm bore

When $Dn < 300{,}000$ — the use of grease is practical
$Dn > 300{,}000$ — use oil

Over 50 mm bore

When $Dn < \dfrac{300{,}000}{\sqrt{D/50}}$ — the use of grease is practical

$Dn > \dfrac{300{,}000}{\sqrt{D/50}}$ — use oil

For tapered and spherical roller bearings of both radial and thrust types having a bore—

Of 50 mm and over

When $Dn < \dfrac{150{,}000}{\sqrt{D/50}}$ — the use of grease is practical

When $Dn > \dfrac{150{,}000}{\sqrt{D/50}}$ — use oil

Some rolling bearing manufacturers recommend a limiting range of 150,000 to 200,000 for grease in all types of rolling bearings except spherical roller bearings, for which the *Dn* should not exceed 100,000.

At ultrahigh speeds, the bearing parts become the critical components due to centrifugal force. Therefore special consideration must be given to the lubrication of rolling bearings operating beyond a *Dn* of 1,000,000, up to which point an oil is essential.

The operating *temperature* for any bearing is said to be normal between 40 and 180°F. Such temperatures are not critical for either the bearing or lubricant, but any excessive departure from this range requires special attention to the oil (Fig. 13-15) or grease.

VISCOSITY. The general oil viscosity recommendations for ball and roller bearings are based on operating temperatures as shown in Tables 13-5 and 13-6.

A ball bearing operating at a temperature of say 140°F, measured at housing surface, would have an oil temperature within the bearing 10° higher, or 150°F. The oil should have a viscosity at 150°F of 70 SUS. Therefore, an oil with a VI of 100, having a viscosity of 160 SUS at 100°F would be required. This would be in the physical property range of a high-quality SAE 10W motor oil, or 155 SUS at 100°F (light) turbine oil.

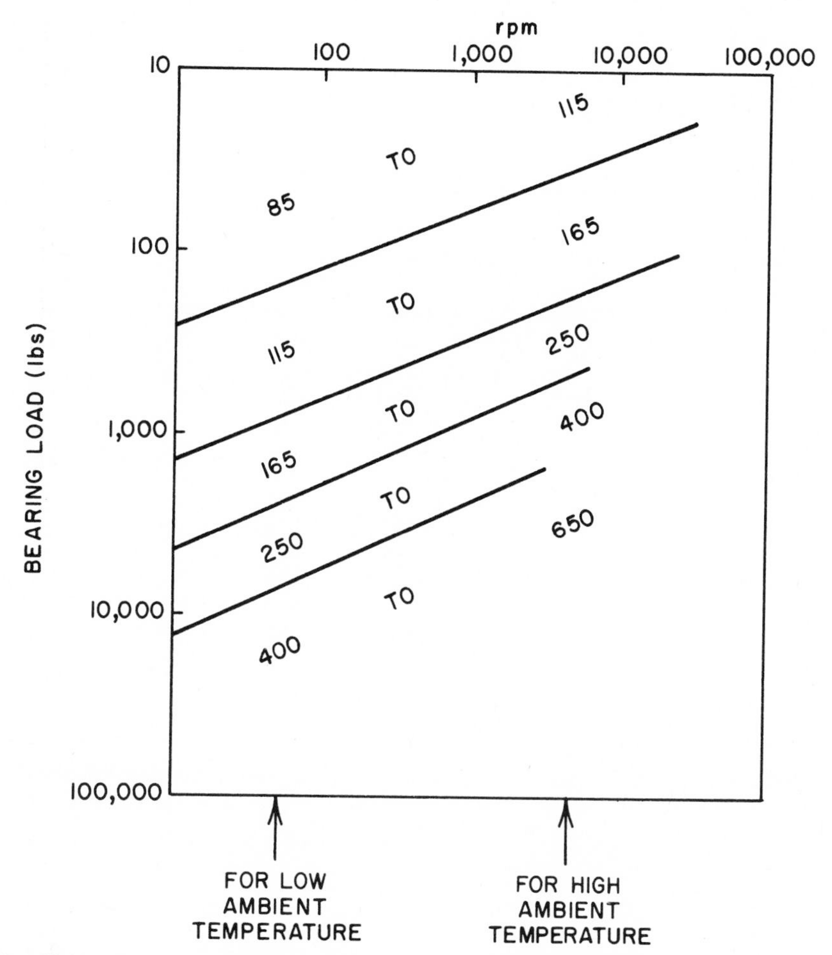

Fig. 13-15. Oil viscosity selection chart when load and speed are known. *Courtesy, SKF Industries, Inc.*

Table 13-5. RECOMMENDED OIL VISCOSITY FOR ROLLING BEARINGS BASED ON OPERATING TEMPERATURES

Type of Bearing	*Viscosity (SUS at Operating Temperature)*
Ball bearings	70
Cylindrical roller bearings	70
Spherical roller bearings	100
Spherical roller thrust bearings	150
Tyson tapered roller bearings	100
Heavily loaded spherical roller thrust bearings	
<20 rpm	1000
<5 rpm	5000

For any type bearing, the oil viscosity is governed by the speed and load. In considering elastohydrodynamic films, the relationship of speed, load, and viscosity must have favorable relationship in order to establish an effective minimum thickness in the areas of surface deformation.

A method of determining the elastohydrodynamic lubricating film parameter is by applying a formula expressing the relationship of all operating factors to bearing geometry:*

$$\wedge = H(\mu_0 \alpha N)^{0.7} P_0^{-0.09}$$

where

$\wedge$ = elastohydrodynamic parameter
H-factor = value pertaining to a given bearing geometry
μ_0 = dynamic viscosity in lb sec/in.^{2}
α = pressure coefficient of viscosity psi^{-1}
N = speed of rotation
P_0 = equivalent load

Table 13-6. RECOMMENDED OIL VISCOSITY FOR RADIALLY LOADED ROLLER BEARINGS WITH LIMITED THRUST, NORMAL LOADS

Operating Temperature	*Speed (rpm)*	*Viscosity (SUS at 100° F)*
Below 150°F	$<$ 750	300
	$>$ 750–3600	150
	3600	60–75
150–210°F	$<$ 750	600
	750–3600	300
	$>$3600	150
Over 210°F	750–3600	2500

Bearings subject to shock or heavy loads use leaded compounds of about 300 SUS at 100°F for each of above conditions, except at temperatures above 210°F, in which case a compounded cylinder oil should be used.

*SKF Industries, "A Guide to Better Bearing Lubrication."

The value of the H-factor is based on the bearing series and bore. The base of lubricant—paraffinic, naphthalenic, or synthetic—governs the value of $(\mu_0\alpha)^{0.7}$ for different viscosities. P_0 is the sum of the radial F_r and axial F_a forces modified according to individual bearings according to $P_0 = X_0F_r + Y_0F_a$ of which the values of X_0 and Y_0 are listed in manufacturers' catalogs.

The values of $\wedge$ fall between 1 and 4 for most rolling bearing applications. A value less than 1 portends a breakdown of the hydrodynamic film resulting in a reduction in life of the bearing. A value of 4 or above assures an uninterrupted effective elastohydrodynamic lubricating film that increases the rated life of the bearing by as much as twice the published value in some instances. The relation of $\wedge$ values to percentage of film is shown in Fig. 13-16.

Spherical roller bearings of both the radial and thrust types require an oil of higher viscosity than other kinds of rolling bearings because of the curvative difference between the end surface of the roller and its contacting shoulder. This variation develops a hydrodynamic film between the two surfaces. Heavily loaded spherical roller thrust bearings at speeds less than 20 rpm should be lubricated with oil having 950–1000 SUS at the operating temperature. At speeds less than 5 rpm use a 5000 SUS oil.

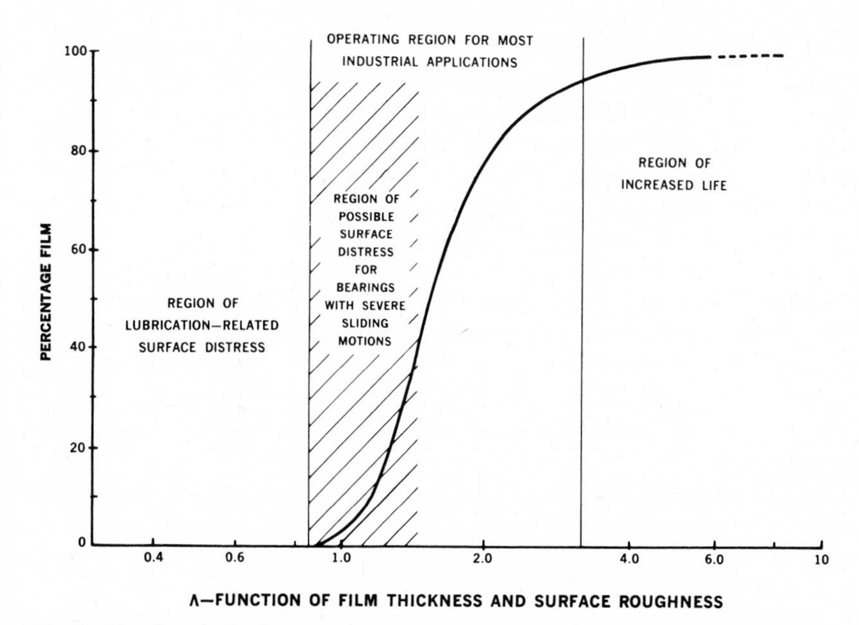

Fig. 13-16. Elastohydrodynamic parameter vs percentage film. *Courtesy, SKF Industries, Inc.*

Oil Application

Commensurate with speed and other factors, the method used to supply oil to a rolling bearing usually depends on convenience of application. Bearings that serve equipment remote from other frictional components, such as those of an independent fan or pump, require a self-contained method of application. Bearings that are part of an integrated mechanical system, as are found in internal combustion engines and machine tools, usually depend on a flow of oil from a central source some distance away from all frictional parts.

Self-contained oil application devices include such all-loss devices as drop-feed and wick-feed lubricators that apply oil directly to the rolling elements and raceways. Also self-contained are reuse devices such as the several types of constant-level oilers that are noncirculatory, providing a bath of oil in which the rolling elements are partially submerged during their passage through the area of maximum load. Constant-level devices include also the flinger type, which provides a spray of oil to the bearings. The flinger rotates with the supported shaft. For vertical shafts, the flinger dips directly into the oil bath, whereas the horizontal shaft flinger is fed by a wick leading to it from the reservoir. Self-contained reuse devices are not recommended for use in locations of high ambient temperatures unless they are cleaned and the oil changed regularly.

Liquid oil is also fed to the bearings via a central distributing system. The oil in such systems is continually recirculated through an endless system so that it is subjected to reuse indefinitely. The circulated oil should enter the bearing at the top of the bearing so it may cascade over the frictional parts. It must also be permitted to drain immediately from the bottom of the bearing so that a dry sump is formed. The dry sump arrangement is particularly essential if the bearings are subjected to elevated temperature as they are in paper machine dryer rolls.

Oil in its vaporous form is also an effective method of lubricating rolling bearings. This method of application is referred to as misting or fogging. The oil fog system is one of all-loss and capable of serving one or more bearings simultaneously and continually. It may be classified as a central system when applied to a bank of bearings served by individual conduits or located within a common case or housing in which an oil mist is maintained.

Oil bath lubrication is satisfactory for low and moderate speed application. Oil should never be higher than the center of the lowest ball in the bearing. The level should be checked regularly. Overfilling results in excessive operating temperature.

Wick-feed lubrication (Fig. 13-17) is one of the older methods used to apply oil to a rolling bearing and still remains effective and inexpensive.

Drip-feed oilers are also used effectively when they are located correctly relative to the rolling elements, as in Fig. 13-18.

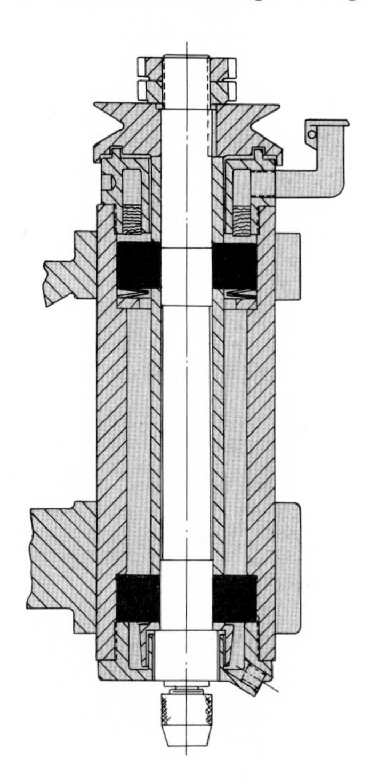

Fig. 13-17. Wick-feed oiler on rolling bearing. *Courtesy, The Fafnir Bearing Co.*

Splash oiling is usually the method chosen for gear shaft bearings in which the bearings and gears are both lubricated with the same lubricant. Figure 13-19 shows a method of leading oil from the case walls into the bearings when normal splash is marginal. Where oil splash is copious, bearings should be equipped with shields to prevent overheating.

Oil jet lubrication (Fig. 13-20) is applied to heavily loaded and high-

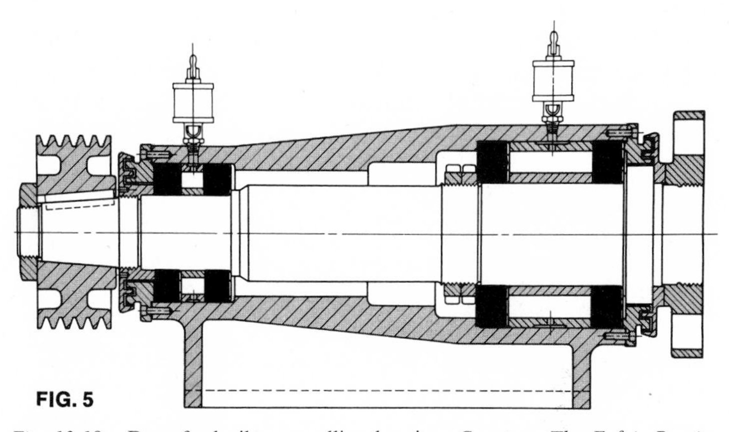

Fig. 13-18. Drop-feed oiler on rolling bearing. *Courtesy, The Fafnir Bearing Co.*

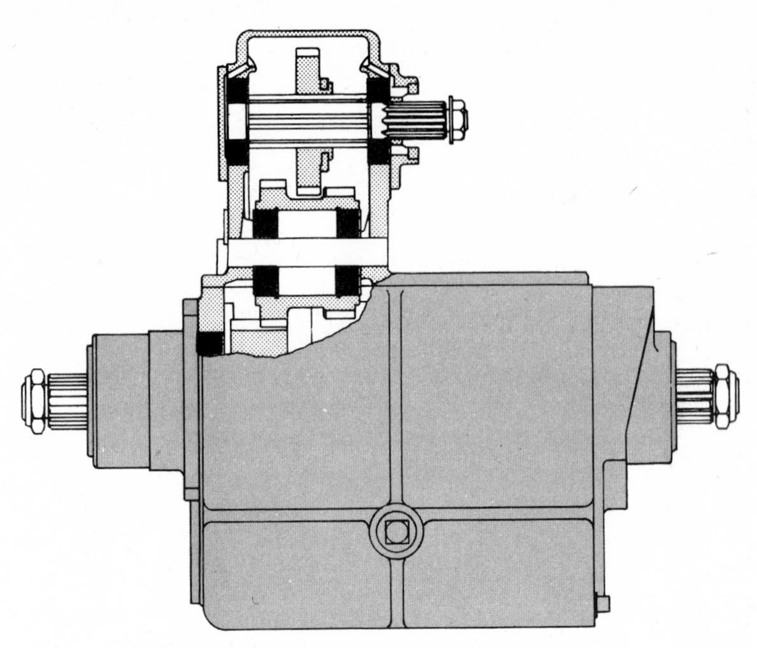

Fig. 13-19. Splash oiling of gear shaft rolling bearings from gear reservoir. *Courtesy, The Fafnir Bearing Co.*

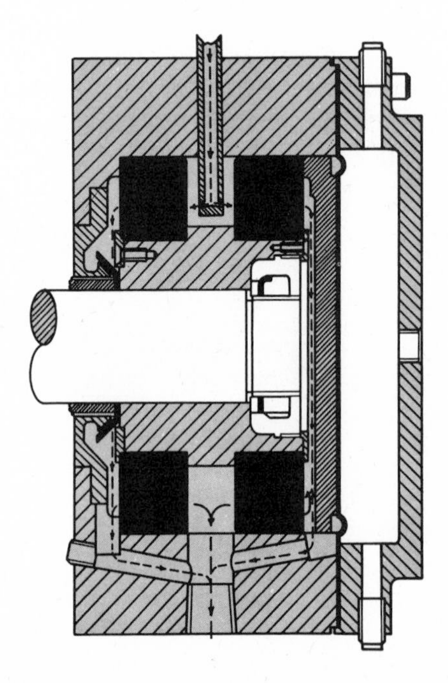

Fig. 13-20. Lubrication of rolling bearings by oil jet. *Courtesy, The Fafnir Bearing Co.*

speed rolling bearings. Oil is introduced through the small jet orifice at the top into the clearance between inner ring and retainer bore and leaves through larger scavenging ports at the bottom. The difference in size between the jet orifice and scavenging port prevents a buildup of oil in the bearing and excessive temperature rise.

Oil-fog lubrication is recommended for high-speed continuous operation of rolling bearings (Fig. 13-21). The success of this method is based on the proper location of oil-air entry ports, correct ratio of oil quantity to air pressure, and adequate exhaust of air-oil mist after lubrication.

All these oil application devices are discussed in detail in Chapter 11.

Rolling Bearings Lubricated With Grease

GREASE CLASSIFICATION

Group 1. *General Purpose Greases.* Greases that are expected to give proper lubrication to bearings whose operating temperatures may vary from −40 to 250°F.

Group 2. *High-Temperature Greases.* Greases that are expected to give proper lubrication to bearings whose operating temperatures may vary from 0°F to 300°F.

Group 3. *Medium-Temperature Greases.* Greases that are expected to give proper lubrication to bearings whose operating temperatures may vary from 32 to 200°F.

Group 4. *Low-Temperature Greases.* Greases that are expected to give proper lubrication to a bearing whose operating temperature may go as low as −67°F or as high as 225°F.

Group 5. *Extreme-High-Temperature Greases.* Greases that are expected to give proper lubrication for comparatively short periods of time where bearing operating temperature may be as high as 450°F.

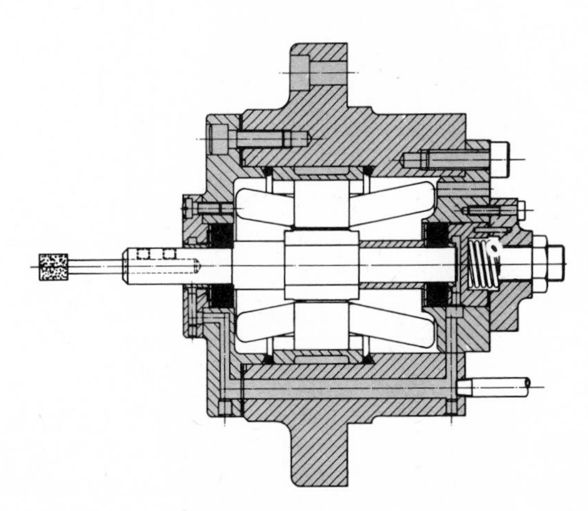

Fig. 13-21. Lubrication of rolling bearings by oil fog or mist. *Courtesy, The Fafnir Bearing Co.*

The requirements of these five groups of greases cover applications where the lubrication is not affected by extremely heavy loads, high speeds, or excessive humidity. The tabulation of Test Requirements, Table 13-7, is intended to establish for the above five groups of greases those characteristics needed to meet the functional requirements of good rolling bearing greases.

For all five groups in Table 13-7, the maximum number of dirt particles per cubic centimeter according to particle size is—over 75 microns, none; 50–75 microns, 50; 20–50 microns, 2000; and 5–20 microns, 5000.

The consistency stability for all groups is measured by the maximum ASTM penetration after 100,000 strokes, which should never increase more than 100 points. For Groups 1, 4, and 5, penetration should never be more than 375; for Groups 2 and 3, never more than 350.

SERVICE CATEGORIES. After determining the feasibility of using a grease in a rolling bearing, a decision must be made as to what grease to use. This will depend on the type of service to which the bearing is exposed. Service categories include the following:

1. *General service.* Ambient conditions normal—atmospheric pressure, temperature between 32 and 200°F, speeds up to 3000 rpm, moderate unit loads, absence of water, and relative absence of radioactivity. A typical example is the electric motor operating a machine tool.

2. *High-temperature service.* An ambient and/or process temperature from 200 to 275°F. Otherwise all other conditions associated with general service prevail. Typical examples are tenting frames used in the textile industry and oven car wheels.

3. *High-speed service.* Speeds that remain within the *Dn* limitations for a grease (0–300,000). Otherwise general service conditions apply. A typical example is the high-speed machine tool spindle.

4. *Subatmospheric pressure service.* Marked by a high evaporation rate. Examples include vacuum chambers, high-flying aircraft and space vehicles.

5. *Low-temperature service.* Ambient and/or process temperatures less than 32°. Typical service includes any equipment subjected to subfreezing temperature including aircraft and food freezing systems.

6. *Extended life service.* Associated with light series bearings packed for bearing life. Typical services would include many hand power tools, home appliances, industrial electric motors, and spindles. The usual life expectancy of the grease is about 44 months, after which the bearing should be cleaned and new grease installed. It is often more economical to replace the old bearing with a new one.

7. *Wet service.* Bearings exposed to water. While proper sealing is essential, the grease used should be impervious to water. Some greases will emulsify to form a slurry which will protect the bearing for a limited period, but if excessive water is present, such grease will eventually wash away, leaving the bearing without proper lubrication. Typical examples are the bearings used on the wet end of paper machines.

Table 13-7. TEST REQUIREMENTS FOR ROLLING BEARING GREASES

Test	*Group 1*	*Group 2*	*Group 3*	*Group 4*	*Group 5*
ASTM penetration, normal worked*	250–350	200–300	220–300	260–320	250–310
Oxidation, maximum drop	10 lb in 500 hr	10 lb in 500 hr	10 lb in 500 hr	5 lb in 500 hr	5 lb in 100 hr at 250°F
Low-temperature torque	1 rev. in max. of 10 sec at −40°F	—	—	1 rev. in max. of 5 sec at −67°F	1 rev. in max. of 10 sec at −40°F
Water resistance, maximum loss	50%	—	50%	20%	50%
Dropping point, minimum	300°F	350°F	300°F	300°F	450°F
Evaporation	—	—	—	1.5% max. in 22 hr at 250°F	4.0% max. in 22 hr at 400°F

* National Lubricating Grease Institute Code: No. 0, 355–385; No. 1, 310–340; No. 2, 265–295; No. 3, 220–250; No. 4, 175–205. Courtesy of SKF Industries, Inc.

8. *Bearing storage.* To protect bearings from corrosion during storage they should be well coated with an anticorrosion material. For nonlubricated bearings, this is done by dipping them in a corrosion-proofing compound containing a preferential wetting agent capable of displacing moisture as well as any acid left on external surfaces during handling. If prelubricated, the grease used should be stable both chemically and physically. Chemical stability precludes early deterioration of the grease; physical stability gives resistance to oil separation or "bleeding." The loss of separated oil from the bearing by leakage can be retarded by means of creep barrier films, described in Chapter 7.

GREASE RECOMMENDATIONS. To meet all the requirements described for the various services outlined above a number of greases are required. General service, category 1, represents the major industrial grease requirement. Combining in one grease other special characteristics often reduces the number of lubricants required in many cases. Other special requirements are not given a specific category because they may be made a normal characteristic of one or more of the greases that represent the categories listed.

One important special requirement for greases used in central distribution systems having long feed conduits is *pumpability*. Combining this feature with the capability to withstand relatively high temperatures (above 250°F and below 500°F) and incorporating them in an otherwise general service grease results in a product possessing wide utility. To add the additional benefits of *high film strength*, chemical stability, and resistance to water, to the same grease would widen its service scope appreciably. Fortunately for all concerned, such improvements have been made in modern greases either by the use of additives or by modifying the molecular structure or both.

GREASE APPLICATION. Application of grease to rolling bearings may be accomplished by all conventional methods such as pressure guns, central station, or hand packing (Figs. 13-22 and 13-23). Regardless of the method used, the important factors are the care and regularity with which it is done. More rolling bearings are destroyed by overlubrication than by underlubrication. A survey of industrial plants showed that the frequency of application to grease-lubricated rolling bearings ranged from every week to once every eight months. It would have been interesting to have had the survey conclude with the relative number of bearing failures in each case as well as at those plants where frequencies numbered various intervals in between. The point is that the frequencies could be correct in all cases. For example, if the bearings were never filled beyond one-third to one-half of their grease capacity by volume, no harm would result in any case. Where rolling bearings are adequately sealed and a proper grease used, the interval between application is indefinite under normal operating conditions. The latter is the basis of packed-for-life lubrication.

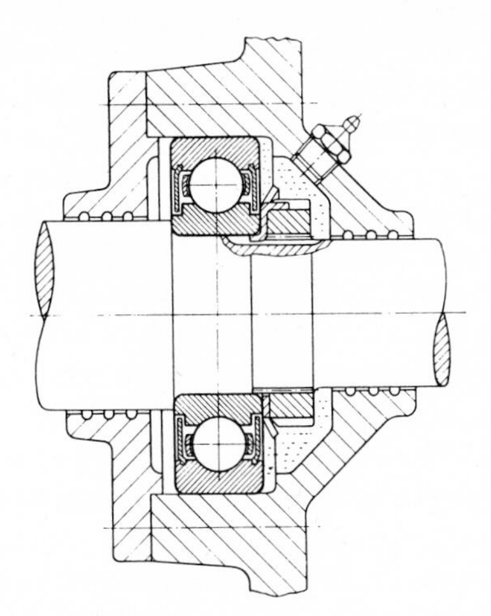

Fig. 13-22. Grease chamber for electric motor with fitting at top. *Courtesy, SKF Industries, Inc.*

There are several rules associated with the grease lubrication of rolling bearings:

1. Use the correct grease according to the type of rolling element and the service in which it is used. The oil used to construct the grease should have about the same viscosity as that if an oil alone were employed for lubrication.

2. Apply a grease of lowest practical consistency commensurate with temperature, speed, leakage, and position of the bearing. Lower consistency greases generally contain relatively more oil, and oil is the ideal lubricant for rolling bearings.

3. Never fill a rolling bearing to the full capacity of its cavity. Otherwise it will run hot and subsequent expansion of the grease will eventually break the seals.

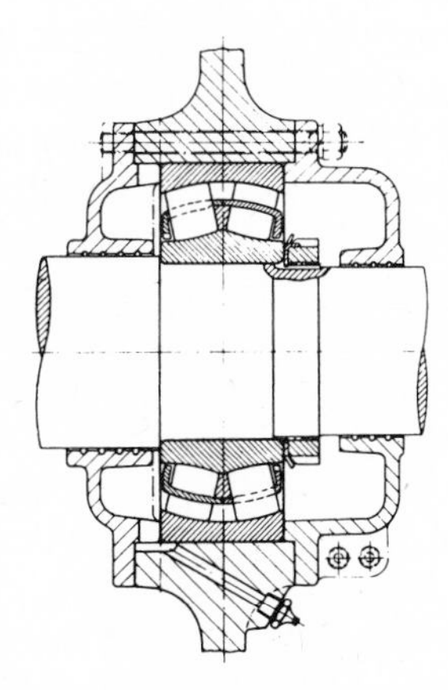

Fig. 13-23. Roller bearing arrangement for large electric motor with fitting at the bottom. *Courtesy, SKF Industries, Inc.*

4. Maintain a volume of grease in the bearing equal to approximately ⅓ to ½ its space capacity. Exceptions to this rule are low-speed bearings, for which the cavity may be nearly filled to capacity.

5. Lubricants should never act as an antidote for mechanical defects for extended periods, but they do have some influence in a number of instances under certain conditions. In this regard, grease offers some damping effect on low-frequency vibration to which rolling bearings are sometimes exposed.

6. Before attaching the nozzle of a pressure gun to a grease fitting, wipe the nozzle clean with a lintless cloth, preferably one dampened with an industrial cleaning solvent. The action is a precaution against forcing foreign matter into the bearing along with the new grease.

The question invariably arises as to how to know when to stop filling a rolling bearing because of the nontransparency of the housing. Except in the case of a newly hand-packed bearing, there is no one simple answer. One way of assuring against an overpacked bearing is to provide a pressure relief opening at the bottom of the housing, fitted with a threaded plug that is temporarily removed during each grease application. After application by, say, a pressure gun through a ball check fitting, the relief opening is allowed to remain open for a short period of bearing operation, to allow any excess grease to be forced out. The flow of grease will finally stop when the pressure in the housing has found its own "comfortable" level. The plug should then be replaced in the opening.

Depending on how much the bearing is exposed to contaminants and water during operation, the application interval should be planned so as to allow one complete change of grease in each bearing on a basis of rotation. In an extremely dirty location, this could mean every application, whereas every sixth application would be sufficient in a clean and dry environment. The change of grease is accomplished in the same manner described above, except that feeding of the grease continues until all the old grease is removed, evidenced by the appearance of new grease at the drain. It is not always necessary to continue feeding new grease until its starts its exit, since drainage of old grease will start when the housing pressure rises sufficiently. If drainage stops before all the old grease is evacuated, another spurt or two of the grease gun may complete the task. The point is not to waste new grease. The old grease can be used to lubricate any rough plain bearings or ways.

FAILURE OF ROLLING BEARINGS

Rolling bearings are subject to damage for various reasons. When damage occurs their complete failure is inevitable either immediately or at some interval thereafter, depending on the kind and extent of the damage.

The cause may be due to fatigue, neglect, mishandling, and exposure to deleterious influences that may or may not be foreseen. The factors leading to ultimate failure can be classified as either mechanical, chemical, or electrical.

Mechanical: abrasion, flaking, spalling, cracking, fracture, fatigue, brinelling, false brinelling, galling (smearing), seizure, wear, glazing (Figs. 13-24 through 13-30).

Chemical: corrosion (due to acids, bases, or salts), rust (corrosion due to moisture), fretting corrosion (Fig. 13-31).

Electrical: etching, pitting, fluting (Figs. 13-32 and 13-33).

Mechanical Damage

FLAKING. The breaking off of minute thin flakes from race surfaces caused by incipient cracks due to continuous deflection of the metal. Initial evidence of metal fatigue due either to overloading, load concentration, or service life limitation.

SPALLING. A rough area on the bearing surface resulting from continuous flaking (Fig. 13-34).

CRACKING. Occurs in depth of the surface material caused by excessive

Fig. 13-24. Advanced abrasive wear. *Courtesy, SKF Industries, Inc.*

Fig. 13-25. Cracks caused by faulty housing fit. *Courtesy, SKF Industries, Inc.*

strain as when the inner race is forced over an arbor (shaft, tapered sleeve, or adapter). Also caused by excessive preloading.

FRACTURE. An extension and aggravation of cracking. A condition of total failure.

BRINELLING. A peening effect. Permanent cavities in the race surface caused by indentation of rolling elements during shock or intermittant heavy

Fig. 13-26. Fatigue caused by impact damage during handling or mounting. *Courtesy, SKF Industries, Inc.*

Fig. 13-27. False brinelling caused by vibration with bearing stationary. *Courtesy, SKF Industries, Inc.*

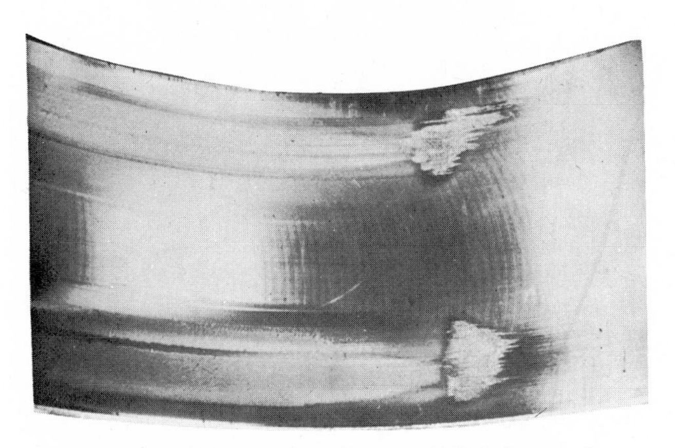

Fig. 13-28. Skid smearing. *Courtesy, SKF Industries, Inc.*

Fig. 13-29. Grooving caused by wear due to inadequate lubrication. *Courtesy, SKF Industries, Inc.*

Fig. 13-30. Glazing caused by inadequate lubrication. *Courtesy, SKF Industries, Inc.*

Fig. 13-31. Corrosion of roller surface caused by formation of acids with some moisture present. *Courtesy, SKF Industries, Inc.*

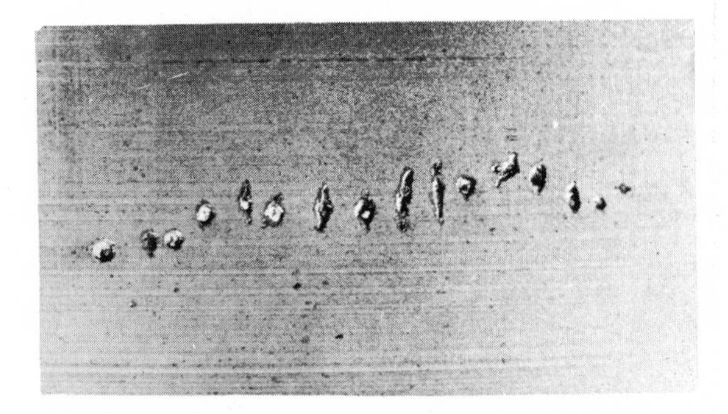

Fig. 13-32. Electric pitting on surface of spherical outer raceway caused by passage of large current. *Courtesy, SKF Industries, Inc.*

Fig. 13-33. Fluting on inner raceway of cylindrical roller bearings caused by prolonged passage of electric current. *Courtesy, SKF Industries, Inc.*

(a) (b) (c) (d)

Fig. 13-34—a, b, c, d. Progressive stages of spalling caused by inadequate lubrication. *Courtesy, SKF Industries, Inc.*

loading. Indentations usually respond in shape to that of the ball or roller when the surface material has been subjected to a force beyond its elastic limit.

FALSE BRINELLING is a corroding action often occuring between the surfaces of two contacting metal bodies which often *appear* to have no relative motion in regard to each other. It leaves an accumulation of powdery debris at the interfaces. In the case of ferrous metals the debris has been analyzed as iron oxide and iron powder. The iron oxide is principally Fe_2O_3 with traces or more of Fe_3O_4. Some tests reveal a substantial proportion of FeO. The debris has some distinct coloring of fairly bright red mixed with rust and brown. Aluminum oxides in the form of fine powder are usually white but are distinctly black when formed by fretting corrosion. Titanium is also subject to fretting.

When a fretted surface is gently cleaned with fine steel wool the affected surfaces appear to be of haphazard roughness with some pitting. In most cases the erosive-like deterioration occurs uniformly from one bearing to another.

Fretting damage occurs when (1) the surfaces of two metal bodies are in close contact, (2) contact is made with normal pressure, (3) some relative motion exists in the form of vibration at the interfaces, (4) the touching surfaces are clean, (5) the atmosphere is dry.

The slight vibration at the contact surfaces is translatory, not spacial. The action is one of slippage, with very little motion. The contact area is disturbed sufficiently to cause the asperities to break off, which not only exposes virgin surfaces but causes the debris to act as an abrasive to expose additional clean metal surfaces and generate localized heating and subsequent welding.

To reduce or eliminate fretting, (1) reduce or eliminate the vibration, (2) limit or remove the translatory slippage, (3) shield the joint against the intrusion of ambient gases that initiate corrosion processes.

The translatory or linear slip may be reduced by applying a coating over one or both surfaces. The purpose is (1) to increase the force of friction, which opposes motion, and (2) to provide a cushion in which the action is transfered to the molecular laminations of the cushioning material. Plating the surfaces with a material such as chrome would increase the friction. Bonding a solid lubricant such as molybdenum disulphide or lead iodide to the contact surfaces provides the cushion.

In chrome plating, the surfaces are ground to about 0.002 in., undersize; this is followed by plating to about 0.003, oversize, and finish by grinding to required size.

Shielding helps but is not a final solution. Smothering the joint in oil or grease helps to exclude air and oxygen, but many greases contain water as a bonding agent, and oils are hygroscopic. Under ordinary condition neither of the latter would be critical, but under the system supporting fretting the corrosion process continues in spite of them even if at a reduced rate.

Interposing a thin shim of PTFE between vapor-blasted surfaces has eliminated fretting activity to a large degree.

Ball and roller bearings, static bearing supports of machine tools, flanges used in various enterprises such as aircraft, butt joints of structural members, face plates of work roll bearing housings (cold rolling mills) and similar joints and pairs are susceptible to fretting corrosion and ultimately fretting fatigue.

GALLING. A blemish caused by rubbing that often leads to *seizure*. Also known as *smearing*. Usually caused in rolling bearings by failure of the rolling elements to rotate, at which time they slide along on their races. It is also caused by any sliding motion of the rolling elements such as swirling or spiraling, during which they roll in one direction and slide in some other diagonal direction. Galling may be axial, circumferential, or cycloidal in attitude. Under normal operating circumstances, the cause of galling may be lack of lubricant or an excessive amount of high-consistency grease.

SEIZURE. The bonding of metal particles from opposing rubbing surfaces due to high local temperatures. It generally results in the displacement of the surface area involved. Seizure develops from galling or rapid wear and often leads to the displacement of large surface areas. It is more prevalent in sliding bearings, however.

WEAR. Not ordinarily associated with ball and roller bearings. A polishing effect does occur on the bore surface when, due to looseness of fit, the inner race stick-slips on its supported shaft. Slip increases as the bore wears to greater diametrical dimension.

MISCELLANEOUS CAUSES OF FAILURE. The markings left on bearing surfaces by hard foreign materials vary considerably. Caught between the rolling elements and their raceways, a particle of sufficient hardness will develop indentations or random shaped blemishes. A broken cage or any particle of it will severely damage a bearing and make it unserviceable. The cause of damage is obvious upon inspection. The characteristic surface markings resulting from all causes should be recognizable in order to prevent further occurrence.

Chemical Damage

CORROSION. Displacement of metal by the reaction of chemical compounds such as acids, alkalines, and salts with exposed surfaces. Any liquid or gas having a high or low pH value is injurious to certain metals, particularly iron and most steels. Many nonferrous metals are also susceptible to attack. Depending on the wetting ability of protective surface coatings, the nature of metal damage may be erratic, or in a form of etching. Carbon dioxide in the atmosphere when dissolved in water will react to produce carbonic acid that will attack metal surfaces.

RUST. A form of corrosion that occurs in ferrous metals when moisture is present. The reaction of ferrous ions with the hydroxyl ions in the water produces ferrous hydroxide, $Fe(OH)_2$, which reduces to a reddish brown residue (rust) that eventually turns to a fine powder that covers the affected area. Rust first appears as spore-like mounds that leave pits when the residue is removed.

Electrical Damage

Eddy currents from a main circuit and those developed statically may enter the frame of any machine and short circuit to the most available ground. If a rolling bearing is located in the short circuit, the spaces between roller and race provide gaps across which the current must jump. The gap is bridged by a film of lubricant of low dielectric strength that enables the current to flow through it, providing the difference in potential between the roller and race is great enough.

Oil is hygroscopic and takes in moisture when exposed to the atmosphere, and most greases contain water as a bonding agent as well as oil, so that neither offers much resistance to current flow across the distance involved. The sparks or arcs so generated heat the spots where they occur, burning a crater in the race surface. A continuous repetition of the arcing process results in a surface of many craters and a noisy bearing that ultimately fails. Proper grounding helps to eliminate this condition.

14

GEARS

The purpose of a gear is to transmit power and motion from a driveshaft to a driven shaft. Gears are one of the basic frictional components of a machine, ranking with bearings and cylinders. They are made in all sizes, from those used in fine instruments to those of large mechanical presses and coal pulverizers that weigh several tons.

TYPES OF GEARS

Gears can be classified according to the relative positions of their shafts:

(a) Spur and herringbone gears transmit power between parallel (nonintersecting, noncrossing) shafts.
(b) Bevel gears (straight tooth and spiral tooth) transmit power between intersecting shafts at any angle relative to each other. However, right-angle drives predominate.
(c) Worm and hypoid gears transmit power between nonintersecting shafts that are at right angles to each other (their axes cross).
(d) Helical gears are designed to transmit small loads between nonintersecting shafts whose axes cross each other at any angle.

A BULL GEAR is the main driving gear of a machine, usually the largest and strongest.

A PINION is a gear with a small number of teeth that usually engages a larger gear.

SPUR GEARS have their teeth cut straight across the face of the gear blank (Fig. 14-1). The profile of spur gear teeth takes the form of an involute,

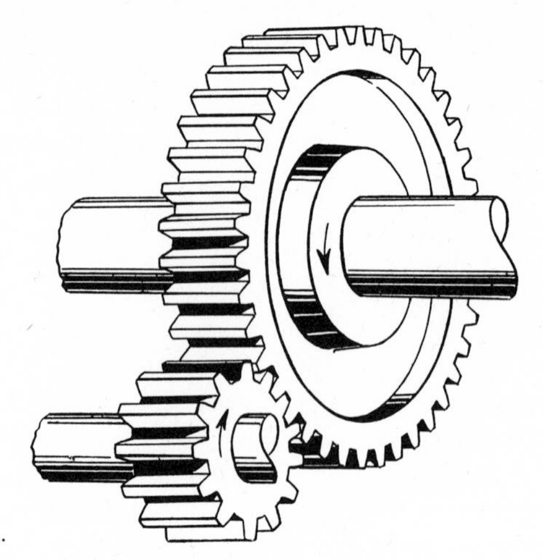

Fig. 14-1. Spur gear and pinion.

scribed by a fixed point on a cord as it unwinds from around a given (base) circle. Spur gears have theoretical line contact straight across the face of meshing teeth. Deformation under load, however, develops area contact. As the teeth pass through mesh, the area of contact sweeps over the working surface of each tooth. This in effect develops a series of axial lines across the working surfaces of meshing teeth (Fig. 14-2a).

As the teeth start to mesh, radial sliding occurs at the initial line of contact, which is at the top of the driven tooth and at the bottom of the driving tooth. As meshing continues, the line sweeping upward on one tooth and downward on the other, sliding diminishes until a point immediately below and approaching the pitch line is reached. At this point, the meshing teeth start to roll over each other and continue the rolling action through the center of mesh (at the pitch line) to a point slightly beyond and above the pitch line. Sliding again commences, but in the reverse direction, increasing in magnitude until the rotation of the gear causes the teeth to break contact. Contact is broken at the tip of the driving tooth and at the root of the driven tooth.

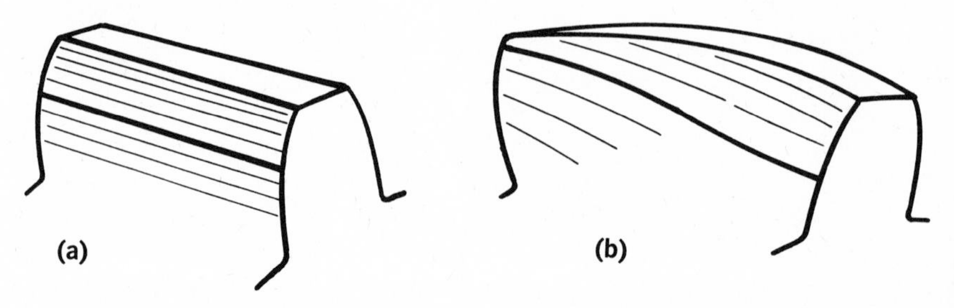

Fig. 14-2. Line of contact between gears in mesh; (a) spur gear, (b) helical gear.

Both sliding and rolling action occur when spur gears are in mesh and operating (Table 14-1).

HERRINGBONE GEARS (double helical gears) have similar frictional characteristics as described for spur gears; i.e., they have both sliding and rolling friction (Fig. 14-3). However, instead of the straight contact line across the tooth face, as with spur gears, the helical design causes contact to be made along a diagonal line (Fig. 14-2b). In other words, the leading end of a helical tooth passes through a portion of mesh before the trailing end enters mesh. This diagonal or slanted line, which becomes steeper as the helix angle becomes greater, crosses the pitch line at the moment the gears reach the center of mesh. This means that the line of contact is shortest (infinitesimal) at the instant mesh commences, becoming increasingly longer as it sweeps upward and downward over the driving and driven teeth, respectively. The length of the diagonal contact line reaches maximum, i.e., across the entire face of the tooth, when the gears are in full mesh. As motion continues, the line of contact diminishes in length until it reaches zero at the instant contact is broken.

Because of the helix angle, herringbone gears are virtually continuous step gears. Step gears are made up of a number of gears placed side by side on the same shaft, each tooth of which is advanced slightly relative to its preceding neighbor to give a steplike appearance, rather than the straight-line effect of the spur gear. The effect of the helix angle allows

Fig. 14-3. Single reduction gear set with double helical gears. *Courtesy, The Falk Corporation, subsidiary of Sundstrand Corp.*

several teeth to be in mesh simultaneously, which distributes load over a greater area and promotes smooth operation. Therefore, the small areas of contact developed during initial stages of approach to and final stages of departure from full mesh are not subjected to any momentary pyramiding of unit pressure.

The helix angle of single helical gears may range up to 45 degrees, imposing axial thrust along the shafts. This load must be carried by the affected bearing. By using two opposing helical gears side by side on the same shaft, the thrust from one counteracts the thrust of the other. The herringbone gear is actually two opposing helical gears cut on the same gear blank, which gives the former a chevron-like or herringbone appearance.

BEVEL GEARS are cone shaped and are designed with either straight (Fig. 14-4) or spiral teeth (Fig. 14-5). Line contact prevails in both types and follows a pattern similar to that described for spur and herringbone gears, respectively. Because the spiral angle is such as to allow one end of a spiral (of a spiral bevel gear) to engage its mate before the end of the preceding spiral has broken contact, smooth operation results.

The frictional load developed by straight bevel gears is due to rolling action. Rolling friction predominates in spiral bevel gears, although some sliding occurs.

WORM GEARS consist of a worm and a worm wheel or gear (Fig. 14-6). The worm is formed by a thread having a definite lead, encircling a worm (blank) shaft, in which it is cut. The thread encircles the relatively small

Fig. 14-4. Straight bevel gears. *Courtesy, Gleason Works.*

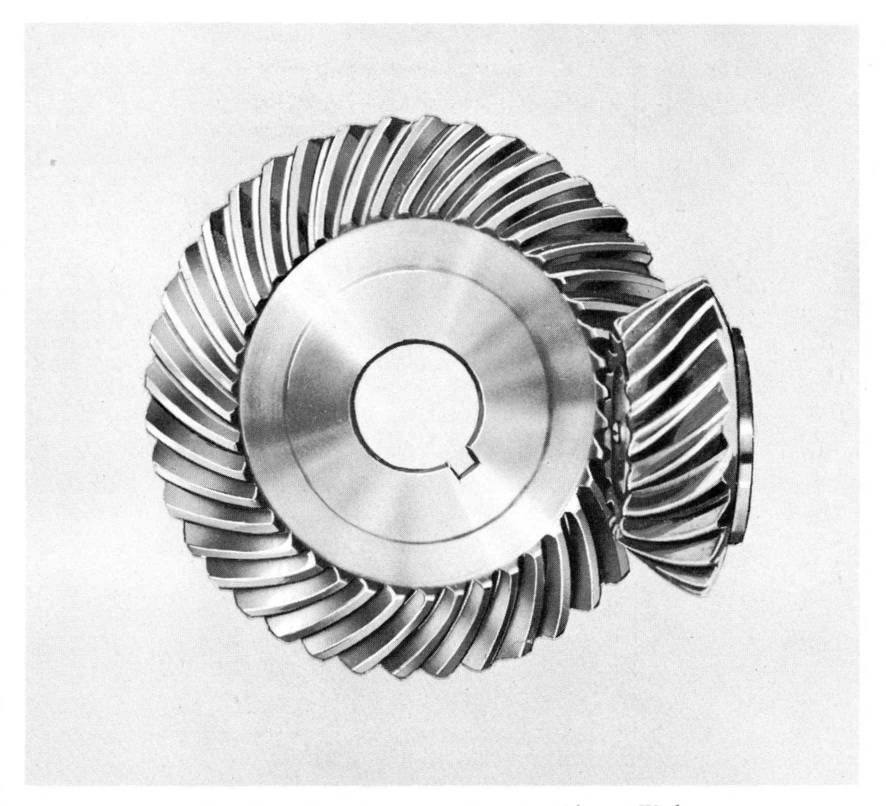

Fig. 14-5. Spiral bevel gears. *Courtesy, Gleason Works.*

diameter blank one or more times. The lead is the distance that the thread advances for each complete turn of the worm. If the diameter of the pitch circle of the worm were increased indefinitely, it would tend to approach a straight line. Under this circumstance, the worm would become a *rack*, which makes the worm a form of endless rack. When the axis of the worm

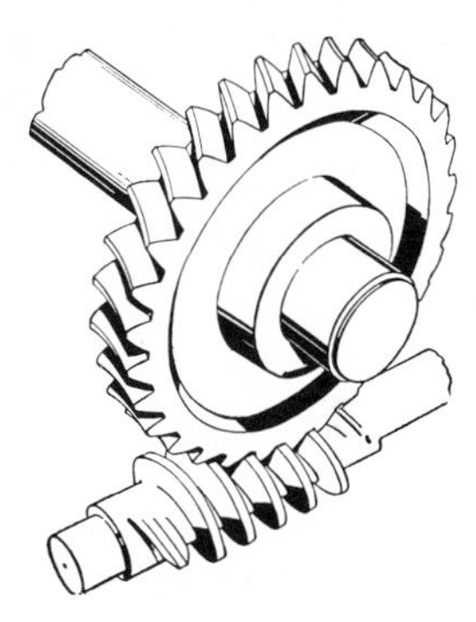

Fig. 14-6. Single-throated worm gear.

and the wheel are at right angles to each other, the teeth of the wheel will mesh with the thread of the worm. Any advance of the thread accomplished by rotating the worm will cause the wheel to turn on its axis. The arc through which the wheel turns for each revolution of the worm depends on the lead or the helix angle. The helix angle is the angle made by the slant of the thread relative to a line perpendicular to the axis of the worm. The teeth of the worm gear are cut diagonally across the blank to conform with the helix angle of the worm.

Worm gears may be single, double, or multiple threaded. The ratio of the number of teeth on the gear to the number of threads on the worm is called the *velocity ratio*, or

$$\text{velocity ratio} = \frac{\text{number of teeth on gear}}{\text{number of threads on worm}}$$

If the wheel has 60 teeth for a triple-threaded worm,

$$\text{velocity ratio} = {}^{60}\!/_{3} = 20 : 1$$

which means that the gear turns once for every 20 revolutions of the worm. For a double-threaded worm,

$$\text{velocity ratio} = {}^{60}\!/_{2} = 30 : 1$$

and for a single-threaded worm the velocity ratio would be 60:1. Thus, the worm gear is capable of large speed reductions. The number of teeth in mesh increases as the number of threads on the worm increases. The length of the worm is usually six times the *pitch,* the latter being the center-to-center distance between threads.

Worm gears may be nonthroated, single-throated, or double-throated. The nonthroated worm gear consists of (*a*) a straight worm, i.e., the outer surfaces of all threads are the same distance from the worm axis, and (*b*) gear teeth that are straight across the outside surface (width) of the gear blank. The single-throated worm gear also has a straight worm, but the outside surface of the gear teeth is curved (concave across the width) to conform with the curvature (circumference) of the worm at the root. The concave design permits the gear teeth to partially wrap around (envelop) the worm. Double-throated worm gears have, in addition to the throated gear teeth, a double-throated worm. Such a worm is designed to permit partial envelopment of the gear by the worm, accomplished by making the worm concave relative to its length. This means that the distance between the outside surface of the threads and the worm axis gradually becomes greater as the threads wind away from the center of the worm in both directions along the axis. Thus, the worm and the gear of a double-throated worm gear possess mutual envelopment. The concave form of the worm allows more teeth to be in mesh at the same time.

Modifying worm gears to obtain partial envelopment changes the frictional relationship between worm and gear teeth in the following manner:

1. Nonthroated worm gears permit point contact only and loads transmitted by such gears act on a series of points as the rotation of the worm progresses through one revolution. For this reason, nonthroated worms are limited in their performance and are seldom used for transmitting power of any appreciable magnitude.

2. Single-throated worm gears attain *line contact*. As the worm revolves, contact is developed on a series of curved lines that run diagonally across the gear tooth. Starting at the tip of the gear tooth, the line of contact progresses radially to the limits of the working depth. The contact line of a worm gear is similar to the slanted line of a helical gear tooth in that it becomes progressively longer as full mesh is approached, after which it diminishes until contact is broken. Simultaneously, the contact line progresses outwardly on the worm threads. During rotation of the worm, considerable sideslip occurs all during the period of contact. With worms of low helix angle the directions of sideslip and contact line are nearly coincident. The line of contact becomes steeper, i.e., tends toward a radial direction, as the helix angle becomes greater.

3. It is probable that the contours of opposing surfaces of double-throated worm gears promote contact in the form of a radial band having an appreciable width rather than a line. As the band would extend simultaneously across the entire width of the gear tooth as in the case of line contact described for single-throated worms, the total bearing area would depend on the width of the band. Frictional relationship between gear and worm in a double-throated worm gear is one of sliding action only.

Because sliding action predominates in worm gear operation, the use of metals possessing a low coefficient of friction is desirable. A combination of a bronze worm gear mated with a steel worm results in a worm gear set with good wearing characteristics, a low coefficient of friction, and a good load-transmitting ability.

Worm gears are subject to high thrust loads. A load that exceeds the worm gear's capacity will result in scoring and peeling.

The *mechanical rating* of a worm gear is the horsepower it can transmit without damage to its worm or gear. Its *thermal rating* is the horsepower it can transmit without reaching a dangerously high temperature. Limited overloads can in some instances be tolerated by the use of a proper lubricant. For example, worm gears driving a bank of mechanical mixers ran hot upon changing the batch ingredients that required additional power over the usual mixture to agitate it. The temperature of the worm set returned to normal when another type of lubricant was introduced.

Experience has shown that a rise in temperature of single-enveloping worm gear lubricants of 90°F above ambient is a satisfactory operating condition. For double-enveloping worm gears, such as the Cone worm gear (manufactured by the Michigan Tool Company), the normal rise in temperature approaches 100°F above ambient without presenting any problem. Temperatures above 200°F cannot be tolerated for any appreciable time. An outboard bearing located too close to this type of worm gear often raises the operating temperature.

The thermal rating is based on the thermal radiating capability of the individual housing, which increases with area and temperature difference between internal and ambient levels. The more efficient the unit, the less thermal energy is generated. Efficiency is dependent on the reduction ratio. The worm gear unit is selected on the basis of lowest permissible rating, whether mechanical or thermal, but the thermal rating should not be exceeded for any gear set.

Worm gears are classified by the American Gear Manufacturers Association according to service conditions.

Class I—Operation does not exceed 8–10 hr out of 24 under normal conditions, and it is free of recurrent shock loads. The service factor is 1.

Class II—8- to 10-hr service, during which shock load is recurrent, or 24-hr service with no shock. The mechanical rating furnished by the manufacturer is divided by a service factor of 1.2, and it should not exceed the thermal rating.

Class III—24 hr; shock load service. Divide mechanical rating by a service factor of 1.333.

Class IV—Intermittent service, where the maximum cycle of operation is not more than 15 min running time in a 2-hr period. Divide mechanical rating by a service factor of 0.7.

Class V—Low-speed service, worm speed less than 100 rpm. Use output torque ratings in inch-pounds at 100 rpm.

HYPOID GEARS may be considered in many respects a compromise between a spiral bevel gear and a worm gear (Fig. 14-7). In spiral bevel gears the axis of the pinion and the gear intersect, while in the worm gear the axis of the worm (pinion) is tangent to the gear. The axis of the hypoid pinion lies in a plane anywhere between the two extremes, but always right angular to the axis of its gear. It resembles the spiral bevel gear in appearance only.

Although rolling friction predominates in spiral bevel gears, some sliding occurs. For all practical considerations, worm gears develop sliding friction only. Hypoid gears, being a combination of worm and spiral bevel, develop both rolling and sliding friction, the latter due principally to sideslip. The ratio of sliding to rolling friction, however, depends on whether the hypoid arrangement (offset of pinion relative to gear axis) approaches that of spiral bevel or worm gear. As the distance between gear and pinion shaft

Fig. 14-7. Hypoid gears. *Courtesy, Gleason Works.*

center lines increases, the ratio of sliding to rolling friction becomes greater. Thus, when heavy loads are transmitted through hypoid gears having a pinion that is extremely offset, tremendous frictional force develops. The pressure developed between the contacting surfaces of hypoid gear and pinion under heavy load is great enough to promote welding and cause scoring of the teeth. To prevent scoring under heavy loading, special ingredients must be added to hypoid gear lubricants.

Hypoid gears may be designed for other than a 90-degree shaft angle, but they are usually employed for right-angle drives.

HELICAL GEARS may be designed to transmit motion between shafts that cross each other at any angle. However, contact between driver and driven gear is made on a point. Both sliding and rolling occur between helical gears, the ratio of sliding to rolling friction increasing as the shaft angularity approaches 90 degrees. When helical gear shafts are at right angles, the action is one of sideslip on a point, a condition not conducive to heavy loading. For this reason, helical gears are seldom used to transmit loads of any appreciable magnitude between nonparallel cross shafts.

Table 14-1 gives a summary of contact and frictional characteristics of various types of gears.

ZEROL BEVEL GEARS. The Zerol bevel gear (trademark name) is similar to the straight bevel gear except that its teeth are curved lengthwise. Unlike the spiral bevel gear, it has a zero degree spiral angle with the axis of the gear. A straight line drawn radially from the center point of the gear would pass through a corresponding point at each end of the tooth to form a chord of the arc made by the intervening length of tooth. The curvature of tooth

Table 14-1. MEANS OF CONTACT BETWEEN GEAR AND PINION, FRICTIONAL CHARACTERISTICS, AND TOOTH DEVELOPMENT OF GEARS

Gear	*Means of Contact**	*Predominant Frictional Characteristic†*	*Pitch Surface on Which Teeth Are Developed*
Spur gears	Line	Sliding and rolling	Cylindrical
Herringbone gears	Line	Sliding and rolling	Cylindrical
Bevel gears:			Conical
Straight tooth	Line	Rolling and sliding	
Spiral bevel	Line	Rolling and sliding	
Worm gears:			Cylindrical
Nonthroated	Point	Sliding	
Single-throated	Line	Sliding	
Double-throated	Band	Sliding	
Hypoid	Line	Sliding	Hyperbolic
Helical (nonparallel shafts)	Point	Sliding	Cylindrical

* Deformation of tooth surface under load creates definite *area* contact in all gears. For teeth forming line contact the line becomes a band probably 1/64 to 1/4 in., depending on the size of gear.

† More or less rolling and sliding occur in nearly all gears; the table lists the predominant frictional characteristics.

develops an overlapping effect, resulting in smoother operation as compared with straight bevel gears. Zerol gears are adaptable to high speeds.

INTERNAL GEARS have teeth cut into the inner surface of a ring, and the gear they mesh with is usually a spur gear, as shown in Fig. 14-8a.

RACK AND GEAR. A rack is a bar with teeth on one of its sides. It meshes with a gear, usually a pinion (Fig. 14-8b). The rack and gear convert rotary motion to linear motion, or vice versa. Their teeth may be straight or helical.

The rack usually moves back and forth on its bed-way as the gear on a fixed shaft changes direction of rotation. The rack may be horizontal, vertical, or angled.

FACE GEARS. Resembling bevel gears, face gears have teeth cut on their face instead of the outer surface. The teeth project out from the blank parallel to the radial axis. They function similarly to a bevel gear. Their mating pinion and face gear are designed with spur teeth and the shafts usually intersect at a 90 degree angle.

Other gears are designed for special purposes (Figs. 14-8 and 14-9). The tooth design usually follows the principles employed by gears already described. Such gears include *elliptical gears* whose teeth are cut on a blank of nonuniform diameter (elliptical). They transmit a rhythmic, pulsating motion. *Intermittent* gears are designed to transmit intermittent motion by a series of nonmeshing surfaces about the gear blank that correspond to similar surfaces on the pinion blank. They are also called *stop-motion* gears. *Equalizer* gears are made to neutralize the effect of large bar links of conveyor side chains.

A GEAR TRAIN is a series of gears mounted on their respective shafts whose only function is to carry the gears or act also as random stations of power takeoff. A gear train is sometimes used to perform a number of

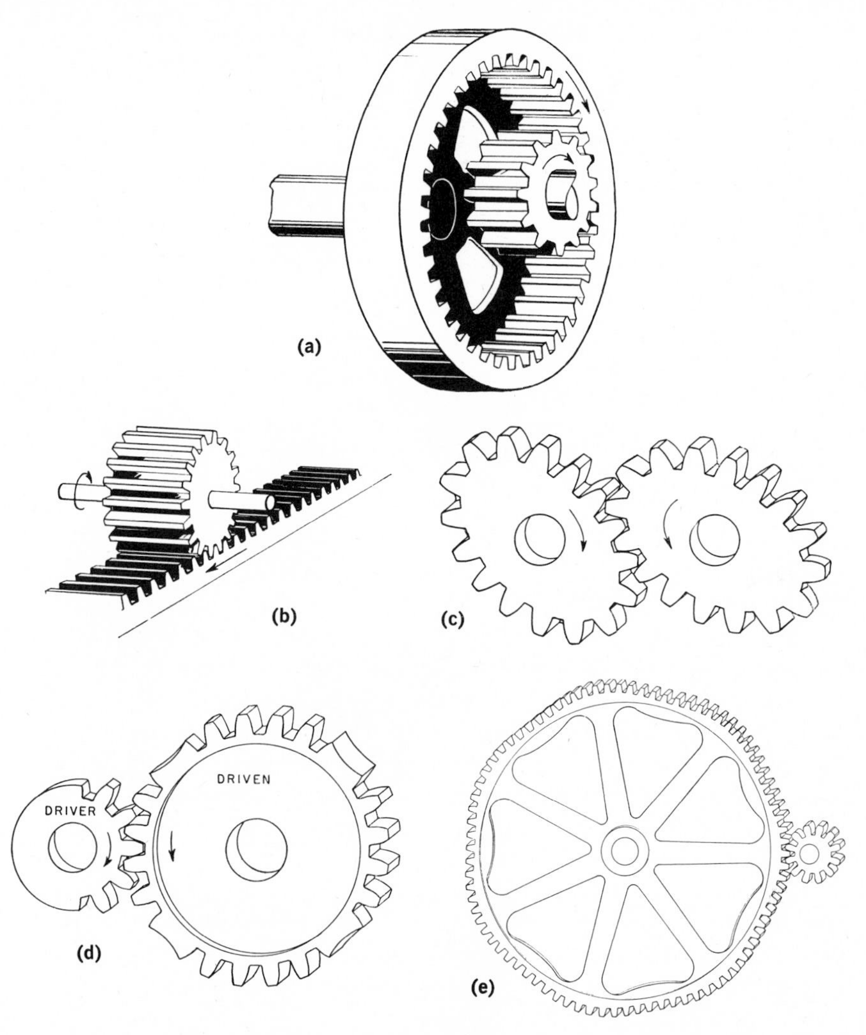

Fig. 14-8. Various types of gears: (a) internal, (b) rack and gear, (c) elliptical, (d) intermittant or stop, (e) equalizer. *Courtesy, Mobil Oil Corp.*

functions by utilizing a common power source. Gear trains also consist of a number of gears of similar size relatively located as to extend and synchronize the movement of a continuing process, as in a paper machine dryer roll drive.

A GEAR SET is a pair or series of pairs of gears having the function of changing some dynamic characteristic in a definite pattern of continuity from power source to power takeoff.

Gear sets are identified with the number of stages required to attain a given ratio of reduction. Any type of gear can do this, but the herringbone,

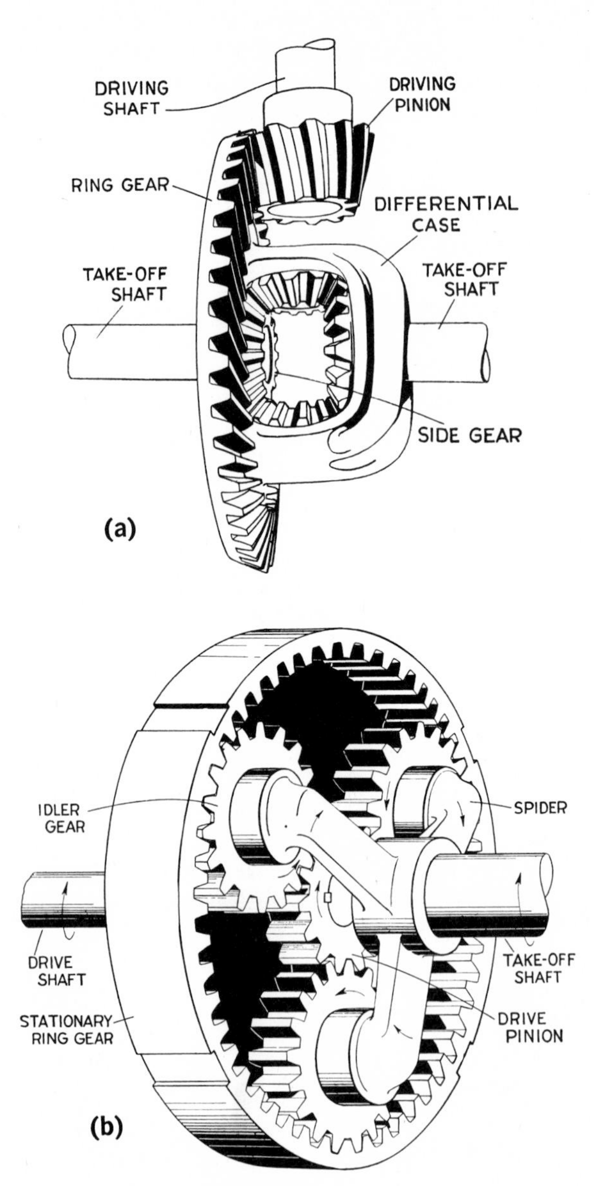

Fig. 14-9. Differential gears (top) and planetary gears (bottom). *Courtesy, Mobil Oil Corp.*

double helical, spur, and bevel gears represent the major portion of industrial and marine enclosed gear sets. According to the number of speed reduction stages, they are called

Pinion stands. Both gears revolve at same speed for purposes of synchronization.

Single-reduction. Ratio accomplished in one stage. (Fig. 14-3).
Double-reduction. Ratio accomplished in two stages. (Fig. 14-10)
Triple-reduction. Ratio accomplished in three stages. (Fig. 14-11)
Quadruple-reduction. Ratio accomplished in four stages.

GEARMOTORS consist of an electric motor connected to a gear set of appropriate reduction ratio, both of which are encased in a common housing through which the power takeoff shaft passes. Gear motors are self-contained in that the necessary bearings, cooling components, and lubrication systems are all included in the one assembly. They are made in various sizes from very small to large multiple-reduction units. In making lubricant recommendations for gearmotors already installed, carefully note the type of gear used, because many of them employ worm gears, especially where the takeoff speed is very low.

SPEED INCREASERS, or step-up gears, are gear sets in which the larger gear does the driving. Lubrication of such gears is the same as for speed reducers. They are used for overdrives in the transmissions of automobiles and to increase the speed of centrifugal compressors in refrigeration systems, particularly for large central air conditioning installations.

Fig. 14-10. Double reduction gear set. *Courtesy, The Falk Corporation, subsidiary of Sundstrand Corp.*

Fig. 14-11. Triple reduction gear set. *Courtesy, The Falk Corporation, subsidiary of Sundstrand Corp.*

GEAR DESIGN

The power transmission ability of gears comes from the meshing of opposing teeth or threads that project from a blank on which they are formed. Most gears are designed and manufactured precisely according to their intended service. In modern machinery, there is no place for a "rough" gear. The contact between gears is positive except for the play, or circumferential freedom, that exists between teeth by design and that is often increased by wear.

GEAR NOMENCLATURE. The various parts of a gear are identified according to their function and design significance as labeled in Fig. 14-12. Not shown in the figure is length of tooth, which is the measure of its extremities on a straight line directed from the paper toward or away from the viewer.

Backlash

Backlash is the clearance, or play, between gear teeth in mesh at the tightest point of engagement. It is required to provide space for the lubrication film, to allow for expansion during temperature changes, and to allow for slight misalignment of shafts. Wear increases backlash. Backlash can be increased or decreased by adjusting the center distances of the pair.

The amount of backlash can be determined with a thickness gauge, or feeler gauge, when the gears are in place. Measure the smallest width between the front and back of opposing teeth when the respective back and front of the same teeth are held tightly together. The desired change in backlash can be obtained by changing the center distances determined as follows:

$$\text{change in center distance} = \frac{\text{change in backlash}}{2 \times \tan \alpha}$$

where α is the pressure angle.

For spur gears with an α equal to $14\frac{1}{2}$, 20, and 25 degrees, multiply the desired change in backlash respectively by 1.933, 1.374, and 1.072 to determine the corresponding change in center line distance.

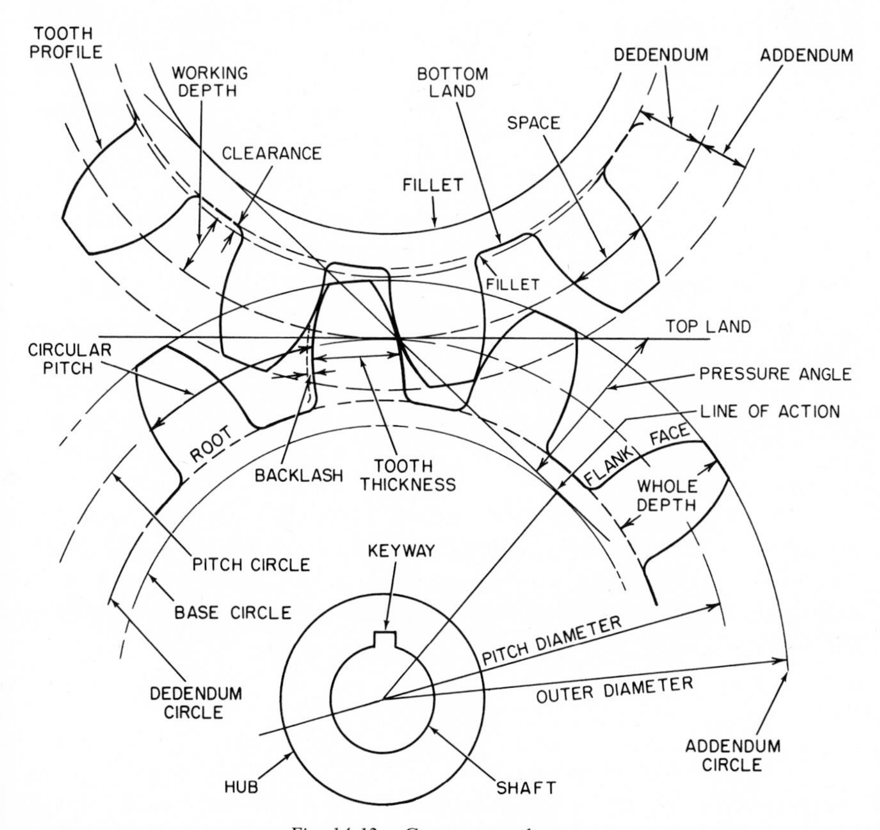

Fig. 14-12. Gear nomenclature.

GEAR SPEED. When two mating gears have the same pitch diameter, their shaft speeds are the same. To increase the speed of the driven gear relative to the driver, the former must be of smaller pitch diameter. To gain mechanical advantage, the pitch diameter of the driven gear must be larger than that of the driver, using the lever principle. The relationship in regard to relative speed is one of inverse proportionality and may be expressed in terms of speed and number of teeth on each gear. Thus,

where

N_1 = speed of drive gear (rpm)
N_2 = speed of driven gear (rpm)
T_1 = number of drive gear teeth
T_2 = number of driven gear teeth

$$\frac{T_1}{T_2} = \frac{N_2}{N_1} \qquad N_2 = \frac{N_1 T_1}{T_2} \qquad N_1 = \frac{N_2 T_2}{T_1}$$

The mechanical advantage (MA) is directly proportional to the relative number of teeth on each gear, so

$$\text{MA} = \frac{T_1}{T_2}$$

The number of gear teeth can be substituted for pitch circle in any case of proportionality, direct or inverse, because the diametral pitch of each mating gear is the same. The gear smaller in pitch diameter does not necessarily have fewer teeth than the larger gear. Examples are the crossed helical and hypoid gears, whose pitch diameters are not proportional to the tooth ratio. This enables the smaller gear sometimes to have more teeth than its larger mate. Thus for hypoid and crossed helical gear sets, it is not always possible to identify the smaller gear of the pair as a pinion.

Gear Functions

Gears are arranged in many ways, depending on the intended service. They are utilized chiefly to change speeds, torque and direction of motion. Some of the more common gear arrangements perform the following functions:

1. Reduce speed
2. Increase speed
3. Transmit power at same speed
4. Increase torque

5. Change angular direction
6. Reverse direction
7. Convert any type of motion to any other—rotary, linear, circular, intermittent, reciprocating, oscillating, planetary, vibratory, and spacial
8. Synchronize shaft speeds in rolling mills (steel, rubber, etc.)

Consideration must be given to the gains and losses that accompany any change in dynamic characteristics of a gear system. The magnitude of gain or loss is a function of ratio or the relative size or number of teeth of the pair of gears involved. Any increase in speed is accompanied by a decrease in torque or mechanical advantage. Conversely, any reduction in speed is accompanied by an increase in torque. In multiple reduction gear sets, the pair of gears having the lowest speed is made stronger in order to transmit the increase in torque. Where it is necessary to reduce speeds and thus increase torque through a large ratio, it is more efficient, both mechanically and economically, to perform the reduction in stages rather than using one pair of gears greatly differing in size. This is analogous to using a low- and high-pressure cylinder of an air compressor to attain a high compression ratio by stages in interest of thermal efficiency and structural economy.

The mechanical rating divided by the service factor represents the actual *service rating,* providing that it never exceeds the thermal rating.

GEAR MATERIALS. Conventional gears used in the industrial, marine, aircraft, and automotive fields are usually steel or cast iron. Many of the worm gears are made of special gear bronze. Steels vary in carbon content and hardness, depending on the intended service. Alloys are also used.

Nonmetallic gears are also used. Some of the better known materials are Formica, Bakelite, Micarta, and Ryertex. Such gears are impervious to petroleum oils and greases so may be lubricated like metallic gears.

Rawhide gears are made to minimize noise. They must be run dry because of the deteriorating effect of petroleum products.

GEAR LUBRICATION

The difference in loading from one pair of gears to another in a multiple-reduction gear set must be considered when selecting a suitable lubricant to meet all conditions of speed and load. Because all frictional surfaces within the gear system, including the gear shaft bearings, are supplied from a common lubricant source, some compromise must be made in the selection of a lubricant. The compromise is usually one of viscosity, and the recommendations shown on lubrication charts usually reflect such considerations.

Integrated gear sets may be fully or partially enclosed within a gear case. Others are open to the atmosphere. Some open gears are surrounded by a metal guard in which a "hand hole" is cut for lubrication and inspection. On other machines, the cover guard is augmented with a *slush pan* that acts as an oil reservoir through which the lower portion of one of the gears passes to provide lubrication. Open gears usually rotate slowly and carry relatively high loads, so that the lubricant selected usually differs from that selected for the faster enclosed gears of the spur, bevel, and helical types. The lubricant for open gears also varies according to the method of its application.

Enclosed gears are usually lubricated with oil that is applied either by splash (Fig. 14-13), scoop (reservoir), internal or external circulation (Fig. 14-14), or a combination of methods. Open gears may be lubricated by spray, slushing, or various hand methods. A spray may be automatic or manually applied by means of a pressurized container equipped with a nozzle. Hand application methods include pouring and swabbing (with a brush).

Lubricating Film Formation on Gear Teeth

Gears, like journal bearings, may be subjected to boundary, mixed and flooded lubrication conditions. Heavily loaded gears of any kind, and hypoid gears particularly, depend on a special type of protection other than the normal hydrodynamic film that is satisfactory for normal loads. This is true

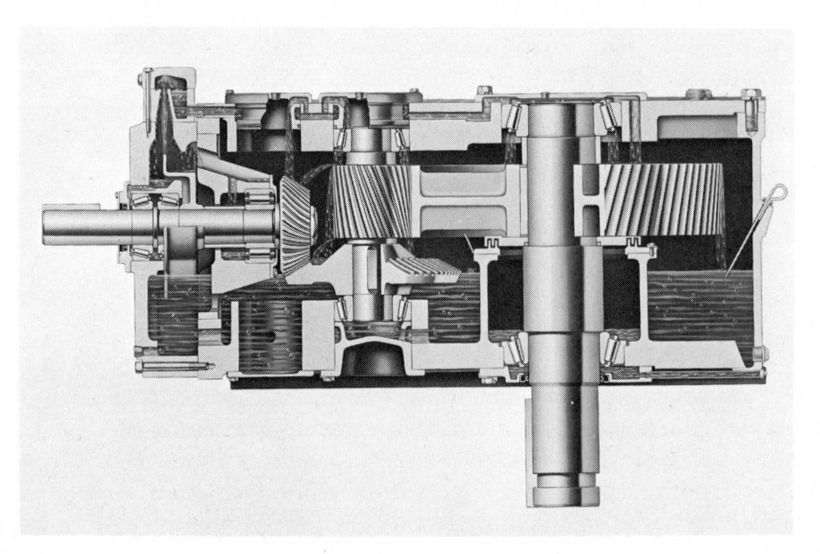

Fig. 14-13. Cross-section of a vertical right-angle speed reducer with downward shaft extension, showing splash lubrication system. Note oil slinger. *Courtesy, The Falk Corporation, subsidiary of Sundstrand Corp.*

Fig. 14-14. Pinion stand equipped with external pressure lubrication system. *Courtesy, The Falk Corporation, subsidiary of Sundstrand Corp.*

especially on gears whose teeth are designed for a predominantly sliding action in contrast to rolling action.

If the load interferes with the smooth formation of a continuous hydrodynamic wedge regardless of viscosity, the condition is one of *extreme pressure.* If the load intermittently threatens the wedge formation, the condition is said to be one of *mild extreme pressure.*

"Mild extreme," although ambiguous, is somewhat established in the vernacular of field engineering. It is loosely applied to bearings with less than 50% of the structural capability to withstand the unit pressures implied. It is more accurate to distinguish pressures according to degree and classify them by codes based on minimum and maximum limits. There is no semblance of quasi-extreme pressure in a gear system of a marine type steam turbine connected to a propeller shaft, or generator, regardless of power developed. An automobile accelerated for less than three seconds at rates associated with hot-rod operation, on the other hand, develops pressure on hypoid gears that could only be expressed by compounding the prefix "extreme" several times in classifying the degree of pressure.

The *line of action* as applied to spur gear teeth is shown in Fig. 14-15.

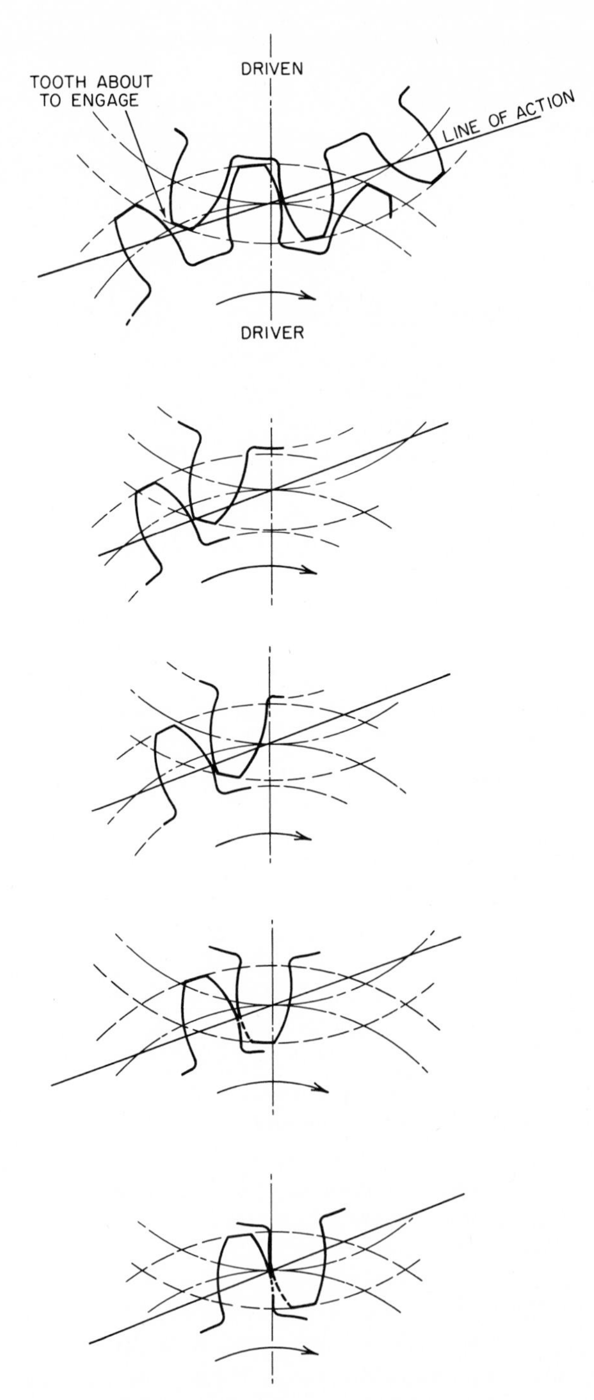

Fig. 14-15. Gear teeth at line of action.

Also shown are the successive points of contact between the meshing teeth as they move along the line of action. Note that the action is first one of sliding, followed at the pitch line by rolling action, and then reverting to sliding action. A similar series of sliding-rolling-sliding actions occurs on all types of gear teeth as they pass through mesh, except that the degree of one or the other varies according to design. As stated earlier, however, the action of hypoid gears approaches complete sliding action in the form of *sideslip*. Worm gears for all practical purposes embraces sliding action only.

As the meshing teeth enter the sliding phase during initial contact, their curvature of contour is conducive to the formation of a hydrodynamic oil wedge. As the teeth proceed into their rolling phase, the forces necessary for hydrodynamic development are no longer present, resulting in a tendency for the surfaces to come together and make contact, or at least revert to thin film lubrication. This is one of the advantages of adequate oil viscosity; the oil film already formed during the sliding action will adhere to the surfaces of the teeth and prolong their separation, providing certain basic rules are followed.

An oil of adequate viscosity commensurate with load and speed must be supplied to prevent rapid oil squeeze-out. The critical factor here is *rate of slide relative to pressure*. Where both sliding velocity and pressure are excessive, the heat generated compels the temperature to rise, leading to other complications, such as the lowering of viscosity and subsequent destruction of the surfaces. The critical factor is a permissible value based on gear geometry and metallurgy, coupled with lubricant capabilities, both as to type and application. If the slide-rate and pressure relationship make the oil inadequate in quantity and viscosity as the rolling phase is initiated at the point (area in practical system due to surface deformation) of maximum pressure along the line of action, metal-to-metal contact may result. The evidence of this lies in a number of observations of spur and helical gear teeth after extended periods of operation with lubricants not fortified with special film strength or extreme pressure additives.

If the oil is viscous enough to meet the operating conditions, the effect of the teeth to force the oil from between them will be retarded long enough for the teeth to pass through the rolling phase and force hydraulic separation. This is called *squeeze-film action*. In the event of extreme or near extreme pressures, neither the hydrodynamic nor hydraulic films may develop sufficiently to prevent metal-to-metal contact. In such case, extreme pressure or film-strength compounds, respectively, carried by the oil are necessary, and initiate their action to eliminate spot welding when a sufficient rise in thermal energy develops as a result of low-level boundary friction.

In the interest of an effective squeeze-film, time and force are important factors, especially during the rolling phase. With relatively high speeds, light loading, and more viscous oils, the side leakage is reduced and hydraulic action retained. Any reduction in speed, increase in load, or lowering of oil viscosity, or any combination of these three tends to encourage side leakage and reduce film thickness correspondingly. Therefore, as with journal bearings, the following statements also become axiomatic in respect to gears:

1. Low speeds and high pressures require a high viscosity oil.
2. Intermediate speeds and pressures require a medium viscosity oil.
3. High speeds and light pressures require a low viscosity oil.

A triple-reduction speed reducer, for example, with herringbone gears in all stages is a unit in which all three conditions are somewhat represented. Consider also a case in which loading is in the high horsepower range, say 300 hp, and the speed reduction is in the range of 18:1, from 1800 rpm to 100 rpm. Gears for this service are necessarily large, both in mass and length of tooth. Tooth length governs the unit tooth pressure measured in pounds per inch of tooth face. The speed-load relationship of the first reduction would require an oil with a viscosity of 300 to 400 SUS at the operating temperature, and the third reduction would require an oil viscosity of 1500 to 1800 SUS at the operating temperature. However, because all gears are lubricated with a common oil, the viscosity of the oil at 100°F would depend on the cooling facilities of the unit. If the oil is circulated through an oil cooler, the viscosity required could be about 600 to 900 SUS at 100°F. If, on the other hand, cooling depended on natural radiation, convection and conduction only, an oil viscosity of about 900 to 1500 SUS at 100°F would be proper. The selection depends on the operating temperature, which will vary in the first case by the volume of oil applied per unit of time (gallons per minute), and in the second case by the area of the gear housing and the capacity of the oil reservoir.

Oil should always be held to the lowest operating level governed by the temperature of the lowest gear teeth in order to develop fluid friction. However, a lower operating temperature can be achieved by raising the oil level and thereby increasing the housing area of heat conduction between oil and the atmosphere.

Open Gears

Open gears may be of any type with the possible exception of hypoids and some worm gears. Spur gears are by far the most prevalent.

The lubrication requirements of open gears are like those of enclosed gears, and when the open gears are sufficiently covered, the same lubricant can be used for similar operating conditions. However, because of the tendency of open gears to throw off conventional oils and the limited suitability of most greases, special residual lubricants or high-viscosity oils are used. These usually require a sump and are very effective for relatively low peripheral velocities.

In determining the lubricant to be used an open gears, several factors must be considered, (1) degree of enclosure, (2) speed (rim velocity), (3) size (pitch diameter), (4) environment, (5) method of lubricant application, and (6) accessibility of gears.

Except when there is some means of lubricant recovery, such as with a sump, open gears must be lubricated on the all-loss principle regardless of the method of application (Figs. 14-16 and 14-17). Any all-loss system ultimately reverts to thin film or boundary lubrication after a new supply of oil or grease is rendered microscopically thin by a combination of run-off, throw-off, and squeeze-off from the pressure area of the teeth.

To counter run-off, a lubricant must possess high viscosity, or consistency, together with persistence to maintain a film on the surfaces of the teeth.

To reduce squeeze-off, the lubricant must possess a strong persistence of film, which means qualities of adhesion and cohesion.

Open gears are exposed to all varieties of environmental conditions. They operate canal locks during northern winters and furnish the train of

Fig. 14-16. Open gear lubricant applied by mechanical force-feed lubricator. *Courtesy, Mobil Oil Corp.*

Fig. 14-17. Open gear lubricant applied by slush pan (bath). *Courtesy, Mobil Oil Corp.*

motion required by dryer rolls of paper machines, where humidity is high and ambient temperature may exceed 140°F. Ring gears rotate cement kilns in heat, rain, and dust. They act as bull gears on large mechanical presses in automobile body production plants where environment is usually not a serious factor, but lubricant throw-off is both highly undesirable and hazardous.

The conventional type of open gear lubricant is an extremely heavy-bodied petroleum residual material having a tar-like appearance being black and tacky. In former years, such materials usually referred to as gear shields, were hard set and required heating to attain sufficient softness in order to apply them. They are now, except for the more liquid gear shields, rendered fluid by the addition of a solvent during processing by the supplier. The solvent, or diluent, is a volatile nonflammable chlorinate-hydrocarbon that evaporates upon application to the gear. The residue, when evaporation is complete, is a plastic rubbery film that covers the teeth to shield them against wear, dust, and water. Some types of gear shield are fortified with extreme pressure antiwear additives. Their rubbery characteristics present a somewhat dry firm surface to the atmosphere, tending to discourage the deposit of dust and to shed water.

Where there is adequate encasement, heavy-bodied oils, preferably leaded compounds or cylinder oils containing 5 to 10% fixed oils, such as lard oil or acidless tallow, may be used. The point to consider in the choice of gear lubricants is that more effective lubrication and greater convenience is obtained with circulating oils than with residual films periodically applied. For small open gears or gear trains a relatively soft grease may be used.

GEAR FAILURE. Improper lubrication, excessive loading, inadequate surface hardness, abrasion, and other factors cause damage to gear teeth leading to their ultimate failure. Some of the surface irregularities related to gear tooth damage appear in Figs. 14-18 through 14-22. The causes of such irregularities on metal surfaces and others not shown are defined and described in Chapter 13.

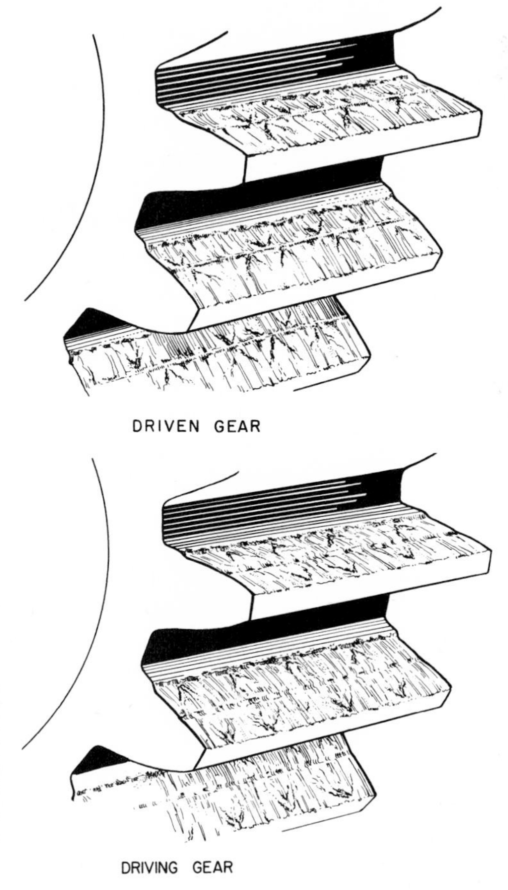

Fig. 14-18. Galling. (Top) The direction of slide on the teeth of a driven gear is always toward the pitch line. (Bottom) On a driving gear, the direction of slide is always away from the pitch line. *Courtesy, Mobil Oil Corp.*

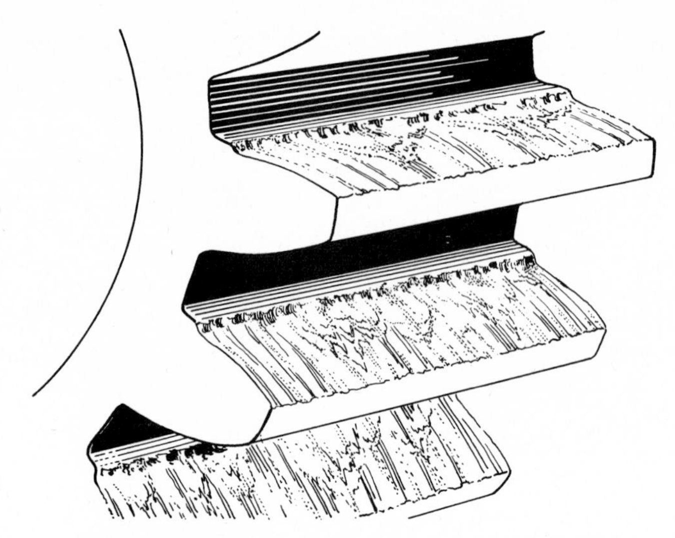

Fig. 14-19. Advanced galling. *Courtesy, Mobil Oil Corp.*

Fig. 14-20. Rippling. Sometimes ripples appear on the surfaces of hypoid gears. *Courtesy, Mobil Oil Corp.*

Fig. 14-21. (Left) Incipient pitting. (Right) Ridging and pitting, which may result from the lack of effective lubrication of hypoid gear teeth. *Courtesy, Mobil Oil Corp.*

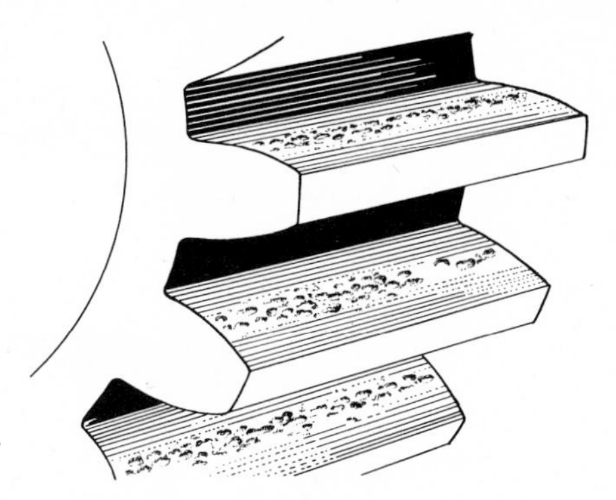

Fig. 14-22. Destructive pitting. *Courtesy, Mobil Oil Corp.*

AGMA STANDARD TABLES

The American Gear Manufacturers Association publishes standard specifications for the selection of industrial gear lubricants. They cover all types of gears, as shown in Tables 14-2 through 14-9. The tables are drawn up according to a code number, each of which designates a lubricant viscosity suggested for various gears and operating factors, and they should be followed for gears designed according to AGMA standards.

The tables are extracted from *AGMA Standard* with the permission of the publisher, the American Gear Manufacturers Association, 1330 Massachusetts Avenue N.W., Washington D.C. 20005. The issues extracted are, for Tables 14-2, 3, and 4, "Lubrication of Industrial Enclosed Gearing" (AGMA 250.02); for Tables 14-5, 6, and 7, "Lubrication of Industrial Gearing" (AGMA 251.01); and for Tables 14-8 and 9, "Mild Extreme Pressure Lubricants for Industrial Enclosed Gearing" (AGMA 252.01).

Oils for AGMA Nos. 1 through 6 should have a viscosity index of 30

Table 14-2. Viscosity Range for Various AGMA Lubricants

AGMA Lubricant Number	*Viscosity Range (SUS)*	
	At 100° F.	*At 210° F.*
1	180–240	
2	280–360	
3	490–700	
4	700–1000	
5		80–105
6		105–125
7 comp*		125–150
8 comp*		150–190
8A comp*		190–250

* The oils marked "comp" are those compounded with 3 to 10% of acidless tallow or other suitable animal fat.

Table 14-3. LUBRICANT RECOMMENDATIONS FOR ENCLOSED GEARING

Type and Size of Unit Main Gear Low Speed Centers	*Ambient Temperature (°F)* 15 to 60 Use AGMA Number	50 to 125 Use AGMA Number
Parallel shaft, (single reduction), up to 8 in.	2	3
Over 8 in. and up to 20 in.	2	4
Over 20 in.	3	4
Parallel shaft, (double reduction), up to 8 in.	2	3
Over 8 in. and up to 20 in.	3	4
Over 20 in.	3	4
Parallel shaft, (triple reduction), up to 8 in.	2	3
Over 8 in. and up to 20 in.	3	4
Over 20 in.	4	5
Planetary gear units		
O.D. housing up to 16 in.	2	3
O.D. housing over 16 in.	3	4
Spiral or straight bevel gear units		
Cone distance up to 12 in.	2	4
Cone distance over 12 in.	3	5
Gearmotors	2	4
High-speed Units*	1	2

*For speeds over 3600 rpm or pitch line velocities over 4000 rpm. It may be desirable to use a lubricant one grade higher or lower than those specified, depending on operating conditions.

Table 14-4. CYLINDRICAL AND DOUBLE-ENVELOPING WORM GEAR UNITS

Worm Centers	*Max. Worm Speed (rpm)*	*Ambient Temperature (°F)*		*Worm Speed** (above rpm)*	*Ambient Temperature (°F)*	
		*15–60**	*50–125*		*15–60**	*50–125*
Up to 6 in. inclusive						
Cylindrical Worms	700	7 comp	8 comp	700	7 comp	8 comp
Double-Enveloping Worm	700	8 comp	8A comp	700	8 comp	8 comp
Over 6-in. centers up to 12-in. centers						
Cylindrical Worms	450	7 comp	8 comp	450	7 comp	7 comp
Double-Enveloping Worms	450	8 comp	8A comp	450	8 comp	8 comp
Over 12-in. centers up to 18-in. centers						
Cylindrical Worms	300	7 comp	8 comp	300	7 comp	7 comp
Double-Enveloping Worms	300	8 comp	8A comp	300	8 comp	8 comp
Over 18-in. centers up to 24-in. centers						
Cylindrical Worms	250	7 comp	8 comp	250	7 comp	7 comp
Double-Enveloping Worms	250	8 comp	8A comp	250	8 comp	8 comp
Over 24-in. centers						
Cylindrical Worms	200	7 comp	8 comp	200	7 comp	7 comp
Double-Enveloping Worms	200	8 comp	8A comp	200	8 comp	8 comp

*Pour point of the oil used should be less than the minimum ambient temperature expected.

**Worm gears of either type operating at speeds above 2400 rpm or 2000 f/min, rubbing speed may require force-feed lubrication. In general, a lubricant of lower viscosity than recommended in the above table may be used with a force-feed system.

minimum for ordinary applications. Where temperatures vary more than 80°F, a VI of at least 60 is desirable. Oils for AGMA compounds Nos. 7, 8, and 8A must have a VI of at least 90. The compounds are animal fats, such as acidless tallow, in amounts of 3 to 10%.

Table 14-5. RECOMMENDED VISCOSITY OF LUBRICANT FOR CONTINUOUS METHODS OF APPLICATION (SUS AT 210°F)

		Pressure Lubrication		*Splash Lubrication*		*Idler Immersion*
		Pitch Line Velocity		*Pitch Line Velocity*		*Pitch Line Velocity*
Ambient Temp.[1] *(°F)*	*Character of Operation*	*Under 1000 ft/min*	*Over 1000 ft/min*	*Under 1000 ft/min*	*1000 to 2000 ft/min*	*Up to 300 ft/min*
15–60[2]	Continuous	80–105	60–80	80–105	60–80	150–250
	Reversing or Frequent "Start-Stop"	80–105	60–80	125–150	105–125	150–250
50–125[2]	Continuous	125–150	105–125	125–150[3]	105–125[4]	400–600
	Reversing or Frequent "Start-Stop"	125–150	105–125	190–250[3]	150–190[4]	400–600

[1] Temperature in vicinity of the operating gears.
[2] When ambient temperatures approach the lower end of the given range, lubrication systems must be equipped with suitable heating units for proper circulation of lubricant and prevention of channeling. Check with lubricant and pump suppliers.
[3] When ambient temperature remains between 90 to 125 °F at all times, use a 400 to 600 SUS viscosity lubricant.
[4] When ambient temperature remains between 90 to 125 °F at all times, use a 190 to 250 SUS viscosity lubricant.

Table 14-6. RECOMMENDED VISCOSITY OF LUBRICANT FOR INTERMITTENT METHODS OF APPLICATION LIMITED TO 1500 FT/MIN PITCH LINE VELOCITY[1] (SUS AT 210°F)[2]

	Mechanical Spray Systems		
Ambient Temperature (°F)[3]	*Mild Extreme Pressure Lubricant*	*Residual Compound*[4]	*Gravity Feed or Forced Drip Method Using Mild Extreme Pressure Lubricant*
15–60	—	1000–3000	—
40–100	500–575	3000–10,000	500–575
70–125	850–1000	3000–10,000	850–1000

[1] Feeder must be capable of handling lubricant selected.
[2] Viscosities given are for lubricants in nondiluted form.
[3] Ambient temperature is temperature in vicinity of the gears.
[4] Diluents must be used to facilitate flow through applicators.

Table 14-7. RECOMMENDED QUANTITIES OF LUBRICANT FOR INTERMITTENT METHODS OF APPLICATION WHERE PITCH LINE VELOCITY DOES NOT EXCEED 1500 FT/MIN. FOR AUTOMATIC, SEMIAUTOMATIC, OR HAND SPRAY SYSTEMS

Gear Diameter (ft)	Ounces per Application at Intervals[1] of: ¼ Hour, Face Width (in.) 8	16	24	32	1 Hour, Face Width (in.) 8	16	24	32	4 Hours[2], Face Width (in.) 8	16	24	32
10	.2	.3	.4	.5	.8	1.2	1.6	2.0	5.0	6.0	8.0	10.0
12	.3	.3	.4	.5	1.2	1.4	1.8	2.2	6.0	7.0	9.0	11.0
14	.3	.4	.5	.6	1.4	1.6	2.0	2.4	7.0	8.0	10.0	12.0
16	.4	.5	.6	.7	1.6	2.0	2.4	2.8	8.0	10.0	12.0	14.0
18	.5	.6	.7	.8	2.0	2.4	2.8	3.2	10.0	12.0	14.0	16.0

[1]The spraying time should equal the time for 1 and preferably 2 revolutions of the gear to insure complete coverage. Periodic inspections should be made to insure that sufficient lubricant is being applied to give proper protection.

[2]Four hours is the maximum interval permitted between applications of lubricant. More frequent application of smaller quantities is preferred.

Table 14-8. VISCOSITY RANGE OF AGMA LUBRICANTS

AGMA Mild EP Lubricant Number	Viscosity Ranges (SUV Seconds) At 100° F	At 210° F
2 EP	280–400	
3 EP	400–700	
4 EP	700–1000	
5 EP		80–105
6 EP	-	105–125
7 EP		125–150
8 EP		150–190

Table 14-9. RECOMMENDED AGMA MILD EP LUBRICANTS FOR INDUSTRIAL ENCLOSED GEARING*

Type of Unit	Size of Unit Main Gear Low-Speed Centers (in.)	AGMA Mild EP Lubricant Number, Ambient Temperature (°F) 15–60	50–125
Parallel shaft—single reduction	Up to 8	2 EP	4 EP
	Over 8–up to 20	2 EP	4 EP
	Over 20	4 EP	4 EP
Parallel shaft—double reduction	Up to 8	2 EP	4 EP
	Over 8–up to 20	4 EP	4 EP
	Over 20	4 EP	4 EP
Parallel shaft—triple reduction	Up to 8	2 EP	4 EP
	Over 8–up to 20	4 EP	4 EP
	Over 20	4 EP	5 EP
Planetary gear	OD of housing—up to 16	2 EP	4 EP
	OD of housing—over 16	4 EP	4 EP
Gearmotors	All sizes	2 EP	4 EP
Spiral or straight bevel gear	Cone distance—up to 12	2 EP	4 EP
	Cone distance—over 12	4 EP	5 EP

* Worm gearing is excluded from these recommendations. Some mild EP oils have been used for worm gear lubrication. For specific recommendations, consult the gearing manufacturer.

15

TORQUE CONVERTERS AND TRANSMISSIONS

Mechanical components designed to transfer power from one mechanism to another by a rotating motion may be simple or complex. A simple process may require only a coupling to transfer power along a common axis without change of torque or angular velocity, as in the case of a motor-generator arrangement for the conversion of alternating current to direct current. A complex process may involve changes in torque, angular velocity, direction, and/or plane.

TORQUE CONVERSION

Torque is the twisting force applied to a rotating member of a machine. It is measured in pounds-feet, where "pounds" is the force applied perpendicularly to an arm (moment arm) extending from the axis of rotation, and "feet" refers to the straight-line distance from the axis to the point on the arm to which the force is applied. Thus if a force of 500 lb is applied to a point on the arm 2 ft from the axis, the torque is said to be $500 \times 2 = 1000$ lb-ft. For a gear, wheel, shaft, or any other circular component the moment arm is represented by a straight line scribed radially anywhere between the axis of rotation (center of the shaft, usually) and the point (action line) to which a drive component transfers its force. The point to which maximum force is applied between a drive gear and driven gear (involute profile) is usually just above the pitch circle. As applied to gears, wheels, and pulleys, the force $\times$ moment arm $=$ moment of force, or torque.

Power is a function of both torque and angular velocity at the point of applied force when the driving component and driven component are circular, regardless of axis relationship. Power is also a function of time, so that the relationship of all factors can be expressed as lb-ft × angular velocity. Thus if a circular component rotates at an angular velocity of 3300 ft/min when a torque of 1000 lb-ft is applied, the power developed would be

$$\frac{\text{torque} \times \text{distance}}{\text{time}}$$

or in one minute the conditions expressed above would be 1000 lb-ft × 3300 ft/min, or 3,300,000 ft-lb. Since 33,000 ft-lb/min is equal to one horsepower, 3,300,000/33,000 = 100 hp. If the 1000 lb-ft torque is the result of applying a force of 1000 lb on a line of action 1 ft from the axis of a wheel, the remaining factor, angular velocity, must correspond to any change in the length of moment arm. This means that by applying the force of 1000 lb to a line of action 0.5 ft from the axis, the torque would become 0.5 × 1000, or 500 lb-ft. Considering that the drive wheel rotates at constant speed, the angular velocity of the driven wheel must respond correspondingly to obey the law of energy conservation: energy can be converted but never gained or lost. The corresponding angular velocity would therefore be

$$\frac{500 \text{ lb-ft} \times \text{angular velocity}}{33{,}000 \text{ ft-lb}} = 100 \text{ hp}$$

and

$$\text{angular velocity} = \frac{100 \times 33{,}000}{500} = 6600 \text{ ft/min}$$

If the moment arm is extended to 2 ft, then the torque would equal 2000 lb-ft, so that

$$\text{angular velocity} = \frac{100 \times 33{,}000}{2000} = 1650 \text{ ft/min}$$

Where the driver and driven components are of the same radius, there will be no torque change relative to each other, such as transmitting power from one gear to another of the same pitch diameter. This is because the moment arms of both gears are equal so that the angular velocities also remain the same. The only changes that occur in this arrangement are (1) a change in rotational direction and (2) some power loss due to friction, which is the efficiency factor and must be shown in the power equation.

Providing the speed and size of the drive component remain constant, any change in the moment arm of the driven component (whether such changes occurs in steps or by modulation) brings about a commensurate change in torque, so that any power transmission that brings about such a change is known as a torque converter, or speed changer.

Rotating members through which such conversions are made may be of the stepped type such as gears, fixed pulleys connected by a belt, and sprockets connected by a chain. Where such changes occur by infinite variation, the rotating members used are cones, disks, adjustable cone pulleys, interlocking dual pulleys, variable-speed pulleys, and various eccentric cam arrangements. The connecting drive elements in variable-speed transmissions are flat belts, V-belts, chains, and steel rings. Compounded arrangements include gear trains and sun gears.

Transmitting components include flexible couplings, gear type, universal joints, slipper joints, mechanical clutches, hydraulic and electrical couplings, variable-speed drives, belt drives, and chain drives.

COUPLINGS

A coupling transmits power from one component to another on a common straight-line axis. Couplings may be rigid or flexible.

RIGID COUPLINGS are primarily a pair of keyed or splined flanges placed on the adjoining ends of coaxial shafts and bolted tightly together. Being rigid, they have no internal relative motion and so cannot compensate for any axial misalignment in respect to adjoining shifts. Rigid couplings have no need for lubrication.

FLEXIBLE COUPLINGS serve the same purpose as rigid couplings but are designed to allow for limited misalignment of the adjoining components. Misalignment is due to one or more of the following: temperature changes, vibration, installation inaccuracies, structural deformation (as from settling, loading, or excessive overhang), and bearing wear.

Flexible couplings can be classified as either (1) *flexing couplings* or (2) *articulated joints*. Regardless of type, they should be able to cope with the misalignment anticipated, angular and/or parallel (Figs. 15-1 and 15-2). The closer they approach the driving characteristics of a rigid coupling, especially at high speeds, the greater the positive response. Another important factor in flexible coupling behavior is free end-float, which allows some axial drift.

Flexing couplings are those that depend entirely on the resiliency of the flexing component to compensate for axial distortion (Figs. 15-1 and 15-2). Nonmetal flexture materials such as rubber and synthetic rubber relieve the stresses that would otherwise develop in the metal parts. Some flexing couplings compensate for angular and parallel misalignment and make end-float allowance. They do not require lubrication.

Articulated joints are those that attain flexibility by mechanical means, some of the more important types of which are gear, grid, pin and bushing, chain, block and jaw, disk, and slider. All require lubrication.

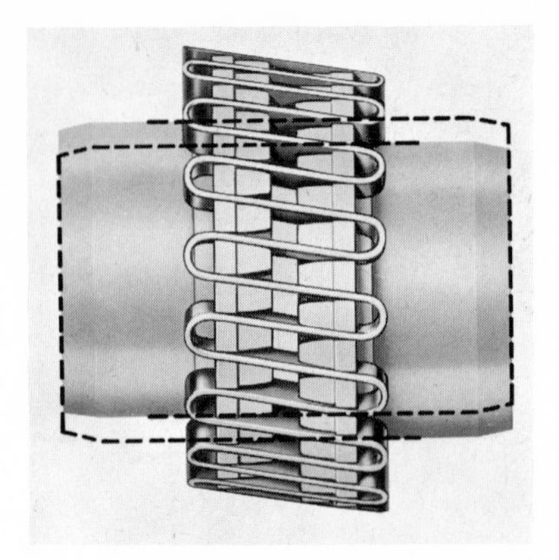

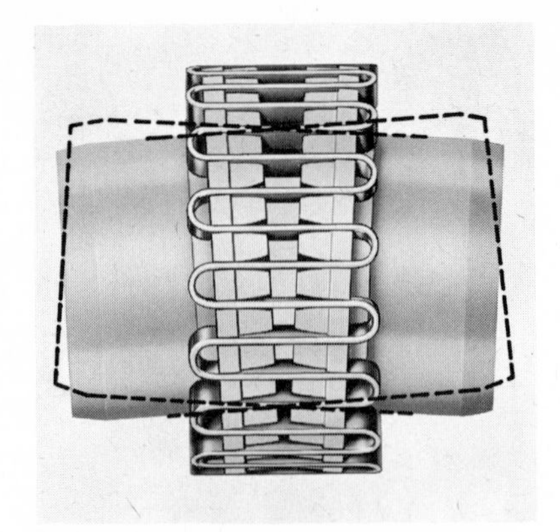

Fig. 15-1. Flexible coupling, articulated joint (grid type). (Left) parallel misalignment; (right) angular misalignment. *Courtesy, The Falk Corp., subsidiary of Sundstrand Corp.*

Couplings are selected on the basis of ability to (1) transmit the required torque and speeds, (2) meet the load characteristics (steady, fluctuating, reversing), (3) compensate for misalignment, and (4) remain rigid and in balance under all operating conditions.

GEAR TYPE COUPLINGS. The conventional gear type coupling has either a straight or crown tooth form (Fig. 15-3) designed to permit limited flexibility. The coupling consists of two identical flanged hubs with external teeth cut on the outer rim. An outer housing has internal teeth which mesh with the external rim teeth. The power is transmitted from one toothed (or splined) flange to the center member and thence to the other toothed (or splined) flange.

Misalignment is compensated for by a sideward spline action between the meshing teeth and not by any relative rotating motion between the gears. The gear teeth are a form of spline drive and are located as far away as

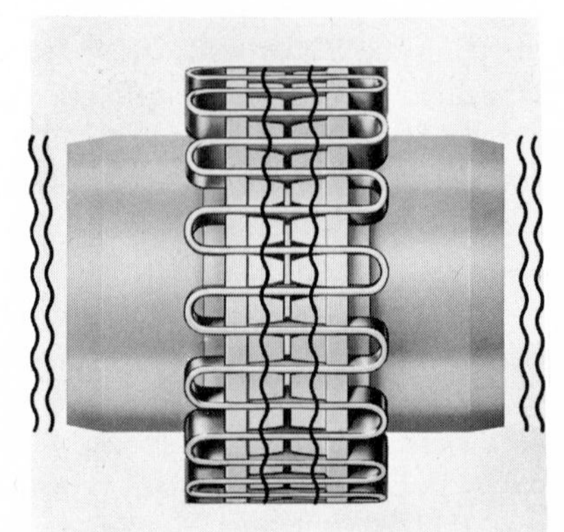

Fig. 15-2. Flexible coupling, free end float. The grid type coupling shown is suitable for most industrial applications. *Courtesy, The Falk Corp., subsidiary of Sundstrand Corp.*

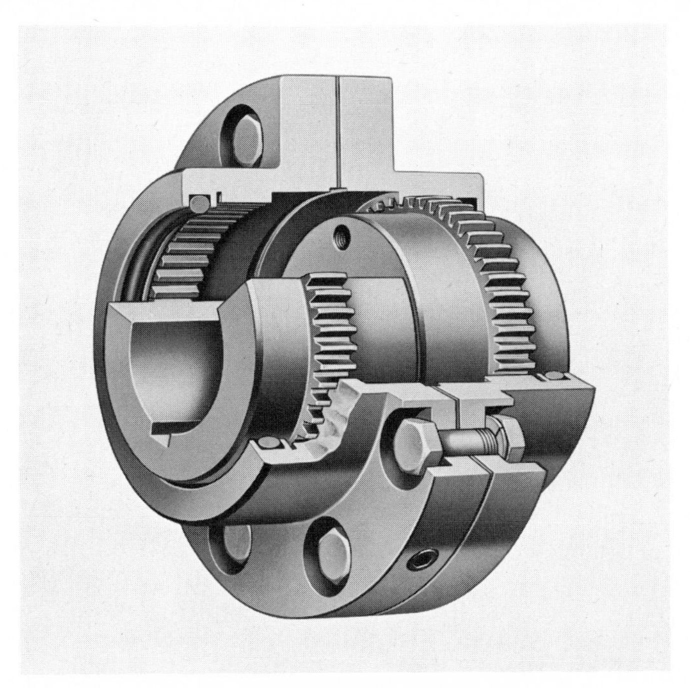

Fig. 15-3. Crowned tooth gear coupling. Lubrication plugs for each sleeve are located 180 degrees apart. One is visible here at bottom. Coupling can be assembled with plugs 90 degrees apart to permit convenient access. *Courtesy, The Falk Corp., subsidiary of Sundstrand Corp.*

possible from the ends of the adjoining shafts to permit maximum flexibility. Couplings with two hubs are double-engagement couplings. There is also a single-engagement coupling which has teeth only on one of the hubs and in the half sleeve used with that hub.

Adequate lubrication of gear couplings is essential for satisfactory operation. They should be lubricated at least every six months, oftener when the coupling is exposed to excessive moisture, extreme temperatures, rapid reversing, shock loads, or excessive misalignment.

Lubricant Specifications. The use of a lubricant specified by the coupling manufacturer, even for normal service conditions, is recommended. Such lubricants should take precedence over those described under "Coupling Lubrication" in the event of any variation.

If the application involves unusual operating conditions (heavy shock loads, frequent axial movement, large speed variation, or extreme temperatures), such information should be submitted to the manufacturer for his special consideration and specific lubricant recommendations. The above rules should apply to all types of equipment, including various type couplings.

GRID COUPLINGS. The grid type flexible coupling consists of two flanged hubs milled axially on the outer edges for the insertion of a snug-fitting grid (Fig. 15-4). The grid is a spring-like component formed from a steel ribbon folded alternately over itself and then formed into a circle and inserted into the axial slots milled in the hubs. The grid is both an interlock for the hubs and the flexible element which allows all four of the shaft distortional influences listed previously. It is the rocking, bending, and sliding action of the grid within the slots that provides flexibility.

A torque control grid type coupling is designed with a friction lining on a torque sleeve to allow continued rotation of the driveshaft in the event of excessive resistance of the driven shaft to turn for any reason. The amount of friction is controlled by springs, which are adjusted by bolts that pass through them from the torque sleeve to a flange on the opposing hub. Another type of grid coupling (spacer) may be installed or removed without disturbing the connecting shafts.

Grease Specifications. The manufacturer of the Steelflex coupling (Fig. 15-4) publishes the following lubrication instructions for their grid coupling:

> Adequate lubrication is essential for satisfactory operation. Lubricate coupling at least once every 12 months. Lubricate oftener when the

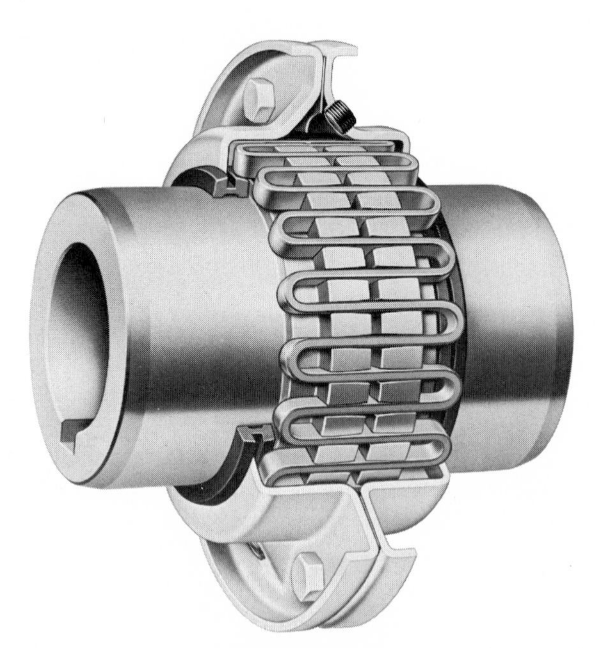

Fig. 15-4. Tapered grid (Steelflex)® coupling with horizontal split cover especially suited to reversing service. Also available with vertical split cover. Note seal. *Courtesy, The Falk Corp., subsidiary of Sundstrand Corp.*

coupling is exposed to excessive moisture, extreme temperatures, rapid reversing, shock loads, or excessive misalignment.

The following specifications are recommended for the Falk Steelflex couplings that are lubricated annually and operate within ambient temperatures of 0 to 150°F (−18 to +66°C). For temperatures beyond this range, consult the factory.

Drop point—300°F (149°C) or higher.
Consistency—NLGI No. 2, with worked penetration value in the range of 250 to 300.
Separation and resistance—Low oil separation rate and high resistance to separation from centrifuging.
Liquid constituent—Should possess good lubricating properties equivalent to a high quality, well-refined petroleum oil.
Inactive—Must not corrode steel or cause swelling or deterioration of neoprene.
Clean—Free from foreign inclusions.

Semipermanent Lubrication. Up to three years between lubrication checks may be obtained through the use of "still-bottom" asphaltic-based lubricants (residuums) with a viscosity of 2000 SUS at 210°F (99°C). At normal temperatures, this lubricant is tacky, and therefore not easily pumped. Couplings should be hand packed, or the grease heated and poured into the coupling. Some of these greases are available with a cutback solvent which facilitates use in a grease gun. These greases, however, are not recommended for applications that exceed 100°F (38°C) ambient.

PIN AND BUSHING COUPLINGS. The pin type coupling is composed of two hubs whose flanges are connected by pins imbedded in rubber-bushed holes in the opposite flange. The rubber acts as a resilient element that can compensate for any limited axial misalignment. Bronze bushing inserts are used to line the rubber surface, which would otherwise be in contact with the pins, to prevent chafing. Synthetic rubber should be used to resist the effect of lubricants. An aluminum soap-base grease of NLGI No. 2 consistancy is again preferred for the lubrication of pin type couplings because of its resistance to oil-soap separation under the centrifugal forces involved. Pin type couplings are variously constructed, the one described above being representative of the others.

CHAIN COUPLINGS. The opposing hubs of a chain type coupling (Fig. 15-5) have sprockets instead of flanges at the outer ends. Around these is wrapped either a single-strand roller chain (with cylindrical rollers) or a silent chain.

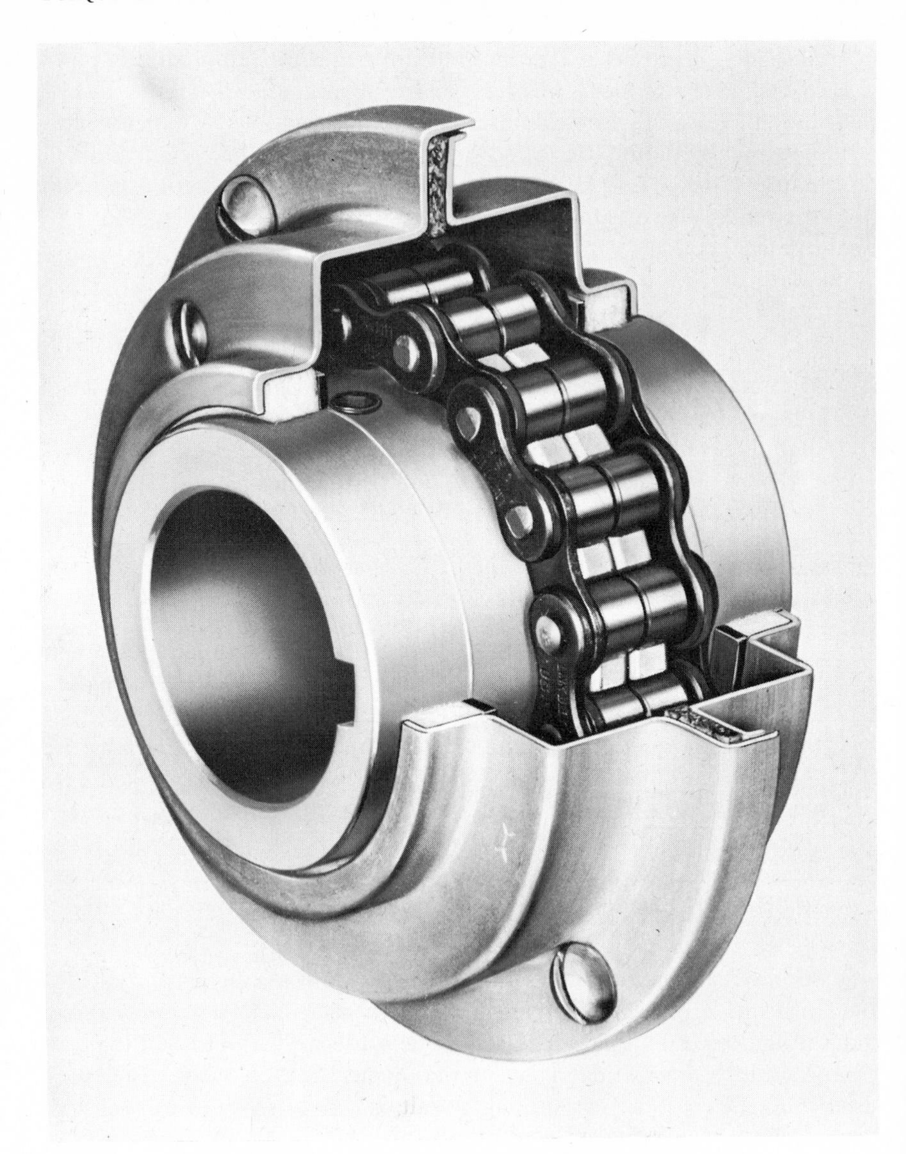

Fig. 15-5. Chain coupling with roller chain. Note access for lubricant. Remove screw and apply grease. Note seals. *Courtesy, Link Belt Chain and Conveyor Components Division, FMC Corp.*

A chain coupling is usually encased to protect it against corrosives, abrasives, and lubricant throw. The casing also provides a safety factor. The flexibility of chain couplings is the combined effect of roller pin clearances and general flexing capabilities of the chain and sprocket arrangement.

Lubrication of Chain Couplings. To obtain maximum operating efficiency and long life, roller chain shaft couplings should be enclosed in a housing and properly lubricated. Lubricate with a noncorrosive, nonseparating grease of medium consistency having a melting point of at least 50°F above surrounding room temperature. Greases used for lubricating universal joints are satisfactory. Do not use ordinary cup grease or greases that separate by centrifugal force. Use NLGI No. 1 grease, preferably of the aluminum soap-base type.

Housings are provided with screw plug openings for grease gun lubrication. Grease should cover the chain and teeth. When housings are completely filled with grease, the action of the coupling will pump the excess lubricant out of the housing during the initial period of operation.

When a casing is not used, pack the coupling chain and teeth with a No. 2 NLGI consistency grease.

Whether the coupling is open or enclosed, periodic oil soaking is advisable in order to provide a protective film on the internal surfaces of links, pins, and rollers. The oil used for soaking should be fairly light (500–700 SUS at 100°F) and should contain a rust inhibitor, a film strength additive, and an adhesive agent. The grease may be an aluminum soap-base grease containing an adhesive agent of No. 2 consistency or a leaded-lime grease of either No. 0 or No. 1 consistency, depending on the ambient temperature, the dividing line of which may be about 140°F.

BLOCK AND JAW COUPLINGS. Another flexible coupling consists of interlocking jaws which embrace a centered square block which acts as the driving member. The jaws and block are drilled to allow passage of the butting shafts. Only sufficient clearance is allowed between all contacting members to allow limited flexing due to angular misalignment without slap or shock. The block and jaw principle allows end float, although its ability to meet parallel misalignment is limited by the clearance between the jaws and block to allow sufficient floating action of the block.

The block and jaw coupling is also known as the Oldham coupling, floating block coupling, and claw type coupling.

Aluminum soap-base grease of No. 2 consistency is recommended for lubrication of this type of coupling; an alternative is a lead-lime soap-base grease of No. 0 consistency.

DISK COUPLINGS. The disk type flexible coupling utilizes laminated metal disk rings between the flanges of two hubs to achieve flexibility. The rings are also bolted to a center member, which may have the form of a circular plate, spool, or floating shaft. Flexibility lies in the bending capabilities of the disks to allow displacement of the center member and permit misalignment of the adjoining shafts. A single type metal disk coupling is made without a center ring in which the disk ring is bolted to one flange and at alternate points to the opposing flange. The rigidity of this type enables it to support shafts of equipment having only one bearing.

MISCELLANEOUS COUPLINGS. The flexible couplings described represent types in more general use. Only their major principles have been covered, since there are many modifications associated with them that offer other benefits in addition to the alignment deformation factors listed previously. Such additional benefits include vertical operation, torque control, overload or shock protection (either by friction or shear), space economy (compactness), bearing substitution (floating shafts), and end float (excessive shaft drift).

Miscellaneous flexible couplings include the double slider, ribbed compression, keyless compression (light loads), cottered joint and shear pin, universal joint, and slipper joint.

Coupling Lubrication

Rigid couplings and flexing couplings do not require lubrication, since there is no relative motion between metal parts. Couplings with preimpregnated frictional parts, so-called "oiless bearings," in continuous extended operation require that the lubricant with which they were initially impregnated be replenished periodically. Parts whose frictional surfaces are fabricated of materials possessing a low coefficient of friction, such as a nylon chain and sprocket, are self-lubricating, as are surfaces overlaid with Teflon.

For all other articulated couplings, the lubrication requirements, including those of the specific applications that have been mentioned, are based on the factors of temperature, angular velocity, load transmitted, radial position and location of frictional parts, effectiveness of seals, coupling enclosed or open, and type of lubricant application facility.

TEMPERATURE. Cold operation, as prevalent in low-temperature climates, cold storage rooms, high altitudes etc., requires an oil with a pour point below 0°F, and a viscosity less than 220 SUS at 100°F.

Intermediate temperature operation, between 0 and 40°F, requires a lubricant viscosity of 225 to 250 SUS at 100°F. Low pour point leaded compounds of appropriate viscosity are also recommended for this environment.

Temperatures between 40 and 120°F require either an oil of 600 to 2500 SUS at 100°F or a grease. A lead-lime-base grease No. 0 or No. 1 (NLGI) consistency or a No. 2 aluminum soap-base grease is preferable. A leaded compound having a viscosity of about 400 SUS at 210°F is also highly satisfactory.

High-temperature conditions describe levels up to about 500°F. Such environments require a grease having excellent oxidation resistance and a drop point approaching the maximum temperature of coupling exposure. High-temperature greases have a nonsoap base and are made with an oil of a viscosity corresponding to that required if an oil were used under the same physical conditions, including temperature, speed, and load. Up to the temperature at which some physical and chemical changes are initiated when continuously exposed (about 150°F), lead compounds are excellent lubricants. That temperature varies according to the quality of the compounds, including the type of lead soap used in their manufacture.

ANGULAR VELOCITY. The peripheral speed of couplings governs the centrifugal forces developed by any fluid contained within them. It also centrifuges the heavier ingredients of a lubricant, including greases, so as to cause separation. A high angular velocity thus presents two important considerations:

1. Couplings with their critical frictional components close to the outer radial extremity require that the lubricant be continuously available in that region. If the design presents tortuous labyrinths through which a lubricant must pass to reach the remote outer parts, then that lubricant must possess adequate fluidity. Otherwise such parts become starved for lubrication. This poses no problem if the lubricant is applied directly to peripheral parts except that the more centralized parts may be neglected. To be effective, the lubricant must circulate through all parts of the coupling. This will not occur if the fluidity of the lubricant requires a centrifugal force magnitude that does not exist. Such a condition develops in a coupling turning at low angular velocity with a grease having an inappropriately high consistency, or with an oil whose pour point is inconsistant with a low temperature. A demarcation level between the use of an oil or a grease is recommended at about 60 ft/min, below which oil should be used. This level is predicated on the alternative use of such greases formed to be otherwise best suited for couplings including those recommended previously in this chapter for the various types. The above conditions also explain the reasons for not using high consistency greases in couplings under any normal operating situation.

2. The centrifuging factor refers chiefly to the use of greases for the lubrication of couplings. The bonding power of the ingredients in some greases is greater than in others. Under the elevated centrifugal forces developed in high-speed couplings, such as used on large turbogenerators, this bond must be exceptionally strong to prevent, primarily, the separation of the soap from the oil. Where this occurs, examination shows a nonlubricating residue remaining in the coupling after relatively short periods of continuous operation.

LOAD TRANSMITTED. Lubrication of heavily loaded couplings, such as those used in steel mill service, must receive the same consideration as that of any heavily loaded plain bearing. This is particularly true of spindle couplings, both gear and slipper types, subject to high unit pressures developed on large rolling mills. This requires lubricants of high film strength, good heat resistance, and adhesiveness. Heavy-bodied leaded compounds, lithium soap base greases containing a lead-sulfur extreme pressure additive plus molybdenum disulphide, and calcium complex greases also containing an extreme pressure additive are usually considered for steel mill coupling service.

Although the lithium type grease appears to maintain a more stable consistency when water leaks into the bearing area, the calcium complex grease shows greater heat resistance. Both types are generally water repellent. This is important in steel mill operation because of exposure to a large volume of water in certain areas. The use of leaded compounds must be considered in view of their temperature limitations and the necessity for continuous (automatic) application. (See slipper couplings, following)

RADIAL LOCATION OF FRICTIONAL PARTS. Lubricant selection in regard to the location and accessibility of frictional parts of couplings is discussed under Angular Velocity above. We merely emphasize the necessity of selecting a lubricant capable of absorbing and dissipating heat generated within the coupling.

EFFECTIVENESS OF SEALS. Oil, being a more effective heat transfer agent than grease, is preferred in some operations. Its use is practical only when the coupling is adequately sealed.

TYPE OF LUBRICANT APPLICATION FACILITY. Some enclosed couplings are equipped with fittings through which grease or oil is applied by pressure gun. Others require periodic grease packing by hand. Large, heavily loaded couplings transmitting hundreds of horsepower may be open or enclosed and operate for many hours between shutdown periods. Grease must be used if there is no provision for continuous automatic application of leaded compounds, which is generally the case. Even where grease is used, inspection and application are standard operating procedures between turns or shifts of steel rolling mill operations.

UNIVERSAL JOINTS AND SLIPPER COUPLINGS

It is often necessary to transmit power at some limited angularity relative to the main power shaft. Beyond the angular limitations allowable with either the universal joint or slipper coupling either one of the other torque converters or transmissions must be used i.e. gear arrangements, chain drives, flexible shafts or belt drives. Where the axial line of power is offset ie. the ultimate drive shaft is parallel but not in alignment with the main power shaft, double universal joints or slipper couplings are used. Angularity may be variable or constant. Shafts may also be made to telescope in order to vary the distance between ends.

Universal Joints

The universal joint (Fig. 15-6), or *Hookes coupling,* consists of a swivel arrangement enabling one shaft to rotate at the same speed as the other while allowing it to turn about the axis of the joint at some desired fixed

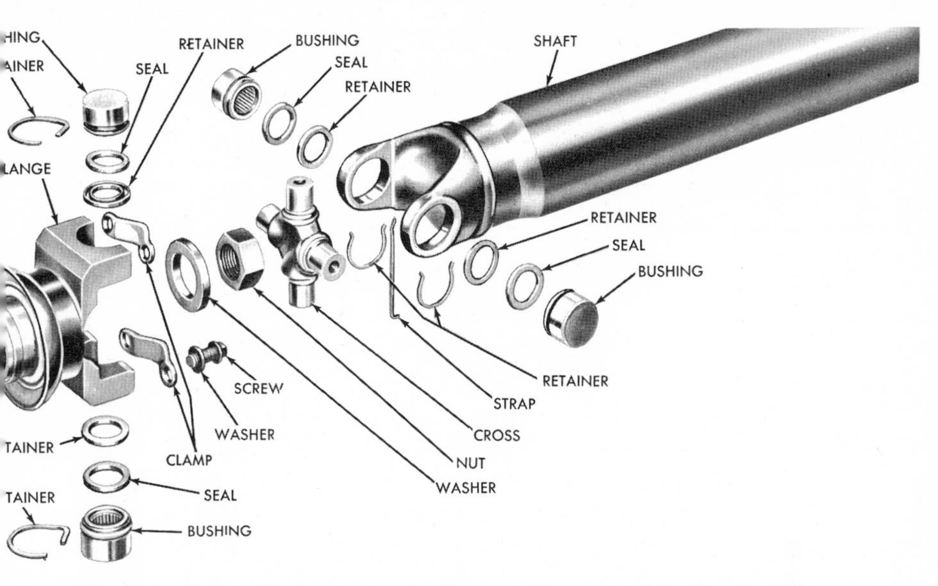

Fig. 15-6. Universal joint, cross and roller type. *Courtesy and ©, Chrysler Corp.*

or variable angle. In the case of universal joints the swivel is fixed by a pin, bolt, or other appropriate fixture so that there is no relative axial motion between connected shafts.

Slipper Couplings

The slipper coupling transmits angular motion as does the universal joint but differs in the relation between the connecting components and the magnitude of power transmitted. The slipper coupling (Fig. 11-35) consists of a jaw and spade (or tongue). The spade, located on the drive shaft, is inserted into the jaw, located on the driven shaft, and the assembly turns as the jaws act in a twisting motion to turn the spade.

Unlike the universal joint, the driven member of a slipper coupling is free to move in a limited axial direction with the jaw while turning about the axis of the slipper. Brasses are attached to the jaws to provide an effective frictional surface. This eliminates the need of a pin, which would be subject to high shearing stress of the thrust loads generated by the metal rolling processes performed on heavy-duty steel mills on which they are used.

CLEARANCE BETWEEN SPADE AND JAWS. The need to move axially requires that sufficient clearance be maintained between the opposing surfaces of the spade and jaw to provide for an adequate lubricating film. If the clearance is not sufficient, high operating temperatures and rapid wear result. The clearance dimension depends on the material of construction, whose expansion rate is a function of conducted heat from the rolling process, the size of the slipper, and the desired temperature of operation predicated on the allowable shock or slap. The latter is relative to the contacting surfaces that are subject to impact during periods of startup, shutdown, or more serious, frequent torque changes during operation.

The method of mathematically determining a practical clearance between the spade and jaw surfaces of a particular slipper coupling used on a steel (slab) mill follows.

Thickness when formed at 80°F:

Clutch jaw lip	10 in.
Brasses (bearing)	2.75 in.
Spade	15.5 in.

Materials:

Clutch jaw	forged steel
Spade	forged steel
Bearings	alloy of copper, aluminum & iron.

Coefficient of expansion of structural members:

Forged steel	0.00000685 max.
Brasses	0.0000095 (calculated according to alloy ingredients)

Spade dimensions at n temperature rise:

$$15.5 \text{ in.} + [(15.5 \times 0.00000685) \times n] = 15.511$$

where

$$n = 100°\text{F rise.}$$

Where $n = 100°$F, 120°F, 130°F, . . . , 180°F, then the spade expansion for each 10°F increment of temperature rise would be as shown in Table 15-1, based on a starting room temperature of 80°F prevailing when the slipper was formed.

Table 15-1. ALLOWANCE FOR SPADE EXPANSION

Temperature of Spade (°F)	*Temperature Rise (°F)*	*Spade Dimension (in.)*	*Spade Expansion (in.)*
80	—	15.5	—
180	100	15.51062	0.011
190	110	15.51170	0.012
200	120	15.51274	0.013
210	130	15.51380	0.014
220	140	15.51486	0.015
230	150	15.516	0.016
240	160	15.517	0.017
250*	170	15.518	0.018
260	180		

* Temperature (bulk) at 0.008-in. clearance which instigated this study and during which slipper life did not extend beyond 40 hr of operation.

Brass dimensions at n temperature rise:

$$2.75 + (2.75 \times 0.0000095) \times n = 2.7526$$

where

$$n = 100°\text{F temperature rise.}$$

With the same procedure as used for the spade, for the same increments of temperature rise the brasses would expand as follows:

Temperature of Brass (°F)	*Temperature Rise Above 80° (°F)*	*Brass Size and Expansion (in.)*
80	0	2.75
190	110	2.752874
200	120	2.753135
210	130	2.75339
220	140	2.75365
230	150	2.75392
240	160	2.75417
250	170	2.75444
260	180	2.75469

Clutch dimensions at n temperature rise, determined as for the above increments:

$$10 \text{ in.} \times 0.00000685 \times n$$

Temperature Rise (°F)	*Brass Expansion (in.)*
100	10.00343
110	10.00377
120	10.0042
130	10.0044
140	10.0048
150	10.00514
160	10.0055
170	10.00583

Assembly dimensions (inches) at n temperature rise
$n = 0$

Spade	15.5
Clutch	10.0
Brass	2.75
Total	28.25

$n = 100°F$ rise

Spade	15.511
Clutch	$10.00 + \frac{0.00685^*}{2} = 10.00343$ in.
Brass	2.7526
Total	28.26703
Original	−28.25
	0.01703 = 0.017-in. clearance

Continuing the same method of calculation for all values of n from 100 to 170, the clearance in each case would be as follows:

Operating Temperature (°F)	*Minimum Clearance Under Maximum Heat Conditions (in.)*
80	0.008*
180	0.017
190	0.019
200	0.021
210	0.022
220	0.024
230	0.025
240	0.027
250	0.029

* Original clearance

Considering the temperature at which an optimum fit would be accomplished to minimize slap and to maintain a reasonably low lubricant oxidizing environment, a clearance of 0.017 in. was selected. This was obtained by removing about 0.0085 in. from each of the brass faces. With the automatic application of a leaded compound with a viscosity of about 1800 SUS at 100°F, the recommended clearance of 0.017 in. resulted in a reduction of bulk temperature to about 185°F and an indefinite extension of slipper life.

The methods of continuous and automatic lubricant application to slipper couplings is described in Chapter 11. When hand lubricated periodically, as during machine downtime, a grease of an NLGI No. 0 or 1 with extreme pressure characteristics should be applied with a pressure gun.

GREASE RECOMMENDATION. A lime-lead base grease of No. 0 or 1 NLGI consistency having a Timken OK load of about 35+ and a drop point of 400°F, or a leaded calcium complex grease having the same specification.

*Divided by 2 here because the clutch jaw expands in all directions.

OIL RECOMMENDATION. A leaded compound with a viscosity of about 150 to 400 SUS at 210°F. Misting systems should include an oil heater to reduce viscosity sufficiently to facilitate application. A temperature of 100 to 120°F is usually maintained but should not be exceeded.

TORQUE CONVERTERS

Many of the components referred to as transmissions are actually torque converters, because they change a given torque and angular velocity to another torque level turning at the same rate. Thus gear sets, chain drives, automotive transmissions, mechanical clutches, variable-speed units, and some types of slip couplings (electric and hydraulic clutches) are all torque converters. Gears are treated in Chapter 14. Planetary gears, sliding gears, constant mesh gears, and differential transmissions are all torque-converting components.

Clutches

There are several types of clutches including jaw clutches, friction clutches, and overrunning clutches. Clutches usually ease or dampen the shock of a load such as that carried by a drive shaft and transmitted to a driven shaft by direct engagement, but not necessarily providing torque conversion.

FRICTION CLUTCHES involve the use of opposing surfaces that mutually engage to transmit power by the interlocking force of friction. The two opposing surfaces may be disks or cones and are known respectively as the *drive plate,* or *pressure plate,* and the *driven plate,* or *cover plate.* The plates are held together during operation by spring pressure and are usually disengaged by pushing a pedal mounted on a shaft (fulcrum), which also carries a throwout yoke that presses against a thrust bearing. When the pedal is pushed forward, the yoke pressing on the thrust bearing moves the sleeve to which the pressure plate is attached, withdrawing that plate from the driven plate against the tension applied by the spring.

OVERRUNNING CLUTCHES transmit rotating motion in one direction only. If for any reason the drive shaft stops, the driven shaft continues to rotate by its own inertia (overrun) until it also comes to a halt. If the direction of the drive shaft is reversed, it will not engage the driven shaft and no power is transmitted. One method of obtaining overrun is to provide a series of wedge-shaped radial protrusions on the drive race (Fig. 15-7). Rollers are inserted in these wedges between the drive race and the driven race located around the drive member. As the drive race rotates in the same direction as the rising slope of the wedge, the rollers are forced upward to

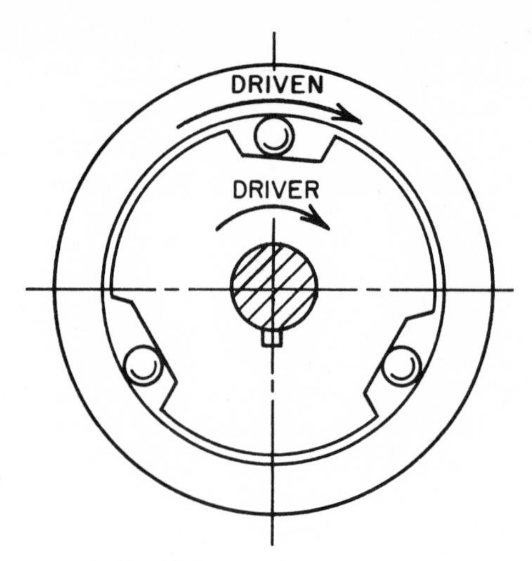

Fig. 15-7. Overrunning clutch.

become wedged between the two races, causing the driven race to rotate with the driving member. Cages are inserted to hold the rollers in some slight contact with both races even when power is not being transmitted.

Overrunning clutches are used to prevent reverse motion upon shutdown as with loaded conveyors and hoisting equipment. In such cases the clutches are in an overrun attitude during normal operation and cannot serve the dual purpose of drive and lock.

The *sprag clutch* is a type of overrunning clutch in which a dragging member, or an inclined sprag, runs free in one direction but binds in the opposite direction. It is held against the surface of the opposing race by a *garter spring*. A triple wrinkle energizing spring is used instead of a garter spring in some drag clutches.

Lubrication of Clutches

Friction clutches, with the exception of those used on pneumatic and electric hoists, require lubrication only on the pilot and release bearings. These bearings are designed either for oil or for grease. If designed for grease, application may be performed by hand (packed) or by pressure gun. The grease preferred is the same one required for wheel bearings, which is a short-fiber soda base grease having an NLGI No. 5 (worked penetration of about 260 at 77°F.). Some manufacturers recommend an NLGI No. 2 grease.

Clutch release bearings designed for oil require one with a viscosity of 500 SUS at 100°F, which corresponds to SAE 30 rating.

Some types of pneumatic hoists require oil to lubricate and cool their friction plates. The demand here is for an oil that can be readily squeezed from between the plates when full power transmission is required but that will otherwise protect them when the maximum pressure is not required and slippage occurs. Successful performance is obtained with an oil having an API gravity of 29.0 to 30.0, a viscosity of 300 SUS at 100°F, and a VI of 90 to 100. It should be free of lubricity additives.

Overrunning clutches of either the roller or sprag types require a lubricant that will not only lubricate the frictional parts but also absorb the heat generated when the binding members phase in (take hold). Roller types require a heavy oil having a viscosity of about 900 to 1200 SUS at 100°F (SAE 50); 1200 SUS is recommended when used in overdrives of older type automobiles. Sprag-types are lubricated by oil or grease, depending on design and purpose. For oil-lubricated sprag type clutches, an oil of 150 to 300 SUS at 100°F is recommended by a major manufacturer. Neither roller or sprag clutches should use an extreme pressure or antiwear lubricant because of their deteriorating tendencies under the high spot temperatures developed. Where grease is called for, use a NLGI No. 2 grease having high resistance to oxidation such as a lithium base grease containing an oxidation inhibitor. When operated below 20°F ambient temperature, a lubricant designed for cold operation should be used, such as a refrigeration type oil with a pour point of about −35°F or lower, corresponding to viscosities in the range of 70 to 90 SUS at 100°F.

Centrifugal Clutches

The centrifugal force developed within a rotating system tends toward a straight line in an outward radial direction in respect to the axis of the system. The automatic centrifugal clutch uses this principle. The driving member and the driven member are not in contact when the driving member is at rest or is turning slowly. In such a situation, the driving member, which is a part of the impeller unit, assumes an attitude at some angularity below 90 degrees along the line of shaft axis. When the drive shaft is brought up to sufficient speed, the centrifugal force on the pivotally mounted parts overcomes the force of spring tension, allowing these parts to move out radially, permitting the face of the impeller to press against the runner face and move the driven shaft.

There are various designs of the mechanism that separates the impeller and runner faces during periods of low or zero centrifugal force. The requirement is usually for more rapid response in both take-hold and release at the faces, which may be positioned anywhere from perpendicular (faces) to parallel (shoes) relative to the axis of rotation.

Centrifugal clutches protect the drive component against excessive overloads and jamming of the driven member for any reason. Both of these conditions cause the driver to slow down sufficiently or stop, allowing spring tension to exceed centrifugal force and cause separation of the clutch faces.

Hydraulic Coupling

Hydraulic couplings (Fig. 15-8) are variously designed to act as slip clutches, speed changers, or direct drive transmissions. One advantage of a fluid coupling is the ease with which power conversion can be accomplished without shock and at a much reduced risk of stalling the drive component. It makes possible the connection of a high-speed power supply, such as a turbine or an internal combustion engine, with relatively slower-speed driven components such as a ship's propeller or automobile drive shaft (although the helical gearing now used for ship propulsion has replaced the hydraulic coupling). Fluid drives have become universally adopted to smooth the accelerating process in automobiles and some industrial equipment. They also act to control torque and provide protection to the driving member in case the driven component becomes jammed.

The principle of hydraulic coupling involves the use of the shearing

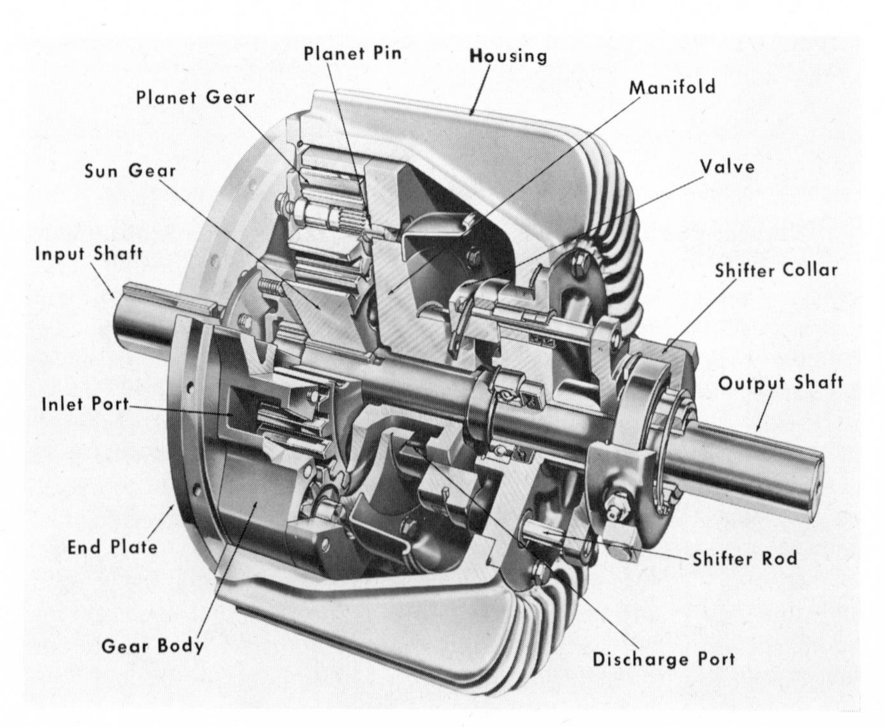

Fig. 15-8. Varidraulic coupling. *Courtesy Dynamatic Division, Eaton Yale & Towne, Inc.*

force of a light-viscosity fluid, usually a mineral oil, to supply the transmitting kinetic energy from a drive member, or *impeller,* to the driven member, or *runner*—both located in a shell containing the fluid. The clutch assembly housed in a casing is never completely filled with fluid. The impeller and runner are vane type rotors attached to opposing but coaxial shafts. When the impeller is rotated the fluid is forced by centrifugal action to flow radially. This sets up a circulating process whereby the fluid enters the runner from the impeller at the outer circumference forcing it toward the inner circumference of the runner, where it reenters the impeller and continues the flow as long as the impeller rotates. As the fluid flows from the impeller under maximum force at the outer circumference, it strikes or rifles through the vanes of the runner, causing the latter to rotate in the same direction but never at the same speed of the impeller. Slippage losses compensate for the difference in speed. The angle of the impeller vanes in respect to the runner vanes relative to the radial plane of the coupling is 90 degrees. Thus if the angle of the runner vane is 50 degrees, the angle of the impeller vane would be $50 + 90 = 140$ degrees, and for 40, 60, 70 degrees, the impeller angle would be 130, 150, 160 degrees, respectively.

To avoid a free flow of the fluid through the runner vanes at some point of revolution, the number of vanes on each rotor must be different. The number of vanes in one rotating element usually corresponds to a ratio of 9.0 or 9.5 to 10, or 0.9 to 0.95 times that of the other. If the number of vanes on each element were equal, the flow of fluid would be unihibited or unobstructed when at some instant during rotation the opposing vane channels were congruent. This would lead to momentary loss in torque conversion accompanied by negative runner velocity impulses.

The heat generated in fluid couplings is often dissipated by circulating the fluid by means of a pump located outside the clutch compartment through a cooling coil. The heat results from the friction from the rapid molecular shear of the fluid.

The slip of hydraulic couplings is a function of impeller speed. It is minimum, about 2.5%, when the impeller is running at full governed speed, and it increases to 100% at some lower governed speed, about 40%.

The hydraulic coupling can be made into a torque converter with high torque ratio at high speeds (strong coupling phase) and low speeds under normal operating conditions (low coupling phase) by the introduction of variable-pitch reactor blades (vanes). In such a component the reactor blades may be operated 75 degrees apart between two extreme positions. The difference in accelerating rate then becomes a function of tangential flow magnitude from one *approaching* that flow angle to one that is *more nearly* tangential flow. The variable pitch converter can change the speed of the runner when the impeller is powered by a constant speed machine such as a synchronous electric motor.

The use of a hydraulic coupling in conjunction with stepped transmissions softens any shock during speed changes. This is due to the smoothing out effect of the slippage that occurs in the coupling.

COUPLING FLUID. Fluids used in hydraulic torque converters are lubricants, but their primary function is that of a power medium. The fluid is in continuous reuse in the hydraulic coupling, and it must be sensitive in its response to torque changes under various environmental conditions, particularly temperature. When internally generated heat is considered as in an automobile engine pulling a trailer up a long grade, the temperature may range from −40 to 300°F. However, because of the mechanical construction of such couplings, lubricating qualities must receive adequate priority.

Desirable characteristics of a hydraulic coupling fluid are long service, maximum torque capacity (which is proportional to fluid weight), noncorroding to parts, nondeteriorating to seals and gaskets, and operating flow capabilities. The approximate specifications shown below are a fair compromise where natural properties are in some conflict with more important required properties. (Some hydraulic converters require a fluid with a lower pour point than −40°.)

Gravity, API	28.0
Pour point, max. (°F)	−40.0
Flash point, min. (°F)	350
Viscosity, SUS at 100°F	200
Viscosity, SUS at 210°F	52
Viscosity index	140
Required additives	Oxidation inhibitor
	Corrosion inhibitor
	Detergent
	Defoaming agent

Because of the expansion properties of oils, the casing of the hydraulic coupling should never be filled completely, especially those used on automobiles to provide automatic transmission.

DRY FLUID COUPLING. (Fig. 15-9) The "fluid" is heat-treated steel shot. This is contained in the housing, which is keyed to the motor shaft. When the motor is started, centrifugal force throws the shot to the perimeter of the housing, packing it between the housing and the rotor. After the starting period of slippage between housing and rotor, the two become locked together to achieve full load speed, without slip and with 100% efficiency. Flexidyne is available in fractional sizes to 1000-hp.

The advantages of dry fluid coupling are as follows: (1) The motor can accelerate quickly and start the load smoothly. (2) Smaller motors can be used, the demand rate is lowered, and the smooth starts prevent breakage and lower maintenance. Starting torque can be tailored to driven load requirements. (3) There is no slip at normal operating speeds, which means

Fig. 15-9. Dry fluid coupling. Dodge Flexidyne. ® *Courtesy, Dodge Manufacturing Corp., division of Reliance Electric Co.*

no wear, no heat, no power loss. (4) Slips at overloads are somewhat greater than the preset starting torque. (5) The current draw is low, usually less than twice the nameplate amperage during starting and overload periods. Reversing loads are easily handled.

A grease similar to that recommended for electric motor bearings is recommended for the ball bearings used for the dry fluid coupling.

Electric Coupling

The electric or eddy-current coupling (Fig. 15-10) consists of a primary rotor and secondary rotor, either of which may be the driving member. The primary rotor is similar to the frame of an electric generator. It contains

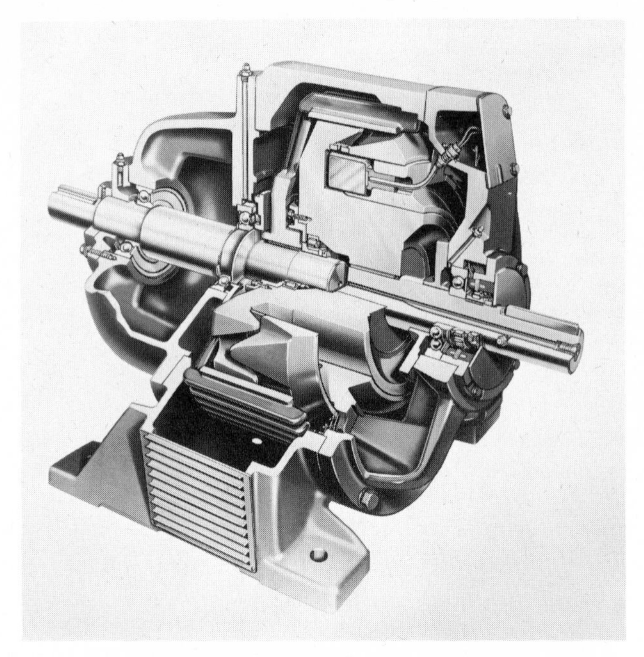

Fig. 15-10. Air-cooled eddy-current coupling. *Courtesy, Dynamatic Division, Eaton Yale & Towne, Inc.*

the windings and inwardly extending poles on which coils are wound that carry the exciting current from a direct current source through a pair of slip rings and brushes. The secondary rotor is an armature having a magnetic core on which are mounted copper bars imbedded in radial insulated slots cut in the outer surface of the rotor. The copper bars are connected to each other by copper rings attached to both ends of each one.

A magnetic flux is developed in the two circuits formed by the two rotors when an electric current is set up through the slip rings from an external source. By rotating one of the rotors the magnetic lines are cut by the copper bars of the secondary rotor to induce an electromotive force. Since these magnetic lines of force are short-circuited, they cause a heavy current to flow, even at relatively slow speeds.

Cutting the lines of force in this manner sets up a reaction between the secondary current and the primary magnetic field causing a magnetic resistance, or drag, to produce torque on the driven rotor. Unlike the hydraulic coupling in which the torque increases continuously with slip throughout the range from minimum to maximum at the point of stalling, the torque of electric couplings increases continuously with slip only up to a point of about 8 to 10% slip, after which it diminishes rapidly. To meet torque requirements of a given operation, a magnetic coupling must be designed for a maximum torque considerably above that of the driving component, or about 70 to 75%, which is attained by increasing the magnetic capacity of the coupling. This means that torque capacity of the coupling be designed within the 0 to 8% slip range.

Because the energy losses within the coupling are converted to thermal energy, the unit must be adequately ventilated or liquid cooled (Fig. 15-11) for heat dissipation.

Lubrication of Electric Couplings. The only frictional parts associated with this coupling are the bearings supporting the overhung rotor shafts. These should be serviced in the manner prescribed for electric motor and generator bearings discussed in Chapter 21.

VARIABLE-SPEED TRANSMISSIONS

Speed changers may be stepped or modulating. Stepped speed changers shift from one combination of torque-speed ratio to another of a higher

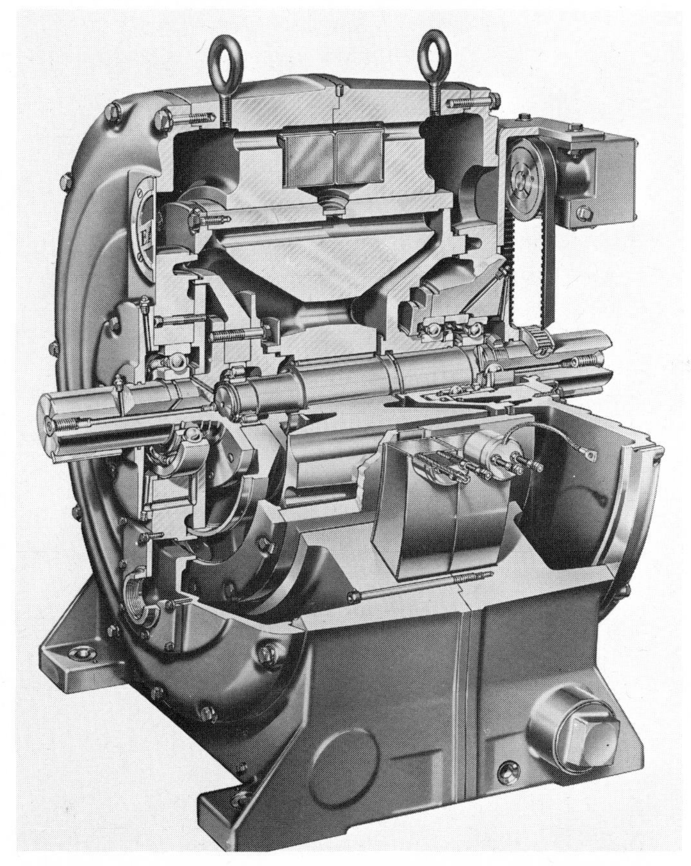

Fig. 15-11. Heavy duty liquid-cooled eddy-current coupling. *Courtesy, Dynamatic Division, Eaton Yale & Towne, Inc.*

or lower ratio in definite steps of, say, from 4 : 1 to 2 : 1 to 1 : 1. This method of changing speed is exemplified by the gear shift mechanism of automobiles. Modulating speed changers, or variable-speed transmissions, change torque-speed ratios gradually from one incremental combination to another. The variation in speed in the latter type is infinite.

Variable-speed drives provide speed control in order to attain synchronization of different mechanical parts, such as maintaining constant tension of material flowing in a manufacturing process, e.g., cold rolling of steel strip and paper drying.

To accomplish variable speed with gears an infinite number of ratios would be required. This in essence would resemble a pair of reversed cones so mounted and manipulated as to allow for the required shaft spacing which would vary as the ratio modulates through the range between minimum and maximum. One of the earlier types of variable-speed transmissions, the *flat belt type,* employed two reversed cones connected by a flat belt. The second cone was guided and shifted by a belt guide (Fig. 15-12). As the belt was shifted along the axis of the cones or pulleys, the speed of the driven cone changed according to the diameter ratio at any point along the line of travel.

The *friction disk type* variable-speed transmission consisted of a driving disk and a driven disk mounted perpendicular to each other (Fig. 15-13).

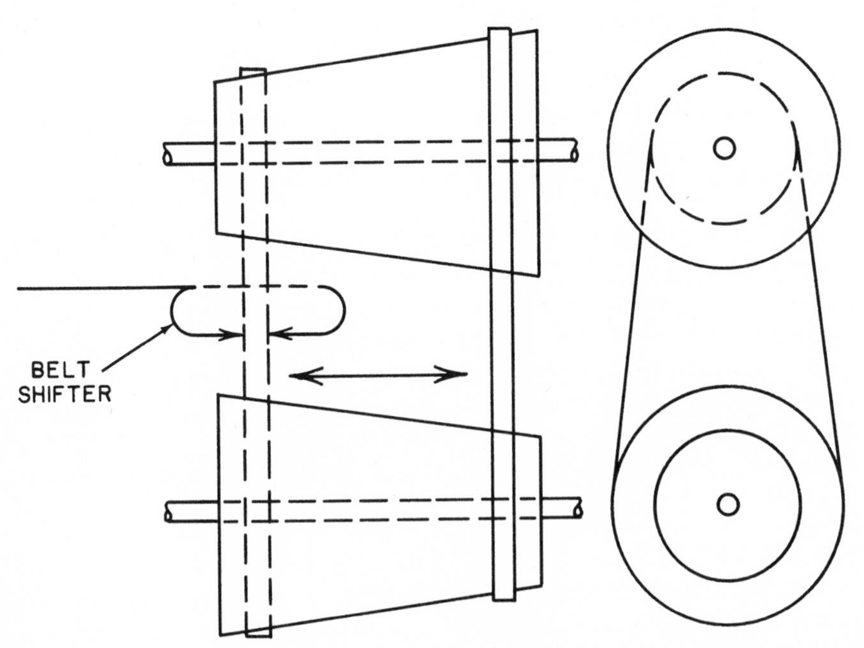

Fig. 15-12. Basic principle of belt-operated variable-speed transmission.

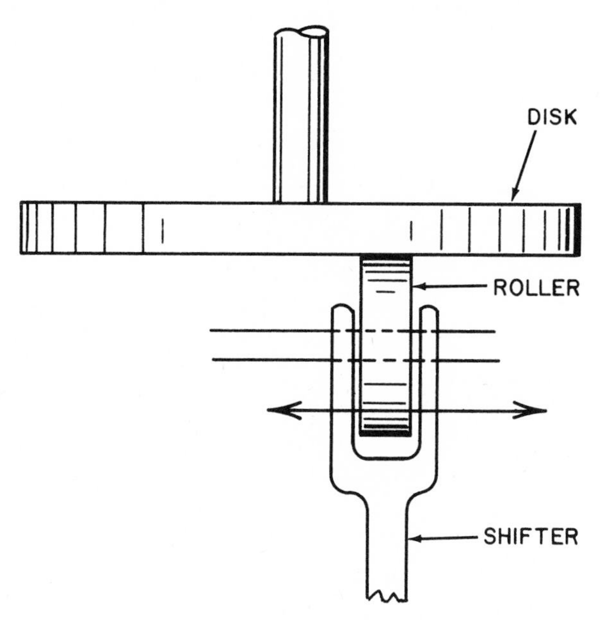

Fig. 15-13. Variable-speed transmission with friction disk.

The driving roller, in frictional contact with the driven disk, can be shifted radially across the face of the disk. The speed of the roller is constant and varies the speed of the driven disk at a ratio commensurate with any given radius of contact.

ADJUSTABLE CONE PULLEYS, PARALLEL SHAFT. Modern variable-speed transmissions employ either a chain, V-belt, or steel ring to connect driving and driven adjustable cone pulleys or V-shaped wheels mounted on parallel shafts.

POSITIVELY, INFINITELY VARIABLE (PIV) TRANSMISSION. The PIV® drive is a positive drive rather than a friction drive. It consists of chain connecting adjustable conical pulleys as in other V-drive types. It differs from the friction conical wheel drives in that the chain and wheel faces are designed to allow positive contact (Fig. 15-14). The chain consists of laminations that slide transversely relative to each other and are tapered at these ends to conform to the cone shape of the wheels. The wheels have alternately staggered radial teeth and grooves in their faces. When the wheels are mounted, the teeth of one are opposite the grooves in the other. As the chain enters the V formed by the conical wheels, the teeth of each force the laminations transversely between the teeth of the opposite wheel so that positive contact is made on both sides of the chain. The teeth are radially oriented to allow the chain to be lifted or lowered away toward the axis, but they prevent it from slipping tangentially.

Fig. 15-14. Cutaway view of Positively, Infinitely Variable (PIV) transmission. Note radial teeth and grooves in the cones. *Courtesy, Link-Belt Chain and Conveyor Components Division, FMC Corp.*

Lubrication of Variable-Speed Drives

Most of the variable-speed drives employing fabric and rubber belts require lubrication only on their shaft bearings, except those equipped with adjusting screws. The bearings are usually equipped with grease fittings. A good grease of NLGI No. 1 consistency is recommended, preferably one possessing extreme pressure characteristics. Screws may be lubricated with either a similar type grease or a fairly heavy oil containing a tackiness additive to prevent drippage. Screws should be run out when they are lubricated to assure complete coverage all along their length. Grease-lubricated worm gears should have all teeth well-coated but not excessively.

Positively, Infinitely Variable (PIV) speed transmissions are oil-lubricated by splash from a bottom reservoir similar to that employed in reduction gear sets. Because of the PIV speed response sensitivity, the oil selected must be of the correct viscosity at any temperature range. The most recent recommendations are as follows:

Below 0°F	150 SUS at 100°F
0 to 40°F	600 SUS at 100°F
40 to 100°F	1800 SUS at 100°F
100 to 150°F	2500 SUS at 100°F

Only a highly refined oil free of impurities should be used. It should have a high viscosity index (over 90) and should contain rust and oxidation inhibitors, a dispersant, and a defoaming agent.

Variable-speed transmissions with steel rings on cast iron pulleys usually require an oil of about 150 SUS at 100°F for best performance, with some allowance made for other than room temperatures. For any make or type of transmission the lubricant should be selected on the basis of the physics involved and the conditions surrounding the installation. Manufacturers' recommendations and instructions usually reflect such factors.

Because of wide variations in design, no general lubricant recommendation can be made to apply to variable-speed drives as a group.

BELT DRIVES

Belt drives are made up of a drive pulley and driven pulley usually mounted on parallel shafts connected by an endless belt that transmits power from one shaft to another. Torque may or may not be converted, depending on the relative diameters of the driving and driven pulleys. Formerly much of the power of industry was transmitted by belt systems. Individual motor drives and more efficient and versatile gearing have reduced the number of flat belt drives, although many still exist on older machines or where their use is desirable or convenient. The flat type belt has been replaced by other types, particularly where the span between shaft centers is relatively short and speed ratios are relatively high. Some of the advantages of friction belt drives include protection against overloading by slippage, no lubrication, and the cushioning effect between driver and driven machines.

Belts vary as to cross section and type and may be flat, round, V-shaped, ribbed, toothed, linked, and cogged.

Flat belts are ribbon-like in cross section and are made in various widths and thicknesses depending on their service. Materials for flat belts include leather, rubberized fabric, canvas, plastic, nylon, and rayon. Leather and rubberized fabrics are the more common materials. Leather requires periodic dressing with a compound that should contain the same currying ingredient used in its preparation.

Compound flat belt power transmission systems include a series of shafting leading from the power source to the final machine. In the order mentioned the system includes—

The head shaft, or drive shaft, leading from the power source usually rotating at about 600 rpm. It transmits power to a line shaft.

The line shaft distributes power at speeds up to about 400 rpm to various jack shafts.

Jack shafts are speed changers and provide desirable angular velocity to countershafts.

Countershafts furnish the transmitted power to individual machines, either directly or through speed reducers, variable-speed drives, clutches, etc.

Flat belts are usually mounted on pulleys that are suitably crowned to keep the belt centered. The principle is the tendency of a moving belt to climb to its largest diameter. Crowns range from less than $\frac{1}{16}$ in. to a maximum of $\frac{1}{8}$ in. per foot of pulley face width, the rule being that the faster the belt speed the less crown required. However some pulleys are flat while others are tapered for speed adjustment.

Flat belts should be made 1 in. narrower than their pulleys up to 7 in. face width and 2 in. narrower for pulleys above 7 in. Reverse direction relative to parallel shafts is accomplished by crossing the belt.

V-belts attain a wedge action (Fig. 15-15) when running in a grooved sheave. They are never given a complete V-shape but are flattened at the bottom so as not to interfere with the wedging action desired. The V-belt is subject (a) to tension in the upper region of cross section when they pass around a sheave and (b) to compression in the lower region. In between the two is a neutral region that carries the load and is reinforced with cords or cables.

V-belts are used singly or in multiples (Fig. 15-16) up to a practical number of about ten, above which fewer belts of greater cross section are recommended. V-belts and sheave grooves are matched to fit each other. The neutral region of the belt represents the pitch line, the speed of which is the pitch line velocity, and the diameter of which is the pitch diameter. V-belts can be used for high-speed ratios and short shaft centers.

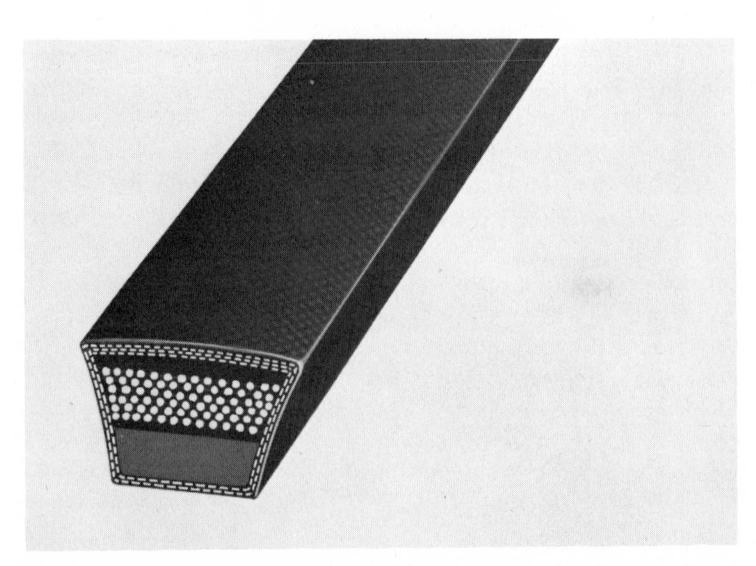

Fig. 15-15. Cross section of V-belt. *Courtesy, Dodge Manufacturing Corp., Mishawaka, Indiana, division of Reliance Electric Co.*

Fig. 15-16. Multiple V-belt drive. *Courtesy, Dodge Manufacturing Corp., Mishawaka, Indiana, Division of Reliance Electric Co.*

Ribbed belts are made with a series of V-ribs running lengthwise with the belt and fit into grooved sheaves as in the manner of V-belts described above. They are usually made of rubber.

Tooth belts are flat belts whose contacting surface has equally spaced teeth that fit a gear-like pulley. It is a positive speed transmission, the meshing teeth preventing slippage between the surfaces in contact.

The transmission of power by belting is a function of friction between belt and pulley with the exception of the ribbed belts. Therefore the arc of contact between the two should be made as great as possible. An accepted value of 165 degrees is assumed to be a fair average for an industrial drive. Where the relative size of pulleys is large in relation to span, an idler pulley is used to depress the belt more snuggly around the pulley. If both pulleys are of the same diameter, i.e., a 1 : 1 ratio, the arc would be 180 degrees, theoretically. To keep belts snug about their pulleys, some takeup arrangement should be provided to eliminate slack due to stretch if an idler is not feasible.

SPEED RATIOS. For any variations between the diameter of a driving pulley and a driven pulley the ratio is found by dividing the diameter of the larger one by the diameter of the smaller one, or

$$\text{speed ratio} = \frac{\text{diam. of larger pulley}}{\text{diam. of smaller pulley}}$$

so that the speed ratio of a combination of a 6-in. and 12-in. pulley is 12/6, or 2 : 1. This means that a 12-in. pulley rotating at 900 rpm will cause the 6-in. pulley to rotate 2 × 900 or 1800 rpm. To determine the speeds or diameter for any combination of driver and driven pulleys use the appropriate equation of those following:

$$\text{rpm of driven} = \frac{\text{diam. of driver} \times \text{rpm of driver}}{\text{diam. of driven}}$$

$$\text{rpm of driver} = \frac{\text{diam. of driven} \times \text{rpm of driven}}{\text{diam. of driver}}$$

$$\text{diam. of driver} = \frac{\text{diam. of driven} \times \text{rpm of driven}}{\text{rpm of driver}}$$

$$\text{diam. of driven} = \frac{\text{diam. of driver} \times \text{rpm of driver}}{\text{rpm of driven}}$$

Line Shaft Lubrication

Periodic application of oil to journal bearings by hand, although not a preferred method, is frequently the only method possible. For a bearing lubricated by hand, the oil must be selected on the basis of three important considerations:

1. Oil life is not a factor in all-loss systems, so that any fairly good, inexpensive, all-purpose oil can be used.
2. Viscosity is an important factor because it governs the leakage rate of oil from a bearing.
3. An oil with strong adhesive properties will maintain a protective film after most of the fluid has leaked from the bearing.

To meet these conditions an oil containing an adhesive agent and having a viscosity of about 400 SUS at 100°F proves highly satisfactory for maintaining normal temperature operation and extending bearing life. An oil with a similar viscosity but with a film strength agent instead of the tackiness additive is recommended for continuous all-loss application, as by bottle oiler or wick-feed cup.

Ring-oiled bearings supporting shafting are of the reuse type and require an oil of high chemical stability, as do all recirculation systems not subject to excessive leakage. A very effective oil is one containing rust and oxidation inhibitors and with a viscosity of 300 to 450 SUS at 100°F.

The selection of bearing lubricants should be based on the speed and loading condition of the head or line shaft in order to simplify lubricant handling and avoid confusion in application.

Bearings are held in place by shaft hangers, or pedestals (pillow blocks), which can support self-aligning bearings. Hangers are equipped with basins for catching oil that drips from the bearings.

Rolling bearings are lubricated with grease applied by pressure gun or central system. A calcium base grease of NLGI No. 2 consistency is commonly used, but a calcium complex or lime-lead grease of the same consistency is preferred in order to provide for the heavier loads of the drive shaft and the high temperatures encountered near the ceiling, where much of the line shafting is located. As with all grease-lubricated rolling bearings, shaft bearings should be filled only to two-thirds capacity and inspected periodically.

OIL. Oil is used on most plain bearings. Shaft bearings may be designed for the all-loss or reuse systems of lubricating. All-loss systems include periodic application by hand as with an oil can, or continuously by means of bottle oilers, wick-feed cups, or drip oilers. However the reuse method employing oil rings is more commonly used, especially on bearings not readily accessible due either to height or moving machine barriers.

CHAIN DRIVES

The many kinds of chains include sling, hoist, crane, conveyor, oven, anchor, tire, and transmission. There are also two or more chain designs that fit into any of the above categories. Of interest here are chains that transmit power from a drive sprocket to a follower sprocket. They are made in different sizes both as to link dimension and chain length. Chains are used in single and multiple strands. Some chains are so long that they must be supported between the drive and follower sprockets. Supports include idler sprockets, ways, and channels, or tracks. All transmission chains require lubrication.

Chains are made of links joined together either by spring clips (for larger pitch chains) or cotter pins to form an endless loop. The type of link identifies the chain. Transmission links are designed to mesh with the teeth of the sprockets over which they travel in either direction. Links are connected at a point called the *joint*. The *pitch* of the chain is the center-to-center distance between two adjacent joints. This corresponds to the pitch of the mating sprockets. The *pitch circle* of a sprocket is the center line distance measured by a circle drawn through the centers of the joints of an endless chain all of whose links will exactly mesh with all of the teeth of the sprocket. *Pitch diameter* is the diameter of the pitch circle. *Pitch line speed* is equal to the number of teeth in the sprocket times the pitch of the chain times the rpm of the sprocket. If the pitch of the chain is expressed in inches, the pitch line speed will be inches per minute.

There are three basic types of transmission chains: (1) enclosed malleable iron chain (pintle), (2) roller chain, and (3) silent chain.

ENCLOSED MALLEABLE IRON CHAINS are heavy-duty chains used in rugged service such as coal drags, elevators, and conveyors (Fig. 15-17). The

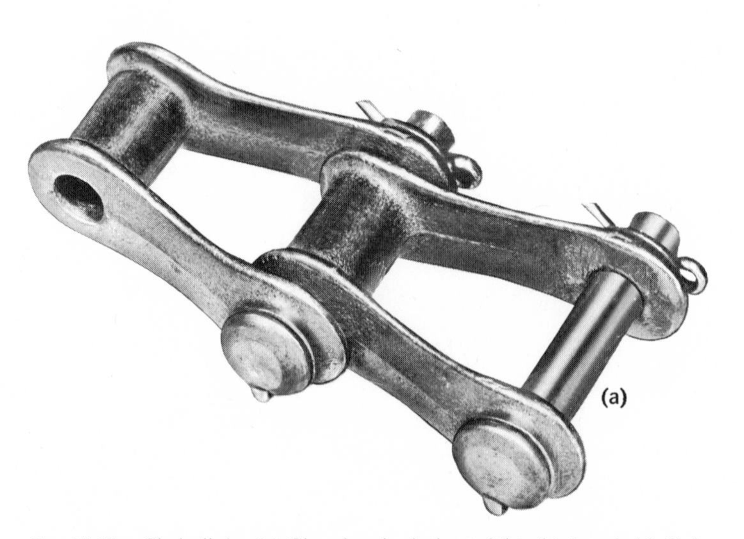

Fig. 15-17. Chain links. (a) Closed end, pintle at right, (b) detachable link, (c) conveyor chain, (d) stamped detachable link. *Courtesy, Link-Belt Chain and Conveyor Components Division, FMC Corp.*

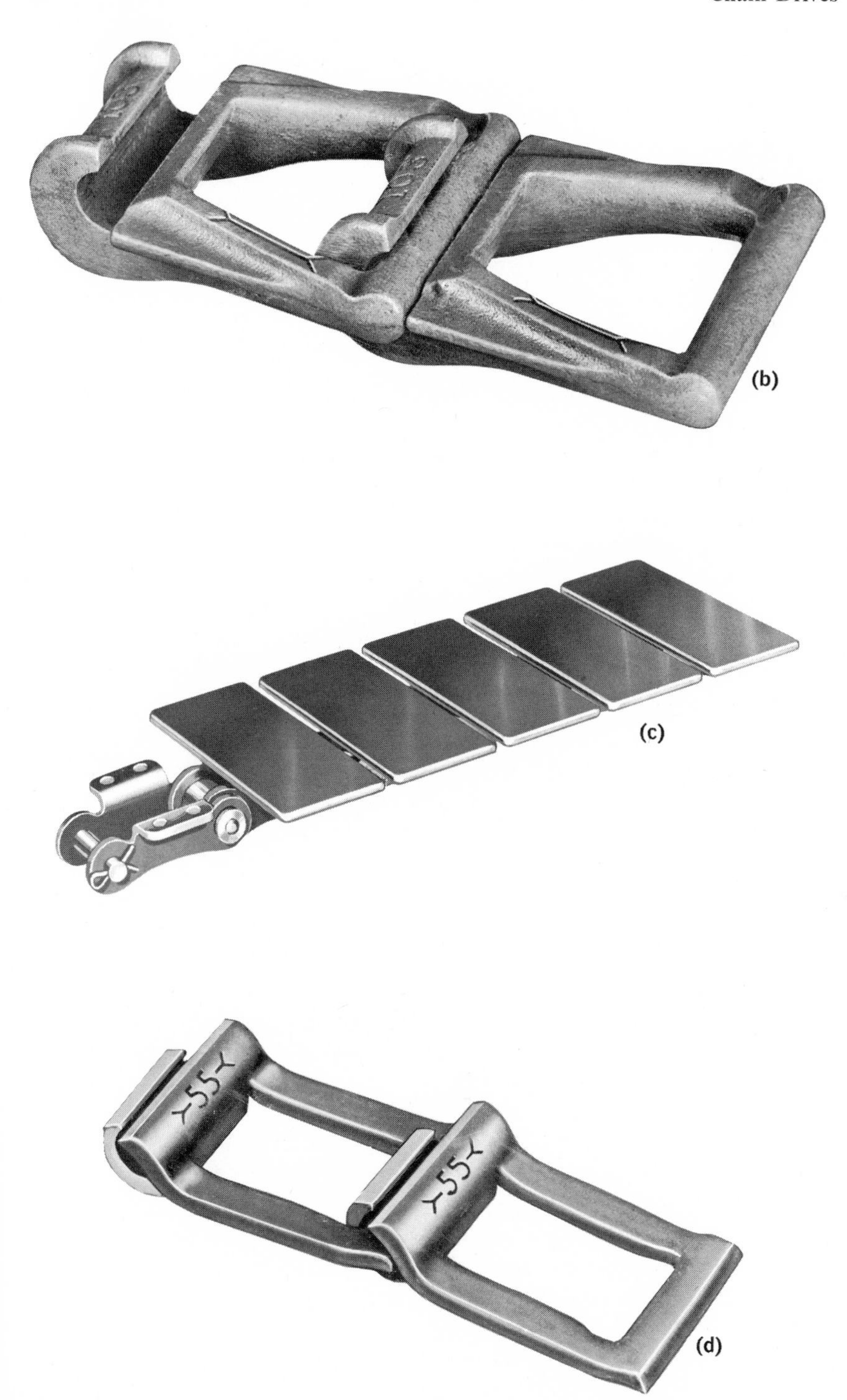
(b)
(c)
(d)

links of this chain are joined with a pin that forms one end of the link, which is inserted in a bushing that is a fixed part of the other end of each link. The Evart or open hook chain is similar, but the hook at one end of each link joins with the cross end of the adjacent link to be made detachable, one link from the other (Fig. 15-17). Detachment will not occur as long as the chain is not bent over on itself and the links subject to a side-sliding action. Open hook chains have a maximum speed of about 500 ft/min and are designed for light loads. They are often made of steel and in some instances used where lubrication is impractical.

ROLLER CHAINS are very flexible and are the type commonly used on bicycles (single stranded), open machine drives, tension controls, and some flexible couplings. Roller chain links are made up of a link pin on each end, a pin link plate that joins the pins of adjoining links together, two rollers supported by the pins, bushings that act as intermediate bearings between pins and rollers, and roller link plates that hold the rollers in place. Rollers perform the dual function of reducing friction when the chain is in rolling contact with the sprockets and separating the end pieces of the link (Fig. 15-18). The permissable speed of roller chains increases as the pitch becomes shorter. The recommended maximum speed of a $\frac{3}{8}$-in. pitch chain running on a sprocket with 24 teeth is about 4200 ft/min. In contrast, the maximum speed recommended for the same sprocket with a $2\frac{1}{2}$-in. pitch is about 300 ft/min. To increase the load capacity of roller chains, multiple strands are used. Ratings of multiple-strand chains are proportional to the number of strands. The tensile strength of a single-strand roller chain with a $\frac{3}{8}$-in. pitch, $\frac{3}{16}$-in. width, and a 0.200-in. roller is about 2100 lb. For a $2\frac{1}{2}$-in. pitch, a roller $1\frac{1}{2}$ in. wide and 1.56 in. in diameter, the tensile strength is about 95,000 lb. per strand. Thus a $\frac{3}{8}$-in. pitch single-strand chain with a 24-tooth sprocket having a speed of say 800 rpm is rated at about 2.2 hp, whereas a $1\frac{1}{4}$-in. pitch chain with a 24-tooth sprocket at the same speed is rated at about 50 hp, providing both chains respectively meet ASA 35 and ASA 100 specifications.

SILENT CHAINS. Silent chains are made up of a number of flat leaves connected by pins to provide a flexible endless rack of various widths (Figs. 15-19 and 15-20). The leaves are contoured to form teeth at each end that mesh with gear-like sprockets with cut teeth. Some silent chains are equipped with side guides; others have middle guides. The guides are straight contour leaves which pass through slots cut in the sprockets. They guide the chain and prevent it from drifting sideways on the chain wheel.

The silent chain drive is quiet, highly efficient, drives positively (without slippage), and is suitable for high speeds, some designs exceeding 5000 ft/min. They operate in short or long center distances between sprockets. For best results smaller sprockets should have at least 21 teeth. Along with finished steel roller chains, silent chains are preferred for speed ratios up to 10:1 and speeds over 700 ft/min, although they are very practical at

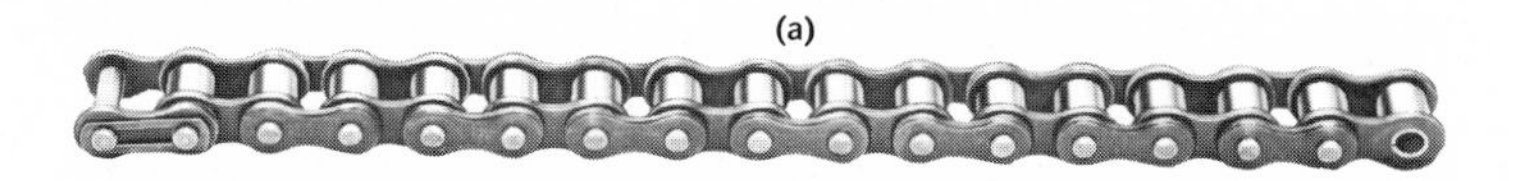

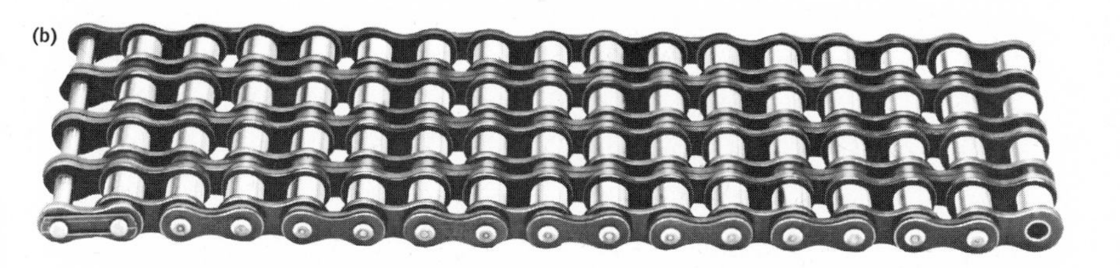

Fig. 15-18. Roller chains. (a) Single strand, (b) multiple strand, (c) sprocket. *Courtesy, Link-Belt Chain and Conveyor Components Division, FMC Corp.*

Fig. 15-19. Silent chain and sprocket. Note guides and guide slots cut in sprocket. *Courtesy, Link-Belt Chain and Conveyor Components Division, FMC Corp.*

lower speeds. Both finished steel roller and silent chains should be operated with cut tooth sprockets rather than cast teeth, which are more suitable for malleable or pressed steel chains of the closed and hook type, respectively. Silent chain drives are very suitable for primary drives of either light or heavy loading, while roller chain drives are used chiefly in secondary drives. Lubrication recommendations for these chains are shown in Table 15-2.

The *speed ratio* of a chain drive is determined by dividing the number of teeth in the larger sprocket by the number of teeth in the smaller. The number of teeth or speed for any combination of three known values relative to driver and follower sprockets can be determined in a manner similar to that shown earlier for belt drives, except that the number of sprocket teeth is substituted for pulley diameter in the various equations.

Fig. 15-20. Large silent chain drive in operation. *Courtesy, Link-Belt Chain and Conveyor Components Division, FMC Corp.*

Table 15-2. OIL VISCOSITIES AND APPLICATION METHODS RECOMMENDED FOR ROLLER AND SILENT ENCLOSED CHAINS

Temperature Range (°F)	*Viscosity SUS at 100°F (minimum)*	*Method of Application According to Speed*
Below 32	175 (−50°F maximum pour point)	
32–50	300 (+20°F maximum pour point)	
50–100	600	
100–150	900–1200	
600 fpm or less		Bath or brush
600 fpm to 1500 fpm		Bath, brush, or drip
1500 fpm or higher		Disk, spray, or circulation

Lubrication of Chain Drives

All transmission chains should be lubricated. The closed malleable iron chains are fabricated of such metal as to combat the abrasive particles associated with the material-handling equipment on which they are used. When application allows a suitable oil to continuously reach the external and internal surfaces of pins, bushings, leaves, and rollers as well as the teeth of sprockets, single- and multiple-strand chains will last for many years. Smearing grease on a roller chain is a waste of time and material, except that it might afford some protection to the sprocket teeth.

Chains should be allowed to soak in oil for a few minutes before installation. The oil should be free of any ingredient that would tend to gum when exposed to air at operating temperature. Under ideal conditions chains should be run in oil fortified with rust and oxidation inhibitors. This is not always practical, especially where a high-speed chain drive is not covered or when its location precludes the use of a sump of any kind. Mist oil application is an excellent substitute for an oil bath. This too requires complete enclosure. Bath, splash, force feed, disk, drip feed, brush, and spray are some of the automatic methods of applying oil discussed in detail in Chapter 11.

Roller and silent chains require similar types of lubricating oil with the viscosity depending on operating temperature and heat transfer capabilities of the system. The oil specifications shown in Table 15-3 apply to both enclosed and open chains. Oven chains require special considerations as shown.

An effective method of applying oil to chains is by brush lubricators (Fig. 15-21). Brush selection may be made from the chart in Fig. 15-22. All these oils should contain rust and oxidation inhibitors. For temperatures below 32°F they should also contain a VI improver and a pour point depressant.

For high-temperature operation as found in ovens, conveyor chains should be lubricated with a solid lubricant, either graphite or molybdenum disulphide. These are deposited on the chain by a diluent, which evaporates

Table 15-3. OIL RECOMMENDATIONS FOR GENERAL OPERATION OF CHAINS ACCORDING TO SAE RATINGS

Ambient Temperature (°F)	*SAE Grade (Viscosity)*
Below 0	*Automatic Transmission Oil (200 SUS at 100°F)*
0–20	10W
20–40	20W
40–100	30W
100–120	40W
120–160	50W

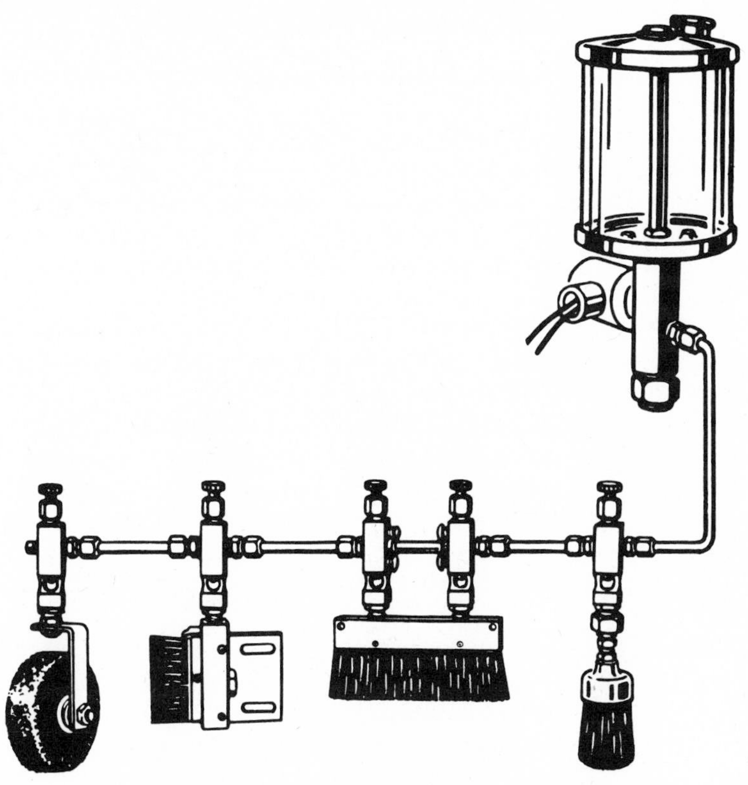

Fig. 15-21. Brush type chain oilers. *Courtesy, Oil-Rite Corp.*

in the hot area. For ovens of 500°F or less, a synthetic fluid such as one of the chlorofluorocarbon polymers may be used. These polymers have good thermal stability below the temperature mentioned. They also exhibit excellent lubrication properties and are nontoxic.

Some improvement has been made in the lubrication of roller chains by the use of sintered steel bushings preimpregnated with oil. Any oil forced out of such bushings during operation is reabsorbed when the chain is idle. To assure continuous protection, however, it is recommended that such chains be subject to periodic oil applications, preferably by bath soaking during downtime periods. A light-bodied oil (about 150 SUS at 100°F) should be used to facilitate reabsorption.

Some manufacturers provide slots in the bushings to facilitate penetration. Such slots enhance the effect of brush, spray, drip, and other drenching methods of oil application.

The oil when applied by drip or spray should be applied to the upper edges of the sideplates on the lower span of roller chains. For single-strand chains the oil should be applied between the pin link plates and roller link

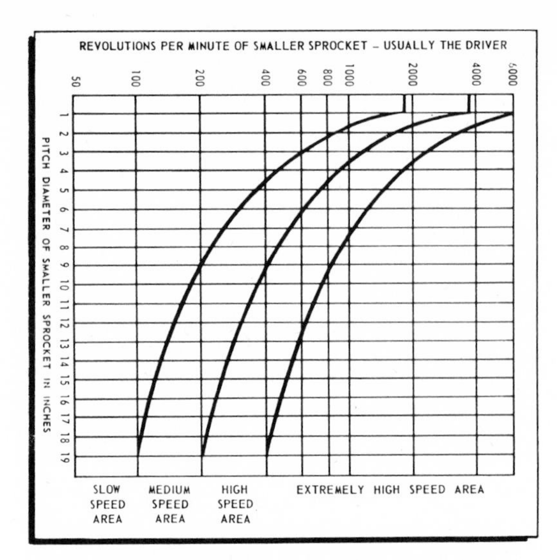

Fig. 15-22. Selection chart for brushes used to lubricate chains. Use bristles on the brush lubricator as follows: for the slow-speed area, use standard bristles, for medium-speed area use standard or nylon, and for high-speed area use nylon bristles. A force-feed oiling system should be used in the extremely high-speed area to carry away heat. Generally, the operating temperature should not exceed 160°F. *Courtesy, Oil-Rite Corp.*

plates, preferably on both sides of the chain. Double-strand chains should be lubricated at the center plates, and multiple strands should be lubricated at alternate rows of plates.

Inadequate lubrication of chains may result in excessive noise, wear, and broken parts. Too heavy a lubricant may cause the chain to cling to the sprocket excessively, and oxidized oil deposits may bring about stiffness at the bushings.

Phase-Shifting Differential

The phase-shifting differential (Fig. 15-23) is a reliable, simple, and precise method for the correction of register or adjustment of phase applicable to conveyor drives, packaging machinery, printing presses, and paper cutoff machines.

OPERATION. Both shafts of PSD units rotate in the same direction, and either may be used as the input to obtain a 2 : 1 speed increase or decrease. Rotation may be in either direction. Wide-span shaft mounting accommodates high overhung loads.

Use of the spider shaft as the input gives an inherent 2 : 1 speed step-up between input and output. Conversely, using the side gear shaft as input results in an inherent 2 : 1 speed reduction. When the spider shaft is the input, each full turn of the correction shaft changes phase relationship by 3.6 degrees. When the side gear shaft is the input, each turn of the correction shaft results in a 1.8-degree change in phase relationship.

Correction is unlimited in either direction and can be applied while the equipment is running or stopped. Correcting torque is low; therefore, the correction shaft may be operated manually or by motor. Once made, the correction is "locked in."

Correction is made by a worm gear fastened rigidly to one of the differential assembly side gears. A worm on the correction shaft drives the combined worm gear and side gear element to introduce a relative speed change between the input and output shafts of the transmission and thereby change their phase relationship.

The high ratio of 100 : 1 between the correction shaft and the worm gear permits extremely fine adjustments to be made with ease. At the same time, phase corrections of considerable magnitude can be accomplished speedily since the low torque needed by the correction shaft permits it to be turned rapidly with a low power motor.

LUBRICATION. A top plug is located at the top of the phase-shifting differential for the introduction of oil. An oil capable of adequately protecting the worm gear is required so that a high grade cylinder oil of about

Fig. 15-23. Specon ® phase-shifting differential (PSD). *Courtesy, Industrial Products Division, Fairchild Hiller.*

2500 SUS at 100°F is recommended for ambient temperatures above 40°F. For ambient temperatures below 40°F, a pneumatic cylinder oil of about 800 to 1000 SUS at 100°F and a pour point of 0°F containing an antiwear agent, adhesive agent, and a small amount of an emulsifier. A cold-weather alternative lubricant is a leaded compound of about 1400 SUS at 100°F and 0°F pour point.

16

HYDRAULICS

Hydraulic systems applied to industrial, marine, and aircraft equipment are designed to perform with a high degree of versatility in controlling the various factors associated with the transmission of energy. Such factors include direction of motion, velocity, force, and the positioning of working components. Hydraulic energy is applied to machine tools, transportation vehicles, lifting devices, and material compacting presses. Hydraulic control systems facilitate the steering of marine vessels and automobiles and actuate aircraft landing gear, among many other tasks. The action of hydraulic systems is free of the inertial forces and the noise associated with mechanical linkages.

A hydraulic circuit is basically a liquid circulation system in some part of which work is performed. Without the work-producing components, the only pressure required would be that necessary to move the liquid against gravity and friction, assuming that it was drawn from and delivered to an environment of equal pressures. To perform work, an additional force is required, which is developed from some external power source. This power is usually in the form of mechanical energy, and because the hydraulic link is fluid, it cannot be paired with another link by a connector, such as a wrist-pin that connects a piston rod to a piston to form a fixed pair. The hydraulic link, therefore, must be held by the pressure of the liquid to form an *open pair*. This takes the form of either (1) a piston in a cylinder to produce lateral motion, (2) a fluid motor consisting of a rotating element within a constraining housing to produce rotary motion, or (3) a diaphragm to produce lateral pulsating motion. The diaphragm is used in some types of controls and in diaphragm pumps.

VERSATILITY OF HYDRAULIC DRIVES. Hydraulic drives have many advantages:

1. Linear or rotary motion
2. A wide range of speeds
3. Infinite speed changes (stepless)
4. Smooth speed transmission
5. Accelerating and decelerating action
6. Operational flexibility, including cycle control
7. Versatility of control methods ranging from manual to fully automatic
8. Reversal of motion direction
9. Compactness
10. Adaptation to other energy controls, including electrical, mechanical, pneumatic, and separate hydraulic systems
11. Inertial damping
12. Intrinsic overload protection

PASCAL'S LAW

Some of the aspects of the hydraulic system are complex, but the principle of operation is founded on a simple law of fluid equilibrium, Pascal's Law. Pascal's Law is based on static fluids, so it does not consider the energy losses inherent in working systems nor the temperature of the fluid. It is stated thus:

When pressure is exerted on any part of a fluid within an enclosed chamber, an equal pressure is transmitted undiminished to all equal areas and in all directions through the fluid within the chamber, regardless of its shape.

To illustrate the principle as set down in Pascal's Law see Fig. 16-1.

HYDRAULIC SYSTEMS

Hydraulic systems applied to industrial, marine, and aircraft equipment are designed to perform with a high degree of versatility in controlling the various factors associated with the transmission of energy. Hydraulic energy is applied to many kinds of production machines, machine tools, vehicles, lifting devices, and material compacting presses. Hydraulic control systems facilitate the steering of marine vessels and automobiles and actuate aircraft landing gear, among many other tasks.

With few special components hydraulic systems are capable of many variations in power conversion and power transmission. Speeds, pressures, and direction can be infinitely variable. Hydraulic systems may be com-

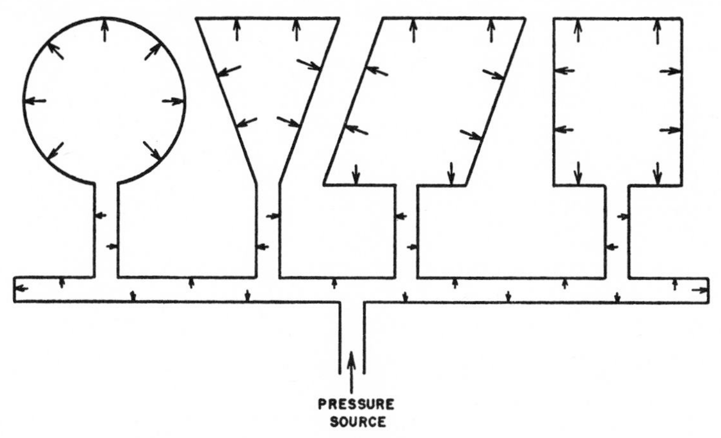

Fig. 16-1. Demonstration of Pascal's Law.

pletely manual or fully automatic. The adoption of electrohydraulic servomechanisms has added greater versatility to the hydraulic systems in their ability to control speed, pressure, positioning, indexing, tension, and other senses of machinery for uses ranging from paper making to steel manufacture.

HYDRAULIC CIRCUIT COMPONENTS (Figs. 16-2 and 16-3). The practical hydraulic circuit consists of seven basic components:

1. *Container or reservoir* for holding and supplying the liquid to the working part of the system.
2. *Pump,* motivated by an external source such as an electric motor, to feed the liquid to the system and convert mechanical energy to hydraulic energy.
3. Hydraulic motor or *actuator,* to convert hydraulic energy back to mechanical energy for the purpose of performing useful work.
4. *Valves,* to control flow according to a desired pattern and regulate the pressure of the liquid as required by the work process.
5. *Accumulator,* to maintain constant pressure.
6. *Piping,* to convey the liquid to and from each of the above components according to the order of the system.
7. *Filters,* to keep the oil clean.
8. *Hydraulic fluid,* may be either a mineral oil or a synthetic liquid.

HYDRAULIC PUMPS

Hydraulic pumps are either positive displacement or nonpositive displacement types. Positive displacement pumps can be further classified as

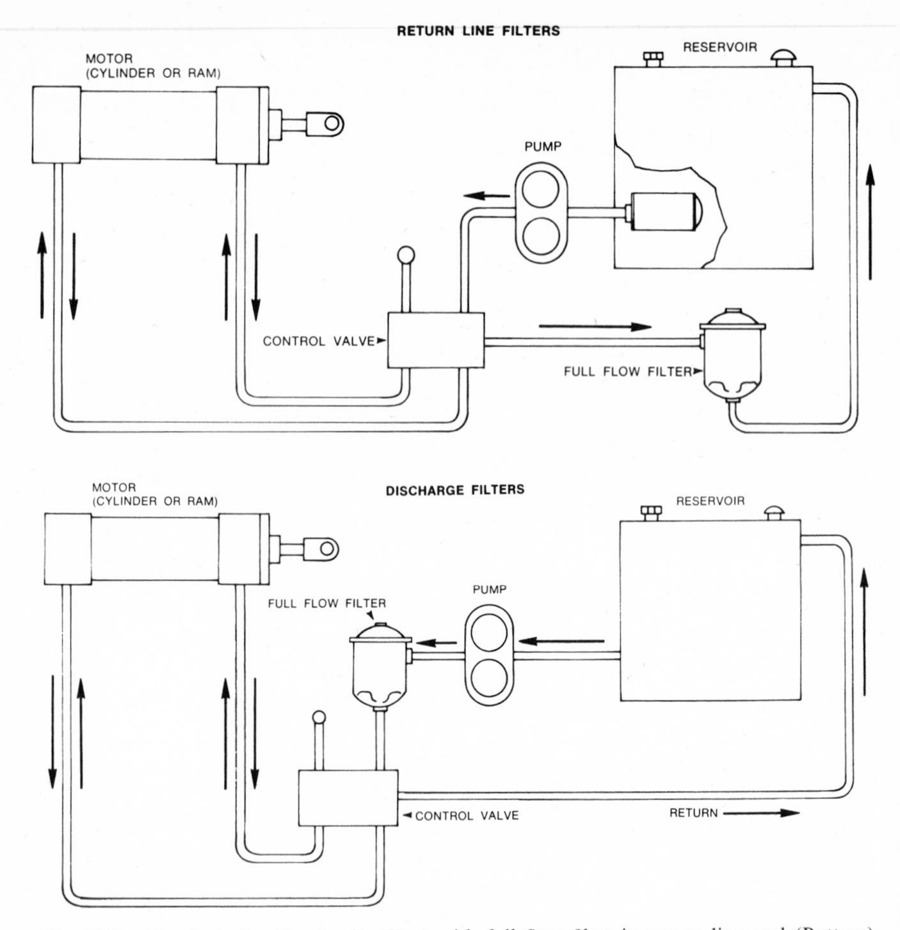

Fig. 16-2. Simple hydraulic circuits (Top) with full flow filter in return line and (Bottom) with full flow filter in discharge line. *Courtesy, Fram Corp.*

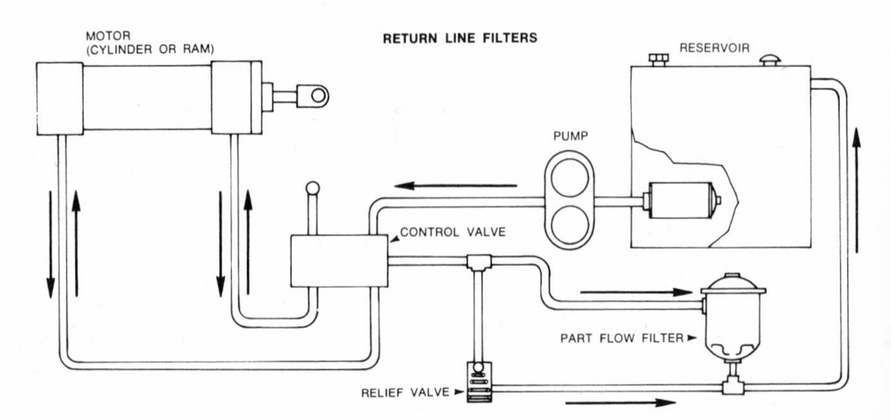

Fig. 16-3. Simple hydraulic circuit with bypass filter. *Courtesy, Fram Corp.*

being fixed displacement or variable displacement pumps. Most hydraulic pumps used for transmission of power are of the rotary, positive displacement type, and they include the gear, vane, and piston pumps.

GEAR PUMPS. The gear type of hydraulic pump includes those designed (1) with both drive and driven gears having external teeth, or (2) with an external tooth gear fitted eccentrically within a larger gear having internal teeth, both of which turn in the same direction. Both internal and external gear pumps are shown in Fig. 19-18. Gear pumps are of the positive fixed-displacement type and are nonpulsating.

VANE PUMPS. In these pumps, a rotor on the pump shaft carries movable vanes in slots in its periphery (Figs. 16-4 and 16-5). Springs push the vanes outward against the contour of a surrounding cam ring. Rotor, vanes, and cam ring form a pumping cartridge and are sandwiched between two port plates. As each spring-loaded vane follows the inner face of the cam ring, it moves outward as it enters the contoured area of the rings, drawing oil from the inlet ports. Having passed the inlet ports, the vane travels up the slope of the discharge portion of the contoured area, pushing oil ahead of it. Here, as the total area for the trapped oil is reduced, pressure is created and the oil is discharged through outlet ports in the cam ring at pressures up to 2500 psi.

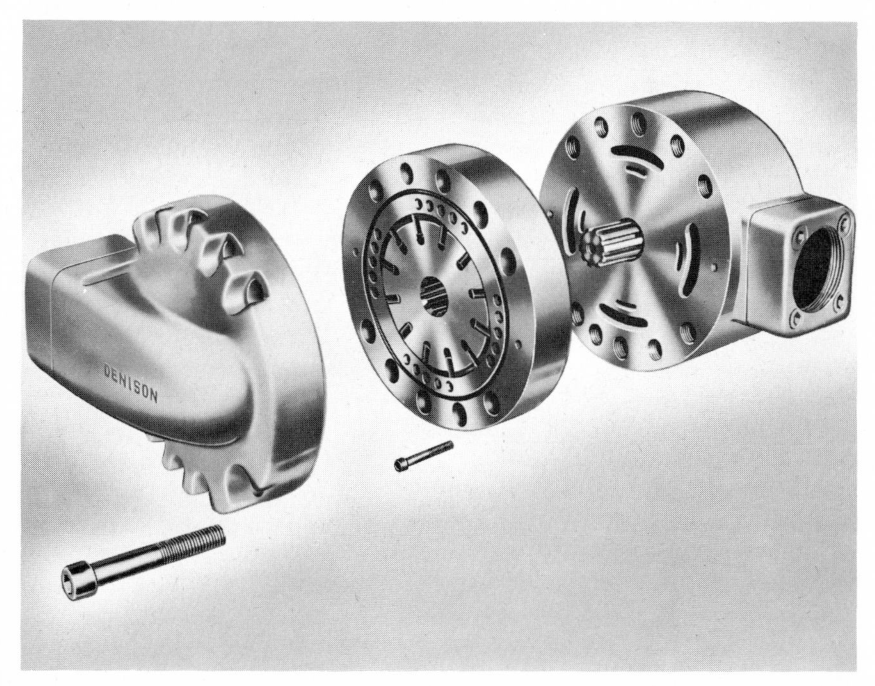

Fig. 16-4. Vane type hydraulic pump motor, rotor in center. *Courtesy, Denison Division, Abex Corp.*

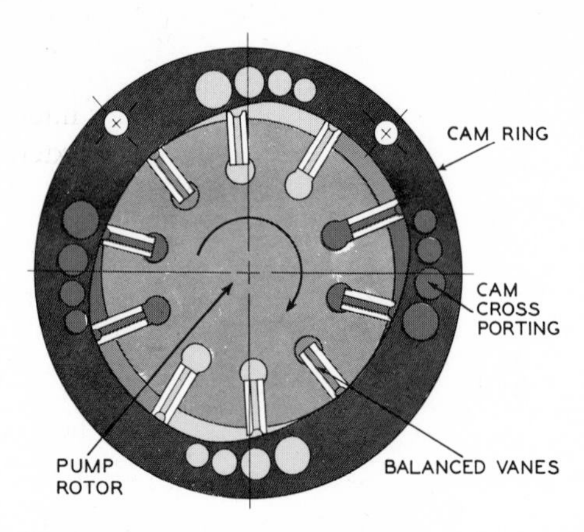

Fig. 16-5. Cross section of vane pump rotor. Capacities up to 147 gpm and 2500 psi. *Courtesy, Denison Division, Abex Corp.*

The pressure limitation of vane pumps is about 1000 psi, but up to 2000 psi can be attained by placing two vane pumps in series. Speeds are 1200 to 1800 rpm according to size. Systems using vane pumps are best served by a hydraulic liquid having a viscosity of 150 to 220 SUS at 100°F.

The amount of fluid handled by any pump is a function of its size and speed. Vane pumps are designed to deliver volumes of 1 gallon per minute and more.

Two vane pumps of different capacity are used so that, when the one of larger volume is unloaded at some intermediate pressure, the smaller one continues at maximum pressure. The purpose of this arrangement is to allow initial rapid advance of an actuator (ram type) followed by a slower advance as required, first, to bring a tool up to the workpiece and then, second, to start cutting at the lower speed. This is common practice on most machine tools.

The vane pump described above is basic for all such pumps. However, such constant-volume pumps are variously designed for variable delivery and adjustable volume control.

PISTON PUMPS. The axial piston pump and the radial piston pump represent the two basic types of hydraulic piston pumps. Both are rotating units and are capable of variable delivery.

RADIAL PISTON PUMPS are positive-displacement types capable of developing high pressures (2500 to 3000 psi) and can be made reversible (Fig. 16-6). Reversibility allows the flow of the hydraulic fluid to be reversed without special valving and without reversing the drive motor. They are capable of performing under heavy loads.

The radial piston pump consists of a rotor or cylinder block driven by a drive shaft off an electric motor, which typically has nine cylinders. The valve spindle on which the cylinders are mounted is axially drilled to allow passage of the hydraulic fluid in a circumferential valve slot that forms an open arc of slightly less than 180 degrees to permit entrance of the fluid into each cylinder in turn as the rotor turns.

The cylinders are closely fitted with free pistons whose outer ends are held in contact with an outer rotating ring by centrifugal force and the pressure of the fluid entering through the slot of the valve. When the ring and spindle are concentric, the rotor rotates about the center of the spindle without causing any radial motion of the pistons, resulting in zero delivery. When the outer ring is moved off center relative to the spindle, the rotation of the rotor causes the pistons to reciprocate radially, drawing fluid into the cylinders during the outward movement and forcing it out of the cylinders during inward movement. The oil forced out of the cylinders enters a discharge line through another set of passages in the valve spindle opposite the suction ports via a discharge port represented by another but similar valve slot. The amount of eccentricity of the outer ring relative to the spindle governs the magnitude of volume delivery, being greatest with maximum offset. This is the basis of variable delivery.

The pistons are equipped with sliding or rolling elements through which their reaction with the outer ring is transmitted.

One of the characteristics of the radial pump is that, by adjusting the eccentricity of the outer ring either to the right or left, the direction of fluid flow is reversed.

AXIAL PISTON PUMPS have a cylinder block bored to provide a number of cylinders (usually nine) in which pistons are free to reciprocate axially relative to the axis of the rotor (Fig. 16-7). The cylinder block is mounted in a yoke that can be adjusted at various angles to the drive shaft. The length of travel of the pistons depends on the angle at which the cylinder block is tilted. The pistons are connected to a drive shaft flange by piston rods with ball joints at each end.

When the block is in line with the drive shaft, the pistons do not reciprocate but merely rotate as one with the assembly. However, by tilting the cylinder block during its rotational motion, the pistons move out as the distance between block and flange increases through 180 degrees, or $\frac{1}{2}$ revolution of the drive shaft, and move in as the distance becomes less during the remaining 180 degrees of rotation. The greater the angularity or tilt of

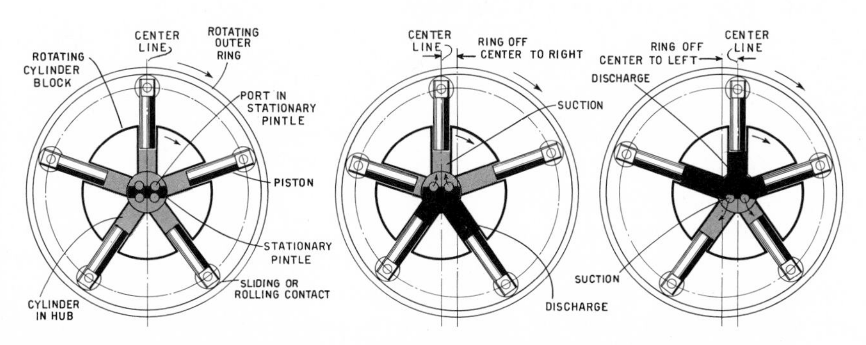

Fig. 16-6. Variable volume radial piston pump. *Courtesy, Mobil Oil Corp.*

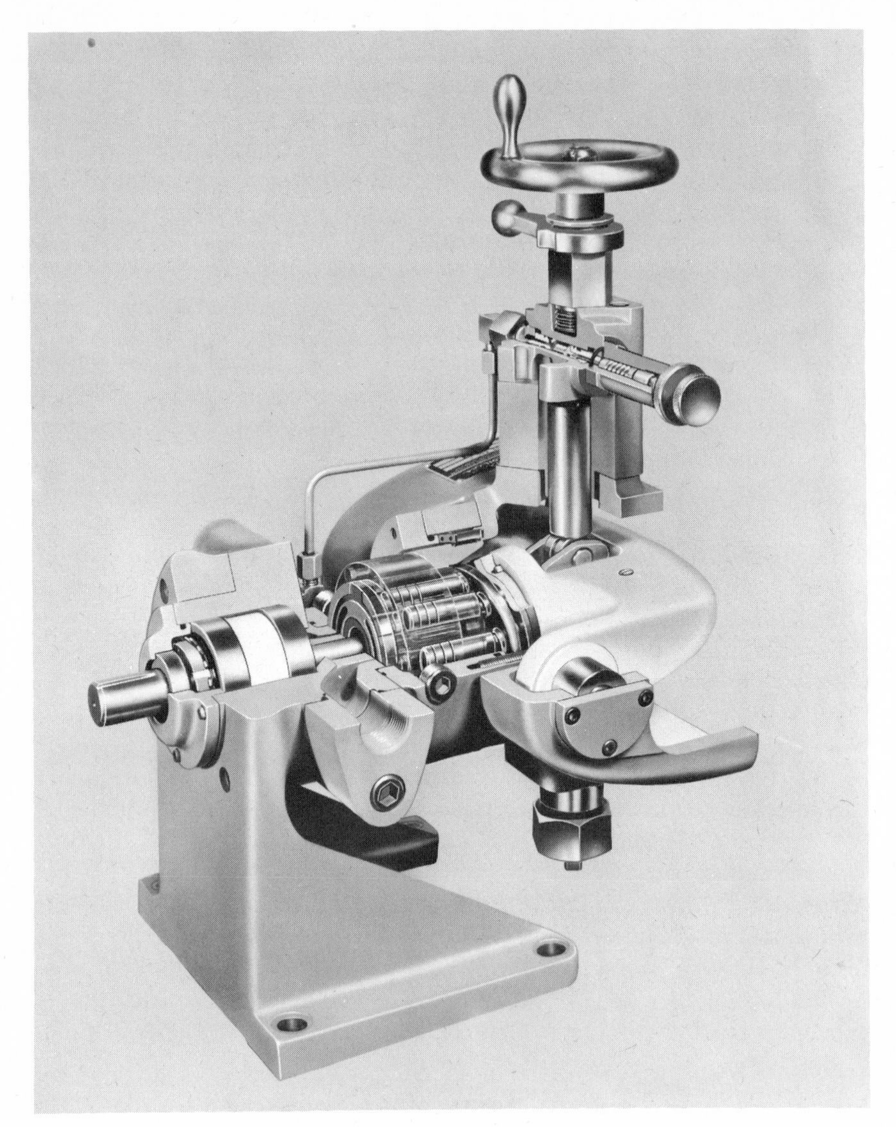

Fig. 16-7. Axial piston, variable volume pump with hand wheel control and pressure compensated. *Courtesy, Denison Division, Abex Corp.*

the cylinder block, the longer becomes the stroke of the pistons in either direction, resulting in increased volume of fluid delivered. The converse is also true.

The working parts (Fig. 16-8) are encased in a housing filled with an oil that fulfills the dual purpose of lubricating the frictional parts and replacing any oil that leaks from the power transmission fluid. The lubricated

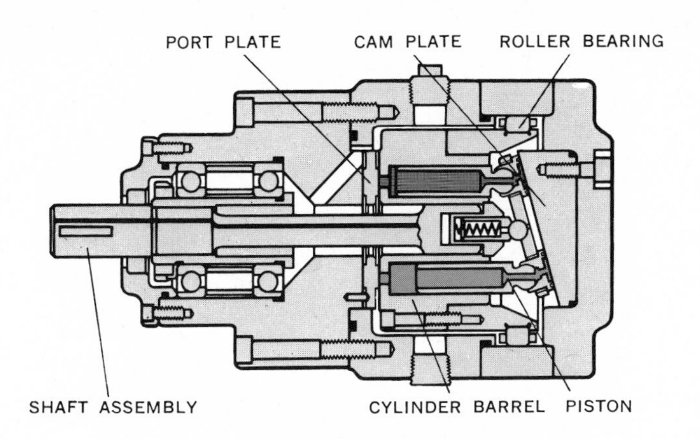

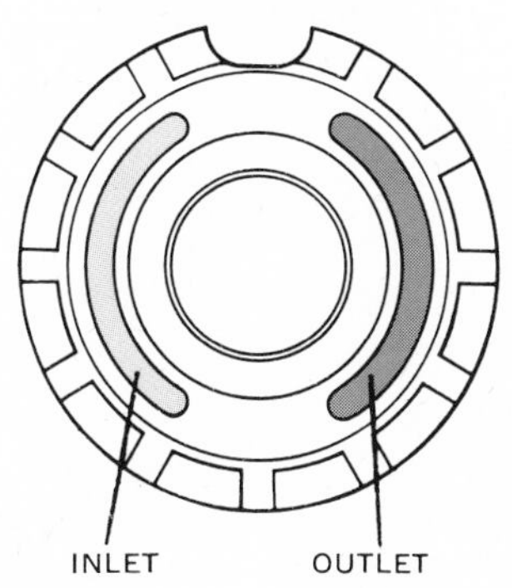

Fig. 16-8. Cutaway of axial piston pump. *Courtesy, Denison Division, Abex Corp.*

parts include the various ball joints and the bearings that support the drive shaft and cylinder block.

Like the radial pump, the axial pump is also equipped with a valve plate cut with a circumferential slot through which the hydraulic fluid leaves the pump from the reservoir and enters into the working circuit.

Piston type pumps are capable of developing very high pressures commensurate with loads (Table 16-1).

Table 16-1. Hydraulic Pump Characteristics

Type of Pump	*Capacity Range*	*Pressure Range*	*Speed*
External gear	Low to high	Low to medium	Low to medium
Internal gear	Very low to medium	Low to medium	Low to medium
Vane type	Low to medium	Medium to high (max. ~1000 psi)	Medium to high
Axial piston	Low to high	Medium to high	Medium to high
Radial piston	Low to high	Medium to high	Medium to high

ACTUATORS

The conversion of hydraulic energy to mechanical energy is accomplished in the actuator in the form of a *rotary fluid motor,* or a *linear ram,* a piston-cylinder arrangement. The actuator is usually connected to the working component of the host machine. Basically, the actuator has the opposite function of the hydraulic pump, but is not necessarily of the same principle. For example, a gear pump may motivate a gear motor, as rotary actuators are commonly called. However, a gear pump may also motivate a vane motor. The combination of one to the other is governed by the intended service of the hydraulic system.

LINEAR ACTUATORS. The ram type actuator is a cylinder-piston arrangement in which the piston moves linearly to produce reciprocation motion as required for shaper tools, grinder tables, etc. It is actuated by introducing the hydraulic fluid under pressure into the cylinder on one side of the piston located on one end of the cylinder and driving the piston to the other end of the cylinder. The piston is returned to its original position either by spring tension (single action) or by reversing the flow of pressurized fluid and allowing it to enter the opposite end of the cylinder to drive the piston back to its original position. The latter method is typical of hydraulic systems used on power equipment and is referred to as a double-acting differential type cylinder.

ROTARY ACTUATORS, OR FLUID MOTORS. These components are similar to the rotary pumps, which can also be used as actuators. In some cases, slight modification of the pump may be necessary to permit their opposite function. The discussion covering rotary pumps of the gear, vane, and piston (both radial and axial) types also applies to the basic designs of rotary actuators operating on the same principles. The major difference lies in the actuating source. The hydraulic pump is actuated by an external source, usually an electric motor. The actuator is motivated by the force of the hydraulic fluid generated by the pump. The functions of the actuator and pump can be demonstrated by the radial piston pump and actuator arrangement. The pump is rotated by an electric motor through a drive shaft connected to the cylinder block. The radial piston type actuator is motivated by the hydraulic fluid under pressure acting on half (or approximately half) the pistons, causing them to move outward and rotate the cylinder block and output shaft. The same approach may be used to comprehend the motivation of the gear and vane type actuator, i.e., the energy of the flowing hydraulic fluid drives the gears and vane rotor.

The reverse function described above may not be possible with some types of vane and piston type components, either because of design or special application. Such conditions are represented by the limited-rotation vane motor and the piston type unit that requires a crank arm to convert linear motion to rotary motion.

HYDRAULIC VALVES

Control Valves

The three factors that can be made variable in hydraulic systems are pressure, direction of flow, and volume per unit of time (minutes). All these factors are controlled by valves. The rate of flow can also be controlled by a variable-volume pump.

In a hydraulic system, pressure is the factor that governs the force or torque exerted by the actuator. By limiting the pressure, the force of the actuation becomes correspondingly limited.

The flexibility of the system in responding to the required changes in direction of the reconverted mechanical motion is dependent on the inherent ability of the fluid to change direction of flow on demand.

The speed of the actuator is determined by the volume of fluid available to it relative to its volumetric displacement. The versatility of a hydraulic circuit, then, is based on the facility with which ultimate force, speed, and direction of motion can be controlled.

PRESSURE CONTROL VALVES. Valves used to control pressure include relief valves, pressure-reducing valves, unloading valves, and foot valves.

Pressure relief valves reduce the pressure in the high-pressure circuits, which they serve. They consist of a spring-loaded ball, or plunger, that normally blocks the entrance to the valve body in which the dynamic components are housed. Pressure on the ball or plunger tends to compress the spring and allow fluid to flow from the high-pressure circuit. The tension exerted by the spring is fixed at a desired magnitude by an adjusting screw supported by the valve body. When the adjusted tension is exceeded by the pressure in the high-pressure circuit, the ball or plunger is unseated, resulting in an opening corresponding to the differential in opposing forces. When the circuit returns to a pressure below that corresponding to spring tension, the ball or plunger is forced against the opening to shut off flow through the relief valve.

A compound relief valve is used in place of the simple type described above to prevent the pressure override prevalent in the pressure relief valve. This is accomplished by providing a light spring combined with a pressure-balanced piston to hold the valve closed. A separate liquid passage is provided to unload the pump when the actuator is stopped for some reason. However, in some circuits, a specially designed unloading valve is used.

Relief valves are connected between the pump line and a line leading back to the reservoir so that any fluid bled from the high-pressure circuit is bypassed around the pump.

DIRECTIONAL FLOW VALVES. There are several kinds of directional flow valves, including the rotary spool, sliding spool (Fig. 16-9), and poppet types. Direction control is necessary for several reasons, the most frequent being

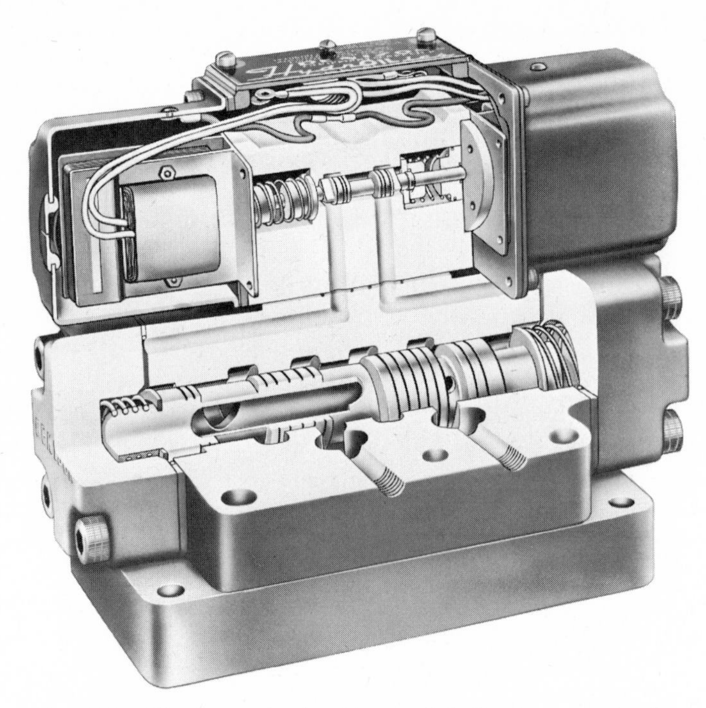

Fig. 16-9. Pilot-operated solenoid-controlled direction control valve. *Courtesy, Denison Division, Abex Corp.*

to maintain a ram type actuator in reciprocating motion. To get such motion, the pressurized fluid from the pump first enters, say, the left end of a cylinder to force the piston to move toward the right end. When the piston reaches its extreme limit of travel in that direction, two energy changes occur simultaneously: (1) the pressure of the fluid on the left side of the piston drops to that in the reservoir, and (2) pressurized fluid enters the right end of the cylinder to force the piston back to its original position, during which the depressurized fluid is swept out of the cylinder through the same port that it entered, to return to the reservoir. When the piston reaches its extreme point of travel to the left, the process is repeated.

Considering that the fluid entering and leaving the cylinder passes through a common control point or valve, such a valve would be designed to have four pathways through which the fluid can flow in and out of each end of the cylinder. This four-way valve is designed to open and close the pump circuit and drain line, both alternately and simultaneously relative to each end of the cylinder.

If the depressurized fluid is to be returned to the reservoir through paths other than those that also feed the pressurized fluid to the cylinder, only a two-way valve would be required. A valve that allows passage of the fluid in only one direction is a check valve, or one-way valve.

Valves are operated electrically, mechanically, hydraulically, and manually (Fig. 16-10). Mechanical operation involves either a cam or trip action, and solenoids are used for electrical operation. Hydraulic actuation can be accomplished by the system's fluid by means of pilot valves, which may be a spool type, moved in one way by the pressurized fluid and returned by spring tension.

The four-way valve of the sliding spool type is one of the more common of the directional valves used on larger systems. The spool in one position (Fig. 16-11) allows the fluid from the pump to enter one end of the cylinder through the annular space, while it permits the fluid from the other end of the cylinder to return to the reservoir. By sliding the spool into the other position (Fig. 16-12), the flow is reversed.

FLOW CONTROLS. In addition to the necessity for applying hydraulic fluid at an appropriate pressure and direction to produce a given workpiece, the factor of speed which the actuator lends to its mechanical linkage, say, the reciprocating table of a planer, the cutting speed is relatively slow compared with the speed at which the worktable is returned to its original position in readiness to repeat the cutting process. The rapid return of the table is a time-saving maneuver.

There are several ways by which the speed of the actuator may be

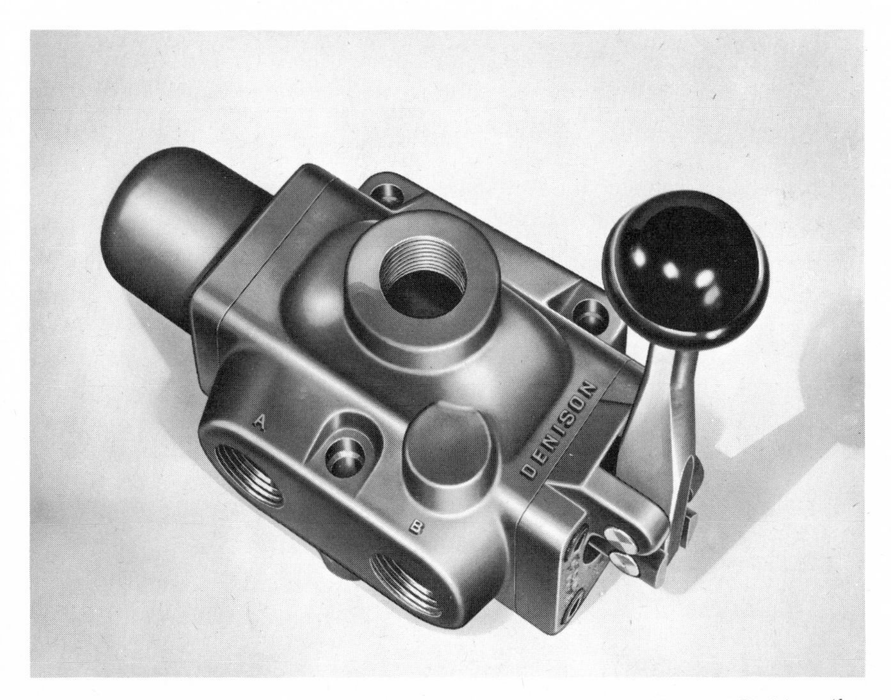

Fig. 16-10. Manual control for directional control valve. *Courtesy, Denison Division, Abex Corp.*

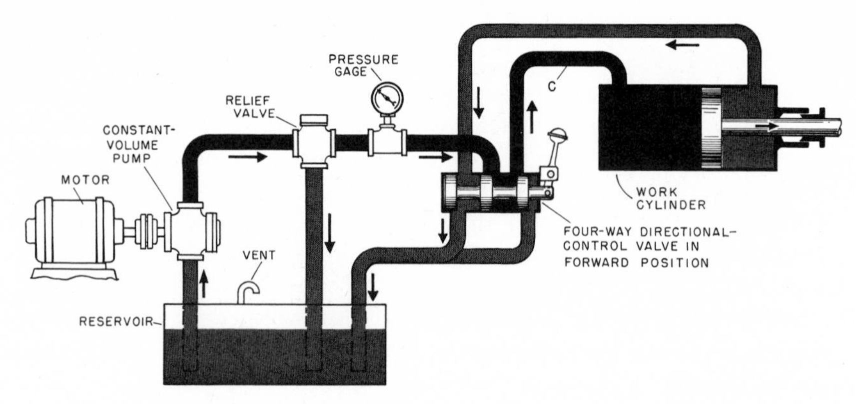

Fig. 16-11. Four-way directional valve, spool forward. *Courtesy, Mobil Oil Corp.*

changed. All involve the *rate* at which the moving element of the actuator is displaced by the pressurized fluid. Speeds then are based on rate of displacement or flow rate. If the pressure can be held constant, the speed of the actuator becomes a function of the volume of fluid that can be made available in a given time.

One way to accomplish the dual speed of the planer table is to provide separate pumps, a small pump for the work stroke and a larger one for the return stroke. Another method would be to use a variable-delivery pump, which would deliver its full capacity (full pump piston stroke) for the return stroke and then adjusted at some shorter stroke to achieve the slower table travel desired for the cutting phase. The third way to attain dual speeds is to use a constant delivery pump supplemented with appropriate flow control valves to regulate the amount of fluid that enters the actuator per unit of time. The third method is the most feasible in that it is more compact than the first method and more flexible than the second, because additional

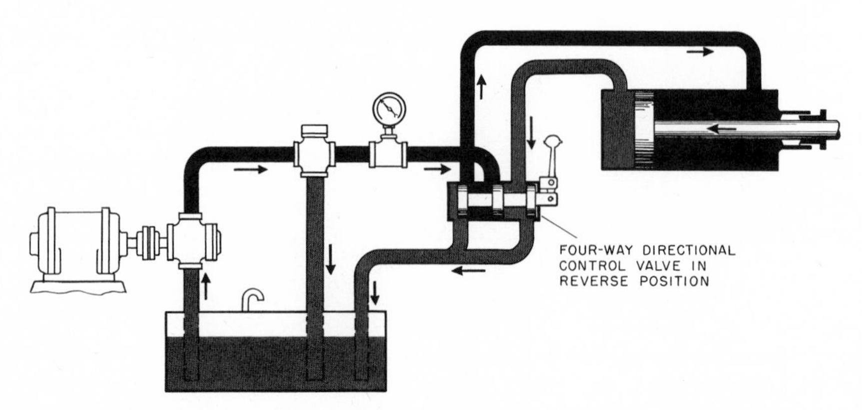

Fig. 16-12. Four-way directional valve, spool reversed. *Courtesy, Mobil Oil Corp.*

actuators may perform other functions. For example, a cross feed mechanism may be driven off the same circuit.

FLOW CONTROL VALVES. The fixed-delivery pump and flow control valve arrangement must be designed not only to change the rate of displacement, but also to maintain constant pressure. The importance of constant pressure is stressed because of the pressure drop across the flow control valve that varies the feed rate. This drop in pressure must be compensated for by any system using either a fixed-delivery or variable-delivery pump.

Flow control valves can serve as metering devices, or as a *bypass* arrangement for returning the excess fluid to the reservoir not required to meet the conditions set by the throttle.

The bypass flow control consists of an orifice for limiting flow to the actuator and a spring-loaded relief valve for returning the excess fluid to the reservoir. A pressure compensator built into the control is made up of a light, spring-loaded spool normally held in a closed position by the fluid pressure beyond the throttle, plus the light spring tension. To allow the work pressure to be exerted on the spool, a bleed line is also built into the control to permit the connection. Thus the pressure ahead of the throttle may be greater than the actuator pressure by the spring tension when the spool is in a closed position. The spool moves to an open position when system pressure is greater than the relief valve setting. Any pressure drop across the throttle orifice is sensed by the compressor spring, which maintains constant flow through the control. Flow control valves are located variously in the circuit depending on the work load characteristics.

1. In the pressure line ahead of the actuator (meter-in) when the work load is opposed to the piston travel as with grinding table feeds.
2. In the cylinder discharge line (meter-out) where the load tends to race or run away, as with drilling operations.
3. In a line connecting the main pump circuit with the reservoir (bleed-off) to allow a given volume of fluid to return to the reservoir at the work load pressure and not the relief valve pressure setting. This is used for reciprocating motions having constant load.

CHECK VALVES permit the flow of fluid in only one direction. Pressure in one direction, plus spring tension, keeps the valve closed, while pressure in the opposite direction acts against the spring to lift the valve off its seat to allow flow. Thus a check valve will feed a pressure circuit but will not allow the fluid to back up into the feeder lines.

Sequence Valves

Machines having two or more motions in sequence, i.e., a second motion starts when the first one ends, require a sequence valve. This valve closes the pressure line to the second motion while the pressure line to the first

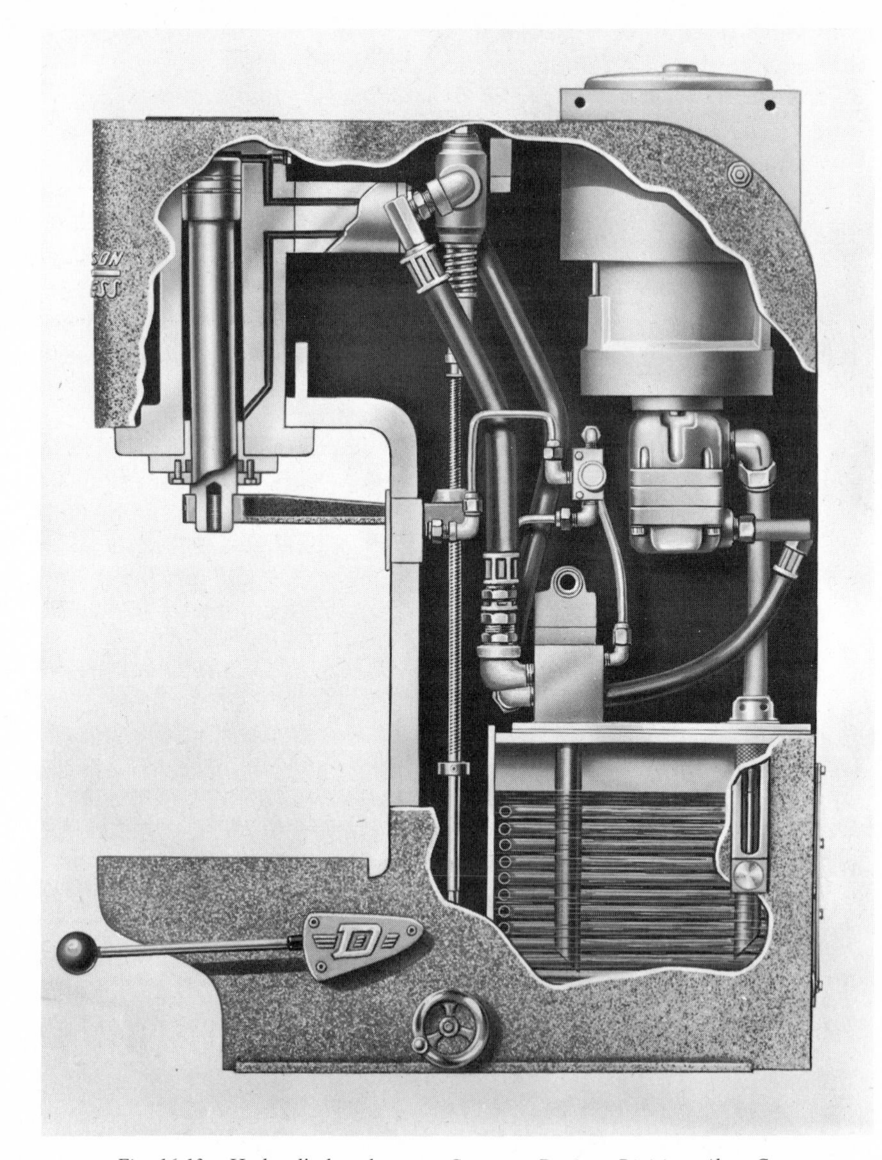

Fig. 16-13. Hydraulic bench press. *Courtesy, Denison Division, Abex Corp.*

motion is open. When the first motion stops, the pressure buildup causes a spring-loaded valve to lift and open the line to the second motion. Both actuators are then under pressure, one static, the second dynamic.

Pilot Valves

Some spool-type directional flow valves are shifted by hydraulic pressure through a pilot valve that may be operated manually, mechanically (with cams or levers), or electrically (with solenoids). The pilot valve is actually a directional control valve that directs fluid from the pump into the main spool-type valve to shift the spool in the manner described for this type of control.

ACCUMULATORS

Accumulators in hydraulic systems have several purposes, the most important of which is to provide increased capacity during extended hold or dwell periods when pressure losses occur due to leakage or branch circuit demands. An accumulator eliminates the necessity of using a larger pumping unit to provide the extra energy during such temporary periods. It also acts to dampen pulsations and eliminate shock during operation of the circuit. Accumulators may be loaded by weight (Fig. 16-18), springs, or pneumatically. The pneumatic type may have the compressed gas in direct contact with the oil or the gas may be separated from the oil by either a piston or a diaphragm (Fig. 16-15). The relative location of an accumulator in a power unit circuit is indicated in Fig. 16-14).

HYDRAULIC FLUID

The connecting link of a hydraulic mechanism is the fluid used to transmit hydraulic energy. Any liquid, including water, will transmit energy, but a number of other considerations require the use of more suitable fluids in hydraulic systems serving modern precision industrial, marine, aircraft, and automotive machinery. Some of these considerations are as follows:

VISCOSITY. The viscosity requirements vary according to operating temperatures and in some instances with the type and size of pump used and the service. Large piston type pumps operating on large presses require fluid viscosities of 600 to 900 SUS at 100°F.

For the majority of industrial requirements, most hydraulic systems respond satisfactorily with oils having viscosities in the range of 150 to 325 SUS at 100°F for operating temperatures up to 165°F. For industrial hydraulic power units powered by either a vane or gear type pump, an oil of 300 to 325 SUS at 100°F is preferred, especially if it is to be used on a variety of machines, as is often the case in large automobile body factories.

Oils increase in viscosity with increasing pressures, so that the viscosity at atmospheric pressures, and a standard test temperature of 100°F is an important consideration. Because the smooth and quick response of coordinated hydraulic components depends on a corresponding flow response of the fluid, it follows that any given combination of dynamic parts is selective as to the viscosity of the fluid that results in their optimum performance. The use of an oil that is too low in viscosity tends to result in increased leakage, loss of volumetric efficiency, reduced pressure, slippage, and the

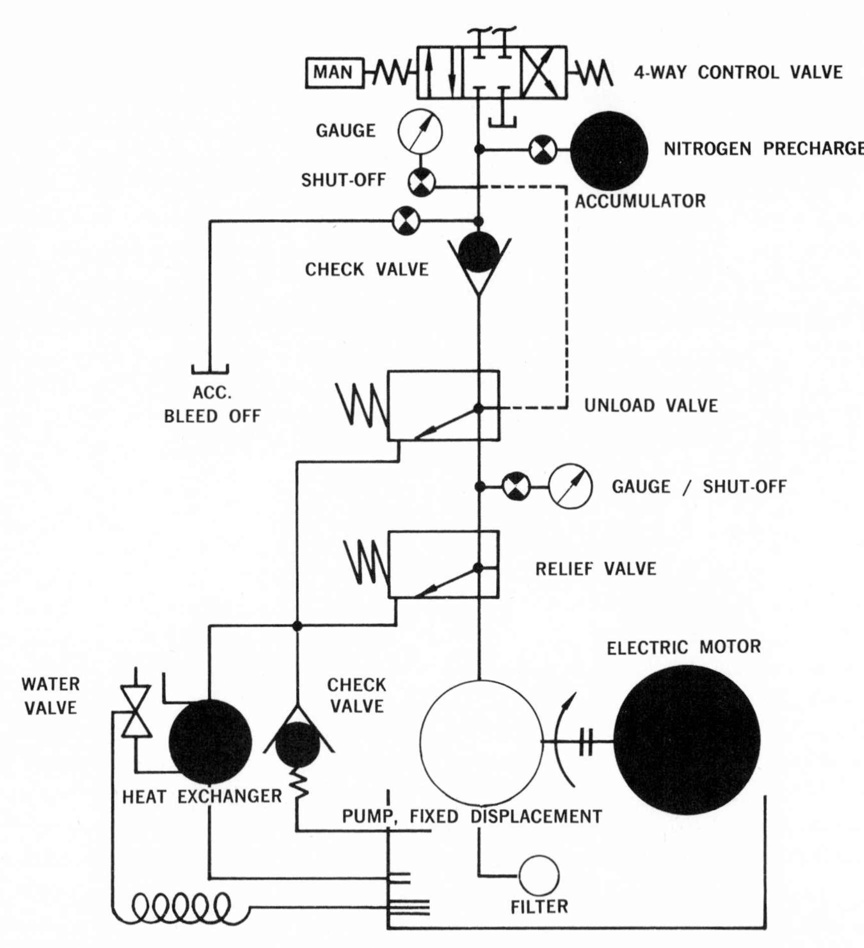

Fig. 16-14. Diagrammatic plan of a fluid power system for industry. Such systems are connected to production machines either as an auxiliary or supplemental power unit. *Courtesy, The Cardwell Machine Co.*

possibility of erratic control. An oil too high in viscosity causes an undue temperature rise, greater pressure drop, slow response to controls, and a higher power requirement.

CHEMICAL STABILITY. As in any other type of lubricant circulating system, the hydraulic fluid must possess high resistance to oxidation that tends to occur in presence of air, heat, and some contaminants. Oxidized materials are acidic and therefore corrosive to the components of hydraulic systems.

WEAR RESISTANCE. The severity of the service to which the hydraulic system will be exposed dictates the antiwear characteristics of the fluid to be used. While the major function of the fluid is to transmit energy, it has the additional task of lubricating the frictional parts of the various moving components. Effective lubrication acts to minimize wear and frictional energy; this energy is absorbed by the fluid as thermal energy, causing a rise in temperature. To meet the more severe conditions prevailing in high-energy regenerating systems, the fluid should be fortified with additives capable of combatting the elevated frictional forces. This means using fluids of greater film strength than possessed by the more conventional type fluids, such as rust and oxidation inhibited turbine type oils of high quality. The discussion of antiwear agents (Chapter 10), described the effectiveness of zinc dithiophosphate in combatting wear and its use as a metal deactivator. Its use in hydraulic fluids subject to severe operating conditions is highly recommended.

LUBRICITY. The property of lubricity is one that promotes motion with minimum effort, which in turn denotes oiliness, or slipperiness. Any hydraulic fluid must possess such a property in order to facilitate motion and reduce the power required by the various frictional components. The property of lubricity may be an adjunct to but not necessarily an ultimate criterion of wear resistance.

VISCOSITY INDEX. Where the temperature of the fluid is subject to wide fluctuation in temperature, it is desirable that a low rate of viscosity change per degree of temperature change be one of the attributes of the hydraulic medium. For systems operating within what is considered to be a normal temperature range of 120 to 140°F, the requirement for a high viscosity index fluid is not important. Because many of the highly refined oils have a VI of over 75, they can serve most industrial hydraulic systems adequately.

FIRE RESISTANCE. Fire resistance is not a factor in the case of conventional industrial equipment, such as machine tools and hoists, and mineral oils are used for most hydraulic fluids. Where there is risk of fire, several

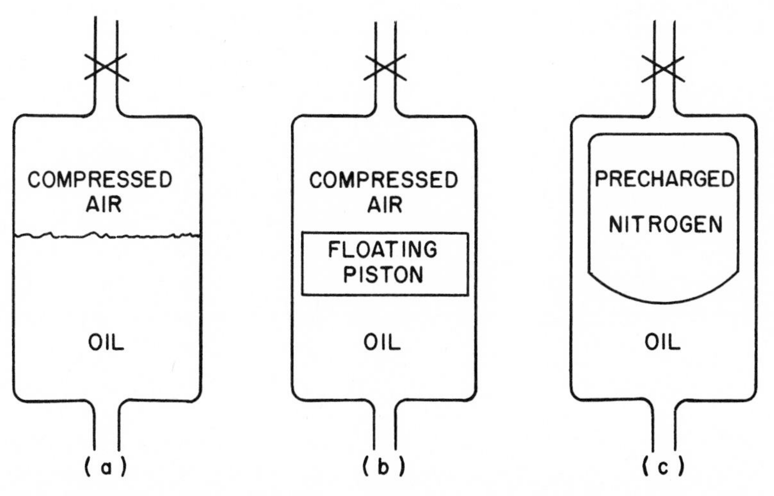

Fig. 16-15. Accumulator, pneumatic. (a) Direct contact of oil with compressed gas, (b) oil separated from gas by piston, (c) flexible bag precharged with nitrogen.

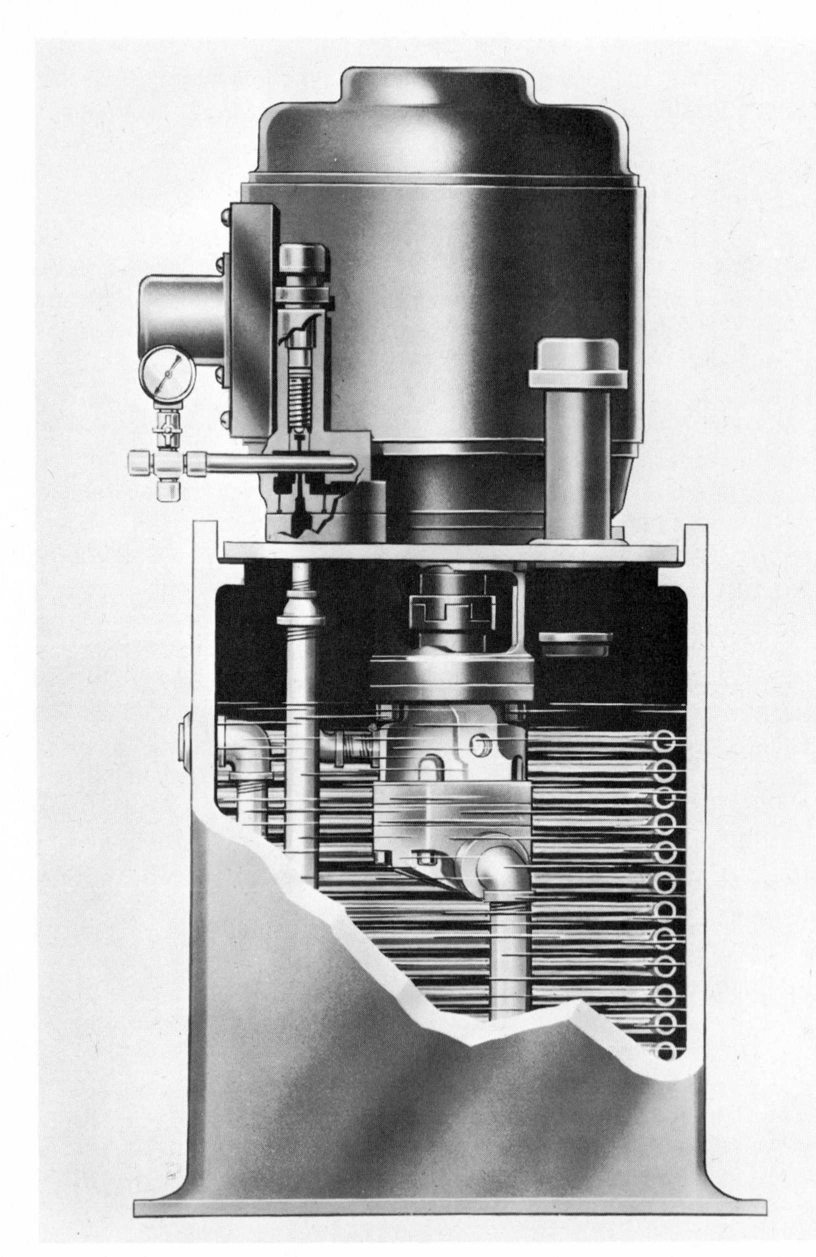

Fig. 16-16. Vertical hydraulic power unit. *Courtesy, Denison Division, Abex Corp.*

synthetic fluids, described in Chapter 6, are available and give excellent performance. They include pure phosphate esters, phosphate ester bases, water glycols, and water-in-oil emulsions. Some synthetic fluids show non-Newtonian characteristics, decreasing in viscosity as the rate of shear in-

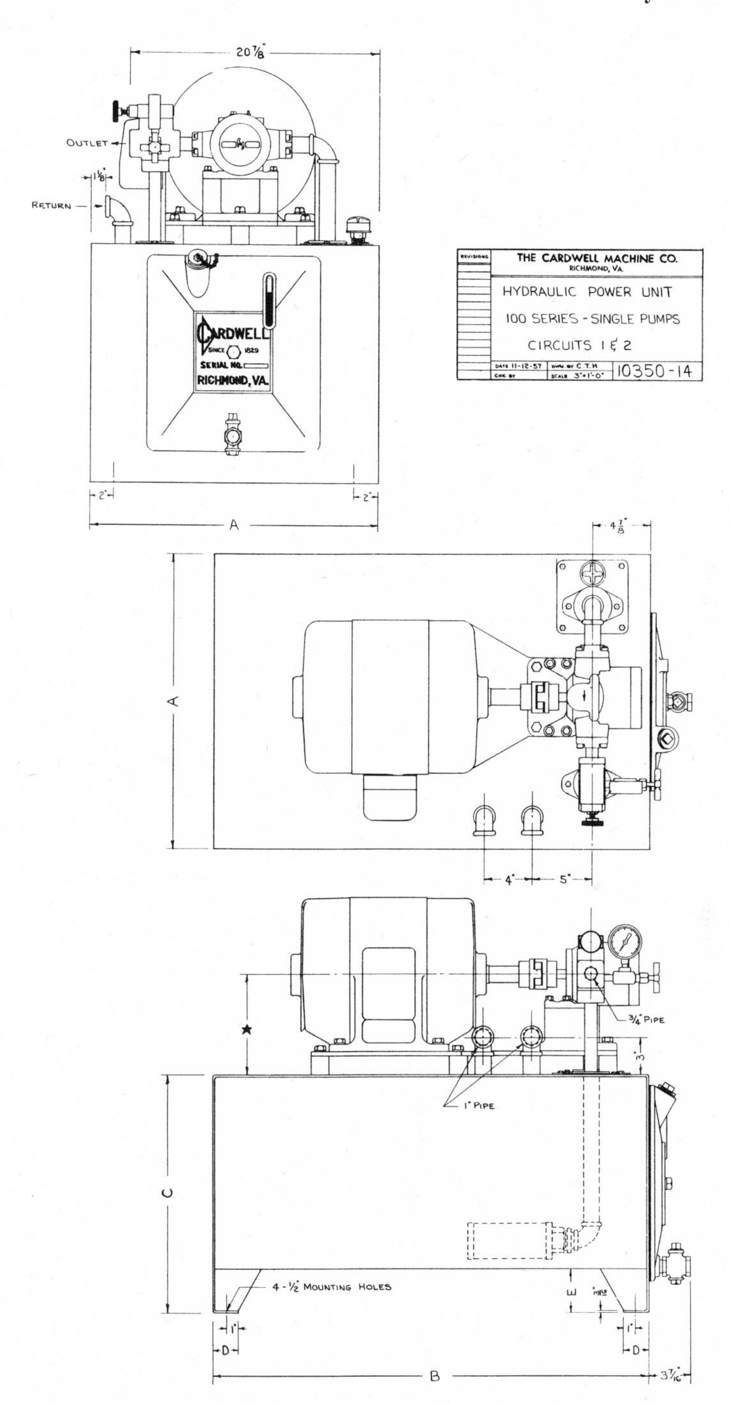

Fig. 16-17. Hydraulic power unit with single pump. *Courtesy, The Cardwell Machine Co.*

creases. Thus adequate size piping must be furnished at the suction side of the pump to prevent cavitation.

Fire-resistant hydraulic fluids must be used in systems that form parts of machines where a source of ignition is ever present, such as the molten metal processed on die casting machines or sparks from hydraulically operated welding equipment. They should also be used in systems operating in areas where the fluid may be exposed to furnaces and where other hot metal surfaces are located, or where molten metals are being poured, as in foundries. Considering that some hydraulic systems operate at pressures from 500 psi to over 5000 psi, any break in the fluid line may cause a stream to shoot up to 60 ft or more. If mineral oil is used, a fire hazard exists wherever a sparking electric motor, soldering or welding operation, open gas flame, etc., exists within the possible radius of oil spray.

The phosphate ester fluids are popular with users of hydraulic systems requiring the safety of fire-resistant properties. They have a flash point of 455°F and a fire point of 685°F. Their fire-resistant properties are best demonstrated by the following test results:

Auto-ignition temperature	1175°F+
Molten metal ignition test (flood)	1700°F
Molten metal ignition test (drip)	1500°F
SAE-AMS hot manifold test	1400°F

Water as a Hydraulic Fluid

Some manufacturing plants use central hydraulic systems for some of their processes, and many of these systems use water as the hydraulic fluid to operate rams that perform pressing, forming, and similar operations. An accumulator (Fig. 16-18) in central systems maintains constant pressure on the fluid. Critical lubrication areas in addition to the pumps (usually reciprocating) are at the rams and the seals through which they pass in their reciprocating motion. Where exposed, such points may be lubricated with water-resistant grease containing graphite or molybdenum disulfide, usually by periodic swabbing.

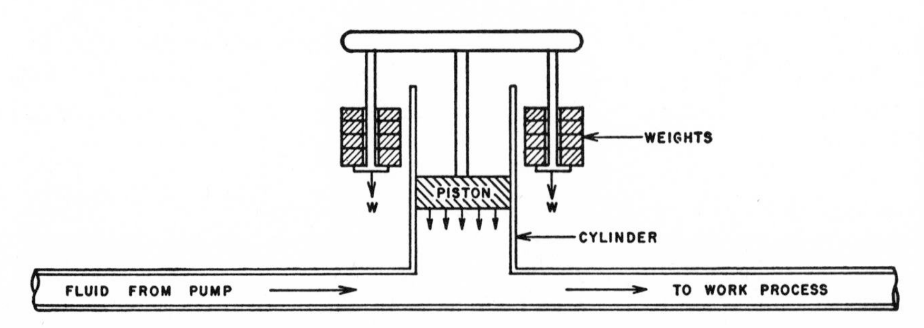

Fig. 16-18. Accumulator, pressure maintained by weight.

Internal parts may be protected against wear and rust by introducing a soluble oil into the water to form an emulsion. If the emulsion is of some noticeable color, such as white, it may seem that the addition of the soluble oil initiated leakage. This is usually not the case. The water that leaked at the same rate previously simply was not noticeable.

OPERATIONAL TROUBLES

Air in a hydraulic system causes slow or erratic response, with loss of motion due to compression of air in the fluid during work stroke. Aeration is due to (1) the level of pump intake being above the fluid level in the reservoir, (2) leaky shaft seals, (3) a loose or defective pump intake pipe, (4) the return line exit being above fluid level.

Cavitation occurs at pump suction due to cavities in the flowing fluid caused by insufficient fluid being fed to the pump (starving), due to (1) viscosity of fluid too high, (2) excessive pump speed, (3) an obstruction in inlet piping or strainer.

High-temperature operation is evidenced by continuous and too early opening of the relief valve. This causes stalling under normal load. Too high fluid viscosity is frequently the cause of heat. An oil cooler may help.

Loss of pressure causes the system to fail. It may be due to (1) a defective relief valve (one that stays open), (2) an inoperative pump, (3) a low oil level, or (4) a bypassed actuator.

Erratic operation may be due to the formation of oxidized oil deposits on valves, pump, and actuator parts. It is usually due to the use of poor quality oils or excessive temperature.

One method of cleaning systems to remove mineral oil deposits is to mix a safety solvent with the oil already in the system and let the system operate for about 48 hr. Then the system is drained and new oil put in. A mixture containing about 5 to 10% solvent is sufficient to clean the system effectively. Special solvents are available from oil companies handling lubricants.

LEAKAGE. Whether fluid leakage is due to low-viscosity oils or loose joints, the loss of fluid is costly. If annual makeup is equivalent to the quantity representing the amount of fluid in the system when filled to the recommended level, it means that the fluid is being changed once per year. Considering that a batch of oil should perform satisfactorily withoug changing from 2 to 10 years, depending on initial quality and operating conditions, leakage may prove costly, regardless of price per gallon of fluid. Where maintenance is improperly handled and joints leak excessively, the cost of fluid may soar to the equivalent of 10 to 40 times that of one complete filling per year, which again means that the fluid is being literally changed 10 to 40 times per year.

Two steel mill scrap-baling presses with a fill capacity of 500 gallons of fluid required 11,000 gallons of makeup in one year. When efforts to have the presses overhauled to stop the leakage failed for production reasons, it was recommended that a lower quality fluid be used to reduce operating cost. The recommendation was adopted and proved successful, but the limit to which this method of reducing costs can be carried is governed by and must be weighed against the effect that any substitute fluid has on the wear rate on the pumps and the valves.

Excessive leakage usually means major seal and joint deficiencies, and only complete replacement will rectify the condition. Leakage of reasonably low magnitude can be reduced by the use of fluids having a limited expanding effect on synthetic rubber seals. This property is described more fully in the discussion of aniline point (Chapter 7).

FIRE RESISTANCE TESTS

The phrase fire resistance is used instead of fire proof, since most hydraulic fluids can be ignited. Fire-resistant fluids are nonflammable under most conditions encountered industrially, and if ignition does occur, the flame is short lived. However, they all have some ignition temperature which varies according to the source of high temperature and method of fluid contact. Tests for fire resistance include these:

1. *Autogenous Ignition Temperature* (*AIT*). This is the minimum temperature to which a vapor-air mixture must be raised to ignite it. It is described under ASTM designation: D 286-30.

2. *Molten Metal Ignition Temperature.* This is the temperature at which ignition occurs when fluid is dripped, flooded, or sprayed onto the surface of a given molten metal.

3. *High-Pressure Ignition Test.* The high-pressure test determines the ignition point of a fluid upon being misted or sprayed through a small orifice under 1000 psi pressure through the flame of an oxy-acetylene torch.

4. *Low-Pressure Ignition Test.* Similar to the high-pressure test, this one has the pressure limited to 100 psi and uses burning materials, such as a cloth saturated with a lubricating oil.

5. *Contact Ignition Test.* This determines the temperature of ignition when a fluid is dripped onto the surface of a hot manifold (SAE-AMS).

6. *Arc Ignition Test.* Flash occurs at a given pressure at which fluid is sprayed through an orifice into the arc across the electrodes of a carbon lamp.

7. *Static Carbon-Arc Test.* To determine flash or burning point by a static carbon arc passing from an electrode to the bottom of a pan holding the test fluid of a given depth, in a given time of sustained arc.

17

MACHINE TOOLS

The designation machine tool may be applied to any productive machine, but its use here describes only those machine tools found in the metal working industry and machine shops. Portable hand tools are not included.

The basic kinds of machine tools are identified by their fundamental type of motion and their metal-removing or reshaping characteristic:

Drilling machines
Turning machines
Milling machines (surface peeling with rotating cutter)
Planers and shapers (flat surface peeling cutting tool)
Broaching machines (surface peeling of irregular shape)
Gear making machines
Grinding machines (surface removal and conditioning by abrasion)
Presses

DRILLING MACHINES

The drill press is a machine on which the drill rotates and is fed along its own axis into a stationary workpiece to produce holes of a given depth and diameter. It also performs tapping (internal threading), reaming (internal surface finishing), and countersinking (recessing for fastener heads) operations. There are five general types of drilling machines: upright, radial, horizontal, multiple spindle, and turret.

DRILLING MACHINE PARTS. The principal working parts of a drilling machine are as follows.

Table—for mounting ordinary sized workpieces. It can be raised, lowered, or swung relative to the column to which it is clamped.

Head—contains the feed mechanism as well as the controls for feed rate and direction. It also supports the spindle, quill, and turnstile.

Spindle—the basic rotating member that provides the tool chuck for securing the tapered shanks of drills, taps, and reamers.

Quill—supports and guides the spindle by means of a rack and pinion through which the feeding force is transferred to the tool.

Turnstile—a one, two, or four-armed (cross-armed) lever for supplying manual pressure to the tool through the pinion and rack arrangement.

Gear drive—mounted on the upper end of the column, supplies the rotating motion to the spindle through a vertical drive shaft. On machines with automatic control, it also supplies the power to feed the tool into the workpiece and bring it back.

Drive shaft—connects the spindle to the drive gears and is splined to allow free vertical motion through the heart of the driven gear in which the spline shaft also acts as a key to provide for its rotating motion.

Electric motor—provides the power to the entire machine. It is usually located at the top and in back of the column to balance the whole assembly, and it is connected to the spindle drive gear by a shaft. In some cases power is transmitted directly through a coupling, or by a belt arrangement. In the latter case, the spline shaft is allowed vertical motion through the hub of the driven pulley.

Jigs have the triple purpose of acting as tool guide, template, and workpiece holder or support. They vary in shape and size depending on the geometry of the workpiece but all permit accurate approach of the drill point to the desired hole center. Drill jig plates or covers are fitted with hardened bushings at the drill entrances, and their internal diameter is equal to the drill size.

Fixtures are devices designed to hold workpieces rigidly, or they may be jigs on the table of the machine. Vices, V-blocks (for cylindrical pieces), and parallels (for flat pieces) are typical fixture forms. Fixtures are usually clamped to the table by bolts whose heads slide within inverted T-slots in the table top.

Drilling machines have been given many fully automatic or semi-automatic innovations involving all aspects of the machinery art. Adjustments, positioning, feeding, reversing, tool changing, indexing, contour tracing, and other operations can be controlled automatically. Most of the manual effort and control has been replaced by automation through the use of mechanical, hydraulic, and electrical energy systems, particularly in competitive, high-production operations. However, many manually or manual-start, semiautomatic operated machines are required by plant maintenance shops, tool rooms, specialty contractors and pattern shops.

The foregoing describes the parts and functions of a basic type of *upright* drilling machine of the belt, gear, and positive drive types that possess great utility and are widely used where workpiece handling and hole production present no specific problems. Recommended lubricants are given in Table 17-1.

Some modifications in the basic design must be made to meet drilling problems associated with heavy or odd-shaped workpieces, mass production, and drilling angle variations. They include the following types of drilling machines.

RADIAL DRILLING MACHINES. Radial drills are designed for downward vertical drilling only (Fig. 17-1). A floor-mounted base acts as its table and on it is mounted an upright column. Clamped to the column is a horizontal arm that can be raised and lowered by means of an elevating screw or swung about the axis of the column. The arm supports the head, which can be moved toward or away from the column and clamped at any distance within the limitation of arm length. The arm length varies from 3 ft to 11 ft, according to the model, as do the elevating and swivel capabilities of the arm. The head is the working component, containing the vertical spindle, feed and speed gearing, and controls. The design of radial drills gives them great versatility, permitting, in addition to large-diameter hole drilling, such operations as reaming, tapping, and boring of heavy and cumbersome workpieces. Lubricants are listed in Table 17-2.

Universal radial drills allow drilling at any desired angle. The head is able to swivel at the end of an arm, which in turn can be raised, lowered, or rotated on or about its column. Such machines can be moved to the workpiece.

A horizontal-spindle radial drill has a head mounted directly on the column, allowing only horizontal operation. The column can be positioned at any point along the length of the base. It can handle large workpieces.

Deep-hole drilling machines are especially made for producing holes of depths many times their diameter. They are usually vertical but their mechanism is similar to that of the upright drill.

Gun-barrel drilling is performed on a horizontal machine. The cutting end of the drill is concave and has ground edges and a hollow shank to allow

Table 17-1. LUBRICANTS FOR DRILLING MACHINES

Part To Be Lubricated	*Type of Lubricant*	*Method of Application*	*Approximate Viscosity (SUS at 100° F) or NLGI Consistency*
Ways and slides	Way oil	Any	300
Spindle bearings	Spindle oil	Any	100–150
Headstock	Turbine oil	Bath	150
Gears, quick return unit	Turbine oil	Circulation or bath	300
Gear cases	Turbine oil	Bath	300
Feed worm gears	Way oil	Bath	300–500
Elevating screws	Way oil	Any	300–500
Elevating gears	Turbine oil	Bath	300
Miscellaneous parts	Machine oil	Hand or cup	150–300
Miscellaneous parts	Grease	Any	No. 2

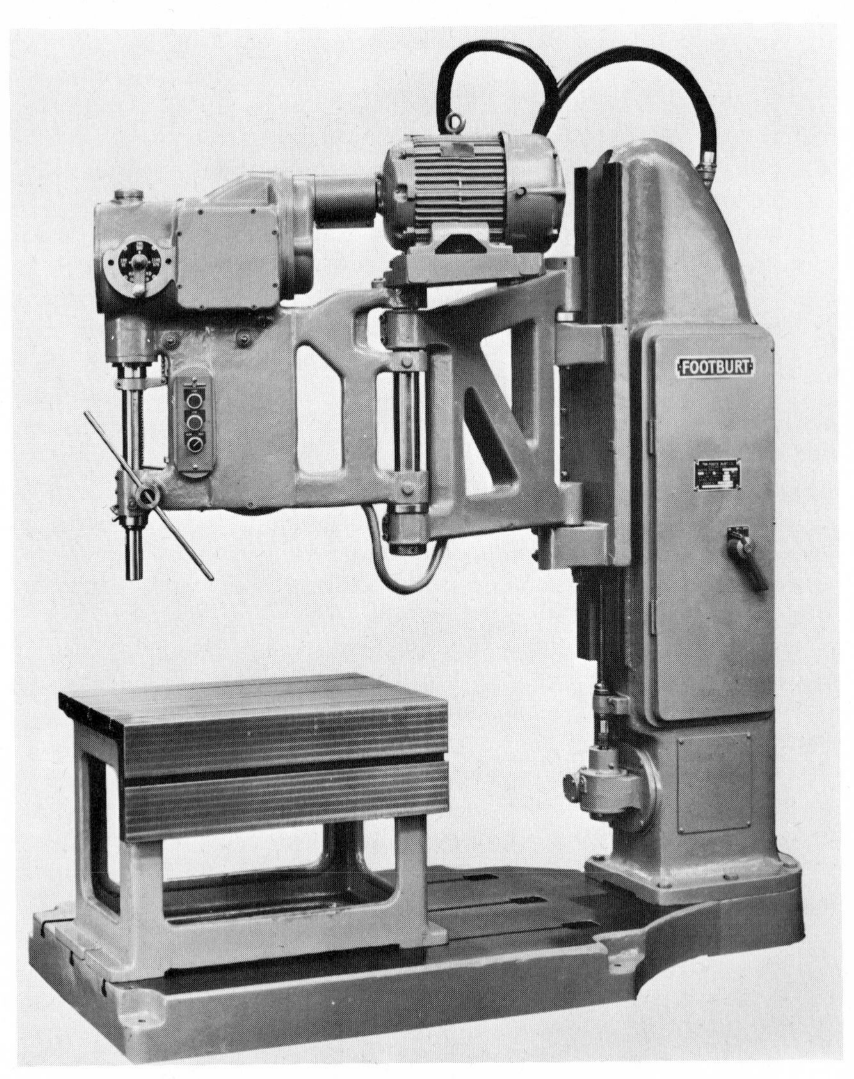

Fig. 17-1. Radial drill. *Courtesy, Footburt-Reynolds Machinery Division, Reynolds Metals Co.*

the passage of cooling fluid. The fluid also washes the chips out of the developing hole. The drill can be automatically withdrawn from the workpiece periodically to relieve the chips.

HORIZONTAL DRILLING MACHINES are built in several varieties. The way type is equipped with multiple spindles centrally driven and set up for a given workpiece in continuous production. The spindle head approaches the workpiece on ways. The table type is similar to the horizontal radial drill.

Table 17-2. LUBRICANTS FOR RADIAL DRILLS

Parts To Be Lubricated	*Type of Lubricant*	*Method of Application*	*Approximate Viscosity (SUS at 100°F) or NLGI Consistency*
Feed worm gear	Cylinder oil way oil	Bath	2500 900
Hydraulic arm and column clasp	Hydraulic oil	Circulation	300
Clutches and hand mechanical gears	Hydraulic oil	Circulation	300
Arm ways	Way oil (preferred)	Any	300
Gibs	Way oil (preferred)	Any	300
Sliding base ways	Way oil (preferred)	Oil pump	300
Elevating screws and column	Way oil	Oil pump	300
Spindle head	Hydraulic oil	Reservoir	300
Hydraulic system	Hydraulic oil	Circulation	300
Miscellaneous parts	Lithium or calcium complex grease	Gun	No. 1

It has a table that can rotate to enable successive drilling on all sides of the workpiece, a function that lends itself well to programming.

GANG DRILLING MACHINES are comprised of a number of upright drill heads separately driven, and the individual columns are mounted in line on a common base and have a common table. They provide a convenient way of drilling various size holes in the same workpiece by transfer along the table.

Another multiple-spindle machine similar to the gang drill is the in-line drilling machine. In-line drills have a common power source that enables a number of them to be rotated and fed simultaneously. The only geometric variation allowed is in the spacing of spindles.

Universal joint drilling machines enable the area placing of a number of spindles anywhere within the limitations of the table. All spindles are driven and fed from a common power source.

THE TURRET DRILLING MACHINE is another departure from the older type gang drill. While the parts remain similar to that of the upright machine, the turret has several spindles that can be rotated to make available any tool change involving drill size or other operation, such as reaming, tapping, counterboring, and trepanning. On some later models, the speed of each spindle is variable and made independent of all others on the same turret, including its opposite number. This makes for an ideal programming arrangement, especially when coupled with some form of automatic workpiece positioning.

Jig Borers

Older methods of guiding a drill include (1) the use of a jig, or bushing, (2) using a punch mark, and (3) counter-boring through an already drilled or reamed hole.

The jig borer (Fig. 17-2) is designed to act as its own jig in accurately locating or spacing holes to be drilled by a single-point tool. Such accuracy lies in the use of a precision spindle and a work-supporting table, both of

Fig. 17-2. Jig borer. *Courtesy, Moore Special Tool Co., Inc.*

which are movable. The spindle is quill mounted to gain rigidity. A spindle sliding head is mounted over a table that can be roughly adjusted both longitudinally and transversely by lead screws and then accurately compensated to within 0.0001 in. by cam action on the lead screw nut. In general, precision adjustment measures include the use of end measuring rods, graduated scale, and lead screw micrometers. Lubricants are listed in Table 17-3.

To provide maximum rigidity and wide speed range while keeping frictional heat to a minimum, jig borers are equipped with preloaded roller bearings.

The boring tools are of the single point variety generally. The machines are heavier than the scope of work performed on them appears to warrant in order to provide the rigid structure required for precision. They are either of the open sided or two-column type. Open sided models include those with horizontal and vertical spindles. Their primary use is the production of precision jigs, dies, and gauges, and they are often used to determine the accuracy of work produced on other type machines. They should be operated in temperature-controlled environments in which air pollution is held to a minimum. The lubricants used on the various frictional parts of jig borers must be selected on the basis of high chemical stability, antiwear qualities and their ability to avoid stickslip action at slow speeds. This is particularly true with production models that are virtually a combination of jig boring and milling machines.

Table 17-3. LUBRICANTS FOR JIG BORERS AND JIG GRINDERS

Part To Be Lubricated	*Type of Lubricant*	*Method of Application*	*Approximate Viscosity (SUS at 100°F) or NLGI Consistency*
Carriage and table ways			
Light work	Way oil	Any	150–300
Heavy work	Way oil	Any	900
Feed control	Hydraulic oil	Circulation	150
Drive gears	Hydraulic oil	Circulation	200–300
	Lithium or calcium complex grease	Gun	No. 1
Lead screws	Way oil (preferred)	Hand	300
Guide rods and slides	Way oil	Any	150
Vertical slides	Way oil	Pump	300
Rotary tables	Grease	Usually prepacked	—
Speed gears (light duty)	Spindle oil	Bath	60
Feed gears (light duty)	Spindle oil	Bath	60
Miscellaneous parts	Way oil	Hand, central oiler	300
Micrometer	Spindle oil	Hand	60
Hydraulic system	Hydraulic oil	Circulation	150
Pneumatic hydraulic system	Hydraulic oil	Circulation	150

TURNING MACHINES

Turning machines can remove material from the external areas of a workpiece axially, from the internal areas of a hole axially (boring), and from external areas of a body radially (facing or squaring). Lathes also perform the following operations: external threading, internal threading (tapping), reaming, drilling and counterboring, knurling, and polishing. Lubricants are listed in Table 17-4.

The basic types are the engine lathe, turret lathe, vertical lathe, each of which can be manually operated or be fully automatic, and the automatic production lathe.

ENGINE LATHE. The engine lathe is so named because it was powered by a steam engine in earlier years. It is an all-purpose lathe adaptable to all kinds of turning work, and it requires skillful handling. Its principal parts are as follows:

Bed—supports all other components and receives all the forces developed during operation. Kinematically it is called the *frame.*

Ways—located on the bed, they act as the tracks of the machine over which certain moving parts travel. Their surfaces must be aligned and leveled with precision.

Carriage—rests on the ways to support the metal-cutting tools. It consists of a saddle and an apron. It carries the tool during the turning operation

Table 17-4. LUBRICANTS FOR TURNING MACHINES

Part To Be Lubricated	*Type of Lubricant*	*Method of Application*	*Approximate Viscosity (SUS at 100° F) or NLGI Consistency*
Ways and slides	Way oil	Any	300
Headstock (engine lathes)	Turbine oil	Any	150–300
	Turbine oil	Spray	150
Tailstock	Hydraulic oil (AW*)	Hand	300
	Way oil	Hand	300
Aprons	Leaded compound or way oil	Plunger pump, bath or spray	300
Saddles	Way oil	Plunger pump	300
	Turbine oil	Any other	150
Quills	Hydraulic oil (AW*)	Any	150
Lead screw	Way oil	Plunger pump	300
	From apron	Splash	Same as apron
	Way oil	Hand	300
Saddle drive gears	Turbine oil	Any	150–300
Tool holders	Way oil (preferred)	Any	300
Feed gears	Turbine or way oil	Splash	300

* AW = Antiwear agent.

between the headstock and tailstock. It is the component that transfers the cutting forces from the tool to the bed ways.

Saddle—part of the carriage that forms the bridge spanning the ways to support the *compound rest* on which the *tool post* (holder) is mounted. It is also fitted with a *cross slide* for positioning the compound rest (slide and swivel) according to the desired tool angle, and it can be manually or automatically (powered from the headstock through the apron) moved back and forth on the carriage to permit cross feeding.

Apron—the front part of the carriage that contains the gears, clutches, and various levers for transmitting and controlling transverse motion to the carriage from the *feed rod,* and the *split-nut* that engages the *lead screw* used when cutting threads.

Tailstock—mounted on the ways in the same manner as the saddle and may be positioned at any desired location along the ways according to the length of the cylindrical workpiece. It contains the *dead center* that acts as a focus and support for the end of the workpiece opposite the headstock (live center) and its adjusting screw and hand wheel.

Headstock—mounted on the bed at the end opposite the tailstock and contains all the components associated with the rotational action of the lathe. These include the usually hollow *spindle,* in which a *live center* is sometimes locked, and which also supports the *face plate* or *chuck* that holds and rotates the workpiece. Chucks with three or four jaws hold the work in the manner of a vise, and a *dog,* clamped to the workpiece and protruding through a channel in the face plate, transmits motion by the face plate to the work, which in turn is held and focused on the *center point* of the spindle. The spindle is generally completely hollow to allow end feeding of bar stock.

The headstock also houses the gears that drive the spindle and supply power to the feed box, and in some instances, to the lead screw and feed rod. It also contains the spindle speed changer and selector with its control levers and mechanism.

The *feed rod* transmits power to the carriage apron gears to cause longitudinal motion.

The *lead screw* also transmits longitudinal motion to the apron, through engaging half-nuts attached to the apron, but is generally used only when a thread-cutting operation is performed on the lathe.

With certain modifications, the engine lathe, or geared-head lathe, is converted or expanded to perform special operations while maintaining, generally, its typical parts composition. The major varieties of the engine lathe are the following:

1. *T-lathe,* whose headstock is positioned at right angles to the bed to facilitate machining of large diameter and ungainly workpieces.

2. *Oil-country lathe,* equipped with both inboard and outboard chucks, i.e., one on either end of the headstock and a large-diameter hollow spindle for threading and cutting off long lengths of pipe.

3. *Tool-maker's lathe,* (Fig. 17-3) with additional controls and attachments to gain greater precision and versatility.

Other modifications include the ability to remove the bed, or slide part of it out of the way at the headstock to allow greater turning radius (swing) of the workpiece. A power-operated chucking and tailstock accelerates loading and unloading operation, which increases production. This is the *manufacturing lathe.* The tracer lathe incorporates controls that can repeat desired contours on a repeat production basis.

TURRET LATHE. A modification of the engine lathe, the turret lathe is equipped with a square, hexagonal, or octagonal turret that replaces the dead center effect of the tailstock. Mounting a different tool on each or any number of the available turret stations enables many different operations to be executed without the necessity of changing the tool by a sequence

Fig. 17-3. Toolmaker's lathe. *Courtesy, Monarch Machine Tool Co.*

Fig. 17-4. Saddle type turret lathe. *Courtesy, Gisholt Machine Co.*

of indexing. Greater versatility may be achieved by adding more tools on each side (front and rear) of the carriage cross slide.

The two fundamental types of turret lathes are identified according to the manner in which the turret is mounted and traverses the workpiece. They are the *saddle type* and *ram type* turret lathes.

The saddle type (Fig. 17-4) is one in which the complete carriage including the turret moves longitudinally on the ways. This construction imparts ruggedness and permits maximum stroke. The turret of this machine is adjustable so that it can perform along a fixed center or provide cross sliding (or side motion) for contour cutting. It also allows tool offsetting as required, for example, for boring.

The ram type machine (Fig. 17-5) is designed to have its saddle carry a ram on which the turret is mounted. The ram slides longitudinally on the saddle, which is clamped in a fixed position on the bed ways at a predetermined point. The turret and ram act as one component in traversing the workpiece. The distance of travel depends on the length of the ram, so that this type of machine is adapted to shorter workpieces.

Both the saddle type and ram type turret lathes may be *bar fed* or set up for *chucking* (Tables 17-5 and 17-6). Bar fed machines are those in which the bar stock is fed through the hollow spindle from the far outer end of the headstock, to protrude into the working area adjacent to the inner end of the headstock. Chucking machines are those in which the unfinished individual workpieces, in any form, are placed and locked in the chuck located at the inner end of the headstock or work area (Fig. 17-6).

AUTOMATIC LATHES. The turret lathe is adaptable to completely automatic operation because of its multiple faces or stations, its cross-fed char-

Fig. 17-5. Ram type turret lathe. *Courtesy, Gisholt Machine Co.*

acteristics, and the feasibility of converting the rotating axis of the turret from a vertical to a horizontal attitude, with the tools arranged parallel or at right angle to the workpiece. It is also possible to incorporate multiple spindles and other more complex mechanisms and controls. Lubricants are listed in Table 17-7.

The single-spindle screw machine is a fully automatic turret lathe that can change the axis about which the tools index from vertical to horizontal. This arrangement, called *toolslides,* allows for the tools to be positioned and

Table 17-5. LUBRICANTS FOR AUTOMATIC CHUCKING AND BAR MACHINES

Part To Be Lubricated	*Type of Lubricant*	*Method of Application*	*Approximate Viscosity (SUS at 100° F) or NLGI Consistency*
Hydraulic system or attached power unit	Hydraulic oil	Circulation	200–300
Pneumatic cylinders	Air line oil	Air line oiler	75–100
Miscellaneous parts	Hydraulic oil	Central system	200–300
Miscellaneous	Lithium or calcium complex grease	Gun	No. 2
Spindle bearing seals	Lithium or calcium complex grease	Gun	No. 2
Slides, camshafts	Lithium or calcium complex grease	Gun	No. 2

Table 17-6. LUBRICANTS FOR SINGLE SPINDLE CHUCKING MACHINES

Part To Be Lubricated	*Type of Lubricant*	*Method of Application*	*Approximate Viscosity (SUS at 100°F) or NLGI Consistency*
Headstock	Hydraulic oil	Circulation or bath	150–200
Gears	Hydraulic oil	Bath	150–200
Cross slides	Lithium or calcium complex grease	Any	No. 2
Air chuck	Lithium or calcium complex grease	Any	No. 2
Air cylinder gland	Hydraulic oil	Hand	200

fed at right angles, while the turret slide or end slide is positioned and feeds parallel, relative to the axis of the workpiece. The production capabilities are further expanded by introducing more than one tool-holding cross slide radially (at right angles) about the axis of the workpiece as well as the tool-holding turret, which travels with, but rotates about, the parallel axis of the workpiece. In some models, the work rotates only while the tools travel traversely as well as normal to the axis of rotation. Where it is necessary to prevent deflection of thin workstock, greater support is provided by

Fig. 17-6. Chucking lathe (Series 220 N/CC chucker) with 6-in. bar feed. *Courtesy, Monarch Machine Tool Co.*

Table 17-7. LUBRICANTS FOR SINGLE AND MULTIPLE SPINDLE AUTOMATICS

Part To Be Lubricated	*Type of Lubricant*	*Method of Application*	*Approximate Viscosity (SUS at 100°F) or NLGI Consistency*
Spindles			
Over 1800 rpm	Hydraulic oil	Circulation	150
Under 1800 rpm	Hydraulic oil	Circulation	300
Ways	Way oil	Central pump	300
Miscellaneous parts on way system	Way oil	Central pump	300
Working tool slides	Way oil	Any	300–900
Circular ends	Way oil	Bath	900

machining close to the fulcrum or spindle. This is done by continuous feeding of the workstock while the machining is done by toolslides that move only normal to the axis of workpiece rotation.

The multispindle automatics (Fig. 17-7) allow several workpieces to be machined simultaneously. The operations involved may require all or

Fig. 17-7. Multispindle automatic screw machine. *Courtesy, Davenport Machine Tool Co., Inc.*

any combination of turning sequences, including forming contours, drilling, reaming, counter-boring, shaving, parting, cutting off, facing, tapping, and threading.

There are several kinds of multiple-spindle automatics, usually referred to as *bar machines.* The two basic models are the *indexing machine* and the *stacked type.* Each spindle is synchronized with its own tool or combination of tools, both end-working and cross-sliding tools as required.

The indexing model has a carrier spool that rotates intermittently about an axis that is horizontal and parallel to the length of the machine. Through the spool protrude four, six, or eight rotating hollow spindles, each of which carries a revolving bar to be machined. The spindles are concentric to the circumference of the carrier spool, and each is equipped with a collet chuck (split tapered bushing) for grasping and rotating the barstock, along with the individual rotating spindles.

The stacked type of multispindle automatic has a series of rotating spindles that remain in a fixed horizontal position, usually one directly above the other. Each spindle is equipped with its own chuck through which barstock feeds simultaneously into separate groups of identical tools. The machine produces the same number of similar finished pieces as there are bars fed. As in the indexing automatic, the tools used either move in to the workpiece at right angles to the axis of rotation of the spindle or from the end and parallel to the same axis. The mechanisms for these two operations are in both single- and multispindle automatics respectively called *working tool slides* and *circular ends.*

Automatic hydraulic production lathe—(Fig. 17-8)—a highly versatile lathe that handles machine parts having wide swing and up to 40 in., lengths.

The automatic chucking machine processes individual forgings and castings rather than barstock. It is equipped with a spindle carrier but has special manual or power chucks to hold the workpieces.

MILLING MACHINES

Because of its versatility and accuracy, the milling machine has become very important in the machining industry. Its basic action is one on which the workpiece moves at a right angle to the axis of a rotary toothed cutting tool. The number of rotary tools used on milling machines are many and varied, each designed to perform particularized functions ranging from simple key way cutting to the milling of spirals, helices, and cams. Lubricants are listed in Table 17-8.

Milling machines are categorized according to their design and versatility. Some manufacturers describe their models as being *plain, universal* (Fig. 17-9), or *vertical,* and others use the terms *knee* and *bed.* However,

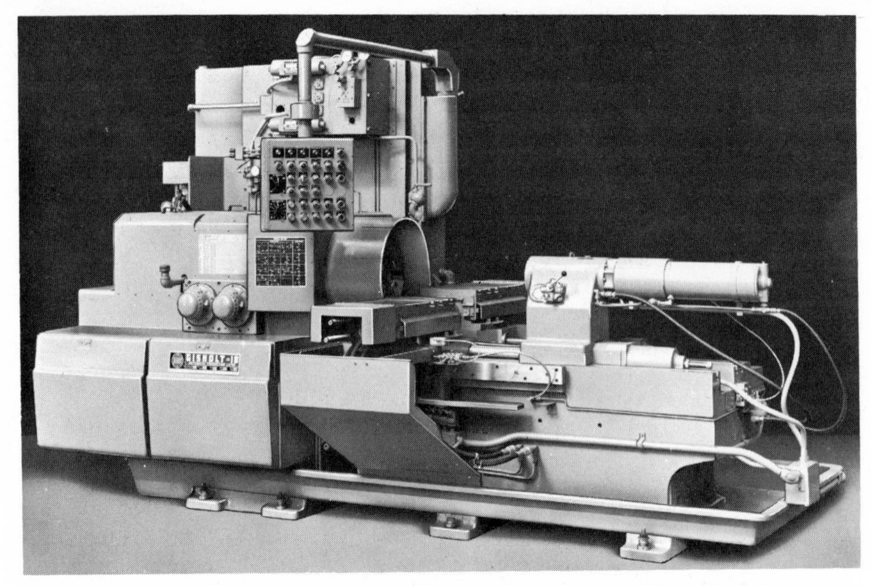

Fig. 17-8. Automatic hydraulic production lathe. *Courtesy, Gisholt Machine Co.*

most standard milling machines are classified as being either *high speed, knee and column, planer,* or *fixed bed* types.

Milling machine components include the following:

Base—the foundation of the machine supporting the other components and also used as a cutting oil reservoir.

Column—supported by the base and supporting the working components described below.

Table 17-8. LUBRICANTS FOR MILLING MACHINES

Part To Be Lubricated	*Type of Lubricant*	*Method of Application*	*Approximate Viscosity (SUS at 100°F) or NLGI Consistency*
Spindle	Spindle oil	Any	150–300
Saddle	Way oil	Hand or pump	300–900
	Turbine or mist oil	Mist	500–1000
Knee (gears)	Hydraulic oil (AW)	Bath	300
Knee (vertical slides)	Way oil	Pump plunger	300–900
Columns	Turbine or way oil	Circulation	300
Ways	Way oil	Hand or pump	300–900
	Mist oil	Mist	300–1000
Gears	Hydraulic oil	Bath or circulation	300
Arbor yoke	Hydraulic oil	Bath or pump	300
Indexing mechanism	Grease	Gun	No. 2

Fig. 17-9. Universal milling machine (Dynamaster). *Courtesy, Brown & Sharpe Manufacturing Co.*

Head—includes the drive mechanism for the rotation and feed of the spindle.

Ram—supported on ways atop the column; it supports and positions the head horizontally.

Spindle—located in the head; it provides mounting for and rotates the cutter. It may be vertical or horizontal, depending on the type of machine.

Table—provides support for the workpiece and rides on the saddle ways, on which it moves horizontally to feed the work to the cutter.

Saddle—provides the ways for the movement of the table and in turn moves on the ways supported by the knee.

Knee—a bracket or knee-like casting that supports the table which in turn is supported by and adjustable on the column to permit raising and lowering to accommodate the workpiece.

Elevating screw—has the dual purpose of supporting and elevating or lowering the knee.

Overarm—acts to support the arbor through an *arbor yoke.*

Arbor—holds the various cutters and is driven by the spindle, into which it is fitted by a tapered shank.

Bed—part of a bed type milling machine. Provides the support for the ways on which the table moves. It also contains the drive mechanism for the table.

Headstock—also a part of the bed type milling machine. It forms part of the bed and contains the drive mechanism of the spindle. It also supports the spindle carrier.

Spindle carrier—supports the spindle and in turn is supported and moves vertically on the headstock ways.

The cutters of milling machines remove metal in two ways. Conventionally the cutters rotate opposite to the direction of the workpiece. With *climb-cut milling,* the cutters rotate in the same direction as the workpiece.

The high-speed milling machine is belt driven, and the spindle may be either vertical or horizontal.

The *knee and column milling machine* may be of the plain, universal,

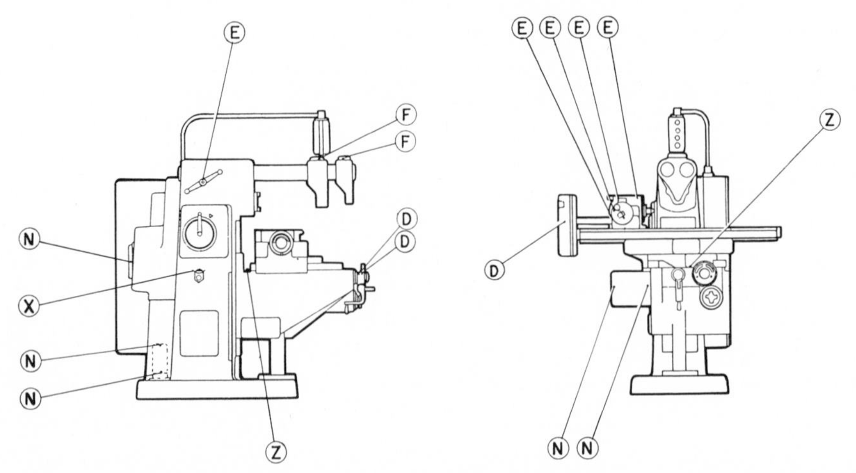

Fig. 17-10. Lubrication points of a universal milling machine, normal use. D, oil daily with good grade machine oil, 300 SUS at 100°F. E, fill when necessary with neutral nonfibrous grease. N, bearings permanently sealed. X, keep filled to gauge (automatic lubrication) with good grade machine oil, 300 SUS at 100°F. F, fill when necessary with good grade machine oil 300 SUS at 100°F. Z, keep filled to gauge (hydraulics system) with good grade hydraulic oil, 150 SUS at 100°F.

or vertical type. The universal has a table that can be swiveled horizontally, permitting a greater range of operations.

The *fixed bed or manufacturing* milling machine is a heavy-duty machine capable of mass production. It has a rigid, solid bed that supports the other components. The single-spindle and two-spindle types are respectively known as simplex and duplex machines. They are highly automatic as required for continuous production.

The planer type milling machine is also of the bed type and is designed for heavy workpieces. It is equipped with a long table that feeds the workpiece to the cutter and is similar to the conventional planer in appearance.

Special types of milling machines include rotary head, ram, and tracer types.

Points of lubrication on a universal milling machine are shown in Fig. 17-10.

SHAPERS

Unlike the milling machine, the shaper has a stationary table on which the workpiece is fastened while being machined to a flat surface at various attitudes by a single, rigid cutting tool passing across it with a reciprocating motion, cutting in just one direction. The tool is held in a *tool block,* which is hinged to allow the tool to ride back over the workpiece without cutting during the reverse stroke.

The construction of the horizontal plain shaper (Fig. 17-11) includes some of the components common to many other machine tools, such as the bed that supports the column, which houses the drive mechanism. The ways for the cross rail are also mounted on the column. The cross rail helps support the table to which the workpiece is clamped and feeds the tool horizontally. A reciprocating ram mounted on top of the column moves the tool alternating away and toward the column. The tool is held in the tool block, which is attached to the head at the front of the ram. The table is supported at one end by the cross rail as mentioned and at the front end by an adjustable *outrider* support. The front end of the shaper is the end at which the table is mounted. Shaper lubricants are listed in Table 17-9.

The plain shaper differs from a universal type shaper in that the universal table can be swiveled about a vertical or horizontal axis as to allow greater versatility in positioning the workpiece without resorting to special fixtures.

PLANERS

The planer has a reciprocating table that moves the workpiece past a stationary tool to produce a flat or contoured surface on the workpiece. Planers are primarily used for machine parts ordinarily too large for shapers

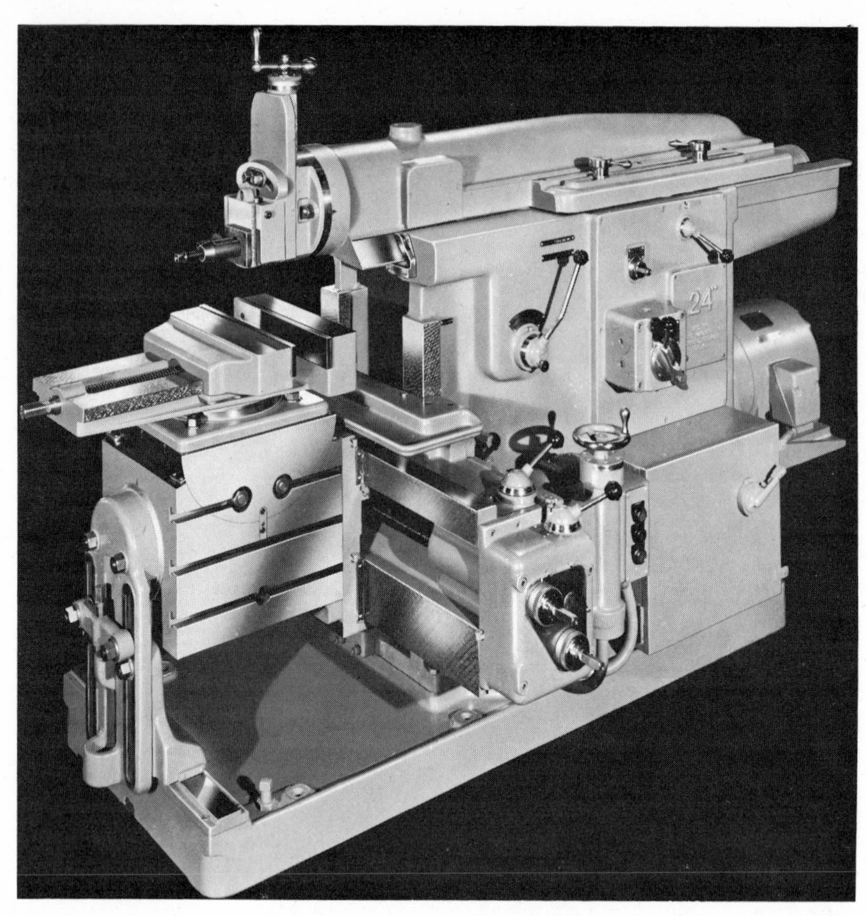

Fig. 17-11. Ram shaper. *Courtesy, Rockford Machine Tool Co.*

Table 17-9. LUBRICANTS FOR SHAPERS

Part To Be Lubricated	*Type of Lubricant*	*Method of Application*	*Approximate Viscosity (SUS at 100° F) or NLGI Consistency*
Hydraulic system	Hydraulic oil	Circulation	300
Ways	Hydraulic oil	Circulation or from hydraulic system	300
Gears	Cylinder oil or leaded compound	Splash	2500
Crossfeed screws	Cylinder oil or leaded compound	Any	2500
Miscellaneous parts	Lithium or calcium complex grease	Gun	No. 2

and milling machines, but they operate similarly in that metal is removed with a series of straight cuts by one or more tools simultaneously. Planer lubricants are listed in Table 17-10.

Planers are designed either with one *open side* or with a *double housing,* the latter of which is bridged on the top by an *arch.* They are equipped with a *bed, table, side head, cross rail, rail head, saddle, elevating screws* for rail and side head, and *housings,* or columns.

The bed supports the table and *table ways* on which the table reciprocates, as well as the housings. The heads carry the tools, which are lifted above the workpiece on the counter-stroke (noncutting) of the table by a clapper box. Rail heads are mounted on the housing ways to feed the tool horizontally, while the side heads similarly mounted feed the tool vertically to allow for side cutting. The cross rail is mounted on the column ways and located above and across the table. It supports the rail heads and provides for the horizontal cross feeding of the tools. Motion is provided from the drive units by a series of shafts variously called horizontal elevating, horizontal rapid traverse, vertical rapid traverse, wheel and feed shafts.

The open-side planer (Fig. 17-12) has only one housing or column, leaving one side open to accommodate wider workpieces than can be

Table 17-10. LUBRICANTS FOR PLANERS

Parts To Be Lubricated	*Type of Lubricant*	*Method of Application*	*Approximate Viscosity (SUS at 100°F) or NLGI Consistency*
Hydraulic system	Hydraulic oil	Circulation	300
Table ways	Hydraulic oil	Off hydraulic system	300
Quill	Hydraulic oil	Oiler	300
Table ways	Way oil	Separate pump	500–900
Elevating screws	Cylinder oil or Leaded compounds	Any	2500
Elevating screw drives	Cylinder oil or Leaded compounds	Any	2500
Drive gears	Leaded compound	Any	2500
Spindle drives	Hydraulic oil	Any	200–300
Rapid traverse drive	Leaded compound	Any	2500
End support drives, gears or rack and pinion	Leaded compound	Any	2500
Cross column traverse screw	Leaded compound	Any	2500
End support column, face and bearing	Leaded compound	Any	2500
Side head gears	Hydraulic oil	Any	300
Miscellaneous parts	Hydraulic oil	Central oiler	300
Rail head screws, ways, guides, screw bearings	Way oil	Any	500–900
Box on rail end	Oil of low paraffinity (refrigerating oil)	Reservoir	200–300

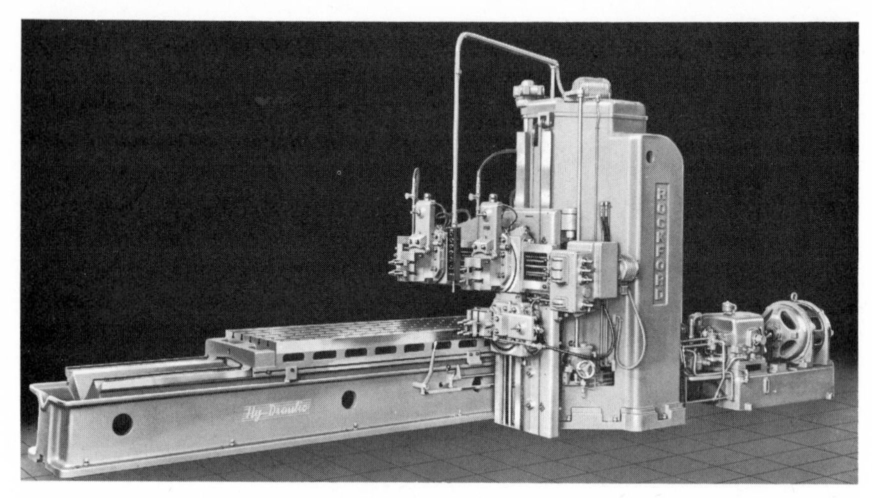

Fig. 17-12. Openside planer. *Courtesy, Rockford Machine Tool Co.*

handled by the double-housing type. Double-housing planers are usually identified by the width between housings.

Planers are usually very large and can be designed with extremely long beds to provide extended table strokes for long workpieces.

BROACHING MACHINES

A broach is a round or flat bar with a series of graduated cutting edges that are either radially or transversely oriented. The broach shapes an outside surface or a hole by being either pushed or pulled translationally over or through the surface. Each succeeding tooth removes an additional amount of metal. In order to gain smooth motion with minimum effort, the dimensions of each succeeding tooth should be gradually increased as the broach moves through or over the workpiece. Also, by diminishing the amount of increase from first to final cut, the machining progresses from a roughing to a finishing operation as the broach moves along its cutting path. Broaching machine lubricants are given in Table 17-11.

The operation is performed on either a vertical or horizontal machine (Fig. 17-13). The horizontal broach has a base that houses the driving equipment and supports the ways and pull mechanism, clamps, and broach pilot. The base of a vertical broaching machine also supports the column and rams. Like most modern machine tools, motion of the various parts is usually hydraulically actuated, although some models are equipped with electro-mechanical drives. Some vertical broaching machines have double rams whereby one ram cuts while the other returns to its original position.

Table 17-11. Lubricants for Broaching Machines

Part To Be Lubricated	*Type of Lubricant*	*Method of Application*	*Approximate Viscosity (SUS at 100°F) or NLGI Consistency*
Hydraulic system (where used)	Hydraulic oil	Circulation	150–300 (depending on pump)
Drive gears	Gear oil	Bath	1800
Miscellaneous parts	Machine oil	Any	300
	Grease	Any	No. 1

Fig. 17-13. Vertical single-slide broaching machine showing broach. *Courtesy, Footburt-Reynolds Machinery Division, Reynolds Metals Co.*

A continuous broaching machine (Fig. 17-14), made for mass production of standard parts, has a stationary tool and a moving workpiece. The continuous motion is provided by a heavy roller chain and sprocket arrangement, which conveys the work through the machining process.

GEAR-MAKING MACHINES

High performance gears are made by cutting or milling spaces from the periphery of a gear blank. Gears can be cut on a broach, miller, or shaper, but special machines are more efficient.

The *gear-tooth forming machine,* or gear cutter, utilizes a rotary milling cutter to produce the given tooth profile. When one space is completed, the machine indexes to position the cutter exactly on center with the location of the next space. The gear is completed when the blank has made one revolution.

The *gear shaper* (Fig. 17-15), or planer, forms the gear teeth by a series of reciprocating strokes of a form tool especially shaped (having the shape of the tooth space) as to remove metal from the blank in the same manner as that described for the rotary milling cutter. As one space is completed, the machine indexes to form the next succeeding space until the gear is completed.

In the *gear-tooth generating* machine, the relationship of two gears in

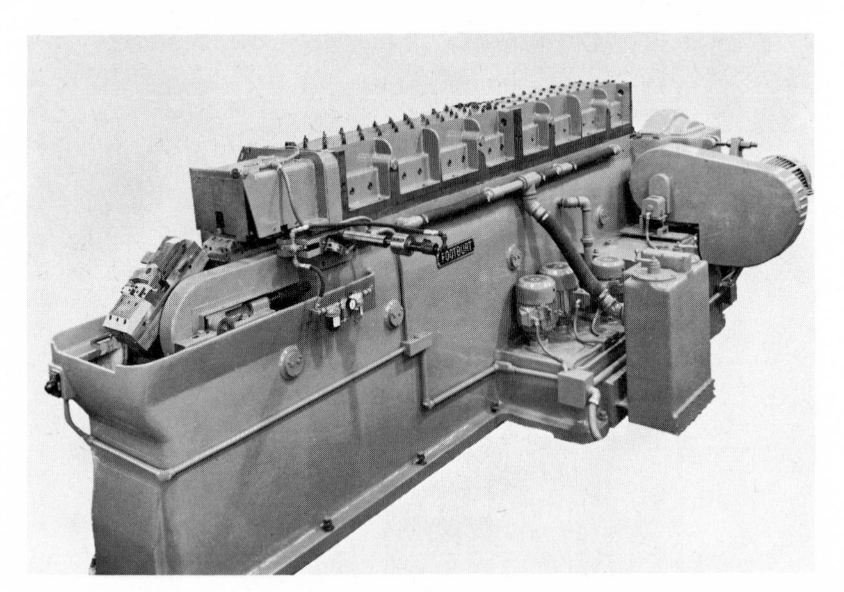

Fig. 17-14. Continuous chain type surface broaching machine. *Courtesy, Footburt-Reynolds Machinery Division, Reynolds Metals Co.*

Fig. 17-15. Gear shaper. *Courtesy, The Fellows Gear Shaper Co.*

working contact is applied to the cutter and blank while both are in motion relative to each other. The cutter may be a *hob,* a rotating worm gashed along its axis to form cutting edges. The rotating hob advances through the rotating blank to form the space between each pair of adjacent teeth. Another type of cutter is in the form of a reciprocating pinion.

The *gear shaver* improves gear tooth surface finish and profile accuracy. It has a rotary shaving cutter with gear teeth ground to a profile congruent with the desired shape of the teeth to be finished. The cutter teeth are serrated to form numerous rectangular notches that scrape off small shavings from the blank. Shaving does not require index gearing, since the cutter either drives or is driven according to the size of the part.

Rack shaving is performed by a rack with serrated teeth that roll back and forth on the gear being shaved. The effect is the same as the cutting action of a rotary shaver.

Ring gears, racks, and gear segments are produced by a broaching machine, which can make spur and helical teeth. Small internal gears can be cut in a single pass of the broach.

A typical automatic oil application device for machine tools is that installed on the hobbing machine shown in Fig. 17-16.

FINI-SHEAR®. A modern method of finishing a wide variety of gears, cams, gerotors and other shapes with precision quality is with the Fini-Shear (Fig. 17-17). The Fini-Shear (made by The Fellows Gear Shaper Company)

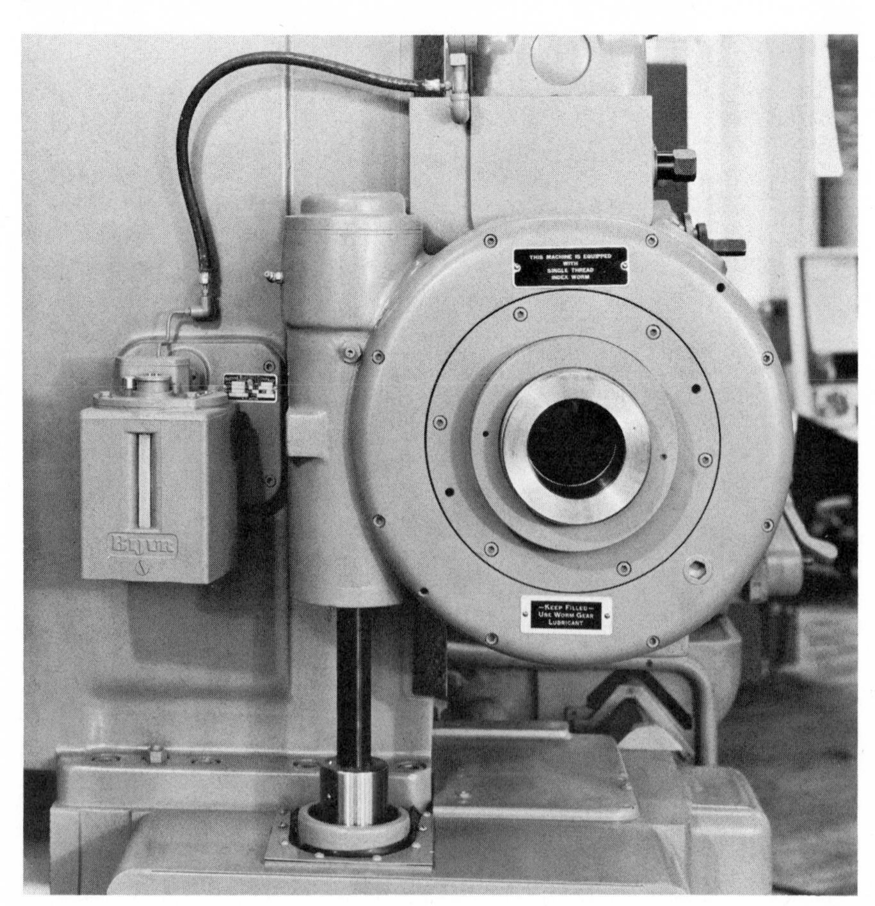

Fig. 17-16. Automatic oiler mounted on a hobbing machine. *Courtesy, Barber-Colman Co.*

Fig. 17-17. Fellows No. 8 Fini-shear®. *Trade name of The Fellows Gear Shaper Co. Reproduced with permission of The Fellows Gear Shaper Co.*

synchronizes the drives of both cutting tool and workpiece through precision-ground timing gears. The cutting tool is essentially a master gear with minimum side clearance and sharpened face. The axis of the cutter is crossed with respect to the workpiece, providing a shearing action as the rotating cutter feeds axially into the synchronously rotating work. The degree of smoothness of the workpiece before finishing has little influence on the final accuracies of the Fini-Shear.

Gear grinders finish already formed gears that have been hardened beyond the practical range of cutting processes. Grinding machines are discussed as the next category.

GRINDING MACHINES

The numerous types of grinding machines range from the simple bench and pedestal grinders to those on mass production lines, such as the grinder for finishing crankshaft journals.

The amount of metal removed by precision grinders is so small that precautions must often be taken to avoid thermal expansion of any of the working and worked parts by providing refrigeration for the grinding coolant. Particle and magnetic filtration of the coolant is also required in some operations to avoid scratching the workpiece.

The conventional grinding machine consists of many of the parts found on other machine tools, but special lubrication considerations involve the wheel spindle bearings, ways, and any hydraulic systems utilized (Table 17-11).

Cylindrical grinders are those on which both the grinding wheel and work rotate on parallel axes (Fig. 17-18). Either the workpiece or the wheel moves along its own axis, and external surfaces are ground.

Center grinding refers to workpieces that are mounted between centers such as the roll grinder in Fig. 17-19.

Centerless grinding refers to workpieces that rotate between the grinding wheel and a regulating or backup wheel and depend on a rest blade which supports and guides the workpiece between the wheels.

Universal grinders (Fig. 17-20), usually referred to as tool-room grinders, are capable of both cylindrical and internal grinding and have great adjustment versatility, both horizontally and vertically. Points of lubrication of a universal grinding machine are shown in Fig. 17-21.

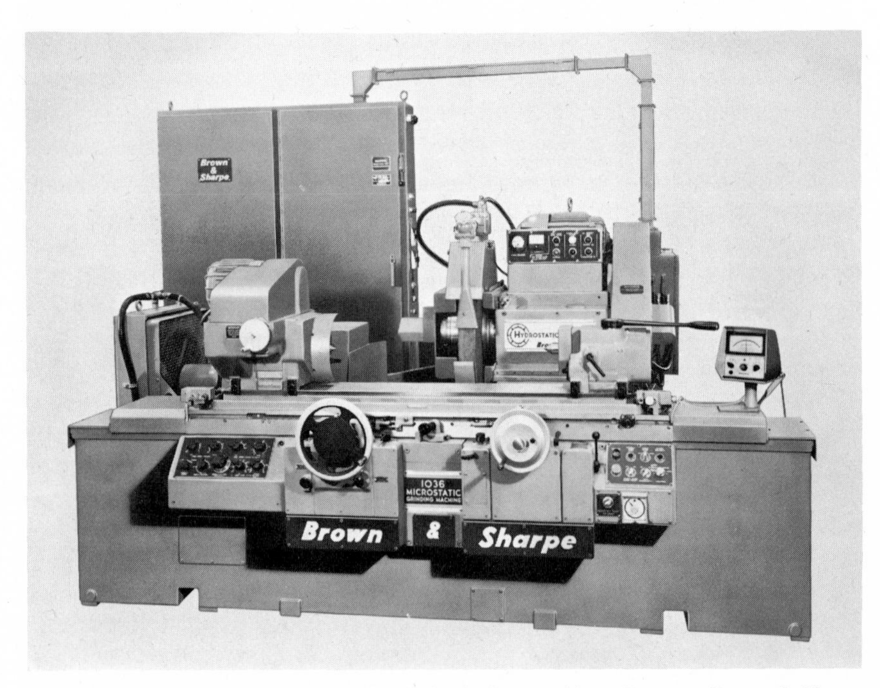

Fig. 17-18. Microstatic hydraulic cylindrical grinding machine. *Courtesy, Brown & Sharpe Manufacturing Co.*

Fig. 17-19. Farrel heavy duty roll grinder shown grinding a work roll to one microinch finish. *Courtesy, Farrel Corp.*

Internal grinders grind the internal surfaces of bushings, bores, sleeves, etc. Internal grinding can be accomplished either by rotating the workpiece on its own axis or by the planetary and rotating motion of the grinding wheel as it moves along the axis of the stationary workpiece.

Fig. 17-20. Universal grinding machine. *Courtesy, Brown & Sharpe Manufacturing Co.*

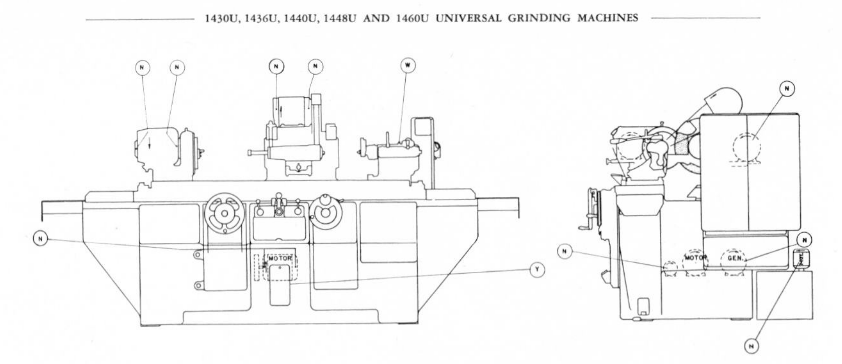

Fig. 17-21. Lubrication diagram for universal grinding machine. W, oil weekly with good grade machine oil, 300 SUS at 100°F. N, bearings permanently sealed. Y, for automatic lubricating and hydraulic systems: Fill tank with good grade, high lubricity, way and hydraulic oil having a viscosity of 150 SUS at 100°F. Renew oil filters annually; maintain oil level. *Courtesy, Brown & Sharpe Manufacturing Co.*

Chucking grinders are those whose workpieces are bored and held in a chuck and rotated while the grinding wheel grinds the internal surface.

Surface grinders (Fig. 17-22) may have a horizontal spindle and present the peripheral surface of the wheel to the work. Others have a vertical spindle and present the side or face of the wheel to the work.

Horizontal spindles are positioned at a right angle to the longitudinal travel of the table and feed vertically.

Vertical spindles revolve about their own axis in a fixed position and feed vertically into the workpiece that is attached to a table that rotates and changes grinding area by reason of wheel offset.

Pedestal and bench grinders, smaller than machine types, consist of an electric motor that carries one or two overhung grinding wheels at one or both ends of its shaft. They are also equipped with a workpiece *steady rest.*

Grinding wheels are made of abrasive particles and crystals bonded together to form a strong, tight solid mass. The most familiar grind wheel

Table 17-12. LUBRICANTS FOR SURFACE GRINDERS

Part To Be Lubricated	*Type of Lubricant*	*Method of Application*	*Approximate Viscosity (SUS at 100°F) or NLGI Consistency*
Hydraulic system	Hydraulic oil	Circulation	150–300
High speed spindle	Spindle oil	Any	100–150
Power feed transmission (where used)	Turbine oil	Bath	150
Ways	Wall oil	Any	300
Miscellaneous parts	Turbine oil	Pump	300

Fig. 17-22. Hydraulic feed surface grinder. *Courtesy, Gallmeyer & Livingston Co.*

is the round, flat-sided wheel with a center hole for shaft mounting. Corundum (aluminum oxide), emery (impure corundum plus iron oxide), and diamond represent the natural abrasives.

PRESSES

Presses cut, coin, stamp, press, extrude, draw, and form materials for various purposes. They range from the simple hand-operated arbor press to those that exert loads measured in tons. The most important types are those used for the mass production of a multiplicity of parts from small washers to bodies of all kinds of air, land, and sea vehicles. Production presses are powered either mechanically or hydraulically.

Mechanical Presses

The types of mechanical presses include the high speed, straight sided, inclinable, gap frame horn, and the major components that are common to most are described here. Their lubricants are listed in Table 17-13.

The mechanical press is usually driven by an electric motor through a bull gear and pinion (or crown gears) to a crankshaft (eccentric) that is connected to slides through a pitman (connecting rod), both of which reciprocate to transfer pressure on the die punch and hence to the workpiece die. A belt-driven flywheel is mounted on the crankshaft to provide inertia to the press motion, but the flywheel can be disconnected from the crankshaft by a clutch which allows it to keep running while the operation of the press is stopped. A brake is provided to stop the crankshaft from turning when necessary. All of these components are mounted on a frame supported on a base. The adjustments and controls include the slide adjustment that controls the low limit of upper die travel and other components that control slide speeds and stroke rates.

Some presses are equipped with pneumatic systems to activate the clutch, brake, and other control devices and to counterbalance the weight of the slide. The straight-side press is constructed with uprights that are held rigid by a bridge, or crown, which also contains the drive mechanism. The uprights are fitted with gibs that guide the reciprocating slides. The whole assembly is held firmly together by vertical tie rods located at each corner of the straight-side press.

Table 17-13. LUBRICANTS FOR MECHANICAL PRESSES

Parts To Be Lubricated	*Type of Lubricant*	*Method of Application*	*Approximate Viscosity (SUS at 100°F) or NLGI Consistency*
Crown gears	Machine or gear oil	Reservoir (circulation)	1800
Counterbalance cylinder	Machine or gear oil	Hand	1800
Hydropneumatic cylinder	Hydraulic oil	Hand	300
Pneumatic cylinders	Air line oil	Air line lubricator	300
Miscellaneous parts	Grease*	Central system	No. 1
Bull gear and pinion	Residuum	Hand or spray	800 SFS† at 210°F (without diluent)
Flywheel shaft bearings	Grease	Central system or gun	No. 1
Clutch shaft bearing	Grease	Central system or gun	No. 1
Connection screw	Cylinder oil or leaded compound	Hand	2500

* Lime, lithium, or calcium EP complex.
† Saybolt Furol Seconds.

Very large presses have their driving mechanism located below the bed (underdrive) rather than in the crown. This arrangement requires the slide to be pulled down by upward extending pitmans, rather than pushed down as in top drive models.

The action of straight-side presses may be single, double, or triple. A single-action press completes an operation with a single stroke of a single slide. A double-acting press is one having an auxiliary outer slide that holds the blank while it is being formed. The outer slide is activated by cams on the crankshaft or by toggles supported by the frame.

Toggles are a series of compound levers linked in tandem that increase the mechanical advantage of a pressure system. They are used on multiaction presses to accomplish an effective dwell (hesitation) of the blank-holding member.

Knuckle-joint presses employ a lever arrangement (made up of two levers of equal length) through force applied at their joints from the crankshaft or eccentric, which provides lost motion to allow a short dwell at the bottom of the slide stroke. They also provide a limited stroke of great force as required by presses used in swaging and embossing.

A *flywheel press* is directly driven (without gears) and is coupled to the crankshaft through a mechanical clutch of the positive engagement type.

All mechanical presses do not have their flywheel mounted on the crankshaft. A type of inclinable press is equipped with an auxiliary back shaft to which the flywheel is coupled through an air-friction clutch.

In place of the bull gear and pinion drive some types of straightside mechanical presses are equipped with crown gears that run in an oil bath and actuate the slides through an eccentric arrangement.

Hydraulic Presses

The hydraulic press differs from the mechanical press chiefly in the use of a pressurized fluid working within an enclosed cylinder and piston arrangement to actuate the slide. A hydraulic fluid such as a highly refined mineral oil transfers the power in modern presses. Hydraulic oils and lubricants are listed in Table 17-14.

The advantage of hydraulic force over mechanical force lies in the uniformity of unit pressure over the entire surface on which it is exerted, and it continues constant over the entire period of piston stroke. This eliminates any lost motion, or spring-back, that is often associated with mechanical systems. The working stroke is usually longer than that of crankshaft or eccentric throw, so that hydraulic presses tend to operate slower than mechanical presses. They are made single-, double-, and triple-acting.

Because they are infinitely variable in all aspects of their operation such as speed, stroke, and pressure, they are ideal for certain operations.

Table 17-14. LUBRICANTS FOR HYDRAULIC PRESSES

Part To Be Lubricated	*Type of Lubricant*	*Method of Application*	*Approximate Viscosity (SUS at 100° F) or NLGI Consistency*
Hydraulic system	Hydraulic oil	Circulation	300
Pull-back or push-up cylinders	Hydraulic oil	Reservoir	300
Stroke-control rack	Hydraulic oil	Reservoir	300
Ram cylinder	Hydraulic oil	Reservoir	300
Safety pawls	Air line oil	Air line lubricator	300

MISCELLANEOUS EQUIPMENT

In addition to the machine tools already described, there are drop hammers, power hacksaws, metal hand saws, welding units both stationary and portable, press brakes, clinchers, portable hydraulic power units, straighteners, shears, cutoff machines, and others, including attachments that broaden the functions of conventional machines.

A modern innovation in machine tools is virtually a machine center. Most of these are designed to produce specialty items, such as pump housings, that are transferred in turn from one operation to another until it is completely or partially machined, drilled, bored, reamed, tapped, milled, or whatever else is required. They are hydraulically powered in all operational phases and controlled by automation (numerical control, Fig. 17-23),

Fig. 17-23. Numerically controlled lathe. *Courtesy, Monarch Machine Tool Co.*

except in cases of bulky workpieces that must be manually handled during some transfer operations.

The selection of lubricants for machine centers follows the same pattern applied to any other machine tool wherein each type of frictional system is analyzed to determine lubricant requirements.

Many integrated machine companies include pattern shops and foundries to support their major production lines. Such enterprises require their own specialized equipment.

Lubrication of Woodworking Machine Tools

Bearings of woodworking machine tools are usually close fit and subject to fairly high speeds (1800 rpm or over). Some of the spindle bearings are prepacked with grease. Oil is usually applied by hand. Spindle type oils should contain an oiliness or film strength agent having a viscosity of 100 to 125 SUS at 100°F, depending on the clearance or age of the machine.

Foundry Equipment

Pig iron is melted in a *cupola.* Other metals, such as brass, are melted in a gas or electric *furnace.* The molten metal is carried from the cupola or furnace to the prepared molds in *ladles* supported by hoists or cranes that travel on overhead rails to the mold area. Pouring is accomplished by tilting the yoke-held ladle with a hand or motorized worm and gear mechanism either directly into large molds or into hand ladles to be relayed into smaller molds.

Upon cooling, the molds are "shaken out," either onto the foundry floor, or onto vibrating shake-out conveyors and screens. This loosens the bulk of the mold sand from the castings. Then they are placed in *tumblers* or *sand blasters* that remove the remaining sand and dust. Iron balls, or "stars," are tumbled with the castings to impart shock.

The final cleaning stage is accomplished in sand blasting machines that remove all traces of sand so as not to dull the finishing tools. In centrifugal blasting, an impeller imparts the required sand or shot velocity. Castings are also cleaned by water blasting, at pressures up to 600 psi.

Finishing machines include power saws, abrasive cut-offs, grinders, shears, chipping hammers, belt sanders, and sledge hammers. Grinders and chippers are usually pneumatically powered. Hand work is held to a minimum.

Radioisotopes may be used to check the interior soundness of castings. This method involves the use of X rays and gamma rays as well as the fluoroscope in detecting any flaws within the body mass. Alternative methods include the use of magnaflux (magnetic powder) on ferrous metals, or by visual inspection of selected samples by destruction.

When hollow parts are cast, the space is filled with a core consisting of sand and an effective binder, such as linseed oil, made firm by baking in a core oven. Mixing of the core materials is accomplished in a mixer called a *mulling machine*. The cores are transported in and out of the oven on core cars. Core oven temperatures range between 300 and 650°F, depending on the binder, and the lubrication of core car wheels may present a problem when mineral oils or greases are employed, although some types of high-temperature greases have proven satisfactory in the lower temperature ranges.

LUBRICANTS FOR FOUNDRIES. Most of the frictional parts of foundry machinery are satisfactorily served by conventional lubricants discussed heretofore for electric motors, various type gears, pneumatic cylinders of molding machines (ramming, squeezing and jarring operations), roll-overs, air tools, air compressors and hydraulic equipment (e.g., combination of jolt, roll-over and draw-molding machines). The latter type machine is often actuated by a hydraulic—pneumatic power piston that separates the liquid from the gaseous fluid.

Special consideration must be given to high temperature greases for core-oven car wheels and stringy type oils and greases for the eccentrics and bearings of shake-out screens and vibrating conveyors. Because of the presence of abrasive foundry dust and sand, adequate sealing must be provided all bearings and gear housing.

Injection Molding Machine

Plastic and rubber articles of an infinite variety of shapes are produced by molding them under hydraulic pressure on the injection molding machine (Fig. 17-24). The lubricants required on such machines include the hydraulic

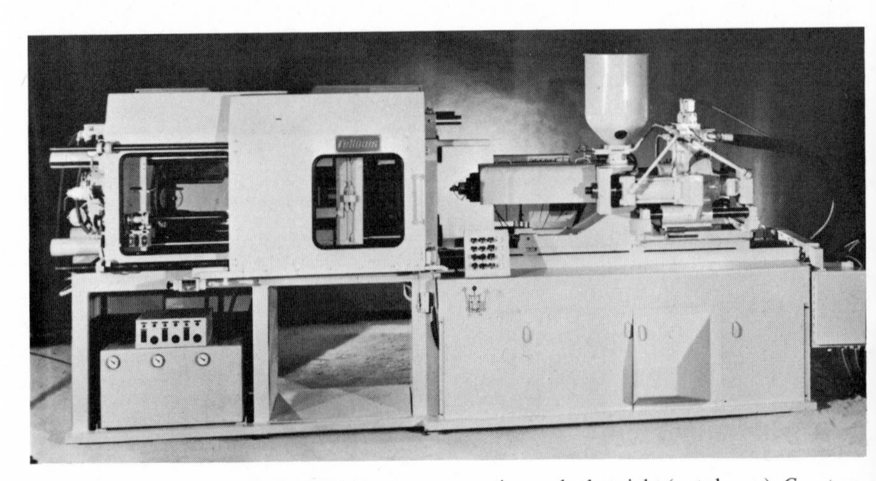

Fig. 17-24. Injection molding machine. A computer is attached at right (not shown). *Courtesy, The Fellows Gear Shaper Co.*

Table 17-15. LUBRICANTS FOR INJECTION MOLDING MACHINES

Part To Be Lubricated	*Type of Lubricant*	*Method of Application*	*Approximate Viscosity (SUS at 100°F) or NLGI Consistency*
Hydraulic system	Hydraulic oil		
Up to 60 hp		Circulation	300
Over 60 hp		Circulation	600
Guide bushing	Lithium or calcium EP complex grease	Gun	No. 1
Injection and plasticizing unit	Lithium or calcium EP complex grease	Gun	No. 1

fluid and the various tie rod bearings, slides, adjusting gears, ways, idler gear bushings, trunnion guide rod bearings, locking pin bearings, slides, and connecting pin bearings. Not all models of injection molding machines contain all the parts mentioned. These components, except the ways, are usually lubricated with grease having a No. 1 NLGI consistency and of lithium or lime-lead base. Application of grease is either by gun or central system. A way oil of 500 to 1000 SUS at 100°F viscosity is recommended. The viscosity of the hydraulic oil depends on the type of pump and pressure with 150 SUS at 100°F used for pressures below 1000 psi and 300 SUS at 100°F for pressures over 1000 psi.

Pneumatic Tools

Pneumatic tools convert the compression and expansion forces of air under fairly high pressure (110 psig) to mechanical work. Conversion is usually accomplished either by a piston-cylinder arrangement or by a rotor having axial vanes or radial blades similar to those of axial- and radial-flow hydraulic pumps. The piston-cylinder type has a valve that directs the high-pressure air to the cylinder, first to one side of the piston and then to the other, in a double-acting sequence.

The reciprocating air tool (Fig. 17-25) is designed for percussion response, such as the repeated blows of a hammer, chisels, riveters, chippers, caulkers, rock drills, and concrete breakers. Rotary air tools (Fig. 17-26) are also well-adapted to turning tools, such as drills, lug and bolt (torque) wrenches, and grinders.

Some air tools, such as drills used in mines, tunnels, and quarries, transmit both reciprocating and rotary motion. Such tools are designed either to allow rotative impulses to emanate from the piston or from an independent air motor located in the front head of the tool. Where an independent air motor is used, it is directly connected to the chuck while the piston operates free-floating.

Some air tools are equipped with an integral oil reservoir that is kept supplied by hand. Care must be taken to prevent abrasive dust and dirt

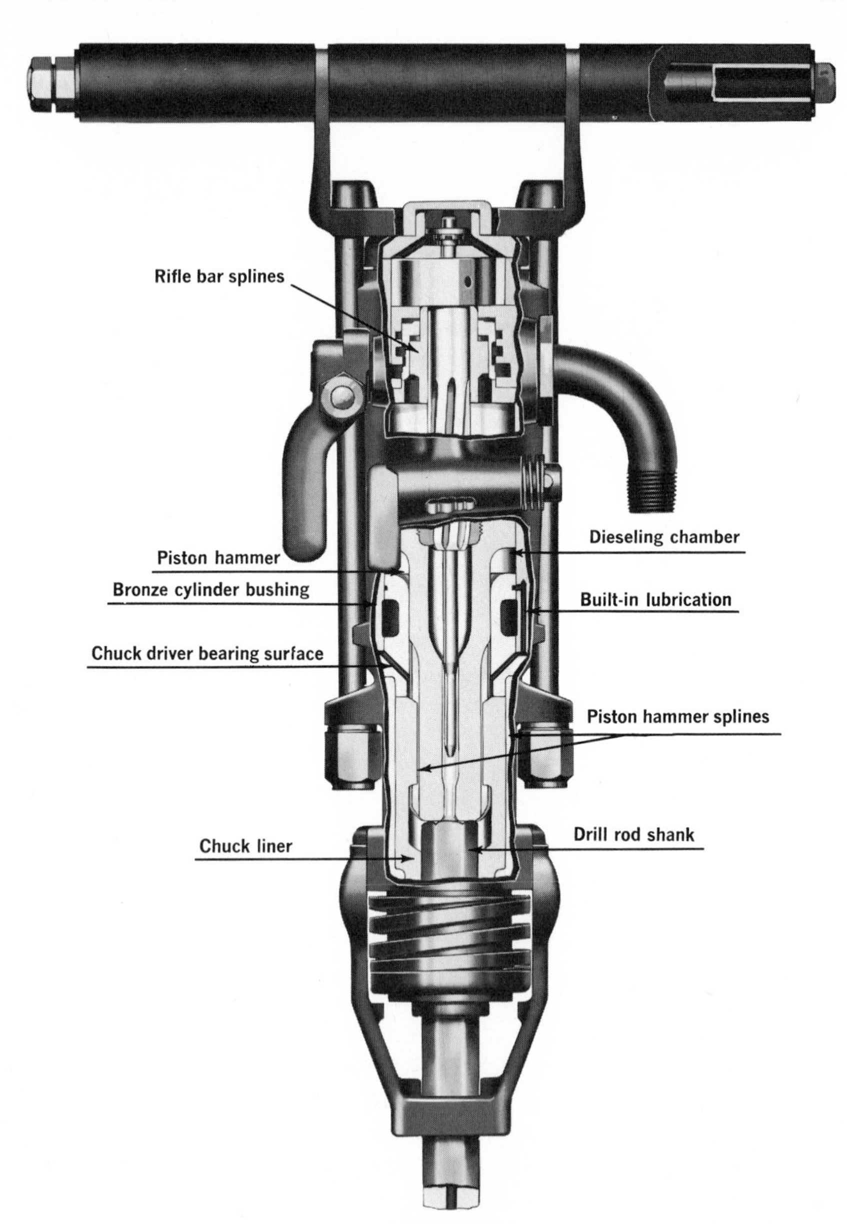

Fig. 17-25. Cutaway view of typical reciprocating sinker rock drill. Note important points of lubrication. *Courtesy, Henry Pohs, Gardner-Denver Co.*

from entering the reservoir through the fill hole during oil replenishment. The only other path through which abrasive materials can enter the air tool is through the air lines along with the compressed air. Otherwise, the tool itself is made substantially dustproof. Other air tools have their lubri-

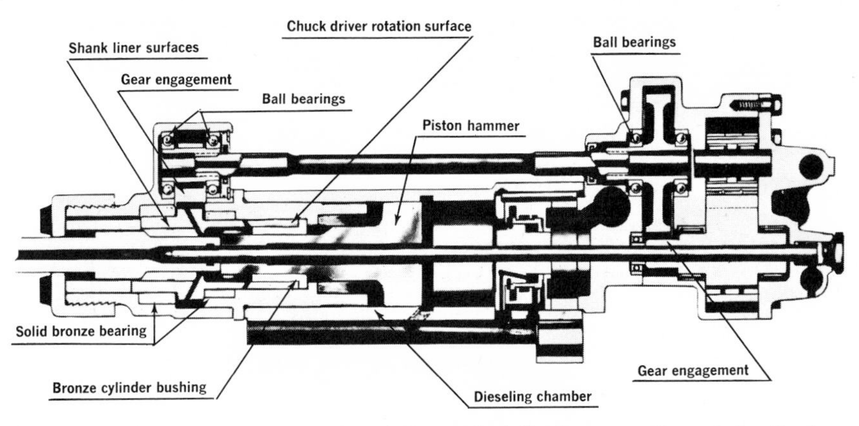

Fig. 17-26. Cutaway view of rotation drifter rock drill. *Courtesy, Henry Pohs, Gardner-Denver Co.*

cating oil supplied through the air line by the use of air line lubrication (Chapter 11).

Another contaminant that enters with the air is moisture. As long as the air remains warm the moisture remains in the form of water vapor, but when the air expands, it cools, and the condensate tends to displace a lubricating film from the surfaces of frictional components such as the pneumatic cylinders. To prevent this, the lubricating oil should contain an emulsifying additive.

With the use of proper lubricants and effective air line filters, pneumatic tools are capable of giving long service life. Correct lubrication of air tools is important because of their high-speed operation and the volume of air passing through them per unit of time. For example, percussion air tools, such as the chipping hammer will deliver upwards of 3000 blows per minute during which time 10 to 12 cu ft of free air will pass through it.

In rotary tools speed depends on the air pressure. Speed control may be attained by the addition of planetary gears driven off the rotor shaft. If their housing is not closed, grease must be applied by a pressure gun using a mixed or lithium base with a No. 2 NLGI consistency. They should not be over-lubricated. For operation at 40°F and below, a No. 1 or lower consistency grease should be used.

Oil, usually applied from an air line oiler, lubricates the frictional parts of pneumatic rock drills. In regions subject to temperature extremes, the oil viscosities listed in Table 17-16 are recommended.

Overlubrication can cause dieseling, during which the oil burns from the drill cylinder. The result is scoring of the cylinder liner (Fig. 17-27) and the formation of abrasive carbon residue and acids that create further abnormal wear and corrosion of the cylinder walls.

Table 17-16. VISCOSITIES OF OIL FOR PNEUMATIC ROCK DRILLS AT VARIOUS TEMPERATURES

SAE Number	*Ambient Temperature Range (°F)*
10	0–40
30	40–80
40	80–110
50	Above 110

Electrical Discharge Machining

Dies are conventionally formed by removing metal from the surface of a blank, or impression. This operation of die sinking is often difficult and time consuming. These forging and casting methods are slowly giving way to electrical discharge machining. The cutting element is a graphite cathode preshaped by conventional machining to the size and profile of the impression desired. The workpiece or blank becomes the anode in the electrical discharge system.

Both the cathode tool and the anode workpiece are immersed in a reservoir of dielectric fluid. Charged with a pulsating direct current of from 0.5 to 300 amp, the graphite tool is moved toward the fixed workpiece until they are 0.0002 to 0.007 in. apart, depending on the quality of surface finish desired. Upon reaching the specified distance or gap between the two elements and commensurate with the magnitude of the impressed current, the dielectric strength of the fluid breaks down and a current jumps across the gap with a flow of sparks. This has an eroding effect on the area of discharge, removing infinitely small grains of metal from the workpiece. As the tool moves slowly over the blank, while the specified gap is maintained, it continues to remove additional grains until an exact reverse image of the

Fig. 17-27. Corrosion due to acids and corrosion by-products developed by dieseling. *Courtesy, Henry Pohs, Gardner-Denver Co.*

Table 17-17. DIELECTRIC OIL PROPERTIES FOR ELECTRICAL DISCHARGE MACHINING

Property	*Finish Cut*	*Intermediate Cut*	*Rough Cut*
API gravity	45.0	42	39
Flash point COC, °F	180.0	270	310
Viscosity, SUS at 100°F	32	45	50
Color*	+25 to +30	+25 to +30	+25 to +30
ASTM rust test	Passes	Passes	Passes

* Color may be ASTM No. 1 maximum in all cases.

tool profile is impressed in the anode or workpiece. The metal particles so displaced are continually flushed away from the eroding area.

Table 17-17 may be used as a guide in selecting a suitable dielectric oil for electric discharge machining.

18

ENGINES AND TURBINES

The oils that lubricate internal combustion engines are variously applied to the frictional components that include the cylinders, valve mechanism, connecting rods and main bearings, cams and camshaft bearings, and wristpins. In small engines, the oil in the crankcase is applied by splash from the crankcase. In automotive engines, including those designed for marine service, the oil is applied by a combination of splash and pressure circulation. Circulation is accomplished by a pump in the crankcase that forces the filtered oil through small tubes and passages, drilled through the shafts and rods, leading to the various frictional parts. In some engines the oil also cools the pistons. The oil serving large stationary and marine engines is often supplied from an external reservoir, circulated through the frictional parts of the engine, drained to the engine crankcase, filtered, cooled, and recirculated.

GASOLINE RECIPROCATING ENGINES

Most gasoline engines are used in automobiles. Their lubricating oils, referred to as motor oils, are of either the straight mineral or additive oil types. The straight mineral oils are primarily employed to lubricate two-cycle engines in which the oil is mixed with the gasoline prior to application (Fig. 18-1). Automobile engine lubricants are required to be highly refined from selected stocks, have a high VI, and be of correct viscosity for their intended service. To perform satisfactorily in service, motor oils must contain such additives as will promote high chemical stability and engine cleanliness and prevent corrosion of engineering metals.

Performance of motor oils is evaluated according to standards based

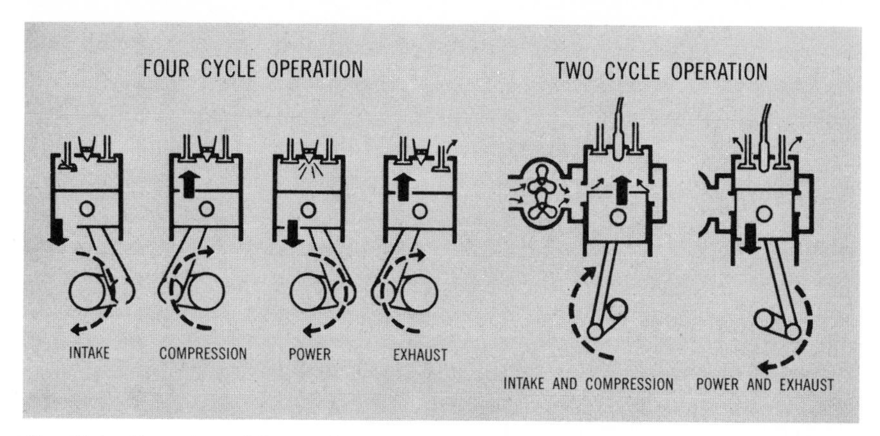

Fig. 18-1. Operation of four-cycle and two-cycle engines. *Courtesy, Cummins Engine Co.*

on severity of service, ranging from prolonged high-speed or high-load operation under high-temperature conditions to idling at subzero temperatures. Their performance is classified according to their capacity to meet various service conditions and is evaluated on the results of API Engine Service Tests. Motor oils are used for all gasoline engines except aircraft engines, and they are designated as Service MS, MM, or ML. Performance specifications and requirements have been prepared by various organizations and engine builders that convey their own methods of determining required standards under certain conditions, including fuel characteristics. One of these is the military specification, designated as Spec. MIL-L-2104A, which covers the specifications required to meet Service MS.

An AMA. Lubricant Evaluation Sequence is widely used to indicate performance of lubricating oils. They are divided into sequences having the following ratings:

Sequence IIA, IIIA: low-temperature deposition, rusting, and high temperature oxidation stability.
Sequence IV: high temperature, high speed, scuffing, and wear.
Sequence V: low temperature, sludge deposition.

An MS Sequence VB Test replaces the original MS Sequence V Test to evaluate how well a motor oil prevents engine deposits of sludge and varnish during stop-and-go operation. The new test also rates the oil on preventing emission valve deposits on a closed PCV system.

An oil may be evaluated, for example, according to Service MS, Sequence IIA-IIIA; or Service MS, Sequence IV; or Service MS meeting the performance specification MIL-L-2104A.

Crankcase Ventilation

Normally, the atmosphere in an internal combustion engine crankcase would be similar to that in any other oil reservoir subject to agitation as by splash, throw, etc. However, because of the leakage of fumes from the combustion chambers by blow-by past the piston rings during the compression and power strokes crankcases become highly contaminated with products of combustion and other debris including some unburned fuel.

Included in the blow-by are carbon monoxide, carbon dioxide, water vapor, unburned fuel, air, oxygenated hydrocarbons, oxides of sulfur and nitrogen, soot, additive salts, and metallic fines resulting from wear of engine parts. All these mix with the oil mist and products of oxidation that are normal to crankcase conditions.

As the cylinder, piston rings, and lands wear, the blow-by volume increases. If allowed to remain in the crankcase, they prove harmful in several ways:

Water condenses and enters the lubricating oil mass.
Raw fuel dilutes the lubricating oil.
Cold sludge forms.
Oxi-materials such as varnish and lacquer form on engine parts.
The lubricating oil becomes acid, leading to an increase in viscosity, corrosiveness, and general chemical and physical deterioration of the oil mass.

The viscosity of a lubricating oil decreases when diluted with raw fuel, but the condition corrects itself to a large extent by evaporation when the temperature of the engine is elevated for an extended period during continuous operation under load. Much of the water may be eliminated during the same period if some means is provided to evacuate the resulting water vapor from the crankcase.

CRANKCASE VENTILATION METHODS. Effective counteraction to the crankcase contaminating blow-by problem lies generally in two methods:

1. Reduction of the blow-by volume through wear retardation of the responsible parts, the cylinder walls, piston rings, and lands. This can best be accomplished by a maintenance program that includes the use of proper lubricating oils of high quality, changed at appropriate intervals and kept free of abrasive particles by efficient filtering.

2. Exhausting the blow-by from the crankcase to the outside atmosphere by some means of ventilation.

The three basic methods of ventilating crankcases are (1) circulation by aspiration, (2) forced draft, and (3) recycling.

Circulation by aspiration is achieved by providing an exhaust line from the interior of the crankcase to an outlet orifice that can develop a draft from the motion of the vehicle. A suction is developed by directing the exhaust end of the exit line toward the rear of the vehicle. The draft tube should be beveled for greatest suction. By this arrangement, air drawn into the crankcase through the oil fill-pipe (breather) replaces the blowby gases exiting through the aspirator line. The crankcase end of the exhaust line should be placed at some point that minimizes the amount of lubricating oil mist allowed to leave with the blow-by gases. The natural circulation method is restricted to speeds over 20–25 mph.

Forced draft. A forced draft augments the suction developed by the aspirator and provides air movement through the crankcase regardless of vehicle speed. A blower driven by a small electric motor will force air through the inlet of the oil fill-pipe. This places a positive pressure within the crankcase when the outlet orifice is dimensionally balanced with the blower capacity to maintain constant flow. The blower is started and stopped according to the on-off position of the ignition switch. This makes a highly effective ventilating system.

With either the aspirator or blower-aspirator type arrangement, some means must be provided to prevent abrasives such as road dust from passing through the breather or oil fill-pipe. For the aspirator, a restricted breather cap over the fill-pipe entrance is effective at conventional speeds. With a blower, an air filter should be attached to the blower outlet and arranged for easy periodic cleaning. The blow-by gases are exhausted to the atmosphere.

Recycling. With the increased efforts to reduce air pollution, the aforementioned crankcase purging systems require some means of recirculating the blow-by gases through the engine, with part of the gases being continually evacuated through the regular exhaust system of the vehicle.

The three recycling methods employed generally all incorporate *positive crankcase ventilation,* or PCV. The principle of the PCV system is to exploit the suction of the intake manifold to draw the blow-by gases from the crankcase. This can be done from any point prior to the entrance of the air-fuel mixture into the valve chamber. The three more common methods respectively connect the crankcase with (1) the intake manifold downstream of the carburetor, using a conventional breather cap, (2) the air cleaner upstream of the carburetor, using a restricted breather cap, and (3) a combination of the first two methods, using a three-way check valve in the line from the crankcase, in which the system is either open or closed (Fig. 18-2). In the open system a breather cap is used to cover the oil fill-pipe; in the closed system the cap is sealed.

Recycling of blow-by gases decreases the amount of unburned hydrocarbons exhausted to the atmosphere due to their re-exposure with air-fuel mixture to the ignition phase of the power stroke. The PCV system is more apropos for gasoline engines than diesels because of the diesels' more efficient fuel burning.

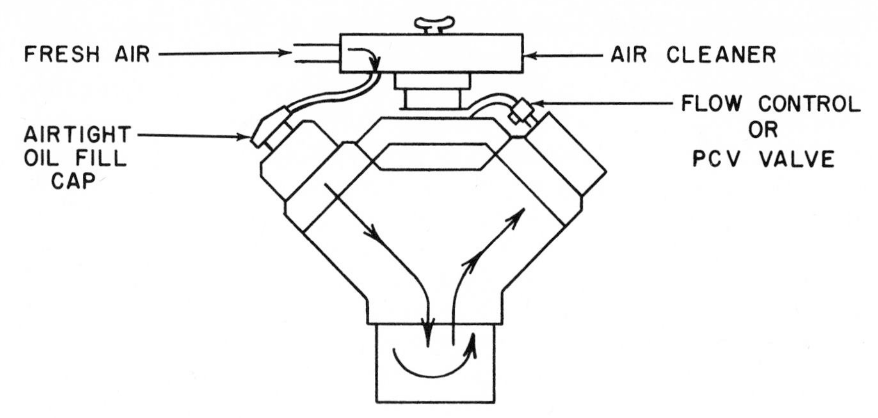

Fig. 18-2. Positive crankcase ventilation system, closed type.

All systems should be so designed or regulated as to prevent too rapid and excessive air flow through the crankcase. This is accomplished by (1) diminishing oil-mist carryover (accompanied by judiciously locating the point of scavenge) and (2) decreasing oil-to-oxygen exposure.

The volume of air flow varies from 2 cu ft/min for small engines to about 7 cu ft/min for larger ones based on engine cubic width displacement.

Lubrication of Outboard Motors

The outboard motor (usually two cycle) is lubricated by oil premixed with the fuel. The oil is distributed to the frictional parts such as the crankshaft bearings, piston rod bearings, and wrist pins, and the lower wall of the cylinders. As the fuel-oil mixture leaves the carburetor, along with air in the proportion of about 15 lb of air per pound of gasoline, it is broken up into a highly combustible mist. Some of the oil remaining in the mixture as it enters the combustion chamber is deposited on the upper wall of the cylinder. Any oil remaining in the mixture upon ignition is burned up along with the charge.

The oil mist forms a coating over all exposed surfaces. This is of concern when excessive oil deposits alter the performance of the spark plug even to the point of preventing its firing. Some engines are provided with crank-case drain pockets from which excess oil, accumulated during low speeds, bleeds off and returns to the intake manifold to be recirculated. Excessive oil here is not a fault because the quantity added to the fuel is based on the more severe operating conditions.

The proportion of oil to gasoline is usually recommended by the manu-facturer of the engine and is usually associated with speed and horsepower under the most severe operation, but the amount of oil is not proportional to the horsepower. Some manufacturers recommend a mixture having a ratio of ½ pint of oil per gallon of gasoline. This is a safe quantity to be used in most outboard two-cycle engines up to about 5 hp if more definite instructions are not available. General recommendations for two-cycle engines of various sizes and services are given in Table 18-1. The proportions given are based on recommendations as determined by a survey of lubrication instructions published by manufacturers.

Since the lubricating oil for the frictional protection of the various moving parts of two-cycle engines is introduced to such members in the form of small droplets transported by the vaporized gasoline, the oil should be of proper viscosity to develop a sufficient thickness of film to protect the bearings and also to provide an effective piston ring seal.

The majority of builders of two-cycle engines recommend a mineral oil with a viscosity meeting the specification SAE 30. The oil should be free of any additives such as detergents that tend to form deposits on the combustion chamber surfaces, including the spark plug. Some oils recommended for two-cycle engines contain a surfactant (surface-active agent) for more improved rust protection. In addition to possessing a reasonable high minimum flash point (above 400°F) and a relatively low maximum pour point (below 20°F) it should have a high viscosity index (VI $\gtreqqless$ 90).

Greater deposit formation is of course associated with straight mineral oils of higher viscosity. This is an important consideration because of the higher piston temperatures prevalent in two-cycle engines as compared with four-cycle engines. To compensate for oils with higher carbon residue, less oil should be used. This is done in the case of two-cycle engines by using oil-fuel mixtures of higher ratios. In some older models of two-cycle engines

Table 18-1. Oil-Gasoline Ratio for Two-Cycle Gasoline Engines

Service	*Horsepower (or rpm)*	*Oil-Gasoline Ratio**
Outboard	½ to 1.5	1:24 (SAE 30)
Outboard	1.5 to 5	1:16 (SAE 30)
Outboard	5 to 15	1:12 (SAE 30)
Outboard	Over 15	1:2 (SAE 30)
Outboard, racing	(5000 to 6000 rpm)	1:6 to 1:4
Rail cars	(1800 to 2000 rpm)	1:16 to 1:12 (SAE 30)
Rail cars	(1000 rpm)	1:20 (SAE 30)
Rail cars	(800 rpm)	1:40 (SAE 60)
Motorcycles		1:25 (SAE 40)
Chain saws		1:16 (SAE 30)
Lawn mowers		1:16 (SAE 30)
Electric generators	1 cyl.—up to 2000 watts	1:16 to 1:12 (SAE 30)
	2 cyl.—3000 watts	1:12 (SAE 30)

* Recommendations are specified either in pints per gallon or by oil-gasoline ratio. To convert ratio to pints per gallon, solve the equation, ratio = $x/8$, where x is the number of pints. For example $1/24 = x/8$, $x = 1/4$ pint.

having a typical rating of about 8 horsepower and a speed of 1750 rpm, used in rail car service, an SAE 60 oil is recommended. Similar type engines of recent manufacture carry an SAE 30 recommendation. However, the effect of the higher viscosity oil relative to carbon residue was offset somewhat by an oil-fuel ratio of 1 : 40 as compared to a ratio of 1 : 12 for some modern engines using SAE 30 oil.

It is of utmost importance that the oil used in two-cycle engines be of sufficient viscosity at the temperatures involved to provide adequate protection to the moving parts against wear. If an adequate amount of oil is continuously supplied to the engine, an SAE 30 or 40 oil has proven highly satisfactory. Most special outboard motor oils are of the SAE 30 rating, which has a typical viscosity of about 475 SUS at 100°F, which for a VI = 95 oil corresponds to about 62 SUS at 210°F.

There are two important rules to follow in connection with the oil-fuel mixture. First, the oil and fuel should always be combined prior to introducing it to the engine fuel tank to obtain more complete mix. The oil should be poured into the gasoline. Second, at altitudes higher than sea level add 10 to 30% more oil than would ordinarily be used to offset the leaner fuel-air ratio.

Supercharging

Supercharging is the act of forcing a greater charge of mixture in to the cylinder than that of the normal suction charge, and to generate a higher initial pressure. This is accomplished by increasing the density of the charge by compressing it at the inlet of the engine.

Supercharging tends to raise the ring zone temperature of internal combustion engines due to the greater burning effect of the fuel. This leads to the requirement for discrete additives to meet the performance tests such as those prescribed for Series 3 oils.

Turbochargers (Fig. 18-3) are similar to superchargers as they are designed to put more air into the cylinders. They differ from the supercharger, however, by utilizing the energy of the exhaust to drive the blower.

DIESEL ENGINES

The performance of diesel engine lubricants must meet exacting specifications according to their service requirements. As with gasoline engine oils, they are classified according to their capacity to meet service conditions based on severity of operation. However, the API Service Classification for diesel engines is Service DG, DM, and DS. For oils that are capable of meeting the requirements of more than one service, say, for severe gasoline engine service and normal diesel engine service, the designation "For Services MS-DG" may be used.

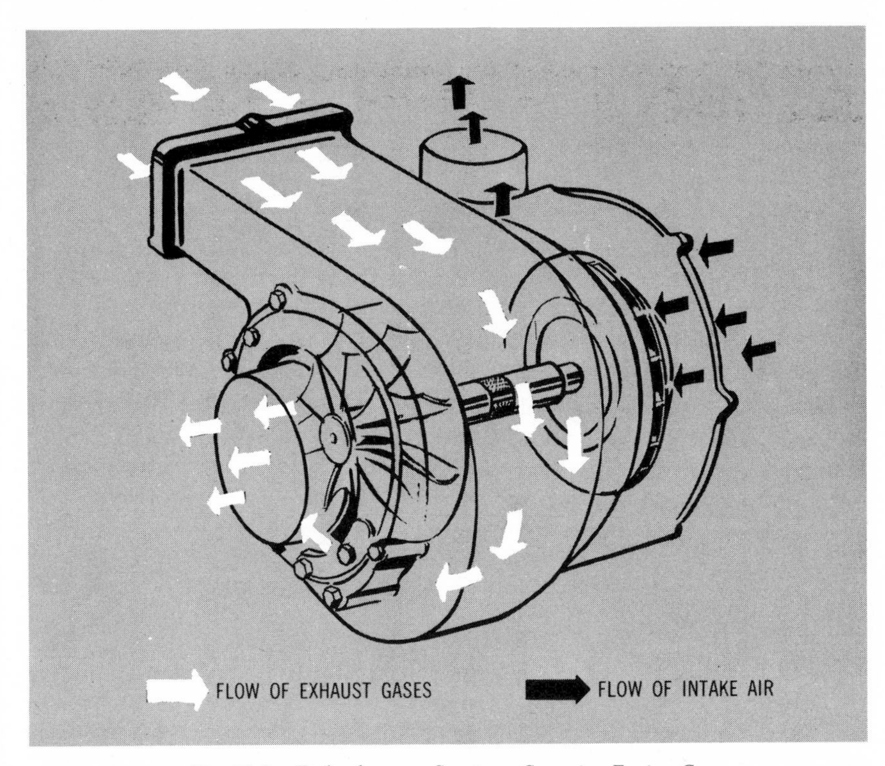

Fig. 18-3. Turbocharger. *Courtesy, Cummins Engine Co.*

Performance specifications may be designated as meeting Specification MIL-L-2104A, or 2-104B. Such specifications may be further designated, for example, as MIL-L-2104A, Supplement 1, for oils meeting MIL-L-2104A, where diesel test fuels contained 0.95 to 1.05% sulfur. A designation of (Series 3) following MIL-L-2104A is used if Supplement (S-1) oils passed a test developed by the Caterpillar Tractor Company. They are then labeled Test No. 1-D and described as Superior Lubricants.

ADDITIVES. The performance of oils rated according to API Service Classification is greatly dependent on additives that resist oxidation, increase detergency, and improve wear characteristics.

Service ML oils are straight mineral oils.

Service MM oils contain oxidation-resistant additives.

Services MS, DG, DM, and DS contain oxidation inhibitors, detergents, and other additives that improve performance; they are described as heavy duty oils.

The desire to reduce the unit weight of diesel engines per horsepower has led to higher speeds and turbocharging. Both factors, plus the increasing use of high sulfur fuels, have presented the need for greater lubrication protection, and better oxidation and thermal stability on the part of the oil, to combat the higher temperatures and pressures involved. New oil

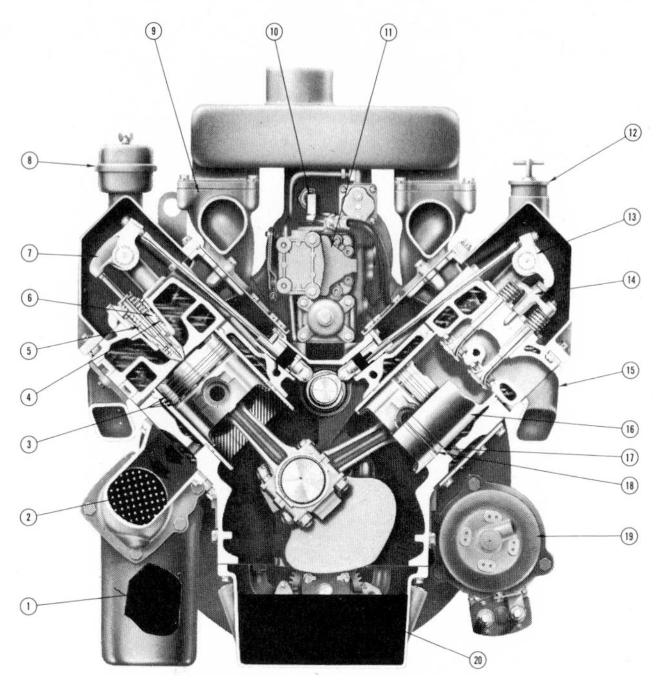

Fig. 18-4. Flow of lubricating oil in diesel engine with oil filter and oil cooler. *Courtesy, Cummins Engine Co.*

1, full-flow lubricating oil filter. 2, oil cooler. 3, piston rings. 4, fuel in. 5, fuel out. 6, injector. 7, injector rocker level. 8, crankcase breather. 9, air intake manifold. 10, fuel suction inlet to pump. 11, fuel pump. 12, oil filler cap. 13, valve rocker lever. 14, valve crosshead. 15, exhaust manifold. 16, cylinder liner. 17, crevice seal. 18, packing ring. 19, cranking motor. 20, lubricating oil pump.

performance specifications have been made necessary, resulting in an increase in severity test levels. This is indicated by a change from MIL-L-2104A to MIL-L-45199 (Series 3), and in some instances beyond the specifications of MIL-L-2104B and Supplement 1 oils. Tests procedures are outlined in tests labeled L-1 and 1-H for Spec. 2104A or Supplement 1 oils or 2104B oil, respectively, and, 1-D and 1-6 for Spec. 45199A oils.

Summary of military specifications for crankcase lubricants.

MIL-L-2104B specification is drawn up for vehicles subject to stop-and-go operation. It supersedes MIL-L-2104B (Amendment 1) and MIL-L-2104B (Supplement 1).

MIL-L-45199B specification provides highest detergent levels to meet the severe service conditions of high output supercharged diesel engines. It is a version of Caterpillar Series 3 Specification.

MIL-L-9000F (*Ships*) is a Navy specification for oils used in diesel engines operating under normal conditions of load and temperature.

MULTIPURPOSE ENGINE OILS. Multipurpose engine oils are capable of meeting to some degree the services of both gasoline and diesel oils. To meet with satisfactory performance in supercharged high-speed diesel engines (700 °F ring zone temperatures), as well as in low-temperature stop-and-start gasoline engines, is a problem met only by selected base stocks fortified with discrete additives. Such oils available are heavy duty lubricants that meet U.S. Spec. MIL-L-2104B, whose detergent-dispersant properties approach a compromise between Supplement 1 and Series 3 oils.

VISCOSITY DESIGNATIONS OF ENGINE OILS. The viscosity of engine lubricating oils is represented by SAE viscosity numbers such as SAE 5W, 10W, 20W through SAE 50. The Saybolt Universal Seconds viscosity corresponding to each number is given in Tables 8-7 and 8-8.

MULTIGRADE ENGINE OILS. Providing viscosities in a single oil to meet seasonal temperatures is one of the purposes of multigrade oil. In order to furnish a single lubricant to eliminate the need for three oils as temperatures change from less than −10 °F to 100 °F or above, an oil that meets the viscosities at the minimum and maximum levels of each is preferred. A multigrade oil is therefore designated SAE 10W-20W-30, and it can be substituted for oils respectively rated SAE 10W, SAE 20W, and SAE 30, each of which must be seasonally replaced.

Fig. 18-5. High-speed diesel engine. *Courtesy, Cummins Engine Co.*

STATIONARY DIESEL AND GAS ENGINE OILS. The capacity of a lubricating oil to perform well is no less important for large stationary engines than for high-speed automotive engines. The deteriorating influences of stop-and-start running of automotive engines remain prevalent in stationary engines subject to intermittent operations, and crankcase ventilation is important in counteracting the problem in both types.

A critical area of carbonaceous buildup in some four-cycle gas engines is the camshaft compartment located in a high-temperature zone above and between V-type cylinders. Proper ventilation and the use of oils having high chemical stability help to improve operating conditions in the camshaft area.

The same considerations given automotive diesel engines operating with high sulfur fuel oil must be given stationary engines using similar fuels.

Diesel lubricating oils are sometimes made with ashless and basic metallic detergent-dispersant additives. Some are made more alkaline to combat the effects of high sulfur fuels. All of them contain antiwear agents in addition to detergent additives. Special lubricants having high detergent qualities are made for diesel engine locomotives where cooling facilities are limited. Moderately alkaline oils are used in marine service employing fuels having about 1 to 1.5% sulfur.

Gas engine lubricants are subject to deterioration due to the effects of nitrogen fixation resulting from improper air-fuel ratios or incorrect ignition adjustment. Oils used under such conditions must be of highest quality, having excellent chemical stability and high resistance to influence of nitrogen products.

LUBRICANT RECOMMENDATIONS BY VISCOSITY. The conventional viscosity for automotive type gasoline and diesel engines is within the range of SAE 30 for temperatures above 32°F and SAE 20 for temperatures below 32°F. For severe temperatures below 0°F, SAE 10W may be used with caution in some carburetor engines in API Service MS and in some diesel engines in API Service DM. The above recommendations are based on constant exposure to the temperatures indicated or not more than about 400°F above them.

Some diesels in tractor service are usually lubricated with an SAE 30 oil for temperatures above freezing and SAE 10W for service below freezing, providing the oil meets Series 3 requirements.

For large industrial stationary diesel engines and those used in marine service, the cylinder temperatures usually are quite high, requiring an SAE 50 or 60 oil. For moderate size engines having lower cylinder wall temperatures, an SAE 40 oil is usually heavy enough. Bearings of such engines, both large and moderately sized, are lubricated from large-capacity circulating systems in which an SAE 30 oil is used. It is general practice, however, to use the same oil for both the cylinders and bearings, in which case an SAE 40 oil is employed even in large engines for both circulation and splash applications.

In any diesel installation, all operating conditions must be considered before a definite oil viscosity can be selected. In railroad service, the limitation of cooling water results in higher operating temperature and is a large factor in viscosity selection. Higher viscosity oils tend toward greater deposit formation, which is somewhat offset by the addition of oxidation inhibitors and detergent-dispersant additives. An SAE 40 oil is usually selected for railroad service.

A summary of lubricant viscosity recommendations for various gasoline, diesel, and gas engine services is given in Table 18-2.

GAS ENGINES

The operation of gas engines is in every way similar to spark-ignited engines operated with liquid fuels such as gasoline. Liquid petroleum gases (LPG) include propane, butane, and a mixture of the two. They differ from the conventional gases such as natural gas, coal gas, blast-furnace gas, producer gas, and coke-oven gas, in that they are maintained as a liquid under pressure prior to their release to the engine fuel system, where they evaporate under lower pressure.

All gases selected for engine fuels have relatively high thermal energy potential (Btu/lb). Except for liquid petroleum gases, all gases used as gas engine fuels require some continuous external bulk source to supply the large volume necessary. This factor limits such engines to stationary service. The ideal arrangements are exemplified by the gas engine compressors operated by natural gas pipeline companies, steel plants that utilize coke

Table 18-2. Recommended SAE Viscosities for Various Internal Combustion Engines

Type of Engine	*Bearing Lubricant Viscosity (Typical SAE No.)*	*Cylinder Lubricant Viscosity (Typical SAE No.)*
Automotive		
Gasoline	30	30 or 10W, 20W, 30W
Diesel engine	30	30
Propane	30	30
Stationary (industrial)		
Diesel engine (4-cycle)	30 or 40	30 or 40
Diesel engine (2-cycle)	30	40
Diesel and dual fuel	30	30
Gas engine	30	30
Locomotive		
Diesel engine	40	40
Marine		
Diesels (4-cycle)	40 or 50	40 or 60
Diesels (2-cycle)	30 or 40	30 or 40
Opposed pistons	30 or 40	40
Supercharged	30	30
Turbocharged	50	60

oven and/or blast furnace gas for their blower engines, and gas producers whose manufactured gas fuels their distributing compressor engines. Such engines operate both four- and two-cycle. The two-cycle engines operate at higher temperatures than the four-cycle engine so that lubricating oil selection and maintenance must be considered accordingly.

FREE PISTON ENGINES

Another method of converting liquid fuel to high-energy gas to drive a gas turbine involves the use of a modified two-stroke-cycle opposed piston diesel engine called a gasifier (Fig. 18-6). The gasifier is a free piston arrangement utilizing intrinsically developed compressed air to return its pistons from the outward extremity of the power stroke or outward dead point back to the inward extremity of the compression stroke, or inner dead point (IDP). The outward dead point (ODP) of the power stroke corresponds

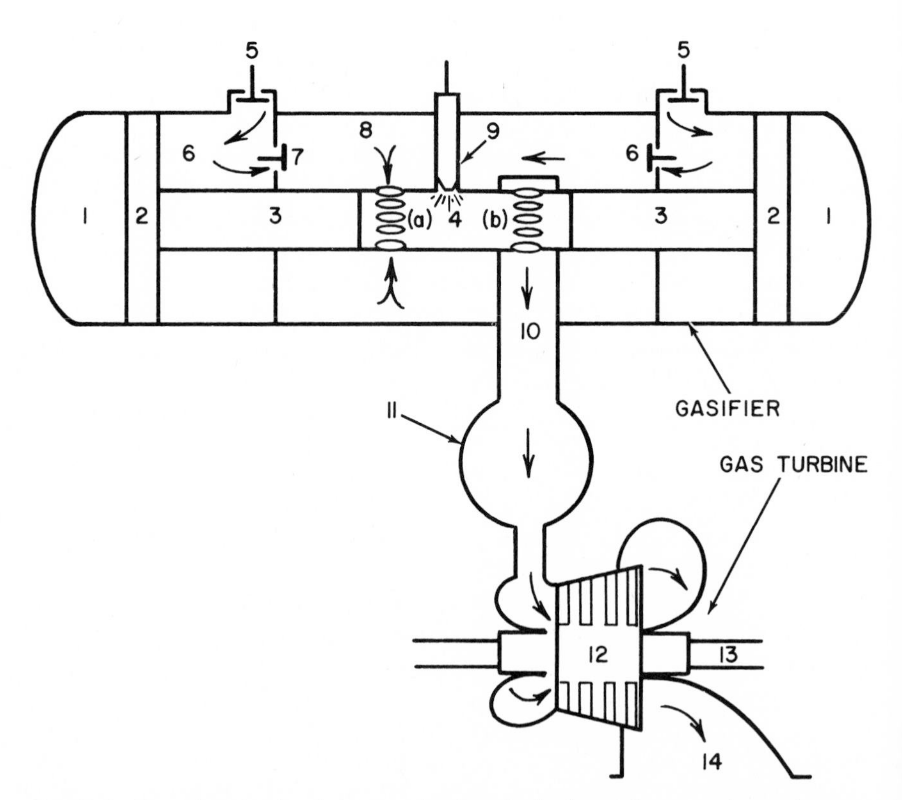

Fig. 18-6. Free piston engine. 1, bounce chamber. 2, air pistons. 3, power pistons. 4, power cylinder. 5, air intake valves. 6, air cylinder. 7, air discharge valves. 8, scavenging air, or air box. 9, fuel injector (multiple). 10, exhaust gas. 11, gas accumulator. 12, gas turbine. 13, gas turbine shaft. 14, gas discharge to atmosphere. (a), scavenging ports. (b), gas discharge ports.

to bottom-dead-center, and the inward dead point of the compression stroke corresponds to the top-dead-center of a conventional two-stroke diesel engine. The compressed air developed during the power stroke, in an enclosed space called the bounce chamber, replaces the crank and flywheel arrangement of the conventional diesel engine to furnish the momentum necessary to return the power pistons to their firing position or point of extreme inward travel. Except for the linkage necessary to maintain their constant relative position, the opposing power pistons are free to reciprocate within the confines of their common cylinder.

Figure 18-6 shows a free-piston gasifier connected to a gas turbine. Using the legend shown with the sketch, the basic operation of the gasifier-turbine may be described as follows:

Startup. Compressed air from an external source enters the bounce chamber (1). Then—

(a) The air piston (2) and the attached power piston (3) move from outward dead point (ODP) toward the center of the unit or inward dead point (IDP).

(b) The fresh air intake valve (5) closes while the air discharge valve (7) opens to allow air into the scavenging or supercharging chamber (8).

(c) The moving power piston covers the scavenger ports (a) and gas discharge ports (b).

(d) The piston continues toward IDP, compressing the entrapped air until the opposing pistons nearly meet head-on, at which point (IDP) the compression ratio reaches up to 50 : 1. This is considerably higher than that attained in a conventional diesel engine. Simultaneously, the pressure in the air box (8) is raised to between 40 to 90 psi by the air piston.

(e) At the ID point, the fuel injector (9) opens to allow fuel to enter between the opposing power pistons, where, because of the high pressure and high temperature, combustion occurs.

(f) The expanding gases formed by the burning fuel drive the piston assembly (power and air pistons) outward to uncover the discharge ports and then to cover the scavenging ports.

(g) The air under pressure from the air box scavenges the power cylinder, driving the burned gases out through the discharge ports into the discharge line (10).

(h) Simultaneously with (f) above, the air piston moving outward with the power piston causes the pressure in the compressor cylinder (6) to drop to just below atmospheric pressure to close the discharge valve (7) and open the air intake valve (5) to allow fresh air to enter the air cylinders.

(i) The pistons (2) and (3), traveling toward ODP at a rapid velocity commensurate with the energy of the expanding combustion gases, are stopped finally at ODP by the air cushions built up in the bounce chambers. The pressure therein reaches a maximum, because of the weight of the piston assembly and its momentum and because the pressure in the power cylinder has dropped to the gas discharge line pressure due to uncovering of the discharge ports. This pressure bounces the pistons back rapidly toward the IDP.

(j) The discharge gases mixed with scavenging air in the discharge line (10) enter the gas collector or accumulator (11), which stores gas acting as an impulse eliminator to smooth out the flow of gas to the gas turbine (12).

Not shown in Fig. 18-6 are the (1) synchronizing rods and linkages that maintain the pistons in synchronism with each other and operate or control certain accessories, including the fuel injector and mechanical force-feed lubricator used to supply oil to the air and power cylinders; (2) starting air tank, which is the "external source" mentioned above that first, furnishes air to a small air cylinder to move the pistons outward to their ODP starting positions during which the air trapped in the bounce chamber is discharged through a relief valve controlled by the starting air pressure and secondly, supplies a measured amount of air to a starting air vessel prior to being admitted to the bounce chambers to start the engine, (3) *balance tube* through which the starting air is led to the bounce chambers from the starting air vessel and to equalize the pressure in the two chambers during operation, (4) *lubricating oil systems* and (5) *pertinent controls*. The cooling water system is also not shown.

The free piston engine is equipped with two basic lubricating systems. These include a mechanical force-feed lubricator that supplies oil to the air compressor cylinders and power cylinders as well as to the balance tube seals, and a separate circulating oil system for the lubrication of accessory equipment such as the fuel pump, controls, and the synchronizing mechanism if the latter is not supplied from the power piston cooling oil system. There is also a circulating oil system that cools the power pistons.

The force-feed lubricator driven off the synchronizing mechanism meters oil to the compressor and power cylinder. The compressor cylinders have a number of ports (usually 8) drilled equidistant around their circumference to which the lubricator is connected via feed lines. The power cylinders also have ports for oil both inward and outward relative to the scavenging and discharge ports, whose feeders are equipped with one-way valves to prevent blow-back to the lubricator.

The power pistons are cooled by a flow of oil supplied to their interior cavities by a central feed line formed by a series of concentric telescoping tubes (to compensate for the reciprocating motion of the power piston) through the bounce chamber. The oil is returned through the bounce chamber within a larger outer tube surrounding the feed tube back to the circulatory oil reservoir. The tubes are sealed to prevent leakage of oil into the bounce chambers. The power cylinders are provided with cooling water jackets as in conventional diesel engines. Only the inboard compressor cylinder heads are provided with water jackets.

Considering that the temperature in the first power piston ring zone ranges from 400°F up to 480°F and that exhaust gas temperatures reach 850°F and above, it is important that the correct oil be used for power piston lubrication. These temperatures exceed those of conventional diesel engines, since the air supplied for combustion enters the cylinder from the air box at about 425°F. Speeds range up to 600 cycles per minute, each cycle representing two piston strokes. Outlet cooling water temperature should average 165°F. The fuel used is generally a residual type or Bunker C fuel oil (viscosity 3000 SUS at 100°F, average), although lighter products including gasoline can be utilized. The type of fuel used is partly a question of economics. Engines using the heavy residual fuels rely on lighter oils for starting. Experience indicates that power cylinder liner wear is greater per 1000 hours of operation with the heavier fuel than with conventional diesel fuels. Wear is reduced in both compressor and power cylinders when the oil feed is increased and also when the number of cylinder application points is increased.

Because of the high temperatures involved in gasifiers, carbonaceous deposits build up rapidly. Consideration of both wear and deposits must guide the selection of a lubricating oil to be used on all parts of the engine. This means, as with all similar situations, that both the physical characteristics and chemical properties of the lubricants must be correct.

For the lubrication of the air compressor and power cylinders an oil having the following approximate specifications is recommended in order (1) to prevent deposit buildup in the air box and piston ring grooves, (2) to retard wear of the piston rings and cylinder liners, (3) to reduce scuffing of pistons, and (4) to prevent fires in air box due to deposit formation.

API gravity	20–22
Pour point, maximum	20–25°F
Flash point, minimum	415°F
Viscosity, SUS at 210°F	90 (minimum), 110 (maximum), SAE 50
Viscosity index	50+
Additives required	Detergent-dispersant Oxidation inhibitor Antiwear agent Corrosion inhibitor

For the cooling of power pistons and the lubrication of accessories, a high quality light-bodied lubricating oil of the so-called "turbine oil type" should have these approximate specifications:

API gravity	28–31
Pour point, maximum	15–20°F
Flash point, minimum	390°F
Viscosity, SUS at 100°F	300 (average)
Viscosity index	85+
Additives required	Rust inhibitor Oxidation inhibitors

The same lubricant is recommended for the gas turbine bearings as for the cooling of gasifier power pistons.

The free piston engine was originally and is still used as an air compressor in single and multistage units. Today it is also used with gas turbines to generate or supply power for public utilities, centrifugal pumps and compressor drives, and ship and locomotive propulsion. When used as an air compressor, part of the air from the air cylinder goes to the power cylinders to supply oxygen to the fuel while the remainder serves the various purposes related to pneumatic systems, such as air-actuated controls and air-driven power tools.

STEAM ENGINES

The frictional components of steam engines include bearings and cylinders with rotating, oscillating, and translatory motions (Table 18-3). Rotating motion occurs on the flywheel shaft, (the main bearing), crankpin bearing, eccentric strap, and the flyball shaft bearings and guides. Oscillating motion is characteristic of the crosshead pin bearing and certain joints of the valve gear of all engines and the valve arm bearings, wrist plate bearings, and the trip mechanism bushings of Corliss engines. Translatory motion occurs on the crosshead guide, or shoe, the valve block and guide, and the cylinder head gland and packing. Included in the latter are the steam cylinders that are especially attended (Fig. 18-7).

Methods of lubricant application vary according to the make and design of the engine. Most of the rotating and oscillating bearings are equipped with drop-feed oilers. Some are hand oiled or equipped with grease fittings or grease cups of various kinds. The main bearings (flywheel shaft) of larger engines are lubricated by chain or ring oilers, sometimes augmented with

Table 18-3. LUBRICANTS FOR STEAM ENGINES

Part To Be Lubricated	*Type of Oil*	*Approximate Viscosity (SUS at 100°F)*
Steam Cylinders		
Steam Temperature (°F)		
Up to 350	Compounded cylinder oil	1800–2500
350–500	Compounded cylinder oil	2500
500–650	Noncompounded cylinder oil	3500–4500
Over 650	Noncompounded cylinder oil	6500
Stationary horizontal engines		
Bearings*	Turbine oil	150–300†
Marine vertical engines		
Bearings	Circulating bearing oil	600

* Including tail rod bearing where used.
† Use the higher viscosity for engines of 20-in. stroke and over.

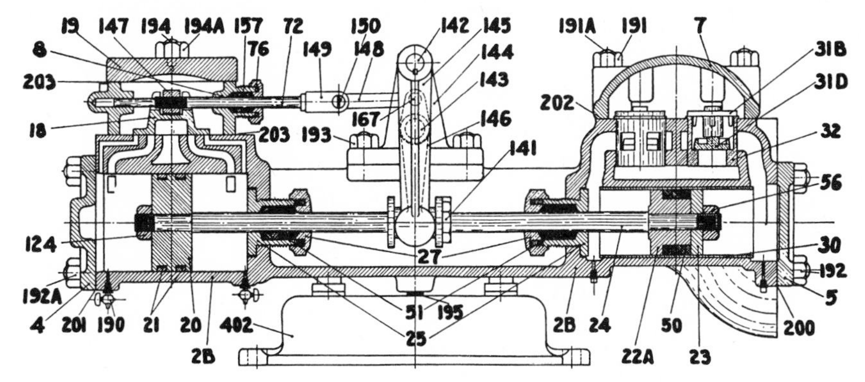

Fig. 18-7. Steam engine cylinder and valve arrangement and connecting fluid pump. Left portion, steam end; right portion, pump end. *Courtesy, American-Marsh Pump Co.*

8, steam chest. 157, packing for valve rod stuffing box. 72, steam valve rod. 31B, discharge valve. 31D, suction valve. 5, cylinder head, fluid. 22A, piston head, fluid. 24, piston rod. 2B, pump body. 51, piston rod packing. 20, steam piston. 4, cylinder head, steam.

drop-feed lubricators. Splash oiling may be used on the crankpin bearing. Crosshead guides or shoes and piston rods are usually lubricated with oil from drop-feed oilers. A 300 SUS at 100°F is used on all oil-lubricated parts except the steam cylinders.

The selection of an oil for steam cylinders is based on the temperature and vapor state of the steam used to drive the engine. The use of highly superheated steam that is exhausted still in a dry state requires an uncompounded oil of high viscosity (4000 to 6500 SUS at 100°F depending on temperature) to combat the elevated temperatures. Superheated steam is steam whose temperature is increased in a superheater at constant pressure. Any condition that causes the steam to condense to any degree during the expansion process necessitates an oil of about 2500 to 3000 SUS at 100°F, compounded with an emulsifying agent such as lard oil or acidless tallow. Steam reduced to a temperature and commensurate pressure at which condensation will occur is said to be saturated. The quality of steam is a measure of the amount of moisture it holds above that required for saturation at a given state of temperature and pressure. Any reduction in the thermal energy content (enthalpy) of steam below that required for saturation at the state pressure results in a decrease in quality, i.e., the saturation point has been lowered and the amount of moisture in the steam increases. Steam having less than 100% quality is "wet steam." Some of this moisture deposits on the surfaces of the expansion chamber, and the rest is carried through the exhaust by the velocity of the steam, whose energy has been correspondingly reduced. The fallout moisture tends to wash a lubricating film from the cylinder and piston ring surfaces, leaving areas of metal-to-metal contact that result in wear. An emulsifying agent is thus required to prevent rapid oil displacement in engines operating under conditions that are conducive to wet steam conditions. The agent may be either the oil in water or water in oil type.

Under substantially high superheated steam conditions, the drop in enthalpy during the expansion process is usually not great enough to cause a water-wet surface condition. However, the use of high-viscosity cylinder oils compensates for the effect of the elevated temperatures involved. The oil film on a steam cylinder surface is very thin, but an effective boundary lubrication condition should prevail.

The force of the steam exerted on the piston is axial (parallel with the piston rod), and any steam entering between the piston and cylinder exerts an equalizing pressure in all directions. This leaves whatever weight remains of the piston allowed by the bending of the piston rod which is the resultant moment of force based on the distance from the supporting rod gland and the stabilizing effect of the crosshead. On engines having exceptionally long strokes or whose pistons are relatively large (as in uniflow engines) and heavy, a tail rod is used to help support and guide the piston. The tail rod extends from the piston in the opposite direction to that of the piston rod through a gland and packing and is supported by an outboard bearing. Tail rods are made slightly bent so that they become straight when loaded.

Oil may be applied to the cylinders directly by mechanical force feed lubricators that feed through lines leading to different points around the cylinder, usually equidistant. The size of the cylinder governs the number of points of application. Oil may also be applied indirectly to the cylinders by hydrostatic lubricators that feed the lubricant into the steam line at a point ahead of the admission valve. Such lubricators operate in conjunction with quills inserted into the steam line, whose ends are cut in a manner best suited for the atomization and uniform distribution within the flow area of the oil applied. These devices also assure oil application to the valves.

Lack of cylinder lubrication is evident by a characteristic groan of the piston as it slides on the surface of the cylinder that is relatively oil free. Groaning will ease when sufficient oil is applied to produce a film of a thickness that causes only a smudge on a piece of tissue paper wiped across the cylinder surface.

GAS TURBINES

Rolling bearings have replaced the plain bearings used in early gas turbines. The major function of the lubricant is to lubricate and cool the low- and high-pressure compressor and turbine bearings of the primary engine. An additional major function is to lubricate the gears of the turboprop that drive the propeller and reduce its speed to practical operating levels. In turboprop engines, it is necessary to specify lubricants based on the demands posed by the gears. Auxiliary components not usually subject to the extreme conditions of the major frictional parts are in general satisfied with the same lubricant. These include the lubricant feed pump, the scavenge pumps, and other accessory gears and bearings. The result is that a central circulating oil system is used to distribute oil to all frictional components. This is not always the case, but the trend toward a one-lubricant system is both general and desirable.

In all circulating oil systems, a filter is essential to catch and hold any sediment. In stationary service, such as engines used as "peak shavers" in large power plants, an auxiliary oil cooler should be used. Not having the advantage of high-altitude conditions experienced by aircraft, some other method of oil cooling should be applied, especially under the heavy loads imposed on stationary units employed in power generation.

The quality requirements placed on lubricating oils for gas turbines are generally similar to those imposed on lubricants applied to any other critical power system. Such requirements include proper viscosity, high viscosity index, volatility commensurate with low evaporation loss, carbon residue, and chemical stability. The gas turbine lubricant must exceed other premium oils in its ability to withstand high temperatures.

THERMAL CONDITIONS OF GAS TURBINES. Gas turbines, including those used in turbojet and turboprop aircraft, develop higher temperatures than other types of turbines and engines. The critical areas are the front and rear bearings of both the low-pressure and high-pressure turbines. The heavy thrust loading and ambient conditions within the gas turbine cause bearing temperatures of about 400°F and bulk oil temperatures of 375°F at the return orifice. If the lubricant is cooled to an inlet temperature of 275°F by means of an oil cooler and fuel oil heat exchanger, the rejected heat load amounts to approximately 3000 to 4000 Btu per minute or higher,

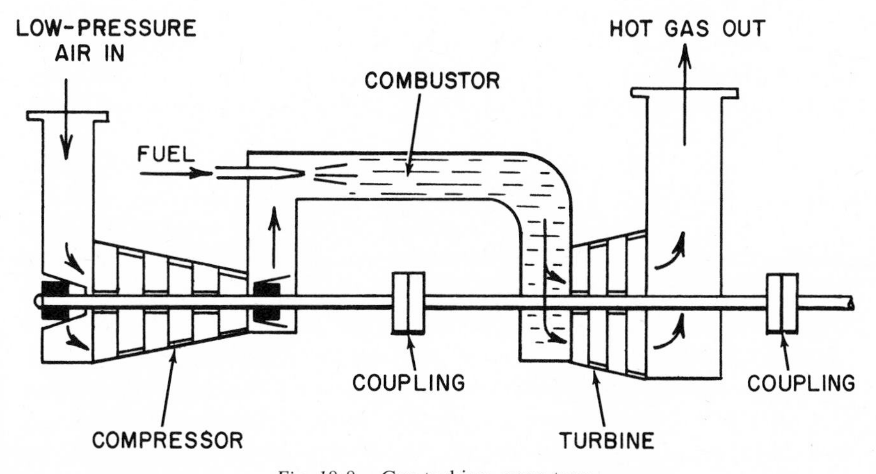

Fig. 18-8. Gas turbine, open type.

depending on the power output of the turbine, which is a function of aircraft speed. The above values pertain to speeds of mach 1 and 2, but they may double at velocities corresponding to mach 4, or four times the speed of sound.

The temperature of turboprop gears reaches 250°F in the mach 1 to mach 2 range.

LUBRICANT SELECTION. The United States Armed Forces is a major consumer of lubricants and has its own physical and chemical standards. These standards are published under identifying code numbers usually having the prefix MIL—the United States Military Specification—which in turn can be divided according to grade. For example, gas turbine lubricants are designated as grade 1005 (5 cSt minimum), or grade 1010 (10 cSt minimum) of MIL-0-6081B. The list of standards is extensive and subject to frequent changes, and in many cases the standards overlap.

The reader should become acquainted with the government system of specifying materials, especially if he is interested in government contracts. However, it is not necessary to rely on this system exclusively, since reputable lubricant manufacturers are quite capable of determining accurate specifications to meet the requirements of existing machines which may or may not compare exactly with the MIL values on every point of reference.

Gas turbine lubricating oils are either mineral oils or synthetic fluids. Stationary gas turbines can use either type, but aircraft and racing cars use synthetic lubricants mostly.

To understand the intricacies presented in the formulation of a satisfactory gas turbine lubricating oil, a résumé of the operating conditions that exist within the engine is in order here. It is given on the basis of the extreme conditions of high-speed, high-flying aircraft.

Elevated bearing temperatures approaching 400°F
High-speed gears and bearings
Heavy loading of gears and bearings
Low starting temperature

To meet these conditions, the lubricating fluid must possess the following characteristics:

Viscosity must be sufficient to provide a protective lubricating film at the high operating temperatures and fluid enough at the minimum temperatures expected to permit relative ease of starting and also to allow adequate flow to the various points of lubrication. Rolling bearings require a lubricant of relatively low viscosity.

Film strength should be great enough to meet the heavy loads (high thrust) imposed on bearings and gears to prevent metal-to-metal contact. Film strength additives are usually used to offset the limited viscosity imposed by cold starting. Where an increase in viscosity is allowed, an improved ZN/P would result, which would preclude the use of film strength additives in some cases.

Chemical stability is essential to withstand the high-temperature environment without chemical change in the lubricant. Such changes include oxidation and subsequent acid formation followed by metal surface corrosion. The general rule is applicable that a mineral lubricating oil deteriorates at a rate that decreases its service life expectancy by one-half for each 18°F rise in operating temperature above a normal of 130°F.

Low volatility is required to minimize evaporation loss. In aircraft operation vapors from the turbine lubricant are carried by air from the compartment pressurizing system and are finally discharged to the atmosphere through pressure-equalizing vents. In other operations, the vapor condenses in the breather conduits. This leads to high acidic aggravation that further carries oil deterioration with accompanying sludge and carbon formation. Traps to eliminate the acid oil vapors are one solution to the problem.

Loss of the lighter (more volatile) ends of the hydrocarbon complex that constitute the lubricating oils leaves the heavier, less volatile ends behind with a subsequent increase in viscosity. Vapors due to evaporation and oil mist from the turbulent effect of high-speed journals form deposits of lacquer and hard carbon on exposed surfaces of housings, seals, and bearing areas. Therefore, lubricants of low volatility are essential.

Nontoxicity, in the interest of human safety.

Rust inhibitors should be contained in gas turbine oils to protect machine surfaces.

Oxidation inhibitors are required under high-temperature operation.

High flash point

High viscosity index

Thermal stability, to meet high temperature operation.

Compatibility with engineering materials used.

NONFOAMING CHARACTERISTICS. Foaming is the result of mixing air and oil under high turbulence. Systems containing large volumes of oil usually break the solution readily enough if the mass is relatively free of strong emulsifiers and if the reservoir settling time is long enough. In systems in which the oil volume is small, foaming cannot be tolerated. Strong emulsifiers in the oil aggravate the foaming problem.

A high degree of foaming (1) causes cavitation at feed pump suction, limiting the amount of oil fed to frictional components; (2) hinders the development of a hydrodynamic wedge; (3) leads to overflow of the froth through vents, fill pipes, and other openings of the system. This results in the loss of oil charge and possible damage to surrounding devices or equipment. The overflow debris left after the emulsion has broken presents a fire and accident hazard. Gas turbine lubricants must be made to resist foaming at any time because the volume, especially of those in mobile service, is relatively small. Even weak solutions are not given adequate settling time in the main reservoirs to break sufficiently.

Mineral Oils. Correctly refined and fortified mineral oil meets a number of the characteristics required for the lubrication of gas turbine engines. One of its disadvantages lies in the volatility factor, where evaporation losses may range from 5 to 10%, depending on its formulation and somewhat by the flash point.

In general, the practice is to specify oils according to SAE number. While the problem of contamination by the combustion gases is absent from gas turbine lubricants, they must necessarily meet the demands already discussed. Viscosity requirements may be generally satisfied with multi-viscosity mineral oils specified as SAE 5-20. Their use should be limited to stationary engines as represented by electric power generators.

Typical Specifications for SAE 5-20 *Motor Oils*

API gravity	29.5
Pour point, maximum	−40°F
Flash point, minimum	325°F
Viscosity, SUS at 0°F	4000 (maximum)
Viscosity, SUS at 100°F	175 (37 cSt approx.*)
Viscosity, SUS at 210°F	49 (7 cSt*)
Additives required	Antioxidant
	Detergent
	Rust inhibitor

Gas turbine engine manufacturers also recommend so-called industrial lubricants, such as those universally used in steam turbines. In addition to the physical properties required as numerically specified, it is essential that they be of higher quality. One criterion is the rate of increase in acidity that occurs when oils are exposed to controlled test conditions at a given high temperature. The stability of an oil may then be estimated by the number of hours it takes before the neutralization number curve changes abruptly from one of slight horizontal grade to one approaching a more vertical attitude, as shown in Fig. 7-1. Highest quality oils will average 4000 hr before the critical point is reached.

Table 18-4 gives typical specifications for industrial type oils recommended for lubrication of direct connected units (bearings) and geared units (bearings and gears).

Gas turbines of the closed or semiclosed types used for ships generally require an oil of higher viscosity than stationary land-based units. Recommended viscosities for the latter are shown in parentheses in the table.

SYNTHETIC LUBRICANTS. This class of lubricants was developed to

*Kinematic viscosity (centistokes)

Table 18-4. TYPICAL PROPERTIES OF GAS TURBINE MINERAL OILS FOR DIRECT CONNECTED AND GEARED UNITS

Property	*Direct Connected Units*	*Geared Units*
API gravity	30	30
Pour point (max.), °F	20	20
Flash point (min.), °F	400	400
Viscosity, SUS at 100°F	150	300 (400)
Viscosity, SUS at 210°F	44	53 (60)
Color, ASTM	1.5	2.5
Viscosity index (min.)	100	90
Rust inhibitor required	Yes	Yes
Antioxidant required	Yes	Yes
Rust test	Passes	Passes

overcome the disadvantages of mineral oils in several respects. One important feature of the synthetic fluids is the decrease in evaporation rate compared with mineral oils when exposed at temperatures of 400°F for a period of 6.5 hr. Minimum flash points of 400–500°F may be obtained with synthetic materials used in gas turbine service. Better high- and low-temperature viscosities are also achieved. Other advantages include good oxidation stability and extended life under the high temperatures encountered. The trend is toward synthetic fluids in all services, including aircraft requirements which initiated their development.

Synthetic lubricants used in gas turbines, particularly in turbojet and turboprop engines, are made up of esters. Esters are compounds made from the reaction of acids and alcohols (Chapter 6), of which *di-ester* and *tetra-ester* are in present use. The latter is a blend of di-esters and polymers (or complex esters) that furnishes, for one thing, the load-carrying ability necessary to meet gear lubricating requirements.

The use of government specifications in connection with synthetic gas turbine lubricants is important. Such specifications report the property limits of such lubricants as determined by extensive testing of present engines. They may also forecast such limits as a result of careful analysis of probable operating conditions in engines that await future development or that are presently in the design or development stage.

Depending on the service intended, of which viscosity at pertinent temperatures is the governing factor, the ester fluids are divided into two major groups, Type I and Type II.

Type I fluids are di-esters and have a maximum viscosity of 13,000 cSt at −65°F and a minimum viscosity of 3.0 cSt at 210°F, according to the specification MIL-L-7808C.

Type II fluids are tetra-esters having a viscosity of less than 10,000 cSt at −40°F and a viscosity of 5 to 7.5 cSt at 210°F.

Table 18-5 shows typical specifications on Type I and Type II lubricants, the latter based on MIL-L-23699.

Table 18-5. TYPICAL SPECIFICATIONS OF TYPE I AND TYPE II GAS TURBINE LUBRICATING FLUIDS

	Type I	*Type II*
Synthetic material	Di-ester	Tetra-ester
Specific gravity	0.92	0.998
Flash point, (min.), °F	400	420
Fire point (min.), °F	450	470
Evaporation % (6½ hr. at 400°F)	5.5	5.5
Viscosity, cSt		
400°F		1.3
210°F	3 (min.)	5.3
100°F	11 (min.)	27 (min.)
−40°F (typical)	—	10,000
−65°F	13,000 (max.)	—
Pour point, °F	−75	−85

The original Type II lubricant was a diester having a minimum viscosity of 7.5 cSt at 210°F. The new tetra-ester type is directed toward better gear protection. Antiwear properties, oxidation stability, and corrosion prevention have been considered in the above specifications. They also possess good viscosity-temperature characteristics to meet the paradoxical conditions of low-temperature starting and high-temperature operation.

The mineral oils discussed above are satisfactory where gas turbine engine bearing temperatures are about 300°F. However, recent development has forced temperatures upward to approach or exceed 400°F. To meet the new demands of both temperature and heavy loading, the development of synthetic fluids became essential.

AUXILIARIES. Equipment required to augment the primary engine is usually present, including starting motors, starting engines, and pumps. Where such equipment is not lubricated from the primary circulatory system, it should be serviced in accordance with the normal procedure for similar components located elsewhere in the plant or ship.

STEAM TURBINES

Steam turbines drive electric generators, centrifugal compressors, pumps, blowers, industrial machinery, such as sugar and paper mills, and propel ships. They operate most efficiently at 1500 to 20,000 rpm, depending on their purpose, and they can be designed to run at variable speeds. The usual speed of industrial turbines ranges from 1800 to 3600 rpm. For many purposes, the single shaft of a steam turbine is directly connected to the driven component, as in electric power generators, or the speed of the shaft may be reduced through gears to drive slower-moving machines.

Steam used to supply energy to the rotating member of steam turbines varies from about 600 psi to above the critical pressure, which for steam

To prevent leakage, stuffing boxes or rotary seals are provided and placed on both sides of the rotor. Any number of blades can be used, usually four or eight. They are made of softer material than that of the cylinder wall, so that any wear will occur on the more easily replaceable member. Larger rotary compressors of the industrial type are jacketed and water cooled.

In smaller rotary compressors, coil springs are placed in the slots behind the blades for the purpose of obtaining sufficient radial thrust to provide a tight seal between the end of the blade and the cylinder wall at all times.

Another type of rotary compressor having a single blade and employed principally in the domestic refrigeration field consists of a steel ring mounted on a camshaft and a steel cylinder, the diameter of which is greater than the ring. As the shaft on which the cam is located rotates concentrically, it causes the ring to revolve eccentrically about the same axis. Because the distance between the center of the concentric shaft and the perimeter of the ring at the point of greatest departure is equal to the radius of the cylinder, the two are in constant contact. The ring, being appreciably smaller in diameter than the cylinder, leaves a crescent-shaped space around the remaining perimeter of the ring between the two. The ring, being free on the cam, rotates as it rolls on the cylinder wall in a gyratory manner. The cylinder is located within a hermetically sealed housing partially filled with oil, which also houses the electric motor that drives the shaft. End plates are installed to enclose the cylinder.

The cylinder wall is drilled in two places in close proximity along its circumference to allow for the entrance of low-pressure gas and its discharge at a higher pressure. Between the two drilled holes or ports a slot is cut across the entire width of the cylinder wall parallel with its axis. From the slot a single blade, backed by a coil spring, emerges to press against the ring. It is free to move in and out with the contour of the ring as the latter revolves. Sometimes the blade is backed by oil pressure.

When the suction port is uncovered, low-pressure gas is drawn in the cylinder within that portion of the crescent-shaped space between the blade and the point where the ring and the cylinder wall are in contact. As the ring continues to revolve, it covers the suction port and traps the gas within the diminishing space between the ever-changing contact point and the blade, forcing it out under higher pressure through the discharge port which is located now on the near side of the blade. At the instant the ring uncovers the suction port, low-pressure gas enters the cylinder and continues to enter until the port is again covered. Similarly, at the instant the ring uncovers the discharge port, high-pressure gas leaves the cylinder and continues to be forced out until the port is again covered.

To prevent the high-pressure gas from flowing back into the cylinder with the uncovering of the discharge port, a check or flapper valve is placed in the discharge line. A check valve is also placed in the suction line to prevent the flow of warm gas back to the low-pressure region from the compressor during periods of shutdown. While the low-pressure gas is drawn directly through a line connected to the suction port, the compressed gas is discharged into the space outside the cylinder but within the sealed housing from where it flows to the receiver.

The drive shaft on which the cam is ground is spirally grooved for the purpose of carrying oil into the cylinder to lubricate the blade and the camshaft bearing. The blade is of composition material.

The rotary compressor is a constant-speed machine. While slight flow interruption occurs at the time of discharge port coverage by the rotor, it is not subject to the more pronounced pulsations of the reciprocating type compressor.

Rotary compressors require lubrication of both bearings and rotor cylinder. For most services, including refrigeration, an oil of 300 SUS at 100°F viscosity is recommended.

Gear Compressors

Not widely used, the gear compressor is the only positive displacement machine free of discharge pulsations. It consists of two meshing gears, mounted and keyed on their own shafts, which rotate within a gas-tight housing. Power is transmitted to one gear, which drives the other. Both gears are fitted closely to the housing so that only slight clearance is maintained between their outermost surface (addendum circle), and the circular contour of the case through approximately 180 degrees. A lubricating oil film completes the seal in that area.

The low-pressure vapor enters the case at the bottom and is carried between the gear teeth to the top of the case, where it is forced out into the discharge line. The vapor is displaced from between the teeth by meshing teeth of the mating gear.

To maintain uniform center and eliminate end thrust, double-helical gears generally are used. The herringbone effect of such gears gives greater strength because of extended line contact between opposing teeth and allows additional teeth to mesh for greater load distribution. Such an arrangement also provides better sealing than would be possible with the usual type of spur gear. A check valve is placed in the suction line to prevent leakage back through the suction valve during periods of shutdown.

Like the rotary and reciprocating types, gear compressors can be arranged to operate as a duplex unit; i.e., two compressors placed on either side and directly connected to an electric motor.

Two-Lobe Compressors

The two-lobe rotary compressor is similar in principle to that of the two-lobe pump shown in Fig. 19-18. Identical mating lobes trap the gas between their outer surfaces and the casing. As the upper arc of the lobes passes the edge of the casing, discharge of gas is initiated while the bottom arc forces it into the discharge line. The lobes are held in a fixed rotating relationship by two identical gears, one mounted on each lobe shaft to provide a link between them. Gears and lobe shaft bearings compose the frictional parts of this type compressor that require lubrication.

Centrifugal Compressors

Centrifugal compressors, employed for refrigeration purposes, are of the kinetic displacement type that functions according to the basic laws of centrifugal force and the effect of that force on gases. The rotating member of the centrifugal compressor is an impeller, consisting of a series of radial vanes through which the low-pressure fluid entering their axial inlet is forced to flow outward toward their peripheral outlet at a higher pressure.

The impeller is mounted and rotated by a shaft to form an integral component that represents the only moving part of the centrifugal compressor. The shaft is supported on bearings that are sealed against the escape of gas and the bearing lubricating oil that continually flows through them. As the impeller rotates, the gas is scooped up by the suction of the blades, whose axial extremities project a short distance into the eye surrounding the shaft and deliver it outward into a voluted or scroll-shaped housing that fully encloses the impeller.

The number of impellers mounted on the shaft represents the number of stages that is required to increase the pressure of the gas from suction to discharge level. To separate the individual stages, a stationary diaphragm is located between each impeller. Most centrifugal compressors are of the multistage type, although a single-stage compressor is available for use in some air conditioning systems.

Centrifugal compressors are relatively large machines capable of developing refrigeration tonnage of from 50 to over 2500 tons, and they operate at speeds approaching 10,000 rpm.

With the exception of auxiliary oil pumps, the impeller shaft is the sole moving part of a centrifugal compressor, whose supporting bearings are the only components requiring lubrication.

LUBRICATION OF CENTRIFUGAL COMPRESSORS. The journal bearings that support the impeller shaft are lubricated with a highly refined oil of 300 SUS at 100°F, usually containing rust and oxidation inhibitors. The oil is recirculated under pressure from a central system and pump that must be put in operation prior to starting the compressor.

Purge-recovering systems are used in large systems to reduce refrigerant losses and cleanse the recovered gas of contaminants, such as noncondensible gases, water, and oil that may leak into the system through the bearing seals or other joints. Such systems involve an auxiliary refrigeration system consisting of a reciprocating compressor, condenser, and evacuator. Oil is removed by an oil separator installed between the auxiliary compressor and its condenser.

Axial Flow Compressors

The principal use for axial flow compressors is in conjunction with gas turbines, described in Chapter 18. The design is similar to that of a reaction turbine in that, if driven in reverse, it not only forces gas along an axial path, but also provides multistaging to gradually increase pressure as the gas passes from each pair of rotating and stationary blades.

The axial flow compressor is a highly efficient machine capable of handling large volumes of air under continuing operation at full capacity. Because it tends to pulsate within normal ranges of output, it is not suitable for general industrial requirements. With a compression ratio of about 4:1, it performs at about 85% efficiency. It is best adapted for conditions having an approximate constancy of flow resistance in order to pass through the region of unstable dip characteristic of propeller fans, especially those having steep pitch. Gas-turbines appear to offer the constant flow resistance compatible with axial flow conditions.

Coupled with diffusion or guide vanes, the axial flow impeller is a compact multistage compressor capable of developing high pressures at speeds not necessarily excessive.

PUMPS

Pumps are used to move liquids from one place to another. Pressures against which pumps operate range from less than one to many thousands of psi. Their volume capacity, usually expressed in gallons per minute, also varies over a wide range. The refrigeration industry uses many pumps to handle liquids, such as water, brine, liquid food products, lubricating oil, and a number of others.

There are three principal kinds of pumps: reciprocating, rotary, and centrifugal.

Reciprocating Pumps

The reciprocating pump is a positive displacement pump. Its principle of operation is similar to that described for the reciprocating compressor having a piston- or plunger-cylinder arrangement as its working element. It is equipped also with suction and discharge valves to allow for entrance and exit of the liquid. Reciprocating pumps may be single- or double-acting, vertical or horizontal, steam- or power-driven, and may or may not be equipped with an air dome.

Direct-acting pumps are so called because the liquid piston is connected directly to a steam-driven piston through a common piston rod. The length of stroke of both pistons of the direct-acting pump is therefore the same, as is their speed. The direct-acting drive possesses the advantage of variable speed, thus providing a variable-capacity pump.

Power pumps are those whose power is derived from an external source, usually through speed-reduction gears and crankshaft arrangement. They are generally constant-speed machines, although speed variation may be obtained by suitable control methods. Power pumps are identified by the number of pistons or cranks, such as simplex (single piston), duplex (two pistons), triplex (three pistons), and quintuplex (five pistons). The throw of the crankshaft governs the length of stroke. Power pumps should be protected by a safety device, such as a relief valve, to avoid damage if subjected to excessive pressure for some reason.

Double-acting pumps have double the capacity of single-acting pumps of comparable size, less the volume taken up by the piston rod when in a position of maximum inward stroke.

Reciprocating pumps are designed to handle liquids having varying degrees of fluidity from low-viscosity solvents to viscous substances such as chocolate syrup and molasses. They are capable of developing tremendous hydraulic pressures, such as required for large presses.

Rotary Pumps

Rotary pumps are also positive-displacement machines. Their principle of operation is similar to that previously described for rotary compressors. The liquid displacement element may be any one of a number of different kinds, including gear, lobe, screw, and vane types (Fig. 19-18). The rotary pump is a constant-flow machine, so that liquid discharge is not subject to the pulsating action experienced with reciprocating pumps. They are used widely in the food industry for handling milk and fruit and vegetable juices of all kinds.

The displacement of rotary pumps is found by determining the volume of the pumping element and multiplying this value with the rpm, as shown for rotary compressors. Some slip is experienced due to pump clearances. Slip loss does not include losses due to inlet conditions, such as the presence of air or other gases to which rotary pumps are sensitive. Wear of the frictional surfaces within a rotary pump is minimized if only those liquids capable of imparting a lubricating effect are handled.

Centrifugal Pumps

The operating principles of the centrifugal pump (Fig. 19-19) are similar to those described previously for centrifugal compressors and fans. Con-

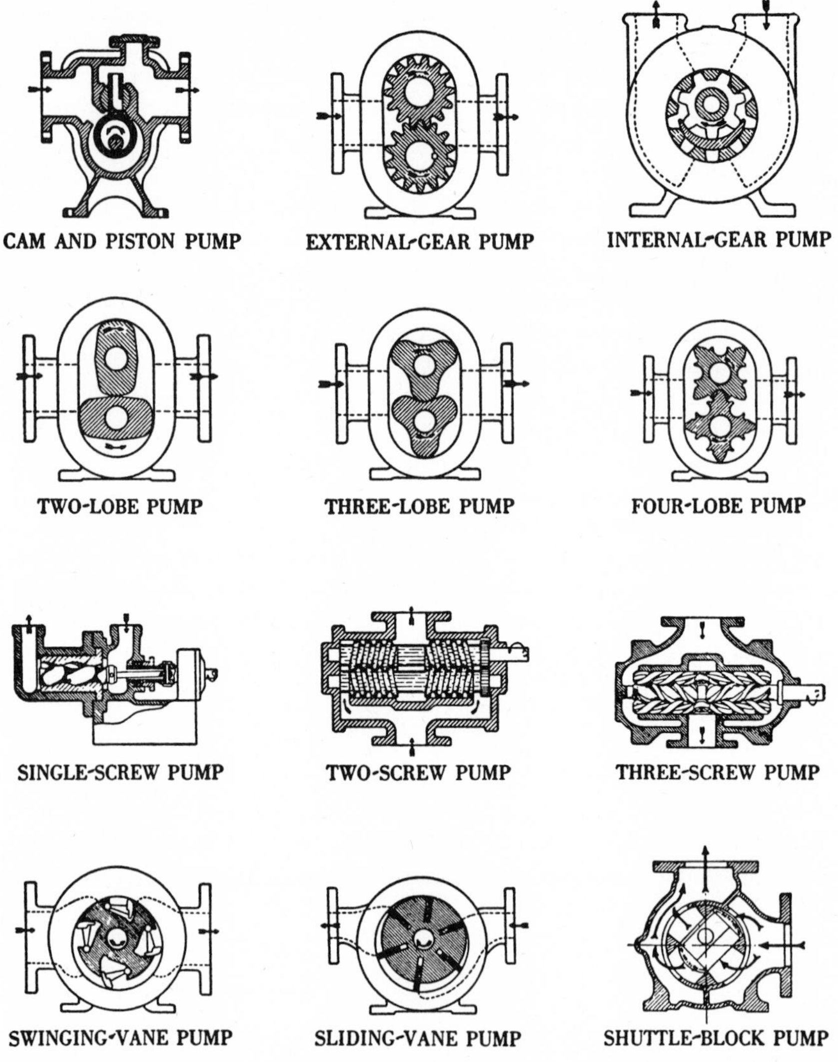

Fig. 19-18. Various types of rotary pumps. *Reprinted, by permission, from the "Standards of the Hydraulic Institute," ed. 10. The Hydraulic Institute, 122 East 42nd St., New York, N.Y.*

sisting of an impeller that carries the vanes, enclosed in a stationary casing, centrifugal pumps are of three general classes, radial flow, mixed flow, and axial flow.

Radial or centrifugal flow pumps force liquid radially outward from the impeller axis by centrifugal force. The pump may be either single suction, in which the liquid enters only one side, or double suction, that allows for liquid entrance on both sides of the impeller.

Mixed flow pumps have their vanes inclined so as to impart a lifting action to the liquid. Lift plus centrifugal force acts to create the velocity head developed by the pump and to discharge the liquid in an axial and radial direction.

Axial flow pumps project the liquid as a fan projects air, i.e., the liquid passes through the pump in a direction essentially parallel to the shaft or impeller axis.

Centrifugal pumps are made with one or more stages according to the desired pressure increase.

Liquid pumps of the horizontal reciprocating types require lubrication as follows. Various methods of application are used.

Crosshead Crank Gears Bearings	Oil—300 SUS at 100°F

Other type pumps—vertical centrifugal pumps	
Upper bearing	Oil—300 SUS at 100°F
Lower bearing (combination guide and thrust bearings)	Grease—No. 2 lime or calcium complex
Horizontal centrifugal pumps	
Rolling bearings	Grease—No. 2 calcium complex or lithium base
Triplex pumps	
Bearings	Oil—300 SUS at 100°F
Stuffing boxes (where used) (packed or compression cup)	Grease—No. 2 calcium complex or No. 3 lime base

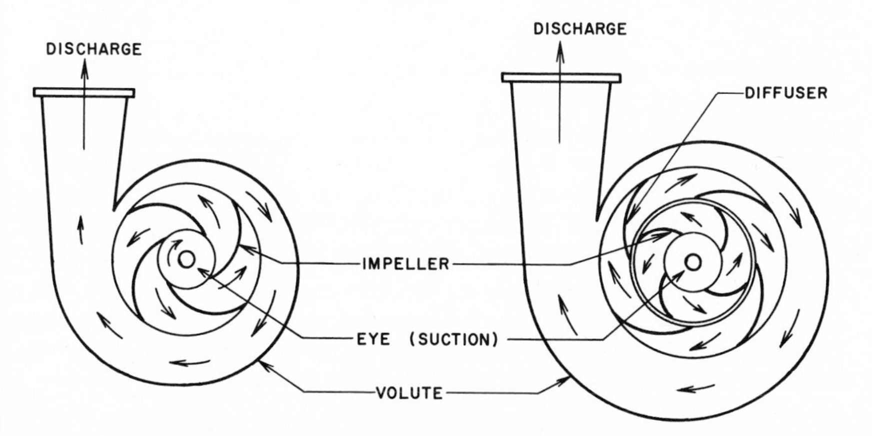

Fig. 19-19. Volute type centrifugal pumps with diffuser at right.

CAVITATION

A pump designed to handle liquids may develop cavitation, a condition that causes a portion of the liquid to vaporize at the point of suction. Such a condition usually involves a myriad of vapor-filled cavities in the liquid.

Cavitation is caused by the loss of pressure within the liquid at various points along the path of normal flow as a result of turbulence. This, in turn, is brought about for various reasons, but chiefly because less liquid is provided to the suction end of the pump than the amount for which the pump is designed and the speed with which it is operated. The lack of sufficient liquid may be the result of undersized piping and valves, incorrect location of vessels relative to head potential, etc. Another factor lies in the vaporizing properties of liquids.

The problem of cavitation is more prevalent with centrifugal pumps, whose low pressure point is at the entrance to the impeller vane or eye. As the fluid is forced through the vane, the increasing pressure causes the vapor cavity to collapse abruptly, resulting in vibration and noise and promoting wear of the impeller metal. Cavitation also interferes with the throughput performance of the pump, which may reduce the production capacity of the manufacturing process served by the pump.

The effect of cavitation is evident by the characteristic of a pump system called the *net positive suction head.* The NPSH is a measurement of energy in feet required to overcome the losses in the suction lines and valves and to allow for the entrance of liquid into the impeller vanes without a loss in performance due to cavitation.

VACUUM PUMPS

The components referred to here are those involved with "high" vacuum, a term used in connection with negative pressures of the magnitude of 1 micron or less.

Methods for "pulling" a vacuum include asperation, condensation, and ejection, but only those using mechanical displacement pumps will be discussed here. Such pumps are not limited to high vacuum uses and may be found in such services as paper mill suction rolls, surface condensers in power plant operation, cigar wrapping machines, and food packaging lines.

RECIPROCATING VACUUM PUMP. In any reciprocating piston-cylinder arrangement, there is always clearance space when the piston reaches its limit of travel. Any gas entrapped in the clearance space re-expands during the suction stroke. The amount of vacuum is decreased by the amount of pressure exerted by the re-expanded gas, which is relatively large for this type of service.

To reduce or eliminate the gas clearance volume, oil is introduced into an inverted cylinder so that it occupies the otherwise gas-filled space when the piston reaches the bottom of its stroke. A bypass from the clearance space to the upper side of the piston eliminates any compression effect on the oil mass underneath.

Some of the gas emulsifies with the oil and, added to the vapor pressure of the oil itself, establishes the final vacuum developed, which may be in the region of 0.1 to 0.05 torr or mm Hg.

ROTARY VACUUM PUMP. The reciprocating pump has been replaced by the more sensitive rotary type, in which oil from a separate reservoir is used both to seal and lubricate the rotor and bearings. The oil so introduced is drawn from the reservoir through the rotor shaft bearings into the cylinder, where it is distributed to the internal frictional parts: the vane-cylinder surfaces and discharge valves.

In an arrangement somewhat similar to booster staging in a refrigeration cycle, higher vacuums can be developed. This is accomplished by linking two or more stages in series, resulting, in limiting levels of 0.01 to 0.001 mm Hg.

Primary or backing pumps are those that discharge directly into the atmosphere; secondary pumps are those that discharge to a lower pressure developed by the primary or backing pump. The latter are capable of developing higher vacuums. Staging can be simply arranged by driving two pumping units on a common shaft and connecting them in series. Both pumping units are entirely submerged in an oil bath. The vacuum within the pumping units draws oil along the drive shaft into the pump cylinders.

Two-stage units employ a pump having an eccentric rotor that maintains contact with and revolves about its cylinder. A vane is kept in positive contact with the revolving rotor by a spring-loaded lever and is reciprocated by the eccentric configuration of the outer surface of the rotor. As the point of contact between the rotor and cylinder approaches the vane, the entrapped gas is discharged through a ball-type discharge valve. When the point of contact passes the discharge port, the vane is pushed up flush with the cylinder, allowing air to enter the free space through an intake port on the side of the vane in the direction of rotation. This air is again compressed by the rotation of the rotor to be discharged as already described and repeated continuously.

LUBRICATION OF VACUUM PUMPS. Vacuum pumps, particularly those that pull a high vacuum (<1 micron of Hg), require an oil for cylinder lubrication having an extremely low vapor pressure to prevent evaporation. An oil of about 300 SUS at 100°F is heavy enough to maintain an effective piston-cylinder seal.

20

SEALS AND PACKING

To prevent leakage between spaces and joints of mechanical work systems, seals, gaskets, seats, and packing are employed. Leakage between isolated spaces may have the following undesirable results: (a) loss or gain of pressure, (b) loss or gain of gas or liquid, (c) contamination of otherwise protected substances.

Leakage may result from defects (sand holes, cracks, etc.) in the space housing material, but interest here is directed toward prevention of leakage at static or dynamic joints. Static joints are those associated with machine components having no relative motion in respect to each other. They are fitted tightly together to form either a continuous conduit (pipes, tubing), or a terminal arrangement in which one component acts as a cover, dome, bounce chamber, pneumatic or hydraulic cushion, engine-head crankcase, or seat. Dynamic joints are those in which one member of a kinematic pair has relative motion in respect to the other. Motion may be reciprocating or rotational. Reciprocating pairs include piston–cylinder, piston rod–stuffing box, or piston rod (plunger)–packing gland arrangements. Rotational parts include journals and rolling bearings.

A type of seal not primarily associated with a joint is the diaphragm, an important component of certain pumps and valves. The diaphragm is a flexible seal, but it may also have dynamic functions.

Leakage in or out of a space is increasingly resisted by tighter fits and greater depth of packing, sealing etc., but some compromise is always acceptable because of the rise in friction that accompanies any rise in interfacial pressures between the opposing surfaces. The art of preventing leakage at any joint is based on the consideration of pressure differentials and the viscosity of the fluids involved. High pressure differentials and low viscosities tend to encourage leakage at a rate commensurate with their values. Where leakage is dangerous, inconvenient or costly, the fluid is often recovered beyond the primary seal before it reaches a region where it is least desired. In many instances, the fluid may be returned to the working system or to some other suitable destination.

Of primary interest here are those seals that are prefabricated in a specific shape, but substances such as pastes, compounds, solders, and packing that confine fluids will also be considered.

Devices employed as leak preventive components differ in their design according to the geometry of the joint and the operating circumstances, hence, the variation in nomenclature mentioned at the beginning of this chapter, although they all act to seal off a given space relative to another. They are defined as follows:

GASKETS

Gaskets are usually installed between two stationary rigid containers to prevent fluid flow in any direction except within the containers or from one to the other (Fig. 20-1). Such containers include those having parallel flanges or concentric cylinders. Leakage from the container is stopped by the application of unit pressure on the gasket greater than the unit pressure of the fluid within the containers. This is usually accomplished by bolting the opposing surfaces tightly together. To meet all the functions intended, gaskets should conform geometrically to the pattern of the flanged surfaces and be constructed of a compressible material that forms an impenetrable barrier to the passage of fluids into or out of the container. The material should also be impervious to the fluids involved and resistant to the operating temperature.

Gaskets are made in sheet, strip, or bulk form. The plain grooved gasket should never be allowed to protrude beyond the height of the groove so as to cause mushrooming upon tightening of the flange (Fig. 20-11).

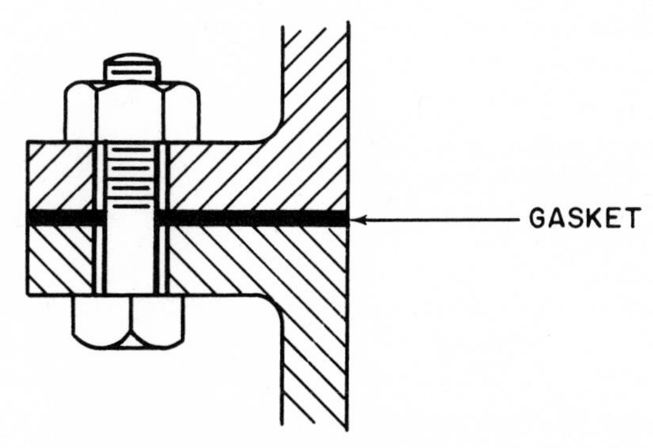

Fig. 20-1. Gasket for joints.

SEALS

Seals here refer to the packings that keep dirt and water out of rolling bearings and lubricant in, and also to rotary shaft seals that prevent the loss of fluid from or the entrance of air into refrigerant compressors of the vertical reciprocating and rotary types, including centrifugal machines. In such compressors, rotary shaft seals are also used to prevent leakage between stages as a result of pressure differentials.

ROLLING BEARING SEALS are made in the form of a ring or washer which is retained in a carrier groove under slight pressure. Felt seals should be made of new, long-fiber wool, free of surface fuzz, and fabricated to a density referred to as medium felt. They should be sized to snug fit the groove and bear lightly on the rotating member. Felt seals should not be used above 200°F, the temperature at which they begin to char. Felt rings are cut with a rectangular cross section, having greater height than width, and are inserted into grooves tapered on one side to provide some compression to the felt and cause sufficient elongation to provide light pressure on the ball or roller. The tapered side of the groove should be nearest the bearing so as to bend the felt fibers inward and return any oil that passes into the seal back to the bearing proper.

Where it is more important to keep abrasive dirt out than to prevent lubricant leakage, slingers are placed outside the housing.

LEATHER SEALS are also used on rolling bearings. Their cross section is usually that of a quarter circle, and they are installed in a groove so that the arc is cupped inward (toward the bearing) for moderate speeds and outward for high speeds to assure some lubrication by leakage in the latter case. Leather seals are retained in their grooves by garter springs (Fig. 20-2) that impose only slight pressure of the seal on the rotating member to avoid burning.

Labyrinth seals, discussed later, and spiral grooves cut into the rotating member are other methods of retaining lubricant in rolling bearings. Inner and outer grooves spiral in opposite directions on the rotating member. The

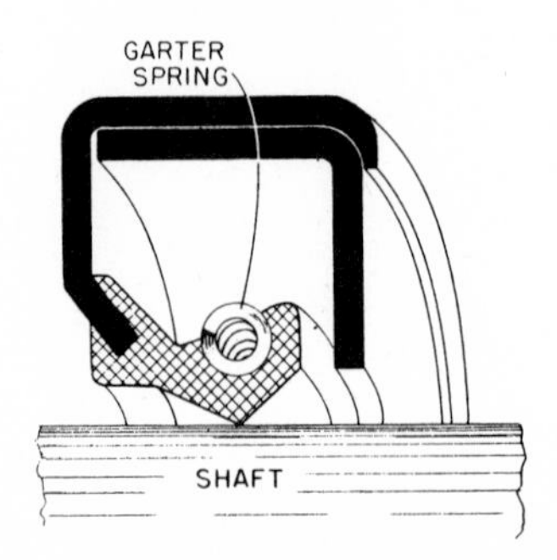

Fig. 20-2. Garter type seal used in hydraulic service.

inner groove returns the lubricant and the outer groove conveys dirt away from the bearing.

ROTARY SHAFT SEALS are of three basic types—simple, bellows and/or spring (Fig. 20-3), and stuffing box (Fig. 20-4). A variation of the simple rotary shaft seal is the positive contact lip seal, which is resilient and shapes itself to the shaft (Fig. 20-5). The stuffing box type is discussed later under the subject of packing.

The bellows type is distinguished by a nosepiece, either rotating or stationary, which further identifies it as a truly mechanical seal. It consists of a sealing ring, a bellows, and a spring, all of which fit around the shaft and are encased within a housing and maintained in position by a cover plate. The sealing ring bears against a shaft shoulder, and the contacting

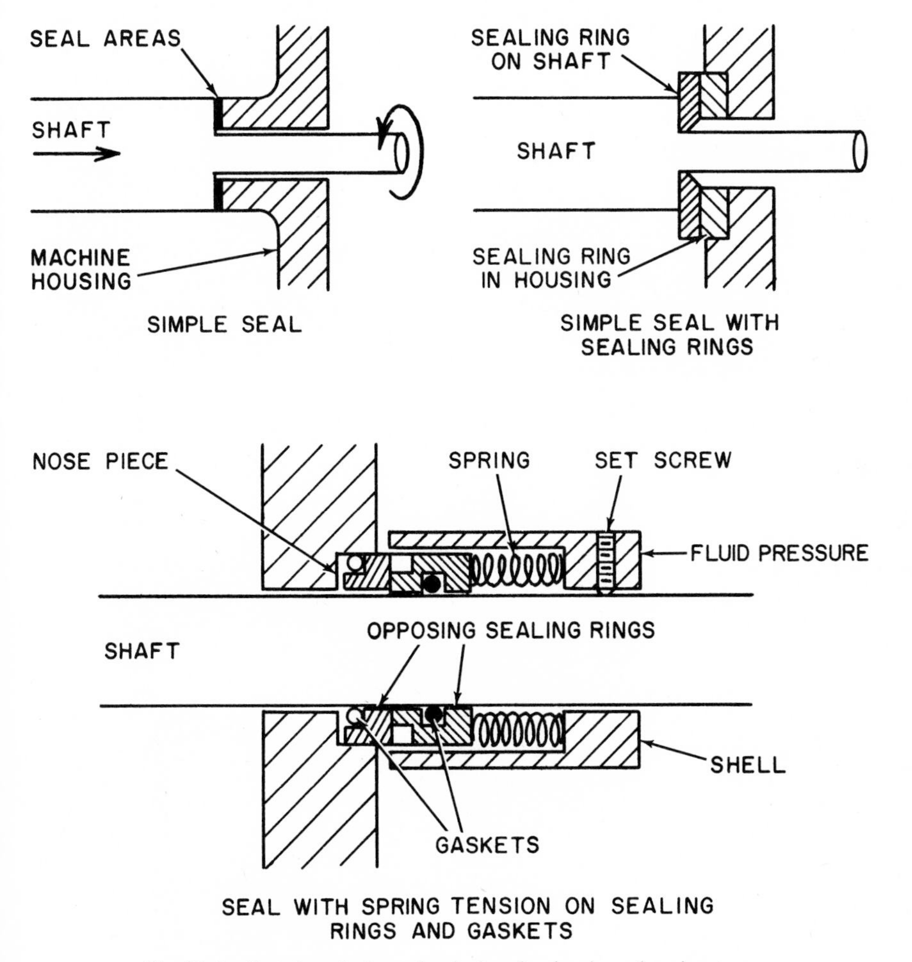

Fig. 20-3. Rotating shaft mechanical seals, simple and spring types.

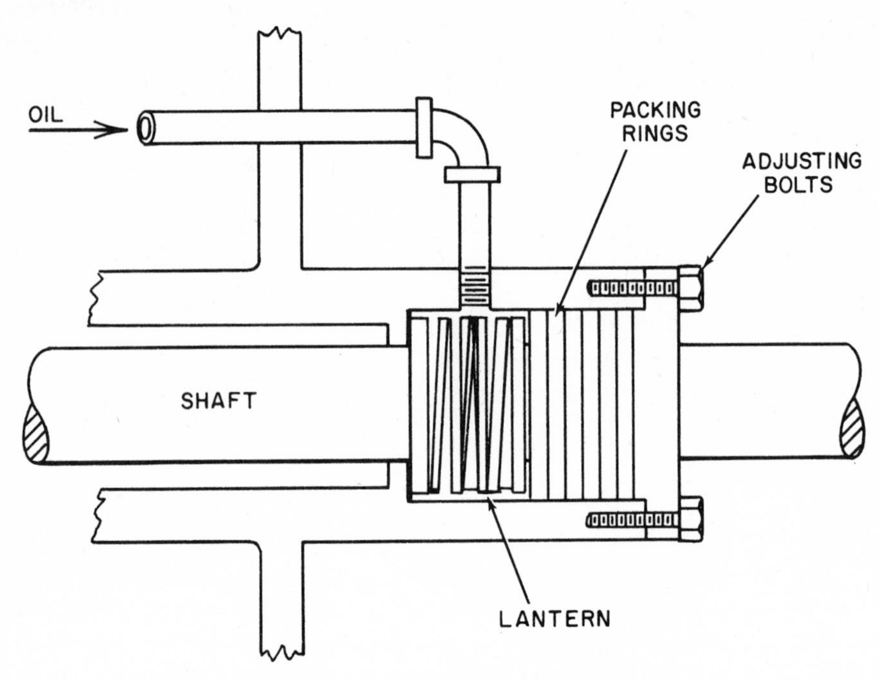

Fig. 20-4. Rotary shaft seal, stuffing box.

surfaces are smoothly finished to form a tight seal. Attached to the sealing ring is a flexible, corrugated metal bellows, the other end of which is fastened securely between the end of the housing and a cover plate. A strong coil spring, which may be located within or outside the bellows, serves the double purpose of keeping the sealing ring tight against the shaft shoulder and

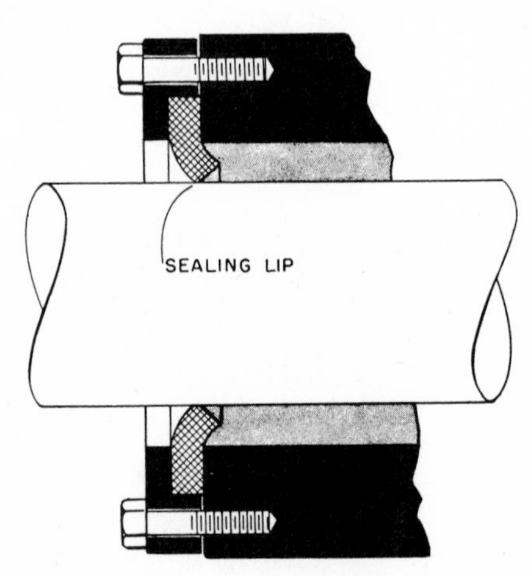

Fig. 20-5. Positive contact lip seal.

expanding the bellows. In this type, the sealing ring acts as the nosepiece and remains stationary. A packed collar that rotates with the shaft and against which the sealing ring bears is a basic modification that distinguishes the two types mentioned above. All seals should be lubricated.

The bellows is necessary to compensate for wear on the bearings or the shaft. Any increase in leakage caused by such wear tends to further expand the bellows, which in turn exerts a correspondingly greater pressure on the sealing surfaces, which are usually bronze. An important consideration in this type of seal concerns the relative position of the center line of the convolutions of the bellows to the center line of the sealing ring.

Because of the difficulty in combining sufficient strength with adequate flexibility, the size of bellows is limited, and bellows are not often used with shafts above 2 in. in diameter. Where bellows are not practical, sealing is accomplished by the use of one or more coil springs to maintain proper contact pressure between the stationary ring and rotating member.

Another type of rotary shaft seal employs oil pressure to maintain sufficient contact on lapped metallic seals in a manner somewhat similar to that described for the bellows type. The oil supplying the pressure also lubricates the rubbing surfaces and acts as a seal during operation. The oil pressure is controlled, and any appreciable drop will automatically shut down the machine on which this type of seal is installed. A shaft seal reservoir located above the shaft provides sufficient oil head to prevent leakage either way through the bearing during shutdown periods.

A labyrinth type seal is used between stages of centrifugal refrigerant compressors. The labyrinths are tortuous passages formed by narrow strips or blades placed concentrically with the shaft and close enough together to form a series of constricted spaces (Figs. 20-6 and 20-7). Because of its tendency to expand, refrigerant gas is subject to a throttling action as it attempts to pass from one space to another. This retards flow sufficiently to prevent leakage, provided enough concentric blades are installed.

CERMET SEALS. Materials composed of a combination of ceramic and

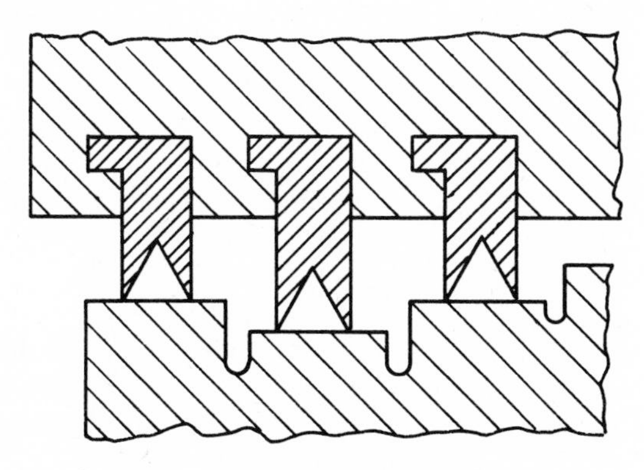

Fig. 20-6. Labyrinth seal.

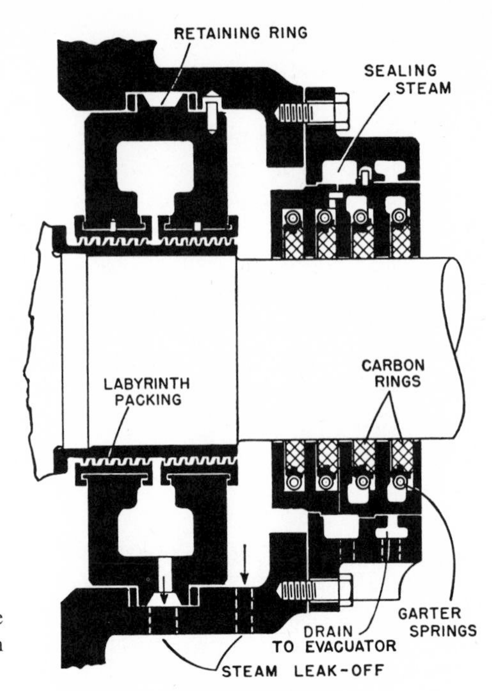

Fig. 20-7. Shaft seal for steam turbine with labyrinth packing and carbon rings.

metal are called cermets. Because of their hardness and resistance to the effect of temperature, cermet shaft face seals are used in pumps handling liquid metal coolants (lithium, NaK) employed in nuclear reactor systems, and cryogenic fluid pumps of high-speed rockets.

Such seals consist of a rotor and stator (male and female) whose sealing surfaces do not wear-in readily so must be lapped flat to within one helium light band and to a finish of 4 microns rms. Lubrication by dry gas depends on a gas film between the rubbing surfaces. Design specifications may include shaft speeds of 5000 to 8000 rpm and an operating life of 10,000 hr. Successful results have been obtained with liquid lithium and sodium-potassium at pump temperatures of 1400°F.

SEAL FACES. Other hard materials besides cermet that bear against a softer rotating material of mechanical seals, such as carbon in most instances, include tungsten carbide, cobalt nickel combinations, nickel iron combinations, and alumina ceramic. Each has its operating limitations, especially in regard to high temperature and other unusual and difficult conditions.

Alumina ceramic, for example, is susceptible to stress shocks and fracture under sudden temperature changes. Heat checking sometimes occurs on cobalt nickel combination materials when subject to high speed, high pressure, or liquids with poor lubricating qualities. Tungsten carbide appears to have some advantage in withstanding high pressure and speed without overheating. Successful results in resisting wear and seal life are reported using tungsten carbide as the stator against a rotating babbitt-filled carbon element under certain conditions which preclude the presence of abrasives.

In any sealing system, including those using rotating versus stationary faces, the important rules to be followed are these:

The fluid in the sealed compartment should be kept free of fines and dirt.

The seal faces should be carefully aligned to assure positive contact at all areas around the periphery of motion.

Surface asperities and other imperfections that may cause scratching or gouging of the seal materials and cause leakage should be eliminated.

Seal distortion, shaft deflection, excessive loading, and insufficient lubrication should be avoided in the interest of effective sealing.

PACKING

Packing refers here to hand-formed or cut material that fits around joints and prevents leakage of confined liquids. Some packings are prefabricated in strands or rings with variously shaped cross sections, such as round, square, rectangular, conical, cupped, O-, V-, and U-shaped (Figs. 20-8 and 20-9). Other packings are bulk materials that require hand shaping within the packing cavity or stuffing box.

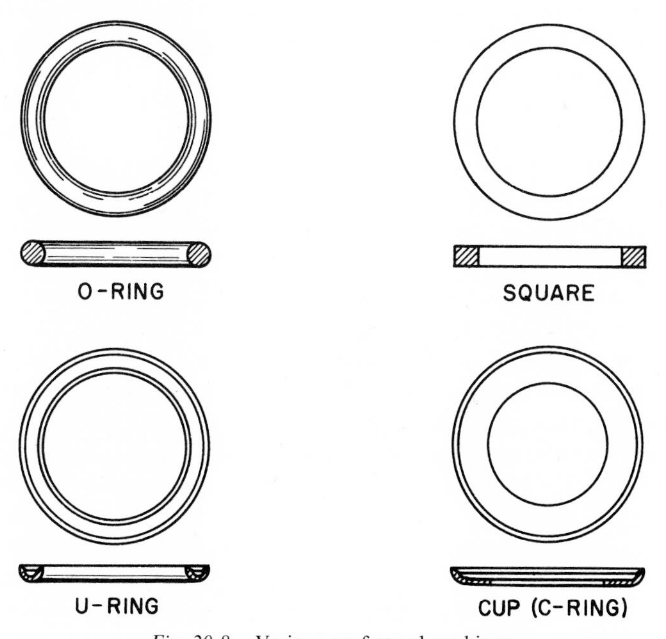

Fig. 20-8. Various preformed packings.

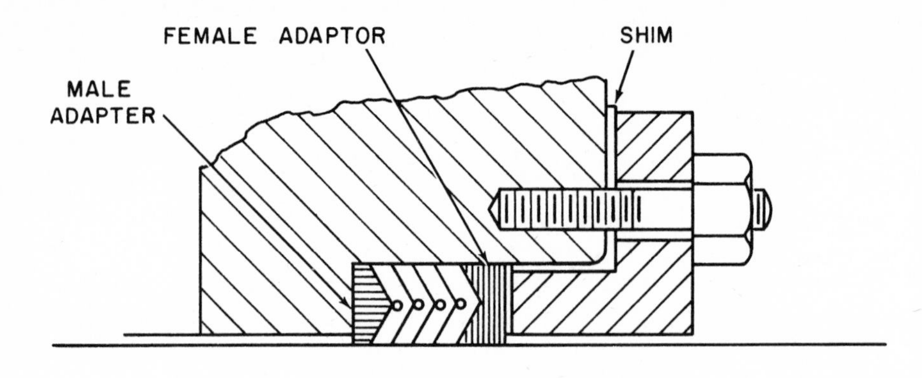

Fig. 20-9. V-ring gland.

THE STUFFING BOX is a type of rotary shaft seal (Fig. 20-4). It consists of a cylindrical housing forming an integral part of the crankcase through which the crankshaft extends. The internal diameter of the stuffing box, being somewhat greater than that of the shaft, allows a series of packing rings to be placed about the shaft and fill the annular space between the shaft and housing. The packing rings are held in place at the compressor end by a shoulder formed on the shaft, and at the flywheel end by a gland secured by adjusting nuts.

The arrangement within the stuffing box usually consists of three sections. The first, nearest the crankcase, is made up of one or more rings. This is followed by a lantern that forms the second section. The third section of five or more packing rings fills out the remainder of the stuffing box, the outer ring contacting the gland. Various forms of lanterns are used, but by the employment of a relatively heavy coil spring, a means of self-adjustment is provided to counteract wear and temperature changes.

The gland should not be drawn up so tight as to cause binding of the packing rings in the shaft, nor should it remain so loose as to preclude sufficient radial expansion to provide effective sealing. As packing wear occurs, a compensating adjustment must be made by tightening the gland nuts. If there is leakage after the gland has been drawn up close to the housing, it is an indication that the packing is worn and must be rebuilt or replaced.

To effect greater sealing and to provide lubrication and cooling, oil may be fed to the stuffing box from the compressor lubricating system by a line introduced at a point where the lantern is located. Using split or sectionalized rings of the correct size and materials assures a close fit on the shaft when only slight pressure is applied at the gland if installed properly. The rings

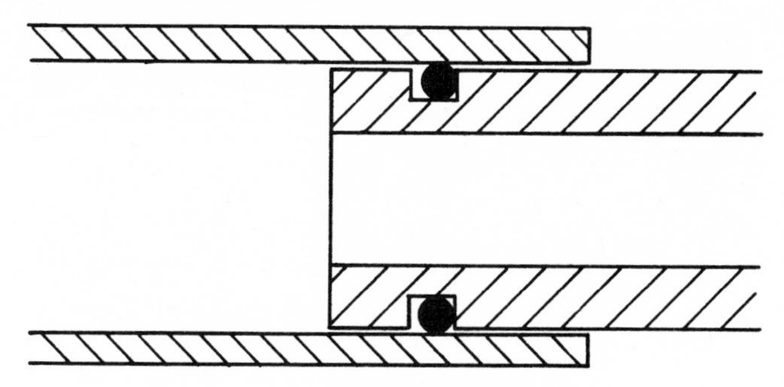

Fig. 20-10. O-ring joint.

should also fit tightly to the wall of the stuffing box. Their necessary contact with the shaft requires materials that are nonabrasive and possess low frictional qualities.

Packings such as the rings described for the stuffing box above are called *radial seals,* because they compress against the shaft and expand against the housing radially. *Axial seals* make contact with flat surfaces that are perpendicular to the axis of the shaft, such as the nosepiece described for the bellows and spring arrangement.

The use of O-rings has increased with the wider employment of hydraulic systems, especially with machine tools and other forms of automatic machinery (Fig. 20-10). O-rings are round and their cross section is also round. They may be used as gaskets (Fig. 20-11) or packing for sliding members, such as stems, rods, and plungers. The materials of which the various forms of packing are made depend primarily on the compatibility of the fluid which they are to restrain.

Diaphragms

The use of flexible diaphragms is two-fold, (1) to limit the lateral movement of fluids, and (2) to absorb such motion by its flexibility. The

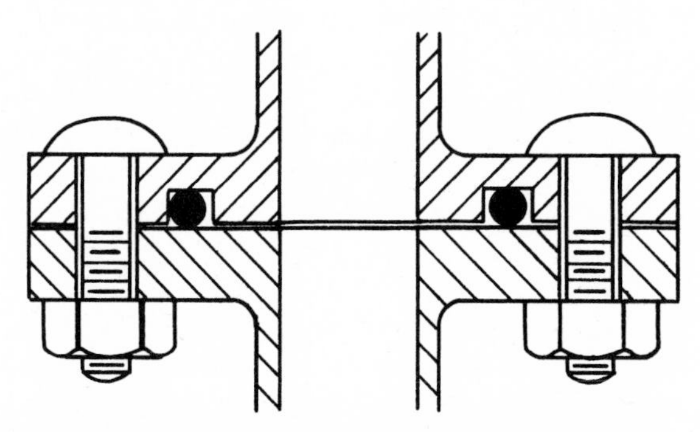

Fig. 20-11. O-ring used on a flange.

bellows described for the rotary seal is a form of diaphragm of the cylindrical type. It can be variously used to restrict the flow of a fluid so as to act as an impulse pump. Corrugated hoses as used with breathing apparatus provide another example of such a diaphragm.

Flat diaphragms of the cylindrical type are used as valves, valve stems, or piston packing. When circumferentially restrained, as exemplified by a drum head, the diaphragm is capable of transmitting limited motion by alternate changes in pressure exerted on one or both sides of a flat sheet or within a bellows.

Piston Rings

To prevent blow-by of gases under high pressure in the combustion or compression chamber of internal combustion engines, gas or air com-

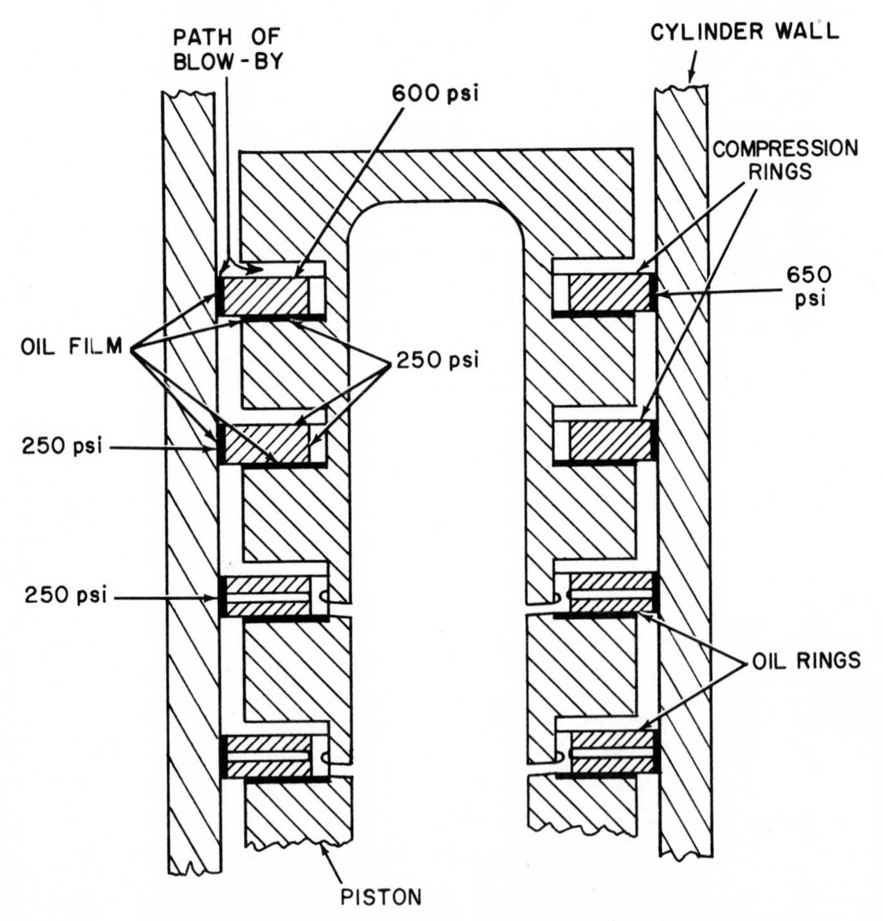

Fig. 20-12. Piston-cylinder arrangement showing rings sealed by lubricating oil to prevent blow-by.

pressors, etc., into the crankcase between the piston and cylinder walls, compression rings are installed. Oil wiper rings are used in conjunction with compression rings to prevent the lubricating oil from being forced into the combustion or compression chamber.

Such rings are not continuous but are split in order to allow for radial dimension (flexibility) changes and to allow their installation into grooves that are cut circumferentially in the piston. Piston rings are described in Chapter 12, and they are mentioned here to emphasize their role as dynamic packing components.

SEALING AND PACKING MATERIALS

Packing materials may be metallic, nonmetallic, or a combination of the two (Table 20-1).

Metallic packings are made up of a number of metals having the qualities required to provide adequate flexibility, wear resistance, strength and low frictional characteristics. The major metals used include iron or its alloys as piston rings, axial seal rings, washers, retainers, shields, etc., bronze base alloys and copper for diaphragm nosepieces, segmental rod packings and lantern rings, and lead and babbitt for gasket materials. Most metals are also used in the form of wire, sheets and strips in conjunction with nonmetallic materials that are fabricated into packing.

Nonmetallic packings are made from many materials, including carbon, leather, felt, cork, rubber, synthetic rubbers gutta-percha, hydrocarbon polymers (plastics), vegetable fibers (flax, wool, cotton, jute, etc.), synthetic

Table 20-1. Compatibility of Seal Materials With Lubricating Oils

Material	*Compatibility*
Natural rubber	NC
Neoprene	FC
Nitrile (Buna N)	C
Styrene (Buna S)	FC
Butyl	NC
Silicon rubber	NC
Viton*	C
PTFE (polytetrafluoroethylene)	C
Cork	C
Leather	C
Vegetable fibers	C
Gasket paper, glue and glycerine	C
Carbon (rings)	C
Felt (synthetic rubber binder)	C
Metallic seals	C
Asbestos	C

Key: C, compatible; NC, noncompatible; FC, fairly compatible.
* Tradename of E. I. duPont deNemours.

cloth (nylon), PTFE (Teflon), and asbestos. Powdered graphite, talc, and mica are used to provide lubrication in some instances or as plunger surface fillers when combined with other soft materials or applied directly to the surface treated.

CARBON. Pistons are equipped with carbon rings when no lubrication is provided. The piston rods are held on center by externally lubricated guides to prevent side thrust against the cylinder walls. Carbon also serves as the material for axial seals and segmental rod packing.

LEATHER is used as oil seals for ball and roller bearings. Usually formed in the familiar cup and V-shapes, it is effective in keeping the bearing lubricant in and foreign matter out when cupped inward or outward, respectively. Leather is used for packing hydraulic rams. At high speeds, some means of lubrication is necessary to prevent burning and to keep the leather pliable.

FELT. In the form of a ring or washer, felt is a common material for sealing ball bearings. Woven felt is stronger than pressed felt, but the latter is satisfactory if not stretched during installation. A synthetic rubber binder or coating is used for felt when used as an oil seal or gasket.

CORK is a widely used gasket material, either in solid form or as corkboard, in which granules are bound together with binders of various kinds, such as glue or polymers.

ASBESTOS is a heat-resisting material used for packing in various forms—sheets, braids, fillers, etc. Yarn types are built up with twisted copper wire to give it strength or saturated with petrolatum. Multiple layers of asbestos sheet are pressed together to form packing of generally used shapes or for gasketing when suitably combined with corrugated or spirally wound metals. It is often used with graphite as a surface smoothing ingredient. A good grade asbestos is recommended for use as packing for oxygen-handling equipment because it in itself does not contain any vegetable ingredients (oils or fibers) that would prove hazardous in the same manner described in the discussion on carbon rings versus lubricants.

SYNTHETIC RUBBER. The various synthetic rubbers all have individual properties. The following list includes the more common types used as packing materials.

Neoprene is a chloroprene, a polymer of chlor-butadiene, called polychloroprene. It has many of the excellent properties of natural rubber but does not have the resistance to oil or heat that some other synthetic rubbers have. It is resistant to many chemicals, abrasives and flame and is used in various forms of packing where compatible with fluids in contact.

Buna S, also known as GR-S, is a styrene and copolymer of butadiene and styrene. It is not as strong as natural rubber but has greater heat resistance. It is made in various forms for packing such as the lip seal shown in Fig. 20-7 having a resiliency that enables it to shape itself to the shaft.

Buna N is an acrylo-type rubber similar to buna S, except that it contains the substituent acrylonitrile in place of styrene. When vulcanized as packing, it possesses high resistance to swelling in petroleum lubricating oils and is fairly heat resistant.

Butyl rubber is an isobutylene type, being a copolymer of isobutylene with isoprene. It does not possess good solvency but is resistant to attack by acids. It is used for gaskets in acid-handling equipment.

PTFE, polytetrafluoroethylene (Teflon) is a thermoplastic material having a relatively high softening point. It has high heat resistance and is chemically inert, making it an ideal packing material where either relatively high temperature or reactive chemicals are exposure factors.

NYLON is a polyamide of various types and structures described as nylon 6,11, 6,6, and 6,10, having different starting materials. They are tough materials having high tensile strength and resistance to abrasion. They possess higher melting points than most other thermoplastics and are light in weight. Because they are subject to moisture regain, the electrical properties deteriorate as the relative humidity increases and improve under dry conditions. They have good mechanical properties when dry and are resistant to high temperature. They are self-lubricating, a property advantageous in any number of packing applications, as well as other frictional components, such as gears, cams, and bearings. However, they become brittle at temperatures above 250°F.

PLANT FIBERS. Flax, jute, and cotton have been used as packing materials for many years. They are effective at normal to low temperatures when used in strands, braids, or cloth sheets. They are usually treated with various binders, such as animal fats and wax.

NATURAL RUBBER is compounded with other materials, such as sulfur, metal particles, graphite, and zinc oxide, and is usually vulcanized at the time or after it is formed or molded into the required shape. It is sometimes strengthened with yarns, as of asbestos or cotton. It is widely used for a packing material. It is subject to swelling and deterioration in the presence of certain lubricants, including petroleum base oils, and it is not used in systems employing such fluids.

OAKUM, untwisted hemp rope, is treated with tar where water or drainage fluids are carried by cast iron pipes. The oakum is used to seal the bell mouth of bell-and-spigot joints peculiar to cast iron pipe.

LITHARGE is a yellow powder of lead monoxide (PbO). When mixed with glycerine, it provides a sealing material for threaded pipes and other joints where it can be conveniently brushed or swabbed on the contacting surfaces. It hardens to a cement-like coating to furnish an effective leakproof seal.

RED LEAD, lead tetroxide (Pb_3O_4), is a bright red powder used principally as a protective coating for ferrous metals. When combined with linseed oil, it provides a material for sealing metal cracks and joints, such as threaded iron or steel pipes.

Solder

Solder is a fusible metal alloy used for joining metals and for sealing metallic joints. Solders should (a) be lower in melting point than the metals to be joined, (b) possess an affinity for the metals on which they are to be applied, (c) have sufficient strength upon solidification commensurate, with the service demands of the junction, (d) be used on clean metal surfaces, even down to the bare metal, (e) be used in conjunction with a flux which promotes melting and offsets the effect of metal impurities, and (f) flow readily when heated.

The ingredients of more common solders are tin and lead, although antimony is used in small amounts to impart hardness. Other alloying materials, such as bismuth, cadmium, sodium, and thallium, are also used either to impart affinity for certain metals or to impart certain properties. Brazing solders that must be applied by a brazing torch include such metals as aluminum, brass, nickel, and silver. They are described as hard solders and possess much higher melting points (about 1200°F to 1600°F) than the plumber's solders, or wiping solders.

Soldering is associated with static joints, particularly seamless tubing made of materials capable of being soldered or brazed. Copper tubes are exceptionally applicable to the soldering process when the joint between two lengths of tubing is augmented with fitted couplings, tees, elbows, etc. Copper tubing is usually selected to transfer lubricants and liquid coolants.

Steel pipe joints are often welded, especially where high pressures and temperatures are involved. Both multilayer gas welding and electric arc welding techniques are employed.

21

GENERAL EQUIPMENT LUBRICATION

Many difficulties experienced with frictional systems are not attributable to the lubricant applied according to the rules. The fault may lie with the individual making the initial lubricant recommendations if he fails to recognize the destructful potentialities associated with a given system. For example, static electricity is generated when paper slides over paper in an atmosphere of low relative humidity, as occurs on some large printing presses. If not properly insulated, grounded (or air conditioned), electrical current passing through any frictional part will often cause metal displacement by electrolysis. Other destructive potentialities are shock loading, overloading, abnormal temperatures, fuel dilution, water washing, inaccessibility of frictional part to lubricant, over-lubrication, contaminated environment, lack of care of lubrication system and lubricant, poor preventive maintenance, restricted ventilation, and inefficient scheduling of inspection.

To assist the reader in recognizing some of the critical operating areas associated with various kinds of machines, industries and processes, several of each are discussed in this chapter. All have their so-called "suffering points" that must be recognized and dealt with. Those discussed here include the following:

Electric motors and generators	Paper mills
Motor-generator sets	Rubber mills
Hydraulic turbines	Sugar mills
Steel mills (strip)	Wire rope

ELECTRIC MOTORS

Electric motors represent an important component in the field of lubrication. They range in size from small fractional horsepower to the large motors used in steel mills of 10,000 horsepower and over.

HOUSINGS. Electric motors may be of the open-frame type or fully enclosed, the latter of which include those rated as explosion proof. The degree of enclosure affects the operating temperature, which in turn affects the temperature of their bearings. Assuming fairly good ventilation, the temperature increase above ambient of open-frame motors is about 70–75 °F, and fully enclosed motors will rise about 100 °F. During startup under cold ambient conditions, the excessive friction and wear problem of inadequate fluidity and consistency of oils and greases, respectively, is greater. This may be rectified by the use of lighter-bodied lubricants or lubricants designed for lower temperatures.

At high-temperature operation, the use of heavier-bodied products to offset the thinning out process is not a critical startup problem. It does, however, present an oxidizing environment unless the lubricants used are properly designed and inhibited to combat high-temperature exposure.

Under any condition of operation, including speed and temperature, lubricants used for electric motor bearings of either plain or rolling bearing types (Fig. 21-1) must meet the requirements described in Chapters 12 and 13. For general applications, the oils and greases recommended meet the viscosity and consistency specifications for both plain and rolling bearings, as shown in Table 21-1. Recommendations are typical but will vary for some makes according to temperature and speed range.

GEARMOTORS. Units that integrally combine an electric motor with speed reduction gears are referred to as gearmotors. Gears employed include helical, herringbone, and worm types. Some units have a planetary arrangement of helical gears. Speeds may be reduced to as low as 2 rpm with motors generally designed for 1750 rpm. Their size ranges from small control units of about $1/_{10}$ hp to large drives of over 100 hp.

The oil used to lubricate the gears of gearmotors also lubricates the bearings. The large units include their own pumps to supply oil to the bearings and gears. However, the type of gears, input speed (motor speed) and horsepower, and operating temperature govern the selection of the oil employed, as shown in Chapter 14. Where the gear-free end bearing of the electric motor is lubricated separately, the recommendations for electric motors should be followed for that bearing.

WINDINGS. Electric motors are usually equipped with an impeller at one end to force air over the windings for ventilation. Precaution must be taken during operation to see that oil is prevented from being drawn into the motor, saturate the windings, interfere with the electrical flow, and cause short circuiting, and ultimately, fire. Slingers help to deflect particles of oil

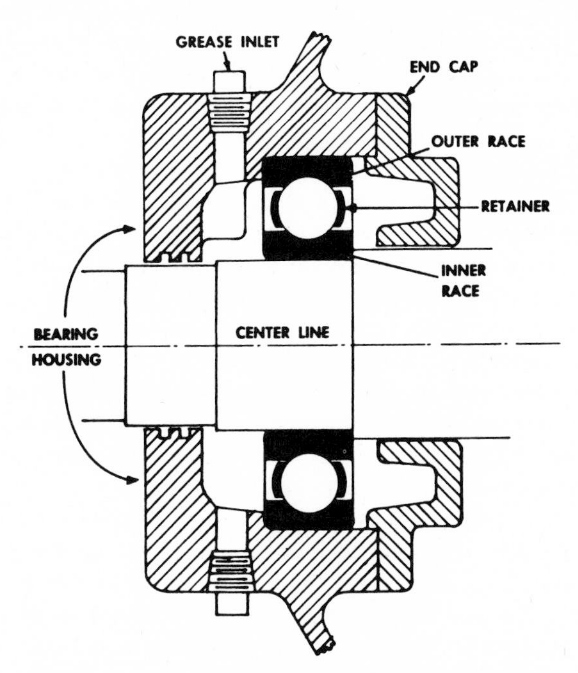

Fig. 21-1. Cross-sectional view of grease-lubricated ball bearing designed for electric motors. Note relief plug at bottom for removal of old grease. The grease content can be brought to a proper balance (excess grease removed) if the plug is left out for a short period after starting. *Courtesy, Allis-Chalmers.*

carried by windage, but they may interfere with air flow. Therefore, filling of the bearing cavity or other application device should be done carefully to avoid spillover or during shutdown.

ELECTRIC GENERATORS. The construction of electrical generators is basically the same as that of electric motors. The difference in operating principle lies in the effect of an externally driven rotor of a generator on the magnetic field as compared with a driving rotor of an electric motor.

Table 21-1. LUBRICANTS FOR ELECTRIC MOTOR BEARINGS

Oil-lubricated bearings (speeds up to 3600 rpm)

Ambient Temperature	*Viscosity* (SUS at 100°F)
Below 32°F	150 to 225 (pour < -15°F)
32 to 200°F	300

Grease-lubricated bearings (speeds up to 1800 rpm)

Ambient Temperature	*Consistency* (NLGI No.)
-10 to 32°F	0
32 to 200°F	2

The generator converts mechanical energy to electrical energy, whereas the motor converts electrical energy (from the generator) to mechanical energy.

Generators may be designed to produce either alternating or direct current, but not both in an individual unit. Horizontal generator bearings, such as those used on steam turbine generator sets, are lubricated with the same oil as the turbine and are generally served by a common pressure circulating system.

MOTOR-GENERATOR SETS. Some industrial processes and controls require direct current. Where alternating current is the universal current used for transmission and distribution, it becomes necessary to convert from ac to dc. While mercury-arc rectifiers are often used, especially in steel mills, the motor-generator set is also widely employed to attain such conversion. Motor-generator sets are also designed to change the frequency of ac, say from 60 cycles to 25 cycles.

Motor-generator sets are arranged by connecting one to the other in a common shaft broken only by the insertion of a flexible coupling between them. They are usually mounted on the same base. Several dc generators may be operated by a common ac motor, whereby the voltage may or may not differ for each generator.

The lubrication of motor-generator sets is similar to that described for electric motor bearings, and the same precautions related to windings should be taken.

Electric generators are also driven by internal combustion engines and hydraulic turbines.

HYDRAULIC TURBINES

Turbines actuated by a head of water are of two basic types: impulse and reactive. The most suitable arrangement for reactive turbines is the vertical shaft which necessitates the use of thrust bearings as well as guide bearings. Some of these bearings include strips of lignum vitae or phenolic materials and are lubricated with water, but a rust and oxidation inhibited oil is also employed, using either the tapered land, tilting pad, or similar type thrust bearing. An emulsifying type of oil should augment water in the lignum vitae bearings.

With another type of bearing designed for hydrostatic lubrication, the rotating member is lifted from the stationary surface by external oil pressure.

Horizontal hydraulic turbines of the impulse type (Pelton wheel) are supported on horizontal bearings that may be considered to require similar oils as those used in horizontal steam turbines.

Hydroturbines actuated by adjustable-blade propeller (reactive) runners have the problem of keeping oil from leaking out of and water leaking into their hub. To counteract this condition, a heavy-bodied oil of about 1800

SUS at 100°F viscosity should be used. The lubricants re
hydraulic turbines and auxiliaries are given in Table 21-2

STEEL MILLS

The rolling of steel is one of the most dynamic processes of industry. Because of the operating conditions mentioned previously, selected lubricants effectively applied and carefully maintained are necessary for the protection of the various frictional components involved.

MILL DRIVES. Power for the roll stand is provided by large electric motors (±5000 hp) through reduction gears to a lead spindle connected to a pinion stand and thence through spindles to the work rolls. Connections to the pinion shafts and roll necks are made with universal joints or slipper couplings.

The lead spindle is connected to the bottom pinion of two-high mills and to the center pinion of three-high mills. The pinions are helical gears that divide the power between the rolls driving adjacent rolls in opposite directions. Table 21-3 lists typical lubricant viscosities and consistencies for mill drive components.

Roll Stand

The heart of the rolling process, the roll stand, consists of two housings or *posts* that support the rolls and *screwdown assembly,* the latter of which controls the space between the work rolls to a dimension of the desired thickness (gauge) of the strip (Fig. 21-2). Also associated with the roll stand are the necessary side guards, guides, roll changer, manipulators, tilting tables (where required) and operating controls. On some types of housings, windows are provided to facilitate roll changing. The major points of lubri-

Table 21-2. LUBRICANTS RECOMMENDED FOR HYDRAULIC TURBINES AND AUXILIARIES

Part To Be Lubricated	*Type of Lubricant*	*Method of Application*	*Approximate Viscosity (SUS) or NLGI Consistency*
Thrust bearings (above 32°F)	Turbine oil	Bath or circulation	300
Guide bearings	Turbine oil	Any	300
Kaplan running hub	Turbine oil	Bath	300
Governors	Turbine oil	Any	300
Wicket gates	Grease (calcium complex or lithium base)	Any	No. 1

Table 21-3. LUBRICATION OF MILL DRIVE COMPONENTS

Frictional Part	*Method of Application*	*Type of Lubricant*	*Lubricant Viscosity (SUS at 100° F) or NLGI Consistency*
Electric motor bearings	Circulation	Oil	300 R & O
Reduction gears			
Gears	Splash	Oil	1800–2200*
	Circulation	Oil	1500–1800*
Bearings	Any	Grease	No. 1 lithium or calcium-lead complex
	Any	Oil	1500–1800*
Pinion stand			
Pinions	Splash	Oil	1800–2200*
	Circulation	Oil	1500–1800*
	Any	(Same oil as reduction gears when combined with pinion)	
Pinion necks	Any	Oil	600–900
Spindle			
Carrier bearing	Gun	Grease	No. 1 lithium or calcium-lead complex
Couplings† and slipper bearings	Any	Grease	No. 1 lithium or calcium-lead complex
	Any	Oil	600–900 leaded compound
	Mist	Oil	600–900 leaded compound

* Leaded compounds preferred.
† Also see discussion of spindle couplings, Chapter 15.

Fig. 21-2. Roll stand of universal slabbing mill. Note vertical rolls and their drive spindles. *Courtesy, Mesta Machine Co.*

cation include the work roll bearings, backup roll bearings (Fig. 21-3) where present, screwdowns, screwdown drives, manipulators, and table bearings.

SCREWDOWNS AND SCREWDOWN DRIVES. The screwdown is a vertical screw and nut arrangement in which the bronze nut, weighing as much as five tons, is electrically driven to position the top roll for each pass through the mill, except where fixed passes are used, such as on a continuous mill. Screwdowns are usually supported on rolling bearings and may be driven hydraulically (older mills), or through a worm gear. On high-speed mills the rate of raising and lowering of the top roll may approach or exceed 40 ft/min, accomplished by an electric motor that drives the nuts up to 60 rpm and handles a load as high as 200,000 lb.

The screw and nut are subjected to the force of the rolling process and require a heavy bodied lubricant having mild extreme pressure characteristics, usually a leaded compound having a viscosity of 400 to 500 SUS at 210°F. It is important that the lubricant be applied in a manner to assure a protective film over the entire thread surface of both screw and nut during the turning operation. For screwdown drives with worm gears, a similar oil as that described for the screwdown is recommended. The screw is steel but the nut is bronze, having a composition of approximately 87% copper, 11% tin, 1.5% nickel, and the remainder of equal parts of phosphorus and lead.

WORK ROLL BEARINGS. The work roll is in direct contact with the piece being rolled and exerts pressure to reduce its thickness. Extending from each

Fig. 21-3. Four-high two-stand cold mill, showing ends of top backup roll. Screwdowns are shown extending vertically above the backup rolls. *Courtesy, Mesta Machine Co.*

end of the roll are necks that act as journals which rest on the bearings that support the roll. The backup rolls carry most of the load pressure, but this is not a major factor in the lubrication of work roll bearings. They are subject to high temperature because they are close to the hot steel passing between the rolls and to contaminants associated with steel mill operation.

Work roll bearings are usually lubricated with grease, which in modern mills is applied from a central system that services a number of other frictional components. The grease must possess high film strength and good thermal stability. The long feed lines associated with central systems or bulk grease stations require the grease to have excellent pumpability. Grease prepared for steel mill central system service should have partial rheopectic properties to enable it to be pumped long distances through restricted feed lines and to provide a seal for bearings during the shearing process to prevent entrance of contaminants, including water. Therefore, the grease should also resist the washing effect of water.

BACKUP ROLLS AND THEIR BEARINGS. One backup roll is located above the top work roll and the other below the bottom work roll of a four-high mill. They are large rolls compared with the work rolls, and their purpose is to prevent deformation of the latter by the screwdown pressure exerted between them during the rolling process, especially on wide pieces to maintain uniform thickness. The backup rolls are rotated by friction developed by contact with the work rolls which are driven by the mill drive. The force exerted on the top backup roll by the screwdowns causes high unit roll neck pressures as well as the continual generation of heat within the backup roll bearing assembly.

Back-up roll bearings are either roller bearings or specially designed sleeve types.

The roller bearing type is lubricated by grease usually applied from the central system that also serves the work rolls. Especially designed sleeve bearings equipped with roller or ball thrust bearings are lubricated with oil supplied from a circulation system. The oil is directed through the bearing assembly in such a manner as to attain maximum cooling and complete distribution to all frictional surfaces. Thus the oil has the dual function of a heat transfer medium and a lubricant.

A popular type of roll bearing is shown in Figs. 21-4 and 21-5, in which the difference in design for the thrust and nonthrust side roll necks is shown. For bearing sizes of less than 21 in., a ball thrust is used in place of the roller thrust used in larger bearings. The type shown includes a rotating sleeve that fits over and is keyed to the tapered roll neck and a nonrotating bushing mounted and secured in a protective chock. The oil film in the bearing is constantly maintained by the hydrodynamic action of the rotating sleeve, to which a surplus of oil is supplied at a controlled temperature. Note the neck seal that keeps the oil in the bearing and the roll coolant and scale out. The flinger directs oil discharged from the sleeve and bushing

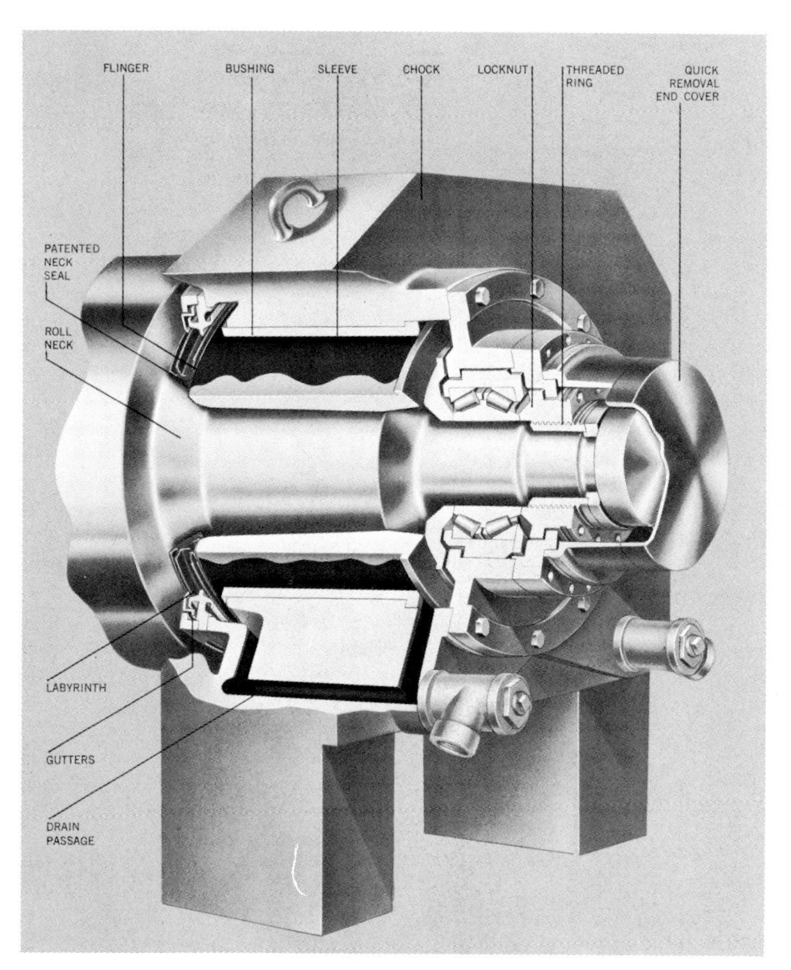

Fig. 21-4. Cutaway view of thrust side of backup roll bearing. *Courtesy, Morgan Construction Co.*

into a deep sump in the chock, from which is passes to the drain. A series of rotating and stationary gutters and deep interlocking labyrinths incorporating twin seals which rub on a chrome-plated surface prevent roll coolant from entering the oil and keep oil in the bearings. If some contaminant does succeed in getting by the seal, it cannot enter the bearing directly because it must first flow to the chock sump, from which it would drain to the receiving tanks of the circulation system and subsequently be filtered out. The oil is delivered to the journal through internal passages within the chock. The radial load is carried entirely by the oil film between the sleeve and bushing. The thrust assembly does not receive any of the load.

The bearing can be removed quickly by backing off the locknut, an action that pulls all components away from the roll neck taper.

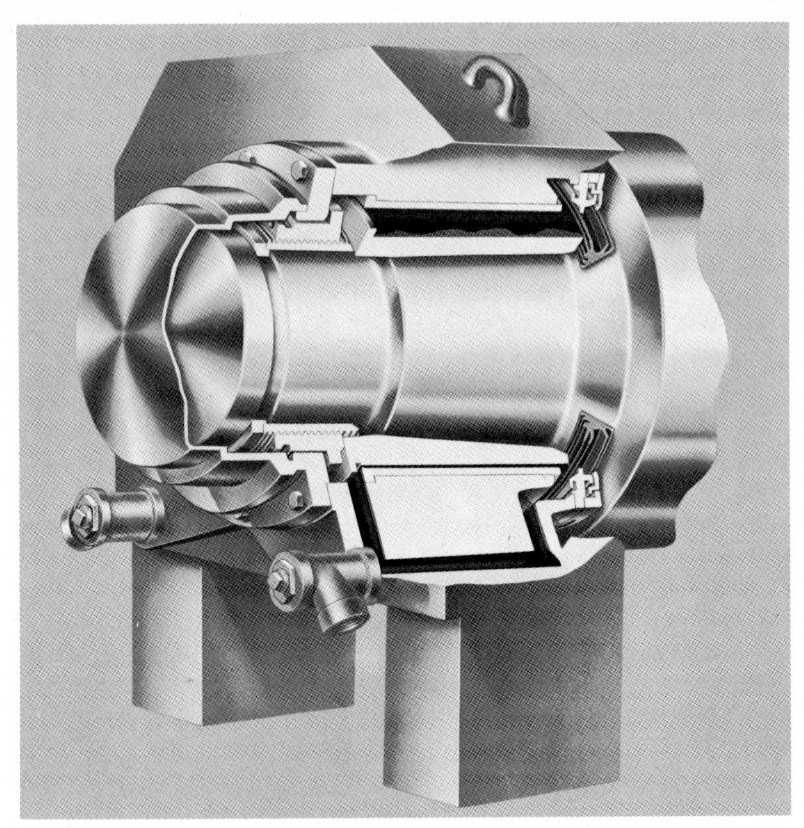

Fig. 21-5. Cutaway view of nonthrust side of backup roll bearing. *Courtesy, Morgan Construction Co.*

Where oil is used it is circulated, and where there are other frictional points being supplied from the same source the system is large, containing hundreds of gallons of oil. The circulating system is equipped with an oil cooler, the usual filters, and centrifuges to remove the water which is used copiously in primary mills to control the temperature of the rolls, or to remove furnace scale from slabs at the scale breaker and roughing stands. To purify the oil system, two tanks are installed to provide alternate batch settling. To maintain a safe water level in the large oil headers (usually located below the mills) and prevent its return to the bearings, the header should be slanted toward the intake and a sight glass installed as a tell-tale for periodic blow-down or line drain. The drain valves should be installed at the bottom end of water legs, several of which should be located at different points of the header. The recommendation concerning the header is important with a system in which the settling and centrifuging capacity are otherwise not as great as they should be.

HYDROSTATIC LUBRICATION OF BACKUP BEARINGS (Fig. 21-6). The control of strip thickness both as to gauge measurement and uniformity throughout its length and width is highly important. This is particularly true during cold reduction that follows hot rolling. Both the work and backup rolls are finished with a slight positive crown in order to improve strip shape uniformity. Another important factor in achieving uniformity is to maintain a stable roll temperature. This can be accomplished when the mill is operating at normal roll speed and roll bearings are effectively lubricated and cooled by the circulating oil under hydrodynamic conditions. Any increase in coefficient of friction due to a substantial reduction in speed or during startup results in an abnormal generation of heat within the bearing, and some of this heat is conducted to the roll body, where it causes dimensional changes. However, the roll returns to its original dimensions with a substantial increase in speed, and hydrodynamic conditions again prevail. These dimensional fluctuations, of course, have an effect on the uniformity of strip gauge.

To eliminate the loss of lubricant film during deceleration, an auxiliary hydrostatic lubrication system automatically becomes active when the roll speed is reduced to 250 ft/min, and it remains active while the mill continues to run below that speed and while it is stopped. When the mill is again brought up to speeds above the cutoff point of 250 ft/min, the hydrostatic system pump automatically shuts off and hydrodynamic forces again take

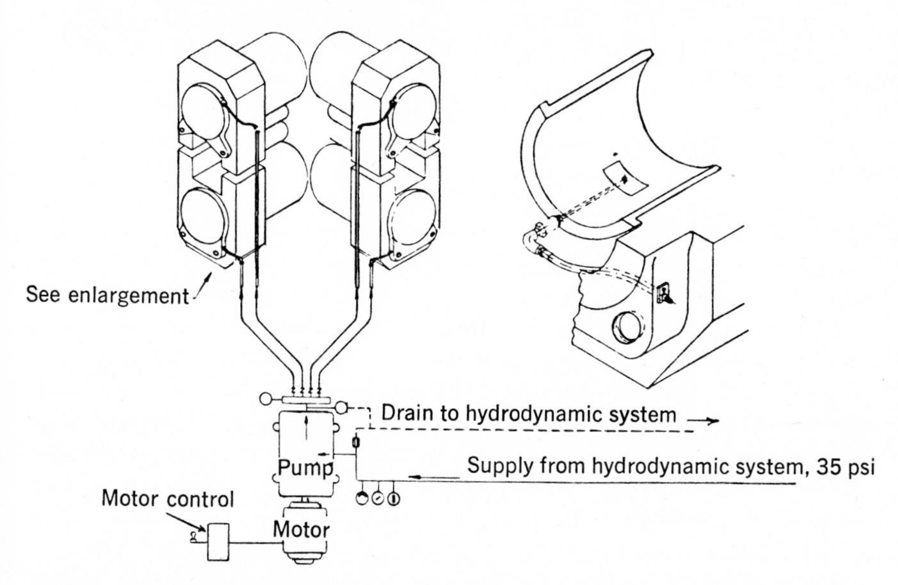

Fig. 21-6. Hydrostatic lubrication system for backup roll bearings. Oil is taken from hydrodynamic lubrication system through pump to load zone of each bearing. *Courtesy, Morgan Construction Co.*

over within the bearing. The hydrostatic system is not a separate entity, but uses the oil from the existing hydrodynamic circulation system by boosting its pressure to the required higher level and returning it to the same system via a common drain line upon leaving the bearing. If for any reason, the hydrostatic system fails, the bearings will continue to be protected in the same manner as they would if their lubrication depended solely on the regular hydrodynamic process. No claim is made that the augmentation of a hydrostatic lubricating system will solve the problem of nonuniformity of strip gauge, but any successful effort to reduce fluctuation in roll temperature would have some favorable contributary result. The major benefit derived is the achievement of lower cost backup bearing maintenance because of the protection of an adequate oil film during the otherwise critical periods of startup and speeds low enough that would otherwise considerably depreciate the bearing characteristic number (ZN/P).

LUBRICATING OIL FOR ROLL NECK BEARINGS. Some water enters the circulating oil system servicing roll neck bearings despite the efforts to keep it out. The oil must therefore possess the ability to separate readily from water either in the centrifuge or the settling stage of the circulation system. To encourage rapid water separation, the oil is free of additives that would cause strong emulsions. New oil should have a minimum Herschel demulsibility of 1620 and contain a film strength additive to meet the high pressures involved, especially with backup roll bearings. To prevent appreciable changes in viscosity, the oil used must possess a high viscosity index, preferably above 90.

Unlike many other frictional components, the viscosity requirements of roll neck bearing oils cannot be standardized. Mills in many cases are designed to meet given rolling requirements, so that the bearing designer or manufacturer is the only one capable of making viscosity recommendations commensurate with size, pressures, cooling capacity, and other operating factors. Although oils for roll neck bearings are made in a wide range of viscosities up to about 3000 SUS at 100°F, the selection of an individual viscosity for a new mill by any one other than the builder cannot be done with precision.

MIST LUBRICATION OF ROLL NECK BEARINGS. The lubrication of tapered roller roll neck bearings of both work and backup rolls requires an oil viscosity of about 750 SUS at 100°F or higher for large bearings and 1000 SUS at 100°F or higher for high-speed bearings. Oil should be of high quality and high viscosity index, derived from solvent-refined petroleum, and should have a Timken OK Load test of at least 35 lb. They must also possess good thermal and oxidation stability and be nonfoaming. Oils designed for mistability for high total density and such anticlogging characteristics are preferred. Such oils with viscosities of 1000 to 1500 SUS at 100°F are available. Those with the higher viscosity have better surface adhering qualities and produce less stray mist. It has been found that by heating the pressurized air better results are obtained.

MANIPULATORS. Straightening, moving, and turning the piece on the roller table to position it for each succeeding pass is the function of the manipulator. This may consist of stationary fingers that appear up through the rollers to engage the piece on the entering side of the rolls as the roller table is lowered. A second type of manipulator consists of arms attached to side guards that reach across the roller table from either or both sides to manipulate the piece and place it in the desired position.

The moving arms are powered by an electric motor through gears of either the open or enclosed type. Some older types are powered hydraulically. Table 21-4 lists the lubricants recommended for the various frictional components of manipulators of the moving type.

Table Rolls

To advance the piece to be rolled to and from the roll stand, tables made with a series of rotating rollers are used. The rollers are supported on their frame by rolling bearings (modern mills), each roller driven by a miter gear mounted on a line shaft turned by an electric motor. The motor gears are lubricated by splash with a leaded compound having a viscosity of 1500 to 2000 SUS at 100°F or, because of water contamination, the same oil used for the backup roll bearing can be used.

One problem associated with the entrance of water into the gear reservoirs is that the oil eventually overflows. Efforts to eliminate overflow include the use of leaded compounds containing fillers, such as clay, to increase the weight of the lubricant above that of water so that it would be the latter that overflows while the lubricant maintains its proper level. The table roller bearings are usually of the rolling type and are lubricated by the same grease used in other grease-lubricated bearings of the mill, applied either by gun or from a central system (Table 21-5).

Runout tables are those that approach the coilers, the final phase of the hot strip rolling process. The rollers of the runout tables are individually

Table 21-4. MANIPULATOR LUBRICANTS

Part To Be Lubricated	*Method of Application*	*Type of Lubricant*	*Viscosity or NLGI Consistency*
Bearings	Gun	Grease	No. 1 or 2 lithium or lime-lead (EP) grease
	Any	Oil	600 SUS at 100°F machine oil
Enclosed gears	Splash	Oil	AGMA No. 6 or 7 leaded compound
Open gears and racks	Brush or spray	Residual	700–900 SFS at 210°F undiluted
	Slush pan	Residual	400 SUS at 100°F undiluted
		Leaded compound	120–130 SUS at 210°F

Table 21-5. ROLLER TABLE LUBRICANTS

Part To Be Lubricated	*Method of Application*	*Type of Lubricant*	*Viscosity (SUS) at 100° F or NLGI Consistency*
Table roller bearings	Any	Lithium or lime-lead grease	No. 0 or 1
Line shaft bearings	Gun	Lithium or lime-lead grease	No. 0 or 1 1500–2000*
	Bath	Leaded compound or straight mineral oil	Same oil used in backup roll bearings
Gears	Bath or circulation	Leaded compound	1500–2000*

* Lower viscosity preferred where fluid is circulated.

driven by variable-speed direct current motors. The strip is cooled with water sprays both above and below the rollers. It is important that the grease used for the bearings of the rollers be resistant to water wash.

COILERS. The coilers automatically reel the strip after it has passed through the last finishing mill and cooling spray. The strip enters between the rotating reel, and expanding segments equipped with rollers guide the strip about the reel. As the strip leaves the last finishing stand, it travels at about a mile a minute on some modern rolling mills. This means that the end of the strip possesses considerable whip as it changes direction upon entering the coiler. Care must be taken to protect or isolate from the flapping end any lubricant (grease) feed lines leading to the various frictional parts of the coiler. Two or more coilers are provided that are alternately used to reel the next piece in turn as each completed coil is unloaded.

Cold Reduction of Strip Steel

After the hot strip is coiled, cooled, and uncoiled, it is passed through a pickling process (thin sheets are annealed) and recoiled. It is then again uncoiled and fed through a series of four-high tandem cold reduction stands (Fig. 21-7) to produce the required gauge, having previously been coated with a palm oil or similar fluid to protect its surface after pickling.

The cold reduction stands are similar to the hot strip finishing stands already described, so that their lubrication need not be repeated here. One possible source of contamination associated with cold reduction is the presence of fine metallic dust in the atmosphere between roll stands. This can be abrasive upon entrance to the work roll and backup roll bearings. One way to prevent this is to install a roving magnetic screen within the space indicated above for the collection and automatic removal of such fines.

Critical areas of roll stands are the friction plate surfaces between the work roll bearing chocks and the housing, where some vibration continually occurs, resulting in wear. The wear causes the formation of metal pieces

Fig. 21-7. Five-stand four-high cold mill. Note coils in foreground. *Courtesy, Mesta Machine Co.*

that break out of the friction plate surfaces and form a rough ball-joint, creating a lock that often prevents easy removal of the work rolls at the time of roll change. The wear is eliminated by swabbing the friction plate surface prior to the installation of a replacement roll with a heavy, black residual type lubricant, such as recommended for open-gear lubrication.

Other Steel Mill Equipment

The integrated steel plant contains a large variety of mechanical equipment, from the skip-hoist that brings up the raw materials to the bell of the blast furnace to the lift-trucks and cranes that carry the finished product to storage. Much of such equipment is an assembly of basic frictional components, such as the bearings, gears, and cylinders described previously. Except for temperature conditions and the presence of water in some areas, most of the lubricants used may be selected using conventional standards. Synthetic lubricants or hydraulic fluids must sometimes be considered because of the temperatures involved, such as for the dolomite and ore pushers in the open hearth. Other frictional components subject to high temperature are the ingot strippers, soaking pit crane tongs, ingot buggies, and coke pushers.

Using the lubricating principles described for the hot strip mill and in previous chapters will assist in understanding many problems associated with other steel product manufacturing processes and equipment not covered here. Other equipment associated with flat steel piece production includes tilting table mechanism, shears, levelers, coil elevators, pallet conveyors, turntables, and reheat furnace components. Other processes that employ frictional equipment include the pickling line, tinning line, and annealing of thin strip.

PAPER MILLS

The functions of the integrated paper mill can be divided as follows:

1. Conversion of raw material to pulp
   (a) from wood (mechanically or chemically)
   (b) from rags
2. Preparation of *stuff* from pulp and other paper ingredients, including clay, sizing, alum, and coloring matter.
   (a) beating
   (b) refining
3. Paper making
   Wet end
      Fourdrinier, or
      Cylinder part
      Press part
   Dryer part
      Dryer rolls
      Dryer felt rolls
   Finishing part
      Calenders
      Reels, winders, slitters, cutters, and lay boys
   Drives

Conversion of Raw Material to Pulp

Before the logs are converted to pulp, their bark is removed and reduced to chips or granules. Rags are freed of loose dust and dirt and cut into small pieces. The equipment for preparing wood includes barkers, slashers, splitters or saws, jackers, chippers, rechippers, hogs, chip crushers, conveyors, deckers, centrifugal screens, shaker screens, and pumps. Thrashers, cutters and secondary dusters are employed in preparing rags.

The frictional components of this machinery are usually of the conventional types. They include plain and rolling bearings, toothed and worm gears, chains and couplings. Any appropriate method of applying oil and grease is employed, including bath, oil cups, wick-feed oilers, ring-oiled systems, compression grease cups, and pressure grease gun.

The selection of lubricants must be made in consideration of the heavy and shock loads prevailing on such equipment as well as the presence of water. To meet the load conditions, intermediate bodied oils (600 to 900 SUS at 100°F) containing antiwear agents should be used, selected on the basis described for the particular component in previous chapters. Oils selected for wet conditions often require an emulsifying additive. Multi-purpose greases (described in Chapter 6) of appropriate consistency are preferred, including those having a lime-lead EP complex or lithium soap base, plus molybdenum disulfide capable of meeting both load and water conditions. Hydraulic grinder feed mechanisms usually employ water-soluble oil having a water-to-oil ratio of about 30:1. Table 21-6 shows typical lubricant recommendations for equipment used in preparing logs for pulp production.

Table 21-6. LUBRICATION OF EQUIPMENT FOR PRODUCING WOOD PULP

Part To Be Lubricated	*Type of Lubricant*	*Approximate Viscosity (SUS at 100°F) or NLGI Consistency*
Barker		
Bearings, rolling	Grease, multi-purpose	No. 1
Chippers		
Bearings, rolling	Oil, leaded compound	600
	Oil, R & O	900–1500
	Grease, multi-purpose	No. 2
Re-chippers		
Bearings, rolling	Oil, R & O	900–1500
	Leaded compound (winter) (low pour point)	600
Chip Crushers		
Bearings	Oil (reuse) R & O	600
	Oil (all-loss) SCO	2500
	Grease, multi-purpose	No. 1
Splitters		
Ways	Oil, way oil	300
Hydraulic system	Oil, R & O	300 or Mfgr. rec.
Screens, chip and knot		
Bearings	Oil, R & O	600
Gears	Oil, R & O	600
Screens, flat		
Bearings	Grease, multi-purpose	600
Screens, centrifugal		
Bearings	Oil, R & O	300
	Grease, multi-purpose	No. 2
Grinders		
Bearings, rolling	Grease, soda-soap (light oil & soap)	No. 4

R & O, rust and oxidation inhibitors; SCO, steam cylinder oil.

Rags are cooked in either a vat (kier) or a globe-shaped rotary digester to produce rag pulp. Special attention must be given the trunnion bearings and girth gear of rotary digesters due to the higher temperatures from the use of steam. The trunnion bearings act also as steam joints and require a compounded steam cylinder oil of about 2500 SUS at 100°F viscosity. They should be drained and cleaned at regular intervals (about every three months of operating time) to prevent the formation of deposits. If a block grease is used in place of oil, its penetration should conform to the steam temperature, which for steam at 20 psi would be about 40 penetration at 77°F. The girth gear is lubricated with a residual type lubricant that can be sprayed on at the pinion mesh point during operation (Table 21-7).

Preparation of Paper Stuff

Raw pulp is processed into fibrous "stuff" ready for the paper machine first by the beater and then by the refiner. The beater has the functions of (1) bringing the pulp to the necessary state of hydration, (2) mixing the required amount of clay, alum, sizing, and coloring pigments with the pulp, and (3) rubbing and pressing the fibers. The beater is an oval vat or tub divided except at both ends by a partition forming two connecting channels. In one channel, a beater roll, mounted in a driveshaft is installed. The roll shaft bearings are the only frictional components on a beater.

Older types of beaters are equipped with half bearings (bottom) to compensate for the lifting effect of the pulp passing between the opposing knives, an action that adds shock to an already heavy load of the roll. To meet this load combination, a grease having a consistency of No. 3 or 4 NLGI and a high drop point should be packed into the bearing so that it is carried to the pressure area with the rise and fall of the shaft. Because some splash prevails, a lime soap or lithium soap-base grease should be used to withstand the water washing effect. Block grease may also be used

Table 21-7. LUBRICATION OF EQUIPMENT FOR PRODUCING RAG PULP

Part To Be Lubricated	*Type of Lubricant*	*Approximate Viscosity (SUS) or NLGI Consistency*
Rag cutter		
Bearings	Grease, multipurpose	No. 1
Gears, open	Residuum	700 at 210°F Furol
Gears, enclosed	Oil, with rust and oxidation inhibitors	600 at 100°F
Digester		
Worm gear and rack	Oil, compounded cylinder	2500 at 100°F
Worm gear bearings	Grease, multipurpose	No. 2
Girth gear and pinion	Residuum	700 at 210°F Furol
Trunnion bearings	Block grease	40 penetration
	Oil (all-loss)	1800–2500 at 100°F

with good results providing it has an appropriate melting point. Later models are equipped with shock absorbers accompanied by full circular plain bearings. These bearings are best lubricated with an oil of about 600 SUS at 100°F viscosity applied by a wick-feed oiler which should be maintained at a high level to assure maximum flow. Rolling bearings are also used, and a No. 2 NLGI consistency lime-lead complex grease is recommended. Where provisions are made for oil, the same type recommended for wick-feed application is preferred. The beater is being gradually replaced by a vertical type pulper.

Vertical beaters. The beating element of the vertical beater is rotated by an electric motor through either bevel gears or hypoid gears located below the vat. This type of beater, identified as a Hydrapulper, is gradually replacing the beater described previously.

(Hydrapulper is a proprietary name of the Black-Clawson Company.)

Refiners. After leaving the beaters, the pulp is further conditioned to a uniform consistency during which the distance between rubbing elements is more accurately controlled and rotating speeds are higher. One of the older such refiners is the Jordan, which is gradually being replaced by the disk type refiner.

The Jordan has a main shaft supported by bearings located at both ends of the machine that are lubricated by wick-feed oilers or compression grease cups. The bearing at the adjusting end is a combination radial and thrust bearing.

The disk type refiner is a horizontal machine consisting of two or more disks, one of which is the floating type. Another type is made up of two disks, one or both of which rotate.

The stuff is transferred from one refining process to another by a series of pumps, during which it is subjected to several screening operations involving rotating cylinder screens, eccentrically vibrated flat screens, and centrifugal screens for removing heavy constituents. The conditioned stock is then introduced to the flow or head box of the fourdrinier paper machine or the vat of the cylinder machine.

Table 21-8 shows the basic types of equipment used in the stuff preparation phase of paper making and their recommended lubricants. Other machines either further refine the pulp or control the consistency of the stuff. Where the stock handled is above 150°F, the viscosity of the oils for refiners, including Jordans, should be increased to about 1500 SUS at 100°F.

Junk removers (listed in the table) operate in a twisting motion to remove wire threads and other debris from waste paper cartons, etc., after they have been broken from their bales and before beating.

Paper Machines

There are two basic types of paper machines, the fourdrinier and cylinder machines. A third type is a short fourdrinier that has a large drying

Table 21-8. LUBRICANTS FOR STUFF PREPARATION EQUIPMENT

Part To Be Lubricated	*Type of Lubricant*	*Method of Application*	*Approximate Viscosity (SUS at 100°F) or NLGI Consistency*
Beater			
Bearings	Oil	Any	600
	Grease	Gun	No. 2
	Grease	Packed	No. 3
Jordans			
Bearings	Oil	Any	600
	Grease	Cup or Gun	No. 2
Vertical beaters			
Gears, bevel	Oil	Circulation	900
Gears, hypoid	Oil	Circulation	SAE 90
Bearings, tailstand	Grease	Gun	No. 2
Pumps, centrifugal	Grease	Gun	No. 2
	Oil	Any	300–400
Screens			
Bearings (eccentrics)	Oil	Bath	400–600
Bearings (shaft)	Grease	Gun	No. 2
	Oil	Any	400–600
Junk removers			
Worm gears	Oil	Bath	2500 (compounded)
Open gears	Residuum	Any	100 at 210°F, Furol
Bearings	Grease	Gun	No. 2

roll about 14 ft in diameter to which the paper from the wire is transferred and pressed against its polished surface until dry. Both the fourdrinier and cylinder machines consist of a wet end and a dryer end. It is the wet end that represents the substantial difference in the two machines so that any description of the dryer end of one will be fairly descriptive of the other. Both the fourdrinier and cylinder machines have common lubrication problems posed by water washing at the wet end and elevated temperatures at the dryer end.

The paper machine is a series of rolls from one end to the other, all of which must be driven either (1) directly from a power source through appropriate gears, variable-speed transmissions, belts and chain drives, (2) by gear trains from a power driven roll shaft, or (3) by friction between adjacent rolls (Fig. 21-8).

Fourdrinier Machine (Fig. 21-9). The name fourdrinier applies to the wet end of the paper machine on which the stuff flows from a *head box* (reservoir) to a wide, endless belt of fine wire mesh driven by a *lower couch roll.* The wire travels upward over a *breast roll* and then forward over a series of small *table rolls* and downward over *the bottom couch roll,* returning to the breast roll over *carrying, guide,* and *stretch* rolls. A deckle strap prevents the stock from running off the side of the wire.

Fig. 21-8. Paper mill drives showing drive shaft, V-belts, variable-speed transmissions, and gear reducers. *Courtesy, Sandy Hill Corp.*

Fig. 21-9. Paper machine, Fourdrinier part (wet end). Head box, table roll, deckel strap and part of the press section are shown. *Courtesy, Sandy Hill Corp.*

The pulp web leaves the fourdrinier part to enter the press section of the wet end, where it is supported on a belt of endless felt that carries it through the first, second, and third set of *press rolls* and on to the dryer rolls. The press rolls squeeze a large amount of the remaining water from the pulp web. The felt is supported and guided through the machine by a series of felt tighteners and *felt rolls.*

After leaving the final set of press rolls, the pulp enters the *dryer section* of the machine, where the remaining water is removed by a series of steam-heated rolls.

Cylinder Machine (Fig. 21-10). The cylinder machine varies from the fourdrinier in that the fine mesh wire is replaced by a series of cylinder vats to which the pulp and water mixture is introduced to the machine. The number of cylinder vats depends on the thickness of the paper being made. A cylinder mold rotates within each vat. The cylinder mold consists of a cylindrical framework over which a coarse wire mesh is wound. This, in turn, is covered with a face wire of fine mesh. The water passes through both wires, while the pulp adheres to the face wire during rotation of the mold.

Lubrication of Wet End. The various rolls of both the fourdrinier and cylinder machines are supported by either plain or rolling bearings. Modern machines employ rolling bearings that are oil or grease lubricated. The trend toward oil prevails with higher paper speeds of fourdrinier machines, but experience has proven that grease gives satisfactory results for paper speeds up to 4000 ft/min on a Yankee machine. The ultimate limit may not be represented by this operation, however. When oil is employed, it is applied from a central circulation system. The viscosity of the oil used for rolling bearings should conform to the values listed in Table 13-5. On machines where the viscosity requirements of wet end bearings are similar to that required for the dryer bearings, a common system may be utilized. However, if a higher viscosity oil is necessary to meet high temperature conditions prevailing at the dryer bearings than that required at the wet end, separate systems must be provided in each case.

Where grease is used, consideration must be given to its resistance to water wash and high chemical stability, both of which are important to provide good performance over extended periods of operation. Effective seals should be used and protected against damage. The author prefers a lithium base or calcium EP complex grease of No. 1 NLGI consistency applied once a week for continuous operation while maintaining the rule of keeping the bearing $\frac{1}{3}$ to $\frac{1}{2}$ filled. Under severe operating conditions, the bearings should be purged regularly.

On older machines using half bearings to support table rolls, a lime-base grease may be applied by hand. However, on such machines that also employ pillow blocks with rolling bearings for other table roll bearings, a lithium or calcium EP complex grease should be applied to all bearings.

The bearings of rolls operating on the wet end of cylinder machines are lubricated with grease applied by pressure gun. The rules applying to grease-lubricated wet end bearings of fourdrinier machines also apply to the cylinder machines.

Fig. 21-10. Cylinder paper machine having multiple cylinders (wet end). *Courtesy, Sandy Hill Corp.*

The bearings of the press roll (Fig. 21-11) may be lubricated with grease in the case of rolling bearings and oil for plain bearings. The same grease recommended for bearings of the fourdrinier part is also recommended for the press roll bearings. For plain press roll bearings, a 700 SUS at 100°F oil containing a small amount of emulsifying agent is recommended. For all-loss systems, oil application is provided by a wick-feed or drop-feed oiler.

The load applied to the press rolls is usually provided by a hydraulic system in which the oil viscosity depends on the type and make of pump used.

Dryer Section (Fig. 21-12). The dryer section of any paper machine consists of a number of steam-heated hollow cylinders so arranged that the pulp web from the press section passes over each roll in turn as it travels toward the finishing end of the machine.

The friction parts of a dryer include the dryer roll bearings located on the front end of the machine, and the steam joints and gear train, both open and enclosed, located at the back of the machine (Figs. 21-13 and 21-14). Because of the steam, all frictional components operate in a high-temperature environment. Because of the high temperature, oil must be used to lubricate the dryer bearings and the steam joints that serve as back roll bearings through which the steam enters the dryer cylinder. The dryer cylinder gears of the open spur type are lubricated with a residium applied either by swab or spray. More modern machines provide oil circulation from a central system for the enclosed gear lubrication. The front roll bearings

Fig. 21-11. Press section of paper machine. *Courtesy, Sandy Hill Corp.*

Fig. 21-12. Dryer section of paper machine. *Courtesy, Sandy Hill Corp.*

are lubricated with oil applied variously by circulation from the same system that serves the enclosed gears or from a separate system when open gears are employed. Oil application to the dryer bearings is also made by mist oil lubricators, each serving 8 to 16 bearings. Older machines employed block grease or oil from a wick-feed cavity forming part of the bearing housing to lubricate dryer bearings.

Steam joints are either grease or oil lubricated. Rolling bearing types are equipped with grease fittings for application by pressure gun, and they require a high-temperature nonsoap grease. If spring-sealed steam joints are used, they are lubricated with the same oil as used for the front dryer cylinder bearings applied by oil cups or some other all-loss arrangement. When served by an oil circulating system, the joint should be equipped with a dry sump. Table 21-9 summarizes the lubricants recommended for the dryer section.

Fig. 21-13. Dryer section, open gear trains. *Courtesy, Sandy Hill Corp.*

Finishing Section. The finishing section of the paper machine follows the dryer section and consists of calender slitters and winders or rolls. If the paper is to be cut to lengths, cutters and lay boys are used. The lubrication of reels, winders, slitters and cutters involves conventional lubrication methods of applying grease or oil.

The *calender* (Fig. 21-15) consists of rolls stacked uniformly, one directly above the other, through which the paper passes to be "ironed" or polished. The bottom roll is driven from an external power source. Each of the remaining rolls is rotated by its adjacent lower roll by friction originating at the power roll. Lubrication of the roll bearings is provided by oil that cascades downward through each bearing in turn from a reservoir located at the top of the calender. The used oil drops into a sump from which it

Fig. 21-14. Dryer section, enclosed gears. Note steam joints. *Courtesy, Sandy Hill Corp.*

is pumped back to the reservoir. A variation of that system provides a separate bypass line to feed individual bearings to eliminate the cascade principle, which tends to carry abrasive particles from one bearing to another. The oil used is generally the same as used for the lubrication of the dryer bearings.

Table 21-9. Lubricants for Dryer Section

Part To Be Lubricated	*Type of Lubricant*	*Method of Application*	*Approximate Viscosity (SUS) or NLGI Consistency*
Dryer bearings			
Steam pressure			
Below 70 psi	Oil	Any	600–650 at 100°F
Above 70 psi	Oil	Any	900–950 at 100°F
Steam joints	Oil	Any	Same as dryer bearings
	Grease, nonsoap	Gun	No. 1
Dryer gears, open	Residuum	Any	700 SS Furol at 210°F (without diluent)
Dryer gears, enclosed	Oil	Circulation	Same as dryer bearings

Fig. 21-15. Paper machine calender. Note lubricant distribution system. *Courtesy, Sandy Hill Corp.*

RUBBER MILLS

Separate processes to convert the natural or synthetic rubber polymers to finished products include mastication, mixing, forming, and curing or vulcanization. Most of the processes are applicable to plastics, linoleum, and tile. The process of breaking down the polymer is called mastication. This breakdown operation is performed on a rubber mill (Fig. 21-16) consisting of two closely spaced parallel rolls operating at different speeds between which the solid polymer passes. Modern methods employ dies through which the polymer is forced to accomplish the same purpose. The major parts of a rubber mill to be lubricated are the roll neck bearings, roll adjustment mechanism, and the gear train.

On most mills, the gear train consists of an open bull gear and pinion driven from an enclosed gear reducer. The bull gear and pinion in turn drive the connecting gears that rotate the rolls. The space between the rolls is controlled by adjusting screws or worm gears.

The bull gear and pinion are lubricated when the lower teeth pass through an oil slush pan. Connecting gears are similarly lubricated on some mills, but on others the oil is circulated. The sleeve type roll neck bearings are either oil lubricated, by mechanical force-feed lubricator or circulation, or grease lubricated, by hand pressure gun or other device. Rolling type roll neck bearings are lubricated by circulating oil. Roll adjustment screws usually require grease applied by pressure gun. Adjustment worm gears are lubricated either with grease, or, when a suitable housing is provided, with oil. Table 21-10 shows the lubricants recommended for rubber mill machinery.

Mixers. The mixer is similar in some respects to the breakdown mill. However, it is in the mixer that the various required ingredients are added and dispersed within the plastic mass. The Banbury Mixer®, (registered trademark of the Farrel Corp.) is a modern machine that performs the same task as the older type mixer in a much shorter time. The Banbury Mixer (Fig. 21-17) consists of the frictional parts shown in Table 21-11.

Forming

From the mixing mill, the rough sheets are variously formed by calendering, extruding, or molding. Extruders form long lengths of any desired cross section by forcing the sheets through suitable dies. Articles are molded into desired shapes by pressing the rough sheets in molds. Rough sheets

Fig. 21-16. Rubber mill with direct drives, adjustable speed and unit base. *Courtesy, Farrel Corp.*

Table 21-10. LUBRICANTS FOR RUBBER MILL EQUIPMENT

Part To Be Lubricated	*Type of Lubricant*	*Method of Application*	*Approximate Viscosity (SUS at 100°F) or NLGI Consistency*
Gear reduction units	Oil	Splash	2500
(enclosed)	Oil	Circulation	1500–2000
Bull gear and pinion			
Under 150°F	Leaded compound	Splash	2500
Over 150°F	Oil	(slush pan)	4000+
Connecting Gears		Splash (slush pan)	
Under 150°F	Leaded Compound		2500
Over 150°F	Oil		4000+
Roll Neck Bearings			
Sleeve type			
Under 150°F	Oil	Any	1800
Over 150°F	Oil	Any	4000+
Any temperature	EP grease (calcium-lead complex or lithium EP)	Any	No. 1
Rolling type			
Under 150°F	Oil	Circulation	2500
Over 150°F	Oil	Circulation	6000
Roll-Adjusting Mechanism			
Screws	Grease	Gun	No. 1
Worm	Grease	Gun	No. 1
	Leaded compound	Bath	2500

are formed into finished sheets of a given dimension by rolling the material between calender rolls.

Extrusion and molding machines are usually powered hydraulically. The oil selected depends on the kind of pump and pressures involved, ranging between a viscosity of 150 to 300 SUS at 100°F. Other parts, including gears, screws, ways and bearings, usually require conventional lubricants as described for such components in previous chapters. The same recommendations generally apply to the friction components of calenders with the exception of sleeve type roll bearings, which require oils having viscosities based on the temperature of the material handled, method of application, and whether or not the bearings are watercooled. Oils for calender roll bearings range from 900 to 6000 SUS at 100°F and are of the turbine type.

A four-roll inverted calender showing lubrication feed lines leading to the roll bearings is shown in Fig. 21-18.

SUGAR MILL

Sugar mill (Fig. 21-19) roll journal bearings present some of the most difficult problems of lubrication. The unit pressures in some mills exceed 1500 psi at rotating speeds as low as 1 rpm, and as a result, their tempera-

Fig. 21-17. Banbury mixer®. *Courtesy, Farrel Corp.*

Table 21-11. LUBRICANTS FOR BANBURY MIXER®

Part To Be Lubricated	*Type of Lubricant*	*Method of Application*	*Approximate Viscosity (SUS at 100°F) or NLGI Consistency*
Combined reduction unit, bull gears, and connecting gears	Oil	Splash or circulated	3000
Rotating weight piston rod	Oil	Air line oiler	300–500
Floating weight piston packing	Oil	Hand or air line oiler	500–700
Discharge door piston packing	Oil	Hand or air line oiler	500–700
Dust stops	Special oil containing plasticizer	Mechanical force-feed lubricator	
Rotor bearings (sleeve)	Grease	Grease lubricator	No. 1
Rotor bearings (rolling)	Grease	Packed	No. 2
Rotor bearings	Oil	Circulation	900

Fig. 21-18. Four-roll inverted calender. *Courtesy, Farrel Corp.*

tures are high. The high pressure on the roll journals is developed by hydraulic power through cylinders located above the top mill bearing. The high pressure is required in order to squeeze the maximum amount of juice from the sugar cane as it passes through the mill.

Modern sugar mills are driven by turbines through a series of gear sets as shown in Fig. 21-20. The gear arrangement consists of a high-speed double reduction unit followed by a second unit containing the intermediate and main gear sets. All gear sets are self-contained enclosed units, completely sealed and lubricated from their own oil reservoirs. Table 21-12 lists the lubricants required by the sugar mill.

WIRE ROPE

Wire rope, or cable, is composed of strands of metal wire laid up evenly around a core. The core supports the strands and in some instances provides internal lubrication. A 6 × 37 wire rope consists of six strands, each strand having 37 wires (Fig. 21-21). If the core is another strand instead of fiber, the first number is increased by one. Flexibility is achieved by increasing the number of wires in a strand or the number of strands in a rope, or both.

The core material is usually sisal (hemp), making up an extremely hard-laid rope. Sisal provides a resiliency that reacts with the motion of the strands as they respond to cycles of stresses from no-load to maximum load and back to no-load. Asbestos cores are used in rope made for high-temperature service, such as ovens. To obtain greater strength a wire rope is used as the core. Absorbent fiber cores provide some lubrication to the inner section of the rope.

Wire Rope Lubrication

The wire rope is a machine with moving parts in the form of wire strands, and they require lubrication in order to retain their strength and flexibility. The lubrication of wire rope is complex and often neglected and

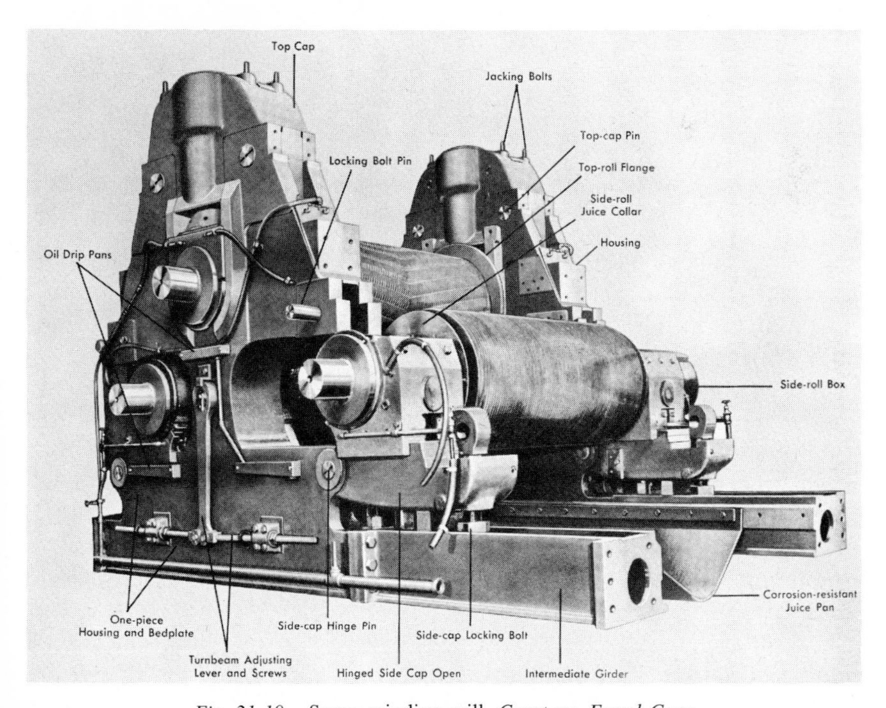

Fig. 21-19. Sugar grinding mill. *Courtesy, Farrel Corp.*

Fig. 21-20. Self-contained gear set for turbine-driven sugar mill. *Courtesy, Farrel Corp.*

widely misunderstood. It may be described according to the several phases involved:

The core is lubricated during its manufacture. Fiber cores are soaked in a chemically stable lubricant possessing penetrating ability. The core also receives additional lubricant during the stranding operation. The wires and strands are automatically lubricated at the die of the stranding machine during fabrication. This initial lubrication serves to protect the wire rope against rust by coating them and filling any spaces in which moisture could settle. It also lubricates the rope during its early use.

To maintain the rope in good condition, a system of intermittent or continuous lubricant application should be followed. Some of the important lubrication rules are as follows:

(1.) Steel strand cores should be bathed in the same lubricant as used on the finished rope during fabrication.

Table 21-12. LUBRICANTS FOR SUGAR MILL AND SUGAR MILL DRIVE

Part To Be Lubricated	*Type of Lubricant*	*Method of Application*	*Approximate Viscosity (SUS) or NLGI Consistency*
Roll journal bearings	Mineral oil preferred	Mechanical Force feed	400 at 210°F
Hydraulic system	Oil	Circulation	300 at 100°F, or as recommended by pump builder
Grease-lubricated parts	Grease, lithium or calcium complex, EP	Gun	No. 2
Turbine bearings and Single-reduction gears on common system	Oil	Circulation	300 at 100°F
Single-reduction gears, self-contained	Oil	Splash	600 at 100°F
Double-reduction gears, self-contained	Oil	Splash	600–900 at 100°F
Main reduction gears, self-contained			
Second reduction	Oil	Splash	700–900 at 100°F
Third reduction	Oil, leaded compound preferred	Splash	400 at 210°F

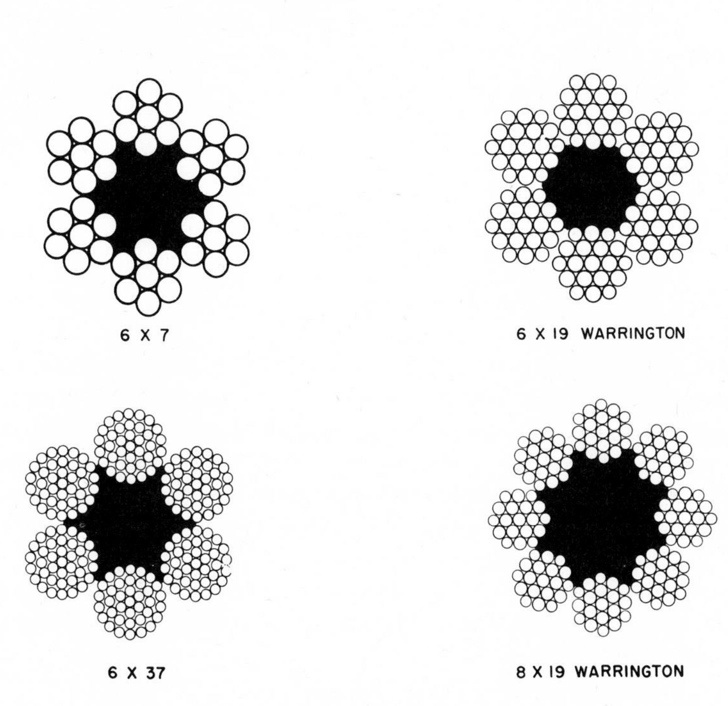

Fig. 21-21. Cross sections of wire rope.

(2.) The lubricant used during the stranding operation must be fluid enough to flow into all spaces to completely cover the wire surfaces. It should have good adhesive and cohesive qualities. The usual practice is to heat the lubricant and maintain the bathing temperature during the stranding and closing processes, wiping the excess lubricant away from the rope as it passes through the die. An important property of wire rope lubricant, especially during fabrication, is its ability to "set up" rapidly so that drippage does not occur during transportation and storage of the coil. Some lubricants, such as petrolatum, have a natural tendency to congeal readily, while other types require an additive, such as a small amount of wax (about 2%).

(3.) Cables in service should be treated periodically with penetrating oil, by immersion, if practical, to assure continuous core lubrication.

(4). The penetrating oil treatment should be followed immediately with a coating of the wire rope lubricant used. This may be accomplished by installing a trough containing the lubricant through which the rope passes as it is being wound on its drum. It is advantageous to use the same lubricant in service as was used during fabrication.

(5.) The lubricant may be applied intermittently or continuously, depending on the type of service.

(6.) Wire rope lubricants must not surface harden, which entraps moisture within the rope and causes corrosion.

(7.) Cables subject to insurance or government inspection are usually lubricated with light-colored lubricants for the sake of observing internal physical conditions.

(8.) Cable should not be dragged over the ground, especially where there is mine debris, sand, etc. Use sheaves, rollers, or lubricated troughs. Metal balls, such as those used in ball mills, placed in a trough make good anti-friction rollers for moving cable provided they are laid the entire length of the trough to support inclined or horizontal slack during off-loading.

(9.) Proper sized sheaves should be used and the rope should not be allowed to kink or be flexed too sharply.

When properly lubricated, the destructive forces tending to shorten the service life of wire rope will be greatly diminished. Such forces include metal fatigue, frictional wear, abrasion, corrosion and cross-nicking due to failure of a deteriorated core to support the strands. Cross-nicking occurs on the outer wires of the strand next to the core and results from overlap of the strands when the core collapses for any reason, such as lack of lubricant. It is commonly believed that the extra strength (about 7.5%) provided by wire cores over fiber cores is responsible for the longer life of the former. The actual reason for longer life of wire core ropes is the added support they give to the strands, which provides greater resistance to crushing.

WIRE ROPE LUBRICANT. The lubricant used for the protection of wire rope should possess the following characteristics:

(1.) Film strength to withstand the high rubbing pressures exerted between the wires in contact under load.

(2.) Adhesiveness, to the point of tackiness. This property not only assures a bond between the lubricant and wire surface, but also provides traction necessary in preventing slippage on the traction or hoist drum. Traction is particulary important for elevator cables.

(3.) Cohesiveness, to prevent drippage during storage, transportation, and service, and to eliminate throw.

(4.) Imperviousness to water to minimize lubricant loss when cable exposed to rain or other wet environment. Cables used in oil field drilling are often subject to attack by the chemicals present in "black water." Lubricants should be resistant to such solutions. Marine cables are usually zinc coated to protect them against rust and salt water attack. Highway guard cables and guy ropes are usually galvanized.

(5.) Fluid enough to be applied cold in the field so that it will creep through the entire wire complex.

(6.) Resistance to 140–160°F heat without change in lubricant structure. Cold weather often requires heating the fluid, and the process is standard on some stranding and closing operations during fabrication. To meet the exigencies of both the manufacturer in the plant and the user in the field, the same brand of lubricant may be used provided one of greater consistency or viscosity is supplied in the field. The difference lies in the prevailing field temperature as against the maximum fabrication temperature of 160°F.

The kind of lubricant chosen depends to a large extent on the service to which the rope will be exposed. The most demanding uses of wire rope are hoisting and hauling. Lubricants for such purposes are as follows:

1. Compounded oils—Heavy-bodied oils compounded with lead soap, preferably lead naphthanate. The viscosity selected depends on the temperature of operation.

2. Grease—Excellent results are obtained with a grease made up of black oil thickened with lime soap. In mines where temperatures are fairly constant at some level between 50 and 70°F, the grease should approach a semifluid state at point of application. If the hoist house is above ground, this means a grease of higher consistency should be used in the summer than during the cold months. When applied during rope fabrication, the heavier product should be heated to 140°F maximum. To assure rapid set, the grease must contain a suitable additive, such as a small amount of paraffin wax or rosin.

3. Petrolatum—A good wire rope lubricant that can be applied by heating to about 150°F in a bath or by spray. It may be brushed on at temperatures above 70°F if occasionally preceeded by the application of a light penetrating oil to protect the core and inner wires and displace any moisture.

4. *Other lubricants*—Because the lubrication of wire rope is so often neglected, especially in industrial plants, almost any lubricant should be used rather than none at all. If the question is one of economy or the reluctance to stock a special lubricant for the purpose, old oils drained from hydraulic

systems, gear cases, etc., make effective wire rope lubricants. If water washed and filtered, such oils are suitable for a number of requirements, such as open chains and sprockets, chair carrier troughs, and rough gears.

PENETRATING OIL. Except when low-viscosity oils (say, under 600 SUS at 100°F) are used regularly, wire ropes should occasionally be treated to a soaking in a light-bodied oil containing an oiliness additive, usually referred to as a penetrating oil. This assures proper protection of the core and internal wires. When this is followed by the continuous application of the grease in a trough, the life of the rope may be tripled. In mines, for example, where the renewal of wornout and frayed cable means the expenditure of several thousand dollars, the extra care is profitable.

METHODS OF APPLICATION. Figure 21-22 shows five methods of applying lubricant to wire rope. The lubricant may be swabbed, brushed, or poured on by hand. Hand pouring or spray application should be done at the beginning of flexing about a sheave, at which point the rope strands tend to spread sufficiently to allow entrance of the lubricant into the inner wires. Swabbing by hand should not be practiced at the entrance of sheaves, drums, etc., because of the danger to personnel.

WIRE ROPE CLEANING. It is often necessary to apply an extra heavy coating of grease or petrolatum to wire rope to protect guide and support rollers against wear, or shredding in the case of wooden supports. If the coating hardens, preventing fresh lubricant from reaching the interior wires, entraps moisture that causes rusting, and interferes with periodic inspection, it should be removed. Such a coating is probably caused by the use of the wrong lubricant or an accumulation of dirt and debris.

The cleaning method selected should be suited to circumstances. Cleaning methods include:

Solvents can soften and loosen petroleum-base coatings. They should possess a high flash point and be nontoxic. The solvent process must be performed in the open air above ground to avoid the accumulation of combustible gases in enclosed spaces, as in mines. Such solvents include kerosene, mixtures of penetrating oil and kerosene (in about 1:4 ratio) and emulsified kerosene.

The softening process is more effective if the cable is allowed to soak. Otherwise, the solvent should be applied several times until the desired results are obtained. The softened coating may then be removed by wiping, by centrifugal force during operation of the rope, or by blasting with compressed air or steam. When steam is used, the rope must be thoroughly dried afterward to remove any condensate. Drying is best accomplished with compressed air at high velocity.

Steam can also be used alone to remove wire rope coatings. Applied by jets, the steam loosens the coating by *heating*.

Plants with a degreasing pit in their production line possess an ideal wire rope cleaner. The rope should be exposed to the trimethylene chloride

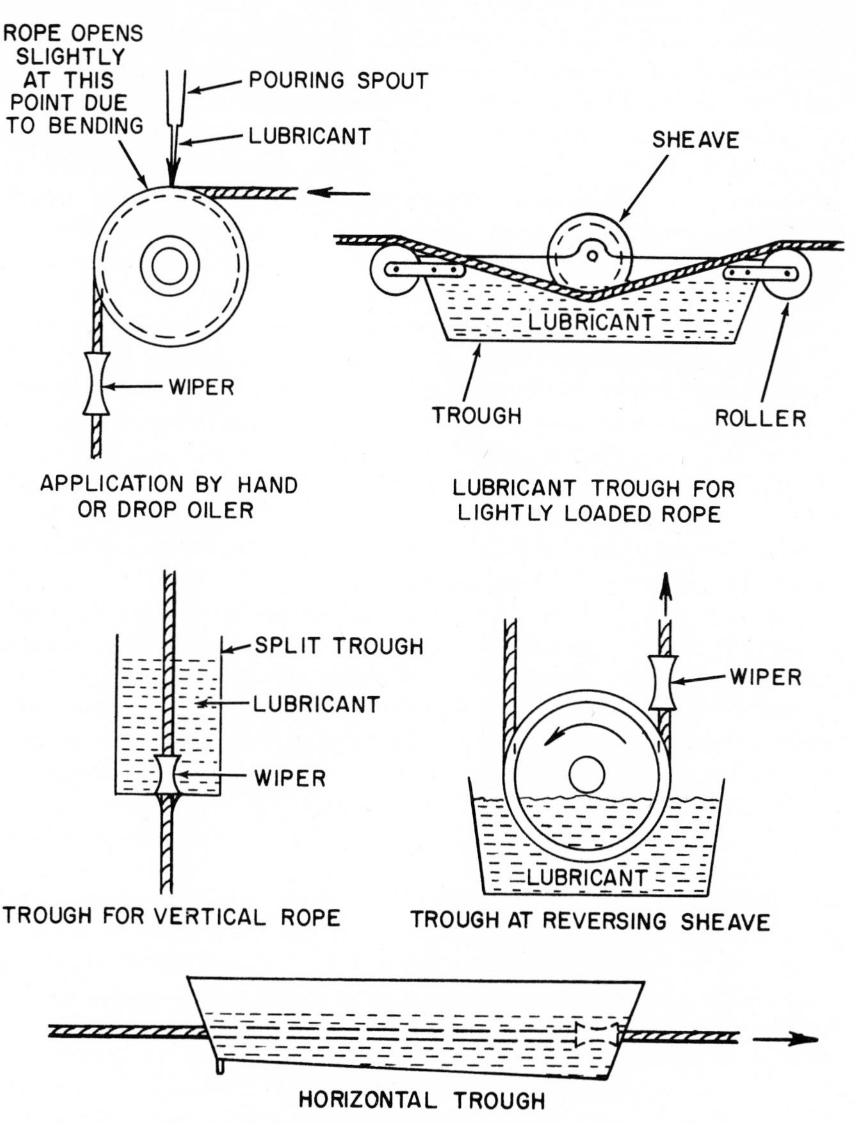

Fig. 21-22. Methods of applying lubricant to wire rope.

vapor only long enough to soften the surface coating to avoid removing the core lubricant. This can be accomplished by a series of dunkings rather than prolonged immersion. The degreasing process must be followed by soaking the rope in a penetrating oil to assure core protection, which in turn is followed by an application of the regular lubricant employed whether a compounded oil, grease, or petrolatum.

Wire brushes can also be used to remove coatings. Best results are obtained by leading the brush or brushes to the cable in line with the strands to remove the debris from the cusps formed by the adjoining strands and wires. Several power-driven brushes may be mounted about a moving cable 120 degrees apart to provide a precise cleaning schedule when operated at specified intervals.

EXTREME ENVIRONMENTS

Technological advances have presented a number of difficult lubrication problems. Lubricants are required to cope with speeds exceeding 100,000 rpm, high vacuum (1×10^{-9} torr), temperatures above 900°F and below −250°F, and nuclear radiation. In some instances, more than one of these conditions exist simultaneously.

In an environment of high vacuum, evaporation is the major factor which eliminates any consideration of conventional oils and greases and synthetic fluids. It also rules out the use of some conventional solid lubricants, such as graphite, which depends on adsorbed moisture for lubricity.

The stability of many lubricants is affected by radiation. Oils change in viscosity, and even molybdenum disulphide is subject to certain characteristic changes.

At high temperatures, conventional lubricants oxidize rapidly, leaving unwanted residues.

Under cryogenic temperature conditions, conventional lubricants solidify and cause mechanical stoppages.

Gears running at ultrahigh speeds are subject to a high rate of attrition when ploughing through an oil bath or striking particles of sprayed lubricant.

Other considerations, especially on manned spacecraft, are lubricant

Table 21-13. PRETREATMENTS FOR METAL SURFACES TO IMPROVE WEAR LIFE OF RESIN-BONDED SOLID LUBRICANTS

Metal	*Pretreatment*
Steel (except stainless)	Grit blasting or other mechanical process Phosphating Sulfiding
Aluminum	Anodizing Chemical coating
Chromium and nickel plate	Vapor, grit, or sandblasting
Titanium	Vapor or grit blasting
Cadmium plate	Phosphating
Zinc plate	Phosphating
Copper and copper alloys	Bright dipping
Magnesium	Dichromate treatment
Stainless steel	Vapor or grit blasting, chemical etching

Courtesy, National Aeronautics and Space Administration.

toxicity and compatibility with various other materials, such as anodic coatings.

Supersonic aircraft, parts of which are constructed of metals having high strength-to-weight ratios, such as titanium, are subjected to fretting corrosion at the joints, which must be treated to reduce or eliminate the damaging process.

The solution to the lubrication problems experienced with extreme environments appears to lie with solid materials in most cases. Such materials take the form of either an applied lubricant such as Teflon (Fig. 21-23), or of a self-lubricated component e.g., carbon rings, carbon and graphite electric motor brushes.

SELF-LUBRICATED COMPONENTS. Rolling bearings are widely used in abnormal environments, particularly ball bearings. Gears of various types are also involved. The cage or separator rubbing on the balls is the major source of friction within the ball bearing, while the faces of the teeth represent the frictional parts of the latter. As a result, these are the areas of greatest probable failure of the two components. By reducing the coefficient of friction between the cage and rolling elements, and between the faces of gear teeth, wear is substantially reduced in both cases.

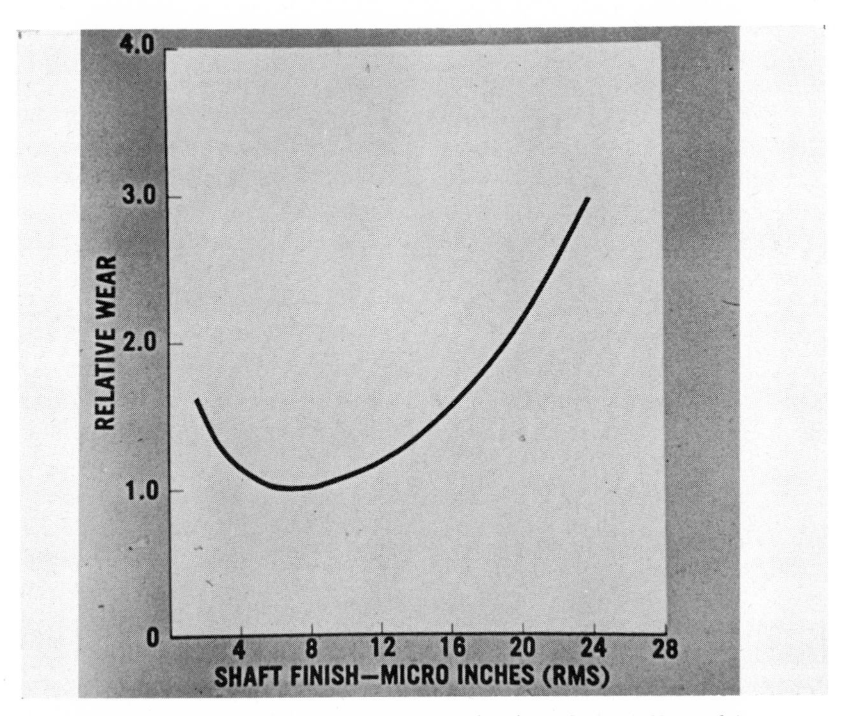

Fig. 21-23. Effect of shaft finish on wear of Teflon bearings. *Courtesy, National Aeronautics and Space Administration.*

Gear teeth, being essentially cantilever beams, must have sufficient load capacity. This requires a material having appropriate stress capability, which eliminates the use of many self-lubricating materials for tooth structure, especially if it is to carry any appreciable load. With such materials as are practical for gear tooth construction only the rim, including the teeth, need be so composed, while the blank to which the rim is bonded or fastened may be of any conventional and appropriate structural material. Another alternative is to shroud the faces of the teeth with a self-lubricating solid in the form of an overlay or inlay, so installed as to prohibit snagging by the opposing teeth as they enter or depart mesh.

Plain bearings may be treated in the same manner as the gear teeth, having only the pressure area protected by a self-lubricating solid overlay much like the thin frictional surface of laminated shell-backed bearings used in internal combustion engines.

Because rolling bearings experience most friction between the rolling elements and cage, the entire cage assembly is constructed of a material with self-lubricating characteristics. It is not feasible to construct the rolling elements of self-lubricating materials because of the heavy load they carry and the low proportion of the total bearing friction they develop.

The construction of supported, self-lubricated, solid, frictional components is, in most cases, the simplest part of the extreme environment problem. The most difficult task is the research and development necessary to provide suitable new materials to meet the conditions.

While the quest proceeds for solids that may be used for the construction of bearing surfaces described above the solid materials now available continue to be effective as lubricants for some of the extreme conditions as well as for those same environments existing under less than critical magnitudes. For example, molybdenum disulfide proves effective and is acceptably stable under relatively high vacuum. It is only when temperatures in vacuum become high enough to promote an excessive rate of outgassing that it reaches a point of useful limitation. Graphite is effective under various environmental conditions not associated with moisture evaporation and where a substratum of oxide film is allowed to exist on frictional surfaces. Silicone greases are effective at temperatures approaching 650°F in the atmosphere.

In self-lubricated sintered bearings, the impregnated lubricant fills the spaces made available by poracity. Such bearings are made up basically of a structural component for supporting a load while acting also as a reservoir for the required lubricant. Under the circumstances in which sintered bearings operate, which could include one or more of the extreme conditions, a liquid lubricant suffices.

Where the lubricant is necessarily a solid, such physics that control the flow of liquids in sintered bearings cannot be utilized, e.g., capillary action, or even squeeze action if such were possible. Therefore, the problem reverts again to the necessity of exposing the solid lubricant to the rubbing components in the form of a rubbing strip, or by proper impregnation, or by alloying one of the opposing surface materials.

Oilless bearings are plain bearings made up of a mixture of lead and powdered PTFE (polytetrafluoroethylene, Teflon), impregnated into a sintered bronze matrix. The impregnated bearing is then mounted on a steel backing to enhance its strength and rigidity. The coefficient of friction of these bearings varies from $f = 0.04$ at high loads and low speeds to $f = 0.16$ for light loads and high speeds. A light film of lead-PTFE mixture is usually applied to the opposing surfaces to protect them during the running-in period.

LUBRICATION AT 600 TO 1000°F. No liquid lubricant or grease is effective at temperatures much above 600°F. Solid lubricants serve satisfactorily up to 1000°F if secure bonding can be accomplished. Good results have been attained with procelain enamel binders for temperatures up to 1000°F if the surfaces are properly prepared before applying the bond.

LUBRICATION FROM −65°F TO 450°F. The requirement related to supersonic aircraft for a lubricant capable of meeting both low and high temperatures is limited to neopentyle and aryl-urea-pentoerythritol greases. Silicones may be used for temperatures ranging from 87°F to over 500°F under some operating conditions.

BASIC LUBRICANTS AND TEMPERATURE LIMITATIONS

Lubricant	*Temperature Range* (°F)
Organic fluids	−70–500
Greases	−100–250
PTFE fluorocarbon	or −65–450
Dry films	−400 to about 500
Liquid metals	below −100 to over 650
Ceramics	0–1100
Ferric iron compositions	1000–2000
	−40–1400

22

NUCLEAR REACTORS

Because of its effect on conventional lubricants, the subject or radioactivity is of interest in this text. This is particularly true in view of the growing use of nuclear energy in commercial enterprises, such as power plants, metallurgical processes, product inspection and control, and research activities, including wear studies of frictional equipment.

ZONES OF RADIOACTIVITY

Power generating plants are designed to convert thermal energy to mechanical energy and thence to electrical energy. In the conventional power plant thermal energy emanates from various kinds of fuel (coal, oil, or gas) burned in the furnace to transform water to high-pressure, superheated steam directly in the boiler. In the nuclear power plant thermal energy is one of the products of nuclear fission reaction that occurs in a *reactor vessel,* which is the furnace of the plant. The heat of fission reaction is removed from the reactor by a circulating *coolant* which, in turn, transfers the heat to water. The water is transformed to high-pressure, superheated steam in a heat exchanger called a *steam generator.* From the steam generator, the steam passes through a turbine, in which it gives up much of its thermal energy to drive an electrical generator. The spent, low-pressure steam from the turbine is condensed and pumped back through the steam generator to repeat the cycle.

For the purpose of lubrication, the nuclear power plant may be divided into two zones according to the intensity of radio-activity as follows:

1. The reactor, or zone of high radioactive intensity.
2. The power plant, or zone of substantially low radioactive intensity.

1. REACTOR. The reactor vessel contains the fuel (an isotope, uranium-235) enclosed in sealed zirconium tubes to form the core. The fuel fissions to produce neutrons and release energy. The reactor also contains boron and cadmium control rods that can be moved in and out of the core to absorb neutrons and regulate the rate of heat generated according to changing power demands. The third essential component of the reactor is the cooling fluid which removes the heat generated during the nuclear reaction within the core. The cooling fluid can be light water (boiling or pressurized), heavy water gas (carbon dioxide or helium), or liquid metal. Liquid metal coolants include mercury, rubidium, lithium, and sodium. Potassium is sometimes alloyed with sodium to keep the coolant liquid at room temperature. The essential reactor components are encased within a containment vessel.

Another essential component is the *moderator,* a material that can slow down neutrons in order to maintain efficient chain reaction within the core. Moderators include light water, heavy water, graphite, beryllium, and some hydrocarbons.

The reactor is surrounded by a shield to prevent the escape of radiation. Shields are usually constructed of concrete, although water supplemented with lead is also used. A reflector surrounds the reactor core within the thermal shielding to return escaping neutrons back to the core.

Reactors differ in their core structure according to the end product desired. Breeders produce fissionable material, and power generators produce electrical energy. Others produce neutrons for nuclear experimentation and research, and another type produces commercially needed isotopes.

The environment within the reactor vessel is highly radioactive. Therefore, the use of any material subject to rapid deterioration by the effect of radiation, including polymers and lubricants containing hydrogen, cannot be considered. Lubricants for coolant pumps or blowers, control rod mechanisms, and other instruments within the high radioactive zone must be especially selected, adequately shielded, or effectively sealed against prolonged exposure.

2. POWER PLANT. The nuclear power plant consists of the same equipment found in conventional power plants, including steam turbines, condensers, condensate pumps, and feed water pumps. This part of the nuclear power plant complex is effectively shielded against radiation from the reactor, and the steam used is generated so as to preclude carry-over of substantial levels of radioactivity into the power-generating components. Thus the radiation in the power-generating area is of low magnitude, usually at levels of about 0.3 rad/hr. or less.

Lubricants used for turbines, generators, pumps, and other mechanical components may be of the conventional types selected according to those listed in this text for similar equipment, including both mineral oils and synthetic fluids. They will not be discussed further here.

POWER PLANT REACTORS

There are several types of reactors used or proposed for producing high-pressure steam to drive turbines. They include the pressurized water reactor, boiling water reactor, liquid metal reactor, organic-cooled reactor, and the gas-cooled, graphite-moderated reactor. The pressurized water reactor is described below.

Pressurized Water Reactor

In the pressurized water reactor, the water acts as a reactor coolant as well as a moderator. The fuel used is enriched uranium. Water is pressurized up to 1200 to 2000 psi and is circulated through the reactor and a heat exchanger (steam generator) by a coolant pump. The high pressure is maintained in the primary circuit (Fig. 22-1) to prevent the water from boiling at the operating temperature, which may reach 540°F at the reactor outlet.

The high-pressure, high-temperature water in the primary circuit (loop) transfers heat through a heat exchanger to water in the secondary circuit at a much lower pressure, about 175 psi, converting it to steam, a process analogous to the indirect expansion refrigeration system. The secondary circuit is composed of the heat sink for the primary system (heat exchanger) steam turbine, condenser, feed water preheaters, feed water and condensate pumps, and necessary piping (Fig. 22-2). Unlike the primary circuit, the secondary circuit is not exposed to core radiation. The water in the primary circuit is continuously purified to minimize radioactive debris and its radioactive effects.

The fuel is usually in the form of water-compatible uranium dioxide clad in a sheath of either zirconium or stainless steel. Low-alloy (mild) steel piping is used in the primary circuit.

PRESSURIZERS. In a pressurized water reactor, the pressure in the primary system must be maintained at a level corresponding to the temperature desired, which is always somewhat below the boiling or saturation point (Fig. 22-3). This means that temperatures for different reactors may be anywhere from 567°F at 1200 psia to 635°F at 2000 psia, the pressure being limited by structural and efficiency considerations. To assure that boiling does not occur, however, the temperature is dropped below the saturation point by as much as 100°F. The fluctuating work demands on the part of the turbine require alterations in the coolant pressure. This is particularly true with ships (Figs. 22-4 and 22-5), in which pressure fluctuations occur according to operational contingencies, during which more or less heat is required to be transferred to the secondary system. This tends to keep the conditions in the primary system in a state of thermodynamic unbalance, even to the point of generating steam in the reactor. This serves

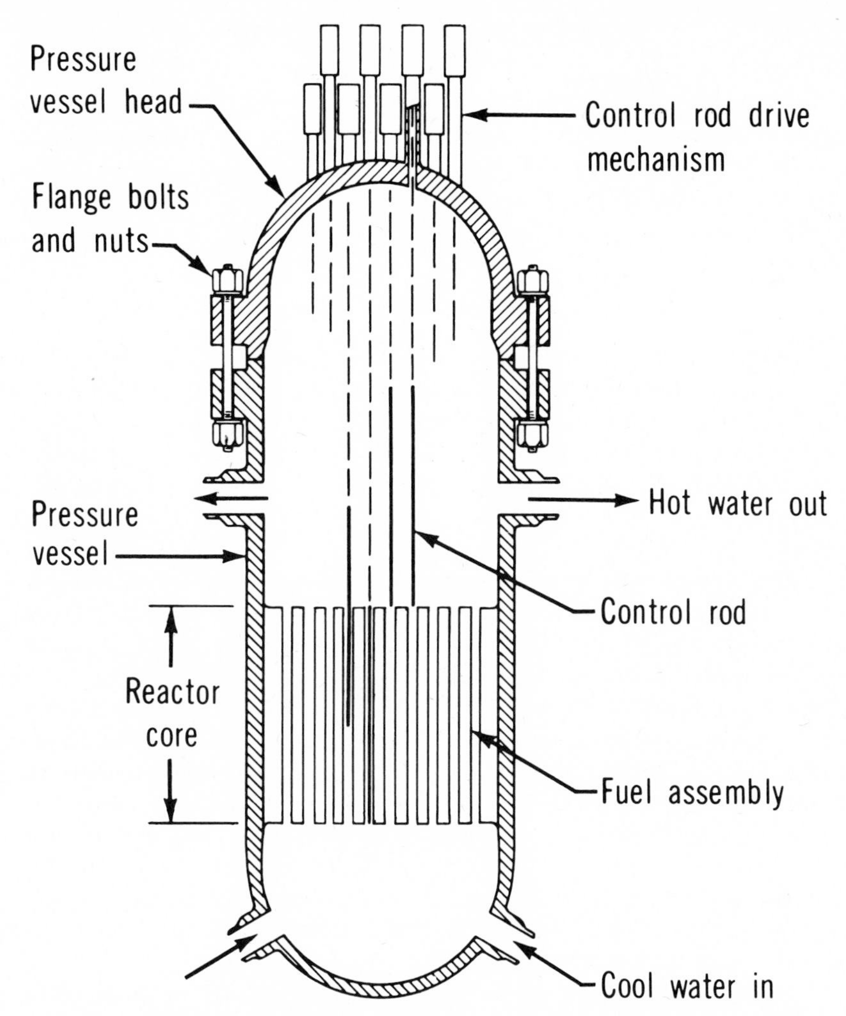

Fig. 22-1. Vertical cross section of a pressurized water reactor vessel (primary circuit). *Courtesy, U.S. Atomic Energy Commission.*

to defeat the purpose of the coolant in its thermal effect in the reactor, accompanied by the possible burnout of the fuel, especially since the liquid water also acts as a moderator to control neutron energy.

To control thermal equilibrium in the primary loop, the pressurizer supplies or reduces heat to form and control a head of steam, which is allowed to form in the pressurizer only, in an auxiliary space located above a controlled liquid coolant water level. Thus only in the pressurizer is the temperature-pressure relationship allowed to reach a level above saturation. It is also provided with a pressure relief valve. Heat is supplied when needed from an external source, such as electrical heaters located in wells below

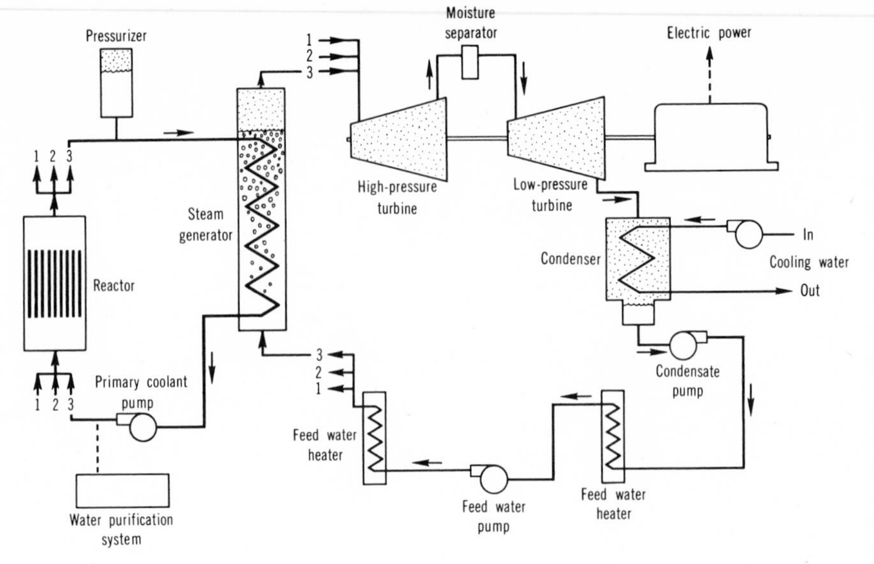

Fig. 22-2. A complete nuclear power plant based on a pressurized water reactor. *Courtesy, U.S. Atomic Energy Commission.*

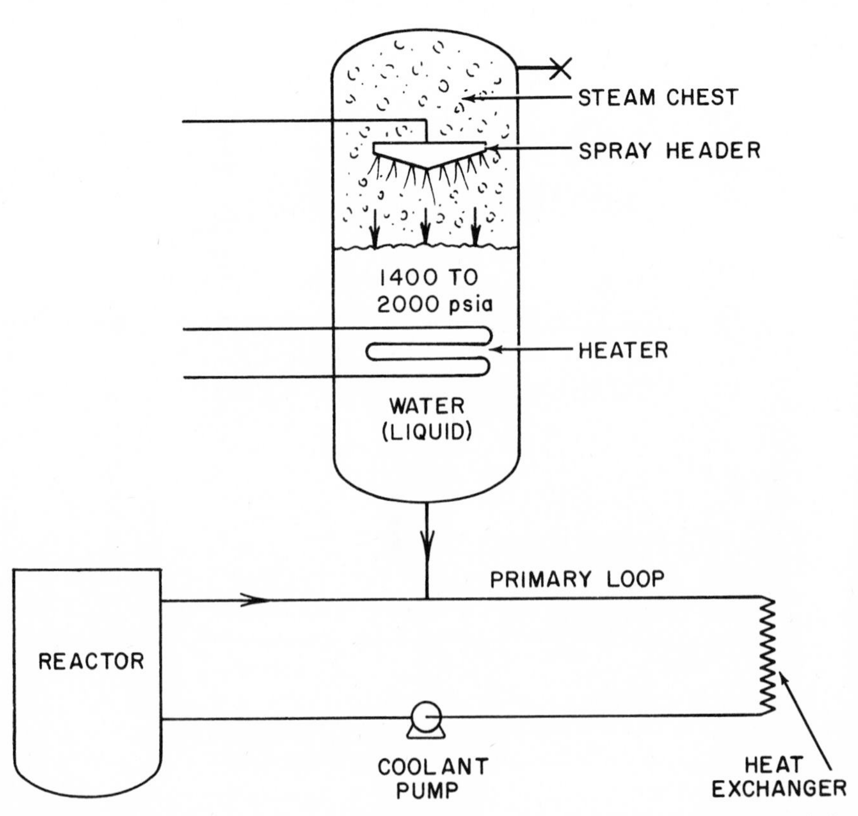

Fig. 22-3. Pressurizer for pressurized water reactor system. Note that steam occurs only in the steam chest.

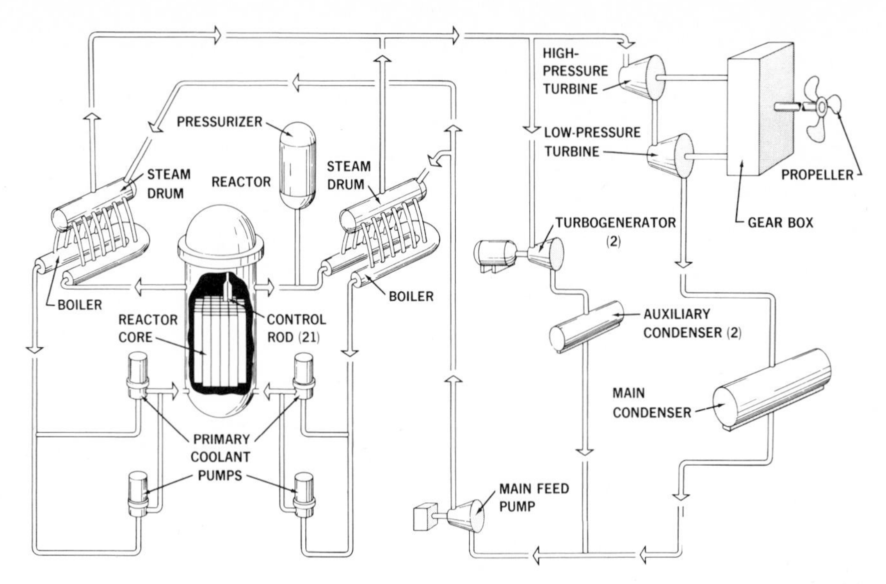

Fig. 22-4. Basic parts of the propulsion system for a nuclear-powered ship, such as the *N.S. Savannah. Courtesy, U.S. Atomic Energy Commission.*

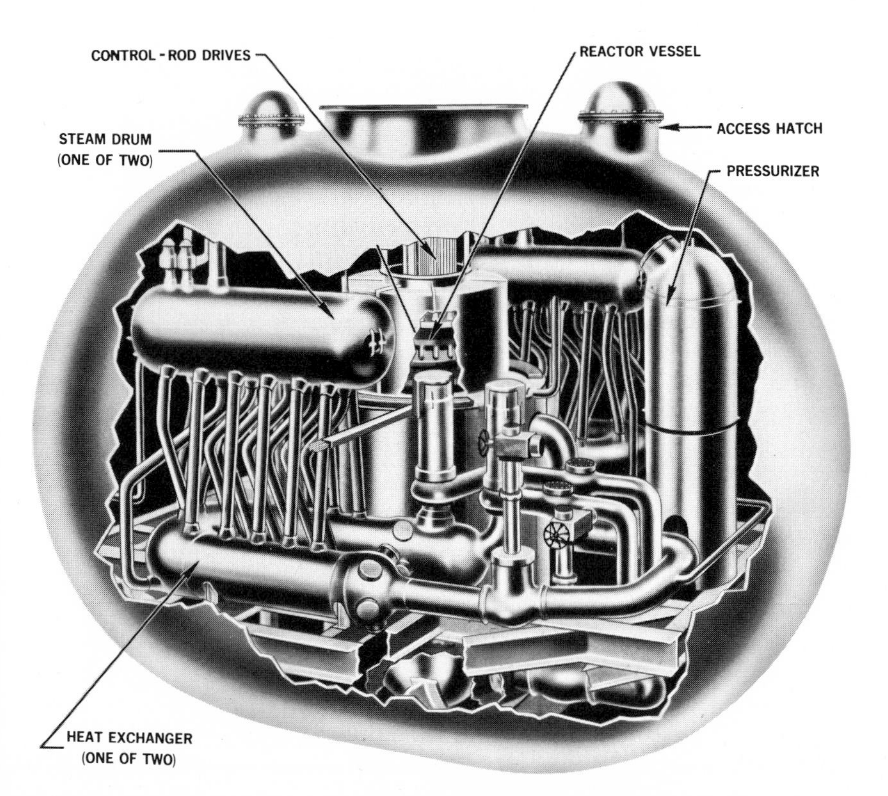

Fig. 22-5. Containment vessel housing the reactor, pumps, and steam generators of the *N.S. Savannah's* nuclear steam generating plant. *Courtesy, U.S. Atomic Energy Commission.*

the water level. Cooling is provided by spray headers located in the steam chest; these reduce the vapor pressure when activated.

The capability of the pressurizer to alternately raise and lower the vapor pressure within the steam chest allows an equilibrium pressure in the primary loop to be maintained.

RADIATION EFFECTS ON ENGINEERING MATERIALS

The physical and chemical changes brought about in target materials (i.e., those materials bombarded by radioactive particles and rays emitted from the nucleus of radioactive nuclei) are functions of the kind of radiation and its intensity. Total effectiveness may be reduced by barriers, such as cladding of fuels, provision of reflectors, and the erection of suitable biological defenses which substantially attenuate the effectiveness of all radiations.

IONIZATION. Nuclear radiations are reduced in force (attenuated) by passing through any barrier mass by an amount depending on the type and thickness of the mass and the form and energy of the radiation. The slowing down process may be accomplished by absorption, or capture of the high energy alpha, beta, or neutron particles or gamma rays. Absorption by or the displacement of electrons of a target atom results in ionization.

When a positively charged alpha particle is near a negatively charged target electron, the attraction between them becomes very strong, and the electron may be forced out of its orbit to be released as a free-moving high-speed particle. The target atom now becomes positively charged because of the new imbalance posed by the loss of a negatively charged electron from the neutral atom. The atom becomes a positive ion by the definition that a positively charged atom is a positive ion by reason of the removal of an electron which carries a negative charge. While an electron is not an ion in the true sense, it may be considered as one in an ionization process because it is a charged particle. An *ion pair* therefore is formed by the combination of the atom that has been converted to an ion and the electron that was removed and which is not really an ion. The offending alpha particle continues on its way to repeat the ionizing process in other atoms, during which it loses energy by imparting a portion to each electron it attracts from orbit. When it finally reaches a state of equilibrium with other atoms of the target material, it picks up two stray electrons to form a neutral helium atom, because the alpha particle is the nucleus of the helium atom.

The beta particle is an electron with a negative charge. When a beta particle passes in the vicinity of a negatively charged target electron, the two particles repel each other. The repelling force may be strong enough to remove the electron from its orbit and, as in the case of the alpha particle, ionization occurs. Primary beta particles of high speed do not exist in the nucleus, but they appear during the transfer of energy when nuclei break down under radioactivity. Secondary or slower beta particles originate in the outer electron shells and not in the nucleus. The beta particle also loses energy as ionization of other atoms occurs and slows down to reach a state of equilibrium with its environment. Secondary ionization by alpha and beta particles is not significant.

The behavior of gamma rays in producing ionization is quite different than that described above for alpha and beta particles. Gamma rays are of zero mass, but they have the effect of a particle having the nature of electromagnetic radiation. They interact variously with other electromagnetic fields. Three major ray interactions are the photoelectric effect, the Compton effect, and pair production.

The major radiations out of the core of the reactor are neutrons and gamma rays. Gamma rays are responsible for the production of ion pairs (ionization), and neutron damage outside the core causes the transmutation of some materials. Selection of materials, such as lubricants and seals, that are least affected by radiation is the most reliable method of assuring extended operating life of the material.

ORGANIC LUBRICANTS. Continued exposure of organic lubricants to radiation results in their ultimate deterioration, which may be described as being similar to that of an electrical discharge. Since most lubricants are either wholly or partially organic, their deterioration is predictable and varies in a given period according to the type of radiation and its intensity. Mineral oils or oils made from petroleum are actually organic and not mineral in nature, because they originate from plant and animal material. Graphite is similarly organic.

Mineral oils and greases made with mineral oils at first become less viscous as a result of ionization due to radiation. Then their viscosity increases to the point of solidity upon continued exposure to radiation. This applies also to most of the synthetic fluids used either as lubricants or hydraulic liquids, especially those made up of polymers. There are some exceptions; polypropene oxide, for example, improves when exposed to radiation.

In general, the effects on organic compounds by reaction with radiations are complex and capable of producing (a) smaller molecules (thinning out process) or (b) larger molecules (thickening process). Such reactions also include adding hydrogen to unsaturated compounds or dehydroginating those compounds with a more stable bond. The products of deterioration include sludges of high molecular weight as well as gases, such as methane and hydrogen, with low molecular weight.

Conventional mineral oils are satisfactory as lubricants for the mechanical equipment of nuclear power plants, except in the reactor zone. For this area, molybdenum disulfide has proved to be relatively stable. There are other proprietary compounds capable of satisfactory lubrication.

Tests show that aromatic compounds improved with additives are less affected by radiation than aliphatic compounds. According to such tests, the terphenyls appear to give the best results.

Lubricants used or considered in highly radioactive environments, such as related to reactor zones, include solids (molybdenum disulfide), liquid metals, water, discrete types of grease (usually proprietary), certain synthetic fluids (polyphenyl ethers), gases (hydrostatic films), and self-lubricating films (various oxides, heterogeneous materials, etc.), and indantherene-base greases.

Concrete is subject to a loss of moisture by radiation exposure resulting in a reduction in strength. This disadvantage is greatly overcome by adequate structural shielding. Water, lead, and iron are effective shielding materials against gamma rays.

Low-carbon steel is free of alloys susceptible to rapid deterioration by irradiation. It is used with some success for pressure vessels and primary circuit piping of the reactor system. Under certain temperature conditions low-carbon steel is subject to more rapid embrittlement by neutron irradiation, leading to brittle fracture and other cumulative effects. Damage to steel in general is due to crystal lattice displacement of its iron constituent. Ionizing radiation does not affect metals but does produce heat.

The boron content of boron-molybdenum welding materials has a highly absorbent cross section and is thus subject to transmutation by neutron bombardment wherein

$$\text{boron-1D} + \text{neutrons} \longrightarrow \text{lithium-7} + \text{helium-4} + 2.8\ \text{MeV}$$

which has a dangerous alpha radiation effect.

High-polymer materials, such as plastics and rubber, suitable for insulating conventional electrical wiring and for seals, are subject to rapid deterioration when irradiated. Being high-polymer materials, their bond is due to a state of ionic balance which is easily disturbed by electron, gamma ray, and neutron bombardment. The result is hardening, embrittlement, and surface oxidation. Specific polymers, such as polyvinylchloride, evolve hydrogen chloride during irradiation. Hydrogen and low-molecular-weight hydrocarbons are also subject to molecular change.

Water decomposes into oxygen and hydrogen gases under intense radiation, causing some dissociation.

Graphite is an effective moderator that slows down neutrons to thermal velocities in a reactor to increase fission probability. The thermal velocity of neutrons is about 2200 meters/sec at normal temperatures. By their comparative energy, neutrons are classified as thermal, intermediate (epithermal), and fast.

Thermal neutrons—less than 100 eV
Intermediate neutrons, also called resonance neutrons—100 eV to 0.1 MeV
Fast neutrons—greater than 0.1 MeV

When graphite is used as layering for reflectors, which surround the core of thermal (power) reactors it serves to slow down most of the neutrons leaving the core.

LUBRICATION OF REACTOR ZONE EQUIPMENT

Hydrocarbon type lubricants can be used only for nuclear power plant equipment protected from radiation and whose frictional parts are designed to prevent such lubricants from entering the reactor, either directly or by way of the coolant. Hydrogen, being a moderator, tends to reduce neutron speeds to levels at which fission occurs. In fast-breeder reactors, the latter action may cause loss of control.

Synthetic fluids, such as polyphenyl ethers and some silicones are considered as being promising liquids in nuclear radiation environments as lubricants, coolants, and hydraulic fluids.

Petroleum-base and synthetic lubricants are not affected by cosmic radiation levels even in space environments, in which evaporation under high vacuum is a limiting factor, but adoption of nuclear power for aircraft and spacecraft may eliminate them entirely.

The frictional components located in the reactor area include the coolant pumps or blowers, equipment for handling the reactor fuel and control rods, and the sliding parts of the reactor. Some of the components are shielded so that conventional petroleum or synthetic lubricants can be used. Where such shielding is not possible, special proprietary greases can be used in some cases. A description of such greases cannot be given here for obvious reasons.

The fluid-piston type bearings of pumps in the primary system of water coolant systems are of the canned rotor type and are lubricated by the water flowing through them. Special consideration must be given the type of materials used as frictional surfaces in rotating bearings in order to resist the galling, wear, and corrosive effect of water. The most satisfactory results are obtained with the use of hard dissimilar high-temperature nonferrous surface metals, such as cobalt-base versus nickel-base materials.

Shielding of hoist drives, including their gears and gear shaft and motor bearings, is provided by a height of water located between the hoist and the top of the reactor. Being over 20 ft high, the water shield reduces radiation to a level sufficient to allow the satisfactory use of conventional petroleum lubricants. The use of molybdenum disulfide in nonsoap-thickened grease has had some success in maintaining stability under radioactive conditions, especially where the cables of the control rod hoist drum enter the reactor zone and become highly radioactive, tending to affect the lubricants of the drive unit.

Bearings of carbon dioxide blowers are protected by shaft seals that keep the gas from escaping into the bearing cavity, so that conventional lubricants can be employed to lubricate such seals. A vacuum pump oil having low vapor pressure is recommended for the purpose.

Sliding components can be isolated from the highly radioactive environment by being located outside the reactor shielding.

The use of liquid metals as lubricants for circulating pumps poses the problem of compatibility with bearing materials. The lubrication properties of liquid metals used primarily as reactor coolants are more fully discussed in Chapter 6.

Table 22-1. LUBRICANTS FOR COOLANT PUMPS, COMPRESSORS, BLOWERS AND DRIVE MOTOR BEARINGS FOR NUCLEAR REACTORS

Equipment	*Type of Reactor*	*Approx. Dosage (rads/hr)*	*Lubricant*
Primary Coolant Pumps	PW	10^3	Water
Centrifugal			
Canned motor bearings			
Pump bearings			
Centrifugal	BW	10^3	
Limited leakage			
Motor bearings			Conventional grease or oil
Pump bearings			Water
Blowers	Gas	1–10	
Motor bearings, shielded			Conventional grease or oil
Blower bearings, shielded			Conventional oil
Canned compressors			Gas
Turbo-compressors			Conventional turbine oil
Sump type propeller pump	LM	10^5	
Drive motor bearings (shielded)			Conventional grease
Pump bearings			Liquid metal
Submerged centrifugal pump	LM		Liquid metal
Standard chemical pump	OM	None	
Drive motor bearings			Conventional oil
Pump bearings			Conventional oil or grease or coolant
Canned rotor centrifugal pump	AH	—	Water

PW, pressurized water. BW, boiling water. LM, liquid metal. OM, organic moderated. AH, aqueous homogeneous.

MEASUREMENT OF RADIATION

Radiation is measured in terms of total dosage, or dose rate. The reporting unit of measurement, the *roentgen,* is based on the extent of ion pair formation, which governs the amount of impingement of gamma radiation.

The roentgen is defined as the amount of radiation resulting in the formation of 2.08×10^9 ion pairs per cubic centimeter of dry air at a temperature of 32°F and a pressure of 1 atm, or standard temperature and pressure conditions, STP. This is equivalent to about 88 ergs of energy when 1 grain of dry air (STP) is exposed to 1 roentgen of gamma radiation. The *rad* is defined as the absorbed dose of nuclear radiation during which 100 ergs of energy is liberated per gram of absorbing material.

Total dosage of radiation is based on the amount of dosage received in roentgens during a given period of exposure.

Dose rate is expressed in roentgens per hour. For small doses, the quantity of radiation is expressed in milliroentgens. Total dosage is equal to the average dose rate per hour multiplied by the number of hours of exposure.

The effect of gamma rays on any material is based upon its ability to produce ion pairs, or ionize. For example, a single exposure dose of less than 25 roentgens will produce no detectable chemical effects in the human body, according to studies made by the United States Department of Defense and published by the Atomic Energy Commission. The dose of radiation received is inversely proportional to the square of the distance from the source. There is also a "thinning down" or attenuation factor, which decreases radiation intensity due to adsorption and scattering of gamma rays by the atmosphere between the source and the material specified. Because of scatter, an object is never fully shielded from the effects of gamma rays unless it is fully surrounded.

HALF-LIFE. The rate of emission of beta particles and gamma rays is expressed in terms of half-life. The half-life varies according to the radioactive substance and is expressed in units of time required for a given element (isotope) to decrease one-half its radioactive value at any given magnitude. The action is one of decay and is continuous.

The presence of radiation is detected by the Geiger counter and the dosimeter, both of which also measure the intensity of gamma rays and other radiations.

23

LUBRICATION MANAGEMENT

A properly managed lubrication program will minimize downtime and its attendant costs. The important factors to be considered in planning a lubrication program are (1) the lubricants: selection, purchase, storage and handling, and application; (2) personnel: organization, training, scheduling, and duty instructions; and (3) reports, or feedback: status of work, lubricant consumption, condition of frictional equipment, and action taken to rectify abnormal frictional conditions.

THE LUBRICANTS

SELECTION. A plant-wide lubrication analysis involving a physical study of individual machines and processes is necessary to determine the kinds and nature of lubricants required for optimum performance. Selections should be made on the basis of operating and ambient conditions related to each machine. Such conditions include temperatures, speeds, loads, moisture or water conditions, dust conditions, method of lubricant application, probable frequency of application or oil change, reuse or all-loss system, filtering facilities, length of feed lines, contaminating influences of processes (e.g., metal-cutting fluids), accessibility of application fittings and appliances, and leakage.

Selection can be made by an experienced plant lubrication supervisor or engineer or by a representative of a potential lubricant supplier. Consideration should be given to machinery builders' instructions and to appropriate performance data offered by suppliers in making a final decision relative to lubricant selection. Where oil reuse systems are involved, the oil selected should have a record of successful performance for extended periods of service in comparable frictional systems. Certain additives, such as rust and oxidation inhibitors, antifoam and antiwear agents, may be required to attain long periods of successful service.

The cost of quality oils containing such additives is usually only slightly higher than that of lower quality oils without additives but with similar physical specifications. The longer life and more effective performance of the better oil results in a much lower overall cost when the lubrication system is properly maintained.

With systems having a high leakage ratio, the cost of high quality oils may not be justified. Systems subject to excessive leakage do not warrant oils with special lubricant characteristics, such as oxidation resistance, when rapid replacement prevents sufficient time for the deterioration of even the lower quality oil. In any case, consideration must be given to the protection of frictional surfaces against wear, which often limits the magnitude to which oil quality can be reduced.

This text has described many additives introduced for specific purposes. For example, a small amount of emulsifying agent is added to a pneumatic tool lubricant to offset the washing effect of condensate from the air cylinders. The substitution in this case of an oil without emulsifying characteristics because of lower price would not be sound because of the more rapid wear that would result and the cost of replacement parts.

An important consideration is the selection of the fewest number of kinds of lubricants commensurate with the protection of all frictional components. Many multipurpose and multigrade oils and greases are available, making possible an appreciable reduction in the number of different brands required in any conventional industrial plant and transportation facility. In addition to the adoption of multipurpose lubricants, additional brand reductions can be made by such practices as selecting multiviscosity fluids, compromising oil viscosities, where possible, and using reuse oils on all-loss components where the loss is reasonably low.

LUBRICANT PURCHASING. It is the responsibility of the purchasing department to procure the required lubricants at the least possible cost. In doing so, the specific lubricants selected by responsible lubrication personnel must be considered. Purchasing solely on the basis of physical specifications is generally risky, because the usual values of specific gravity, pour point, viscosity, and color are not criteria for performance.

Price alone must not be the governing factor in the purchase of lubricants. It costs approximately six dollars to apply a dollar's worth of grease. Unscheduled downtime attributable to improper lubrication can cost many times the annual cost of lubricants required for operation. There are numerous cases in which the expenditure of an additional 10 to 20% for a better lubricant or more effective application would have reduced plant maintenance costs by a much higher sum.

Savings in the cost of lubricants can be made by larger volume purchasing. This can be done by consolidating requisitions so that a number of drums can be made by a single delivery from the supplier. Not only do some suppliers offer a discount for deliveries based on a minimum number of drums, but savings in the cost of accounting procedures can be made.

Purchasing lubricants in bulk, with deliveries made into large storage tanks by tank wagon (up to 1500 gallons of oil) or more economically by tank transport (5500 gallons or over), also reduces the unit cost of oils. Bulk grease delivery is also cost-reducing when practical. On the other hand, maintaining excessive inventories tends to increase cost because of the large idle investment involved. A minimum-maximum inventory based on rate of turnover coupled with single delivery orders appears to be the most economical and convenient purchasing procedure for the average size plant.

Minimum and maximum inventory levels are practical bases for purchasing lubricants, with quantity dependent on the rate of turnover. The practice assures an adequate supply at all times and compensates for the time elapse between ordering and receiving shipment.

STORAGE AND HANDLING. Oils and greases are delivered either in bulk or in 400-lb (55-gallon) drums. Both are available in smaller packages (5-gallon cans or 35-lb pails), the handling of which poses very few problems and is often convenient for plant distribution.

For *bulk delivery of oil,* a tank is usually installed in a protected location as close as possible to the unloading point so that the oil can be received by gravity. If the tank is located above the level of the delivery vehicle, a transfer pump is necessary. If two tanks are used, they must be connected by crossover piping equipped with valves. The tanks must be vented at the top to allow air passage in and out when oil is entering or leaving the tank (Fig. 23-1). Air cleaners should be installed at such vents. Each tank must be furnished with drain lines and valves as well as a level gauge. A pump is required to transfer oil from the tank to containers in which it is distributed, usually drums that must be kept free of all contaminants. A strainer should always be used during the transfer of oil from one container to another.

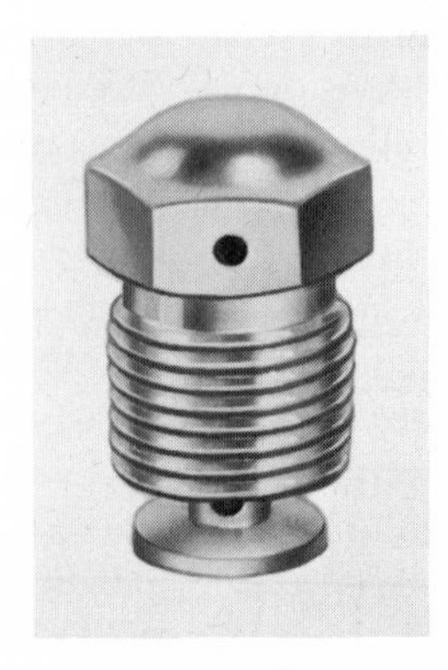

Fig. 23-1. Reservoir vent plug with deflectors.

Bulk purchase of oil has the advantage over purchasing in drums because of lower price per gallon and greater convenience of storage. An additional advantage is gained when the oil can be distributed through piping to various areas of the plant or directly into large machines. Some advantage is lost, however, in the cost of the labor required to fill and clean the containers required for plant distribution. The maintenance of the storage tanks is another expense item, since they must be periodically inspected and kept free of water, which is drained from the bottom if found in appreciable quantity. The presence of water is determined either from the level gauge or by a dip stick, the end of which is smeared with an adhesive material that changes color in water.

Periodic flushing of small tanks and cleaning of all oil storage tanks is necessary. Large tanks are equipped with manholes for cleaning or repair. Working in a tank can be hazardous and should always be done with at least two men, one to act as an aide for the other to prevent an accident, especially suffocation. A gas mask must be worn if the tank is not properly ventilated.

A storage tank is reserved for the same kind of oil; otherwise cleaning is necessary before filling with the new type. To avoid an oil shortage within the plant, the capacity of the tank should be larger than the volume of oil customarily delivered. Thus when the level falls to a certain level an order is placed with the supplier for a new filling with the assurance that the tank will take the load. Any quantity of oil returned to the supplier because the tank is unable to take the entire load is an expense to the buyer.

Bulk delivery of grease has been adopted by large steel plants, replacing the practice of receiving it in 400-lb or larger drums. The delivery of bulk grease has been variously made in huge plastic bags, tank transport, and, even more conveniently, in large, portable, reuseable steel containers or tanks. The steel tank acts as a central station from which the grease is distributed directly to the various frictional components throughout the plant or a given area of a large plant (Fig. 11-66).

The advantages of using bulk grease are the lower price per pound, reduction in loss of grease that hangs up in drums, lower cost of dispensing within the plant, less storage space required compared with barrels, and greater cleanliness. A second tank makes possible continuous plant supply during container replacement periods.

Delivery of oil and grease in 400-lb or 55-gallon drums is convenient for plants where bulk deliveries are not feasible. Unloading is accomplished by rolling the drums onto a receiving platform level with the tailgate of a delivery truck or floor of a box car, or by means of an automatic tailgate that lowers the drums gently to the driveway. Removal to storage is then made by lift truck or hand barrel truck. Upending a drum requires the careful lifting of two men. However, for moving drums short distances, rolling is preferred.

Drums are preferably stored indoors and on their ends. When stored outdoors, they should stand on a dry surface and be covered. If upended and exposed to rain, tilting the drum slightly prevents entrance of water through the bungs due to breathing during a reduction in temperature.

Racks with one or more tiers are a convenient method of storing drums indoors or under cover. A traveling chain hoist is required to move the drums.

When pouring oil from a drum, it should be placed on its side on a sleigh-like cradle with the fill bung down and equipped with a spring-actuated faucet. A drip pan placed on the floor below the faucet catches the drippage. Remove the vent plug when drawing oil. Another method of dispensing oil from a drum is by means of a hand pump mounted over the fill bung with a drain funnel toward which the pump nozzle can be swiveled to direct drippage back to the drum without contamination.

Grease in barrels may be dispensed by a hand- or motor-driven pump, air pressure, or scoop. A follower plate is usually required to force the grease toward the pump suction to avoid cavitation. The barrel should be scraped down when it is almost empty to reduce loss of grease adhering to the sides. A wooden paddle is recommended in dispensing grease by hand because of the danger of damaged frictional parts if for some reason a metal scoop is accidentally dropped. Grease barrels should be kept covered, except when grease is being extracted by hand.

APPLICATION AND MAINTENANCE OF LUBRICANTS. Methods of lubricant application are described in Chapter 11. The method of application selected depends on the geometry and characteristics of the frictional part involved. Automatic devices reduce labor costs and provide positive methods of both oil and grease application in measured amounts. Bottle oilers and wick-feed and drop-feed cups should be used to replace hand oiling where practical. Central oil and grease systems are preferred for many installations, and their use is recommended where practical to assure a constant and convenient lubricant supply to frictional parts serviced. Frictional components that are inaccessible during machine operation should be equipped with either (1) feed lines whose points of application are brought out to a more convenient location or (2) devices that can provide a lubricant for extended periods, such as automatic oilers, constant level oilers, and spring-loaded grease cups. Many components, such as open gears, chains, conveyor parts, and wire ropes are effectively lubricated by spraying. Trip lubricators, mechanical force-feed lubricators, and oil-fed brushes can also be used to eliminate hand methods.

The care of circulating systems and their oils is important in obtaining trouble-free operation and longer life of the lubricant. Leaks should be eliminated to reduce oil consumption and assure an adequate supply of lubricant to bearings and gears at all times. Return line outlets located beneath the reservoir oil surface eliminate aeration that causes foaming and breaks the oil into small droplets that are more susceptible to oxidation than oil maintained as a mass. To assure proper oil levels, all reservoirs should be equipped with oil level indicators, such as window sights (Fig. 23-2) or level gauges.

Rapid deterioration of oil is avoided by keeping temperatures within safe operating limits by the use of oil coolers.

Filters are always necessary in circulating systems to remove contaminants. Adsorption filters for the removal of corroding substances, depth and edge filters to remove solid particles, and centrifuges to separate water and magnetic separators to remove ferrous particles, used when appropriate, assist in protecting frictional parts against wear and corrosion. They also help to extend the service life of the oil. Filters must be cleaned or changed regularly. A substantial pressure drop across a filter is indicative that cleaning or change is necessary.

Extended life is obtained with rust- and oxidation-inhibited oils used in circulating, splash, and bath systems.

Periodic analysis of oils to determine any changes in viscosity, rise in neutralization number, and sediment content can be made to establish oil drain periods. With motor oils, infrared analysis to ascertain the development of excess nitrogen products is another method of establishing oil drain frequency. Lubricating oil analysis is usually provided by oil suppliers to larger customers. They will also analyze oil from large systems (such as

Fig. 23-2. Window sight for determining safe oil level in reservoir.

turbines) once a year, and the results may be kept as a running record of the batch. The report of analysis also includes recommendations for improving the oil condition if necessary.

PERSONNEL

ORGANIZATION. As with all other plant activities, lubrication must be done in an orderly and efficient manner. To carry out his duties completely and effectively, each individual should be instructed as to what he is to do and when it must be done. It is also important that management be certain that all assigned tasks are accomplished or be advised why they are not. Adequate supervision is required to direct and instruct each individual as to what is expected of him. This is the responsibility of the supervisor, whether it be the plant engineer, maintenance superintendent, lubrication engineer, or lubrication foreman. In a small plant, the supervision of lubrication activities is a simple task, usually handled by an individual, such as a foreman, who has additional responsibilities. With large plants, the lubrication organization is more complex, but its functions can be simplified by coordination.

A basic lubrication organization includes the following personnel:

Supervisor—responsible for all lubrication activity, including the installation and care of appliances and the assignment, instruction, and training of personnel under his jurisdiction. He acts on lubrication problems reported to him and maintains liaison with maintenance supervisor, lubricant stores, purchasing department, chemical laboratory, lubricant supplier, representative or engineer, and all departments served by his organization.

Oilerman—(1) applies lubricants to assigned equipment and reports its completion or noncompletion, (2) inspects systems regularly to determine operating temperatures, oil levels, filters, and operating conditions, (3) collects oil samples periodically for laboratory analysis or visual inspection, (4) cleans oil systems and changes the lubricant when necessary, (5) reports abnormal frictional conditions, and (6) performs special lubrication tasks assigned to him by the supervisor.

Mechanic—acts as a troubleshooter pertaining to lubrication problems, repairs lubrication systems and appliances, and installs new appliances or modifies older equipment as directed.

Clerk—distributes assignments to the work force according to schedule, acts on reported lubrication problems through supervisor or as appropriate, reports any unfinished task assigned to oilerman or mechanic, maintains records of oil changes, lubricant consumption, oil losses due to line failures, etc., maintains file of machine builders instructions, lubrication manuals, catalogues, reports, and machine data, and performs all other clerical duties assigned him by the supervisor.

TRAINING OF PERSONNEL. The oilermen and mechanics are required to know the techniques of lubricant application, operating principles and maintenance of application devices, importance of maintaining proper oil levels and lubricant feed rates, limitation of grease application to rolling bearings, etc., a basic knowledge of lubricants and their uses, and the necessity of applying each lubricant to its proper frictional component.

An initial period of on-the-job training with an experienced oilerman followed by attendance at related clinics is an effective method of preparing new lubrication personnel.

The clinics are held by representatives of lubricant suppliers and machinery manufacturers, complemented by instruction from plant personnel. Such clinics can be divided into a number of weekly sessions of about one hour each to provide a progressive training program. Clinics can also serve as refresher courses for experienced personnel and can be informative to members of related departments, such as purchasing and maintenance personnel.

SCHEDULING. Lubricants are consumed or deteriorate in use. Consumption is due to leakage, evaporation, blow-by and burning. Deterioration is caused primarily by heat, which, with the catalysts of water and air, promotes oxidation and forms acids.

All-loss lubrication systems and defective piping, joints, and seals of circulation systems represent the areas of highest lubricant consumption.

Deterioration of lubricants occurs most rapidly in internal combustion engines and within systems not properly vented or sufficiently cooled. Grease tends to harden when exposed to high temperature and pressure and allowed to remain static for extended periods.

In the case of all-loss oil and grease application systems, the rate of lubricant consumption under normal operating conditions is fairly regular. This means that, on the basis of probability, such appliances will require additional lubricant at predictable intervals, whether in terms of hours, days, or months, depending on the appliance.

Similarly, the life of a lubricant in a reuse system is a function of operating conditions, and where such conditions remain relatively constant, the frequency of oil change of that system is predictable within safe limits. Again, it requires experience, guided by periodic laboratory analysis, to determine the probable life span of a given oil for each individual system. The time between introduction of the oil to a given system and the point when it has reached a specific level of deterioration, as determined by laboratory analysis, establishes the frequency of drain periods for that system and for the particular oil. Oil change can be correctly scheduled when the interval is determined.

Manual application by oil can is usually done by the machine operator, as is the operation of a central manual oil or grease pump serving an individual machine. The reason for this is that application is usually required at least twice per shift. The regular oilerman, however, is responsible for refilling the oil cans and central reservoirs.

Ball and roller bearings of electric motors and machine spindles are often over-lubricated. The trend is toward prepacking and adequately sealing such bearings, but they should still be inspected periodically. This is done by checking their operating temperature by hand.

DUTY INSTRUCTIONS. Scheduling the work of an oilerman is done by written instructions handed to him each morning or the morning of a given day of the week, usually Monday. The instructions may consist of a sheet covering a full day's work or a series of cards, one for each machine to be attended to that day.

The instructions given the oilermen are made up by the clerk from a tickler file set up according to the weeks of the year or some other appropriate timetable. Included may be special instructions to investigate and act on requests or complaints received from plant operating personnel.

The instructions to oilermen should be specific. They should include the identification of the machines to be serviced, either by name or number or both, location by department, and tasks to be performed on each machine. Space on the instruction form should be provided for the oilerman to denote completion of each task and a remarks section for explanation of any negative report and reasons. The amount of oil added to each reservoir is an important part of the remarks. The remarks section also provides space for reporting abnormal frictional conditions and their location. The instruction sheet is complete only when dated and signed by the oilerman.

Oilermen become familiar with all the lubricants used and where each is to be applied. Such information is found on the lubrication charts drawn up as a result of the plant-wide analysis mentioned earlier in this chapter. Copies of such charts are usually made available to the oilerman.

There are several methods of guiding lubrication personnel as to the correct point of application of each lubricant. Tags placed on frictional parts or application points are usually not practical, since they are easily detached or torn off. Brass plates, unless installed by the machine builder, require the drilling and tapping of holes for screws to hold them in place. They are convenient and practical. A successful method of identifying lubricants with their points of application and frequency of application is that of a coding system.

Lubricant Application Coding

In manufacturing plants having a wide diversification of machinery, the application of lubricants to the various frictional parts may be greatly simplified by practical guidelines. The goal must always be the safe minimum number of lubricant brands and types, but there always remain the factors of system capacity and frequency of lubricant application and change. The identification of individual lubricants is always a problem in plants

LUBRICATION INSTRUCTIONS

Oilerman John Allen No. 369 Date Sept. 23, 1974

General Instructions	Completed	Remarks
Routine Oiling, Dept. C	✓	
Change Oil, machine 33	No	Machine being rebuilt
Attend lubrication clinic 11 AM	✓	
Check and add oil to machines 78, 12, 91 and 4	✓	12 gal oil #3 to bring all machines to proper level 3gal - No. 78 4 gal - No. 12 2 gal - No. 91 3gal - No. 4

COMMENTS:

Schedule machine 33 for 9/26.

JHAllen
Oilerman's Signature

Fig- 23-2a.

where purchasing is done on the basis of periodic specification bidding unless there is a coding system. The problem is more serious when there is rapid turnover of lubrication personnel.

Codes may be developed according to numbers, colors, symbols, or letters, or some combination thereof. Numbers often identify the lubricant as to service and viscosity. For example, if the number 10 is selected for circulating oils, then viscosity can be represented by either the comparable SAE number or the SUS viscosity at 100°F. Thus a circulating oil having a viscosity of about 300 SUS at 100°F would be coded by the number 10-300.

If identified by SAE numbers, the same oil would be coded as 10-20. However, only one of the two identifying methods should be used in any given plant. Circulating oils are used in most reuse systems, including gear cases, compressors, crankcases, hydraulic systems, electric motors, chain cases, turbine systems, and many other central systems, all of which require an oil of high chemical stability and other mutually desirable qualities. Thus the number 10 would identify a substantial proportion of plant lubricants. Other categories would include the following oils and their suggested basic code numbers and colors:

Spindle oils	Code 2	Blue
General purpose oils	Code 3	Yellow
Way oils	Code 4	Purple
Steam cylinder oils	Code 5	Brown
Air line (pneumatic) oil	Code 6	White
Hypoid lubricant	Code 7	Green
Leaded compounds	Code 8	Black
Adhesive oils	Code 9	Orange
Circulating oils	Code 10	Red
(Other lubricants)	Code	(Two colors)

The number 1 was not used because of its possible confusion with the number 10.

Greases may be coded similarly, but the wide use of multiple purpose types has practically eliminated the necessity. When special greases are used, such as those required for extreme temperatures, they may be specifically identified.

Charts showing the recommended lubricants for each machine and attached thereto have proven unsatisfactory, since they become obliterated or destroyed in time. To indicate type of lubricant and frequency of application simple symbols are practical. For example, a yellow dot at the point of lubrication may represent a general purpose grease to be applied every three months, and a red diamond may signify the use of a high-temperature grease and that the bearing is to be flushed and relubricated at two-week intervals. The diamond shape may be a reminder that all the bases (details) must be touched before the chore of lubrication is complete, particularly on bearings operating at elevated temperatures.

Intervals of application or oil changes may be specified with the following symbols:

Daily	—
Weekly	=
Monthly	×
Quarterly	●
Semiannually	▲
Annually	~

Each interval is based on a work shift of eight hours, so that daily actually means every 8 hr of operation; weekly represents 40 hr; monthly is 160 hr, etc. No modification is necessary for a plant normally operating eight hours per day and five days per week. Even when frequency of lubrication application is specified by a competent engineer, it is an unwritten law of good preventive maintenance that all frictional components be inspected continually. Thus any operating abnormalties, such as an increase in temperature or noise, will be detected early enough to prevent an unscheduled interruption in production.

DISPENSING LUBRICANTS. In carrying out his routine duties of dispensing lubricants, the oilerman requires the necessary tools, oils, and greases. The tools include screwdrivers, wrenches, pliers, bung opener, flashlight, and wiping cloths. A hand-operated grease gun, oil cans, and an oil measuring can are also important items. His stock of lubricants may include one or two greases and several oils, including gear, spindle, air line, way, and machine oils.

To facilitate transportation of the required supplies and to eliminate lost time in frequent trips back and forth to the oil house or room, a lubrication cart is usually necessary. It may be made at the plant or purchased, hand drawn or motorized. Special tank carts equipped with a pump and hose are used to transport new oil from the oil house to a machine reservoir for the purpose of changing oil. Portable filters and their pumping units are also transported similarly. All containers, including tank carts, must be labeled as to its oil content and never used for a different oil unless thoroughly cleaned and flushed with the new oil.

REPORTS (FEEDBACK)

STATUS OF WORK SCHEDULED is reported by the oilerman on his instruction form. If the work is completed, he so states. In the event one or more tasks cannot be accomplished, he should also advise and give the reasons. If the reason is serious, such as a mechanical fault, it should be reported immediately to his supervisor and noted on the instruction form. If the condition is not serious, such as a temporary obstruction, the details should be briefly reported and noted as to when the work can be done or rescheduled.

In large plants, the details of oiler instructions and feedback reports become cumbersome when performed by manual clerical procedures. Where numerous departments and hundreds of machines are involved, the clerical costs increase substantially, especially if a system of higher management information is effectively maintained. The latter must be maintained if *marginal* machines are to be uncovered and either overhauled or replaced.

The problem of an overburdening manual clerical system can be solved by the adoption of an automatic data processing (ADP) system, also referred to as an electronic data processing (EDP) system. Lubrication information can be combined with plant maintenance data to provide a wider range of maintenance information.

The plant lubrication analysis, which includes the name and location of machines and their points of lubrication, lubricants required, methods and frequency of lubricant application, is the foundation on which a data processing system can be installed. The data from the analysis is key punched into standard data processing cards for each machine to produce a master deck. From the master deck of cards, a lubrication schedule and a lubrication data card for each plant lubrication requirement are made available for every week of the year. The cards are color coded according to plant area or department and arranged by week and sequence of lubrication.

The lubrication data cards serve as work orders for the oilerman, who checks off the work as it is completed. If the punched card indicates incompleted work, it is used to print a delinquency report so that corrective action can be taken.

In addition to scheduling and the issuance of work orders, information relative to labor time required, replacement parts, costs, machine downtime, lubricant consumption, and other pertinent information can be assembled and recorded to assist management in controlling costs. A monthly summary of work orders or job tickets lists the cost of lubrication and maintenance of each machine and enables management to spot trouble areas immediately.

LUBRICANT CONSUMPTION can be determined from the work orders of the oilerman, who reports the amount of oil added to a system as makeup or for a complete change of batch.

OIL HOUSE

A centrally located oil house or room serves a number of purposes convenient to all lubrication personnel if designed properly (Fig. 23-3). Opened containers of lubricants can be stored there from which oilermen can fill their oil cans and grease guns or obtain small amounts of oil for makeup purposes. It provides a working area in which the mechanic can repair lubrication equipment that can be moved. Oilermen's schedules can be distributed at the oil house and their work forms returned there when their tasks have been completed. It should have a bulletin board where special notices or work orders can be posted, and a file cabinet containing lubrication bulletins, machine builders' instructions, equipment catalogues, and copies of plant lubrication analyses. Lockers for the storage of tools and lubrication equipment, a desk with a telephone, and repair benches with appropriate vises should also be furnished. Sufficient floor space should be provided for parking lubrication carts, filter and tank carts, dollies, and barrel trucks.

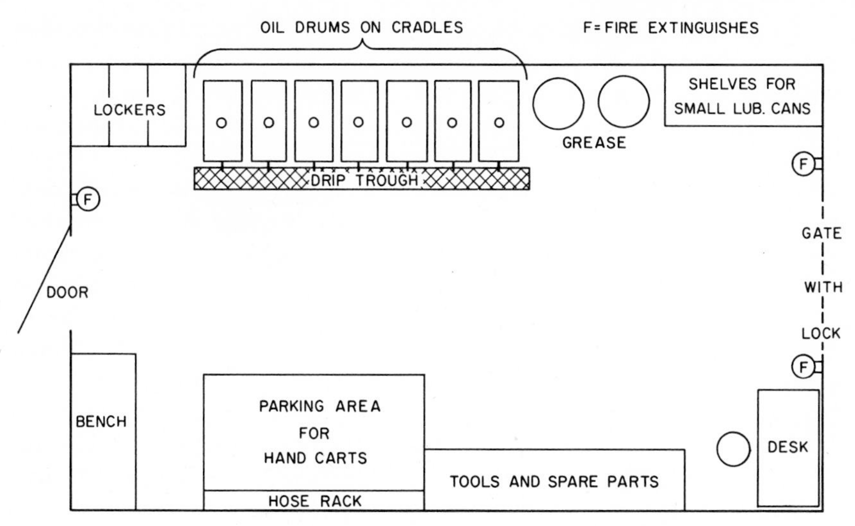

Fig. 23-3. Plan for oil house.

SAFETY

Much time, effort and money is spent by industry each year to provide safety equipment and educating personnel in accident prevention. The lubrication clinics mentioned earlier provide an excellent opportunity to promote safe practices. Safety clinics can be conducted by a representative of an insurance company or a member of the plant safety department. Safety not only to the oilerman, but also to persons near him at all times, should be emphasized.

Hazards related to industrial accidents may be categorized as follows:

1. Mechanical hazards involving the handling of heavy objects, such as full lubricant drums, working near moving machines, use of ladders, spillage of lubricants on floors, loose clothing, use of tools, misuse of compressed air, handling pressurized tools and equipment, and working with safety glasses, hard hats, and safety shoes.

2. Electrical hazards are prevalent when working around live wires, motors, generators, starters, transformers, and control instruments, when handling extension cords while standing on wet floors or in puddles of water, and when using portable electrical tools. Working within an idle machine constitutes a mechanical hazard, but it becomes an electrical function if the starter is not locked in an open position to prevent the machine from being started unsuspectedly by an operator.

3. Fire hazards are created when welding, brazing, or flame cutting is carried on in the vicinity of lubricant drums, feed and return lines of

circulating systems, grease feed lines, and oil storage tanks. Oil or grease on floors and other surfaces constitutes a fire hazard from a flame, spark, lighted cigarette, broken or a bare electric wire, and hot metal. Spontaneous combustion occurs in oily rags piled in a bin or other container, especially if linseed oil is present.

4. Asphyxiation is a hazard associated with prolonged inhalation of certain gases and mists present in the air at a level of concentration injurious to the lungs. Such gases and mists may emanate from chlorinated synthetic fluids, from leaded compounds, and even from straight mineral oils if the fumes are inhaled for extended periods.

5. Suffocation can occur in empty and poorly ventilated storage tanks used for chemicals and petroleum products. Men entering tanks to clean them should be on guard and leave periodically for fresh air.

6. Inflammation of the skin is brought about by constant contact of uncovered parts of the body with petroleum products. Entrance of bacteria into inflamed areas results in dermatitis.

APPENDICES

1. Temperature Conversion, Celsius to Fahrenheit

°F	°C	°F	°C	°F	°C
−60	−51.11	160	72.22	380	193.3
−50	−45.56	170	76.67	390	198.9
−40	−40.00	180	82.22	400	204.4
−30	−34.44	190	87.78	410	210.0
−20	−27.78	200	93.33	420	215.6
−10	−23.33	210	98.89	430	222.2
0	−17.78	220	104.4	440	226.7
10	−12.22	230	109.0	450	232.2
20	− 6.667	240	105.6	460	237.8
30	− 1.111	250	112.2	470	243.3
40	4.444	260	126.7	480	248.9
50	10.00	270	132.2	490	254.4
60	15.56	280	137.8	500	260.0
70	21.22	290	143.3	510	265.6
80	26.67	300	148.9	520	272.2
90	32.22	310	154.4	530	276.7
100	37.78	320	160.0	540	282.2
110	43.33	330	165.6	550	287.8
120	48.89	340	172.2	560	293.3
130	54.44	350	176.7	570	298.9
140	60.00	360	182.2	580	304.4
150	65.56	370	187.8	590	310.0

Interpolation

Degrees Fahrenheit	1	2	3	4	5	6	7	8	9
Degrees Celsius	0.67	1.33	2.00	2.67	3.33	4.00	4.67	5.33	6.00

To convert Celsius to Fahrenheit

$$°F = (\tfrac{9}{5} \times °C) + 32$$

To convert Fahrenheit to Celsius

$$°C = \tfrac{5}{9}(°F - 32)$$

To convert minus values of Celsius to Fahrenheit

$$°F = (\tfrac{9}{5} \times -35) + 32$$

To convert minus values of Fahrenheit to Celsius

$$°C = \tfrac{5}{9}(-40^{*} - (+32))$$

* −40° is the only temperature at which Fahrenheit and Celsius are the same.

2. Properties of Saturated Steam*

Abs. Press. psi (p)	Temp. °F (t)	*Specific Volume* (V)		*Enthalpy*			*Entropy* (s)			*Internal Energy*
		Liquid	Vapor	Liquid (h)	Evap. (L)	Vapor (H)	Liquid	Evap.	Vapor	Evap. (h_1)
1.0	101.74	0.01614	333.6	69.70	1036.3	1106.0	0.1326	1.8456	1.9782	974.6
2.0	126.08	0.01623	173.73	93.99	1022.2	1116.2	0.1749	1.7451	1.9200	957.9
3.0	141.48	0.01630	118.71	109.37	1013.2	1122.6	0.2008	1.6855	1.8863	947.3
4.0	152.97	0.01636	90.63	120.86	1006.4	1127.3	0.2198	1.6427	1.8625	939.3
5.0	162.24	0.01640	73.52	130.13	1001.0	1131.1	0.2347	1.6094	1.8441	933.0
6.0	170.06	0.01645	61.98	137.96	996.2	1134.2	0.2472	1.5820	1.8292	927.5
7.0	176.85	0.01649	53.64	144.76	992.1	1136.9	0.2581	1.5586	1.8167	922.7
8.0	182.86	0.01653	47.34	150.79	988.5	1139.3	0.2674	1.5383	1.8057	918.4
9.0	188.28	0.01656	42.40	156.22	985.2	1141.4	0.2759	1.5203	1.7962	914.6
10	193.21	0.01659	38.42	161.17	982.1	1143.3	0.2835	1.5041	1.7876	911.1
14.696	212.00	0.01672	26.80	180.07	970.3	1150.4	0.3120	1.4446	1.7566	897.5
15	213.03	0.01672	26.29	181.11	969.7	1150.8	0.3135	1.4415	1.7549	896.7
20	227.96	0.01683	20.089	196.16	960.1	1156.3	0.3356	1.3962	1.7319	885.8
25	240.07	0.01692	16.303	208.42	952.1	1160.6	0.3533	1.3606	1.7139	876.8
30	250.33	0.01701	13.746	218.82	945.3	1164.1	0.3680	1.3313	1.6993	869.1
35	259.28	0.01708	11.898	227.91	939.2	1167.1	0.3807	1.3063	1.6870	862.3
40	267.25	0.01715	10.498	236.03	933.7	1169.7	0.3919	1.2844	1.6763	856.1
45	274.44	0.01721	9.401	243.36	928.6	1172.0	0.4019	1.2650	1.6669	850.5
50	281.01	0.01727	8.515	250.09	924.0	1174.1	0.4110	1.2474	1.6585	845.4
55	287.07	0.01732	7.787	256.30	919.6	1175.9	0.4193	1.2316	1.6509	840.6
60	292.71	0.01738	7.175	262.09	915.5	1177.6	0.4270	1.2168	1.6438	836.0
65	297.97	0.01743	6.655	267.50	911.6	1179.1	0.4342	1.2032	1.6374	831.8
70	302.92	0.01748	6.206	272.61	907.9	1180.6	0.4409	1.1906	1.6315	827.8
75	307.60	0.01753	5.816	277.43	904.5	1181.9	0.4472	1.1787	1.6259	824.0
80	312.03	0.01757	5.472	282.02	901.1	1183.1	0.4531	1.1676	1.6207	820.3
85	316.25	0.01761	5.168	286.39	897.8	1184.2	0.4587	1.1571	1.6158	816.8
90	320.27	0.01766	4.896	290.56	894.7	1185.3	0.4641	1.1471	1.6112	813.4
95	324.12	0.01770	4.652	294.56	891.7	1186.2	0.4692	1.1376	1.6068	810.2
100	327.81	0.01774	4.432	298.40	888.8	1187.2	0.4740	1.1286	1.6026	807.1
110	334.77	0.01782	4.049	305.66	883.2	1188.9	0.4832	1.1117	1.5948	801.2
120	341.25	0.01789	3.728	312.44	877.9	1190.4	0.4916	1.0962	1.5878	795.6
130	347.32	0.01796	3.455	318.81	872.9	1191.7	0.4995	1.0817	1.5812	790.2
140	353.02	0.01802	3.220	324.82	868.2	1193.0	0.5069	1.0682	1.5751	785.2
150	358.42	0.01809	3.015	330.51	863.6	1194.1	0.5138	1.0556	1.5694	780.5
160	363.53	0.01815	2.834	335.93	859.2	1195.1	0.5204	1.0436	1.5640	775.8
170	368.41	0.01822	2.675	341.09	854.9	1196.0	0.5266	1.0324	1.5590	771.4
180	373.06	0.01827	2.532	346.03	850.8	1196.9	0.5325	1.0217	1.5542	767.1
190	377.51	0.01833	2.404	350.79	846.8	1197.6	0.5381	1.0116	1.5497	763.0
200	381.79	0.01839	2.288	355.36	843.0	1198.4	0.5435	1.0018	1.5453	759.0
210	385.90	0.01844	2.183	359.77	839.2	1199.0	0.5487	0.9925	1.5412	755.2
220	389.86	0.01850	2.087	364.02	835.6	1199.6	0.5537	0.9835	1.5372	751.3
230	393.68	0.01854	1.9992	368.13	832.0	1200.1	0.5585	0.9750	1.5334	747.7
240	397.37	0.01860	1.9183	372.12	828.5	1200.6	0.5631	0.9667	1.5298	744.1
250	400.95	0.01865	1.8438	376.00	825.1	1201.1	0.5675	0.9588	1.5263	740.7

* Extracted from "Thermodynamic Properties of Steam," by Keenan and Keyes, published by John Wiley & Sons, Inc., New York.

3. Blending Chart for Petroleum Oils

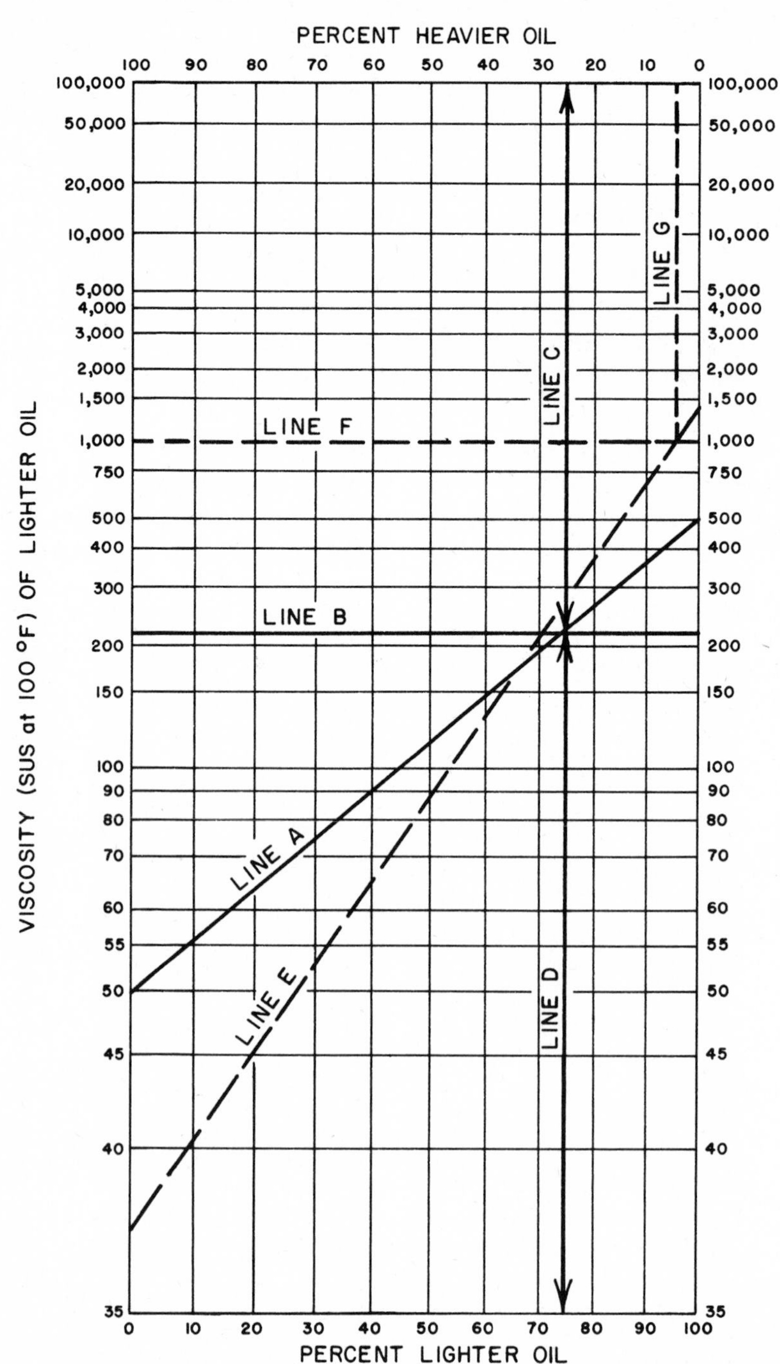

How To Use the Blending Chart

For Blending

Example. To obtain a viscosity of 220 SUS at 100°F by blending a lighter oil of 50 SUS at 100°F with a heavier oil of 500 SUS at 100°F.

1. Connect the viscosities of both oils and draw Line A.
2. Read vertically upward from the intersection of Line A and the required viscosity Line B, or 220 SUS, as shown on the chart, and establish Line C.
3. It is found that 25% heavier oil is required (500 SUS at 100°F).
4. Read vertically downward from the same intersection (Line A with Line B), and establish Line D.
5. It is found that 75% of the lighter oil (50 SUS at 100°F) is required.

Therefore, blending 25% of an oil of 500 SUS and 75% of an oil of 50 SUS oil will result in an oil of 220 SUS, all at 100°F.

To Determine Fuel Oil Dilution

Example. To determine percent dilution of diesel fuel (37 SUS at 100°F) from the blending chart. Assume new motor oil to have a viscosity of 1200 SUS at 100°F, and used oil, 1000 SUS at 100°F, as determined by analysis.

1. Draw Line E connecting 37 SUS (viscosity of lighter oil) with 1200 SUS (viscosity of new oil).
2. Draw Line F on 1200 SUS at 100°F line.
3. Read vertically upward from intersection of Lines E and F on Line G and establish percent dilution, approximately 4%. Condemnation level is about 6% and above.

4. VISCOSITY SELECTION FOR GIVEN LOAD AND SPEED

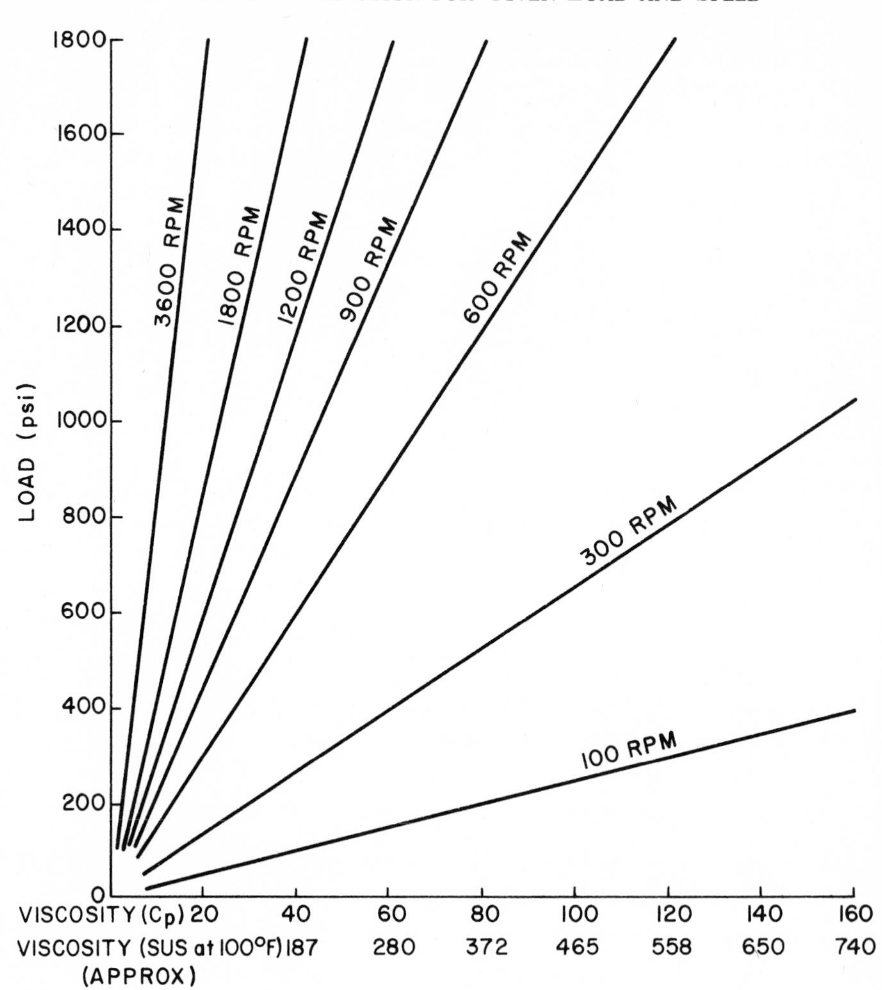

5. Viscosity Conversion Table

Kinematic Viscosity (Centistokes = K)	*Saybolt Universal Seconds*	*Saybolt Furol Seconds*	*Redwood Seconds*	*Redwood Admiralty Seconds*	*Engler Degrees*	*Barbey Degrees*
1.00	31		29		1.00	6200
2.56	35		32.1		1.16	2420
4.30	40		36.2	5.10	1.31	1440
5.90	45		40.3	5.52	1.46	1050
7.40	50		44.3	5.83	1.58	838
8.83	55		48.5	6.35	1.73	702
10.20	60		52.3	6.77	1.88	618
11.53	65		56.7	7.17	2.03	538
12.83	70	12.95	60.9	7.60	2.17	483
14.10	75	13.33	65.0	8.00	2.31	440
15.35	80	13.70	69.2	8.44	2.45	404
16.58	85	14.10	73.3	8.86	2.59	374
17.80	90	14.44	77.6	9.30	2.73	348
19.00	95	14.85	81.5	9.70	2.88	326
20.20	100	15.24	85.6	10.12	3.02	307
31.80	150	19.3	128	14.48	4.48	195
43.10	200	23.5	170	18.90	5.92	144
54.30	250	28.0	212	23.45	7.35	114
65.40	300	32.5	254	28.0	8.79	95
76.50	350	35.1	296	32.5	10.25	81
87.60	400	41.9	338	37.1	11.70	70.8
98.60	450	46.8	381	41.7	13.15	62.9
110.	500	51.6	423	46.2	14.60	56.4
121.	550	56.6	465	50.8	16.05	51.3
132.	600	61.4	508	55.4	17.50	47.0
143	650	66.2	550	60.1	19.00	43.4
154	700	71.1	592	64.6	20.45	40.3
165	750	76.0	635	69.2	21.90	37.6
176	800	81.0	677	73.8	23.35	35.2
187	850	86.0	719	78.4	24.80	33.2
198	900	91.0	762	83.0	26.30	31.3
209	950	95.8	804	87.6	27.70	29.7
220	1000	100.7	846	92.2	29.20	28.2
330	1500	150	1270	138.2	43.80	18.7
440	2000	200	1690	184.2	58.40	14.1
550	2500	250	2120	230	73.00	11.3
660	3000	300	2540	276	87.60	9.4

5. VISCOSITY CONVERSION TABLE (*Continued*)

Kinematic Viscosity (Centistokes = K)	*Saybolt Universal Seconds*	*Saybolt Furol Seconds*	*Redwood Seconds*	*Redwood Admiralty Seconds*	*Engler Degrees*	*Barbey Degrees*
770	3500	350	2960	322	100.20	8.05
880	4000	400	3380	368	117.00	7.05
990	4500	450	3810	414	131.50	6.26
1100	5000	500	4230	461	146.00	5.64
1210	5500	550	4650	507	160.50	5.13
1320	6000	600	5080	553	175.00	4.70
1430	6500	650	5500	559	190.00	4.34
1540	7000	700	5920	645	204.50	4.03
1650	7500	750	6350	691	219.00	3.76
1760	8000	800	6770	737	233.50	3.52
1870	8500	850	7190	783	248.00	3.32
1980	9000	900	7620	829	263.00	3.13
2090	9500	950	8040	875	277.00	2.97
2200	10000	1000	8460	921	292.00	2.82

The viscosity is often expressed in terms of viscosimeters other than the Saybolt Universal. The formulas for the various viscosimeters are as follows:

Redwood	$K = 0.26t - 188/t$	(British)
Redwood Admiralty	$K = 2.396 - 40.3/t$	(British)
Saybolt Universal	$K = 0.22t - 180/t$	(American)
Saybolt Furol	$K = 2.2t - 203/t$	(American)
Engler	$K = 0.147t - 374/t$	(German)
	$= \text{Engler degrees} \times 51.3$	

6. VISCOSITY DECREASE OF LUBRICATING OIL DUE TO DIESEL FUEL* OIL DILUTION

Percent Fuel Oil Dilution	*Initial Viscosity*									
	300	400	500	600	700	800	900	1000	1100	1200
1	285	385	475	570	660	760	850	940	1025	1125
2	270	365	450	540	625	715	800	880	965	1050
3	260	350	430	510	590	670	730	830	900	980
4	250	335	410	485	555	635	705	780	850	910
5	245	320	390	460	520	600	660	725	800	850
6	235	305	370	435	490	560	625	685	745	800

* Viscosity of diesel fuel assumed to be 37 SUS at 100°F.

7. Comparable Viscosity Scales with SAE Numbers

SAE No.	SUS 100° F	SUS 210° F	Centistokes 100° F	Centistokes 210° F
10	150–250	40–45	32–54	4.2–5.7
20	250–400	45–58	54–87	5.7–9.6
30	400–600	58–70	87–130	9.6–13.0
40	600–850	70–85	130–185	13.0–16.8
50	850–1500	85–110	185–320	16.8–22.7

8. Storage Tank Sizes and Capacities

Capacity (gallons)	Size (diameter × length)	Gauge or Dimension	Approximate Weight (lb)
550	48″ × 6′10″	7	800
1000	48″ × 10′8″	7	1250
2000	64″ × 12′0″	7	1995
3000	64″ × 17′11″	7	2800
4000	64″ × 23′10″	7	3580
5000	96″ × 13′4″	¼″	4800
6000	96″ × 16′2″	¼″	5600
8000	96″ × 21′4″	¼″	7000
10000	96″ × 26′7″	¼″	8400

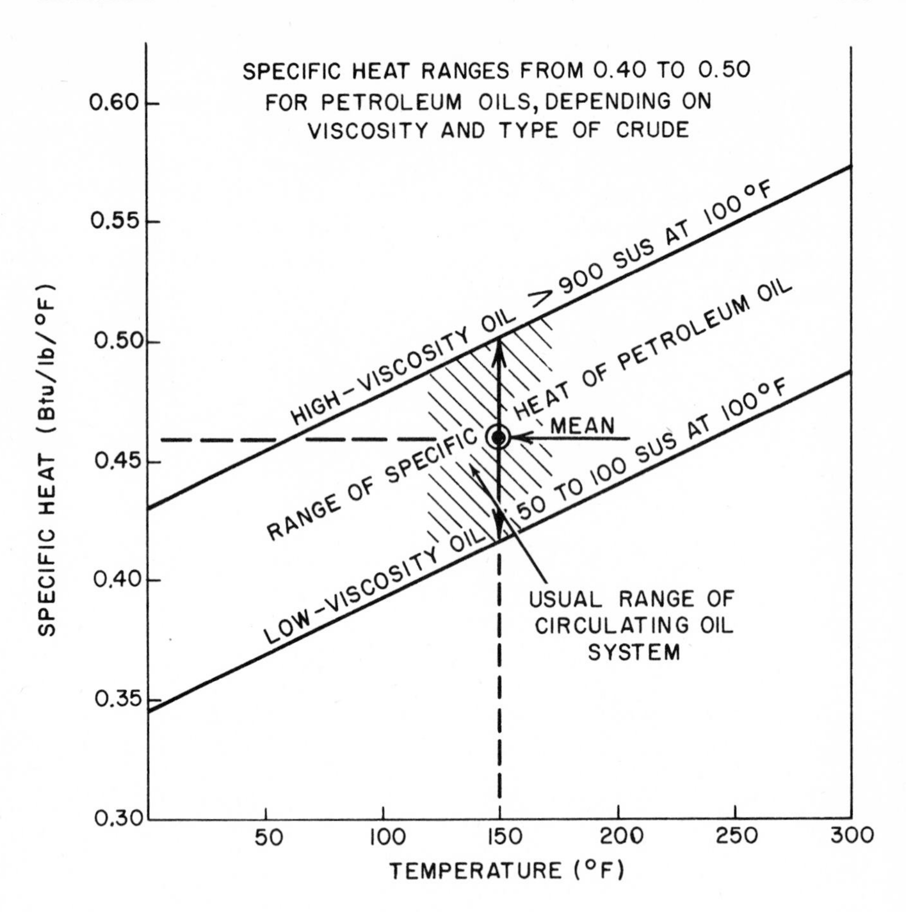

9. SPECIFIC HEAT OF PETROLEUM OILS

Value Varies With Viscosity and Type of Crude.
Mean Value of 0.46 May Be Used for Oil Cooling.

10. Hertz Stresses

Stresses in structural metals of two bodies of various geometrical shapes when forced together can be determined by formulae developed by Heinrich Hertz, a German physicist. Such formulae are used in the design of rolling bearings and gears. Both the contact band and stress pattern can be found by considering the curvature and length of bodies in contact.

In the case of two cylinders Fig. (a) the width of band B is determined by the equation:

$$B = \frac{16W(k_1 + k_2)r_1 r_2}{L(r_1 + r_2)}$$

when

$$k_1 = \frac{1 - \mu_1^2}{\pi E_2} \quad \text{and} \quad k_2 = \frac{1 - \mu_2^2}{\pi E_2}$$

where

W = force (lb)
L = length of cylinders
μ = Poisson's ratio
E = modulus of elasticity

$$\text{maximum compressive stress} = \frac{4F}{L\pi B} \text{ psi}$$

For the geometric shape in Fig. (b), the maximum Hertzian stress S is determined by the equation in which d_1 and d_2 are the diameters of the cylinders in (b).

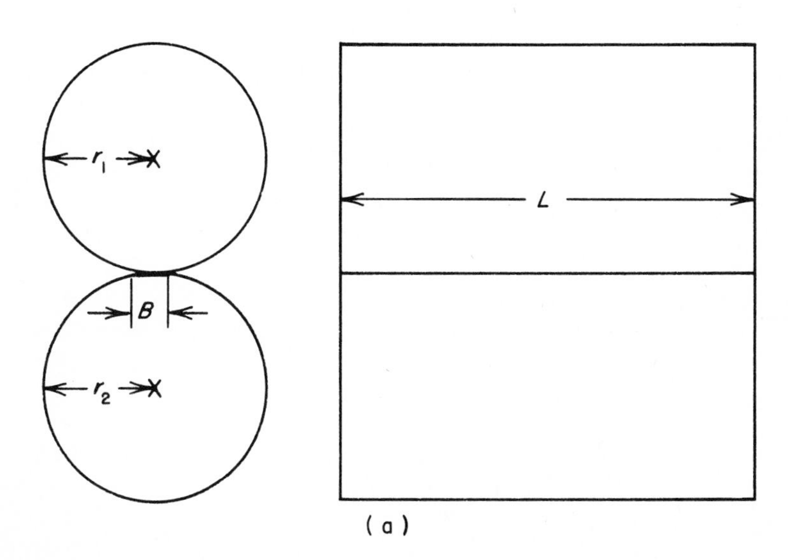

(a)

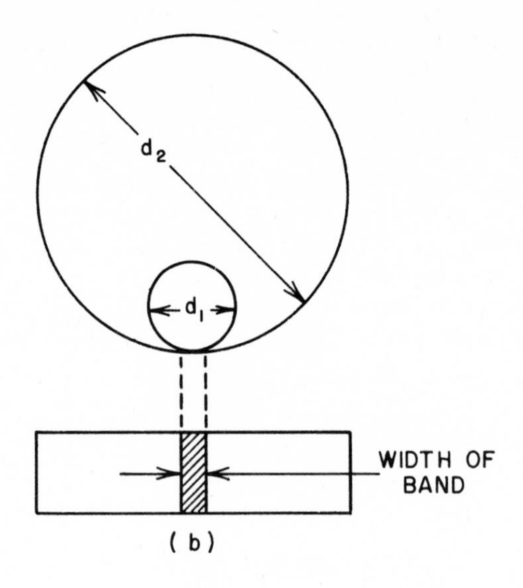

(b)

$$S_{max} = 0.591 \frac{WE}{L} \left(\frac{d_2 - d_1}{d_2 d_1} \right) \text{psi}$$

and the mean stress by the equation

$$S = 0.786 S_{max}$$

11. THE pH SCALE

The term pH is an abbreviation for "potential of electricity for positive hydrogen ions." Its value is expressed by a number denoting the degree of acidity or alkalinity of a substance. The value of pH does not in any way indicate the quantity of acids or alkalines in a solution as found by titration methods.

The pH value is derived by measuring the amount of hydrogen ions (H^+) in grams per liter liberated during ionization of acid or alkaline solutions at various stages of intensities. For example, pure water ionizes to produce 0.0000001 gram of hydrogen or acid ions per liter. Therefore, any substance producing 0.0000001 gram of acid ions (H^+) per liter is considered to be a neutral solution. The greater the number of hydrogen ions present in a solution, the stronger becomes its acid reaction. Pure water, being neutral, is established as a relative point by which intensity comparisons can be made. It follows that (a) any solution producing more hydrogen ions than pure water is acid and to the degree represented by the difference, and (b) any solution producing less hydrogen ions than pure water is alkaline to the degree represented by the difference.

A substance producing 0.000001 gram of hydrogen ions per liter will be slightly acid. A substance producing 0.00000001 gram of hydrogen ions per liter will be slightly alkaline. Note that the fluctuations in measurements are increased or decreased in multiples of 10. This means that for each whole graduation in the pH scale intensity of acidity and alkalinity is raised or lowered 10 times its preceding number. Because hydrogen ion concentration is measured in grams per liter, it requires numerous decimal places for exact recording. The scale is therefore simplified by the use of whole numbers from 1 to 14, which represent the number of decimal places required to record the weight of hydrogen ions in grams per liter of solutions at various levels of acidity and alkalinity.

THE pH SCALE*

pH Value	*Hydrogen Ion (H^+) Concentration Grams Ionizable H^+ per Liter*		*No. of Times Acidity or Alkalinity Exceeds that of Pure Water (pH 7)*	
0.0	1.0	Increasingly Acid ↑	10,000,000	10
1.0	0.1		1,000,000	10^{-1}
2.0	0.01		100,000	10^{-2}
3.0	0.001		10,000	10^{-3}
4.0	0.0001		1,000	10^{-4}
5.0	0.00001		100	10^{-5}
6.0	0.000001		10	10^{-6}
7.0	0.0000001	Neutral	1	10^{-7}
8.0	0.00000001	Increasingly Alkaline ↓	10	10^{-8}
9.0	0.000000001		100	10^{-9}
10.0	0.0000000001		1,000	10^{-10}
11.0	0.00000000001		10,000	10^{-11}
12.0	0.000000000001		100,000	10^{-12}
13.0	0.0000000000001		1,000,000	10^{-13}
14.0	0.00000000000001		10,000,000	10^{-14}

* The pH scale value is equivalent to the number of decimal places required to denote the weight of H^+ in grams per liter.

12. OIL REFINING

A complete discussion of the various processes for refining lubricating oils is not within the scope of this text. However, a brief résumé concerning the division of petroleum crudes into its various important components and some methods of oil refining may lead to a better understanding of the problems associated with lubrication of mechanical equipment.

Distillation

Crude petroleum is what its name implies insofar as lubricating oils are concerned. Crude oils are not made up of a given substance possessing a common boiling point, but of numerous compounds having individual boiling points that enable them to be separated by heating. The process is called fractionating and is accomplished in a fractionating tower that has trays installed at various levels to catch the condensating fluids.

Initial fractionating is called crude reduction, or topping. In the topping tower the light, low-boiling-point products are separated from the heavier, higher-boiling-point products. The remaining substance is made up of that material containing low viscosity lubricating oils down to heavy residuums. The crude is heated by steam before entering the tower. Vaporization occurs immediately. The vapor rises in the tower in which temperature and pressure from bottom to top are carefully controlled by live steam. As the heavier and higher boiling point vapors reach a certain level, they pass through bubble caps. The bubble caps that form part of the trays direct the vapors through a liquid body of the same fraction that has already condensed in the trays under ambient temperatures and pressure at that level. These, in turn, condense in the tray, and the overflow runs to collection tanks outside the tower.

All the products resulting from boiling and consequent condensing produced in a distillation unit are called distillates. Thus, there are gasoline distillates, naphtha distillates, kerosene distillates, gas oil distillates, etc.,

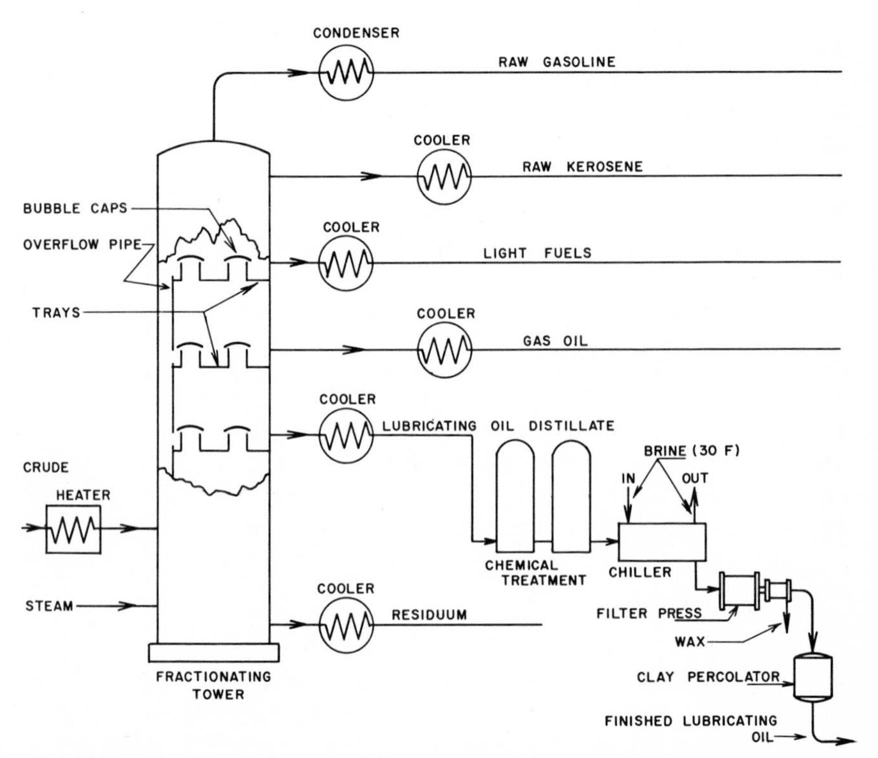

Appendix 12. Typical flow of petroleum from crude (well) to finished lubricating oil.

named in the order of ascending boiling points. Light-bodied fuel oils and kerosene are representative intermediate boiling point fractions in the topping tower.

The residue from the crude still, or topping tower, composed of fractions having higher boiling points than the temperatures carried for the topping process, is pumped to another tower after it is preheated. The distillation process is repeated at suitable temperatures with the light-bodied lubricating oils being drawn off at the top and the relatively heavier products at the bottommost tray level. It is at this point in the oil preparation process that the flash point and the viscosity for any fixed combination of conditions carried on in subsequent refining operations are determined. The residue (residuums) left in the extreme bottom of the still may be heavy lubricating oils or asphalt, depending on the base of the crude. Intermediate levels produce lubricating oil stocks having relative viscosities between the two extremes (Appendix Table 12-1).

The process of distillation is a continuous operation, and not one of intermittent charging.

High temperatures promote the decomposition of hydrocarbon molecules, causing undesirable chemical changes. Such high temperatures are produced in the fire still, where heating the charge is accomplished by direct contact heating as carried on in the old batch process. To reduce the boiling point of the charge during distillation, several methods are employed.

One method is known as *steam distillation.* Steam is mixed with the crude to raise the vapor pressure of the mass above that of the surrounding atmosphere within the still. This lowers the boiling point below what it would be if all the pressure were to be developed within the charge by direct contact heating.

Another method used to reduce the boiling point of the charge is by *direct contact heating,* i.e., in a manner similar to boiling water over a gas flame.

Still another method used, particularly in distilling lubricating oils of the lighter viscosities, is a modification of the above. The boiling point of distillation is reduced by *vacuum distillation.* This method is similar to that described for steam distillation except that a partial vacuum is maintained within the tower during the fractionating process. Vacuum is developed by rapidly condensing light vapors as they emerge from the top of the tower. The action is much the same as that found in steam turbine condensers.

This description of the distillation processes is general and varies in many details for reasons dictated by types of crudes, end results, etc. Fractionating between relative ranges of temperature results in individual products having definite minimum and maximum boiling points. This means that under closer control they, too, can be further divided (cut) into products having closer end points. The substances within a given mass of a distilled product having the lowest boiling point (most volatile) are referred to as "light ends." The less volatile substances in the same mass are "heavy ends."

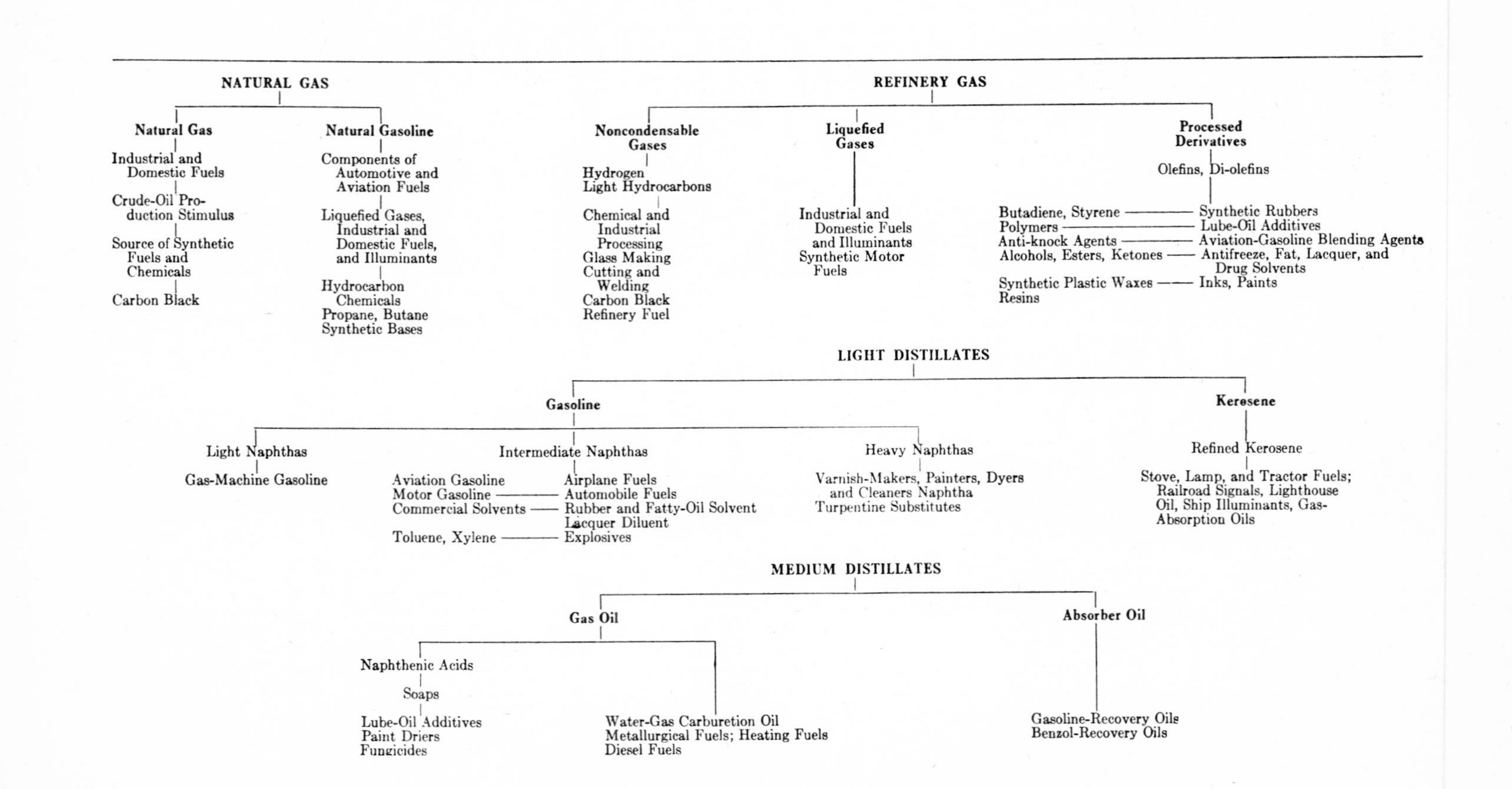
NATURAL GAS
Natural Gas
Industrial and Domestic Fuels
Crude-Oil Production Stimulus
Source of Synthetic Fuels and Chemicals
Carbon Black
Natural Gasoline
Components of Automotive and Aviation Fuels
Liquefied Gases, Industrial and Domestic Fuels, and Illuminants
Hydrocarbon Chemicals
Propane, Butane
Synthetic Bases
REFINERY GAS
Noncondensable Gases
Hydrogen
Light Hydrocarbons
Chemical and Industrial Processing
Glass Making
Cutting and Welding
Carbon Black
Refinery Fuel
Liquefied Gases
Industrial and Domestic Fuels and Illuminants
Synthetic Motor Fuels
Processed Derivatives
Olefins, Di-olefins
Butadiene, Styrene
Synthetic Rubbers
Polymers
Lube-Oil Additives
Anti-knock Agents
Aviation-Gasoline Blending Agents
Alcohols, Esters, Ketones
Antifreeze, Fat, Lacquer, and Drug Solvents
Synthetic Plastic Waxes
Inks, Paints
Resins
LIGHT DISTILLATES
Gasoline
Light Naphthas
Gas-Machine Gasoline
Intermediate Naphthas
Aviation Gasoline
Airplane Fuels
Motor Gasoline
Automobile Fuels
Commercial Solvents
Rubber and Fatty-Oil Solvent
Lacquer Diluent
Toluene, Xylene
Explosives
Heavy Naphthas
Varnish-Makers, Painters, Dyers and Cleaners Naphtha
Turpentine Substitutes
Kerosene
Refined Kerosene
Stove, Lamp, and Tractor Fuels; Railroad Signals, Lighthouse Oil, Ship Illuminants, Gas-Absorption Oils
MEDIUM DISTILLATES
Gas Oil
Naphthenic Acids
Soaps
Lube-Oil Additives
Paint Driers
Fungicides
Water-Gas Carburetion Oil
Metallurgical Fuels; Heating Fuels
Diesel Fuels
Absorber Oil
Gasoline-Recovery Oils
Benzol-Recovery Oils

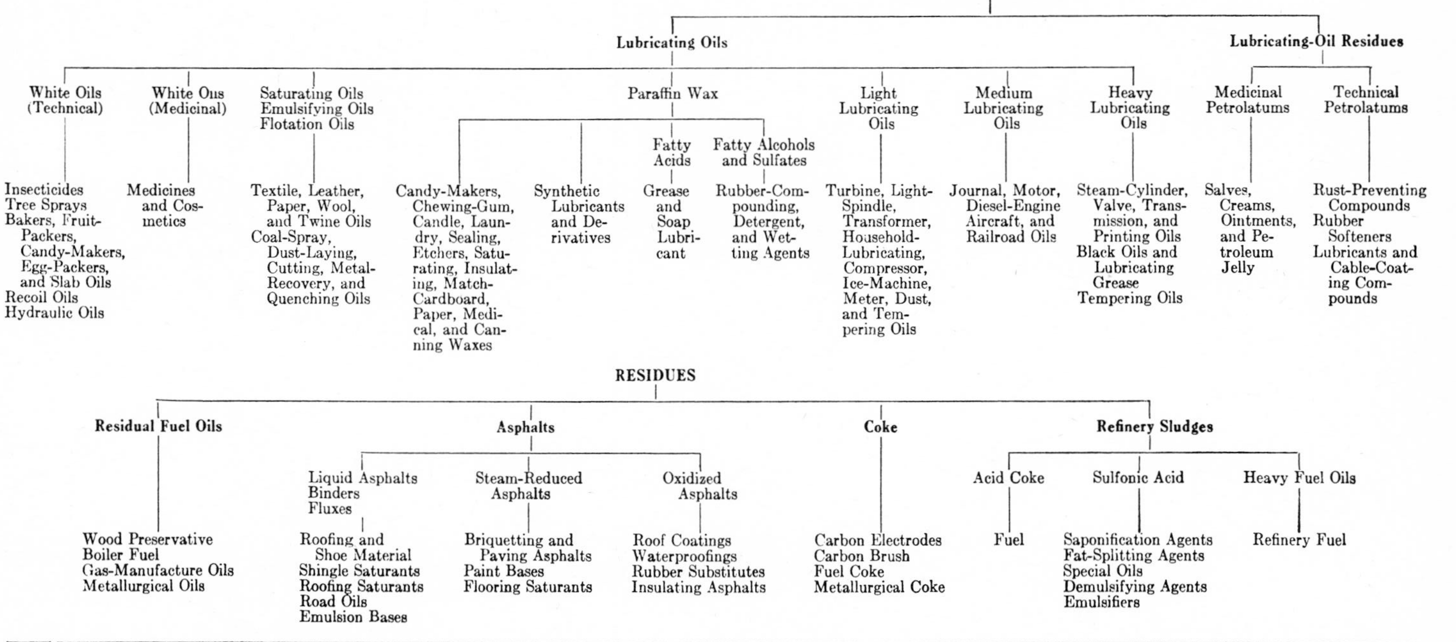

* Courtesy, American Petroleum Institute.

Table A-12. PETROLEUM PRODUCTS

BLENDING. A finished lubricating oil made up by blending two oils of different fractionations (i.e., one of lower viscosity than the other, but of the same base crude) may have a range of volatility not compatible with some critical lubricating requirements. In other words, the "ends" may be too far apart for an express purpose. However, where such practices are kept within reasonable limits, satisfactory results can be obtained in most mechanical operations. Blending oils of different viscosities to obtain some desired compromise is widely practiced with excellent results. Some high-grade motor and industrial oils of certain intermediate viscosities are produced by this method.

REFINING WITH SOLVENTS. During the refining process following distillation, unsaturates, asphalt color bodies, and waxes are removed from the lubricating oil distillates and residuums. This is accomplished by treating the unrefined oil with solvents, of which there are several kinds. Some of the solvents used are furfural (a distillate of oat hulls treated with sulfuric acid), sulfur dioxide, cresylic acid, and propane. All act to dissolve certain impurities in the oil, after which the solvents are recovered and reused. Mixed directly in the oil, solvents remove the impurities for which they are especially selected. The action is one of a solvent rather than a chemical, which accounts for solvent recoverability. Solvent refining may be accomplished variously by using one or two different solvents, processes known as single solvent and duosolvent refining systems, respectively. Sulfur dioxide in its liquid form removes unsaturates. Upon reduction of pressure on the solvent-unsaturate mixture, the sulfur dioxide evaporates, leaving the unsaturates behind. The sulfur dioxide is then recovered in a manner similar to that of any mechanical refrigeration system.

One method of double solvent treatment employs propane and cresylic acid, which act together to remove unsaturates, asphalts, and waxes. In the first stage, liquid propane is mixed with the oil under pressure. Asphalt is not solvent in propane and therefore precipitates. A reduction of pressure on the oil-propane mixture causes propane to evaporate, lowering the temperature, which causes the formation of wax crystals in the oil. A further reduction of oil temperature followed by filtration through cloth filters removes the wax crystals. In the second stage, the oil is scrubbed by cresylic acid. Cresylic acid, being the heavier of the two, percolates down through the oil and removes the unsaturates.

Clay Filtering. Some lubricating oils are refined further by filtering through fuller's earth, a natural claylike substance. Artificial activated clays are also used to a large extent as a substitute for fuller's earth. Clay filtering improves the water-separating characteristics of oil and also acts as a decolorizing medium. The clay may be added directly to the oil in controlled quantities and then filtered out, or the process may be carried out by allowing the oil to percolate down through the clay. A petroleum solvent is added to heavier lubricating oils to increase flow rate during the clay filtering process. The process is one of adsorption.

Acid Treatment. An old method of refining lubricating oils and one still used in special cases is that of sulfuric acid treating. The reaction of sulfuric acid is both chemical and physical.

Acid is added to the oil and the mixture is agitated. Unsaturated hydrocarbon molecules react with the acid forming a sludge which precipitates and settles to the bottom of the lead-lined mixing tank. The acid sludge is then drawn from the tank. The process may be repeated several times, depending on the type of oil and the desired quality of the finished product.

Acid treating is followed by a neutralizing process accomplished by transferring the oil to another tank and treating with caustic soda. The caustic soda reacts with any acid remaining in the oil, the residue of which settles in the same manner as the acid sludge. After the caustic soda treatment, the oil is washed with water to remove all traces of the alkali.

Dewaxing. Wax from petroleum sources may be crystalline, resembling beeswax, or it may be formless, resembling jelly. The latter is called petrolatum. The amount of wax contained in a crude depends on the base. Paraffins have more wax than any other base.

Waxes within a given crude congeal at different temperatures (melting point). For this reason, the temperature of the lubricating oil must be reduced to a level where a sufficient amount of wax can be removed to meet the conditions at which the oil will be used. Wax is the congealing agent in oils and will reduce or prevent flow when the temperature of the oil is reduced to a temperature commensurate with the amount of wax it contains. Therefore, wax content is an important consideration where the lubrication service requires oils to be exposed to low temperatures. Automotive engines and transmission components are affected during winter operation when the oil-congealing temperatures are not sufficiently low. Channeling results, leading to rapid wear. Hard starting is another disadvantage of using high wax content oils. Oils used to lubricate refrigeration compressors, especially in low-temperature systems, are particularly required to be of low wax content. Otherwise, wax separation will occur in the evaporators and cause restricted flow, reduce heat transfer, and create an undesirable condition generally.

To meet special service conditions, some lubricating oils must be dewaxed beyond the stage afforded during solvent treatment. This is accomplished by chilling the oil to a certain temperature in a brine chiller. If drastic dewaxing is to be accomplished, chilling must be done in stages to facilitate filtration, which otherwise would be hampered by excessive wax precipitation.

The chiller is a long cylinder containing a spiral baffle through which the oil is pumped. Cold brine is circulated in a jacket surrounding the cylinder. After the oil is chilled to the desired temperature, say $-40°F$, it is forced through a filter press. The filter press is made up of a series of cloth filters backed by perforated metal plates. The wax is held by the filters and the dewaxed oil withdrawn from the press.

INDEX